# Chemical Biology: Methodology and Applications

# Chemical Biology: Methodology and Applications

Edited by
Oliver Stone

WILLFORD PRESS

www.willfordpress.com

Published by Willford Press,
118-35 Queens Blvd., Suite 400,
Forest Hills, NY 11375, USA

ISBN: 978-1-68285-368-9

**Cataloging-in-Publication Data**

Chemical biology : methodology and applications / edited by Oliver Stone.
p. cm.
Includes bibliographical references and index.
ISBN 978-1-68285-368-9
1. Biochemistry. 2. Biology. 3. Chemistry. I. Stone, Oliver.
QH345 .C44 2017
572--dc23

For information on all Willford Press publications
visit our website at www.willfordpress.com

WILLFORD PRESS

Printed in the United States of America.

# Contents

# Preface

The main aim of this book is to educate learners and enhance their research focus by presenting diverse topics covering this vast field. This is an advanced book which compiles significant studies by distinguished experts. This book addresses successive solutions to the challenges arising in the area of application, along with it; the book provides scope for future developments.

Chemical biology is a fast growing inter-disciplinary field, with contributions from the fields of chemistry, biology, and physics. It involves the study and use of chemical principles to modulate biological systems. The methodology and applications presented in this book provide information to the readers about the practical aspects of chemical biology. From theories to research to practical applications, case studies related to all contemporary topics of relevance to this field have been included in this book. It aims to serve as a resource guide for students and experts alike and contribute to the growth of chemical biology.

It was a great honour to edit this book, though there were challenges, as it involved a lot of communication and networking between me and the editorial team. However, the end result was this all-inclusive book covering diverse themes in the field.

Finally, it is important to acknowledge the efforts of the contributors for their excellent chapters, through which a wide variety of issues have been addressed. I would also like to thank my colleagues for their valuable feedback during the making of this book.

**Editor**

# Revisiting the role of histo-blood group antigens in rotavirus host-cell invasion

Raphael Böhm[1], Fiona E. Fleming[2], Andrea Maggioni[1], Vi T. Dang[2], Gavan Holloway[2], Barbara S. Coulson[2,*], Mark von Itzstein[1,*] & Thomas Haselhorst[1,*]

Histo-blood group antigens (HBGAs) have been proposed as rotavirus receptors. H type-1 and Lewis[b] antigens have been reported to bind VP8* from major human rotavirus genotypes P[4], P[6] and P[8], while VP8* from a rarer P[14] rotavirus recognizes A-type HBGAs. However, the role and significance of HBGA receptors in rotavirus pathogenesis remains uncertain. Here we report that P[14] rotavirus HAL1166 and the related P[9] human rotavirus K8 bind to A-type HBGAs, although neither virus engages the HBGA-specific $\alpha$1,2-linked fucose moiety. Notably, human rotaviruses DS-1 (P[4]) and RV-3 (P[6]) also use A-type HBGAs for infection, with fucose involvement. However, human P[8] rotavirus Wa does not recognize A-type HBGAs. Furthermore, the common human rotaviruses that we have investigated do not use Lewis[b] and H type-1 antigens. Our results indicate that A-type HBGAs are receptors for human rotaviruses, although rotavirus strains vary in their ability to recognize these antigens.

[1]Institute for Glycomics, Griffith University, Gold Coast Campus, Southport, Queensland 4222, Australia. [2]Department of Microbiology and Immunology, The University of Melbourne at the Peter Doherty Institute for Infection and Immunity, Melbourne, Victoria 3000, Australia. * These authors contributed equally to this work. Correspondence and requests for materials should be addressed to T.H. (email: t.haselhorst@griffith.edu.au) or to M.v.I. (email: m.vonitzstein@griffith.edu.au).

Rotaviruses, members of the *Reoviridae* family, are the most important cause of severe diarrhoeal disease and dehydration in young children with around 500,000 deaths each year[1]. The introduction of two vaccines licensed in 2006 and 2008 has been successful in reducing disease severity and death rates[2]. However, several recent studies have revealed a significant increase in intussusception after the administration of the first or second rotavirus vaccine dose[3,4], and vaccine-associated hospitalizations for severe rotavirus gastroenteritis have been reported[5]. These risks must be weighed up against the benefits of preventing rotavirus-associated illness. Rotaviruses with the P[8] genotype of the spike protein VP4 are predominantly associated with severe disease globally, with the P[4] and P[6] genotypes being substantially less common[6]. Virus infectivity is activated by trypsin cleavage of VP4 to produce the VP5* and VP8* subunits. Although both subunits remain associated on the virion surface, initial cell attachment to host cell membrane glycoconjugates is mediated by the VP8* head of the spike protein[7,8]. Gangliosides are glycosphingolipids with one or more sialic acid residues that are key receptor molecules for rotavirus infection[9–14]. Rotaviruses have been classified as sialidase-sensitive or sialidase-insensitive, based on the effect on their infectivity of host cell pre-treatment with sialidase. However, commonly used bacterial and viral sialidases remove only terminal sialic acids from sialylglycoconjugates. Sialic acid recognition by sialidase-sensitive animal rotaviruses via VP8* is structurally well characterized through a series of crystallographic studies[13,15–21]. Our recent studies using nuclear magnetic resonance (NMR) spectroscopy and cell-based assays identified GM1 as a ganglioside receptor for the human rotavirus strains Wa (P[8]) and RV-3 (P[6]), and revealed that the internal *N*-acetylneuraminic acid (sialic acid) residue of GM1 plays a key role in VP8* recognition[12,22]. We concluded that human sialidase-insensitive strains such as Wa and RV-3, and most probably other human strains, require the presence of sialic acid-containing receptors for efficient infection[12,22].

A new paradigm of human P[14] rotavirus binding to A-type HBGA has been recently suggested, related to decreased infectivity of HAL1166 in HT-29 cells following anti-A-type antibody treatment and increased infectivity in CHO cells expressing A-type HBGA (ref. 23). The X-ray crystal structure of HAL1166 VP8* revealed a typical galectin-like fold with the twisted β-sheets separated by a shallow cleft as observed in all available VP8* structures, including those of the sialidase-insensitive human rotaviruses Wa and DS-1 (P[4]), and the sialidase-sensitive animal strains RRV and CRW-8 VP8* (refs 16,23). Notably, the width of the binding site in HAL1166 VP8* appears narrower than in Wa and DS-1, resembling the binding site of RRV and CRW-8. Interestingly, sialic acids bind the VP8* of such animal rotaviruses at the same location within the narrow cleft as HBGA A-type trisaccharide (A-tri) and tetrasaccharide (A-tetra) binds to HAL1166 VP8*, although the key amino acids that interact with sialic acids and HBGA are different. Based on VP8* sequence alignments and rotavirus infectivity assays, it was predicted that the human P[9] rotavirus K8 could also bind to A-type HBGA, although no structural evidence has been reported. The VP8* of P[9], P[14] and P[25] genotype rotaviruses within P genogroup III bind A-type HBGAs in haemagglutination inhibition and synthetic oligosaccharide-based assays[24]. In addition, binding assays with P genogroup II rotaviruses of P genotypes [4], [6] and [8] showed interaction of H type-1 and Lewis[b] antigens with the VP8* of P[4] and P[8] human rotaviruses, while P[6] rotavirus VP8* bound only to H type-1 antigen[25,26].

In the current study, we sought to determine the importance of HBGA receptors for several major human rotavirus genotypes.

Our results demonstrate that A-type HBGAs are important receptors for human rotaviruses, although major variations between virus strains were detected. Using a multidisciplinary approach, we show that P[14] rotavirus HAL1166 and the related P[9] human rotavirus K8 bind A-type HBGAs without engaging the HBGA-specific α1,2-linked fucose moiety. Importantly, human rotaviruses DS-1 (P[4]) and RV-3 (P[6]) also use A-type HBGAs for infection, with fucose involvement. In contrast, the human P[8] rotavirus Wa does not recognize A-type HBGAs and no rotavirus strain we tested uses H-type-1 and Lewis[b] antigens.

## Results

**Binding of HBGAs to rotavirus VP8*.** The evidence supporting the notion that HBGAs are common cell receptors for human rotaviruses remains incomplete and unclear. To address this issue, we report here an NMR solution study of the interactions between A-tri and A-tetra (Fig. 1a), and the VP8* carbohydrate-recognizing domain from human rotaviruses HAL1166 (P[14], III]), K8 (P[9, III]), Wa (P[8, II]), RV-3 (P[6, II]) and DS-1 (P[4, II]). The $^1$H NMR spectrum of A-tri under our experimental conditions (Fig. 1b) revealed existence of α- and β-pyranose (six-membered ring) and -furanose (five-membered ring) forms in aqueous solution that are populated as follows (Fig. 1c): 52% GalNAcα1,3[Fucα1,2]Gal$p$α (A-tri$^{p\alpha}$), 24% GalNAcα1,3[Fucα1,2]Gal$p$β (A-tri$^{p\beta}$), 18% GalNAcα1,3[Fucα1,2]Gal$f$α(A-tri$^{f\alpha}$) and 6% GalNAcα1,3[Fucα1,2]Gal$f$β (A-tri$^{f\beta}$), consistent with a previous report[27].

We have used Saturation Transfer Difference (STD) NMR spectroscopy to study the interaction of rotavirus VP8* protein and A-type HBGAs. This NMR method is based on the transfer of saturation from the protein to bound ligands that are in fast exchange with a binding site, and therefore allows the discrimination of binding and non-binding ligands. The degree of saturation a ligand proton receives from the target protein can be directly translated into a ligand-binding epitope. Ligand protons that are in close contact to the protein surface receive larger saturation and will therefore produce larger STD NMR effects compared to protons that are more solvent exposed[28,29].

We obtained the $^1$H NMR spectrum of A-tri (Fig. 2a) and the control STD NMR spectra of A-tri with glutathione-*S*-transferase (GST) (Fig. 2b) and A-tri in complex with recombinantly expressed GST-tagged VP8* from human rotavirus strains Wa, K8 and HAL1166. The STD NMR spectrum of GST-Wa VP8* in complex with A-tri showed no recognition of A-tri (Fig. 2c). However, the STD NMR spectra of GST-K8 VP8* (Fig. 2d) and GST-HAL1166 VP8* (Fig. 2e) in complex with A-tri unequivocally show that both proteins bind to A-tri. The control STD NMR spectrum of the GST fusion tag alone with A-tri shows no significant signal intensities, confirming that A-tri does not interact with GST. It is worth noting that this important control STD NMR spectrum demonstrates that all observed signals in the spectra of HAL1166 and K8 VP8* originate from binding of A-tri to VP8*. The strongest interaction of A-tri was observed for the methyl protons of the GalNAc *N*-acetamido group (NHAc, $\delta = 1.73$ p.p.m.) for both Hal1166 and K8 VP8* proteins. No STD NMR signal was observed for the H-disaccharide (Fucα1,2Galβ), emphasizing the significance of the GalNAc moiety for A-tri binding to HAL1166 (Supplementary Fig. 1). It is most striking that STD NMR signals were also observed for the GalNAc H2 proton of two A-tri species: A-tri$^{p\alpha}$ and A-tri$^{p\beta}$. In addition, the H4 proton of the galactose moiety also received saturation for A-tri$^{p\alpha}$ and A-tri$^{p\beta}$. However, the X-ray crystal structure of A-tri with HAL1166 VP8* revealed that only the pyranose α-anomer, a biologically irrelevant configuration, binds to the protein[23]. In HBGA precursors, the glycosidic linkage between Gal and

**a**

GalNAcα1-3[Fucα1-2]Gal,
blood group A-Triaose (A-tri)

GalNAcα1-3[Fucα1-2]Galb(1,4)GlcNAc,
blood group A-Tetraose type II (A-tetra)

**b**

H1-GalNAc(A-Tri$^{pα}$)

H1-Gal(A-Tri$^{pα}$)

H1-Fuc(A-Tri$^{pα}$)

H1-GalNAc(A-Tri$^{fα}$)

H1-Gal(A-Tri$^{fα}$)        H1-Fuc(A-Tri$^{pβ}$)

H1-Fuc(A-Tri$^{fα}$)

5.5        5.3        5.1        p.p.m.

**c**

GalNAcα1-3[Fucα1-2]Galpα
(A-tri$^{pα}$, 52%)

GalNAcα1-3[Fucα1-2]Galpβ
(A-tri$^{pβ}$, 24%)

GalNAcα1-3[Fucα1-2]Galfα
(A-tri$^{fα}$, 18%)

GalNAcα1-3[Fucα1-2]Galfβ
(A-tri$^{fβ}$, 6%)

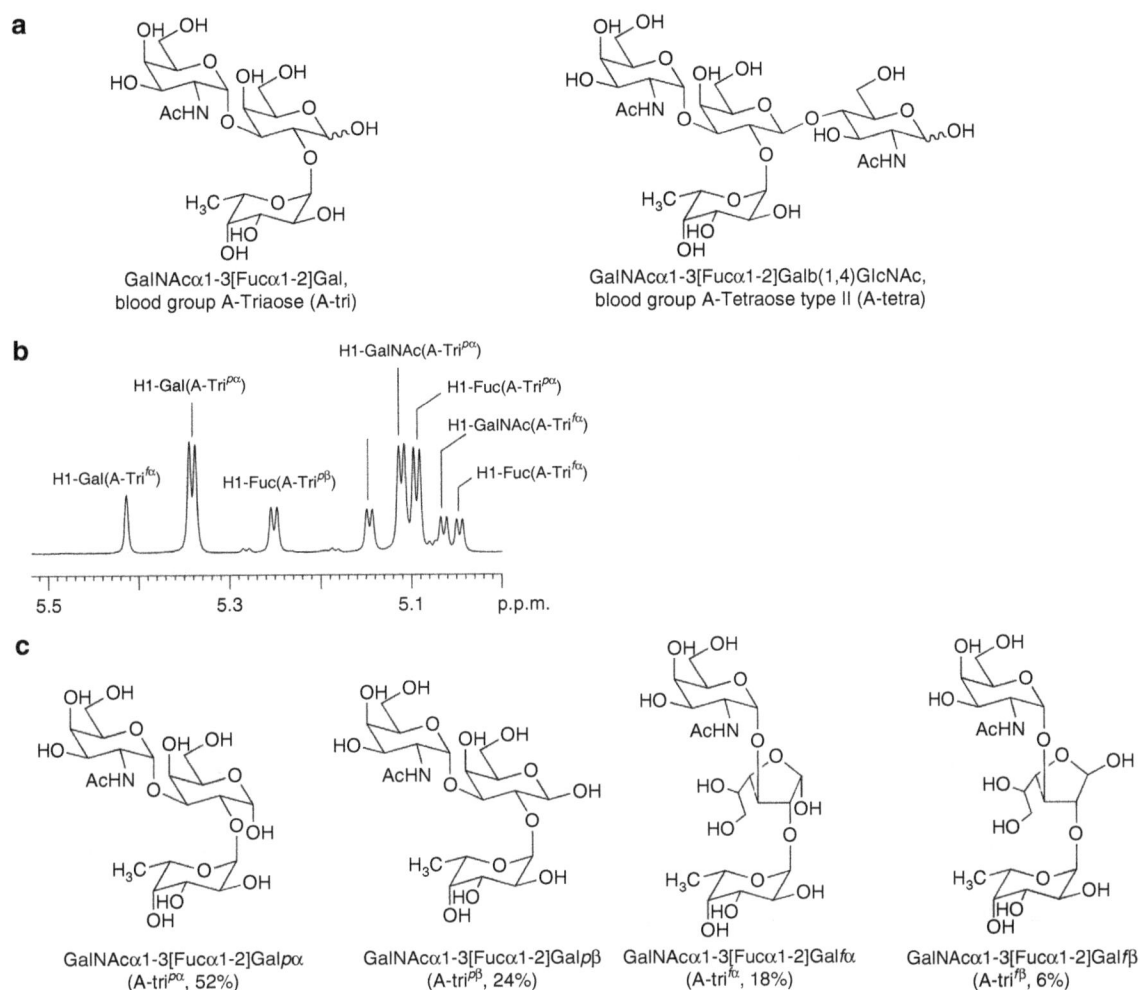

**Figure 1 | Structures and NMR spectra of A-tri and A-tetra glycans.** Structures of A-type HBGAs trisaccharide (A-tri, GalNAcα1,3[Fucα1,2]Gal) and blood group A type-2 tetrasaccharide (A-tetra, GalNAcα1,3[Fucα1,2]Galβ1,4GlcNAc) (**a**). [1]H NMR spectrum shows the anomeric protons of A-tri at 305 K (**b**). Structures of the different A-tri forms and their distribution in solution at 305 K (**c**).

GlcNAc is always a β-configuration. Our solution binding study demonstrates that A-tri$^{pα}$ and A-tri$^{pβ}$ are able to bind to K8 and HAL1166 VP8*, and that the binding epitope of A-tri for both proteins is identical, suggesting the same binding mode of A-tri, irrespective of the reducing end (Supplementary Fig. 2a,b). A homology model of K8 VP8* was generated using the HAL1166 VP8* X-ray crystal structure (pdb accession code: 4DRV[23]) (Supplementary Fig. 3) with a sequence identity of 83.9% and sequence similarity of 90.7%. The binding-site residues show minor variations supporting the identical binding epitope of A-tri determined by STD NMR experiments.

We have also analysed interactions of A-tri with VP8* proteins from P[6] (RV-3) and P[4] (DS-1) human rotaviruses. A-tri binding affinity to RV-3 and DS-1 was identified (Fig. 3b,c, respectively), with a strong STD NMR signal observable for the N-acetamido group of the GalNAc residue. Strikingly, the methyl group of the fucose moiety (CH$_3$ Fuc, ~1 p.p.m.) shows the most predominant STD NMR signal intensity, leading to the conclusion that the fucose moiety of A-tri does interact with the VP8* protein of DS-1 and RV-3. Our NMR study provides the first detection of a direct interaction of the HBGA fucose moiety with any rotavirus VP8* protein. However, absolute STD NMR signal intensities were substantially lower compared with the spectra acquired for A-tri in complex with HAL1166 and K8. To validate protein activity, an equimolar concentration of GM1a glycan, a

confirmed receptor for the DS-1-like P[4] rotavirus strain RV-5 and for RV-3 (ref. 22) was added to the A-tri complex of VP8* protein from DS-1 or RV-3.

Quantitative analysis of the HAL1166 VP8*-A-tri STD NMR spectra revealed that the fucose residue of A-tri does not receive any saturation via the protein. This solution-based observation is in excellent agreement with the reported X-ray crystal structure[23], where the fucose moiety does not make intimate contact with the protein. However, this finding raises the question of how important HBGAs are as receptors for P[III] rotavirus infection, as the fucose residue is the carbohydrate determinant that defines the ABH and Lewis antigens[30]. An interesting and relevant comparison is with another clinically significant human gastrointestinal virus, norovirus. All major genogroups of human norovirus have a clear dependence on HBGAs for infectivity and exhibit a relatively broad binding specificity for A, B, H or Lewis antigens[31,32]. Furthermore, a recent study has led to the proposal that HBGA-like substances associated with human enteric bacteria bind human norovirus and may have an impact on transmission and infection of this virus[33]. Interestingly, crystallography studies revealed that norovirus predominantly interacts with the fucose moiety of the blood group antigens[34-37]. In fact, the α1,3- and α1,4-fucose residues of Lewis HBGAs are claimed to be essential for binding to GII.4 noroviruses[36,37]. STD NMR studies on GII.4 virus-like particles derived from norovirus

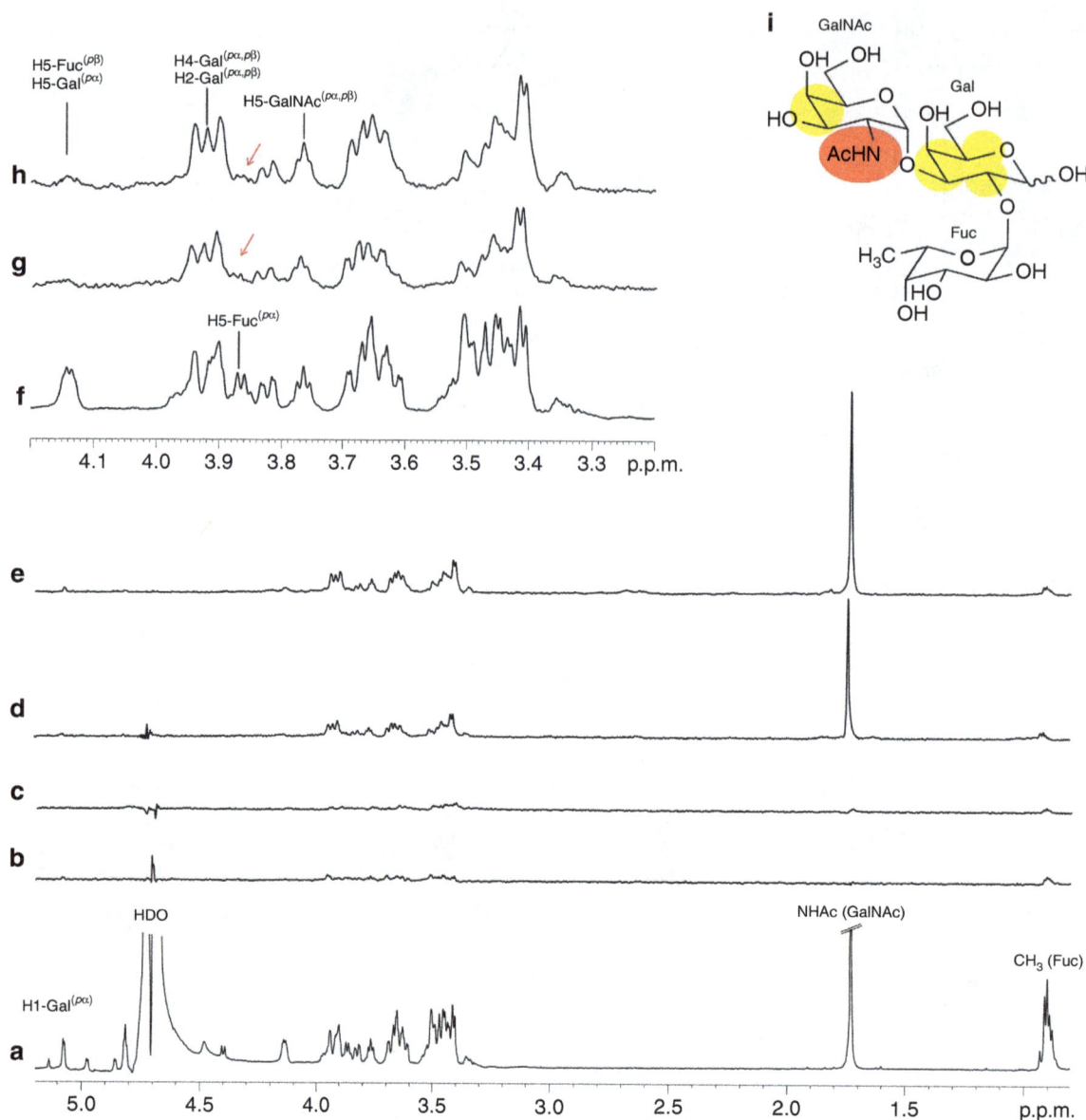

**Figure 2 | NMR spectra of A-tri complexed with K8, HAL1166 and Wa VP8\*.** [1]H NMR (**a**) and STD NMR (**b–f**) of human VP8\* with A-type HBGA trisaccharide (A-tri, GalNAcα1,3[Fucα1,2]Gal) in 70 mM phosphate buffer pH 7.1 and 50 mM NaCl at 280 K. In detail, STD NMR spectra are shown for 24 μM GST and 2.4 mM A-tri (**b**), 18 μM GST-Wa VP8\* and 1.8 mM A-tri (**c**), 36 μM GST-K8 VP8\* and 2.4 mM A-tri (**d**), 24 μM GST-HAL1166 VP8\* and 2.4 mM A-tri (**e**). Expansion of the spectrum from 4.2 to 3.2 p.p.m. is shown for the [1]H NMR (**f**) and STD NMR spectra of 36 μM GST-K8 VP8\* and 2.4 mM A-tri (**g**) and 24 μM GST-HAL1166 VP8\* and 2.4 mM A-tri (**h**). Epitope map of A-tri when bound to K8 and HAL1166 VP8\*: red, strong STD NMR effects; yellow weak STD NMR effects (**i**).

Ast6139 underlined the broad HBGA-binding profile of GII.4 noroviruses. A, B, H and Lewis epitopes, with and without a sialic acid, are recognized by norovirus-like particles, leading to the conclusion that the minimal recognition unit for norovirus was α-L-fucose[38]. Our study herein shows that the fucose moiety is irrelevant for blood group antigen binding to the investigated P[III] rotavirus spike proteins. Clearly, if P[III] rotavirus infection depended on the A, B, H and Lewis blood groups, an engagement of the key sugar residue, fucose, with the viral protein would be anticipated. Although our study indicates that fucose plays no direct role in the recognition event, it does not rule out that it may still play an indirect role in stabilizing the blood group antigen through non-conventional hydrogen bonds as has recently been described for Lewis[x] (ref. 39).

Our study also demonstrates a lack of A-tri recognition by VP8\* from the clinically relevant human P[8, II] rotavirus Wa. We identified that GM1a glycan bound as expected to the A-tri complex of Wa VP8\* and the Wa VP8\* competed with Wa virus for infectivity, confirming protein function. Moreover, using STD NMR spectroscopy no recognition of H type-1 and Lewis[b] antigens by the P[8] Wa VP8\* was observed (Supplementary Fig. 4). In contrast, a recent study based on assays for VP8\* binding to human saliva and synthetic oligosaccharides led to the conclusion that H type-1 and Lewis[b] antigens can interact with Wa VP8\* (ref. 25). Notably, the Wa strain represents an important current and past lineage of the most common human rotavirus serotype worldwide[40] but shows only 43% VP8\* sequence identity with HAL1166, and most of the amino acids in the glycan-binding site differ substantially (Supplementary Fig. 5). Importantly, our present study demonstrates for the first time that VP8\* of DS-1 and RV-3 rotaviruses recognize A-tri. After P[8,II], these viruses represent the next most common genotypes P[4,II] and P[6,II],

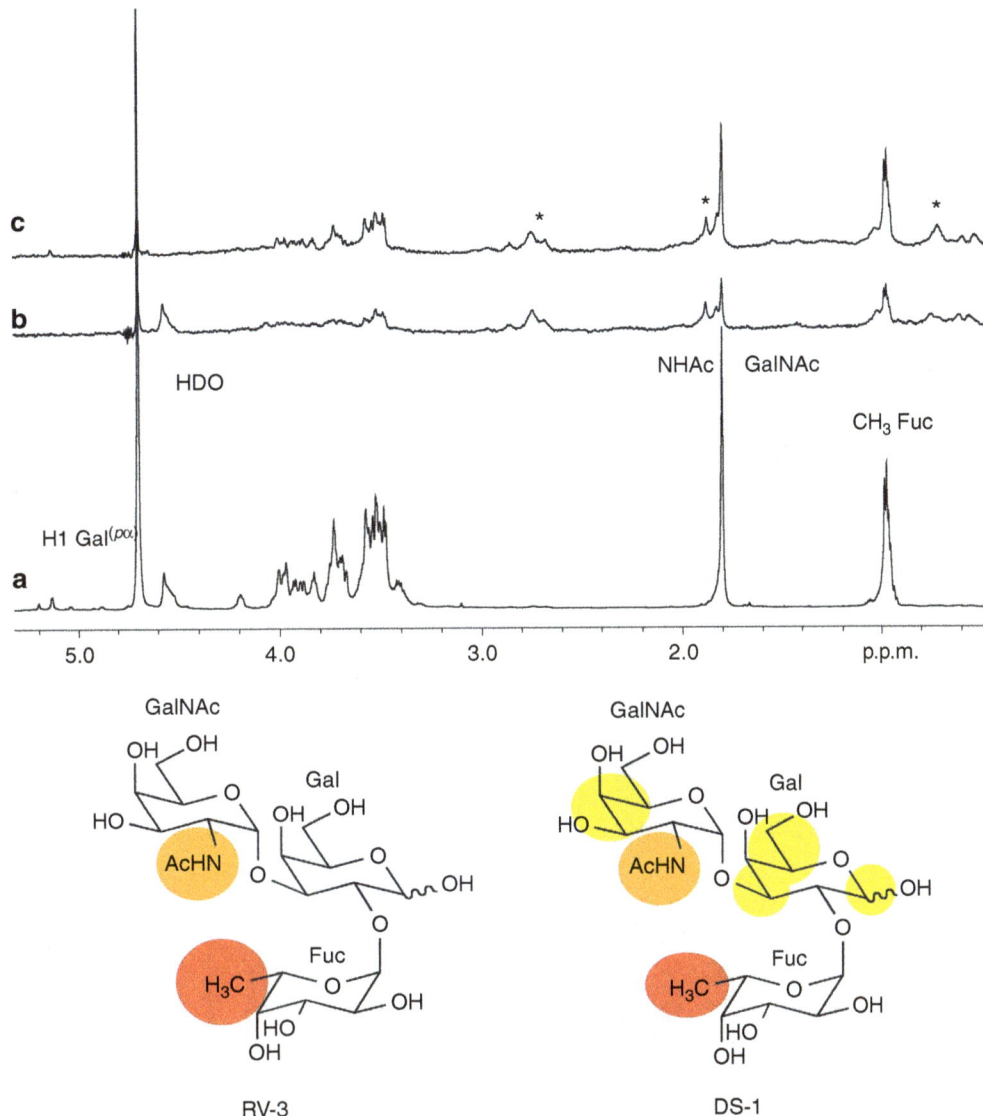

**Figure 3 | NMR spectra of A-tri complexed with RV-3 and DS-1 VP8\*.** [1]H NMR (**a**) and STD NMR (**b,c**) of human VP8\* with A-type HBGA trisaccharide (A-tri, GalNAcα1,3[Fucα1,2]Gal) in 70 mM phosphate buffer pH 7.1 and 50 mM NaCl at 280 K. In detail, STD NMR spectra are shown for 24 μM GST-RV-3 and 2.4 mM A-tri (**b**), and 24 μM GST-DS-1 VP8\* and 2.4 mM A-tri (**c**). Epitope map of A-tri when bound to RV-3 and DS-1 VP8\* are shown in the lower panel VP8\*: red, strong STD NMR effects; orange, medium STD NMR effects; yellow, weak STD NMR effects (\* = protein background signals).

and show relatively low VP8\* sequence identity to HAL1166 (44 and 47%, respectively). Significantly, the A-tri engagement with VP8\* from both DS-1 and RV-3 involves the fucose moiety. This fucose recognition is consistent with a recent study in children from Burkina Faso, which showed an association of P[6] rotavirus disease predominantly with a Lewis-negative phenotype, independent of secretor status[41].

**A-tetra and A-tri have an identical binding epitope for K8 and HAL1166 VP8\*.** We have also used STD NMR spectroscopy to investigate the VP8\* interaction of the biologically relevant A type-2 HBGAs tetrasaccharide (A-tetra), which has an additional GlcNAc residue (Fig. 4a–g). The initial [1]H NMR spectrum of A-tetra shows no distinct forms as in the case of A-tri (Fig. 4a). A control STD NMR experiment of the GST fusion tag in the presence of A-tetra shows no significant STD NMR signals (Fig. 4b), confirming that all STD NMR signals for A-tetra in complex with GST-K8 VP8\* (Fig. 4c,f) and GST-HAL1166 VP8\* (Fig 4d,g) are genuine and are derived from the binding of A-tetra to VP8\* and not GST. Similar to A-tri recognition, the

$N$-acetamido group's methyl protons of the crucial A-tetra GalNAc moiety (NHAc, $\delta = 1.72$ p.p.m.) received the highest saturation, implying a very close contact with the protein surface. However, our STD NMR studies demonstrated that the additional GlcNAc residue $N$-acetamido group methyl protons do not engage with either HAL1166 VP8\* (Fig. 4g) or K8 (Fig. 4f). Furthermore, the GalNAc, and Gal H2 and H3 protons also show strong STD NMR signals, whereas the ring protons of the GlcNAc residue do not produce significant STD NMR signals. The fucose residue is entirely solvent exposed and does not interact with the protein. Therefore, we conclude that the binding epitope of A-tetra is identical to that of A-tri when complexed with HAL1166 VP8\*. Finally, from this study we have discovered that the binding epitopes of A-tri and A-tetra are also identical when bound to K8 VP8\*, with only very small differences observed in STD NMR signal intensities (Supplementary Fig. 2).

**A-tri adopts a different conformation when bound to K8 VP8\*.** The conformation of glycans when bound to proteins (bioactive

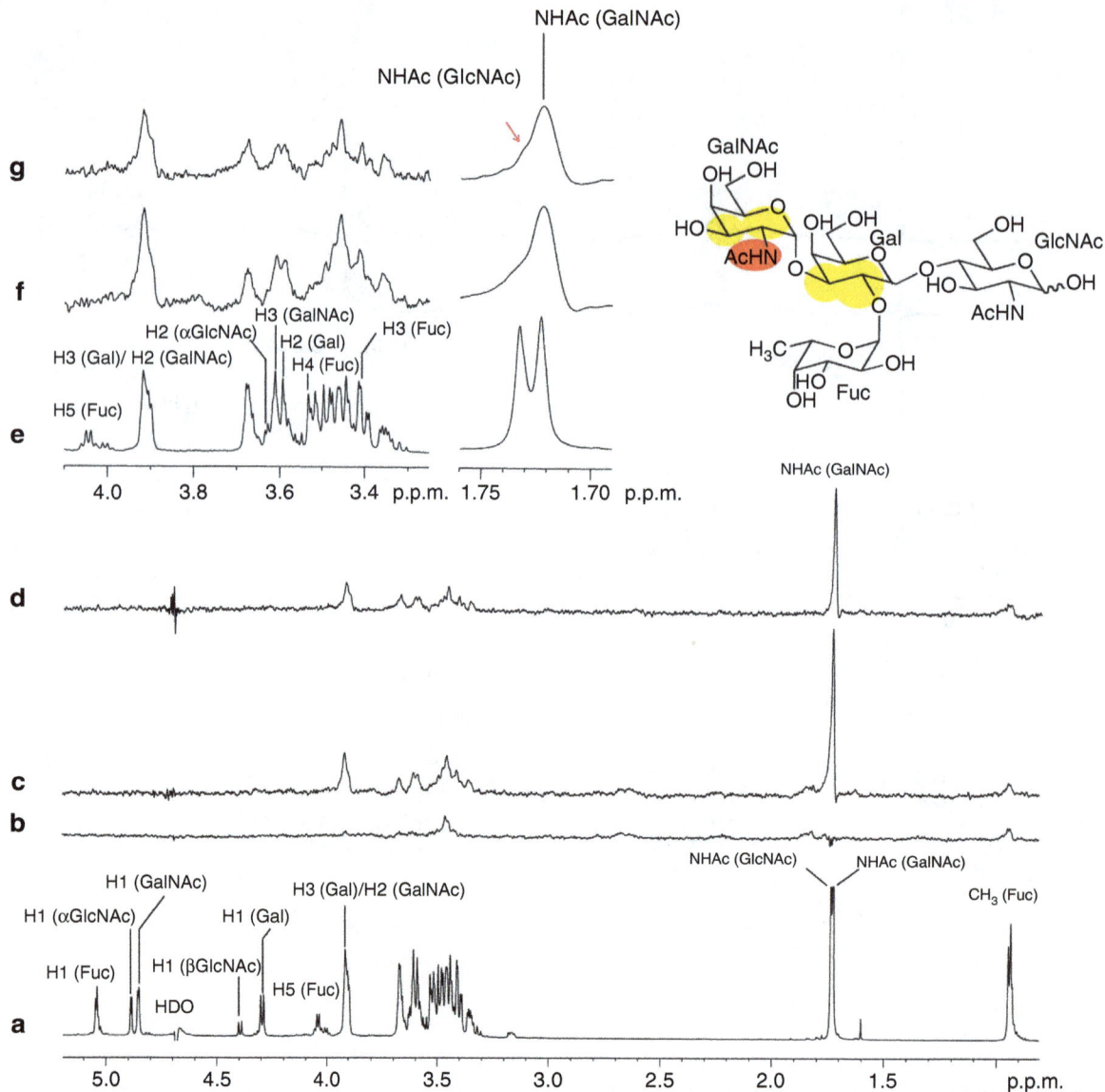

**Figure 4 | NMR spectra of A-tetra complexed with K8 and HAL1166 VP8\*.** [1]H NMR (**a**) and STD NMR (**b–d**) of human VP8\* with A HBGA type-2 tetrasaccharide (A-tetra, GalNAcα1,3[Fucα1,2]Galβ1,4GlcNAc) in 70 mM phosphate buffer pH 7.1 and 50 mM NaCl at 280 K. In detail, STD NMR spectra are shown for 24 μM GST and 2.4 mM A-tetra (**b**), 24 μM GST-K8 VP8\* and 2.4 mM A-tetra (**c**), 24 μM GST-HAL1166 VP8\* and 2.4 mM A-tetra (**d**). Expansion of the spectrum from 4.1 to 3.3 p.p.m. is shown for the [1]H NMR (**e**) and STD NMR of 24 μM GST-K8 VP8\* and 2.4 mM A-tetra (**f**), and 24 μM GST-HAL1166 VP8\* and 2.4 mM A-tetra (**g**). Epitope map of A-tetra when bound to K8 and HAL1166 VP8\*: red, strong STD NMR effects; yellow, weak STD NMR effects (**h**).

conformation) can significantly vary in solution compared with a single conformation obtained with X-ray crystallography. To determine the bioactive conformation of A-tri when bound to VP8\* in solution, we acquired Nuclear Overhauser Enhancement spectroscopy (NOE) NMR spectra of free A-tri in solution and transferred NOE (trNOE) NMR experiments of A-tri in complex with K8 VP8\* following a similar approach to that previously reported[11]. The NOE NMR spectrum shows that A-tri has positive NOEs (diagonal and cross-signals of opposite sign) at a temperature of 305 K (Fig. 5a). On the contrary, the trNOE experiment of a VP8\* K8:A-tri-complex (Fig. 5b) at a 1:10 (protein:ligand) molar ratio clearly indicate that the NOEs are negative, leading to the assumption that A-tri has adopted the rotational correlation time ($\tau_c$) of the VP8\* due to association with the protein. Conversely, the trNOE experiment of A-tri in complex with Wa VP8\* does not result in a change of the sign of NOE signals, indicating that A-tri has no binding affinity for Wa

VP8\* (Fig. 5c). The observed reduced intensity of the NOE signal (Fig. 5c) can be explained by a substantially lower ligand concentration in the trNOE compared with the NOE experiment of the free ligand in solution. A more detailed comparison between the NOE and trNOE spectra indicates significant differences in the NOE pattern. Most striking is the disappearance of the interglycosidic NOE between the fucose methyl group, the CH₃ (Fuc) residue and the H1 (Gal) proton of bound A-tri. This particular NOE shows a strong signal in the uncomplexed NOE spectrum (Fig. 5a), indicating that the bioactive conformation of A-tri must be different to the predominant solution conformation adopted in the absence of the protein. The disappearance of a crucial NOE can be explained by a conformational rearrangement of the ligand on binding to the protein, most probably around the α1,2-glycosidic linkage between the fucose and galactose residue. This is in excellent agreement with the binding epitope obtained from STD NMR

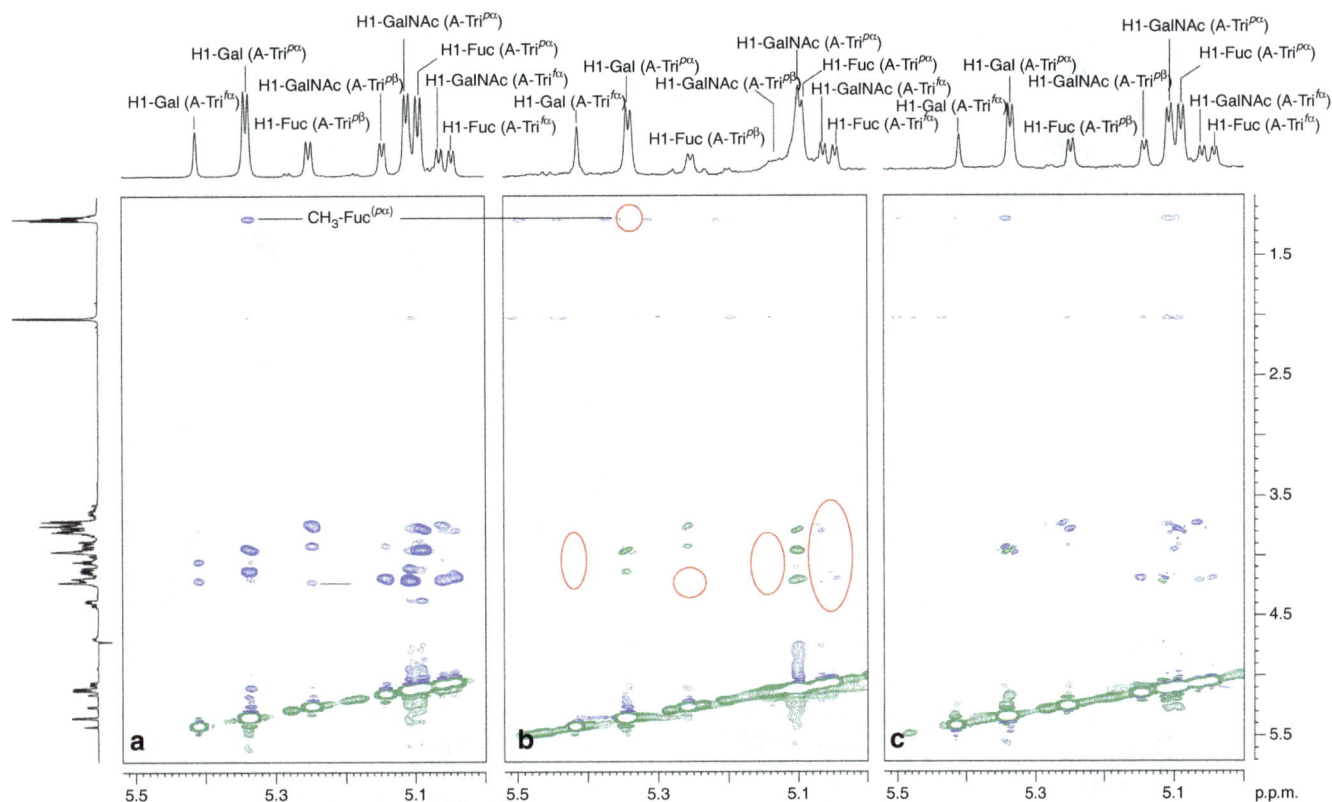

**Figure 5 | NOE of A-tri and trNOE spectra of A-tri in complex with K8 VP8\*.** NOE of A-type HBGA trisaccharide (A-tri, GalNAcα1,3[Fucα1,2]Gal) in aqueous solution at a mixing time $\tau_m$ of 800 ms (**a**). Transferred NOE (trNOE) of A-tri in a complex with rotavirus K8 VP8\* (**b**) and Wa VP8\* (**c**), respectively, at a ratio of 1:12 (protein:ligand) and a mixing time $\tau_m$ of 150 ms. NOE spectrum (**b**) indicates binding due to strong negative NOE's, whereas spectrum (**c**) indicates no binding due to the occurrence of small positive NOEs.

experiments, revealing that the fucose is not engaged with the protein and remains solvent exposed, resulting in a higher degree of flexibility.

**Molecular dynamics simulation reveals flexibility of the fucose residue.** To further confirm the flexibility of the fucose residue, molecular dynamics (MD) simulation of A-tri$^{p\alpha}$ in complex with HAL1166 (Supplementary Fig 6) was conducted based on the X-ray crystal structure (pdb 4DRV)[23]. Our MD simulation reveals that the α1,2-glycosidic linkage between the fucose and galactose residue is highly flexible. Furthermore, our simulation indicates that this linkage can easily adopt a conformation where the intermolecular distance between the H1 (Gal) and CH$_3$ (Fuc) is >4.0 Å. A-tri conformations (63.4%) before reaching energy equilibrium (∼1 ns) show a shorter distance between H1 (Gal) and CH$_3$ (Fuc) (1.8–4.0 Å). However, after reaching equilibrium the MD simulation revealed that only 9.4% of all A-tri conformations indicate a shorter distance (1.8–4.0 Å) and the majority (90.6%) of all conformations reveal a larger distance between the H1 Gal and CH$_3$ Fuc (>4.0 Å) that would not result in the detection of an NOE. This distance calculation is in excellent agreement with the trNOE experimental outcomes that show a complete disappearance of this important NOE. The VP8\* engagement of A-tri$^{p\alpha}$ is mainly through the GalNAc moiety and the non-binding fucose residue appears to have a high degree of rotational freedom. Another important observation is that negative NOEs, clear evidence of binding, could not be detected for A-tri with the galactose adopting a furanose form (A-tri$^{f\alpha}$); hence, it has no binding affinity. This result is not surprising as hexofuranosyl residues are absent in mammalian glycoconjugates

and only found in oligo- and polysaccharides of bacteria, protozoa, fungi, plants and archaebacteria[42]. This finding is further supported by STD NMR experimental analysis that shows VP8\* can only accommodate A-tri when the internal galactose residue adopts a biologically relevant pyranose form. Interestingly, the observed severe line broadening for the H1 (GalNAc) signals for both pyranose α- and β-anomers (A-tri$^{p\alpha}$ and A-tri$^{p\beta}$) suggests close protein interaction and is in excellent agreement with the determined A-tri glycointeractome by STD NMR spectroscopy (see Fig. 2).

**Specificity and affinity of A-type HBGA binding to VP8\*.** To quantify the affinity of A-tri and A-tetra when bound to rotavirus VP8\* protein, we have determined $K_D$-values using $^1$H–$^{15}$N heteronuclear single quantum correlation (HSQC) chemical shift perturbation NMR experiments[43,44]. Upon addition of the ligand to uniformly $^{15}$N-labelled VP8\* protein, changes in $^1$H and $^{15}$N chemical shifts can only be detected for those backbone amides that are in close contact with the ligand. A $K_D$-value of 1.2 mM has been previously reported using the same approach for the sialidase-sensitive rhesus rotavirus (RRV) VP8\* protein and 3'-sialyllactose as ligand[14]. In the current study, we have determined the $K_D$-values for A-tri and A-tetra following the glycan induced chemical shift perturbations of the backbone amides of K8 and HAL1166 VP8\* for ten residues that show the strongest changes (Fig. 6). A complete $^1$H–$^{15}$N HSQC spectral analysis of these perturbations was undertaken (Supplementary Figs 7–9). We found that the affinity of A-tri is significantly stronger for HAL1166 VP8\* ($K_D$ 1.5 mM) compared to K8 VP8\* ($K_D$ 3.4 mM). Interestingly, the blood group-A tetrasaccharide (A-tetra) showed

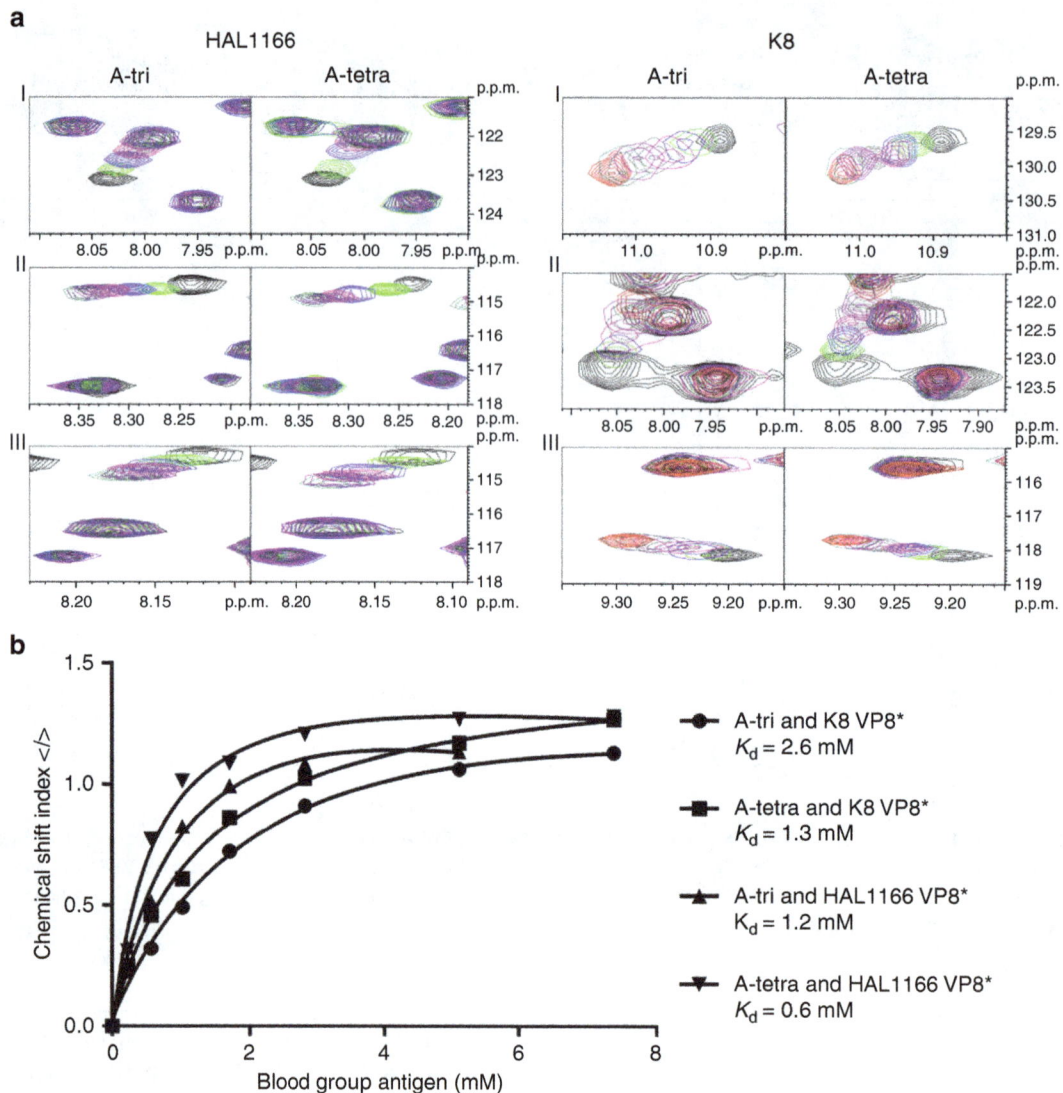

**Figure 6 | $^{15}$N HSQC NMR spectra of A-tri in complex with HAL1166 and K8 VP8\*.** $^{1}$H–$^{15}$N HSQC NMR titration of A-type HBGA trisaccharide (A-tri) and A- type-2 HBGA tetrasaccharide (A-tetra) binding to 0.4 mM HAL1166 and 0.4 mM K8 VP8\*. Overlayed spectra of all titration steps are shown for the three strongest amide shifts (I, II, III; **a**). Dissociation constants ($K_d$) were obtained by nonlinear regression based on a single binding site model fitting the chemical shift index $</>$ as an average of the ten strongest amide shifts (**b**).

lower $K_D$ values when bound to HAL1166 and K8 VP8\*, of 0.8 and 1.7 mM, respectively (Fig. 6). This can be explained by the fact that A-tetra does not adopt α,β-anomeric mixtures of both the pyranose and furanose forms in solution as does A-tri where only the α- and β-pyranose forms of Gal showed binding affinity to VP8\* and not the α- and β-furanose forms. Nevertheless, A-tetra also shows a higher affinity for HAL1166 VP8\* over K8 VP8\*. No changes in the total number of chemical shift perturbations were identified for protein complexes with A-tetra compared to A-tri. This confirms the STD NMR outcomes that the additional GlcNAc residue does *not* interact with the protein and that the binding epitopes of A-tri and A-tetra are identical (Supplementary Figs 8 & 9). No significant chemical shift perturbations were observed for human rotavirus Wa VP8\* in the presence of A-tri (58 mM), providing further evidence that A type HBGAs do *not* bind to Wa VP8\* (Supplementary Fig. 10). Interestingly, addition of Lewis[b] even at a very high concentration (19.7 mM), to $^{15}$N-labelled Wa VP8\* also did not result in *any* chemical shift perturbations. These results are in agreement with our STD NMR results and indicate that Lewis[b] does not bind to VP8\*.

**A-Tri blocks cell binding by VP8\* of HAL1166 but not Wa.** To corroborate our NMR and molecular modeling results, we have investigated if A-tri blocks VP8\* binding to highly rotavirus-permissive MA104 cells (Fig. 7a, Supplementary Fig. 11). Using flow cytometry, we have demonstrated significant levels of cell binding by VP8\* of HAL1166 ($P = 0.0005$), K8 ($P = 0.0060$) and Wa ($t$-test, $P < 0.0001$) over the background (control) binding. Treatment with 2 mM and 10 mM A-tri reduced the specific cell binding of HAL1166 VP8\* by 53% ($t$-test, $P = 0.0074$) and 69% ($t$-test, $P = 0.021$), respectively, but had no effect on specific cell binding by K8 VP8\* ($t$-test, $P > 0.22$) or Wa VP8\* ($t$-test, $P > 0.05$). The increased level of HAL1166 VP8\* cell binding compared with K8 VP8\*, and the dose-dependent A-tri blockade of HAL1166 but not K8 VP8\* binding, can be explained by the higher $K_D$-value of A-tri for K8 VP8\* than HAL1166 VP8\* (Fig. 6).

**A-tri inhibits infection by K8, HAL1166, RV-3, ST-3 and DS-1 but not Wa.** The effect of A-tri treatment on MA104 cell infection by a range of human rotaviruses was determined. All of these

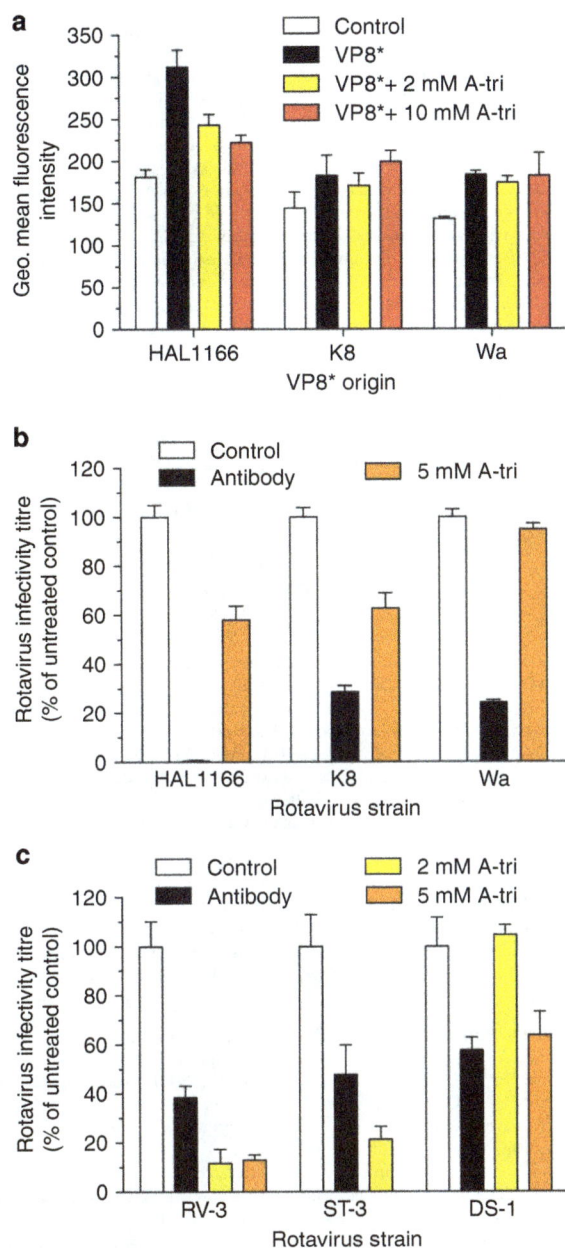

**Figure 7 | A-tri competition with cell binding by rotavirus VP8\* and rotavirus infection.** Effect of A-type HBGA trisaccharide (A-tri, GalNAcα1,3[Fucα1,2]Gal) on MA104 cell binding by VP8\* of rotaviruses HAL1166, K8 and Wa (**a**), the infectivity of these rotaviruses (**b**) and the infectivity of P[6] rotaviruses RV-3 and ST-3, and P[4] rotavirus DS-1 (**c**). The effect of rotavirus treatment with 2 mM or 5 mM A-tri on infectivity was determined. Neutralizing antibodies (Antibody) used as positive controls were B37:1 (HAL1166), RV-4:3 (K8), 1A10 (Wa), RV-3:1 (RV-3), ST-3:3 (ST-3) and RV-5:2 (DS-1). Three replicates were analysed for each condition. The data shown are representative of the findings of two independent experiments. Bars indicate the s.d.

not alter Wa infectivity (t-test, $P = 0.09$; Fig. 7b), in excellent agreement with our NMR results. Overall, we have observed that K8 rotavirus is less dependent on A-type antigen for VP8\*-cell binding than HAL1166, and shows a similar dependence on this antigen to HAL1166 for infectivity. In contrast, Wa rotavirus does not recognize this antigen. Consistent with the A-tri recognition of the VP8\* of DS-1 and RV-3 detected by STD NMR (Fig. 3), A-tri at 5 mM inhibited the infectivity of DS-1 and RV-3 rotaviruses, by 36% (t-test, $P = 0.001$) and 87% (t-test, $P < 0.0001$), respectively (Fig. 7c). Infectivity blockade was still evident at 2 mM A-tri for RV-3 (88%; t-test, $P < 0.0001$) and another P[6] human rotavirus ST-3 (79%; t-test, $P < 0.0001$), but not DS-1 (Fig. 7c).

**H type-1 and Lewis[b] HBGA do not inhibit P[II]/[III] viruses.** We could not detect any blockade of HAL1166, K8 or Wa rotavirus infectivity by H type-1 and Lewis[b] antigen at a 10 mM concentration (Supplementary Fig. 12). This is supported by both STD NMR and $^{1}H$–$^{15}N$ HSQC chemical shift perturbation NMR experiments, which revealed no H type-1 and Lewis[b] antigen recognition by HAL1166, K8 or Wa rotavirus VP8\* (Supplementary Figs 1, 4 and 10). Similarly, 10 mM H type-1 or Lewis[b] antigen did not affect the infectivity of RV-3, ST-3 or DS-1 (Supplementary Fig. 12).

**Discussion**

From the presented data determined by our multidisciplinary receptor studies, using recombinant VP8\* and whole viruses, we conclude that HBGAs play a role in human rotavirus infection of host cells but substantial virus strain-specific variations in HBGA usage are evident. The A-type HBGAs show some interaction with the P[III] HAL1166 strain but may be less important for P[III] K8 rotavirus. For these rotaviruses, the histo-blood group-specific fucose moiety did not interact with VP8\*. Furthermore, other related fucosylated glycans such as Lewis[b] and H type-1 antigen neither bind to HAL1166 and K8 VP8\* nor facilitate infection by these viruses. We have also demonstrated that A-type HBGAs, Lewis[b] antigen and H type-1 antigen neither bind to Wa VP8\* nor facilitate infection by Wa, which is a representative of an important current and past lineage within the most common disease-causing P[II] human rotaviruses[40]. It is a reasonable assumption that a rotavirus strain such as Wa, representing a major P[8] rotavirus lineage, might become prevalent partly because it does not discriminate between hosts according to their secretor status.

We also found that Lewis[b] and H type-1 antigens do not inhibit infection by other P[II] human rotaviruses, including DS-1 (P[4]) and ST-3 (P[6]), whose VP8\* were previously reported to bind one or both of these antigens[25]. We have discovered that A-type HBGA interacts with the VP8\* of human P[4] and P[6] rotaviruses. In these complexes, we determined, for the first time, an involvement of the fucose moiety with the protein surface. This engagement may help explain why A-tri reduces infection by P[6] rotavirus RV-3 more substantially than the P[14] HAL1166 and P[9] K8 strains. However, as the DS-1 infectivity reduction by A-tri was similar to that of HAL1166 and K8, other factors are also likely to be involved, including the difference in usage of other receptors such as integrins between DS-1 and RV-3 (ref. 45).

Taken together, our data revealed that HBGAs are used as rotavirus receptors, consistent with their abundance on the surface of mucosal epithelia of the gastrointestinal tract. However, it is notable that major rotavirus strain variations in HBGA recognition were detected. The relevant human clinical strain Wa did not recognize HBGAs, which is consistent with the observed

viruses were passaged to a similar low extent. Infectivity measurements were more sensitive than VP8\*-cell binding assays for the detection of A-tri use, as a 37% decrease in K8 infectivity following exposure to 5 mM A-tri was observed ($P = 0.0009$; Fig. 7b). As expected from the increased HAL1166 infectivity in cells genetically modified to express A-tri (ref. 23), the infectivity of HAL1166 in MA104 cells was inhibited by 42% after A-tri exposure (t-test, $P < 0.0001$). In contrast, the presence of A-tri did lack of influence of genetically controlled expression of different HBGAs on pediatric rotavirus gastroenteritis[46,47]. Interestingly, two recent studies suggested that non-secretor individuals (null homozygote fucosyltransferase, FUT2) are not susceptible to infection with the most common rotaviruses[48,49]. However, this finding appears inconsistent with the well-established fact that by 2–3 years of age, essentially all children excrete rotavirus at least once and/or seroconvert to rotavirus regardless of their secretor status[50,51]. Although only a small number of children were analysed, a further recent study[41] may shed light on this discrepancy with its finding that non-secretors may be preferentially infected by P[6] rather than P[8] rotaviruses. This also indicates that all children are susceptible to rotavirus infection, independent of their blood group. Consequently, the interdependence of rotavirus infection and disease on secretor status remains questionable. The possibility that rotaviruses, such as noroviruses, may bind HBGA-like substances on intestinal bacteria also should be considered[33,52], especially in the light of recent evidence that antibiotic treatment reduces rotavirus infection[53]. The complex regulation of fucosylation by FUT2 on intestinal epithelium, proposed to be controlled by innate lymphoid cells and involve IL-22 induction by commensal bacteria[54,55] also may contribute to the apparent inconsistencies between studies of the roles of HBGA in rotavirus receptor usage and disease. Our study indicates that HBGA play a role in host cell invasion by several human rotavirus genotypes. However, the clear differences we found between rotavirus strains in their ability to use HBGA show that HBGA may not be essential in this process. In this context, it is important to note that FUT2 activity is not restricted to the biosynthesis of H type-1 precursors (Gal$\beta$1,3GlcNAc). Glycolipids in the epithelial tissues of the gastrointestinal tract are also fucosylated[56] by FUT2. Clearly, secretor status of the individual and FUT2 enzyme activity are not correlated with blood group specificity. More extensive structural, virological and epidemiological studies are required to shed light on the exact role played by HBGAs in rotavirus infection and disease.

## Methods

**Viruses and cells.** The origins of human rotaviruses HAL1166, K8, Wa, RV-3, ST-3 and DS-1 have been described previously[45,57,58]. These viruses were cultivated in the MA104 monkey kidney cell line (ATCC CRL-2378) using DMEM containing 2 mM L-glutamine (Sigma-Aldrich), 20 mM HEPES (Roche) and antibiotics (DMEM) supplemented with 10% (v/v) fetal bovine serum. Rotavirus infectivity was activated with porcine pancreatic trypsin (Sigma) at 20 µg ml$^{-1}$ for 20 min at 37 °C, and infectivity assays were performed as before[45,57].

**Carbohydrates and reagents.** Deuterium oxide was purchased from Sigma Aldrich (Australia) and all glycans, that is A antigen trisaccharide (A-tri), A antigen tetrasaccharide (A-tetra), Lewis b and H type-1 antigens, were purchased from Elicityl (France).

**VP8\* protein production.** Plasmids expressing the gene encoding the VP8\* core proteins of HAL1166 (VP8\*$_{64-224}$), K8 (VP8\*$_{64-224}$), Wa (VP8\*$_{64-223}$), RV-3 (VP8\*$_{64-223}$) and DS-1 (VP8\*$_{64-223}$) for NMR and rotavirus-binding assays were synthesized (HAL1166) or cloned from virus (Wa, K8, RV-3 and DS-1) as described previously for Wa (ref. 12). In brief, complementary DNA produced from viral dsRNA by reverse transcription was used as template for PCR with specific oligonucleotide primers (Supplementary Table 1) to produce the VP8\* gene fragment, which was cloned into the pGEX-4T-1 vector (GE Healthcare). DNA sequencing indicated that the predicted amino acid sequence of pGEX-Wa-VP8\* was identical to the published Wa VP4 sequence[59] (GenBank accession code L34161), except for leucine substituting a phenylalanine as position 163; pGEX-K8-VP8\* amino acid sequence was identical to the published sequence (GenBank accession code Q01641), except for a change from tyrosine to aspartic acid at amino acid position 149; and the sequence of pGEX-RV-3-VP8\*[22] was identical to the most recent database entry for RV-3 VP4 (GenBank accession code FJ998273). Our sequence and FJ998273 both differ from the original published sequence (GenBank accession code U16299) in showing serine rather than proline at aa position 71 (refs 60,61). The pGEX-DS-1-VP8\* amino acid sequence was identical to that published (GenBank accession code HQ650119). The VP8\* gene fragment

of HAL1166 was synthesized (GenScript) with 5'-BamH1 and 3'-EcoR1 restriction sites and ligated into the pUC57-Kan cloning vector through a single blunt-end EcoRV restriction site. Following digestion with BamH1 and EcoR1 of the pUC57-HAL1166-VP8\* plasmid, the VP8\* gene fragment was ligated into pGEX-4t-1, yielding pGEX-4t-1-HAL1166-VP8\*. The final expression plasmid encoded for amino acids 64 to 224 of the VP4 protein and was fused to an N-terminal GST protein. For STD NMR, trNOE and cell-binding assays, the GST portion was not cleaved from the VP8\* proteins. Protein expression was performed as described previously[19]. To prevent the formation of insoluble products, inclusion bodies and aggregates, the expression protocol was optimized for RV-3 VP8\* by adding 20 ml l$^{-1}$ of a 50 × stock solution of Augmedium (Athena Environmental Sciences, Inc., Baltimore, MD, USA), to precondition the cells and induce expression of chaperone proteins, followed by 30 min incubation. Isopropylthiogalactoside was added to reach a final concentration of 1 mM, to induce protein overexpression. Frozen cell pellets were thawed in PBS (137 mM NaCl, 2.7 mM KCl, 10 mM Na$_2$HPO$_4$, 2 mM KH$_2$PO$_4$ pH 7.3) supplemented with 1 mM phenylmethylsulfonyl fluoride (Roche Diagnostics). The cells were lysed with 1 mg ml$^{-1}$ lysozyme supplemented with 1% (w/v) Triton X-100 and 20 µg ml$^{-1}$ DNaseI. Cell debris was removed by centrifugation at 20,000g for 30 min and the supernatant was passed over a glutathione-Sepharose column (Amersham-Pharmacia Biotech). The column was washed with PBS-binding buffer and bound GST fusion protein was eluted with 50 mM Tris-HCl pH 7.4 supplemented with 10 mM glutathione.

**Expression of uniformly $^{15}$N-labelled VP8\* protein.** For the acquisition of $^{15}$N-$^1$H HSQC NMR experiments, HAL1166, K8 and Wa VP8\* core proteins were expressed using $^{15}$N-minimal growth medium. Cells were grown to OD$_{600nm}$ = 0.6 at 37 °C in common LB medium. Cells were pelleted and washed with 1 × M9 salts (22 mM KH$_2$PO$_4$, 45 mM Na$_2$HPO$_4$ × H$_2$O, 8.6 mM NaCl, pH 7.2) and then resuspended four times concentrated in $^{15}$N-minimal growth medium (22 mM KH$_2$PO$_4$, 45 mM Na$_2$HPO$_4$ × 7H$_2$O, 8.6 mM NaCl, 18 mM $^{15}$NH$_4$Cl, 22 mM glucose, 2 mM MgSO$_4$, 0.1 mM CaCl$_2$, 1 × BME Vitamins pH 7.2) and incubated for $^1$H at 25 °C before expression was induced by the addition of 0.1 mM isopropylthiogalactoside. Cells were harvested after 4 h. Purification including removal of the GST fusion protein was accomplished as previously described[19].

**Standard NMR experiments.** Chemical shift assignment of A-Tri and A-Tetra in 20 mM deuterated phosphate buffer and 70 mM NaCl, pH 7.1, was achieved by COSY, TOCSY, HSQC, 1D NOESY and $^1$H NMR experiments at 280 and 305 K. Chemical shift assignments were in agreement with published values[27].

**STD NMR experiments.** All STD NMR spectra were acquired in Shigemi Tubes (Shigemi, USA) with a Bruker 600 MHz Avance spectrometer at 280 K using a conventional $^1$H/$^{13}$C/$^{15}$N gradient cryoprobe system. NMR experiments of A-tri in complex with VP8\*-GST protein and GST as a control experiment were prepared using a protein concentration of 24 or 36 µM and a ligand concentration of 2.4 mM, resulting in a total protein–ligand ratio of 1:100 and 1:67, respectively. The final volume was 200 µl containing 20 mM deuterated phosphate buffer and 70 mM NaCl, pH 7.1. The protein was saturated with a cascade of 40 Gaussian-shaped pulses with duration of 50 ms each at − 0.1 p.p.m., resulting in a total saturation time of 2 s. The off-resonance was set to 33 p.p.m. and 1,024 experiments were acquired. A WATERGATE sequence was used to suppress the residual HDO signal. A spinlock filter with a strength of 5 kHz and duration of 10 ms was applied to suppress the protein background.

**NOE and trNOE NMR experiments.** NOE NMR experiments of free A-tri were acquired by diluting 1 mg of A-tri in 50 µl of 20 mM deuterated phosphate buffer and 70 mM NaCl, pH 7.1, at 305 K on a Bruker 600 MHz Avance spectrometer. The mixing time ($\tau_m$) was set to 800 ms and the relaxation delay to 2 s. trNOE NMR experiments were carried out using 1 mg GST-Wa or GST-K8, with a total molecular weight of 44 kDa and 145 µg of A-Tri that equals a protein:ligand ratio of 1:12 in 20 mM deuterated phosphate and 70 mM NaCl, pH 7.1, at 305 K and 600 MHz. The mixing time ($\tau_m$) for VP8\*-GST protein was set to 150 ms to avoid false-positive NOEs due to spin diffusion effects. A trNOE NMR experiment of GST-K8 protein using the identical experimental setup but without ligand was acquired to subtract it from the trNOE with ligand to remove water background.

**$^1$H-$^{15}$N-HSQC NMR experiments.** To calculate solution dissociation constants ($K_D$) of ligand, we used $^1$H-$^{15}$N-HSQC titration experiments, as was previously employed[15]. Using a standard $^1$H-$^{15}$N-HSQC protocol, NMR spectra of 0.8 mM $^{15}$N-labelled K8, HAL1166 and Wa VP8\* were first acquired in the absence of ligand at 300 K on a 600-MHz NMR spectrometer equipped with a cryoprobe. The NMR buffer consisted of 20 mM Na$_2$HPO$_4$/NaH$_2$PO$_4$ and 70 mM NaCl in 10% D$_2$O. A-tri or A-tetra was titrated to the $^{15}$N-labelled proteins in seven steps with 7.4 mM as the highest ligand concentration. $^1$H-$^{15}$N-HSQC spectra were acquired after each titration step, to follow the ligand-induced changes in the chemical shift perturbations of the protein backbone amide nitrogen and hydrogen.

In a first analysis step, the amide shifts were calculated for the highest ligand concentration weighing proton shifts over nitrogen by a factor of five $\delta(^1H, {}^{15}N) = \delta(^1H) + \delta(^{15}N)/5$ (Supplementary Fig. 7). A threshold value of $\Delta\delta(^1H, {}^{15}N) = 0.04$ p.p.m. is considered as binding of the corresponding amino acid. Alternatively, this threshold is set to the s.d. of all chemical shifts. However, the full analysis shows that both alternative threshold values are very similar for the binding of A-tri and A-tetra to the VP8* proteins of the RV strains K8 and HAL1166 (s.d. for HAL1166$_{A\text{-tri}} = 0.036$, HAL1166$_{A\text{-tetra}} = 0.042$, K8$_{A\text{-tri}} = 0.037$, K8$_{A\text{-tetra}} = 0.040$; Supplementary Figs 8 and 9). Notably, the s.d. for the binding of A-tetra is slightly higher compared with A-tri. Binding affinities were calculated on the basis of the ten strongest backbone amide chemical shifts, which all showed a characteristic pattern of shift movements confirming the specificity of those changes (Supplementary Figs 8 and 9). The chemical shifts were quantified by the chemical shift index for those strongly affected backbone amides according to the following formula: $<I> = \Sigma\{[(N_0 - N)/0.5]^2 + [(H_0 - H)/0.1]^2\}^{1/2}$, where $N_0$ and $N$ are the amide $^{15}N$ chemical shifts in the absence and presence of ligand, respectively; $H_0$ and $H$ are the backbone amide hydrogen chemical shifts in the absence and presence of ligand, respectively[11]. Calculations were performed using GraphPad Prism 6.0, by using nonlinear regression following a one-site total binding model according to $<I> = B_{max}*[A\text{-tri}]/(K_d + [A\text{-tri}]) + NS*[A\text{-tri}] + Background$, with $B_{max}$ the maximum specificity binding and NS the slope of nonspecific binding.

**Homology modelling of K8 VP8\*.** A homology model of K8 VP8* protein was built using the X-ray crystal structure HAL1166 X (pdb accession code: 4DS0) as template using YASARA Structure molecular modelling package (Ver. 13.9.8)[62] (Supplementary Fig. 5). The HAL1166 structure (pdb code: 4DRV[23]) was chosen as a suitable template, as it has a sequence identity of 83.9% and sequence similarity (BLOSUM62 score is $>0$) of 90.7%. The subsequent, fully unrestrained simulated annealing minimization, run for the entire model, resulted in an overall quality Z-score of $-0.081$ and indicates a high-quality homology model. Both proteins share the amino acid sequence (Ser187-Tyr188-Leu189-Leu190-Thr191) that has been identified as containing key residues for HBGA binding in the case of HAL1166 VP8*[23]. After the side chains had been built, optimized and fine-tuned, all newly modelled parts were subjected to a combined steepest descent and simulated annealing minimization (that is, the backbone atoms of aligned residues were kept fixed to avoid potential damage. The hm_build.mcr macro of the YASARA package with default parameters was used.

**MD simulations.** To analyse the flexibility of the $\alpha(1,2)$-glycosidic linkage of A-tri, the three-dimensional structure HAL1166 VP8* protein in complex (pdb 4DRV)[23] a 2.8Ns MD calculation was performed using the YASARA Structure molecular modeling package (Ver. 13.9.8). The model was subjected to further refinement using the md_refine.mcr macro using the AMBER99 force field. Simulation parameters were kept at the values defined by the macro.

**Flow cytometric analysis of VP8\*-cell binding.** Binding by GST-VP8* from HAL1166, K8 and Wa rotaviruses to MA104 cells was assayed in the same experiment at 4 °C as previously described[22]. The optimal (saturating) concentration of each GST-VP8* was initially determined from its dose-dependent cell-binding curve. For A-Tri blockade studies, A-Tri (2 mM or 10 mM final conc.) or DMEM was reacted for $^1$H at 4 °C with the optimal level of GST-VP8* for HAL1166 (150 µg ml$^{-1}$), K8 (600 µg ml$^{-1}$) or Wa (300 µg ml$^{-1}$), then added to cells for 45 min. Stained single cells were analysed by flow cytometry as previously described[22]. The results provided are representative of those obtained in three independent experiments.

**Assays for inhibition of rotavirus infectivity.** A-tri (2, 5 or 10 mM), H type-1 antigen (10 mM), Lewis$^b$ antigen (10 mM), rotavirus-neutralizing monoclonal antibody (as mouse ascites fluid) or DMEM diluent were incubated with trypsin-activated rotavirus for 1 h at 37 °C. The antibodies used were directed either to VP4 or to rotavirus outer capsid protein VP7, which confers glycoprotein (G) serotype specificity. Antibodies comprises the G8-specific B37:1 (ref. 63) (dilution: 1 in 1,000) for HAL1166, the G1-reactive RV-4:3 (ref. 64) (1 in 2,000) for K8, the G1-reactive IA10 (ref. 65) (1 in 10,000) for Wa, the G3-specific RV-3:1 (ref. 57) (1 in 20,000) for RV-3, the ST-3- and P[6]-specific ST-3:3 (ref. 66) (1 in 50,000) for ST-3 and the P[4]-specific RV-5:2 (ref. 67) (1 in 20,000) for DS-1. Confluent MA104 cell monolayers containing $4 \times 10^4$ cells were washed before virus-ligand mixtures or DMEM were transferred onto cells and reacted for 1 h at 37 °C at the optimum multiplicity of infection of 0.02. Following inoculum removal and washing as above, infected cells were incubated for 15 h at 37 °C in 95% (vol/vol) air with 5% (vol/vol) CO$_2$. Acetone-fixed cell monolayers were sequentially stained with rabbit antiserum to rotavirus (1 in 1,000) and fluorescein isothiocyanate-conjugated goat anti-rabbit IgG (Invitrogen; 1 in 80), and virus titres determined by fluorescence microscopy as described previously[68]. For all assays, data represent the mean of triplicate samples and two independent experiments.

**Statistical analysis.** Student's $t$-test was used, with significance set at the 95% level. Error bars on graphs indicate the s.d.

# References

1. Tate, J. E. et al. 2008 estimate of worldwide rotavirus-associated mortality in children younger than 5 years before the introduction of universal rotavirus vaccination programmes: a systematic review and meta-analysis. *Lancet Infect. Dis.* **12,** 136–141 (2012).
2. Richardson, V. et al. Effect of rotavirus vaccination on death from childhood diarrhea in Mexico. *N. Engl. J. Med.* **362,** 299–305 (2010).
3. Weintraub, E. S. et al. Risk of intussusception after monovalent rotavirus vaccination. *N. Engl. J. Med.* **370,** 513–519 (2014).
4. Yih, W. K. et al. Intussusception risk after rotavirus vaccination in U.S. infants. *N. Engl. J. Med.* **370,** 503–512 (2014).
5. Bowen, M. D. & Payne, D. C. Rotavirus vaccine-derived shedding and viral reassortants. *Expert Rev. Vaccines* **11,** 1311–1314 (2012).
6. Banyai, K. et al. Systematic review of regional and temporal trends in global rotavirus strain diversity in the pre rotavirus vaccine era: insights for understanding the impact of rotavirus vaccination programs. *Vaccine* **30**(Suppl 1): A122–A130 (2012).
7. Fiore, L., Greenberg, H. B. & Mackow, E. R. The VP8 fragment of VP4 is the rhesus rotavirus hemagglutinin. *Virology* **181,** 553–563 (1991).
8. Ciarlet, M. et al. Initial interaction of rotavirus strains with N-acetylneuraminic (sialic) acid residues on the cell surface correlates with VP4 genotype, not species of origin. *J. Virol.* **76,** 4087–4095 (2002).
9. Guo, C. T. et al. Ganglioside GM(1a) on the cell surface is involved in the infection of human rotavirus KUN and MO strains. *J. Biochem.* **126,** 683–688 (1999).
10. Delorme, C. et al. Glycosphingolipid binding specificities of rotavirus: identification of a sialic acid-binding epitope. *J. Virol.* **75,** 2276–2287 (2001).
11. Haselhorst, T. et al. Recognition of the GM3 ganglioside glycan by Rhesus rotavirus particles. *Angew. Chem. Int. Ed. Engl.* **50,** 1055–1058 (2011).
12. Haselhorst, T. et al. Sialic acid dependence in rotavirus host cell invasion. *Nat. Chem. Biol.* **5,** 91–93 (2009).
13. Yu, X. et al. Novel structural insights into rotavirus recognition of ganglioside glycan receptors. *J. Mol. Biol.* **413,** 929–939 (2011).
14. Dormitzer, P. R. et al. Specificity and affinity of sialic acid binding by the rhesus rotavirus VP8* core. *J. Virol.* **76,** 10512–10517 (2002).
15. Dormitzer, P. R., Sun, Z. Y., Wagner, G. & Harrison, S. C. The rhesus rotavirus VP4 sialic acid binding domain has a galectin fold with a novel carbohydrate binding site. *EMBO J.* **21,** 885–897 (2002).
16. Blanchard, H., Yu, X., Coulson, B. S. & von Itzstein, M. Insight into host cell carbohydrate-recognition by human and porcine rotavirus from crystal structures of the virion spike associated carbohydrate-binding domain (VP8*). *J. Mol. Biol.* **367,** 1215–1226 (2007).
17. Kraschnefski, M. J. et al. Effects on sialic acid recognition of amino acid mutations in the carbohydrate-binding cleft of the rotavirus spike protein. *Glycobiology* **19,** 194–200 (2009).
18. Kraschnefski, M. J. et al. Cloning, expression, purification, crystallization and preliminary X-ray diffraction analysis of the VP8* carbohydrate-binding protein of the human rotavirus strain Wa. *Acta Crystallogr. Sect. F. Struct. Biol. Cryst. Commun.* **61,** 989–993 (2005).
19. Scott, S. A. et al. Crystallization and preliminary X-ray diffraction analysis of the sialic acid-binding domain (VP8*) of porcine rotavirus strain CRW-8. *Acta Crystallogr. Sect. F. Struct. Biol. Cryst. Commun.* **61,** 617–620 (2005).
20. Yu, X. et al. Crystallization and preliminary X-ray diffraction analysis of the carbohydrate-recognizing domain (VP8*) of bovine rotavirus strain NCDV. *Acta Crystallogr. Sect. F. Struct. Biol. Cryst. Commun.* **64,** 509–511 (2008).
21. Zhang, Y. D., Li, H., Liu, H. & Pan, Y. F. Expression, purification, crystallization and preliminary X-ray diffraction analysis of the VP8* sialic acid-binding domain of porcine rotavirus strain OSU. *Acta Crystallogr. Sect. F. Struct. Biol. Cryst. Commun.* **63,** 93–95 (2007).
22. Fleming, F. E. et al. Relative roles of GM1 ganglioside, N-acylneuraminic acids and α2β1 integrin in mediating rotavirus infection. *J. Virol.* **88,** 4558–4571 (2014).
23. Hu, L. et al. Cell attachment protein VP8* of a human rotavirus specifically interacts with A-type histo-blood group antigen. *Nature* **485,** 256–259 (2012).
24. Liu, Y. et al. Rotavirus VP8*: phylogeny, host range, and interaction with histo-blood group antigens. *J. Virol.* **86,** 9899–9910 (2012).
25. Huang, P. et al. Spike protein VP8* of human rotavirus recognizes histo-blood group antigens in a type-specific manner. *J. Virol.* **86,** 4833–4843 (2012).
26. Ying, Y. et al. Human milk contains novel glycans that are potential decoy receptors for neonatal rotaviruses. *Mol. Cell Proteomics* **13,** 2944–2960 (2014).
27. Strecker, G., Wieruszeski, J. M., Michalski, J. C. & Montreuil, J. Complete analysis of the $^1$H- and $^{13}$C-NMR spectra of four blood-group A active oligosaccharides. *Glycoconj. J.* **6,** 271–284 (1989).

28. Mayer, M. & Meyer, B. Group epitope mapping by saturation transfer difference NMR to identify segments of a ligand in direct contact with a protein receptor. *J. Am. Chem. Soc.* **123**, 6108–6117 (2001).

29. Mayer, M. & Meyer, B. Characterization of ligand binding by saturation transfer difference NMR spectroscopy. *Angew. Chem. Int. Ed. Engl.* **38**, 1784–1788 (1999).

30. Marionneau, S. *et al.* ABH and Lewis histo-blood group antigens, a model for the meaning of oligosaccharide diversity in the face of a changing world. *Biochimie* **83**, 565–573 (2001).

31. Huang, P. *et al.* Noroviruses bind to human ABO, Lewis, and secretor histo-blood group antigens: identification of 4 distinct strain-specific patterns. *J. Infect. Dis.* **188**, 19–31 (2003).

32. Huang, P. *et al.* Norovirus and histo-blood group antigens: demonstration of a wide spectrum of strain specificities and classification of two major binding groups among multiple binding patterns. *J. Virol.* **79**, 6714–6722 (2005).

33. Miura, T. *et al.* Histo-blood group antigen-like substances of human enteric bacteria as specific adsorbents for human noroviruses. *J. Virol.* **87**, 9441 (2013).

34. Cao, S. *et al.* Structural basis for the recognition of blood group trisaccharides by norovirus. *J. Virol.* **81**, 5949–5957 (2007).

35. Choi, J. M., Hutson, A. M., Estes, M. K. & Prasad, B. V. Atomic resolution structural characterization of recognition of histo-blood group antigens by Norwalk virus. *Proc. Natl Acad. Sci. USA* **105**, 9175–9180 (2008).

36. Shanker, S. *et al.* Structural analysis of histo-blood group antigen binding specificity in a norovirus GII.4 epidemic variant: implications for epochal evolution. *J. Virol.* **85**, 8635–8645 (2011).

37. Chen, Y. *et al.* Crystallography of a Lewis-binding norovirus, elucidation of strain-specificity to the polymorphic human histo-blood group antigens. *PLoS Pathog.* **7**, e1002152 (2011).

38. Fiege, B. *et al.* Molecular details of the recognition of blood group antigens by a human norovirus as determined by STD NMR spectroscopy. *Angew. Chem. Int. Ed. Engl.* **51**, 928–932 (2012).

39. Zierke, M. *et al.* Stabilization of branched oligosaccharides: Lewis(x) benefits from a nonconventional C-H...O hydrogen bond. *J. Am. Chem. Soc.* **135**, 13464–13472 (2013).

40. McDonald, S. *et al.* Diversity and relationships of co-circulating modern human rotaviruses revealed using large-scale comparative genomics. *J. Virol.* **86**, 9148–9162 (2012).

41. Nordgren, J. *et al.* Both Lewis and secretor status mediate susceptibility to rotavirus infections in a rotavirus genotype dependent manner. *Clin. Infect. Dis.* **59**, 1567–1573 (2014).

42. Peltier, P., Euzen, R., Daniellou, R., Nugier-Chauvin, C. & Ferrieres, V. Recent knowledge and innovations related to hexofuranosides: structure, synthesis and applications. *Carbohydr. Res.* **343**, 1897–1923 (2008).

43. Meyer, B. & Peters, T. NMR spectroscopy techniques for screening and identifying ligand binding to protein receptors. *Angew. Chem. Int. Ed. Engl.* **42**, 864–890 (2003).

44. Shuker, S. B., Hajduk, P. J., Meadows, R. P. & Fesik, S. W. Discovering high-affinity ligands for proteins: SAR by NMR. *Science* **274**, 1531–1534 (1996).

45. Graham, K. L. *et al.* Integrin-using rotaviruses bind α2β1 integrin α2 I domain via VP4 DGE sequence and recognize αXβ2 and αVβ3 by using VP7 during cell entry. *J. Virol.* **77**, 9969–9978 (2003).

46. Ahmed, T. *et al.* Children with the Le(a + b-) blood group have increased susceptibility to diarrhea caused by enterotoxigenic *Escherichia coli* expressing colonization factor I group fimbriae. *Infect. Immun.* **77**, 2059–2064 (2009).

47. Yazgan, H. *et al.* Blood groups and rotavirus gastroenteritis. *Pediatr. Infect. Dis. J.* **32**, 705–706 (2013).

48. Trang, N. V. *et al.* Association between norovirus and rotavirus Infection and histo-blood group antigen types in Vietnamese children. *J. Clin. Microbiol.* **52**, 1366–1374 (2014).

49. Imbert-Marcille, B. M. *et al.* A FUT2 gene common polymorphism determines resistance to rotavirus A of the P[8] genotype. *J. Infect. Dis.* **209**, 1227–1230 (2013).

50. Velazquez, F. R. *et al.* Rotavirus infections in infants as protection against subsequent infections. *N. Engl. J. Med.* **335**, 1022–1028 (1996).

51. Paul, A., Gladstone, B. P., Mukhopadhya, I. & Kang, G. Rotavirus infections in a community based cohort in Vellore, India. *Vaccine* **32**(Suppl 1): A49–A54 (2014).

52. Jones, M. K. *et al.* Enteric bacteria promote human and mouse norovirus infection of B cells. *Science* **346**, 755–759 (2014).

53. Uchiyama, R. *et al.* Antibiotic treatment suppresses rotavirus infection and enhances specific humoral immunity. *J. Infect. Dis.* **210**, 171–182 (2014).

54. Goto, Y. *et al.* Innate lymphoid cells regulate intestinal epithelial cell glycosylation. *Science* **345**, 1254009 (2014).

55. Pickard, J. M. *et al.* Rapid fucosylation of intestinal epithelium sustains host–commensal symbiosis in sickness. *Nature* 638–641 (2014).

56. Iwamori, M. & Domino, S. E. Tissue-specific loss of fucosylated glycolipids in mice with targeted deletion of α(1,2)fucosyltransferase genes. *Biochem. J.* **380**, 75–81 (2004).

57. Coulson, B. S. *et al.* Neutralizing monoclonal antibodies to human rotavirus and indications of antigenic drift among strains from neonates. *J. Virol.* **54**, 14–20 (1985).

58. Browning, G. F. *et al.* Human and bovine serotype G8 rotaviruses may be derived by reassortment. *Arch. Virol.* **125**, 121–128 (1992).

59. van Doorn, L. J. *et al.* Detection and genotyping of human rotavirus VP4 and VP7 genes by reverse transcriptase PCR and reverse hybridization. *J. Clin. Microbiol.* **47**, 2704–2712 (2009).

60. Rippinger, C. M. Complete genome sequence analysis of candidate human rotavirus vaccine strains RV3 and 116E. *Virology* **405**, 201–213 (2010).

61. Kirkwood, C. D., Bishop, R. F. & Coulson, B. S. Attachment and growth of human rotaviruses RV-3 and S12/85 in Caco-2 cells depend on VP4. *J. Virol.* **72**, 9348–9352 (1998).

62. Krieger, E., Koraimann, G. & Vriend, G. Increasing the precision of comparative models with YASARA NOVA--a self-parameterizing force field. *Proteins* **47**, 393–402 (2002).

63. Tursi, J. M. *et al.* Production and characterization of neutralizing monoclonal antibody to a human rotavirus strain with a "super-short" RNA pattern. *J. Clin. Microbiol.* **25**, 2426–2427 (1987).

64. Coulson, B. S. & Kirkwood, C. Relation of VP7 amino acid sequence to monoclonal antibody neutralization of rotavirus and rotavirus monotype. *J. Virol.* **65**, 5968–5974 (1991).

65. Fleming, F. E. *et al.* Rotavirus-neutralizing antibodies inhibit virus binding to integrins α2β1 and α4β1. *Arch. Virol.* **152**, 1087–1101 (2007).

66. Kirkwood, C. *et al.* Human rotavirus VP4 contains strain-specific, serotype-specific and cross-reactive neutralization sites. *Arch. Virol.* **141**, 587–600 (1996).

67. Coulson, B. S. *et al.* Derivation of neutralizing monoclonal antibodies to human rotaviruses and evidence that an immunodominant neutralization site is shared between serotypes 1 and 3. *Virology* **154**, 302–312 (1986).

68. Kiefel, M. J. *et al.* Synthesis and biological evaluation of *N*-acetylneuraminic acid-based rotavirus inhibitors. *J. Med. Chem.* **39**, 1314–1320 (1996).

## Acknowledgements

We are grateful to Enzo Palombo for the provision of HAL1166 rotavirus, Harry Greenberg for 1A10 antibody and Renee Winzar for technical assistance in protein production. T.H. thanks the Australian Research Council for the award of an Australian Future Fellowship (FT120100419). M.v.I. and B.S.C. also acknowledge the financial support of the Australian Research Council (DP1094393). B.S.C. is grateful to the National Health and Medical Research Council of Australia for the award of a Senior Research Fellowship (ID628319). M.v.I, B.S.C. and T.H. thank the National Health and Medical Research Council of Australia for a Project Grant (ID597439).

## Author contributions

All of the authors contributed to various aspects of the design, experimental, analysis and discussion of the research. R.B. and T.H. performed the NMR experiments. F.E.F. and V.T.D. performed the virological assays. R.B., B.S.C, M.v.I. and T.H. wrote the manuscript.

## Additional information

# 2

# Cellular delivery and photochemical release of a caged inositol-pyrophosphate induces PH-domain translocation *in cellulo*

Igor Pavlovic[1], Divyeshsinh T. Thakor[1], Jessica R. Vargas[2], Colin J. McKinlay[2], Sebastian Hauke[3], Philipp Anstaett[1], Rafael C. Camuña[4], Laurent Bigler[1], Gilles Gasser[1], Carsten Schultz[3], Paul A. Wender[2] & Henning J. Jessen[5]

Inositol pyrophosphates, such as diphospho-myo-inositol pentakisphosphates (InsP$_7$), are an important family of signalling molecules, implicated in many cellular processes and therapeutic indications including insulin secretion, glucose homeostasis and weight gain. To understand their cellular functions, chemical tools such as photocaged analogues for their real-time modulation in cells are required. Here we describe a concise, modular synthesis of InsP$_7$ and caged InsP$_7$. The caged molecule is stable and releases InsP$_7$ only on irradiation. While photocaged InsP$_7$ does not enter cells, its cellular uptake is achieved using nanoparticles formed by association with a guanidinium-rich molecular transporter. This novel synthesis and unprecedented polyphosphate delivery strategy enable the first studies required to understand InsP$_7$ signalling in cells with controlled spatiotemporal resolution. It is shown herein that cytoplasmic photouncaging of InsP$_7$ leads to translocation of the PH-domain of Akt, an important signalling-node kinase involved in glucose homeostasis, from the membrane into the cytoplasm.

[1] Department of Chemistry, University of Zurich, Winterthurerstrasse 190, Zurich 8057, Switzerland. [2] Departments of Chemistry and Chemical and Systems Biology, Stanford University, Stanford, California 94305, USA. [3] European Molecular Biology Laboratory (EMBL), Cell Biology & Biophysics Unit, Meyerhofstrasse 1, 69117 Heidelberg, Germany. [4] Departamento de Química Orgánica, Facultad de Ciencias, Universidad de Málaga, Malaga 29071, Spain. [5] Department of Chemistry and Pharmacy, Albert-Ludwigs University Freiburg, Albertstrasse 21, 79104 Freiburg, Germany. Correspondence and requests for materials should be addressed to P.A.W. (email: Wenderp@stanford.edu) or to H.J.J. (email: henning.jessen@oc.uni-freiburg.de).

**D**iphospho-inositol polyphosphates (InsP$_7$) are second messengers involved in essential cell signalling pathways[1–4]. A distinct difference of InsP$_7$ compared with other inositol polyphosphates is the presence of a phosphoanhydride bond in, for example, the 5-position (5-InsP$_7$, Fig. 1), rendering them a structurally unique class of second messengers. This special feature is also the reason for their nickname 'inositol pyrophosphates'. InsP$_7$ are implicated in the regulation of diverse cellular and metabolic functions in different kingdoms of life[1–8]. It has been proposed that InsP$_7$ bind to the pleckstrin homology (PH) domain of protein kinase B (Akt), and competitively suppress its specific phosphatidylinositol 3,4,5-trisphosphate (PIP$_3$) association at the plasma membrane, thereby inhibiting phosphoinositide-dependent kinase 1 (PDK1)-mediated phosphorylation of Akt[9,10]. However, there remains uncertainty as to whether the reduced phosphorylation of Akt is a result of the inhibition of its membrane association via its PH-domain, since the *in vitro* assays that have been performed do not contain any membrane or membrane mimics. In addition, InsP$_7$ might act either as allosteric inhibitors or as non-enzymatic phosphorylating agents or both[3,11]. Notwithstanding, inhibition of the Akt pathway by InsP$_7$ has an impact on glucose uptake and insulin sensitivity, as exemplified by a mouse model that lacks inositol hexakisphosphate-kinase 1 (IP6K1). These knockout mice have reduced levels of InsP$_7$ and show a lean phenotype on high-fat diet concomitant with increased insulin sensitivity[9]. As a consequence, IP6K1 has recently been proposed as a novel target in the treatment of diabetes and obesity[12]. To address fundamental questions about the mechanism of action of these potent signalling molecules and their subcellular localization, the development of new chemical tools is required.

To understand cellular signalling mediated by second messengers, photocaged analogues that can be activated on demand inside living cells with spatiotemporal resolution have attracted great interest[13]. Unfortunately, preparation of such analogues often requires lengthy synthetic sequences. Phosphorylated second messengers derived from *myo*-inositol such as *myo*-inositol 1,4,5-trisphosphate (Fig. 1) present additional challenges as their polyanionic nature precludes efficient cellular uptake[14]. For example, while cell-permeable and photocaged analogues of different *myo*-inositol polyphosphates (InsP$_x$) and phosphatidyl *myo*-inositol polyphosphates (PtdIns-P$_x$) have been reported[15–18], the phosphate groups typically need to be reversibly masked. This complicates their use, as multiple intracellular hydrolysis events must occur before the free polyphosphate is formed[19]. Even so, no such photocaged derivatives are currently available for the more complex diphospho-*myo*-inositol pentakisphosphates as, for example, 5-InsP$_7$ (Fig. 1)[20].

Here we report the design, step-economical synthesis, photophysical and metabolic evaluation of photocaged 5-InsP$_7$, and significantly, a general solution to the delivery of unmodified polyphosphate probes into cells using guanidinium-rich molecular transporters[21]. On cytoplasmic uncaging, 5-InsP$_7$-mediated PH-domain translocation from the membrane into the cytosol in living cells is demonstrated for the first time on a 15-min timescale.

## Results

**Synthesis.** All current synthetic approaches to access any InsP$_7$ isomer rely on a global hydrogenation in the last step, during which up to 13 protecting groups need to be removed[22–30]. Significantly, however, hydrogenation is incompatible with photocaging groups and many other functional moieties like, for example, fluorophores. To address this problem, a novel strategy based on the development of a levulinate benzyl ester adaptor (LevB) is described. The introduction of this new protecting group and its combination with fluorenylmethyl (Fm) protection[31–33] and photocage introduction enables the previously inaccessible synthesis of the first photoactivatable diphospho-inositol InsP$_7$ probe **9** equipped with a [7-(diethylamino)-coumarin-4-yl]methyl (DEACM) photocage (Fig. 2a). It is noteworthy that this strategy potentially facilitates the introduction of other tags, such as, for example, photoaffinity labels and fluorophores.

The synthesis commenced with benzylidene protected **2** prepared as previously described (Fig. 2a)[23]. The 5-OH position of **2** is available for phosphitylation, allowing virtually any protected phosphate to be introduced. However, none of the existing protecting groups are compatible with the subsequent introduction of the coumarin cage. Such protecting groups would need to be stable under acidic and basic conditions and must enable double deprotection under very mild conditions. To meet these stringent requirements, a new phosphate-protecting group is required. The approach described herein is based on an Umpolung strategy that had been exploited in prodrug design for nucleotides[34,35]. Conceptually, this strategy is useful to generally couple phenol or alcohol protecting groups to phosphates via a benzyl adaptor, greatly enhancing the available protecting group strategies for phosphates (Fig. 2b). A novel P-amidite **3** was developed (Fig. 2, Supplementary Fig. 1), which is connected via a benzyl ester to a levulinate group (LevB). After oxidation to the phosphate triester **6**, the LevB group can be cleaved by hydrazone formation, initiating cyclization and finally a Grob-type fragmentation to give the unprotected phosphates (Fig. 2b).

After coupling P-amidite **3** to alcohol **2** and oxidation, all inositol-protecting groups were cleaved with TFA. The resultant protected monophosphate **4** was phosphitylated with (Fm)$_2$-P-amidite **5** and oxidized. Notwithstanding the significant molecular crowding of hexakisphosphate **6**, both LevB groups were efficiently cleaved under mild conditions (hydrazine*AcOH/ TFA), releasing phosphate **7**. Next, the P-anhydride was formed using Fm-protected photocaged P-amidite **8** (ref. 33). All 11 Fm-protecting groups were then cleaved with piperidine resulting in highly pure photocaged 5-InsP$_7$ **9** in 45% yield over 2 steps (11% overall yield from **1**) on precipitation of the compound as the dodeca-piperidinium salt. The piperidinium ions can also be exchanged with sodium ions by precipitation. In addition, natural 5-InsP$_7$ **10** can be prepared following the same strategy by using (Fm)$_2$-P-amidite **5** in the anhydride forming reaction (8% overall yield from **1**)[36,37]. This eight-step synthesis represents a general strategy to access 5-InsP$_7$ **10** and caged analogues in scalable amounts (30 mg of **9** have been prepared) and very high quality

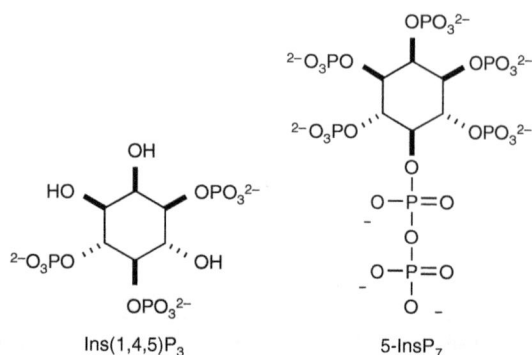

**Figure 1 | Phosphorylated second messengers derived from *myo*-inositol.** Chemical structures of *myo*-Inositol 1,4,5-trisphosphate and 5-diphospho-*myo*-inositol pentakisphosphate (5-InsP$_7$).

**a**

**b**

Relay
Phenol protecting group,
e.g., levulinate (Lev)

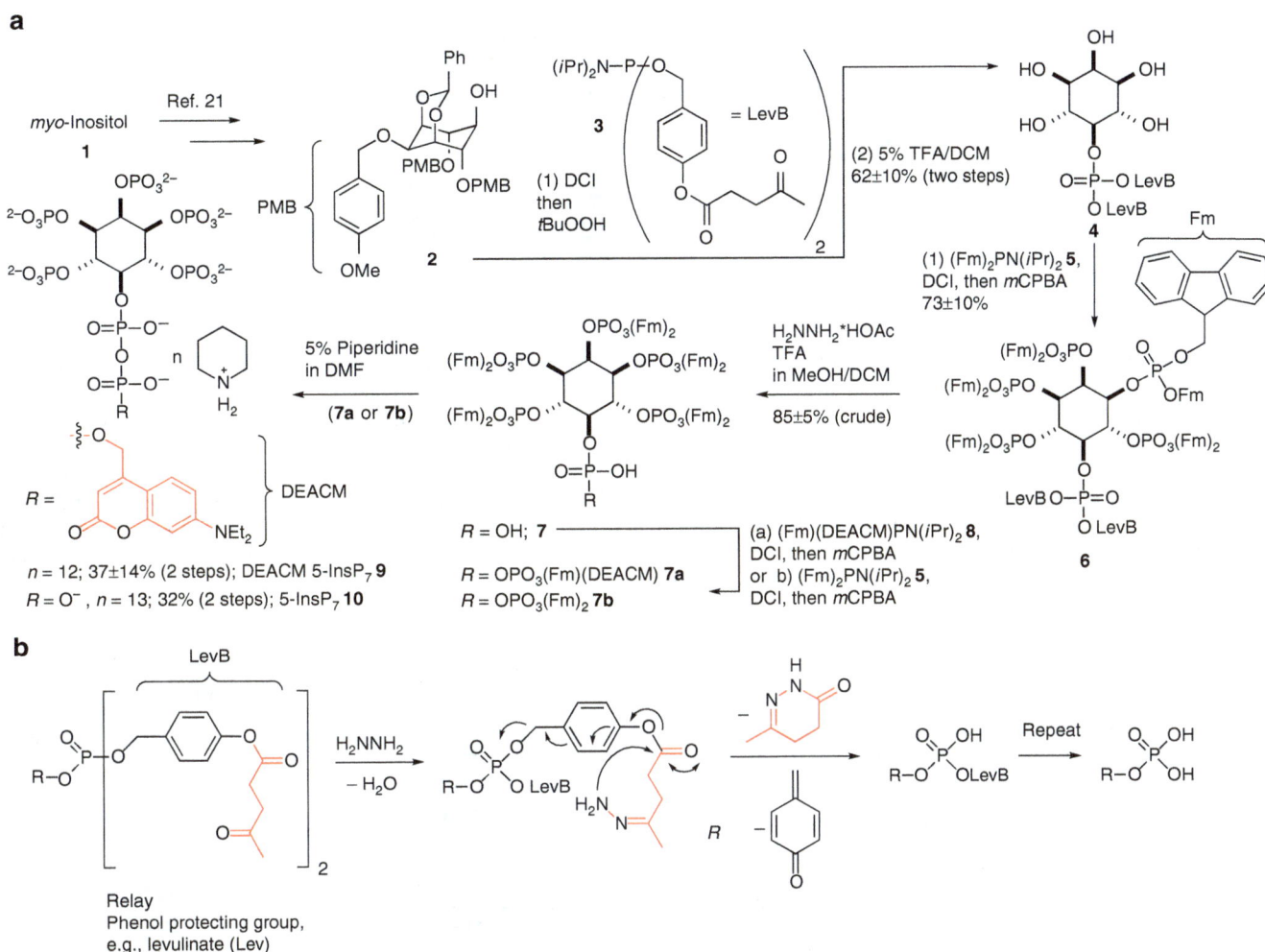

**Figure 2 | Synthesis of photocaged 5–InsP₇ and mechanism of LevB cleavage.** (**a**) Synthesis of DEACM 5–InsP₇ **9** and 5–InsP₇ **10** based on fluorenylmethyl (Fm) protection and a novel phosphate-protecting group (LevB). DCI, 4,5-dicyanoimidazole; DEACM, [7-(diethylamino)-coumarin-4-yl]methyl; Fm, fluorenylmethyl; mCPBA, metachloro perbenzoic acid; TFA, trifluoroacetic acid. (**b**) An adaptor strategy for phosphate release: hydrazine triggers levulinate (red) cleavage and 1,6-elimination (blue) to release free phosphate. Generally, levulinate could also be replaced with other phenol protecting groups.

without the need for a final hydrogenation under aqueous conditions.

***In vitro* stability and photophysical properties.** To serve as a useful tool, DEACM 5–InsP₇ **9** must be stable towards enzymatic digestion to enable cellular uptake and release only on photolysis. To test its stability, **9** was incubated in tissue homogenate (brain, Fig. 3a; liver, Supplementary Fig. 3 and Supplementary Methods) and cell extract (Supplementary Figs 4–5 and Supplementary Methods). Readout was achieved by resolution on polyacrylamide gels (35%, Fig. 3 and Supplementary Methods)[38]. DEACM 5–InsP₇ **9** did not decompose under these conditions over incubation times up to 5 h (Fig. 3a, Lanes III–V and Supplementary Figs 3–5). Thus, **9** is a probe that has the potential to be broadly applied in different cell and tissue types. Importantly, on exposure to ultraviolet light (366 nm, 4 W, distance 10 cm) in extracts, it was cleanly converted into 5–InsP₇ **10**, as verified by PAGE (Fig. 3a, Lanes VI and VII and Supplementary Figs 3–4) and HPLC analysis (Supplementary Fig. 6) with **10** as a standard.

Next, the photophysical properties of DEACM 5–InsP₇ **9** were characterized. The quantum yield for the disappearance of **9** $\Delta\varphi_{chem}$ is 0.71% at 355 nm as determined by actinometry following a novel protocol (Supplementary Methods)[39–41]. The fluorescence quantum yield $\varphi_f$ is 6.2% and the lifetime $\tau_f$ is 1.2 ns. Notably, **9** also exhibits typical coumarin fluorescence at 500 nm (excitation at 386 nm; Supplementary Figs 7–10 and Supplementary Methods)[33,42].

**Cellular delivery and uncaging.** Notwithstanding the efficiency of this synthesis, it was found as expected that DEACM 5–InsP₇ **9**, like other polyanions[11,14], does not readily cross the non-polar membrane of a cell (Fig. 4b,c). To address this problem, its non-covalent complexation, cell uptake and release using guanidinium-rich molecular transporters were studied[43]. **9** was mixed with amphipathic, guanidinium-rich transporter **11** (Fig. 4a), in an equimolar ratio to form nanoparticles. HeLa cells were treated with these complexes and analysed for coumarin fluorescence by flow cytometry. Significantly, while DEACM 5–InsP₇ **9** itself does not appreciably enter cells, the transporter-complexed **9** does, as demonstrated by a 10-fold increase in intracellular fluorescence (Fig. 4b). According to flow cytometry analysis, over 99% of cells display increased levels of **9** following treatment with the transporter/DEACM–InsP₇ complex (Fig. 4c). It is important to note that other transfection reagents like Lipofectamine 2000 do not efficiently deliver **9** into cells (Fig. 4b).

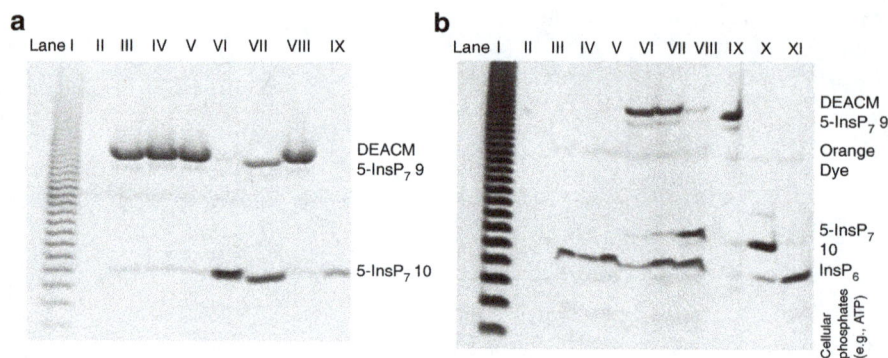

**Figure 3 | *In vitro* and *in cellulo* release of 5-InsP₇.** (a) Analysis of DEACM 5-InsP₇ **9** by gel electrophoresis (PAGE) and toluidine blue staining. **9** is stable for hours in rat brain extract (lanes III–V) and can be uncaged by ultraviolet irradiation (lane VI). Lane I: poly-P marker. Lane II: empty. Lane III: **9** in brain extract (3 h). Lane IV: **9** in brain extract (2 h). Lane V: **9** in brain extract (1 h). Lane VI: **9** in brain extract (1 h), then ultraviolet irradiation (15 min). Lane VII: **9** in distilled water, then ultraviolet irradiation (15 min). Lane VIII: **9**. Lane IX: 5-InsP₇ **10**. (b) Analysis of cellular uptake and *in cellulo* photouncaging with and without MoTr **11** after TiO₂ microsphere extraction followed by gel electrophoresis (PAGE). Bands containing **9** and **10** were additionally extracted and analysed by MALDI mass spectrometry. **9** only enters cells in the presence of MoTr **11** (lanes VI, VII) and can be uncaged in living cells (lane VIII). Lane I: Poly-P marker to assess quality of separation. Lane II: empty. Lane III: HeLa cells (control). Lane IV: HeLa cells + **11** (control). Lane V: HeLa cells + **9** (5 h). Lane VI: HeLa cells + **9** + **11** (5 h), Lane VII: HeLa cells + **9** + **11** (16 h), Lane VIII: HeLa cells + **9** + **11** (16 h, then 10 min irradiation 366 nm, 4W). Lane IX: DEACM 5-InsP₇ **9** (control). Lane X: 5-InsP₇ **10** (control). Lane XI: InsP₆ (control).

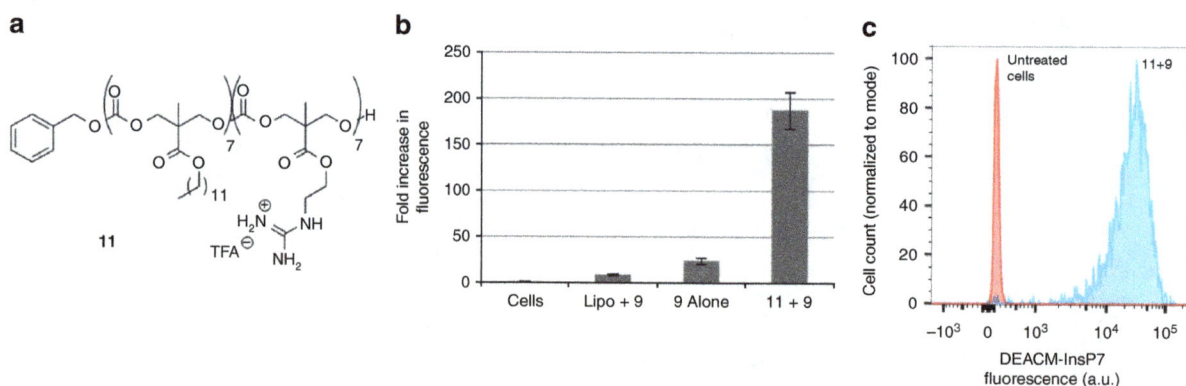

**Figure 4 | Intracellular delivery of photocaged 5-InsP₇ to HeLa cells with a guanidinium-rich transporter.** (a) Structure of amphipathic oligocarbonate transporter, **11**. (b) Cellular uptake of **9** as determined by flow cytometry. Complexes were formulated at a 1:1 mole ratio of **9** (5 µM) to **11**. Values reported are normalized to the autofluorescence of untreated cells. (c) Histogram plot of intracellular fluorescence demonstrates >99% delivery efficiency to cells.

Delivery and intracellular distribution of **9** were further analysed in HeLa cells by confocal microscopy after 4 and 16 h (Fig. 5). The z-stack analysis shows DEACM 5–InsP₇ **9** distributed throughout the cytoplasm at both time points (Fig. 5, single z-slice shown). Both diffuse fluorescence and fluorescent puncta are observed, consistent with mixed diffusion or endosomal uptake and release[44–46]. This is additionally supported by a 65% reduction in cellular uptake when cells were treated at 4 °C, a condition known to inhibit most endocytotic processes (Supplementary Fig. 11).

Cellular uptake, stability and efficient uncaging in living cells was additionally verified by extraction of diphospho-inositol polyphosphates and other cellular phosphates based on a recently published TiO₂ microsphere enrichment method[47]. Here it is shown that this method can also be used to extract analogues such as **9** from complex cell and tissue lysates (Fig. 3b and Supplementary Methods) enabling studies concerning its intracellular fate after delivery. After incubation of DEACM 5–InsP₇ **9** with HeLa cells in the presence or absence of MoTr **11** and repeated washings to remove external **9**, the extracts prepared from those cells (1 million cells) clearly showed a distinct novel band corresponding to **9** in the PAGE analysis after enrichment with TiO₂ and elution (Fig. 3b, lanes VI–VII), whereas no such

uptake could be detected in the control experiment without transporter (Fig. 3b, lane V). To verify its identity, the band corresponding to DEACM 5–InsP₇ **9** was extracted from the gel and analysed by MALDI mass spectrometry, demonstrating its intracellular stability (Supplementary Fig. 12 and Supplementary Methods) for multiple hours. Moreover, efficient intracellular uncaging by irradiation at 366 nm was proven using the same extraction and resolution method (TiO₂ enrichment, then PAGE) in combination with mass spectrometry after extraction of Lane VIII (Fig. 3b). These conditions were found to be of no immediate toxicity (Supplementary Fig. 13 and Supplementary Methods). In summary, the photocaged molecule **9** is efficiently taken up by cells in the presence of MoTr **11**, evenly distributed throughout the cytoplasm, stable for multiple hours in its caged form and can be selectively uncaged to 5-InsP₇ **10**, thus fulfilling the stringent requirements imposed on an intracellular signalling probe.

**PH-domain translocation.** To determine the suitability of the combined delivery and uncaging strategy for a deeper understanding of the effect of InsP₇ fluctuations, PH-domain translocation on cytoplasmic InsP₇ release was studied. The rationale for this experiment is provided by the lean phenotype displayed by IP6K1 knockout mice on high-fat diet and the observation that

|  | Bright field | DEACM-InsP$_7$ | Merge |
|---|---|---|---|

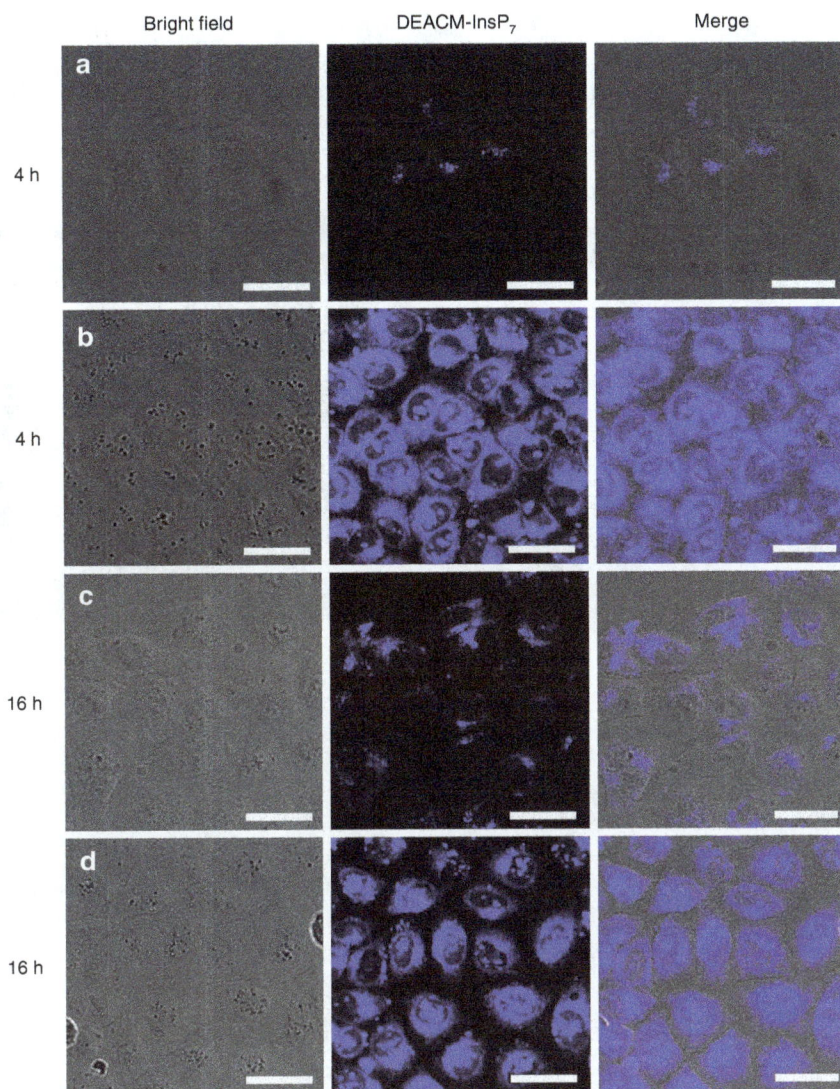

**Figure 5 | The molecular transporter 11 delivers photocaged 5-InsP$_7$9 into the cytoplasm.** Confocal microscopy analysis of HeLa cells incubated with 5 μM DEACM–InsP$_7$ **9** in the presence or absence of transporter **11** after 4 and 16 h of incubation. Efficient uptake of **9** is demonstrated by blue coumarin fluorescence emitted from the compound. Cells treated with (**a**) **9** alone for 4 h, (**b**) **9** + **11** for 4 h, (**c**) **9** alone for 16 h and (**d**) **9** + **11** for 16 h. Scale bars, 35 μm.

InsP$_7$ inhibit Akt phosphorylation *in vitro* and *in vivo* by binding to the PH-domain[9]. Collectively, these findings suggest an effect of 5-InsP$_7$ on membrane localization of Akt. However, no tool to augment any InsP$_7$ within seconds in living cells was previously available. With the new tools in hand, HeLa cells were transiently transfected with a plasmid expressing the PH-domain of Akt fused to an enhanced green fluorescent reporter protein (eGFP)[48,49]. Cells were serum-starved to induce cytoplasmic localization of the PH-domain due to absence of growth factors and therefore inactivation of the PI3K/Akt/mTOR pathway[50]. PH–eGFP plasma membrane association was then efficiently induced within 10 min on external addition of a combination of growth factors (insulin-like growth factor (IGF); endothelial growth factor (EGF)) into the medium. During the starvation period, cells were loaded with caged InsP$_7$ **9**/MoTr **11** nanoparticles for 4 h. This treatment alone had no effect on PH-domain localization. Next, cells were irradiated under a confocal laser-scanning microscope with short laser pulses (375 nm, 10 MHz, 30 s) in different areas (Fig. 6 dotted circle, and Supplementary Figs 14–23), and PH-domain localization was traced using the green channel. After photouncaging, a delayed

but complete PH-domain translocation from the plasma membrane into the cytoplasm was observed, and these results were repeated several times ($n = 4$). Significantly, translocation did not occur when cells were incubated with photocaged InsP$_7$ **9** or MoTr **11** only (Supplementary Figs 18–23). In these cases, the PH-domains remained localized on the membrane for several hours, demonstrating the need for the presence of all components and ruling out photobleaching of eGFP in the irradiated areas. A detailed analysis of additional micrographs in pseudo-colour with ratiometric changes is shown in the Supplementary Information (Supplementary Figs 14–23). This is the first example of controlled 5-InsP$_7$ **10** augmentation inside of a living cell within a few seconds timeframe coupled to a microscopic readout on the single cell level. We posit that this strategy will be useful to understand InsP$_7$ signalling in more detail as previously possible, as evidenced by the delayed PH-domain translocation observed for the first time in our experiments.

## Discussion

This study provides a new strategy to synthesize InsP$_7$ that enables introduction of caging subunits. The potential utility of

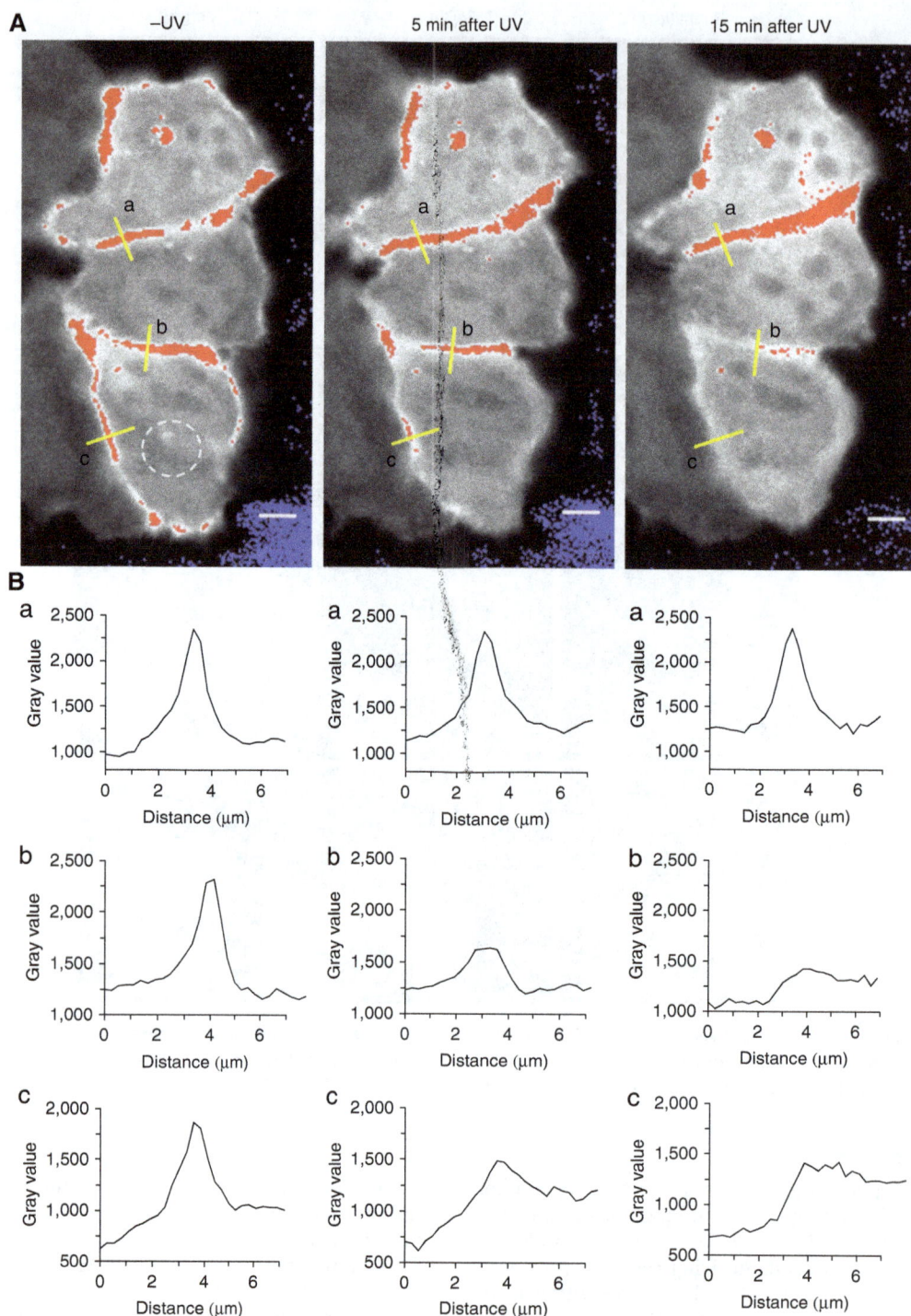

**Figure 6 | PH-domain translocation in irradiated and control cells.** Confocal fluorescence microscopy analysis of PH–eGFP translocation in HeLa Kyoto cells after photouncaging in defined areas (dotted circle). (**A**) Serum-starved cells were loaded with 5 μM **9** + **11** for 4 h and then stimulated with IGF and EGF (100 ng ml$^{-1}$). Robust recruitment of the PH-domain to the membrane is observed. Photouncaging in the dotted area (white circle) is achieved by short ultraviolet laser pulses and the change of fluorescence intensity followed over time (0, 5 and 15 min). (**B**) Development of the fluorescence intensity (indicated as gray value) over time (0, 5, 15 min) in three different membrane sections (a, b, c; distance in μm). Photouncaging leads to translocation of the PH-eGFP construct into the cytoplasm after 5 min from the membrane of the irradiated cell. After 15 min, complete translocation of the PH-domain into the cytoplasm is observed (**B**, b and c), whereas in the non-irradiated control cell the PH–eGFP construct remains localized on the membrane (**B**, a). Images are presented in pseudo-colour, normalized over time. Intensities were acquired pre-saturated, with the entire dynamic range of intensity available. Scale bars, 5 μM.

photocaged 5-InsP$_7$ **9** was demonstrated by photon-triggered uncaging in rat brain homogenate and other cell extracts. A complex of **9** with molecular transporter **11** was then shown to efficiently enter cells after non-covalent nanoparticle assembly.

Collectively, these results provide the first example of the synthesis of a photocaged analogue of InsP$_7$ and of its subsequent delivery into cells using non-covalent complexation with a guanidinium-rich molecular transporter. A recently developed

TiO$_2$ microsphere enrichment method was applied to study the *in cellulo* stability of 5-InsP$_7$ analogues and their efficient photochemical release in combination with MALDI mass spectrometry. We expect that this combined synthesis, delivery and analytical strategy will find widespread and general application in cell signalling studies as a convenient way to rapidly augment 5-InsP$_7$ **10** with spatiotemporal resolution. Along these lines, it was shown that cytoplasmic release of 5-InsP$_7$ triggers delayed but complete membrane desorption of the PH-domain of Akt within 15 min. In the human proteome, PH-domains are the 11th most common domain[51], and the new approach described in this publication will enable a systematic understanding of the effect of inositol-pyrophosphate augmentation on protein localization.

## Methods

**Experimental data of synthetic compounds.** For $^1$H, $^{13}$C and $^{31}$P NMR spectra of compounds and MALDI and HR-ESI MS spectra see Supplementary Figs 24–70. $^1$H NMR spectra were recorded on Bruker 400 MHz spectrometers or Bruker 500 MHz spectrometers (equipped with a cryo platform) at 298 K in the indicated deuterated solvent. $^{31}$P[$^1$H]-NMR spectra and $^{31}$P NMR spectra were recorded with $^1$H-decoupling or $^1$H coupling on Bruker 162 MHz or Bruker 202 MHz spectrometers (equipped with a cryo platform) at 298 K in the indicated deuterated solvent. All signals were referenced to an internal standard (PPP). $^{13}$C[$^1$H]-NMR spectra were recorded with $^1$H-decoupling on Bruker 101 MHz or Bruker 125 MHz spectrometers (equipped with a cryo platform) at 298 K in the indicated deuterated solvent. All signals were referenced to the internal solvent signal as standard (CDCl$_3$, $\delta$ 77.0; CD$_3$OD, $\delta$ 49.0; DMSO-d$_6$, $\delta$ 39.5).

Detailed synthetic procedures for all new compounds are provided in the Supplementary Information (see Supplementary Methods, chemical synthesis).

**Cellular uptake by flow cytometry.** Caged-IP7 **9** and oligomer **11** were brought up in pH 7.4 PBS buffer at 1 mM concentrations. HeLa cells were seeded at 40,000 cells per well in a 24-well plate and allowed to adhere overnight. The IP7:co-oligomer complexes were formed at a 1:1 molar ratio by mixing 8 μl of 1 mM oligomer stock with 8 μl of 1 mM caged-IP7 **9** stock in 184 μl PBS pH 7.4. For conditions with caged-IP7 **9** alone, 8 μl of 1 mM caged-IP7 **9** stock was added to 192 μl PBS pH 7.4. The complexes were allowed to incubate for 30 min at room temperature. The Lipofectamine 2000 control was prepared in OptiMEM according to the manufacturer's instructions (0.75 μl Lipofectamine into 62.5 μl. OptiMEM, 8 μl caged-IP7 **9** stock into 54.5 μl OptiMEM). The cells were washed with ~0.5 ml serum-free DMEM medium, then 400 μl serum-free DMEM was added to wells with untreated cells; 368.75 μl to wells treated with Lipofectamine 2000; 350 μl to treated wells. Then 31.25 μl Lipofectamine:caged-IP7 **9** and 50 μl of the caged-IP7:co-oligomer complexes were added to each respective well for a final caged-IP7 **9** concentration of 5 μM; all conditions were performed in triplicate. The cells were incubated at 37 °C for 3 h. The medium was then removed and the cells were washed with 1.0 ml PBS. About 0.4 ml EDTA trypsin was added, and the cells incubated for 8 min at 37 °C. Next, 0.6 ml of serum-containing DMEM medium was added, and the contents of each well were transferred to a 15 ml centrifuge tube and centrifuged (1,200 r.p.m. for 5 min). The cells were collected and re-dispersed in 200 μl PBS, transferred to FACS tubes, and read on a flow cytometry analyser. Results were analysed using FlowJo software. The data presented are the mean fluorescent signals from 10,000 cells analysed.

**Fluorescence microscopy.** Confocal microscopy was conducted on a CLSM SP5 Mid ultraviolet–visible Leica inverted confocal laser-scanning microscope equipped with a 15 W MaiTai DeepSee twophoton laser (Stanford Cell Sciences Imaging Facility, Award #S10RR02557401 from the National Center for Research Resources). HeLa cells were incubated in serum-free DMEM with 5 μM **9** or **9** + **11** (equimolar), for 4 or 16 h. After incubation, media was removed and cells washed 3 × with 1.0 mg ml$^{-1}$ heparin (180 U mg$^{-1}$, Aldrich) solution in PBS, then imaged in clear serum-free DMEM. Pictures were recorded with a HCX APO L × 20/1.00 water immersion objective and photomultiplier tube detector.

**eGFP–Akt-PH-domain translocation.** Mammalian cells (HeLa Kyoto) were grown up to 50–60% confluence (eight wells Lab-Tek Chamber Slide) in 250 μl DMEM (10% FBS, 10% Pen/Strep, high glucose (4.5 g l$^{-1}$)) at 37 °C, 5% CO$_2$ for 16 h. Cells were then washed with 250 μl PBS followed by a wash with serum-free DMEM (-FBS, -Pen/Step). About 230 μl serum-free DMEM medium were then added to each well.

Transfection of HELA cells with the eGFP Akt-PH-domain for imaging: The transfection mixture was prepared for eight wells. About 200 ng DNA (eGFP–Akt-PH) in 150 μl DMEM (-FBS) were incubated for 10 min at room temperature. Then, 10 μl FuGENE transfecting reagent were added to DNA containing DMEM (-FBS) medium and incubated for 20 min at room temperature. About 20 μl of the transfection mixture were then added to each well. Cells were incubated for 6–8 h at 37 °C (5% CO$_2$.). After this period, the transfection mixture was removed by washing the cells with 250 μl PBS and DMEM (-FBS). Cells were starved by adding 250 μl DMEM (-FBS) and incubated for 14–16 h at 37 °C (5% CO$_2$).

After overnight starvation, the cells were washed with PBS (250 μl) and 230 μl DMEM (-FBS) were added. Afterwards, 20 μl sample mixture (consisting of caged InsP7:Co-oligomer (D:G7:7) complexes) were added to each well and distributed equally. The final concentration of the complex was 5 μM for each well. Cells were then incubated at 37 °C (5% CO$_2$) for 4 h.

**Imaging of Akt-PH translocation.** After 4-h incubation with **9** and **11**, HeLa cells were washed with 250 μl (2 ×) imaging buffer (115 mM NaCl, 1.2 mM CaCl$_2$, 1.2 mM MgCl$_2$, 1.2 mM K$_2$HPO$_4$, 20 mM HEPES). About 200 μl imaging buffer were added to the wells. Translocation of eGFP–Akt-PH to the membrane was stimulated by adding a mixture of growth factors EGF and IGF (100 ng ml$^{-1}$ each). Translocation was monitored over 10 min at 37 °C (5% CO$_2$).

Cells that had responded to treatment with the growth factors and that had the PH-domain localized at the membrane were used in the uncaging experiments by confocal laser-scanning microscopy. Non-illuminated neighboring cells were used as controls.

**Confocal laser-scanning microscopy.** Imaging was performed on an Olympus IX83 confocal laser-scanning microscope at 37 °C in a 5% CO$_2$ high humidity atmosphere (EMBL incubation box). Imaging was performed using an Olympus Plan-APON × 60 (numerical aperture 1.4, oil) objective. The images were acquired utilizing a Hamamatsu C9100-50 EM CCD camera. Image acquisition was performed via FluoView imaging software, version 4.2. The green channel was imaged using the 488 nm laser line (120 mW cm$^{-2}$) at 3% laser power and a 525/50 emission mirror. The red channel was imaged using the 559 nm laser (120 mW cm$^{-2}$) at 2.0% laser power and a 643/50 emission filter. A pulsed 375 nm laser line (10 MHz) was applied for uncaging experiments. For uncaging experiments, circular regions of interest of 4–10 μm diameter were pre-defined. Pre-activation images were captured for five frames (5 s per frame), followed by 30 s of activation within the regions of interest. Recovery images were captured for 35 min at a frame rate of 5 s per frame.

**Image analysis.** Image analysis was conducted utilizing Fiji open source image analysis software tool[52]. Lookup tables were applied to match the colour within the recorded image with the wavelengths of detected light. For comparability, the lookup tables of pre- and postactivated images were set the same weighting.

## References

1. Bennett, M., Onnebo, S. M. N., Azevedo, C. & Saiardi, A. Inositol pyrophosphates: metabolism and signalling. *Cell. Mol. Life Sci.* **63**, 552–564 (2006).
2. Monserrate, J. P. & York, J. D. Inositol phosphate synthesis and the nuclear processes they affect. *Curr. Opin. Cell Biol.* **22**, 365–373 (2010).
3. Shears, S. B., Weaver, J. D. & Wang, H. Structural insight into inositol pyrophosphate turnover. *Adv. Biol. Regul.* **53**, 19–27 (2013).
4. Wilson, M. S. C., Livermore, T. M. & Saiardi, A. Inositol pyrophosphates: between signalling and metabolism. *Biochem. J.* **452**, 369–379 (2013).
5. Rao, F. *et al.* Inositol pyrophosphates promote tumor growth and metastasis by antagonizing liver kinase B1. *Proc. Natl Acad. Sci. USA* **112**, 1773–1778 (2015).
6. Rao, F. *et al.* Inositol hexakisphosphate kinase-1 mediates assembly/disassembly of the CRL4-signalosome complex to regulate DNA repair and cell death. *Proc. Natl Acad. Sci. USA* **111**, 16005–16010 (2014).
7. Laha, D. *et al.* VIH2 regulates the synthesis of inositol pyrophosphate InsP8 and jasmonate-dependent defenses in Arabidopsis. *Plant Cell* **27**, 1082–1097 (2015).
8. Illies, C. *et al.* Requirement of inositol pyrophosphates for full exocytotic capacity in pancreatic beta cells. *Science* **318**, 1299–1302 (2007).
9. Chakraborty, A. *et al.* Inositol pyrophosphates inhibit akt signalling, thereby regulating insulin sensitivity and weight gain. *Cell* **143**, 897–910 (2010).
10. Luo, H. R. *et al.* Inositol pyrophosphates mediate chemotaxis in dictyostelium via pleckstrin homology domain-PtdIns(3,4,5)P3 interactions. *Cell* **114**, 559–572 (2003).
11. Wu, M. X. *et al.* Elucidating diphosphoinositol polyphosphate function with nonhydrolyzable analogues. *Angew. Chem. Int. Ed.* **53**, 7192–7197 (2014).
12. Mackenzie, R. W. & Elliott, B. T. Akt/PKB activation and insulin signalling: a novel insulin signalling pathway in the treatment of type 2 diabetes. *Diabetes Metab. Syndr. Obes.* **7**, 55–64 (2014).
13. Yu, H. T., Li, J. B., Wu, D. D., Qiu, Z. J. & Zhang, Y. Chemistry and biological applications of photo-labile organic molecules. *Chem. Soc. Rev.* **39**, 464–473 (2010).

14. Ozaki, S., DeWald, D. B., Shope, J. C., Chen, J. & Prestwich, G. D. Intracellular delivery of phosphoinositides and inositol phosphates using polyamine carriers. *Proc. Natl Acad. Sci. USA* **97**, 11286–11291 (2000).

15. Best, M. D., Zhang, H. L. & Prestwich, G. D. Inositol polyphosphates, diphosphoinositol polyphosphates and phosphatidylinositol polyphosphate lipids: Structure, synthesis, and development of probes for studying biological activity. *Nat. Prod. Rep.* **27**, 1403–1430 (2010).

16. Hoglinger, D., Nadler, A. & Schultz, C. Caged lipids as tools for investigating cellular signalling. *Biochim. Biophys. Acta* **1841**, 1085–1096 (2014).

17. Kantevari, S., Gordon, G. R. J., MacVicar, B. A. & Ellis-Davies, G. C. R. A practical guide to the synthesis and use of membrane-permeant acetoxymethyl esters of caged inositol polyphosphates. *Nat. Protoc.* **6**, 327–337 (2011).

18. Li, W. H., Llopis, J., Whitney, M., Zlokarnik, G. & Tsien, R. Y. Cell-permeant caged InsP(3) ester shows that Ca$^{2+}$ spike frequency can optimize gene expression. *Nature* **392**, 936–941 (1998).

19. Pavlovic, I. et al. Prometabolites of 5-Diphospho-myo-inositol Pentakisphosphate. *Angew. Chem. Int. Ed.* **54**, 9622–9626 (2015).

20. Glennon, M. C. & Shears, S. B. Turnover of inositol pentakisphosphates, inositol hexakisphosphate and diphosphoinositol polyphosphates in primary cultured-hepatocytes. *Biochem. J.* **293**, 583–590 (1993).

21. Geihe, E. I. et al. Designed guanidinium-rich amphipathic oligocarbonate molecular transporters complex, deliver and release siRNA in cells. *Proc. Natl Acad. Sci. USA* **109**, 13171–13176 (2012).

22. Albert, C. et al. Biological variability in the structures of diphosphoinositol polyphosphates in Dictyostelium discoideum and mammalian cells. *Biochem. J.* **327**, 553–560 (1997).

23. Capolicchio, S., Thakor, D. T., Linden, A. & Jessen, H. J. Synthesis of unsymmetric diphospho-inositol polyphosphates. *Angew. Chem. Int. Ed.* **52**, 6912–6916 (2013).

24. Capolicchio, S., Wang, H. C., Thakor, D. T., Shears, S. B. & Jessen, H. J. Synthesis of densely phosphorylated Bis-1,5-diphospho-myo-inositol tetrakisphosphate and its enantiomer by bidirectional P-anhydride formation. *Angew. Chem. Int. Ed.* **53**, 9508–9511 (2014).

25. Falck, J. R. et al. Synthesis and structure of cellular mediators—inositol polyphosphate diphosphates. *J. Am. Chem. Soc.* **117**, 12172–12175 (1995).

26. Laussmann, T., Reddy, K. M., Reddy, K. K., Falck, J. R. & Vogel, G. Diphospho-myo-inositol phosphates from Dictyostelium identified as D-6-diphospho-myo-inositol pentakisphosphate and D-5,6-bisdiphospho-myo-inositol tetrakisphosphate. *Biochem. J.* **322**, 31–33 (1997).

27. Reddy, K. M., Reddy, K. K. & Falck, J. R. Synthesis of 2- and 5-diphospho-myo-inositol pentakisphosphate (2- and 5-PP-InsP(5)), intracellular mediators. *Tetrahedron Lett.* **38**, 4951–4952 (1997).

28. Wang, H. C. et al. Synthetic inositol phosphate analogues reveal that PPIP5K2 has a surface-mounted substrate capture site that is a target for drug discovery. *Chem. Biol.* **21**, 689–699 (2014).

29. Wu, M., Dul, B. E., Trevisan, A. J. & Fiedler, D. Synthesis and characterization of non-hydrolysable diphosphoinositol polyphosphate messengers. *Chem. Sci.* **4**, 405–410 (2013).

30. Zhang, H., Thompson, J. & Prestwich, G. D. A scalable synthesis of the IP7Isomer, 5-PP-Ins(1,2,3,4,6)P5. *Org. Lett.* **11**, 1551–1554 (2009).

31. Watanabe, Y., Nakamura, T. & Mitsumoto, H. Protection of phosphate with the 9-fluorenylmethyl group. Synthesis of unsaturated-acyl phosphatidylinositol 4,5-bisphosphate. *Tetrahedron Lett.* **38**, 7407–7410 (1997).

32. Mentel, M., Laketa, V., Subramanian, D., Gillandt, H. & Schultz, C. Photoactivatable and cell-membrane-permeable phosphatidylinositol 3,4,5-trisphosphate. *Angew. Chem. Int. Ed.* **50**, 3811–3814 (2011).

33. Subramanian, D. et al. Activation of membrane-permeant caged PtdIns(3)P induces endosomal fusion in cells. *Nat. Chem. Biol.* **6**, 324–326 (2010).

34. Jessen, H. J., Schulz, T., Balzarini, J. & Meier, C. Bioreversible protection of nucleoside diphosphates. *Angew. Chem. Int. Ed.* **47**, 8719–8722 (2008).

35. Thomson, W. et al. Synthesis, bioactivation and anti-hiv activity of the Bis (4-Acyloxybenzyl) and Mono(4-Acyloxybenzyl) esters of the 5'-monophosphate of Azt. *J. Chem. Soc. Perkin Trans.* **1**, 1239–1245 (1993).

36. Cremosnik, G. S., Hofer, A. & Jessen, H. J. Iterative synthesis of nucleoside oligophosphates with phosphoramidites. *Angew. Chem. Int. Ed.* **53**, 286–289 (2014).

37. Hofer, A. et al. A Modular synthesis of modified phosphoanhydrides. *Chem. Eur. J.* **21**, 10116–10122 (2015).

38. Losito, O., Szijgyarto, Z., Resnick, A. C. & Saiardi, A. Inositol pyrophosphates and their unique metabolic complexity: analysis by gel electrophoresis. *PLoS ONE* **4**, e5580 (2009).

39. Anstaett, P., Leonidova, A. & Gasser, G. Caged phosphate and the slips and misses in determination of quantum yields for ultraviolet-A-induced photouncaging. *ChemPhysChem* **16**, 1857–1860 (2015).

40. Anstaett, P., Leonidova, A., Janett, E., Bochet, C. G. & Gasser, G. Reply to commentary by trentham et al. on "caged phosphate and the slips and misses in

determination of quantum yields for ultraviolet-a-induced photouncaging" by Gasser et al. *ChemPhysChem* **16**, 1863–1866 (2015).

41. Corrie, J. E., Kaplan, J. H., Forbush, B., Ogden, D. C. & Trentham, D. R. Commentary on "caged phosphate and the slips and misses in determination of quantum yields for ultraviolet-a-induced photouncaging" by G. Gasser and Co-Workers. *ChemPhysChem* **16**, 1861–1862 (2015).

42. Schonleber, R. O., Bendig, J., Hagen, V. & Giese, B. Rapid photolytic release of cytidine 5-diphosphate from a coumarin derivative: A new tool for the investigation of ribonucleotide reductases. *Bioorg. Med. Chem.* **10**, 97–101 (2002).

43. Stanzl, E. G., Trantow, B. M., Vargas, J. R. & Wender, P. A. Fifteen years of cell-penetrating guanidinium-rich molecular transporters: basic science, research tools, and clinical applications. *Acc. Chem. Res.* **46**, 2944–2954 (2013).

44. Lee, H. L. et al. Single-molecule motions of oligoarginine transporter conjugates on the plasma membrane of Chinese hamster ovary cells. *J. Am. Chem. Soc.* **130**, 9364–9370 (2008).

45. Rothbard, J. B., Jessop, T. C., Lewis, R. S., Murray, B. A. & Wender, P. A. Role of membrane potential and hydrogen bonding in the mechanism of translocation of guanidinium-rich peptides into cells. *J. Am. Chem. Soc.* **126**, 9506–9507 (2004).

46. Wender, P. A., Galliher, W. C., Goun, E. A., Jones, L. R. & Pillow, T. H. The design of guanidinium-rich transporters and their internalization mechanisms. *Adv. Drug Deliv. Rev.* **60**, 452–472 (2008).

47. Wilson, M. S. C., Bulley, S. J., Pisani, F., Irvine, R. F. & Saiardi, A. A novel method for the purification of inositol phosphates from biological samples reveals that no phytate is present in human plasma or urine. *Open Biol.* **5**, 150014 (2015).

48. Cormack, B. P., Valdivia, R. H. & Falkow, S. FACS-optimized mutants of the green fluorescent protein (GFP). *Gene* **173**, 33–38 (1996).

49. Laketa, V. et al. PIP(3) induces the recycling of receptor tyrosine kinases. *Sci. Signal.* **7**, ra5 (2014).

50. Jo, H. et al. Small molecule-induced cytosolic activation of protein kinase Akt rescues ischemia-elicited neuronal death. *Proc. Natl Acad. Sci. USA* **109**, 10581–10586 (2012).

51. Lemmon, M. A. Pleckstrin homology (PH) domains and phosphoinositides. *Biochem. Soc. Symp.* **74**, 81–93 (2007).

52. Schindelin, J. et al. Fiji: an open-source platform for biological-image analysis. *Nat. Methods* **9**, 676–682 (2012).

## Acknowledgements

Imaging was performed with support of the Center for Microscopy and Image Analysis at UZH and the Stanford Shared FACS Facility. We thank Professors Jay Siegel, John Robinson, Adolfo Saiardi, Robert Waymouth, Chris Contag, Lynette Cegelski, Dr. Vanessa Pierroz, Dr. Riccardo Rubbiani and Dr. Vibor Laketa for discussions, materials and procedures. This work was supported by The Swiss National Science Foundation (PP00P2_157607 to H.J.J. and PP00P2_133568 & PP00P2_157545 to G.G.), the National Institutes of Health (NIH-CA031841, NIH-S10RR027431-01 and NIH-CA031845 to P.A.W.), and the Deutsche Forschungsgemeinschaft (DFG, TRR83 to C.S.). Fellowships: National Science Foundation and the Stanford Center for Molecular Analysis and Design.

## Author contributions

I.P., D.T.T., R.C.C., J.R.V., C.J.M., P.A. and S.H. conducted the experiments. H.J.J., P.A.W., C.S., L.B. and G.G. planned the experiments. H.J.J., P.A.W., J.R.V. and C.J.M. wrote the manuscript. All authors discussed the results and revised the manuscript.

## Additional information

# 3

# Transfer hydrogenation catalysis in cells as a new approach to anticancer drug design

Joan J. Soldevila-Barreda[1,*], Isolda Romero-Canelón[1,*], Abraha Habtemariam[1] & Peter J. Sadler[1]

Organometallic complexes are effective hydrogenation catalysts for organic reactions. For example, Noyori-type ruthenium complexes catalyse reduction of ketones by transfer of hydride from formate. Here we show that such catalytic reactions can be achieved in cancer cells, offering a new strategy for the design of safe metal-based anticancer drugs. The activity of ruthenium(II) sulfonamido ethyleneamine complexes towards human ovarian cancer cells is enhanced by up to $50 \times$ in the presence of low non-toxic doses of formate. The extent of conversion of coenzyme $NAD^+$ to NADH in cells is dependent on formate concentration. This novel reductive stress mechanism of cell death does not involve apoptosis or perturbation of mitochondrial membrane potentials. In contrast, iridium cyclopentadienyl catalysts cause cancer cell death by oxidative stress. Organometallic complexes therefore have an extraordinary ability to modulate the redox status of cancer cells.

[1]Department of Chemistry, University of Warwick, Gibbet Hill Road, Coventry CV4 7AL, UK. * These authors contributed equally to this work. Correspondence and requests for materials should be addressed to P.J.S. (email: P.J.Sadler@warwick.ac.uk).

Organometallic complexes are well known as catalysts for organic chemical reactions, for example, olefin metathesis by Grubbs' ruthenium carbene complexes[1,2] and asymmetric hydrogenation of ketones by Noyori's ruthenium arene complexes[3,4]. The prospect of using organometallic complexes as catalytic drugs is attractive since this might allow low safe non-toxic doses of transition metals to be administered, and furthermore introduce novel mechanisms of action that can overcome resistance to current widely used platinum anticancer drugs[5–7]. However, achieving catalytic activity in cells is challenging on account of the presence of many nucleophilic biomolecules, which might act as catalyst poisons. Here we show for the first time that Noyori-type ruthenium complexes can catalytically reduce coenzyme NAD$^+$ in human ovarian cancer cells using non-toxic concentrations of formate as a hydride donor. Moreover, such catalysis greatly enhances the potency of the complexes and increases selectivity towards cancer cells *versus* normal cells. The mechanism of cancer cell death involves the generation of reductive stress, in contrast to organometallic iridium cyclopentadienyl anticancer catalysts, which induce oxidative stress[8]. It is apparent that organometallic complexes have a unique ability to modulate the redox status of cells.

First we considered the choice of organometallic transfer hydrogenation catalysts for cell studies. Several Ru$^{II}$, Rh$^{III}$ and Ir$^{III}$ complexes have previously been reported to catalyse the regioselective reduction of NAD$^+$ to 1,4-NADH in water, using formate as a hydride source[9–11]. Stekhan and Fish *et al.* elucidated the mechanism of reduction of NAD$^+$ using Cp*Rh$^{III}$ bipyridine complexes (formate coordination, loss of CO$_2$, hydride transfer to Rh$^{III}$, then to NAD$^+$ in a weak-association complex)[12–16]. Süss-Fink *et al.*[17] reported the catalytic reduction of NAD$^+$ using a series of Ru$^{II}$, Rh$^{III}$ and Ir$^{III}$ phenanthroline catalysts. [(Cp*)Rh(phen)Cl]$^+$ exhibits turnover frequencies (TOFs) of up to twice those of [(Cp*)Rh(bipy)Cl]$^+$ in aqueous media[17]. Rh$^{III}$ and Ru$^{II}$ catalysts bearing dipyridyl amine ligands functionalized with maleimide are also active, but not as active as (Cp*)Rh(bipy)Cl]$^+$ (ref. 18). In 2012, Hollmann *et al.*[19] reported a tethered Rh$^{III}$ complex containing an analogue of the Noyori ligand TsDPEN immobilized in a poly(ethylene) polymer. The catalytic activity of the heterogeneous catalyst was lower than that of other soluble Rh$^{III}$ complexes.

Initially, we investigated the use of Rh$^{III}$ sulfonamido ethyleneamine complexes for catalytic hydrogenation of NAD$^+$ in cancer cells but found them to be less effective than the Ru$^{II}$ analogues, which were therefore explored in more detailed studies. We showed previously that the efficiency of [(η$^6$-arene)Ru(R-SO$_2$-En)Cl] catalysts depends on both the nature of the arene and on the sulfonamide substituent (R)[20]. For cell work, we compared complexes 1–4 with *p*-cymene (*p*-cym) as the arene and the sulfonamide substituent R = methyl (MsEn, 1), *p*-methylbenzene (TsEn, 2), *p*-trifluoromethylbenzene (TfEn, 3) and *p*-nitrobenzene (NbEn, 4), with analogues 5–7 containing *o*-terphenyl (*o*-terp) as the arene and R = methyl (MsEn, 5), *p*-methylbenzene (TsEn, 6), *p*-trifluoromethylbenzene (TfEn, 7), Fig. 1. The *o*-terp arene was chosen on account of its increased hydrophobicity (Supplementary Table 1) to enhance uptake into cells. The Ru$^{II}$-ethylenediamine complex 8 was studied for comparison since this compound has poor catalytic activity, but good anticancer activity *in vitro* and *in vivo*[21].

## Results
Regioselective catalytic reduction of NAD$^+$ (2 mol equiv) to 1,4-NADH was observed in MeOH-$d_4$/D$_2$O (2:9 v/v) and D$_2$O alone using complexes 1–4 and formate as the hydride source (25 mol equiv) by $^1$H nuclear magnetic resonance (NMR) at

310 K, pH* 7.2 ± 0.1. The TOFs for the reactions (Table 1, determined as described in the Methods section) showed a general trend in which the more electron-withdrawing sulfonamides gave higher catalytic activity: NbEn > TfEn > TsEn > MsEn. Such a trend has been previously reported[20,22,23]. The complexes studied in the present work show comparable TOFs (*ca.* 0.2–7 h$^{-1}$) for the reduction of NAD$^+$ to those reported for [(η$^6$-arene)Ru(N,N′)Cl] complexes (*ca.* 0.006–10 h$^{-1}$), although much lower than those obtained using Rh$^{III}$ or Ir$^{III}$ complexes[17,18].

The catalytic reduction of NAD$^+$ using complexes 5–7 under the same conditions was more rapid, being complete by the time the first $^1$H NMR spectrum was recorded. The reaction was therefore then carried out using only 6 mol equiv of NAD$^+$, and the catalytic reactions were then complete within the first 10 min (5–7). Subsequently, formation of a brown precipitate was observed and, as a consequence, the TOF of these complexes could not be determined. However, it was apparent that the *o*-terp complexes were more active as catalysts than the *p*-cym complexes (for which the highest activity was observed for 4, 20.5 min for completion).

The catalytic mechanism was expected to involve initial binding of formate to Ru followed by transfer of hydride and (irreversible) release of CO$_2$. Evidence for this was provided by complex 5 for which a Ru-H $^1$H NMR peak was detected at −5.5 p.p.m. (Supplementary Fig. 1) as well as mass spectrometry (MS) peaks assignable to both the formate and hydride adducts (Supplementary Table 2).

We then investigated the antiproliferative activity of complexes 1–7 in A2780 human ovarian cancer cells, and compared them with both complex 8 (RM175) and the clinically approved drug cisplatin. Complexes 1–4 were moderately active with half-maximal inhibitory concentration (IC$_{50}$) values ranging from 11.9 to 14.7 μM (Fig. 2; Supplementary Table 3), some 6–12 × less potent than complex 8 (2.2 μM) and cisplatin (1.2 μM). The *o*-terp complexes 5–7 were slightly less active, with IC$_{50}$ values in the range 12.4–21.2 μM. The potency order in both series is MsEn > TsEn > TfEn, the reverse of the order seen for catalytic activity above. However, no correlation would be expected since these antiproliferative experiments were not done under catalytic conditions.

The most notable difference in the properties of the ethylenediamine complex 8 and the sulfonamido ethyleneamine complexes 1–7 is the ability of the sulfonamide substituent to stabilize an adjacent deprotonated Ru-bound N and in turn 16e$^-$ intermediates in the catalytic cycle. We thought that this change in electronic distribution would affect DNA binding, which is thought to play a major role in the activity of complex 8. Complex 8 binds strongly to guanine bases in DNA after initial aquation[24–31].

$^1$H NMR studies showed that aquation of complexes 1–7 was complete in <5 min at 298 K, and therefore faster and more favourable than for 8. As might be expected from the increased charge density on Ru in the sulfonamide complexes, the p$K_a$ values of the aqua adducts of complexes 1–7 were higher (by about two units, range 9.5–9.9, Table 1; Supplementary Fig. 2) than for aquated 8 (p$K_a$ 7.7). The sulfonamido complexes would therefore exist mainly as aqua adducts over the pH range used in the present experiments (close to physiological, pH 7).

Perhaps surprisingly, the model base 9-ethylguanine had a similar affinity for complexes 1–4 as complex 8 (as determined by $^1$H NMR, equilibrium constants 60–105 mM$^{-1}$ compared with 60 mM$^{-1}$ for 8, Supplementary Table 4), and binding (to N7, Supplementary Fig. 3) was rapid (pH* 7.2, 310 K). The adducts were also characterized by electrospray ionization–MS (Supplementary Table 5). No binding to 9-methyladenine was

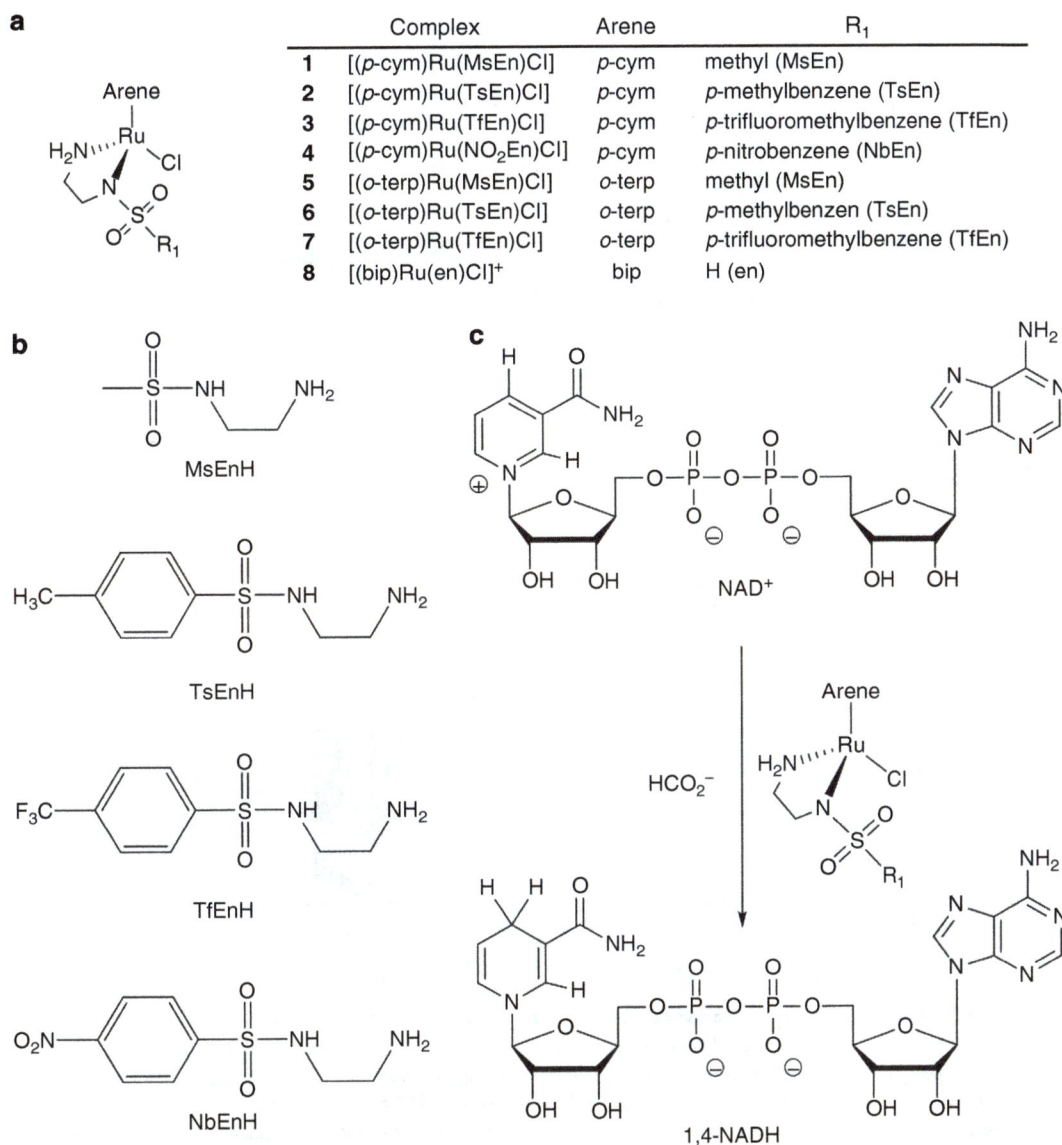

| | Complex | Arene | $R_1$ |
|---|---|---|---|
| 1 | [(p-cym)Ru(MsEn)Cl] | p-cym | methyl (MsEn) |
| 2 | [(p-cym)Ru(TsEn)Cl] | p-cym | p-methylbenzene (TsEn) |
| 3 | [(p-cym)Ru(TfEn)Cl] | p-cym | p-trifluoromethylbenzene (TfEn) |
| 4 | [(p-cym)Ru(NO₂En)Cl] | p-cym | p-nitrobenzene (NbEn) |
| 5 | [(o-terp)Ru(MsEn)Cl] | o-terp | methyl (MsEn) |
| 6 | [(o-terp)Ru(TsEn)Cl] | o-terp | p-methylbenzen (TsEn) |
| 7 | [(o-terp)Ru(TfEn)Cl] | o-terp | p-trifluoromethylbenzene (TfEn) |
| 8 | [(bip)Ru(en)Cl]⁺ | bip | H (en) |

**Figure 1 | Complexes studied in this work and the catalytic reaction.** (**a**) Structure of Ru complexes, (**b**) structures of the ligands and (**c**) catalytic conversion of $NAD^+$ into 1,4-NADH mediated by formate.

**Table 1 | pK$_a$* values for aqua adducts of catalysts 1–4 (and 8 for comparison) and turnover frequencies for transfer hydrogenation of $NAD^+$ using formate.**

| Complexes | | pK$_a$* | TOF ($h^{-1}$) D₂O/MeOH-$d_4$# | TOF ($h^{-1}$) D₂O |
|---|---|---|---|---|
| 1 | [(p-cym)Ru(MsEn)Cl] | 9.86 ± 0.01 | 1.25 ± 0.03 | 1.11 ± 0.02 |
| 2 | [(p-cym)Ru(TsEn)Cl] | 9.78 ± 0.06 | 2.88 ± 0.06 | 1.58 ± 0.04 |
| 3 | [(p-cym)Ru(TfEn)Cl] | 9.71 ± 0.01 | 5.7 ± 0.3 | 3.06 ± 0.05 |
| 4 | [(p-cym)Ru(NO₂En)Cl] | 9.56 ± 0.04 | 9.6 ± 0.3 | 4.1 ± 0.1 |
| 8 | [(bip)Ru(en)Cl]⁺ | 7.71 ± 0.01† | — | — |

pK$_a$*, pKa value determined in deuterated solvent; TOF, turnover frequency.
#23% MeOH-$d_4$/77% D₂O.
†ref. 31, pK$_a$ value (pK$_a$* 7.85).

detected. However, no binding of complex 2 (as a representative of the series) to calf thymus DNA was detected by inductively coupled plasma-mass spectrometry (ICP-MS) measurements of bound Ru even after a 27-h incubation at 310 K (mol ratio 1:3 complex:DNA base pairs, 2 mM cacodylate buffer pH 7.4, 2 mM NaCl). The absence of strong binding to calf thymus DNA was also apparent from circular and linear dichroism and melting temperature studies (Supplementary Table 6; Supplementary

Figs 4 and 5). This weak binding is probably a consequence of unfavourable steric and electronic interactions between the DNA double helix and the sulfonamido side chain in comparison with the unsubstituted ethylenediamine analogue 8. Hence, we can conclude that DNA is not likely to be a target for the sulfonamide complexes.

Next, we investigated the antiproliferative activity of the complexes towards A2780 human ovarian cancer cells in

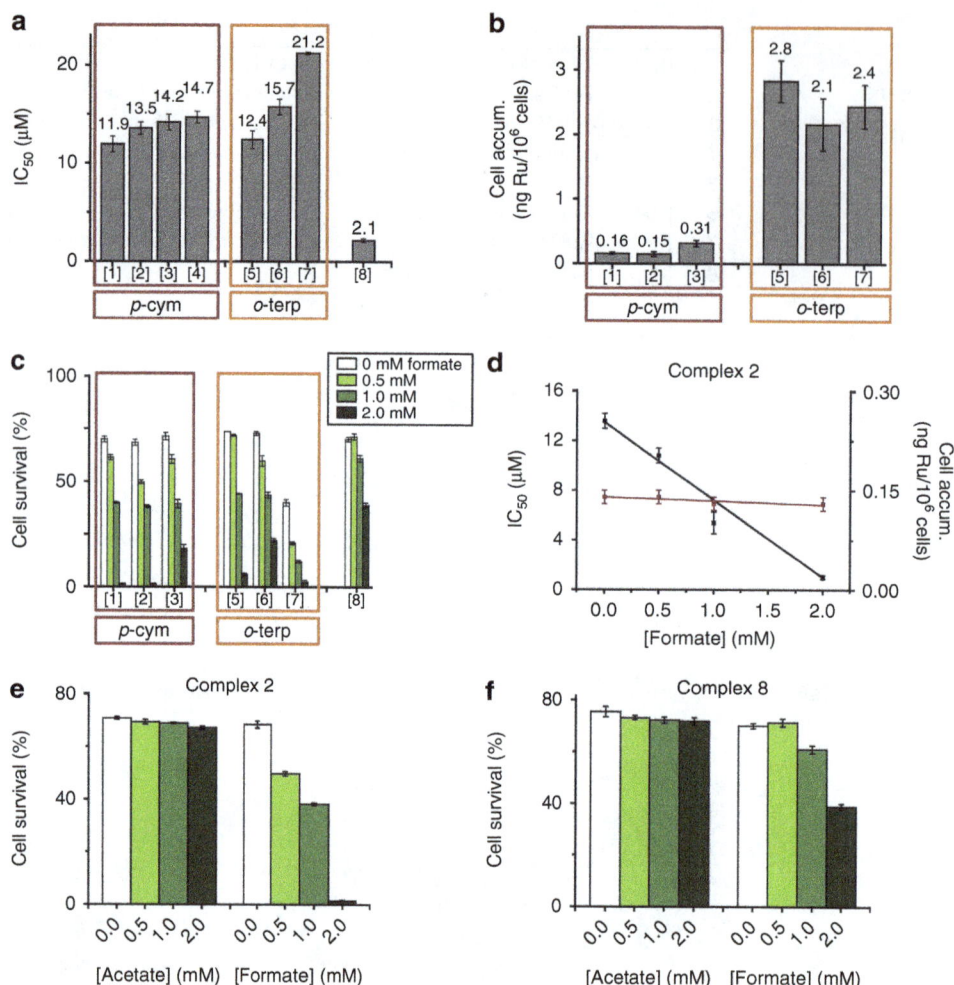

**Figure 2 | Antiproliferative activity in A2780 human ovarian cancer cells.** (**a**) $IC_{50}$ values for complexes **1–8**. (**b**) Cellular accumulation of Ru from $p$-cym complexes **1–3** and their $o$-terp analogues **5–7** after 24 h. (**c**) Percent cell survival when equipotent concentrations of complexes **1–3** and **5–8** (1/3 × $IC_{50}$) were co-administered with different concentrations of sodium formate. (**d**) $IC_{50}$ values for complex **2** when co-administered with various concentrations of sodium formate and the cellular accumulation of Ru under similar conditions. (**e**) and (**f**) Comparison of cell survival (%) for complexes **2** and **8** when co-administered with various concentrations of sodium acetate or sodium formate. All experiments included 24 h of drug exposure. All the experiments were performed as duplicates of triplicates in independent experiments and the error bars were calculated as the s.d. from the mean.

the presence of formate, conditions under which catalytic intracellular conversion of $NAD^+$ to NADH might occur. First we showed that formate itself at concentrations from 0 to 2 mM is non-toxic to the cells (Fig. 3). Then A2780 cells were co-incubated with equipotent concentrations of complexes **1–8** (1/3 × $IC_{50}$) and three different concentrations of sodium formate (0.5, 1 and 2 mM). The antiproliferative activity of complexes **1–7** was enhanced by co-administration with formate (Supplementary Tables 7 and 8; Fig. 2). The degree of enhancement of activity was directly proportional to the formate concentration (Fig. 2), consistent with a direct contribution to the activity from catalytic transfer hydrogenation. The antiproliferative activity of complex **8** was also enhanced by the co-administration of formate, but to a much lower extent (Fig. 2). The largest decrease in cell survival, of *ca.* 50 ×, was observed for complexes [($p$-cym)Ru(MsEn)Cl] (**1**) and [($p$-cym)Ru(TsEn)Cl] (**2**), dropping from *ca.* 69% to 1% when the concentration of formate was increased from 0 to 2 mM (Fig. 2). Complexes containing $o$-terp also showed a significant enhancement of antiproliferative activity, with complex **7** giving the greatest decrease in cell survival (*ca.* 16-fold). In contrast, the presence of formate only decreased cell survival induced by complex **8** (RM175) from *ca.* 70 to 39%.

Next, we investigated whether the formate-induced increase in activity was accompanied by intracellular conversion of $NAD^+$ to NADH using the most active complex **2**. Cellular accumulation of ruthenium in A2780 cells exposed to complex **2** (4.5 μM) showed no significant dependence on the amount of formate added (Supplementary Table 8; Fig. 2), consistent with a direct relationship between the potentiation of activity and the ability of the formate to act as a hydride source. Furthermore, the cellular distribution of Ru from complex **2** (4.5 μM) was very similar in the presence and absence of formate (2 mM), being, in both cases, greatest in the cytosolic fraction (51%) > membrane and organelle fraction (38%) >> nuclear fraction (9%). There was little Ru in the cytoskeletal fraction (Supplementary Table 9).

The $IC_{50}$ of complex **2** in A2780 cells decreased from 13.6 ± 0.6 (no formate) to 1.0 ± 0.2 μM in the presence of 2 mM formate (Fig. 2), notably achieving the potency of the clinical drug cisplatin, but having a different mechanism of action. Importantly, co-incubation with formate seems to increase the selectivity factor of complex **2** (from 3.6 ± 0.2 to 5 ± 1, Supplementary Table 10). Selectivity factors were calculated as the ratio between the antiproliferative activity towards normal cells (MRC5 fibroblasts) and the activity in A2780 ovarian cancer cells.

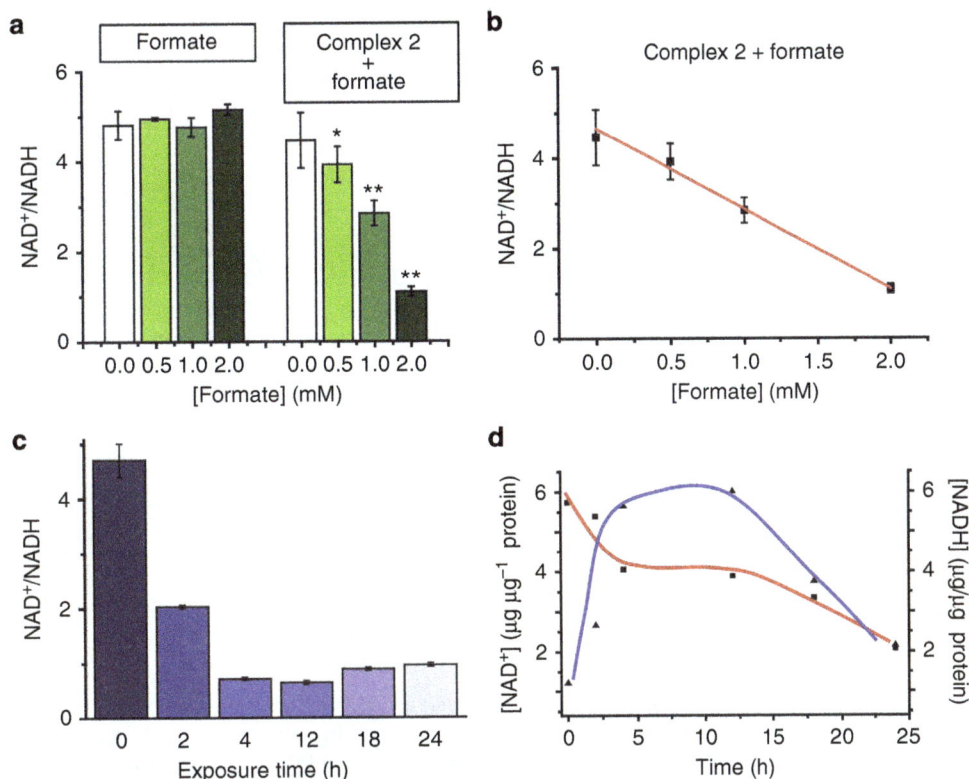

**Figure 3 | Perturbation of the NAD$^+$/NADH ratio in A2780 cells exposed to 1/3 $\times$ IC$_{50}$ of complex 2 and sodium formate.** (**a**) Variation in the concentration of formate (0, 0.5, 1 and 2 mM) in the absence (left) and presence (right) of complex **2**, 24 h exposure. (**b**) Graph showing the linear correlation between the concentration of sodium formate and the NAD$^+$/NADH ratio. (**c**) Effect of exposure time (0, 2, 4, 12, 18 and 24 h) on the NAD$^+$/NADH ratio. (**d**) Variation in the concentration of NAD$^+$ and NADH with time. All the experiments were performed as duplicates of triplicates in independent experiments and the error bars were calculated as the s.d. from the mean. An independent two-sample $t$-test with unequal variances, Welch's test, was used to define the statistical difference between the values obtained for the NAD$^+$/NADH ratio experiment. *$P < 0.05$, **$P < 0.01$.

Cells, and living organisms in general, maintain a tight balance for their redox system by controlling the levels of oxidants and antioxidants. In particular, cancer cells are under constant oxidative stress, and high levels of reactive oxygen species such as OH, O$_2^-$ and H$_2$O$_2$ are common as a result of disturbed mitochondrial function[32,33]. As a consequence, neoplastic tissues are especially sensitive to further changes in the redox balance, for example, those caused by changes in the NAD$^+$/NADH ratio. In contrast, non-cancerous cells have normal-functioning mitochondria that allow them to adjust to changes in their redox balance. This difference provides a strategy for conferring selectivity on anticancer agents that attack cancer cell metabolism[34–36].

The specific role of formate was confirmed by comparison with acetate. The activity of neither complex **2** nor **8** towards human ovarian cancer cells was affected significantly by acetate under similar conditions (Fig. 2; Supplementary Table 11). Acetate, unlike formate, cannot act as a source of hydride for catalytic hydrogenation reactions.

We then investigated whether complex **2** in combination with formate could convert NAD$^+$ to NADH in A2780 human ovarian cancer cells. In the first experiment, A2780 cells were incubated for 24 h with complex **2** and three different concentrations of sodium formate (0.5, 1 and 2 mM). Neither the complex nor formate alone had a significant effect on the intracellular NAD$^+$/NADH ratio (Fig. 3). However, the NAD$^+$/NADH ratio decreased significantly after co-administration of **2** and increasing amounts of formate (from 4.5 to 1.1 with 2 mM formate, Fig. 3).

An apparent TOF (TOF$_{ap}$) of $0.19 \pm 0.01$ h$^{-1}$ was determined for the conversion of NAD$^+$ to NADH in A2780 cancer cells

treated with 4.5 μM complex **2** and 2 mM sodium formate for various times (0, 2, 4, 12, 18 and 24 h). However, the interpretation of this value is not straightforward since many other processes in cells may influence the concentrations of NAD$^+$ and NADH. The NAD$^+$/NADH ratio decreased rapidly for the first 4 h, from 4.78 ($t = 0$ h, untreated) to 0.83 ($t = 4$ h), followed by a 4-h period in which the decrease was less pronounced, reaching a final ratio of 0.74 after 12 h (Fig. 3). After 12 h, a slight recovery was observed, with the ratio reaching 0.98 after 24 h. This might be due to gradual poisoning of the catalyst (reaction with other biomolecules) or degradation.

Flow cytometry analysis of A2780 cells exposed to complex **2** and 2 mM sodium formate showed that the levels of reactive oxygen species (ROS) in treated cells are comparable to those arising from treatment with $N$-acetyl-L-cysteine, a well-known reductant (Fig. 4a). The intracellular generation of ruthenium hydride species may lead not only to the reduction of NAD$^+$ (and NADP$^+$), but also to the reduction of other biomolecules such as ketones and imines[20,37,38].

Induction of apoptosis in A2780 cells exposed to complex **2** and sodium formate was investigated using flow cytometry. Annexin V-fluorescein isothiocyanate (FITC) and propidium iodide (PI) dual-staining allowed the detection of four different populations: viable cells, non-viable cells and early-stage and late-stage apoptosis. Early-stage apoptosis is characterized by changes in the symmetry of the phospholipid membrane, and late stage by further disruption of the integrity of the cell membrane so it becomes permeable to PI. In the case of cells exposed to complex **2** and 2 mM of sodium formate, no population of cells in either of these two stages of apoptosis was detected after 24 h (Fig. 4b), as was the case after treatment with sodium formate alone.

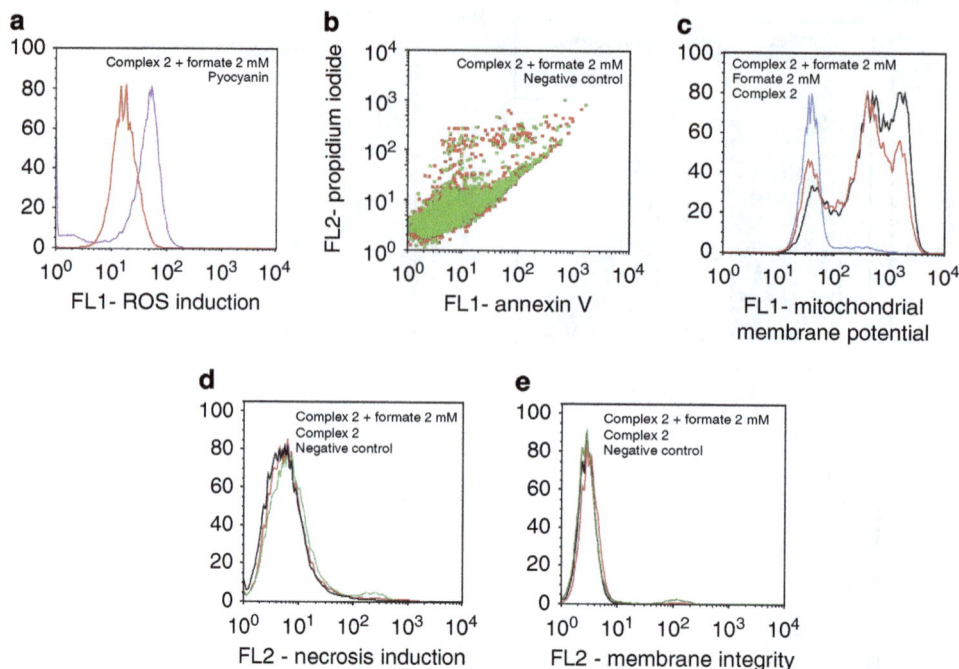

**Figure 4 | Mechanism of action studies on A2780 ovarian cancer cells.** All experiments used 24 h of drug exposure time, fixed drug concentrations (13.6 μM complex **2**, 2 mM sodium formate) and no recovery time. (**a**) Induction of ROS in cells, pyocyanin was used as a positive control; (**b**) induction of apoptosis; (**c**) changes in the mitochondrial membrane potential; (**d**) induction of necrosis and (**e**) changes in cellular membrane integrity.

We also used flow cytometry to investigate whether exposure of cells to complex **2** and sodium formate induced variations in the mitochondrial membrane potential. The induction of apoptosis is closely related to mitochondrial changes. In the intrinsic pathway of apoptosis activation, changes in the mitochondrial membrane potential allow release of cytochrome *c* into the cytosol, which promotes the subsequent activation of the caspase cascade[39]. No changes in the mitochondrial membrane potential were observed (Fig. 4c). This is consistent with a mechanism of action based on the perturbation of the redox state of cells, and the lack of apoptosis.

In the absence of apoptosis after 24 h of drug exposure, we investigated the induction of necrosis by flow cytometry. A2780 cells exposed to complex **2** in the presence and absence of 2 mM sodium formate tested negative for increased fluorescence of the 7-ADD stain in the FL2-red channel, indicating that, under these experimental conditions, there is no cellular necrosis either (Fig. 4d). Furthermore, the same technique was used to confirm cellular membrane integrity. In this case, cells exposed to complex **2** alone or in combination with 2 mM sodium formate showed low fluorescence in the FL2 channel reading for CytoPainter red, which reacts with exposed amines in compromised membranes (Fig. 4e). Other mechanisms that could be considered in future work include the induction of autophagy[40] and pathogenic mitochondrial oxidation[41], both of which have been linked to reductive stress, but it is also possible that the cell death observed here involves new pathways and cannot be simply mapped onto known mechanisms.

## Discussion

It is challenging to design metal complexes that might exhibit catalytic activity in living cells[5-7]. Catalysts for chemical transformations work efficiently under well-defined conditions, whereas cells contain a wide range of nucleophiles, as well as oxidants and reductants, which might readily poison the catalyst *via* substitution or redox reactions. Nevertheless, some success

has been reported[42-45]. For example, manganese macrocycles such as M40403 can act as superoxide dismutase mimics and decompose superoxide into $O_2$ and $H_2O_2$ in cells[46,47]. Iron porphyrin complexes can catalyse the reduction of azides to imines[48], and copper peptide complexes can catalyse the degradation of RNA in hepatitis models[42,49]. Cobalt complexes, which cleave peptide bonds[50-52], can decompose amyloids present in diseases such as Alzheimer's or Parkinson's[53,54]. In particular, organometallic compounds have shown success in the last few years. For example, Ru complexes such as [($\eta^6$-arene) Ru(azpy)I]$^+$ (azpy = *N*,*N*-dimethylphenyl- or hydroxyphenyl-azopyridine) catalytically oxidize glutathione (GSH) to glutathione disulfide (GSSG), increasing the levels of ROS in cells[55].

Meggers *et al.*[56,57] have also shown that organoruthenium complexes such as [(Cp*)Ru(COD)Cl] (COD = cyclooctadiene) or the photoactivatable complex [(Cp*)Ru($\eta^6$-pyrene)]PF$_6$ can catalyse the cleavage of allylcarbamates from protected imines. More recently, the catalytic cleavage of allylcarbamates in cells has been greatly improved by the use of complexes of the type [(Cp)Ru(QA-R)($\eta^3$-allyl)]PF$_6$ (QA = 2-quinolinecarboxylate; R = π-donating groups). The latter complexes can activate protected fluorophores or even protected anticancer drugs in cells[58], an approach that could also provide reduction in toxicity for anticancer drugs. An interesting strategy for avoiding poisoning and increasing the stereoselectivity of organometallic catalysts is to incorporate them in the pocket of a host protein, as illustrated by Ward *et al.* using biotin-labelled Noyori-type Ir$^{III}$ complexes and streptavidin as the protein host[59].

In this work, we have shown for the first time that Noyori-type catalytic transfer hydrogenation can be achieved in living cells, using Ru$^{II}$ arene complexes with a chelated sulfonylethylamine ligand and formate as the hydride donor to convert coenzyme NAD$^+$ into NADH, and thereby modulate the NAD$^+$/NADH redox couple.

Formate at non-toxic concentrations not only initiates the catalysis, but also enhances the anticancer potency of the

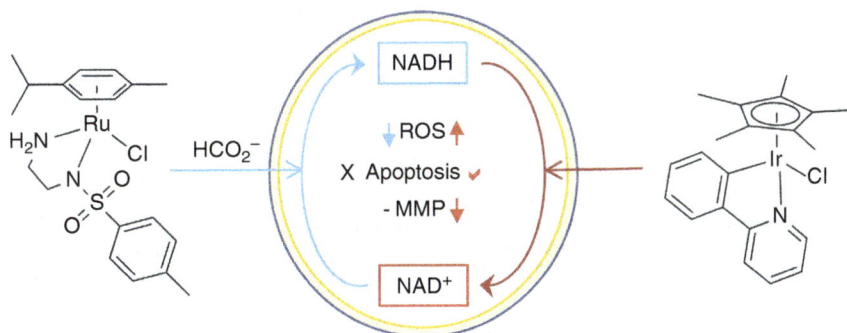

**Figure 5 | The contrasting mechanisms of action of two classes of organometallic anticancer catalysts.** Illustrated is the induction of reductive stress in cancer cells by the $Ru^{II}$ arene complexes reported here, and oxidative stress by $Ir^{III}$ cyclopentadienyl complexes[8]. Both processes are mediated by changes in the $NAD^+/NADH$ redox couple. $Ru^{II}$/formate donates hydride to $NAD^+$, lowers the level of reactive oxygen species (ROS), but does not trigger changes in mitochondrial membrane potential (MMP) nor apoptosis. In contrast, the $Ir^{III}$ complex accepts hydride from NADH, causes apoptosis, alters the MMP and generates oxidative stress via ROS production.

complexes by a factor of up to *ca.* 14. Furthermore, this seems to be accompanied by an increase in selectivity, which for complex **2** rises from $3.6 \pm 0.1$ to $5 \pm 1$ in the presence of formate ($P < 0.01$ in a Welch's test). This introduces a new design concept for transition metal complexes, one which may allow therapeutic doses to be kept low and safe with minimal side effects. Moreover, the mechanism of cell death, reductive stress through conversion of coenzyme $NAD^+$ to NADH, is unusual and different from that of cisplatin and most other anticancer agents. Hence, this new approach should be effective for treatment of cisplatin resistance, which has become a major clinical problem. Apoptosis is usually thought to be the main mechanism of cisplatin-induced cell death, but flow cytometry experiments on A2780 human ovarian cancer cells, together with the lack of changes in the mitochondrial membrane potential, ruled out the induction of apoptosis as the mechanism of cell death for the catalytic Ru complexes studied here.

Whether such a combination of a catalyst and hydride transfer agent (formate) could be useful in therapy remains to be further investigated, but the importance of this work is more fundamental. There now exist organometallic catalysts that can perturb the $NAD^+/NADH$ ratio in cells in either direction: towards NADH using the $Ru^{II}$ arene sulfonamido ethyleneamine/formate system reported here, or towards $NAD^+$ using $Ir^{III}$ cyclopentadienyl complexes (Fig. 5)[8]. Organometallic complexes are unique in their ability to achieve such redox modulation in living cells.

## Methods

**Transfer hydrogenation of $NAD^+$.** Complexes **1–7** were dissolved in MeOH-$d_4$/$D_2O$ (5:1 *v/v*, 1.4 mM, 4 ml). Solutions of sodium formate (35 mM, 4 ml) and $NAD^+$ in $D_2O$ (2.8 mM, 2 ml) were also prepared and incubated at 310 K. In a typical experiment, 200-μl aliquots of each solution were added to a 5-mm NMR tube and the pH* adjusted to $7.2 \pm 0.1$ (total volume 0.635 ml, final concentrations: Ru complex 0.44 mM; $NAD^+$ 0.88 mM; formate 11.02 mM; molar ratio 1:2:25). $^1H$ NMR spectra were recorded at 310 K every 162 s until the completion of the reaction.

Molar ratios of $NAD^+$ and NADH were determined by integrating the peaks at 9.33 and 6.96 p.p.m. ($NAD^+$ and 1,4-NADH, respectively). The turnover number for the reaction was calculated as follows:

$$\text{TON} = \frac{I_{6.96}}{I_{6.96} + I_{9.93}} \frac{[NAD^+]_0}{[\text{Catalyst}]}$$

where $I_n$ is the integral of the signal at $n$ p.p.m. and $[NAD^+]_0$ is the concentration of $NAD^+$ at the start of the reaction.

**Cell culture.** A2780 human ovarian carcinoma and MRC5 human fetal lung fibroblasts were obtained from the European Collection of Cell Cultures. Both cell lines were grown in Roswell Park Memorial Institute medium (RPMI-1640) supplemented with 10% of fetal calf serum, 1% of 2 mM glutamine and 1% penicillin/streptomycin. All cells were grown as adherent monolayers at 310 K in a 5% $CO_2$-humidified atmosphere and passaged at ca. 70–80% confluency.

***In vitro* growth inhibition assays.** The antiproliferative activity of complexes **1–8** was determined in A2780 ovarian cancer cells and MRC5 fibroblasts. Briefly, 96-well plates were used to seed 5,000 cells per well. The plates were left to pre-incubate with drug-free medium at 310 K for 48 h before adding different concentrations of the compounds to be tested. A drug exposure period of 24 h was allowed. After this, supernatants were removed by suction and each well was washed with PBS. A further 72 h was allowed for the cells to recover in drug-free medium at 310 K. The sulforhodamine B (SRB) assay was used to determine cell viability. $IC_{50}$ values, as the concentration that causes 50% cell death, were determined as duplicates of triplicates in two independent sets of experiments and their s.d. were calculated.

**Co-administration of Ru complexes with formate or acetate.** Cell viability assays were carried out with complexes **1–8** in A2780 ovarian cancer cells. These experiments were carried out as described above with the following modifications: a fixed concentration of each Ru complex equal to $1/3 \times IC_{50}$ was used in co-administration with three different concentrations of sodium formate or sodium acetate (0.5, 1.0 and 2.0 mM). To prepare the stock solution of the drug, the complex was dissolved in 5% dimethylsulfoxide and diluted in a 1:1 mixture of 0.9% saline:cell culture medium. This stock was further diluted using RPMI-1640 until working concentrations were achieved. Separately, stock solutions of sodium formate or sodium acetate were prepared in saline. The complex and formate were added to each well independently, but within 5 min of each other.

**$IC_{50}$ of complex 2 on co-administration with formate.** The antiproliferative activity towards A2780 ovarian cancer cells of complex **2** when co-administered with 2 mM sodium formate was determined. These experiments were performed as described above, using a fixed concentration of formate and variable concentrations of the Ru complex. The SRB assay was used to determine cell viability. $IC_{50}$ values were determined as duplicates of triplicates in two independent sets of experiments and their s.d. were calculated.

**Ruthenium accumulation in cancer cells.** Briefly, $1.5 \times 10^6$ cells per well were seeded on a six-well plate. After 24 h of pre-incubation, the complexes were added to give final concentrations equal to $IC_{50}/3$ and a further 24 h of drug exposure was allowed. After this time, cells were washed, treated with trypsin-EDTA, counted and cell pellets were collected. Each pellet was digested overnight in concentrated nitric acid (73%) at 353 K; the resulting solutions were diluted using double-distilled water to give a final concentration of 5% $HNO_3$ and the amount of Ru taken up by the cells was determined by ICP–MS. These experiments did not include any cell recovery time in drug-free media; they were all carried out as duplicates of triplicates and the s.d. were calculated. This experiment was also carried out using complex **2** and co-administering 2 mM sodium formate.

**Ruthenium distribution in cancer cells.** Cell pellets were obtained as described above, and were fractionated using the Fraction PREP kit from BioVision according to the supplier's instructions. Each sample was digested overnight in concentrated nitric acid (73%) and the amount of Ru taken up by the cells was determined by ICP–MS. These experiments were all carried out in triplicate and the s.d. were calculated.

**$NAD^+/NADH$ determination.** Experiments to determine the $NAD^+/NADH$ ratio in A2780 ovarian cancer cells exposed to complex **2** were carried out using the $NAD^+/NADH$ assay kit from Abcam (ab65348) according to the manufacturer's instructions. Briefly, A2780 cells were seeded in six-well plates at a density of

$1.0 \times 10^6$ cells per well. After 24 h of pre-incubation, cells were treated with a fixed concentration of **2** equal to its $IC_{50}$ in co-administration with three different concentrations of sodium formate (0, 0.5, 1 and 2 mM). These concentrations of sodium formate were also used on their own as a second set of negative controls. Drug exposure was allowed for 24 h, after which all supernatants were removed by suction and wells were washed with PBS. Cells were treated with trypsin, detached and counted before being pelleted by centrifugation. Cell pellets were extracted using the NAD/NADH extraction buffer and filtered using a 10-kDa-molecular weight cutoff filter. Samples were split into two, the first half was used to determine total NADt (NAD + NADH), and the second half to determine NADH after heating the samples at 333 K for 30 min. The absorbance at 450 nm was normalized against the protein content in each sample determined using the Bradford assay.

**Time dependence of NAD$^+$/NADH ratio.** Experiments to determine the time dependence of the NAD/NADH ratio in A2780 ovarian cancer cells exposed to complex **2** were carried out using the NAD$^+$/NADH assay kit as described above with the following modifications. The drug exposure time was variable and included time points at 0, 2, 4, 12, 18 and 24 h; the concentration of complex **2** was fixed at $IC_{50}$ levels, and the concentration of formate was 2 mM.

**ROS determination.** Flow cytometry analysis of total induction of ROS in A2780 cells caused by exposure to complex **2** and sodium formate was carried out using the Total ROS detection kit (Enzo Life Sciences) according to the supplier's instructions. Briefly, $1.0 \times 10^6$ A2780 cells per well were seeded in a six-well plate. Cells were pre-incubated in drug-free medium at 310 K for 24 h in a 5% $CO_2$-humidified atmosphere, after which they were exposed to either sodium formate (2 mM), complex **2** (concentration equal to $IC_{50}$) or a combination of both. After 24 h of drug exposure, supernatants were removed by suction and cells were washed and harvested. Staining was achieved by resuspending the cell pellets in buffer containing the green fluorescent reagent. Cells were analysed in a Becton Dickinson FACScan Flow Cytometer using Ex/Em: 490/525 nm for the oxidative stress detection. Data were processed using Flowjo software. The experiment included cells treated with pyocyanin for 30 min as positive control.

**Mitochondrial membrane assay.** Analysis of the changes of mitochondrial potential in A2780 cells after exposure to complex **2** and sodium formate was carried out using the Abcam, JC-10 Mitochondrial Membrane Potential Assay kit according to the manufacturer's instructions. Briefly, $1.0 \times 10^6$ cells were seeded in six-well plates and left to incubate for 24 h in drug-free medium at 310 K in a humidified atmosphere. Drug solutions were added (sodium formate 2 mM, complex **2** at a concentration equal to $IC_{50}$ or a combination of both, in triplicate experiments) and the cells were left to incubate for a further 24 h under similar conditions. Supernatants were removed by suction and each well was washed with PBS before detaching the cells using trypsin-EDTA. Staining of the samples was done in flow cytometry tubes protected from light, incubating for 30 min at room temperature. Samples were immediately analysed on a Beckton Dickinson FAC Scan with fluorescence detection. Data were processed using FlowJo software. This experiment included carbonyl cyanide $m$-chlorophenyl hydrazine as positive control.

**Induction of apoptosis.** Flow cytometry analysis of apoptosis in A2780 cells caused by exposure to complex **2** and sodium formate were carried out using the Annexin V-FITC Apoptosis Detection Kit (Sigma Aldrich) according to the manufacturer's instructions. Briefly, A2780 cells were seeded in six-well plates ($1.0 \times 10^6$ cells per well), pre-incubated for 24 h in drug-free media at 310 K, after which they were exposed to either sodium formate (2 mM), complex **2** (concentration equal $IC_{50}$) or a combination of both. Cells were harvested using trypsin and stained using PI/Annexin V-FITC. After staining, cell pellets were analysed in a Becton Dickinson FACScan Flow Cytometer. For positive-apoptosis controls, A2780 cells were exposed for 2 h to staurosporine ($1 \mu g \, ml^{-1}$). Cells for apoptosis studies were used with no previous fixing procedure as to avoid nonspecific binding of the annexin V-FITC conjugate.

**Induction of necrosis.** Flow cytometry analysis of necrosis in A2780 cells caused by exposure to complex **2** and sodium formate was carried out using a Apotosis/Necrosis detection kit (Abcam) according to the manufacturer's instructions. Briefly, A2780 cells were seeded in six-well plates ($1.0 \times 10^6$ cells per well), pre-incubated for 24 h in drug-free media at 310 K, after which they were exposed to either sodium formate (2 mM), complex **2** (concentration equal $IC_{50}$) or a combination of both. Cells were harvested using trypsin and stained using the membrane-impermeable nuclear dye 7-ADD (Ex/Em 546/647 nm). After staining in the dark, cell pellets were analysed in a Becton Dickinson FACScan Flow Cytometer. For positive-necrosis controls, A2780 cells were exposed to methanol.

**Cellular membrane integrity.** Flow cytometry analysis of cellular membrane integrity of A2780 cells caused by exposure to complex **2** and sodium formate were carried out using the CytoPainter assay (Abcam) according to the manufacturer's

instructions. Briefly, A2780 cells were seeded in six-well plates ($1.0 \times 10^6$ cells per well), pre-incubated for 24 h in drug-free media at 310 K, after which they were exposed to either sodium formate (2 mM), complex **2** (concentration equal $IC_{50}$) or a combination of both. Cells were harvested using trypsin and stained in the dark using CytoPainter Red (Ex/Em 583/603 nm), which reacts with cell surface amines in compromised membranes. After staining, cell pellets were analysed in a Becton Dickinson FACScan Flow Cytometer.

## References

1. Schwab, P., Grubbs, R. H. & Ziller, J. W. Synthesis and applications of RuCl$_2$('CHR')(PR$_3$)$_2$: the influence of the alkylidene moiety on metathesis activity. *J. Am. Chem. Soc.* **118**, 100–110 (1996).
2. Schwab, P., France, M. B., Ziller, J. W. & Grubbs, R. H. A series of well-defined metathesis catalysts–synthesis of [RuCl$_2$('CHR')(PR$_3$)$_2$] and its reactions. *Angew. Chem. Int. Ed.* **34**, 2039–2041 (1995).
3. Hashiguchi, S. *et al.* Kinetic resolution of racemic secondary alcohols by Ru"-catalyzed hydrogen transfer. *Angew. Chem. Int. Ed.* **36**, 288–290 (1997).
4. Haack, K. J., Hashiguchi, S., Fujii, A., Ikariya, T. & Noyori, R. The catalyst precursor, catalyst, and intermediate in the Ru"-promoted asymmetric hydrogen transfer between alcohols and ketones. *Angew. Chem. Int. Ed.* **36**, 285–288 (1997).
5. Shen, D. W., Pouliot, L. M., Hall, M. D. & Gottesman, M. M. Cisplatin resistance: a cellular self-defense mechanism resulting from multiple epigenetic and genetic changes. *Pharmacol. Rev.* **64**, 706–721 (2012).
6. Abu-Surrah, A. S. & Kettunen, M. Platinum group antitumor chemistry: design and development of new anticancer drugs complementary to cisplatin. *Cur. Med. Chem.* **13**, 1337–1357 (2006).
7. Furrer, J., Smith, G. S. & Therrien, B. in *Inorganic Chemical Biology: Principles, Techniques and Applications* (ed. Gasser, G.) 341–364 (Wiley, 2014).
8. Liu, Z. *et al.* The potent oxidant anticancer activity of organoiridium catalysts. *Angew. Chem. Int. Ed.* **53**, 3941–3946 (2014).
9. Chenault, H. K. & Whitesides, G. M. Regeneration of nicotinamide cofactors for use in organic synthesis. *Appl. Biochem. Biotechnol.* **14**, 147–197 (1987).
10. Wu, H. *et al.* Methods for the regeneration on NAD coenzymes. *Green Chem.* **15**, 1773–1789 (2013).
11. Quinto, T., Köhler, V. & Ward, T. R. Recent trends in biomimetic NADH regeneration. *Top. Catal.* **57**, 321–331 (2014).
12. Buriez, O., Kerr, J. & Fish, R. Regioselective reduction of NAD$^+$ models with [Cp*Rh(Bpy)H]$^+$: structure–activity relationships and mechanistic aspects in the formation of the 1, 4-NADH derivatives. *Angew. Chem. Int. Ed.* **38**, 1997–2000 (1999).
13. Lo, H. C. & Fish, R. H. Biomimetic NAD$^+$ models for tandem cofactor regeneration, horse liver alcohol dehydrogenase recognition of 1,4-NADH derivatives, and chiral synthesis. *Angew. Chem. Int. Ed.* **41**, 478–481 (2002).
14. Lo, H. C. *et al.* Bioorganometallic chemistry. 13. Regioselective reduction of NAD$^+$ models, 1-benzylnicotinamde triflate and beta-nicotinamide ribose-5'-methyl phosphate, with in situ generated [CpRh(Bpy)H]$^+$: structure-activity relationships, kinetics, and mechanist. *Inorg. Chem.* **40**, 6705–6716 (2001).
15. Leiva, C., Lo, H. C. & Fish, R. H. Aqueous organometallic chemistry. 3. Catalytic hydride transfer reactions with ketones and aldehydes using [Cp*Rh(bpy)(H$_2$O)](OTf)$_2$ as the precatalyst and sodium formate as the hydride source: Kinetic and activation parameters, and the significance of steric and electronic effects. *J. Organomet. Chem.* **695**, 145–150 (2010).
16. Lutz, J. *et al.* Bioorganometallic chemistry: biocatalytic oxidation reactions with biomimetic NAD$^+$/NADH co-factors and [Cp*Rh(bpy)H]$^+$ for selective organic synthesis. *J. Organomet. Chem.* **689**, 4783–4790 (2004).
17. Canivet, J., Süss-Fink, G. & Štěpnička, P. Water-soluble phenanthroline complexes of rhodium, iridium and ruthenium for the regeneration of NADH in the enzymatic reduction of ketones. *Eur. J. Inorg. Chem.* **2007**, 4736–4742 (2007).
18. Haquette, P. *et al.* Chemically engineered papain as artificial formate dehydrogenase for NAD(P)H regeneration. *Org. Biomol. Chem.* **9**, 5720–5727 (2011).
19. de Torres, M., Dimroth, J., Arends, I. W. C. E., Keilitz, J. & Hollmann, F. Towards recyclable NAD(P)H regeneration catalysts. *Molecules* **17**, 9835–9841 (2012).
20. Soldevila-Barreda, J. J. *et al.* Improved catalytic activity of ruthenium − arene complexes in the reduction of NAD$^+$. *Organometallics* **31**, 5958–5967 (2012).
21. Yan, Y. K., Melchart, M., Habtemariam, A., Peacock, A. F. & Sadler, P. J. Catalysis of regioselective reduction of NAD$^+$ by ruthenium(II) arene complexes under biologically relevant conditions. *J. Biol. Inorg. Chem.* **11**, 483–488 (2006).
22. Wu, X. *et al.* On water and in air: fast and highly chemoselective transfer hydrogenation of aldehydes with iridium catalysts. *Angew. Chem. Int. Ed.* **45**, 6718–6722 (2006).
23. Tan, J. *et al.* pH-regulated transfer hydrogenation of quinoxalines with a Cp*Ir–diamine catalyst in aqueous media. *Tetrahedron* **67**, 6206–6213 (2011).

24. Bugarcic, T. *et al.* Cytotoxicity, cellular uptake, and DNA interactions of new monodentate ruthenium(II) complexes containing terphenyl arenes. *J. Med. Chem.* **51**, 5310–5319 (2008).

25. Chen, H., Parkinson, J. A., Morris, R. E. & Sadler, P. J. Highly selective binding of organometallic ruthenium ethylenediamine complexes to nucleic acids: novel recognition mechanisms. *J. Am. Chem. Soc.* 173–186 (2003).

26. Chen, H. *et al.* Organometallic ruthenium(II) diamine anticancer complexes: arene-nucleobase stacking and stereospecific hydrogen-bonding in guanine adducts. *J. Am. Chem. Soc.* **124**, 3064–3082 (2002).

27. Habtemariam, A. *et al.* Structure-activity relationships for cytotoxic ruthenium(II) arene complexes containing N,N-, N,O-, and O,O-chelating ligands. *J. Med. Chem.* **49**, 6858–6868 (2006).

28. Wang, F., Xu, J., Habtemariam, A. & Sadler, P. J. Competition between glutathione and guanine for a ruthenium(II) arene anticancer complex: detection of a sulfenato intermediate. *J. Am. Chem. Soc.* **127**, 17734–17743 (2005).

29. Wang, F. *et al.* Competition between glutathione and DNA oligonucleotides for ruthenium(II) arene anticancer complexes. *Dalton Trans.* **42**, 3188–3195 (2013).

30. Aird, R. E. *et al.* In vitro and in vivo activity and cross resistance profiles of novel ruthenium (II) organometallic arene complexes in human ovarian cancer. *Br. J. Cancer* **86**, 1652–1657 (2002).

31. Wang, F. *et al.* Kinetics of aquation and anation of ruthenium(II) arene anticancer complexes, acidity and X-ray structures of aqua adducts. *Chem. Eur. J.* **9**, 5810–5820 (2003).

32. Diehn, M. *et al.* Association of reactive oxygen species levels and radioresistance in cancer stem cells. *Nature* **458**, 780–783 (2009).

33. Jungwirth, U. *et al.* Anticancer activity of metal complexes: involvement of redox processes. *Antioxid. Redox Signal.* **15**, 1085–1127 (2011).

34. Romero-Canelón, I. & Sadler, P. J. Next-generation metal anticancer complexes: multitargeting via redox modulation. *Inorg. Chem.* **52**, 12276–12291 (2013).

35. Watson, J. Oxidants, antioxidants and the current incurability of metastatic cancers. *Open Biol.* **3**, 1–9 (2013).

36. Trachootham, D., Alexandre, J. & Huang, P. Targeting cancer cells by ROS-mediated mechanisms: a radical therapeutic approach? *Nat. Rev. Drug. Discov.* **8**, 579–591 (2009).

37. Betanzos-Lara, S. *et al.* Organometallic ruthenium and iridium transfer-hydrogenation catalysts using coenzyme NADH as a cofactor. *Angew. Chem. Int. Ed.* **51**, 3897–3900 (2012).

38. Liu, Z. *et al.* Reduction of quinones by NADH catalyzed by organoiridium complexes. *Angew. Chem. Int. Ed.* **52**, 4194–4197 (2013).

39. Bayir, H. *et al.* Apoptotic interactions of cytochrome c: redox flirting with anionic phospholipids within and outside of mitochondria. *Biochim. Biophys. Acta* **1757**, 648–659 (2006).

40. Galluzzi, L. *et al.* Molecular definitions of cell death subroutines: recommendations of the Nomenclature Committee on Cell Death 2012. *Cell Death Differ.* **19**, 107–120 (2012).

41. Zhang, H. *et al.* Glutathione-dependent reductive stress triggers mitochondrial oxidation and cytotoxicity. *FASEB J.* **26**, 1442–1451 (2012).

42. Hocharoen, L. & Cowan, J. A. Metallotherapeutics: novel strategies in drug design. *Chem. Eur. J.* **15**, 8670–8676 (2009).

43. Li, J. & Chen, P. R. Moving Pd-mediated protein cross coupling to living systems. *Chembiochem* **13**, 1728–1731 (2012).

44. Noffke, A. L., Habtemariam, A., Pizarro, A. M. & Sadler, P. J. Designing organometallic compounds for catalysis and therapy. *Chem. Commun.* **48**, 5219–5246 (2012).

45. Sasmal, P. K., Streu, C. N. & Meggers, E. Metal complex catalysis in living biological systems. *Chem. Commun.* **49**, 1581–1587 (2013).

46. Salvemini, D. *et al.* A nonpeptidyl mimic of superoxide dismutase with therapeutic activity in rats. *Science* **286**, 304–306 (1999).

47. Filipović, M. R. *et al.* Striking inflammation from both sides: manganese(II) pentaazamacrocyclic SOD mimics act also as nitric oxide dismutases: a single-cell study. *Angew. Chem. Int. Ed.* **49**, 4228–4232 (2010).

48. Sasmal, P. K. *et al.* Catalytic azide reduction in biological environments. *Chem. Bio. Chem.* **13**, 1116–1120 (2012).

49. Bradford, S. & Cowan, J. A. Catalytic metallodrugs targeting HCV IRES RNA. *Chem. Commun.* **48**, 3118–3120 (2012).

50. Lee, J. *et al.* Cleavage agents for α-synuclein. *Bull. Korean Chem. Soc.* **29**, 882–884 (2008).

51. Suh, J. *et al.* Cleavage agents for soluble oligomers of human islet amyloid polypeptide. *J. Biol. Inorg. Chem.* **13**, 693–701 (2008).

52. Suh, J. *et al.* Cleavage agents for soluble oligomers of amyloid beta peptides. *Angew. Chem. Int. Ed.* **46**, 7064–7067 (2007).

53. Lee, T. Y. & Suh, J. Target-selective peptide-cleaving catalysts as a new paradigm in drug design. *Chem. Soc. Rev.* **38**, 1949–1957 (2009).

54. Kayed, R. *et al.* Common structure of soluble amyloid oligomers implies common mechanism of pathogenesis. *Science* **300**, 486–489 (2003).

55. Dougan, S. J., Habtemariam, A., McHale, S. E., Parsons, S. & Sadler, P. J. Catalytic organometallic anticancer complexes. *Proc. Natl Acad. Sci. USA* **105**, 11628–11633 (2008).

56. Streu, C. & Meggers, E. Ruthenium-induced allylcarbamate cleavage in living cells. *Angew. Chem. Int. Ed.* **45**, 5645–5648 (2006).

57. Sasmal, P. K., Carregal-Romero, S., Parak, W. J. & Meggers, E. Light-triggered ruthenium-catalyzed allylcarbamate cleavage in biological environments. *Organometallics* **31**, 5968–5970 (2012).

58. Völker, T., Dempwolff, F., Graumann, P. L. & Meggers, E. Progress towards bioorthogonal catalysis with organometallic compounds. *Angew. Chem. Int. Ed.* **53**, 10536–10540 (2014).

59. Köhler, V. *et al.* Synthetic cascades are enabled by combining biocatalysts with artificial metalloenzymes. *Nat. Chem.* **5**, 93–99 (2013).

## Acknowledgements

This research was supported by the ERC (grant no. 247450), EPSRC (grant no. EP/F034210/1), University of Warwick IAS (fellowship for J.J.S.-B.) and Science City (ERDF/AWM). We thank Bushra Qamar and Dr Magdalena Moss for technical assistance with cell culture, Professor Alison Rodger for access to CD and LD equipment and Dr Nikola Chmel, Dr Maria Romero and Feng Chen for assistance with its use and interpretation. We also acknowledge participation in the EU COST Action CM1105.

## Author contributions

J.J.S.-B., I.R.-C., A.H. and P.J.S. designed the research. J.J.S.-B., I.R.-C. and A.H. performed research. J.J.S.-B., I.R.-C., A.H. and P.J.S. analysed data and J.J.S.-B., I.R.-C. and P.J.S. wrote the paper.

## Additional information

# 4

# Chemical basis for the recognition of trimethyllysine by epigenetic reader proteins

Jos J.A.G. Kamps[1,*], Jiaxin Huang[2,*], Jordi Poater[3,*], Chao Xu[4], Bas J.G.E. Pieters[1], Aiping Dong[4], Jinrong Min[4], Woody Sherman[5], Thijs Beuming[5], F. Matthias Bickelhaupt[1,3], Haitao Li[2] & Jasmin Mecinović[1]

A large number of structurally diverse epigenetic reader proteins specifically recognize methylated lysine residues on histone proteins. Here we describe comparative thermodynamic, structural and computational studies on recognition of the positively charged natural trimethyllysine and its neutral analogues by reader proteins. This work provides experimental and theoretical evidence that reader proteins predominantly recognize trimethyllysine via a combination of favourable cation-$\pi$ interactions and the release of the high-energy water molecules that occupy the aromatic cage of reader proteins on the association with the trimethyllysine side chain. These results have implications in rational drug design by specifically targeting the aromatic cage of readers of trimethyllysine.

[1] Institute for Molecules and Materials, Radboud University, Heyendaalseweg 135 , 6525 AJNijmegen, The Netherlands. [2] Department of Basic Medical Sciences, Center for Structural Biology, School of Medicine, Tsinghua University, Beijing 100084, China. [3] Department of Theoretical Chemistry and Amsterdam Center for Multiscale Modeling, VU University, De Boelelaan 1083, 1081 HV Amsterdam, The Netherlands. [4] Structural Genomics Consortium, University of Toronto, 101 College Street, Toronto, Ontario, Canada M5G 1L7. [5] Schrödinger, Inc., 120 West 45th Street, New York, New York 10036 USA. * These authors contributed equally to this work. Correspondence and requests for materials should be addressed to J.M. (email: j.mecinovic@science.ru.nl).

The positioning and chemical diversity of post-translational modifications on histone proteins orchestrate the structure and function of the eukaryotic chromatin[1-3]. One such modification is lysine methylation, which is associated both with gene activation and repression, depending on the type of histone and details of the sequence site[4]. The methylation of lysine residues of histone proteins is a dynamic process that is regulated by SAM-dependent histone lysine methyltransferases, FAD- or Fe(II)/2OG-dependent histone demethylases, and reader proteins (also known as effector proteins) that specifically recognize post-translationally modified lysines in histones and affect the downstream cellular processes[5-7]. Enzymatic lysine methylation can lead to the formation of monomethyllysine (Kme1), dimethyllysine (Kme2) and trimethyllysine (Kme3), with each methylation mark being specifically recognized by different classes of the interacting reader proteins[8]. Lower methylation states Kme1 and Kme2 are specifically read by 53BP1 tandem tudor domains, L3MBTL1 MBT repeats, G9a ankyrin repeats and ORC1 BAH domain, primarily via the cavity-insertion binding mode[9-13]. The constitution of the Kme1/Kme2 recognition site enables the specificity in two ways: the methylammonium group forms the energetically favourable hydrogen bonding and electrostatic interactions with the negatively charged aspartate or glutamate, allowing the methyl group to position towards the aromatic residues, and the narrow binding pocket sterically prevents the access of the bulkier Kme3. The highest methylation state Kme3 is specifically recognized by a structurally diverse class of reader proteins, including plant homeodomain (PHD) zinc fingers, WD40 repeats and members of the Royal superfamily (tudor domain, chromodomain and PWWP domain), in the surface-groove binding mode[8,10]. For the Kme3 reading modules, binding studies of histone peptides showed that binding affinities typically follow the trend Kme3 > Kme2 > Kme1 > K (ref. 14). With the exception of ATRX ADD domain, most characterized reader proteins specifically recognize Kme3 through an aromatic cage that consists of 1–4 aromatic amino acids (Phe, Tyr and Trp) and/or one methionine[15]. Aromatic cages of several reader modules also contain negatively charged Asp or Glu residues. The positioning of the quaternary ammonium (Kme3) group inside the aromatic cage, as demonstrated by structural determination of several reader–Kme3 complexes, suggests that the specific readout process is primarily driven by cation–$\pi$ interactions, although charge-independent interactions may also contribute to the overall binding[8,16-18]. Herein we report clear experimental and computational support for the chemical basis for the recognition of Kme3-containing histones by reader proteins. Our study reveals that the association between trimethyllysine and the aromatic cage of reader proteins is driven by energetically favourable cation–$\pi$ interactions between the positively charged trimethyllysine and the electron-rich aromatic cage, and the trimethyllysine-mediated release of non-optimally structured water molecules that occupy the aromatic cages of reader proteins.

## Results

**Physical–organic chemistry approach.** Specific favourable binding of the positively charged side chain of Kme3 to the aromatic cage of reader proteins could, in principle, be a result of (i) favourable solute–solute interactions (cation–$\pi$ and CH–$\pi$ interactions), (ii) partial desolvation of the Kme3 side chain of histone tails (via the hydrophobic effect), and/or (iii) desolvation of the aromatic cage of reader proteins. To elucidate the underlying chemical basis for the recognition of natural Kme3 by reader proteins, we have carried out detailed comparative studies for binding of 10-mer histone peptides that contain the positively

charged Kme3, its neutral carba analogue Cme3, and the glycine residue that lacks the entire side chain at the fourth position of histone 3 (that is, H3K4me3, H3C4me3 and H3G4; Fig. 1a). We have chosen the simplest uncharged Cme3 analogue to directly probe the involvement of the proposed cation–$\pi$ interactions in reader–histone associations, because it has virtually the same size, shape and polarizability as the positively charged Kme3, but lacks the presence of the fixed positive charge[16]. Values for volumes of Kme3 (160.2 Å$^3$) and Cme3 (158.2 Å$^3$) indicate that, in the case that the binding mode is the same for both side chains, they should displace the same amount of water molecules from the protein site on binding. We have chosen the 10-mer H3G4 peptide to explore the importance of the entire side chain of Kme3 on association with reader proteins. The Kme3/Cme3 → G substitution directly probes the significance of the potential displacement of water molecules that are localized inside the aromatic cage of reader proteins.

**Thermodynamic analyses of reader–histone association.** We use isothermal titration calorimetry (ITC) to obtain full thermodynamic descriptions for binding of H3K4me3, H3C4me3 and H3G4 peptides to five representative reader proteins that specifically recognize H3K4me3 (the PHD zinc fingers of JARID1A, BPTF, TAF3 and the tudor domains of the Royal family of SGF29 and JMJD2A) (Table 1 and Fig. 1b,c)[19-23]. The five reader pockets are different in the aromatic cage composition and architecture, which allows us to examine the effect of individual constitution of the aromatic cage on binding differences. Comparative ITC experiments for the associations of H3K4me3 and H3C4me3 showed that: (i) the positively charged H3K4me3 binds 2–33-fold stronger than the neutral H3C4me3 to 4 out of 5 reader proteins that contain Trp as part of the aromatic cage (JARID1A, TAF3, BPTF and JMJD2A; Table 1); (ii) association of the Kme3 side chain with the aromatic cage is on average about 4.3 kcal mol$^{-1}$ more favourable in enthalpy than the association of the neutral Cme3 group to the same cage; and (iii) association of the Kme3 side chain is about 3.1 kcal mol$^{-1}$ less favourable in entropy than the association of the Cme3 group to the same aromatic pocket. Collectively, these data provide evidence for the presence of the favourable cation–$\pi$ interactions in the natural readout process, as exemplified by the enthalpy-driven association of the naturally occurring Kme3 with the electron-rich aromatic cage of reader proteins. In contrast to other readers that contain at least one Trp residue, H3K4me3 and H3C4me3 bound to the tandem tudor domain of SGF29 with virtually indistinguishable thermodynamics of associations, indicating the lack (or at least a minor contribution) of cation–$\pi$ interactions in the association of Kme3 by the Tyr/Phe-containing half aromatic cage of SGF29 (Table 1). This result is consistent with the well-established observation that the strength of cation–$\pi$ interactions depends on the nature of the aromatic ring[24-32]. Studies on the related protein systems showed that Trp forms significantly stronger cation–$\pi$ interactions with quaternary ammonium ions than do Phe or Tyr residues[24,25]. For SGF29, the electrostatic interactions between Kme3 and D266, and between the positively charged $\alpha$-amino group of A1 and the H3A1 binding pocket importantly contribute to the overall binding affinity of H3K4me3 (refs 22,33).

Negative values of the heat capacities ($\Delta C_p$) for binding of H3K4me3 and H3C4me3 to reader proteins were also determined by ITC. In all the cases examined, we observed more negative values for H3C4me3 than for H3K4me3: JARID1A–H3K4me3 $-162 \pm 4$ cal mol$^{-1}$ K$^{-1}$, JARID1A–H3C4me3 $-182 \pm 3$ cal mol$^{-1}$ K$^{-1}$; TAF3–H3K4me3 $-142 \pm 7$ cal mol$^{-1}$ K$^{-1}$, TAF3–H3C4me3 $-171 \pm 8$ cal mol$^{-1}$ K$^{-1}$; BPTF–H3K4me3

$-103 \pm 6 \, \text{cal} \, \text{mol}^{-1} \, \text{K}^{-1}$, BPTF–H3C4me3 $-145 \pm 7 \, \text{cal} \, \text{mol}^{-1} \, \text{K}^{-1}$ (Supplementary Fig. 1). These results are in agreement with the involvement of the classical hydrophobic interactions for binding of H3C4me3 to the aromatic cage of reader proteins; this suggests that entropy-driven (partial) desolvation of the Cme3 side chain contributes favourably to the binding affinity[34–37]. Binding of the uncharged Cme3 to the aromatic cage can additionally be attributed to the energetically favourable CH–$\pi$ hydrogen bonding with a strong polarization component[38,39].

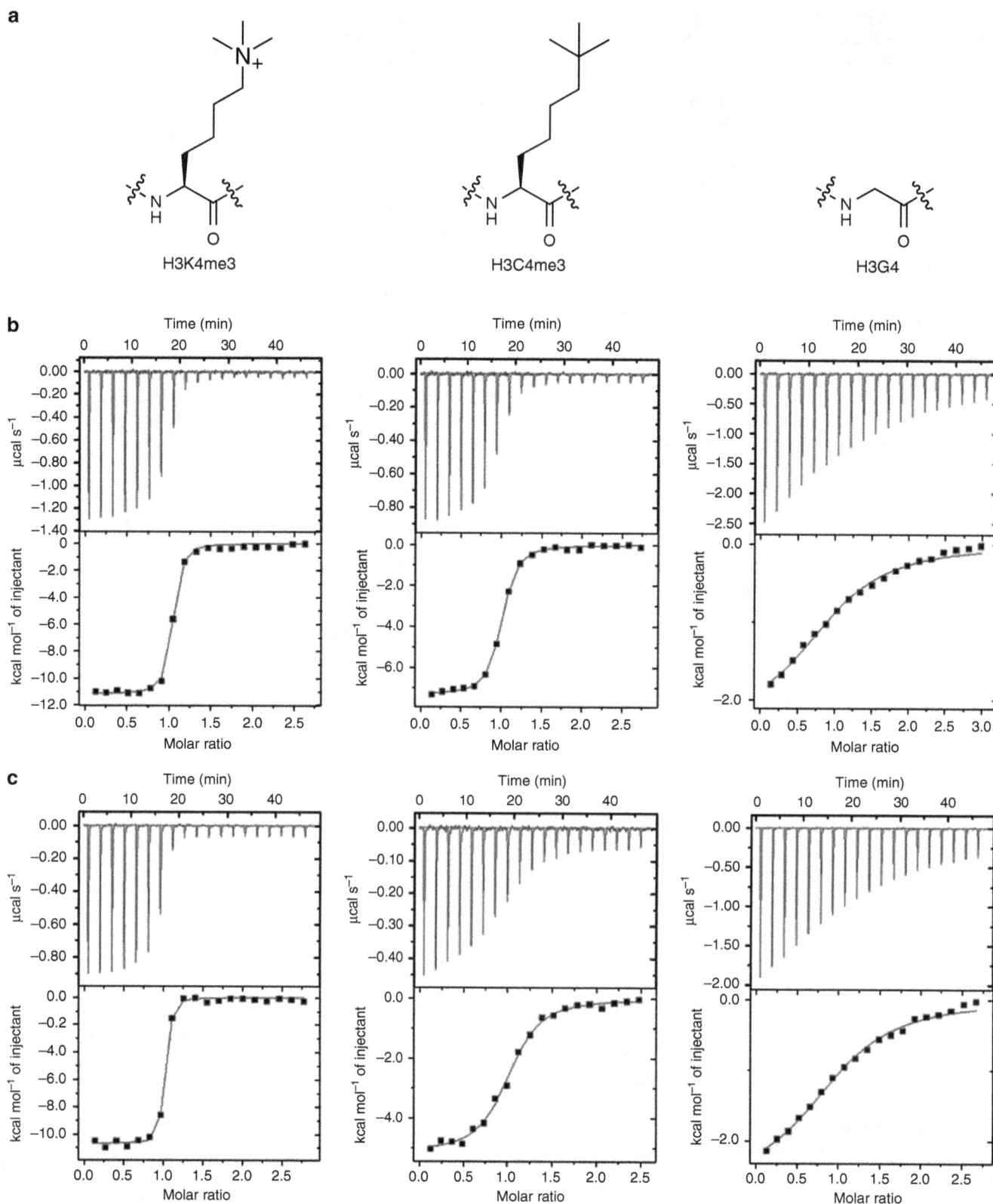

Figure 1 | Thermodynamic analyses of binding. (a) Structures of the positively charged Kme3 and neutral Cme3 and G analogues; (b) ITC curves of 10-mer H3K4me3, H3C4me3 and H3G4 histone peptides binding to the JARID1A PHD3 domain; (c) ITC curves of 10-mer H3K4me3, H3C4me3 and H3G4 histone peptides binding to the TAF3 PHD domain.

We further examined the contribution of the entire Kme3 side chain to the overall binding associations with the aromatic cage of reader proteins. ITC data showed that binding of 10-mer H3G4 to all five reader proteins was dramatically reduced ($>500$-fold) when compared with binding of the H3K4me3 counterpart, highlighting the importance of the entire side chain in the complexation process. More detailed thermodynamic analyses were only possible with JARID1A and TAF3, because both proteins bind to the reference H3K4me3 peptide with $K_d$ values in submicromolar range and the H3G4 peptide had sufficient residual affinity for ITC characterization (Fig. 1b,c): JARID1A–H3G4 ($K_d = 88 \, \mu M$, $\Delta G° = -5.5 \, kcal \, mol^{-1}$, $\Delta H° = -2.1 \, kcal \, mol^{-1}$, $-T\Delta S° = -3.4 \, kcal \, mol^{-1}$) and TAF3–H3G4 ($K_d = 36 \, \mu M$, $\Delta G° = -6.1 \, kcal \, mol^{-1}$, $\Delta H° = -2.5 \, kcal \, mol^{-1}$, $-T\Delta S° = -3.6 \, kcal \, mol^{-1}$). Overall, thermodynamic data revealed that (i) binding of the entire side chain of the Kme3 contributes about $-4 \, kcal \, mol^{-1}$ (that is, about 40%) to the overall Gibbs binding free energy ($\Delta G°$); (ii) favourable enthalpy provides a dominant contribution ($\sim -8.5 \, kcal \, mol^{-1}$) to the binding of the entire Kme3 side chain to the aromatic cage; and (iii) entropy of binding becomes more favourable ($-T\Delta\Delta S° = -4.5 \, kcal \, mol^{-1}$) for H3G4 relative to H3K4me3. In addition to thermodynamics results on H3C4me3, these results indicate that favourable cation–$\pi$ interactions are not solely responsible for strong binding affinity of H3K4me3, but that other types of solute–solute interactions and reader/histone desolvation could also play an important role in the specific readout of Kme3.

**Structural determination of reader–H3C4me3 complexes.** Having shown that the removal of the positive charge in Kme3 (as in the neutral H3C4me3) resulted in reduced binding affinity for most reader proteins due to less favourable enthalpy of binding, we aimed to rationalize these results in conjunction with structural analyses for reader–H3C4me3 complexes. We solved three X-ray crystal structures for complexes with JARID1A, TAF3 and SGF29 at 1.6–2.8 Å resolution (Fig. 2 and Table 2). All three reader–H3C4me3 structures clearly illustrated that the uncharged side chain of C4me3 is positioned well inside the aromatic cages of JARID1A, TAF3 and SGF29, virtually in the same binding mode as the positively charged Kme3 (Fig. 2a–c). The calculated average values of the root-mean-squared deviation for binding of 'Cme3' and 'Kme3'–aromatic cage pairs were: 0.124 Å for JARID1A, 0.261 Å for TAF3 and 0.108 Å for SGF29, respectively, suggesting essentially the same complexation mode engaging in aromatic pocket residues upon binding of neutral C4me3 (Fig. 2d–f). In all three complexes, the carba histone peptide binds to an electrostatically negative surface with the long C4me3 side chain positioned in a surface groove formed by the caging residues (Fig. 2g–i and Supplementary Fig. 2). On the formation of the JARID1A–H3C4me3 complex, the buried solvent accessible

surface area (SASA) of C4me3 (hydrogen atoms added) is 160.6 Å$^2$, which accounts for 38.8% of the total SASA of C4me3, as compared with Kme3 binding to JARID1A with a buried SASA of 163.8 Å$^2$, which equals 39.5% of total SASA. Similar features have also been observed for binding of H3C4me3 and H3K4me3 to TAF3 with 48.3% buried SASA for H3C4me3 and 50.1% buried SASA for H3K4me3, and to SGF29 with 48.6 and 47.3% buried SASA for H3C4me3 and H3K4me3, respectively.

**Computational analyses in the gas and aqueous phase.** Our aim is to elucidate the nature and selectivity of the non-covalent interactions between the aromatic cage that consists of two tryptophan residues of JARID1A (hereafter designated as TRP2 fragment) and the Kme3 versus Cme3 side chain of the histone peptide. To this end, we have quantum chemically characterized the energetics and bonding mechanism in two model complexes, using dispersion-corrected density functional theory at BLYP-D3BJ/TZ2P and COSMO for simulating aqueous solution, as implemented in the Amsterdam Density Functional (ADF) program[40,41]. The model complexes consist of those moieties of the JARID1A–H3K4me3 and JARID1A–H3C4me3 X-ray structures that give rise to the intermolecular interaction in the full reader–histone complexes (Supplementary Table 1). The chosen subsystems were terminated with one hydrogen at $C_\beta$ of the Kme3 or Cme3 side chain and one hydrogen at each $C_\beta$ of the TRP2 fragment. Thus, Kme3 and Cme3 fragments are fully optimized, both as isolated molecules and as molecular fragments in the complex with TRP2. To simulate the structural rigidity that is imposed by the protein backbone in the full protein system, the TRP2 fragment is kept frozen to the X-ray structure, both as a separate fragment and in the complexes. Geometries of the optimized model systems differ only very slightly from the X-ray structures.

Our computations show that, in line with experimental data, there is an energetic preference of $\sim 2 \, kcal \, mol^{-1}$ for the JARID1A–Kme3 over the JARID1A–Cme3 model complex with bond energies $\Delta E(aq)$ of $-10.2$ and $-8.4 \, kcal \, mol^{-1}$, respectively (Table 3). The geometries of the two model complexes are similar, but $NMe_3^+$ in the JARID1A–Kme3 model is somewhat closer to the TRP2 tryptophan cage than $CMe_3$ in the JARID1A–Cme3 model. The closest H–C distances between an $NMe_3^+$ H atom and a C atom of a tryptophan in the JARID1A–Kme3 model is 2.78 Å, while the same H atom is 3.38 Å away from the closest C atom of the other tryptophan. For comparison, the corresponding H–C distances in the JARID1A–Cme3 model are 3.16 and 3.15 Å (Table 3 and Supplementary Fig. 3). A characteristic difference in geometries comes from the conformation of Kme3 and Cme3. In the former, the chain of four carbon atoms has a zigzag conformation whereas, in the latter, this chain is U shaped.

Our bonding analyses reveal that the bond energies $\Delta E(aq)$ associated with the molecular recognition processes of Kme3

**Table 1 | Thermodynamic parameters for the associations of 10-mer H3K4me3 and H3C4me3 peptides (ART(Kme3/Cme3)QTARKS) to five reader proteins\*.**

| | H3K4me3 | | | | H3C4me3 | | | |
|---|---|---|---|---|---|---|---|---|
| | $K_d$ ($\mu M$) | $\Delta G°$ (kcal mol$^{-1}$) | $\Delta H°$ (kcal mol$^{-1}$) | $-T\Delta S°$ (kcal mol$^{-1}$) | $K_d$ ($\mu M$) | $\Delta G°$ (kcal mol$^{-1}$) | $\Delta H°$ (kcal mol$^{-1}$) | $-T\Delta S°$ (kcal mol$^{-1}$) |
| JARID1A | 0.094 | $-9.6 \pm 0.1$ | $-11.0 \pm 0.1$ | $1.4 \pm 0.1$ | 0.34 | $-8.8 \pm 0.1$ | $-7.4 \pm 0.1$ | $-1.4 \pm 0.1$ |
| TAF3 | 0.024 | $-10.4 \pm 0.1$ | $-10.9 \pm 0.1$ | $0.5 \pm 0.2$ | 0.79 | $-8.3 \pm 0.1$ | $-5.2 \pm 0.1$ | $-3.1 \pm 0.2$ |
| BPTF | 0.49 | $-8.6 \pm 0.1$ | $-13.1 \pm 0.1$ | $4.5 \pm 0.1$ | 0.76 | $-8.3 \pm 0.1$ | $-10.0 \pm 0.2$ | $1.7 \pm 0.2$ |
| SGF29 | 1.7 | $-7.9 \pm 0.1$ | $-7.7 \pm 0.1$ | $-0.2 \pm 0.1$ | 1.4 | $-8.0 \pm 0.1$ | $-7.9 \pm 0.1$ | $-0.1 \pm 0.1$ |
| JMJD2A | 0.94 | $-8.2 \pm 0.1$ | $-13.1 \pm 0.2$ | $4.9 \pm 0.2$ | 16 | $-6.5 \pm 0.1$ | $-8.1 \pm 0.1$ | $1.6 \pm 0.1$ |

\*Values obtained from 5–7 repeated ITC experiments. The stoichiometry (histone peptide:reader protein, $n$) = 0.95–1.05.

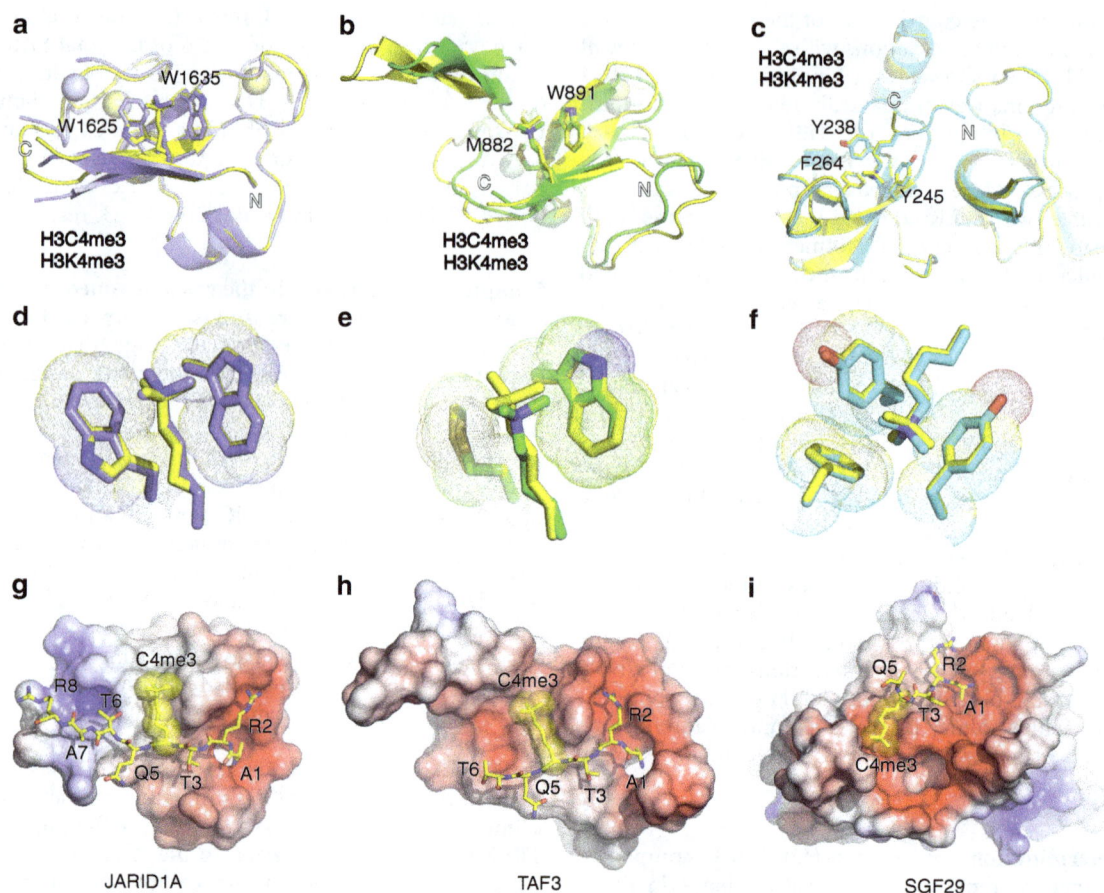

**Figure 2 | Structural analyses of reader–histone interactions.** Structural superimposition of H3C4me3 and H3K4me3–bound complexes of (**a**) the JARID1A PHD finger, (**b**) the TAF3 PHD finger and (**c**) the SGF29 tandem tudor domains. Overall structures are represented in ribbon view with key residues highlighted in stick. In all panels, the H3C4me3 peptides and their complexes are coloured yellow, and the H3K4me3 counterparts are colour coded blue for JARID1A, green for TAF3, and cyan for SGF29. Small spheres, zinc ions. Close-up view of the reader pockets are shown in **d** for JARID1A, **e** for TAF3 and **f** for SGF29. Van der Waals surfaces of caging residues are depicted as dots. Electrostatic surface view of H3C4me3 complexes of (**g**) JARID1A, (**h**) TAF3 and (**i**) SGF29. Red and blue colours indicate negative and positive electrostatic potential, respectively. H3 peptides are shown in stick mode with C4me3 side chain overlaid with dotted van der Waals surfaces.

versus Cme3 in water are essentially identical with the corresponding instantaneous interaction energies $\Delta E_{int}(aq)$ of $-10.3$ and $-8.7\,kcal\,mol^{-1}$, respectively. The reason is that complexation only very slightly changes the geometry of the Kme3 and Cme3 side chains as a result of which the associated deformation strain is negligible, that is, 0.1 and $0.3\,kcal\,mol^{-1}$, respectively. The intrinsic preference for Kme3 over Cme3, that is, the interaction energy $\Delta E_{int}$ between the same structures but in the absence of the solvent, is even more in favour of the former with values of $-27.6$ and $-10.9\,kcal\,mol^{-1}$, respectively (Table 3). The significantly stronger interaction energy of Kme3 is, however, strongly attenuated by the desolvation incurred on binding, which is significantly more unfavourable for Kme3. Thus, solvent effects destabilize the JARID1A–Kme3 complex by $+17.3$, whereas the desolvation penalty in the JARID1A–Cme3 complex is only $+2.2\,kcal\,mol^{-1}$. The reason for this large difference can be attributed to the removal of solvent (desolvation) around the positive charge of the Kme3 side chain ammonium group. Note that the stronger binding in JARID1A–Kme3 causes a reduction in the bond distances (see above), resulting in a computed Pauli repulsion energy between closed shells that is $+6.7\,kcal\,mol^{-1}$ more repulsive for this more stable JARID1A–Kme3 complex.

The reason why the TRP2 unit interacts more favourably with Kme3 than with Cme3 becomes clear from our quantitative Kohn–Sham molecular orbital and energy decomposition analyses (EDA) of the interaction energy $\Delta E_{int}$ (Table 3)[42]. Interestingly, although dispersion $\Delta E_{disp}$ is the largest contributor to the reader–histone interaction, it contributes only $4.4\,kcal\,mol^{-1}$ to the $16.7\,kcal\,mol^{-1}$ difference in $\Delta E_{int}$ between JARID1A–Kme3 ($-27.6\,kcal\,mol^{-1}$) and JARID1A–Cme3 ($-10.9\,kcal\,mol^{-1}$; Table 3). Instead, the difference in stability between JARID1A–Kme3 and JARID1A–Cme3 mainly originates from the electrostatic ($\Delta V_{elstat}$) and orbital interaction ($\Delta E_{oi}$) terms that favour the complex with Kme3 by 9.6 and $9.4\,kcal\,mol^{-1}$, respectively.

The more attractive $\Delta V_{elstat}$ in case of Kme3 goes hand in hand with the significantly more positive charge on all atoms in the Kme3 ammonium, as inferred from our Voronoi deformation density (VDD) atomic charges[43] (Fig. 3a). The nitrogen atom in Kme3 carries a positive charge of $+59$ mili-a.u., which has to be compared with the negative charge of $-40$ mili-a.u. on the structurally analogous carbon atom in the overall neutral Cme3. Importantly, the hydrogen atoms of the trimethylammonium group of Kme3 are also significantly more positively charged than the corresponding ones of the tert-butyl group in Cme3. For example, the hydrogen atom closest to the reader's TRP2 fragment has an atomic charge of $+84$ and $+29$ mili-a.u. in Kme3 and Cme3, respectively (Fig. 3a).

Our Kohn–Sham molecular orbital analyses show that the enhanced orbital interactions $\Delta E_{oi}$ in JARID1A–Kme3 result

**Table 2 | Data collection and refinement statistics.**

| | JARID1A-H3C4me3 | TAF3-H3C4me3 | SGF29-H3C4me3 |
|---|---|---|---|
| *Data collection* | | | |
| Space group | I432 | P2$_1$ | P2$_1$2$_1$2$_1$ |
| Cell dimensions | | | |
| $a, b, c$ (Å) | 108.9, 108.9, 108.9 | 30.2, 50.1, 85.9 | 50.1, 65.2, 105.2 |
| $\alpha, \beta, \gamma$ (°) | 90, 90, 90 | 90, 90, 90 | 90, 90, 90 |
| Resolution (Å) | 50-2.8 (2.87-2.80)* | 50-2.1 (2.14-2.10) | 37.2-1.60 (1.63-1.60) |
| $R_{merge}$ | 6.5 (79.8) | 12.3 (66.9) | 7.3 (77.5) |
| $I / \sigma I$ | 64.9 (3.3) | 17.1 (2.8) | 17.6 (2.6) |
| Completeness (%) | 99.5 (100) | 99.6 (100) | 99.1 (92.1) |
| Redundancy | 17.1 (14.4) | 3.7 (3.7) | 6.8 (6.6) |
| *Refinement* | | | |
| Resolution (Å) | 50-2.8 | 32.6-2.1 | 37.2-1.60 |
| No. reflections | 2,920 | 15,001 | 44,342 |
| $R_{work}/R_{free}$ | 24.6/27.9 | 22.2/28.0 | 20.4/23.9 |
| No. atoms | | | |
| Protein | 399 | 2,064 | 2,865 |
| Ligand/ion | 56/2 | 228/8 | 84/10 |
| Water | 0† | 78 | 381 |
| *B*-factors | | | |
| Protein | 97.3 | 34.9 | 16.8 |
| Ligand/ion | 80.3/86.9 | 29.2/25.8 | 19.7/34.6 |
| Water | | 37.7 | 25.9 |
| R.m.s. deviations | | | |
| Bond lengths (Å) | 0.003 | 0.014 | 0.009 |
| Bond angles (°) | 0.698 | 1.58 | 1.384 |

*Values in parentheses are for highest-resolution shell.
†No water molecules are modelled due to high *B*-factor of the complex structure.

**Table 3 | Quantum-chemical bonding analysis (energies in kcal mol⁻¹, distances in Å) in TRP2–Kme3 and TRP2–Cme3 systems in aqueous solution*.**

| | TRP2-Kme3 | TRP2-Cme3 |
|---|---|---|
| $\Delta E$(aq) | -10.2 | -8.4 |
| $\Delta E$(aq)$_{strain}$ | 0.1 | 0.3 |
| $\Delta E$(aq)$_{int}$ | -10.3 | -8.7 |
| $\Delta E$(desolv)$_{int}$ | +17.3 | +2.2 |
| $\Delta E_{int}$ | -27.6 | -10.9 |
| $\Delta E_{Pauli}$ | 20.8 | 14.1 |
| $\Delta V_{elstat}$ | -15.0 | -5.4 |
| $\Delta E_{oi}$ | -13.0 | -3.6 |
| $\Delta E_{disp}$ | -20.4 | -16.0 |
| d(H$_{Me}$-C$_{TRP-6MR}$) | 3.38 | 3.15 |
| d(H$_{Me}$-C$_{TRP-5MR}$) | 2.78 | 3.16 |

*Computed at BLYP-D3BJ/TZ2P with COSMO to simulate aqueous solution. Structural rigidity imposed by the protein backbone is simulated through constraint geometry optimizations. See also equations (1)-(3) in Methods section.

from both, stronger donor–acceptor orbital interactions and stronger polarization of the TRP2 fragment in the presence of the positively charged Kme3 than in the case of the neutral Cme3. Thus, the VDD analyses based on the two molecular fragments[43] reveal a small but significant charge transfer of 0.04 electrons from the occupied π fragment molecular orbitals (FMOs) on TRP2 to virtual σ*$_{C-N}$ and σ*$_{C-H}$ type FMOs on Kme3 whereas essentially no charge is transferred to FMOs on Cme3. One reason is the much lower energy of the acceptor orbitals in the positively charged Kme3 (Fig. 3b). Another reason is the better overlap between TRP2 π orbitals and the acceptor orbitals of Kme3. This originates from the fact that the low-energy virtual orbitals of Kme3 are mainly localized on the positive

trimethylammonium group through which Kme3 binds to TRP2, as can be seen in the realistic three-dimensional plots of relevant FMOs in Fig. 3b. The low-energy orbitals of Cme3 are more delocalized with less amplitude on the tert-butyl group close to TRP2. Consequently, in most cases TRP2–Kme3 overlaps are significantly larger than TRP2–Cme3 overlaps, as shown for the TRP2 highest occupied molecular orbital (HOMO) and HOMO-1 and the Kme3 or Cme3 lowest unoccupied molecular orbital (LUMO) and LUMO + 1 (Supplementary Table 2).

**WaterMap calculations.** Next, we ran WaterMap calculations for all five systems to evaluate the contribution of aromatic cage desolvation to the affinity of Kme3 and Cme3 for reader proteins. WaterMap computes thermodynamic quantities (free energy, enthalpy and entropy) for simulated water molecules around a protein-binding site using explicit solvent molecular dynamics simulation and thermodynamic characterization. In short, regions of high solvent density from the molecular dynamic simulations are clustered into 'hydration sites', and thermodynamic quantities for these sites are calculated using inhomogeneous solvation theory[44,45]. For all five reader proteins, two to four high-energy hydration sites were identified within the aromatic cage (Fig. 4a–e). These hydration sites are displaced from the aromatic cage by both the Kme3 and Cme3 side chain, but not by the H3G4 peptide. The total free energy contributed by desolvating the aromatic cage (determined as the difference in WaterMap scores between Kme3 and Gly) ranges from 4.3 kcal mol⁻¹ for JARID1A to 8.7 kcal mol⁻¹ for SGF29. Depending on the composition of the cage, this free energy reward can be both entropically and enthalpically driven (Fig. 4f and Supplementary Figs 4 and 5). For example, both TAF3 and JMJD2A contain an Asp residue that can form hydrogen bonds with the binding site water molecules, resulting in more

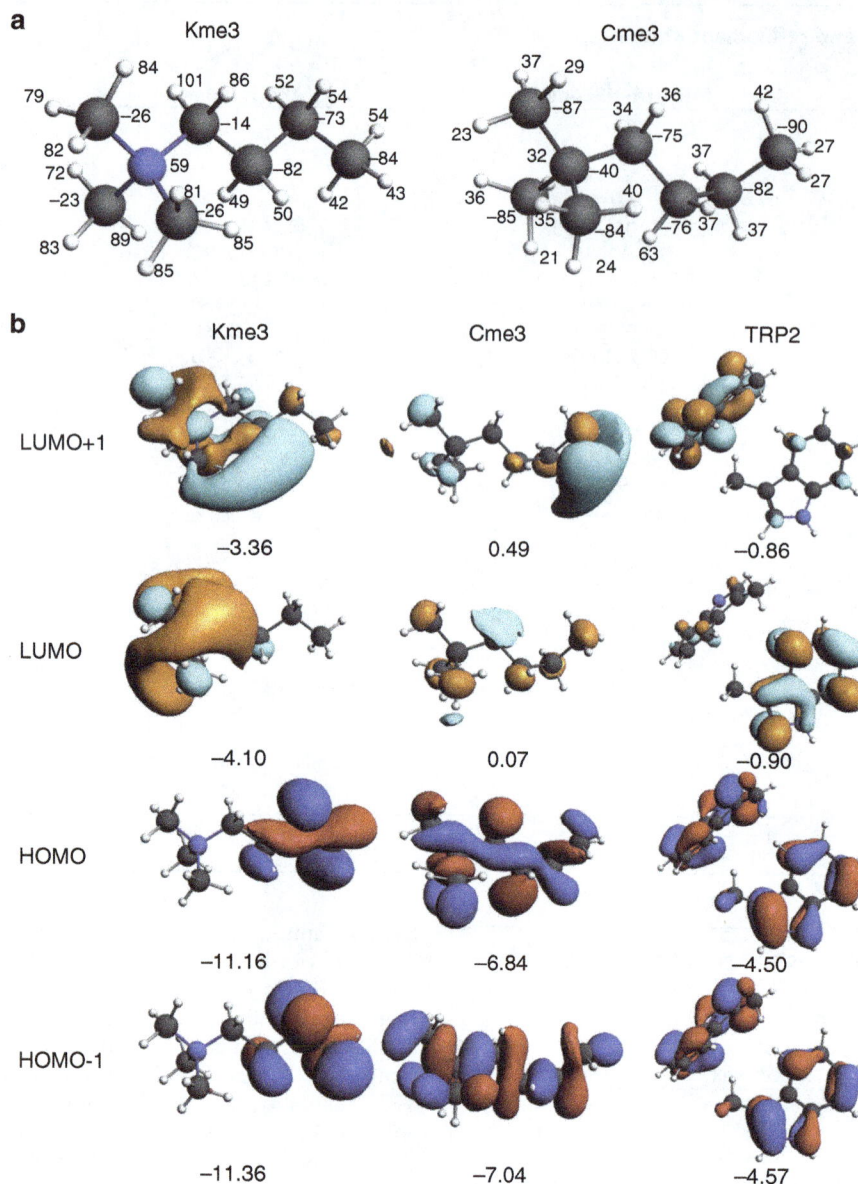

**Figure 3 | Computational analysis of TRP2–Kme3 and TRP2–Cme3 interactions.** (**a**) VDD atomic charges (in mili-a.u.) of Kme3 and Cme3, computed at BLYP-D3BJ/TZ2P using X-ray structures of the full systems (red, negative; blue, positive); (**b**) Frontier orbitals (with orbital energies in eV) of Kme3, Cme3 and TRP2, computed at BLYP-D3BJ/TZ2P using X-ray structures of the full systems (isosurface drawn at 0.03).

favourable enthalpy of the hydration sites in the cage, hence more unfavourable change in enthalpy on displacing those waters on Kme3/Cme3 binding. On the other hand, the BPTF cage is completely surrounded by aromatic residues, producing an enthalpically unfavourable environment for water and therefore a favourable free energy change from water displacement on Kme3/Cme3 binding.

## Discussion
The advances of experimental and theoretical tools developed in the past decade have enabled more extensive analysis of the origins of some genuinely important biomolecular recognition phenomena, including the molecular basis of the hydrophobic effect(s) in protein–ligand interactions and the fundamentals of the receptor–neurotransmitter interactions in neurochemistry[27,46]. This study comprehensively examines the origin of the biomolecular recognition between naturally occurring trimethyllysine-containing histone proteins and their interacting reader proteins that are involved in epigenetic gene regulation

processes. We use the physical–organic chemistry approaches, supported by high-resolution structural analyses of reader–histone interactions, to elucidate the molecular/chemical basis of one of the fundamental non-covalent interactions in epigenetics. Analyses of crystal and solution structures of free (unbound) reader proteins and reader–Kme3 complexes have illustrated that the reader's aromatic cage is largely preformed and does not undergo induced fit for binding of histone substrates (Supplementary Fig. 6). The predominantly static nature of the aromatic cage has an advantage over a more flexible recognition site because it minimizes the loss of conformational entropy of the protein on ligand binding[8]. Binding of the flexible and highly unstructured histone to reader proteins, however, results in a significant conformational change of the histone resulting in a more unfavourable entropy of binding for longer histone peptides relative to shorter histone counterparts[33].

On the basis of the studies of the related proteins that possess the aromatic cages for the recognition of positively charged methylammonium groups, it has been suggested that epigenetic readers recognize Kme3 via cation–π interactions[16,27,37,47]. Our

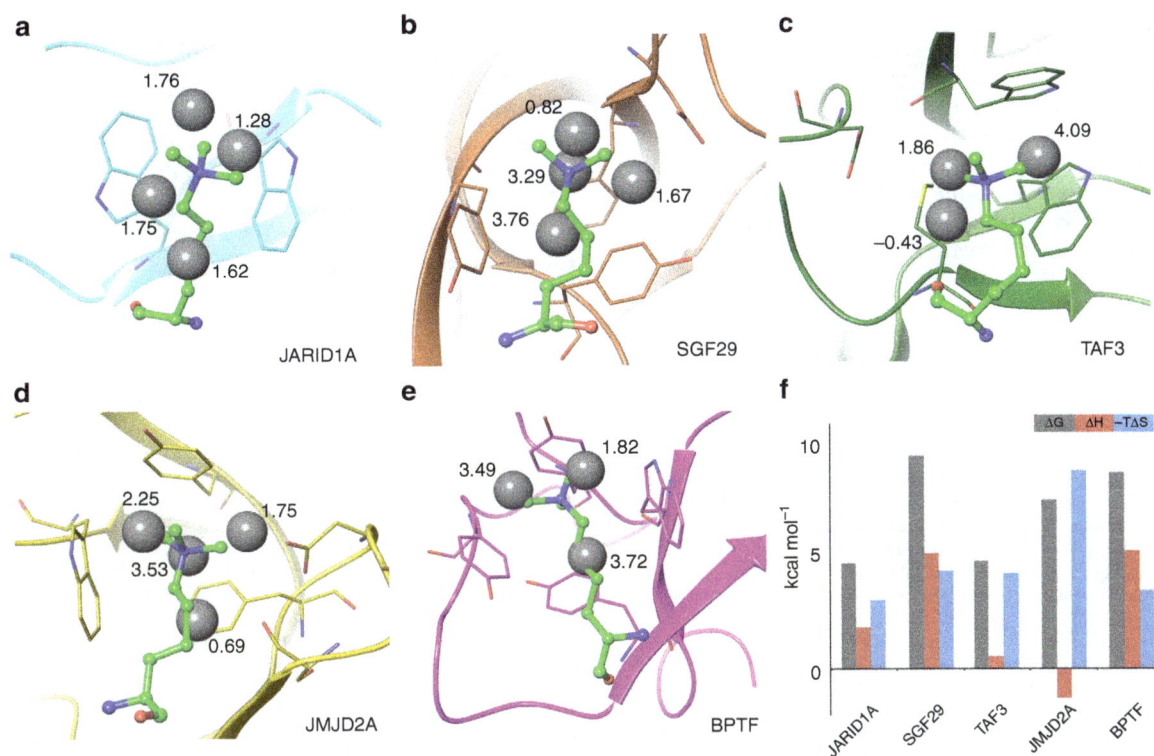

**Figure 4 | WaterMap calculations for the solvation of aromatic cages of reader proteins.** (**a**) JARID1A; (**b**) SGF29; (**c**) TAF3; (**d**) JMJD2A; (**e**) BPTF. Superimposed Kme3 side chain and water molecules are presented as green stick and grey spheres. Numbers adjacent to grey spheres represent the value of the free energy ($\Delta G$) for individual water molecule; (**f**) Thermodynamic parameters for the solvation of the aromatic cages of five reader proteins used in this study.

integrated thermodynamic, structural and computational studies clearly confirm the presence of favourable cation–$\pi$ interactions in the readout of H3K4me3 by reader domains of JARID1A, TAF3, BPTF and JMJD2A. Previous examination of the recognition of neutral Cme3 by HP1 chromodomain, a reader of H3K9me3 that contains an aromatic cage comprising two tyrosine and one tryptophan residues, revealed that HP1 binds to H3C9me3 with substantially lower affinity than H3K9me3, thus suggesting that the positive charge of Kme3 is crucial for the association of HP1–H3K9me3 (ref. 16). Comprehensive structural data on JARID1A, TAF3 and SGF29 in complex with H3C4me3, as described in this work, provide clear evidence that the Cme3 side chain is well positioned inside the aromatic cages of these three reader proteins in the same manner as the positively charged Kme3 (Fig. 2) and thus enable us to interpret the binding calorimetric data (Table 1). Out of three possible mechanisms (that should always be considered in the interpretation of any protein–ligand system), that is, solute–solute interactions, desolvation of ligand (in this case Kme3) and desolvation of protein (in this case aromatic cage), that govern the recognition of Kme3 by reader proteins, we can exclude desolvation of the Kme3 side chain, because charged residues are highly soluble in aqueous media and have to pay a big desolvation penalty to become desolvated. In this regard, it is essential that the energetically unfavourable desolvation of Kme3 is fully compensated (or more correctly overcompensated) by energetically favourable protein–ligand interactions and protein desolvation to provide a strong binding force for the specific recognition of Kme3 by reader proteins. Based on ITC experiments, our observed enthalpy-driven association of positively charged Kme3 (relative to Cme3) to the electron-rich aromatic cage of several reader proteins has its molecular origin in strong cation–$\pi$ interactions. In addition, the methylene groups

of the side chain of Kme3 located within van der Waals distance of the aromatic cages, contribute to the overall binding affinity via weaker, but still favourable, CH–$\pi$ interactions[38,39]. Our quantum mechanical studies, furthermore, reveal that reader–Kme3 association has the strongest dispersion contribution (similar to reader–Cme3), but that the differences in binding affinities between Kme3 and Cme3 are primarily a result of disparities in electrostatic interactions and orbital interactions (Table 3).

Despite the universally recognized phenomenon that biomolecular processes take place in aqueous media and that the hydrophobic effect is a primary determinant of biomolecular association, the role of explicit water molecules has often been ignored in analyses of biomolecular recognition events[48,49], although recent advances have enabled more detailed analysis of the role of water molecules in binding[46,50,51]. Energetically favourable desolvation of protein-binding sites, however, often determines the magnitude of protein–ligand association[52,53]. Our observations that binding affinities of H3G4 with JARID1A and TAF3 are drastically reduced when compared with H3K4me3 led to the hypothesis that the aromatic cages are occupied by high-energy water molecules. Although difficult to confirm experimentally, WaterMap calculations performed on five representative reader proteins (both in apo and holo forms) provided evidence that water molecules located inside the aromatic cages exhibit significant unfavourable free energy (Fig. 4 and Supplementary Table 3). These high-energy water molecules are displaced by Kme3 side chain on binding, which consequently provide a substantial favourable contribution to Kme3 binding.

Collectively, the experimental and computational work presented here suggests that the association between trimethyllysine-containing histones and epigenetic reader domain proteins is driven by favourable cation–$\pi$ interactions and the favourable release of high-energy structured water molecules that occupy the aromatic cages of reader proteins. Our study highlights the hitherto neglected, yet essential contribution of water in a molecular readout process in the established area of epigenetics. This study, furthermore, sheds light on the design of small molecule probes that specifically recognize readers of trimethyl-lysine. In comparison with the advances in development of inhibitors of other epigenetic targets, including bromodomains and various eraser/writer enzymes, there has been very limited success in identification of probes for readers of Kme3 (refs 54,55). Towards this aim, our study provides valuable experimental and computational data needed for the medicinal chemistry community to design and develop potent and selective small molecule inhibitors with therapeutic potential.

## Methods

**General experimental procedures.** All experiments were conducted under the following conditions, unless stated otherwise. Commercially available compounds were supplied by commercial sources and used without any further purification. Dry solvents were obtained by purification of HPLC grade solvents over activated alumina column using an MBraun SPS800 solvent purification system. When stated, degassing of solvents was performed for each reaction individually by passing through $N_2$ (g) for a period of at least 30 min before use. Compound purification done by column chromatography, was carried out using Silica gel, MerckTM grade (pore size 60 Å; particle size 230–400 mesh, 40–63 μm). Reaction progress was monitored by glass thin-layer chromatography plates (TLC Silica gel 60G, F254, Merck, Germany) and observed by ultraviolet light and/or by staining in ninhydrin or permanganate. Compound analyses done by $^1$H NMR, were recorded on a Varian Inova 400 at 400 MHz. $^{13}$C NMR data were either recorded using a Bruker Avance III 500 MHz at 125 MHz or a Varian Inova 400 at 101 MHz. Reported chemical shifts are in p.p.m., moving from high to low frequency and referenced to the residual solvent resonance. Reported coupling constants ($J$) are noted in hertz (Hz). To assign multiplicity of signals the following standard abbreviations were used: s, singlet; d, doublet; t, triplet; q, quartet; quint, quintet; m, multiplet; and br, broad. When possible, $^1$H assignments were made using appropriate two-dimensional NMR methods, such as correlation spectroscopy, heteronuclear single-quantum correlation spectroscopy and heteronuclear multiple-bond correlation spectroscopy. Mass spectrometry and chromatography analysis were done using a Shimadzu UFLC LC-20AD liquid chromatography/mass spectrometry system, equipped with a RPC18 200 × 2 guard column. Typical conditions for a run are: 157 bar, mobile phase; 2 min 5% MeCN 95% $H_2O$, in 16 min decreasing polarity to 100% MeCN, 5 min of 100% MeCN, in 2 min increasing polarity to 95% $H_2O$ for 5 min. Ultraviolet/visible detection of this machine was done by Ultraviolet Visible Shimadzu SPD-M20A (200–600 nm), while mass spectrometry analyses was done using the Thermo scientific LCQ Fleet. HPLC trace analyses were done on a Shimadzu liquid chromatography system; DGU 20A5, using a SPD 20A ultraviolet detector at 214 nm. The machine is equipped with a Gemini-NX 3 C18 column. Typical conditions for a run are: 1 min at 5% MeCN in 95% $H_2O$ (with 0.1% trifluoroacetic acid (TFA)), increase over 30 min to 100%, keep this for 5 min, then over 5 min the concentration is decreased to 5% MeCN in 95% $H_2O$ (with 0.1% TFA).

**Synthesis of Fmoc-L-Cme3.** Supplementary Fig. 7 shows the schematic presentation of the synthetic protocol for the preparation of Fmoc-L-Cme3 (**6**).

**Synthesis of (1).** Boc-Asp(OH)-OtBu (5.81 g, 20 mmol, 1 equivalent), 4-dimethylaminopyridine (223.8 mg, 2 mmol, 0.1 equivalents) and *N*,*N*′-dicyclohexylcarbodiimide (4.95 g, 24 mmol, 1.2 equivalents) were dissolved in dry $CH_2Cl_2$ (40 ml) under $N_2$ atmosphere. To this solution was added ethanethiol (4.7 ml, 64 mmol 3.2 equivalents). After 4 h of stirring the solvent was removed under reduced pressure. The crude product was purified by column chromatography ($SiO_2$, EtOAc in *n*-pentane 5–20%). This yielded thioester **1** (6.26 g, 18.8 mmol, 94%) as a pale yellow oil: $[\alpha]^{25}_D$ +43.4 (c 1.00, $CH_3Cl$). FT-IR $v_{max}$ (cm$^{-1}$): 3,436, 2,980, 2,932, 1,715, 1,688, 1,495, 1,367, 1,250, 1,150, 1,059, 1,023 and 847. $^1$H NMR (400 MHz, CDCl$_3$) δ: 5.42 (d, $J$ = 8.0 Hz, 1H, NH), 4.48–4.36 (m, 1H, αCH), 3.10 (dq, $J$ = 17.0, 5.0 Hz, 2H, βCH$_2$), 2.96–2.78 (m, 2H, SCH$_2$), 1.46 (s, 9H, C(CH$_3$)$_3$), 1.44 (s, 9H, C(CH$_3$)$_3$) and 1.29–1.22 (m, 3H, CH$_3$). $^{13}$C NMR (101 MHz, CDCl$_3$) δ: 196.9, 169.6, 155.2, 82.3, 79.7, 50.8, 45.5, 28.2, 27.8, 23.4 and 14.6. HRMS, calculated for $C_{15}H_{27}NO_5SNa$ [M + Na]$^+$ 356.1508, found 356.1511.

**Synthesis of (2).** To a suspension of Pd/C (375 mg, 10% Pd on activated carbon, 6 wt%) and thioester **1** (6.26 g, 18.8 mmol, 1 equivalent) in degassed dry $CH_2Cl_2$ (40 ml) was added triethylsilane (9 ml, 56.3 mmol, 3 equivalents). The solution was stirred for 90 min, while cooling on a water bath. The black suspension was filtered through celite, concentrated and purified by column chromatography ($SiO_2$, EtOAc in *n*-heptane 5–25%). This eventually yielded aldehyde **2** (4.84 g, 17.7 mmol, 95%) as a clear colourless oil, which solidified over time: $[\alpha]^{25}_D$ −24.2 (c 1.50, EtOH). FT-IR $v_{max}$ (cm$^{-1}$): 3,370, 2,980, 2,935, 1,714, 1,501, 1,368, 1,251, 1,151, 1,054 and 847. $^1$H NMR (400 MHz, CDCl$_3$) δ: 9.74 (s, 1H, C(O)H), 5.34 (d, $J$ = 7.5 Hz, 1H, NH), 4.58–4.39 (m, 1H, αCH), 2.98 (qd, $J$ = 18.0, 5.0 Hz, 2H, βCH$_2$), 1.47 (s, 9H, C(CH$_3$)$_3$) and 1.44 (s, 9H, C(CH$_3$)$_3$). $^{13}$C NMR (101 MHz, CDCl$_3$) δ: 199.3, 169.8, 155.3, 82.4, 79.8, 49.2, 46.1, 28.2 and 27.7. HRMS, calculated for $C_{13}H_{23}NO_5Na$ [M + Na]$^+$ 296.1474, found 296.1471.

**Synthesis of (3).** To a suspension of methyltriphenylphosphonium bromide (2.23 g, 6.16 mmol 1.1 equivalents) in dry tetrahydrofuran (THF; 30 ml) under $N_2$ atmosphere, was added NaHMDS (3.1 ml, 6.16 mmol, 2.0 M in THF, 1.1 equivalents). Aldehyde **2** (1.08 g, 3.66 mmol, 1 equivalent) was dissolved in dry THF (15 ml) and added to the solution after 30 min of stirring. Subsequently, the reaction mixture was stirred for 20 h and then quenched by the addition of KHSO$_4$ (aq) (60 ml, 1 M). The aqueous layer was extracted with EtOAc (3 × 25 ml) and the combined organic extracts were washed with $H_2O$ (50 ml) and brine (50 ml). The organic layer was dried over Na$_2$SO$_4$, filtered and evaporated under vacuum. The crude product was purified by silica column chromatography ($SiO_2$, EtOAc in *n*-heptane 5–20%), affording **3** (919 mg, 3.385 mmol, 60%) as a clear colourless oil. $[\alpha]^{25}_D$ +10.3 (c 0.84, MeOH). FT-IR $v_{max}$ (cm$^{-1}$): 3,352, 2,980, 2,933, 1,715, 1,496, 1,367, 1,251, 1,154, 918 and 847. $^1$H NMR (400 MHz, CDCl$_3$) δ: 5.79–5.63 (m, 1H, CH$_2$ = CH), 5.16–5.09 (m, 2H, CH$_2$ = CH), 5.05 (d, $J$ = 7.5 Hz, 1H, NH), 4.25 (dd, $J$ = 19.0, 8.5 Hz 1H, αCH), 2.63–2.39 (m, 2H, βCH$_2$), 1.46 (s, 9H, C(CH$_3$)$_3$) and 1.44 (s, 9H, C(CH$_3$)$_3$). $^{13}$C NMR (101 MHz, CDCl$_3$) δ: 171.1, 155.2, 132.5, 118.7, 81.9, 79.6, 53.3, 37.0, 28.3 and 28.0. HRMS, calculated for $C_{14}H_{25}NO_4Na$ [M + Na]$^+$ 294.1681, found 294.1683.

**Synthesis of (4).** To a solution of **3** (918 mg, 3.39 mmol, 1 equivalent) in dry $CH_2Cl_2$ (30 ml) under $N_2$ atmosphere, were added second generation Grubbs catalyst (434 mg, 0.51 mmol, 0.15 equivalents) and 4,4-dimethyl-1-pentene (1,860 μl, 15.54 mmol, 4 equivalents). This solution was stirred for 24 h at 50 °C. After cooling down, the solvent was evaporated under reduced pressure. The crude product was purified by column chromatography ($SiO_2$, EtOAc in *n*-heptane 0–10%), affording **4** (650 mg, 1.9 mmol, 56%). $[\alpha]^{25}_D$ −5.4 (c 0.93, MeOH). FT-IR $v_{max}$ (cm$^{-1}$): 3,337, 2,954, 1,716, 1,495, 1,365, 1,248, 1,153, 970 and 847. $^1$H NMR (400 MHz, CDCl$_3$) (Z: E ratio 1: 4.7, most abundant isomer) δ: 5.49–5.58 (m, 1H, CH = CH), 5.32–5.20 (m, 1H, CH = CH), 5.01 (d, $J$ = 8.0 Hz, 1H, NH), 4.29–4.16 (m, 1H, αCH), 2.54–2.33 (m, 2H, βCH$_2$), 1.88 (dd, $J$ = 7.5, 1.0 Hz, 2H, εCH$_2$), 1.47 (s, 9H, C(CH$_3$)$_3$), 1.44 (s, 9H, C(CH$_3$)$_3$) and 0.87 (s, 9H, C(CH$_3$)$_3$). $^{13}$C NMR (126 MHz, CDCl$_3$) δ: 171.4, 155.1, 132.2, 125.8, 81.7, 79.5, 53.6, 47.1, 35.8, 30.8, 29.3, 28.3 and 28.1. HRMS, calculated for $C_{19}H_{35}NO_4Na$ [M + Na]$^+$ 364.2464, found 364.2478.

**Synthesis of (5).** To a suspension of Pd/C (140 mg, 10% Pd on activated carbon, 25 wt%) in dry $CH_2Cl_2$ (20 ml), was added **4** (558 mg, 1.63 mmol, 1 equivalent). The solution was vigorously stirred under $H_2$ atmosphere for 24 h. The black suspension was filtered through celite and washed with $CH_2Cl_2$ (3 × 25 ml). The filtrate was concentrated under reduced pressure yielding **5** (530 mg, 1.54 mmol, 95%) as a slightly brown oil. $[\alpha]^{25}_D$ −14.0 (c 1.00, MeOH). FT-IR $v_{max}$ (cm$^{-1}$): 3,350, 2,954, 2,865, 1,770, 1,498, 1,392, 1,366, 1,249, 1,154 and 849. $^1$H NMR (500 MHz, CDCl$_3$) δ: 4.92 (d, $J$ = 8.0 Hz, 1H, NH), 4.09 (dd, $J$ = 13.0, 7.0 Hz, 1H, αCH), 1.60–1.48 (m, 2H, βCH$_2$), 1.40 (s, 9H, C(CH$_3$)$_3$), 1.37 (s, 9H, C(CH$_3$)$_3$), 1.30–1.13 (m, 4H, γCH$_2$ and δCH$_2$), 1.12–1.04 (m, 2H, εCH$_2$) and 0.79 (s, 9H, C(CH$_3$)$_3$). $^{13}$C NMR (126 MHz, CDCl$_3$) δ: 172.2, 155.4, 81.6, 79.5, 54.0, 44.0, 33.0, 30.3, 29.4, 28.4, 28.0, 26.1 and 24.3. HRMS, calculated for $C_{19}H_{37}NO_4Na$ [M + Na]$^+$ 366.2620, found 366.2619.

**Synthesis of (6).** Protected **5** (295 mg, 0.86 mmol, 1 equivalent) was dissolved in a mixture of TFA: dichloromethane (30 ml, 2:1) and left stirring for 5 h. The solvent was removed under vacuum and the resulting crude product was redissolved in $H_2O$: dioxane (30 ml, 1:1) and the pH of the solution was adjusted to pH 8–9 by the addition of NaHCO$_3$. Subsequently, Fmoc-OSu (435 mg, 1.29 mmol, 1.5 equivalents) was added to the solution. After stirring for 16 h the solution was acidified to pH 3 by addition of HCl (aq) (1 M) and extracted with EtOAc (5 × 20 ml). The combined organic extracts were washed with brine (50 ml), dried over Na$_2$SO$_4$, filtered and concentrated under reduced pressure. The crude oil was purified by column chromatography ($SiO_2$, MeOH in $CH_2Cl_2$ and a few drops of AcOH, 1–4%), affording **6** (295 mg, 0.72 mmol, 84%) as a clear viscous oil. $[\alpha]^{25}_D$ −2.5 (c 0.16, MeOH). FT-IR $v_{max}$ (cm$^{-1}$): 3,326, 2,952, 2,862, 1,710, 1,520, 1,451, 1,214, 1,079, 758 and 739. $^1$H NMR (400 MHz, CDCl$_3$) δ: 7.77 (d, $J$ = 6.0 Hz, 2H, 2 × ArCH), 7.64–7.50 (m, 2H, 2 × ArCH), 7.40 (t, $J$ = 7.4 Hz, 2H, 2 × ArCH), 7.31 (t, $J$ = 7.0, 2H, 2 × ArCH), 5.37–5.20 (m, 1H, NH), 4.54–4.33 (m, 3H, αCH and OCH$_2$), 4.28–4.17 (m, 1H, CH), 1.97–1.82 (m, 1H, βCH), 1.78–1.66 (m, 1H, βCH), 1.44–1.10 (m, 6H, γCH$_2$ and δCH$_2$ and εCH$_2$) and 0.86 (s, 9H, C(CH$_3$)$_3$). $^{13}$C NMR (126 MHz, CDCl$_3$) δ: 177.5, 156.1, 143.9, 141.3, 127.7, 127.1, 124.9, 120.0, 67.1, 53.9, 47.2, 43.9, 32.4, 30.3, 29.4, 26.1 and 24.2. HRMS, calculated for $C_{25}H_{31}NO_4Na$ [M + Na]$^+$ 432.2151, found 432.2153.

**Solid-phase peptide synthesis.** Ten mer histone peptides were synthesized by solid-phase peptide synthesis applying Fmoc chemistry. Peptides contain a carboxylic acid at the C terminus and were made on Wang resin and couplings were done in dimethylformamide (DMF) with Fmoc-protected amino acid (3.0 equivalents), diisopropylcarbodiimide (3.3 equiv.) and hydroxybenzotriazole (3.6 equivalents). Completion of the reaction was determined with the Kaiser test, and removal of Fmoc was achieved by treatment with a large excess of piperidine (20%) in DMF for 20–30 min. Every wash step was performed with 3 × DMF and after building completion the Fmoc was removed followed by wash 3 × DMF and 3 × Et$_2$O continued by drying of the resin in vacuo. The peptides were cleaved from the resin by a mixture of TFA (92.5%), H$_2$O (2.5%), tri-isopropylsilane (2.5%) and ethane-1,2-dithiol (2.5%). After mixing and shaking for 4–5 h, the product peptide was precipitated in Et$_2$O, and the Et$_2$O was decanted after centrifugation (3,500 r.p.m., 3 min, Hermle 220.72 v04). Histone peptides were analysed by liquid chromatography–mass spectrometry and purified by preparative HPLC (Supplementary Figs 8–13). Purified histone peptides were analysed by $^{19}$F NMR spectroscopy, which provided evidence that they appear as TFA salts.

**Preparation and purification of reader proteins.** Reader proteins were prepared and purified following the previously reported procedure[33]. Briefly, the reader domains of BPTF, JMJD2A, JARID1A, TAF3 and SGF29 were expressed in *Escherichia coli* Rosetta BL21 DE3 pLysS hosts, using Terrific Broth medium. The bacteria were cultured to OD600 ~0.6 at 37 °C after which they were induced with 0.5 mM isopropyl-b-D-thiogalactoside overnight at 16 °C. Proteins were purified using Ni-NTA beads for 6xHis-tagged proteins or glutathione sepharose beads for GST tagged proteins, respectively. After purification, the 6xHis tag was cleaved from JMJD2A and SGF29 using TEV-protease and the GST tag was cleaved from TAF3 using thrombin. Protein were purified by size-exclusion chromatography using a Superdex 75 column (GE Healthcare). SGF29 was eluted in 25 mM Tris, 50 mM NaCl, 1 mM dithiothreitol at pH 7.5; JMJD2A and TAF3 were eluted in 50 mM Tris at pH 7.5; BPTF and JARID1A were eluted in 50 mM Tris, 20 mM NaCl at pH 7.5. All proteins were made filter sterile and stored at 4 °C until further use.

**Isothermal titration calorimetry.** Concentrations of histone peptides were measured by ultraviolet spectroscopy at 205 nm, following the previously reported method[56]. All histone peptides were titrated to the same batch of reader proteins. Generally, 350–600 μM of H3K4me3 or H3C4me3 peptides were titrated to 25–40 μM of protein, except for JMJD2A–H3C4me3 (200 μM JMJD2A, 3 mM H3C4me3). H3G4 (5 mM) was titrated to JARID1A (330 μM) and H3G4 (3 mM) was titrated to TAF3 (200 μM). Each ITC titration consisted of 19 injections. ITC experiments were performed on the fully automated Microcal Auto-iTC200 (GE Healthcare Life Sciences, USA). Heats of dilution for histone peptides were determined in control experiments, and were subtracted from the titration binding data before curve fitting. Curve fitting was performed by Origin 6.0 (Microcal Inc., USA) using one set of sites binding model. For each reader–histone system, 5–7 independent ITC experiments were carried out. Measurements of heat capacities were typically done in the interval of 10–30 °C, in triplicate at each temperature.

**X-ray crystallography.** The tandem tudor domain of human SGF29 (residues 115–293) was cloned into a pET-28a-MHL vector, and is expressed, purified as described before[22]. The purified SGF29 is concentrated to 20 mg ml$^{-1}$ as a stock and frozen at −80 °C for future use. Purified SGF29 (15 mg ml$^{-1}$) was mixed with histone peptide H3C4me3 in a molecular ratio of 1:3, and the complex was crystallized in a buffer containing 0.1 M Bis-Tris, pH 5.5, 27% PEG3350, 200 mM ammonium sulphate and 5 mM strontium chloride. Before flash-frozen in liquid nitrogen, the crystals were soaked in a cryoprotectant buffer containing 88% reservoir solution and 12% glycerol.

Human JARID1A PHD finger (aa 330–380) was PCR amplified, and cloned into a modified pET28b vector (Novagen) with an N-terminal 10xHis-SUMO tandem tag. JARID1A PHD finger used for crystallization was expressed in the *E. coli* BL21 (Novagen) induced overnight by 0.2 mM isopropyl β-D-thiogalactoside at 25 °C in the LB medium supplemented with 0.1 mM ZnCl$_2$. The collected cells were suspended in 500 mM NaCl, 20 mM Tris, pH 8.5. After cell lysis and centrifugation, the supernatant was applied to a HisTrap (GE Healthcare) column and the protein was eluted with a linear imidazole gradient from 20 mM to 500 mM, followed by tag cleavage using ULP1. A HisTrap column was used to remove the cleaved 10xHis-SUMO tag after removal of imidazole by desalting. The JARID1A PHD sample flow-through was then pooled, concentrated and polished by size-exclusion chromatography on a Superdex 75 16/60 column (GE Healthcare) under the elution buffer: 150 mM NaCl, 20 mM Tris, pH 8.5. The resultant peak of JARID1A PHD finger was then concentrated to ~17 mg ml$^{-1}$, split into small aliquots and frozen in liquid nitrogen for future use.

As for human TAF3, the PHD finger construct 885–915 was cloned, expressed and purified using essentially the same strategy as JARID1A PHD finger. TAF3 PHD finger was concentrated to ~25 mg ml$^{-1}$ and aliquoted for future use.

Crystallization was performed via the sitting drop vapour diffusion method under 4 °C by mixing equal volume (0.2–1.0 μl) of JARID1A PHD-H3C4me3 complex (1:1.8 molar ratio, 14–16 mg ml$^{-1}$) and reservoir solution containing 0.02 M sodium-l-glutamate, 0.02 M DL-alanine, 0.02 M glycine, 0.02 M DL-lysine HCl, 0.02 M DL-serine, 0.1 M Tris, 0.1 M Bicine, pH 8.5, 12.5% MPD, 12.5% PEG 1 K, 12.5% PEG3350. As for TAF3 PHD-H3C4me3 complex (1:1.4 molar ratio, 22–24 mg ml$^{-1}$), the crystal was grown in the reservoir solution containing 0.03 M magnesium chloride, 0.03 M calcium chloride, 0.1 M MES, 0.1 M imidazole, pH 6.5, 15% PEGMME 550, 15% PEG 20 K. The complex crystals were directly flash-frozen in liquid nitrogen with reservoir solution as cryoprotectant for data collection. The diffraction data were collected at the beamline BL17U of the Shanghai Synchrotron Radiation Facility at 0.9793 Å. All diffraction images were indexed, integrated and merged using HKL2000 (ref. 57). The structure was determined by molecular replacement using MOLREP[58] with the free JARID1A PHD finger (PDB ID: 2KGG) and free TAF3 PHD finger (PDB ID: 2K16) as the search model. Structural refinement was carried out using PHENIX[59], and iterative model building was performed with COOT[60]. Detailed data collection and refinement statistics are summarized in Table 2. Structural figures were created using the PYMOL (http://www.pymol.org/) program.

**Quantum-chemical analyses.** All calculations for TRP2-Kme3 and TRP2-Cme3 complexes were carried out with the ADF program using dispersion-corrected density functional theory at the BLYP-D3BJ/TZ2P level of theory[40,41]. The effect of solvation was simulated by means of the Conductor like Screening Model (COSMO) of solvation as implemented in ADF. The approach has been benchmarked against highly correlated post-Hartree–Fock methods and experimental data and was found to work reliably[61–63].

The bonding mechanism in our model complexes have been further analysed using quantitative (Kohn–Sham) molecular orbital theory in combination with an EDA[42,64]. The bond energy in aqueous solution $\Delta E$(aq) consists of two major components, namely the strain energy $\Delta E_{strain}$(aq) associated with deforming the Kme3 (or Cme3) and the reader from their own equilibrium structure to the geometry they adopt in the complex, plus the interaction energy $\Delta E_{int}$(aq) between these deformed solutes in the complex (see equation (1)):

$$\Delta E(aq) = \Delta E_{strain}(aq) + \Delta E_{int}(aq). \qquad (1)$$

To arrive at an understanding of the importance of desolvation phenomena during the complexation process, we separate the interaction energy $\Delta E_{int}$(aq) into the effect caused by the change in solvation $\Delta E$(desolv) and the remaining intrinsic solute-solute interaction $\Delta E_{int}$ between the unsolvated fragments in vacuum:

$$\Delta E_{int}(aq) = \Delta E_{int}(desolv) + \Delta E_{int}. \qquad (2)$$

In the EDA, the intrinsic interaction energy $\Delta E_{int}$ can be further decomposed as shown in equation (3):

$$\Delta E_{int} = \Delta V_{elstat} + \Delta E_{Pauli} + \Delta E_{oi} + \Delta E_{disp}. \qquad (3)$$

Here $\Delta V_{elstat}$ corresponds to the classical electrostatic interaction between the unperturbed charge distributions of the deformed fragments that is usually attractive. The Pauli repulsion $\Delta E_{Pauli}$ comprises the destabilizing interactions between occupied orbitals and is responsible for the steric repulsions. The orbital interaction $\Delta E_{oi}$ accounts for charge transfer (donor–acceptor interactions between occupied orbitals on one moiety with unoccupied orbitals of the other, including the HOMO–LUMO interactions) and polarization (empty/occupied orbital mixing on one fragment due to the presence of another fragment). Finally, the $\Delta E_{disp}$ term accounts for the dispersion interactions based on Grimme's DFT-D3BJ correction. Furthermore, the charge distribution has been analysed using the VDD method[43].

**WaterMap calculations.** WaterMap has been described in detail in previous works[52,65]. All calculations were run in with default settings. In brief, a 2 ns molecular dynamic simulation of the reader proteins with the peptide removed, is performed using the Desmond molecular dynamic engine[66,67] with the OPLS2.1 force field[68,69]. Protein atoms are constrained throughout the simulation. Water molecules from the simulation are then clustered into distinct hydration sites. Enthalpy values for each hydration site are obtained by averaging over the non-bonded interaction for each water molecule in the cluster. Entropy values are calculated using a numerical integration of a local expansion of the entropy in terms of spatial and orientational correlation functions[44,45]. The contribution of water-free energy to the binding free energy of the peptide is approximated by the sum of the free energies of hydration sites displaced by the ligand on binding.

## References

1. Allis, C. D., Jenuwein, T. & Reinberg, D. *Epigenetics* (Cold Spring Harbor Laboratory Press, 2007).
2. Kouzarides, T. Chromatin modifications and their function. *Cell* **128**, 693–705 (2007).
3. Strahl, B. D. & Allis, C. D. The language of covalent histone modifications. *Nature* **403**, 41–45 (2000).

4.  Zhang, Y. & Reinberg, D. Transcription regulation by histone methylation: Interplay between different covalent modifications of the core histone tails. *Genes Dev.* **15**, 2343–2360 (2001).

5.  Martin, C. & Zhang, Y. The diverse functions of histone lysine methylation. *Nat. Rev. Mol. Cell Biol.* **6**, 838–849 (2005).

6.  Cloos, P. A. C., Christensen, J., Agger, K. & Helin, K. Erasing the methyl mark: Histone demethylases at the center of cellular differentiation and disease. *Genes Dev.* **22**, 1115–1140 (2008).

7.  Yun, M., Wu, J., Workman, J. L. & Li, B. Readers of histone modifications. *Cell Res.* **21**, 564–578 (2011).

8.  Taverna, S. D., Li, H., Ruthenburg, A. J., Allis, C. D. & Patel, D. J. How chromatin-binding modules interpret histone modifications: lessons from professional pocket pickers. *Nat. Struct. Mol. Biol.* **14**, 1025–1040 (2007).

9.  Min, J. *et al.* L3MBTL1 recognition of mono- and dimethylated histones. *Nat. Struct. Mol. Biol.* **14**, 1229–1230 (2007).

10. Li, H. *et al.* Structural basis for lower lysine methylation state-specific readout by MBT repeats of L3MBTL1 and an engineered PHD finger. *Mol. Cell* **28**, 677–691 (2007).

11. Botuyan, M. V. *et al.* Structural basis for the methylation state-specific recognition of histone H4-K20 by 53BP1 and Crb2 in DNA repair. *Cell* **127**, 1361–1373 (2006).

12. Kuo, A. J. *et al.* The BAH domain of ORC1 links H4K20me2 to DNA replication licensing and Meier–Gorlin syndrome. *Nature* **484**, 115–119 (2012).

13. Collins, R. E. *et al.* The ankyrin repeats of G9a and GLP histone methyltransferases are mono- and dimethyllysine binding modules. *Nat. Struct. Mol. Biol.* **15**, 245–250 (2008).

14. Sims, R. J. & Reinberg, D. Histone H3 Lys 4 methylation: caught in a bind? *Genes Dev.* **20**, 2779–2786 (2006).

15. Iwase, S. *et al.* ATRX ADD domain links an atypical histone methylation recognition mechanism to human mental-retardation syndrome. *Nat. Struct. Mol. Biol.* **18**, 769–776 (2011).

16. Hughes, R. M., Wiggins, K. R., Khorasanizadeh, S. & Waters, M. L. Recognition of trimethyllysine by a chromodomain is not driven by the hydrophobic effect. *Proc. Natl Acad. Sci. USA* **104**, 11184–11188 (2007).

17. Ruthenburg, A. J., Allis, C. D. & Wysocka, J. Methylation of lysine 4 on histone H3: intricacy of writing and reading a single epigenetic mark. *Mol. Cell* **25**, 15–30 (2007).

18. Zhenyu, L., Lai, J. & Yingkai, Z. Importance of charge independent effects in readout of the trimethyllysine mark by HP1 chromodomain. *J. Am. Chem. Soc.* **131**, 14928–14931 (2009).

19. Wang, G. G. *et al.* Haematopoietic malignancies caused by dysregulation of a chromatin-binding PHD finger. *Nature* **459**, 847–851 (2009).

20. Li, H. *et al.* Molecular basis for site-specific read-out of histone H3K4me3 by the BPTF PHD finger of NURF. *Nature* **442**, 91–95 (2006).

21. van Ingen, H. *et al.* Structural insight into the recognition of the H3K4me3 mark by the TFIID subunit TAF3. *Structure* **16**, 1245–1256 (2008).

22. Bian, C. *et al.* Sgf29 binds histone H3K4me2/3 and is required for SAGA complex recruitment and histone H3 acetylation. *EMBO J.* **30**, 2829–2842 (2011).

23. Lee, J., Thompson, J. R., Botuyan, M. V. & Mer, G. Distinct binding modes specify the recognition of methylated histones H3K4 and H4K20 by JMJD2A-tudor. *Nat. Struct. Mol. Biol.* **15**, 109–111 (2008).

24. Ma, J. C. & Dougherty, D. A. The Cation – π Interaction. *Chem. Rev.* **97**, 1303–1324 (1997).

25. Dougherty, D. A. Cation-π interactions in chemistry and biology: a new view of benzene, Phe, Tyr, and Trp. *Science* **271**, 163–168 (1996).

26. Gallivan, J. P. & Dougherty, D. A. Cation-π interactions in structural biology. *Proc. Natl Acad. Sci. USA* **96**, 9459–9464 (1999).

27. Dougherty, D. A. The cation-π interaction. *Acc. Chem. Res.* **46**, 885–893 (2013).

28. Nagy, G. N. *et al.* Composite aromatic boxes for enzymatic transformations of quaternary ammonium substrates. *Angew. Chem. Int. Ed.* **53**, 13471–13476 (2014).

29. Cubero, E., Luque, F. J. & Orozco, M. Is polarization important in cation-pi interactions? *Proc. Natl Acad. Sci. USA* **95**, 5976–5980 (1998).

30. Hunter, C. A., Low, C. M. R., Rotger, C., Vinter, J. G. & Zonta, C. Substituent effects on cation-π interactions: a quantitative study. *Proc. Natl Acad. Sci. USA* **99**, 4873–4876 (2002).

31. Wheeler, S. E. & Houk, K. N. Substituent effects in cation/π interactions and electrostatic potentials above the centers of substituted benzenes are due primarily to through-space effects of the substituents. *J. Am. Chem. Soc.* **131**, 3126–3127 (2009).

32. Zhong, W. *et al.* From ab initio quantum mechanics to molecular neurobiology: a cation-π binding site in the nicotinic receptor. *Proc. Natl Acad. Sci. USA* **95**, 12088–12093 (1998).

33. Pieters, B., Belle, R. & Mecinović, J. The effect of the length of histone H3K4me3 on recognition by reader proteins. *Chembiochem* **14**, 2408–2412 (2013).

34. Southall, N. T., Dill, K. A. & Haymet, A. D. J. A view of the hydrophobic effect. *J. Phys. Chem. B* **106**, 521–533 (2002).

35. Chandler, D. Interfaces and the driving force of hydrophobic assembly. *Nature* **437**, 640–647 (2005).

36. Blokzijl, W. & Engberts, J. B. F. N. Hydrophobic Effects. Opinions and Facts. *Angew. Chem. Int. Ed.* **32**, 1545–1579 (1993).

37. Salonen, L. M., Ellermann, M. & Diederich, F. Aromatic rings in chemical and biological recognition: energetics and structures. *Angew. Chem. Int. Ed.* **50**, 4808–4842 (2011).

38. Nishio, M., Umezawa, Y., Fantini, J., Weiss, M. S. & Chakrabarti, P. CH-π hydrogen bonds in biological macromolecules. *Phys. Chem. Chem. Phys.* **16**, 12648–12683 (2014).

39. Takahashi, O., Kohno, Y. & Nishio, M. Relevance of weak hydrogen bonds in the conformation of organic compounds and bioconjugates: Evidence from recent experimental data and high-level ab initio MO calculations. *Chem. Rev.* **110**, 6049–6076 (2010).

40. te Velde, G. *et al.* Chemistry with ADF. *J. Comput. Chem.* **22**, 931–967 (2001).

41. Becke, A. D. Density-functional exchange-energy approximation with correct asymptotic behavior. *Phys. Rev. A* **38**, 3098–3100 (1988).

42. Bickelhaupt, F. M. & Baerends, E. J. *Reviews in Computational Chemistry* Vol. 15, 1–86 (Wiley, 2000).

43. Fonseca Guerra, C., Handgraaf, J. W., Baerends, E. J. & Bickelhaupt, F. M. Voronoi deformation density (VDD) charges: assessment of the Mulliken, Bader, Hirshfeld, Weinhold, and VDD methods for charge analysis. *J. Comput. Chem.* **25**, 189–210 (2004).

44. Lazaridis, T. Inhomogeneous fluid approach to solvation thermodynamics. 1. Theory. *J. Phys. Chem. B* **102**, 3531–3541 (1998).

45. Lazaridis, T. Inhomogeneous fluid approach to solvation thermodynamics. 2. applications to simple fluids. *J. Phys. Chem. B* **102**, 3542–3550 (1998).

46. Snyder, P. W. *et al.* Mechanism of the hydrophobic effect in the biomolecular recognition of arylsulfonamides by carbonic anhydrase. *Proc. Natl Acad. Sci. USA* **108**, 17889–17894 (2011).

47. Persch, E., Dumele, O. & Diederich, F. Molecular recognition in chemical and biological systems. *Angew. Chem. Int. Ed.* **54**, 3290–3327 (2015).

48. Snyder, P. W., Lockett, M. R., Moustakas, D. T. & Whitesides, G. M. Is it the shape of the cavity, or the shape of the water in the cavity? *Eur. Phys. J. Spec. Top.* **223**, 853–891 (2014).

49. Ball, P. Water as an active constituent in cell biology. *Chem. Rev.* **108**, 74–108 (2008).

50. Krimmer, S. G., Betz, M., Heine, A. & Klebe, G. Methyl, ethyl, propyl, butyl: Futile but not for water, as the correlation of structure and thermodynamic signature shows in a congeneric series of thermolysin inhibitors. *ChemMedChem* **9**, 833–846 (2014).

51. Breiten, B. *et al.* Water networks contribute to enthalpy/entropy compensation in protein-ligand binding. *J. Am. Chem. Soc.* **135**, 15579–15584 (2013).

52. Beuming, T. *et al.* Thermodynamic analysis of water molecules at the surface of proteins and applications to binding site prediction and characterization. *Proteins* **80**, 871–883 (2012).

53. Sirin, S., Pearlman, D. A. & Sherman, W. Physics-based enzyme design: predicting binding affinity and catalytic activity. *Proteins* **82**, 3397–3409 (2014).

54. Arrowsmith, C. H., Bountra, C., Fish, P. V., Lee, K. & Schapira, M. Epigenetic protein families: a new frontier for drug discovery. *Nat. Rev. Drug Discov.* **11**, 384–400 (2012).

55. Sippl, W. & Jung, M. *Epigenetic Targets in Drug Discovery* Vol. 42 (Wiley-VCH Verlag GmbH, 2009).

56. Anthis, N. J. & Clore, G. M. Sequence-specific determination of protein and peptide concentrations by absorbance at 205 nm. *Protein Sci.* **22**, 851–858 (2013).

57. Otwinowski, Z. & Minor, W. *Macromolecular Crystallography, part A* Vol. 276 (Academic Press, 1997).

58. Vagin, A. & Teplyakov, A. Molecular replacement with MOLREP. *Acta Crystallogr. Sect. D Biol. Crystallogr.* **66**, 22–25 (2010).

59. Adams, P. D. *et al.* PHENIX: A comprehensive Python-based system for macromolecular structure solution. *Acta Crystallogr. Sect. D Biol. Crystallogr.* **66**, 213–221 (2010).

60. Emsley, P. & Cowtan, K. Coot: Model-building tools for molecular graphics. *Acta Crystallogr. Sect. D Biol. Crystallogr.* **60**, 2126–2132 (2004).

61. Fonseca Guerra, C., van der Wijst, T., Poater, J., Swart, M. & Bickelhaupt, F. M. Adenine versus guanine quartets in aqueous solution: dispersion-corrected DFT study on the differences in π-stacking and hydrogen-bonding behavior. *Theor. Chem. Acc.* **125**, 245–252 (2010).

62. van der Wijst, T., Fonseca Guerra, C., Swart, M., Bickelhaupt, F. M. & Lippert, B. A ditopic ion-pair receptor based on stacked nucleobase quartets. *Angew. Chem. Int. Ed.* **48**, 3285–3287 (2009).

63. Padial, J. S., de Gelder, R., Fonseca Guerra, C., Bickelhaupt, F. M. & Mecinović, J. Stabilisation of 2,6-diarylpyridinium cation by through-space polar-π interactions. *Chem. Eur. J.* **20**, 6268–6271 (2014).

64. Baerends, E. J., Gritsenko, O. V. & van Meer, R. The Kohn-Sham gap, the fundamental gap and the optical gap: the physical meaning of occupied and virtual Kohn-Sham orbital energies. *Phys. Chem. Chem. Phys.* **15**, 16408–16425 (2013).

65. Abel, R., Young, T., Farid, R., Berne, B. J. & Friesner, R. A. Role of the active-site solvent in the thermodynamics of factor Xa ligand binding. *J. Am. Chem. Soc.* **130,** 2817–2831 (2008).

66. Maestro-desmond interoperability tools, version 4.1 (Schrödinger, New York, NY, USA, 2015).

67. Desmond molecular dynamics system, version 4.1 (D. E. Shaw Research, New York, NY, USA, 2015).

68. OPLS2.1. (Schrodinger Inc.; New York, NY, USA, 2015).

69. Wang, L. *et al.* Accurate and reliable prediction of relative ligand binding potency in prospective drug discovery by way of a modern free-energy calculation protocol and force field. *J. Am. Chem. Soc.* **137,** 2695–2703 (2015).

## Acknowledgements

We thank the Netherlands Research School for Chemical Biology (J.M.), the National Natural Science Foundation of China program 31270763 and The Major State Basic Research Development Program in China 2015CB910503 (H.L.), and the Netherlands Organization for Scientific Research (NWO-ALW, NWO-CW and NWO-EW, F.M.B.) for financial support. The SGC is a registered charity (number 1097737) that receives funds from AbbVie, Boehringer Ingelheim, the Canada Foundation for Innovation, the Canadian Institutes for Health Research, Genome Canada through the Ontario Genomics Institute (OGI-055), GlaxoSmithKline, Janssen, Lilly Canada, the Novartis Research Foundation, the Ontario Ministry of Economic Development and Innovation, Pfizer, Takeda and the Wellcome Trust (092809/Z/10/Z to J.Min). We thank W. Tempel for the data collection of SGF29 and the staff members at beamlines BL17U of the Shanghai Synchrotron Radiation Facility for their assistance in data collection of JARID1A and TAF3.

## Author contributions

J.M. conceived and supervised the project. J.J.A.G.K. synthesized Fmoc-Cme3 and prepared histone peptides. J.J.A.G.K. and J.M. carried out thermodynamic studies and analysed the data. J.H. and H.L. performed structural experiments with JARID1A and TAF3. J.P. and F.M.B. carried out quantum-chemical analyses. C.X., A.D. and J.Min. performed structural experiments with SGF29. B.J.G.E.P. expressed and purified proteins for thermodynamic analyses. W.S. and T.B. carried out WaterMap calculations and analysed the results. J.M. wrote the manuscript with contributions from W.S., T.B., F.M.B. and H.L. All authors contributed in editing the manuscript.

## Additional information

**Accession codes:** Coordinates of JARID1A PHD–H3(1-10)C4me3, TAF3 PHD–H3(1-10)C4me3 and SGF29 tandem tudor–H3(1-10)C4me3 complexes have been deposited into Protein Data Bank under accession codes 5C11, 5C13 and 5C0M, respectively.

**Competing financial interests:** The authors declare no competing financial interests.

# 5

# Modelling the Tox21 10 K chemical profiles for *in vivo* toxicity prediction and mechanism characterization

Ruili Huang[1], Menghang Xia[1], Srilatha Sakamuru[1], Jinghua Zhao[1], Sampada A. Shahane[1], Matias Attene-Ramos[1], Tongan Zhao[1], Christopher P. Austin[1] & Anton Simeonov[1]

Target-specific, mechanism-oriented *in vitro* assays post a promising alternative to traditional animal toxicology studies. Here we report the first comprehensive analysis of the Tox21 effort, a large-scale *in vitro* toxicity screening of chemicals. We test ∼10,000 chemicals in triplicates at 15 concentrations against a panel of nuclear receptor and stress response pathway assays, producing more than 50 million data points. Compound clustering by structure similarity and activity profile similarity across the assays reveals structure–activity relationships that are useful for the generation of mechanistic hypotheses. We apply structural information and activity data to build predictive models for 72 *in vivo* toxicity end points using a cluster-based approach. Models based on *in vitro* assay data perform better in predicting human toxicity end points than animal toxicity, while a combination of structural and activity data results in better models than using structure or activity data alone. Our results suggest that *in vitro* activity profiles can be applied as signatures of compound mechanism of toxicity and used in prioritization for more in-depth toxicological testing.

[1] Division of Pre-clinical Innovation, National Center for Advancing Translational Sciences, National Institutes of Health, 9800 Medical Center Drive, Rockville, Maryland 20850, USA. Correspondence and requests for materials should be addressed to R.H. (email: huangru@mail.nih.gov).

Thousands of chemicals to which humans are exposed have inadequate data on which to predict their potential for toxicological effects. Traditional toxicity testing conducted *in vivo* using animal models provides chemical safety reference to humans, but these methods are expensive and low throughput, and it is often difficult to extrapolate the test results to human health effect because of species differences. High-throughput screening (HTS) techniques are now routinely used in conjunction with computational methods and information technology to probe how chemicals interact with biological systems, both *in vitro* and *in vivo*. Progress is being made in recognizing the patterns of response in genes and pathways induced by certain chemicals or chemical classes that might be predictive of adverse health outcomes in humans. However, as with any new technology, both the reliability and the relevance of the approach need to be demonstrated in the context of current knowledge and practice.

The Tox21 programme[1-4], a collaboration between the National Institute of Environmental Health Sciences/National Toxicology Program, the US Environmental Protection Agency/National Center for Computational Toxicology, the National Institutes of Health Chemical Genomics Center (now within the National Center for Advancing Translational Sciences) and the US Food and Drug Administration, aims to identify chemical structure–activity signatures derived through *in vitro* testing that could act as predictive surrogates for *in vivo* toxicity. During the production phase of the Tox21 programme, the Tox21 10 K compound library has been screened against 30 cell-based assays, including nuclear receptors[5] and stress response pathways[6], in a quantitative HTS (qHTS) format in triplicate[7-10].

Here we review the performance of these assays and the data quality, summarize the activities observed from these assays and evaluate the utility of the data towards achieving the Tox21 goals. We find the *in vitro* assay activity profiles useful for hypotheses generation on compound mechanism of toxicity. These data can be applied, together with chemical structure information, to build predictive models for *in vivo* toxicity and prioritize chemicals for more advanced toxicological tests.

## Results

**Assay performance and activity distribution summary.** Twelve of the thirty assays screened performed well in the qHTS format with performance statistics[11] including signal-to-background ratios $\geq 3$-fold, coefficient of variances $\leq 10\%$ and Z′ factors $\geq 0.5$ (Table 1). The other 18 assays, for example, the AR-bla antagonist mode assay, with poorer performance in one or two metrics, for example, lower signal-to-background ratio ($<3$), were compensated by better performance in other metrics, for example, extremely small coefficient of variance ($<5\%$), such that the overall assay performance still withheld as measured by data reproducibility as described below. The positive control titrations embedded in every plate replicated well across the entire screen (Fig. 1) with variations in $AC_{50}$s $<3$-fold for 89% of the assays and $<4$-fold for all assays (Table 1). A more direct measure of assay performance is data reproducibility. Reproducibility[9] as represented by active match, inactive match, inconclusive and mismatch rates was calculated for all assays (Table 2) screened against the three copies of the 10 K library with compounds plated in different well locations in each copy. Seventeen of the

---

**Table 1 | Tox21 10 K qHTS assay summary statistics*.**

| Assay | S/B | Z′ factor | CV | Positive control | Control AC$_{50}$ | Control AC$_{50}$ fold change |
|---|---|---|---|---|---|---|
| AhR-luc | 8 ± 4 | 0.3 ± 0.2 | 16 ± 15 | Omeprazole | 49.5 μM | 3.14 |
| AR-bla agonist | 1.9 ± 0.2 | 0.2 ± 0.1 | 5 ± 1 | R1881 | 1.21 nM | 2.36 |
| AR-bla antagonist | 2.5 ± 0.3 | 0.7 ± 0.2 | 4 ± 1 | Cyproterone acetate | 4.68 μM | 2.51 |
| ARE-bla | 2.1 ± 0.3 | 0.70 ± 0.06 | 5 ± 2 | β-Naphthoflavone | 1.95 μM | 1.29 |
| AR-MDA-luc agonist | 6.6 ± 0.6 | 0.68 ± 0.06 | 15 ± 2 | R1881 | 14.3 pM | 1.52 |
| AR-MDA-luc antagonist | 17 ± 10 | 0.67 ± 0.07 | 8 ± 2 | Nilutamide | 15.3 μM | 1.49 |
| Aromatase | 6.2 ± 0.8 | 0.80 ± 0.03 | 4.7 ± 0.9 | Letrozole | 6.70 nM | 1.32 |
| DT40 *Rad54/Ku70* | 40 ± 4 | 0.8 ± 0.2 | 6 ± 2 | Tetra-octyl ammonium bromide | 416 nM | 1.63 |
| DT40 WT | 40 ± 7 | 0.79 ± 0.08 | 7 ± 3 | Tetra-octyl ammonium bromide | 594 nM | 1.79 |
| DT40 *Rev3* | 40 ± 9 | 0.79 ± 0.09 | 6 ± 2 | Tetra-octyl ammonium bromide | 440 nM | 1.5 |
| ATAD5 | 6.0 ± 0.9 | 0.73 ± 0.04 | 14 ± 2 | 5-Fluorouridine | 2.12 μM | 2.11 |
| ER-bla agonist | 4.7 ± 0.6 | 0.53 ± 0.09 | 4 ± 2 | β-Estradiol | 332 pM | 1.51 |
| ER-bla antagonist | 3.3 ± 0.8 | 0.4 ± 0.1 | 11 ± 3 | 4-Hydroxy tamoxifen | 5.13 nM | 1.58 |
| ER-BG1-luc agonist | 2.5 ± 0.3 | 0.5 ± 0.2 | 10 ± 5 | β-Estradiol | 29.0 pM | 3.75 |
| ER-BG1-luc antagonist | 8.0 ± 0.9 | 0.77 ± 0.07 | 6 ± 2 | 4-Hydroxy tamoxifen | 73.2 nM | 1.28 |
| FXR-bla agonist | 4.0 ± 0.7 | 0.3 ± 0.2 | 7 ± 1 | Chenodeoxycholic acid | 29.8 μM | 1.29 |
| FXR-bla antagonist | 4.4 ± 0.9 | 0.67 ± 0.09 | 3 ± 1 | Guggulsterone | 36.7 μM | 1.3 |
| TR-beta-luc agonist | 9 ± 2 | 0.63 ± 0.07 | 12 ± 3 | T3 | 41.9 pM | 2.14 |
| TR-beta-luc antagonist | 5.0 ± 0.9 | 0.7 ± 0.1 | 6 ± 3 | NA | NA | NA |
| GR-bla agonist | 3.0 ± 0.2 | 0.73 ± 0.05 | 3.3 ± 0.8 | Dexamethasone | 3.64 nM | 1.36 |
| GR-bla antagonist | 1.9 ± 0.1 | 0.5 ± 0.1 | 4.8 ± 0.9 | Mifeprostone | 1.71 nM | 2.16 |
| HSE-bla | 3.9 ± 0.5 | 0.46 ± 0.08 | 4 ± 1 | 17-AAG | 45.3 nM | 1.7 |
| Mitochondria toxicity | 6 ± 3 | 0.6 ± 0.2 | 8 ± 3 | FCCP | 96.2 nM | 2.79 |
| P53-bla | 3.1 ± 0.4 | 0.6 ± 0.2 | 6 ± 2 | Mitomycin C | 1.53 μM | 1.62 |
| PPAR-delta-bla agonist | 2.5 ± 0.3 | 0.67 ± 0.06 | 6 ± 1 | L-165,041 | 36.4 nM | 1.8 |
| PPAR-delta-bla antagonist | 2.2 ± 0.2 | 0.57 ± 0.06 | 4.3 ± 0.6 | MK886 | 38.5 μM | 1.95 |
| PPAR-gamma-bla agonist | 2.4 ± 0.1 | 0.73 ± 0.04 | 5 ± 2 | Rosiglitazone | 11.8 nM | 1.85 |
| PPAR-gamma-bla antagonist | 2.1 ± 0.2 | 0.6 ± 0.3 | 5 ± 2 | GW9662 | 2.47 nM | 3.3 |
| VDR-bla agonist | 1.9 ± 0.2 | 0.5 ± 0.1 | 5 ± 1 | 1α, 25-Dihydroxy vitamin D3 | 35.0 pM | 1.97 |
| VDR-bla antagonist | 2.7 ± 0.2 | 0.55 ± 0.06 | 7 ± 1 | NA | NA | NA |

AC$_{50}$, concentration at 50% activity, AC$_{50}$ fold change $= 10^{SD(\log AC50)}$; CV, coefficient of Variance (derived from the negative control wells); NA, not applicable; qHTS, quantitative high-throughput screening; S/B, signal to background.
*Data are derived from the positive and negative control wells on each plate and presented as mean ± standard deviation ($n = 408$)

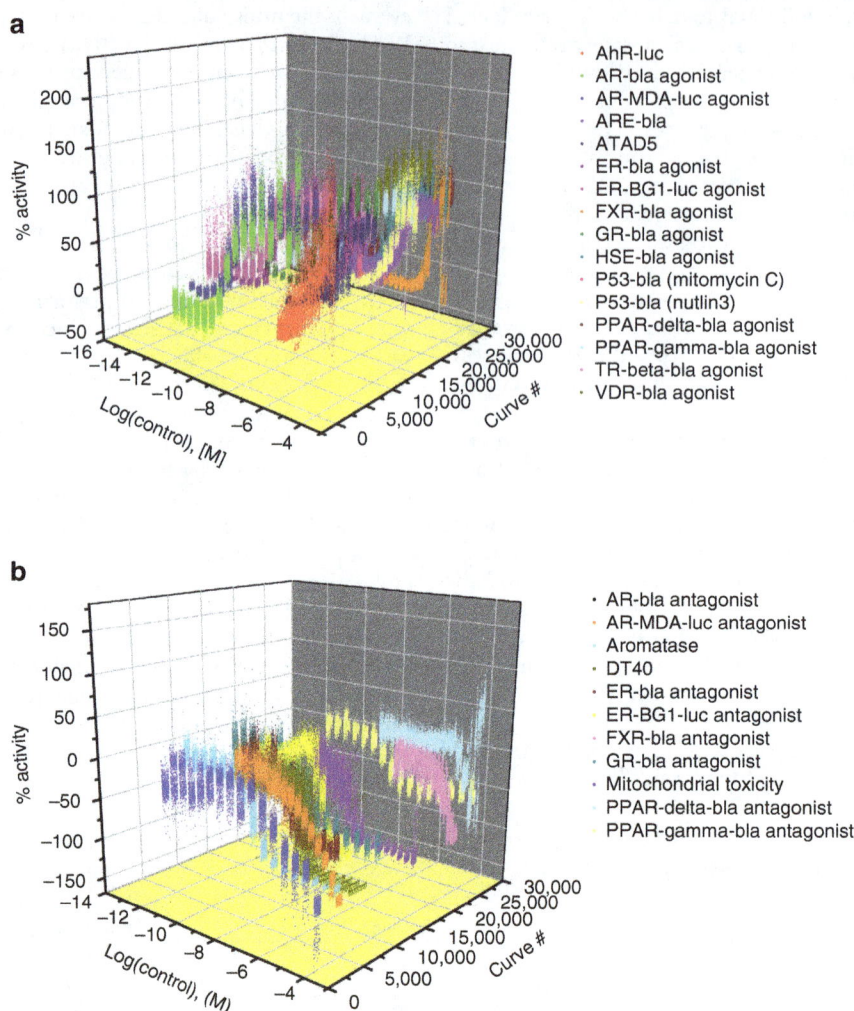

**Figure 1 | Concentration response data of the positive control compounds for the 30 Tox21 phase II assays. (a)** Agonist mode assays; (**b**) antagonist mode assays. The positive control compound is plated as 16-pt. titrations in duplicate in the control columns of every assay plate. In the figure, each concentration response curve is from one plate with a total of 408 plates per assay. The consistency of the control response curves is an indicator of good assay performance.

thirty assays scored (score = 2 × %active match + %inactive match – %inconclusive - 2 × %mismatch) > 90 (grade A) in terms of reproducibility with < 0.5% mismatches in activity (Table 2). Eleven assays had reproducibility scores between 80 and 90 (grade B) with mismatch rates < 1%. Only two assays, the wild-type DT40 and the GR-bla antagonist mode assay, scored below 80, but still above 75, with 1–2% mismatch rates. For the same sample, the average $AC_{50}$ differences between the three runs were < 2-fold for all the assays (Table 2).

The 30 assays screened against the Tox21 10 K collection showed a wide spectrum of activities (Fig. 2) with active rates ranging from 0.43% (VDR-bla agonist mode assay) to 27.4% (DT40 *Rad54/Ku70* mutant assay; Fig. 2a) and potencies ranging from subnanomolar to tens of micromolar (Fig. 2b). The average active rate of the 30 assays was 6.5%. The AR-MDA-luc agonist mode assay had the largest fraction of potent compounds (33.3% of actives had $AC_{50} < 1 \mu M$), whereas the FXR-bla agonist mode assay had no active compound with $AC_{50} < 1 \mu M$.

**Clustering compounds by activity profile.** The 10 K compounds were grouped into 610 clusters by their activity profile

similarity and each cluster was examined for enriched Medical Subject Headings (MeSH; http://www.ncbi.nlm.nih.gov/mesh) pharmacological action (PA) terms (see Online Methods for details). Figure 3 shows the clustered activity profiles and the most significantly enriched MeSH PA term in each cluster (Supplementary Data 1). Of the 553 clusters that contain at least one compound with known MeSH PA (Supplementary Data 1), 544 clusters have significantly enriched terms with $P < 0.05$ (Fisher's exact test), 362 clusters with $P < 0.01$ (Fisher's exact test) and 89 clusters with $P < 0.001$ (Fisher's exact test). This result indicates that compounds with similar activity profiles as determined in the Tox21 screens tend to share similar annotated modes of action (MOAs). For example, all compounds in cluster k36.15 are annotated as cardiotonic agents, including deslanoside, digitoxin, digoxin, ouabain and proscillaridin, which are cardiac glycosides that act by inhibiting Na/K channels[12]. Cluster k30.6 is enriched with statins, including atorvastatin, fluvastatin and cerivastatin, which are hydroxymethylglutaryl-CoA reductase inhibitors[13]. Seven of the eight compounds in cluster k30.9 are antineoplastic antimetabolites, with the only exception of *N*-butyl-*N'*-nitro-*N*-nitrosoguanidine, which does not have a MeSH PA annotation but is a known DNA alkylating agent[14].

**Table 2 | Assay performances measured by reproducibility of the Tox21 10 K triplicate runs\*.**

| Assay | Active match (%) | Inactive match (%) | Inconclusive (%) | Mismatch (%) | AC$_{50}$ fold change | Score$^\dagger$ |
|---|---|---|---|---|---|---|
| ATAD5 | 5.34 | 91.78 | 2.84 | 0.05 | 1.30 | 99.51 |
| DT40 Rad54/Ku70 | 23.92 | 59.47 | 16.50 | 0.10 | 1.32 | 90.61 |
| DT40 Rev3 | 22.38 | 58.07 | 19.29 | 0.26 | 1.40 | 83.02 |
| DT40 WT | 21.07 | 58.89 | 18.65 | 1.39 | 1.49 | 79.58 |
| P53-bla | 11.10 | 84.86 | 4.04 | 0.00 | 1.29 | 103.02 |
| ARE-bla | 15.29 | 68.34 | 15.65 | 0.70 | 1.76 | 81.85 |
| HSE-bla | 7.57 | 86.38 | 6.05 | 0.00 | 1.45 | 95.46 |
| Aromatase | 15.94 | 73.08 | 10.66 | 0.30 | 1.44 | 93.66 |
| Mitochondria toxicity | 17.57 | 67.52 | 14.33 | 0.55 | 1.53 | 87.20 |
| AhR-luc | 8.86 | 78.78 | 12.24 | 0.10 | 1.82 | 84.05 |
| AR-bla agonist | 5.36 | 86.88 | 7.44 | 0.30 | 1.77 | 89.55 |
| AR-bla antagonist | 15.46 | 75.07 | 9.38 | 0.09 | 1.35 | 96.44 |
| AR-MDA-luc agonist | 4.14 | 93.65 | 2.20 | 0.00 | 1.36 | 99.74 |
| AR-MDA-luc antagonist | 13.22 | 76.44 | 10.07 | 0.27 | 1.48 | 92.28 |
| ER-BG1-luc agonist | 16.43 | 71.22 | 12.05 | 0.28 | 1.52 | 91.46 |
| ER-BG1-luc antagonist | 12.03 | 79.72 | 7.96 | 0.29 | 1.48 | 95.25 |
| ER-bla agonist | 7.01 | 87.11 | 5.87 | 0.01 | 1.36 | 95.25 |
| ER-bla antagonist | 9.84 | 77.86 | 11.95 | 0.34 | 1.49 | 84.90 |
| FXR-bla agonist | 2.46 | 93.87 | 3.65 | 0.02 | 1.53 | 95.09 |
| FXR-bla antagonist | 7.50 | 83.04 | 9.35 | 0.11 | 1.73 | 88.48 |
| GR-bla agonist | 6.67 | 87.49 | 5.75 | 0.10 | 1.37 | 94.89 |
| GR-bla antagonist | 9.73 | 75.11 | 13.13 | 2.01 | 1.81 | 77.40 |
| PPAR-delta-bla agonist | 3.46 | 90.79 | 5.71 | 0.04 | 1.71 | 91.91 |
| PPAR-delta-bla antagonist | 5.67 | 86.73 | 7.49 | 0.10 | 1.73 | 90.37 |
| PPAR-gamma-bla agonist | 8.79 | 83.87 | 7.02 | 0.31 | 1.61 | 93.81 |
| PPAR-gamma-bla antagonist | 8.61 | 79.88 | 11.05 | 0.44 | 1.90 | 85.15 |
| TR-beta-luc agonist | 2.13 | 90.72 | 7.15 | 0.00 | 1.38 | 87.84 |
| TR-beta-luc antagonist | 17.15 | 68.35 | 14.18 | 0.32 | 1.39 | 87.82 |
| VDR-bla agonist | 2.25 | 92.53 | 5.19 | 0.03 | 1.62 | 91.79 |
| VDR-bla antagonist | 5.41 | 86.41 | 8.11 | 0.08 | 1.53 | 88.97 |

\*Active match is the percentage of compounds that were reproducibly active, inactive match is the percentage of compounds that were reproducibly inactive, and mismatch is the percentage of compounds that showed conflicting activities in the triplicate runs. A compound is assigned inconclusive if its activity in the triplicate runs was not clearly active or inactive to make a conclusive reproducibility call. Detailed definitions of the reproducibility calls can be found in our previous report[9]. AC$_{50}$ fold change is the average AC$_{50}$ differences in fold of the active compounds in the triplicate runs.
$^\dagger$Score = 2 × active match + inactive match – inconclusive - 2 × mismatch.

Another example is cluster k41.4, which is enriched with oestrogenic compounds, such as 17alpha-ethinylestradiol and non-steroidal oestrogens such as diethylstilbestrol and zearalenone (Fig. 3b). A neighbouring cluster, k41.3, contains bisphenol type compounds such as bisphenols A, B, Z and AF. Even though most of the bisphenols do not have a MeSH PA assigned, they are known oestrogenic compounds, as well[15]. These compounds are not only structurally similar but also share similar activity profiles (Fig. 3b).

Moreover, 537 clusters contain both compounds with known MeSH PA and compounds with no MeSH PA annotation available. If most compounds in the same cluster share similar annotated MOA, then we can use this information to hypothesize on the MOA of the unknown compounds. For example, the pesticide fludioxonil does not have a MeSH PA assigned but co-clustered with the oestrogenic compounds in k41.4, consistent with recent reports on its endocrine disrupting activity[16]. Similarly, the MeSH term 'anti-inflammatory agents' is significantly enriched in cluster k22.1 (Fisher's exact test: $P = 4.90 \times 10^{-12}$). Most of the compounds in this cluster are glucocorticoids with similar steroid type structure. Six out of the sixteen compounds in this cluster do not have MeSH PA annotations, including metformin, a diabetes drug with a distinct structure. Metformin decreases hyperglycemia primarily by suppressing glucose production in the liver (hepatic gluconeogenesis)[17]; however, the molecular target of metformin is not clearly understood. The present co-clustering of metformin with the anti-inflammatory glucocorticoids indicates that it may act in a manner similar to that of the glucocorticoids. Metformin

was identified as an active agonist in both the glucocorticoid receptor (GR) and androgen receptor (AR) assays. Ampiroxicam is another non-steroidal drug found in k22.1 without a MeSH PA annotation. A literature search on ampiroxicam revealed that it is also an anti-inflammatory drug[18] supporting the utility of clustering by activity profile as an indicator for compound MOA.

**Modelling *in vivo* toxicity with *in vitro* and structural data.** For benchmarking our models against available *in vivo* data, we used 72 end points derived from the Registry of Toxic Effects of Chemical Substances database as described in the Methods section. These models are based on compound assay activity profiles clusters (activity-based models) or structure clusters (structure-based models) or both (combined models). The detailed model construction process is described in the Online Methods. The premise for these models is that compounds that share similar *in vitro* signatures and/or structure features are likely to show similar *in vivo* effects. We first attempted to build models for each of the 72 end points using the Tox21 phase II assay activity profiles. The average area under the receiver operating characteristic (ROC) curve (AUC-ROC) values from the 100 randomizations of each model are shown in Fig. 4 and listed in Supplementary Table 1. The AUC-ROC values for the 72 toxicity end points ranged from 0.50 (rat multiple dose inhalation) to 0.90 (rat multiple dose intramuscular injection) with an average of 0.64. In all, 7 of the 72 end points had good predictive models with average AUC-ROC values >0.75. There are five human toxicity end points, including standard Draize test

**a**

**b**

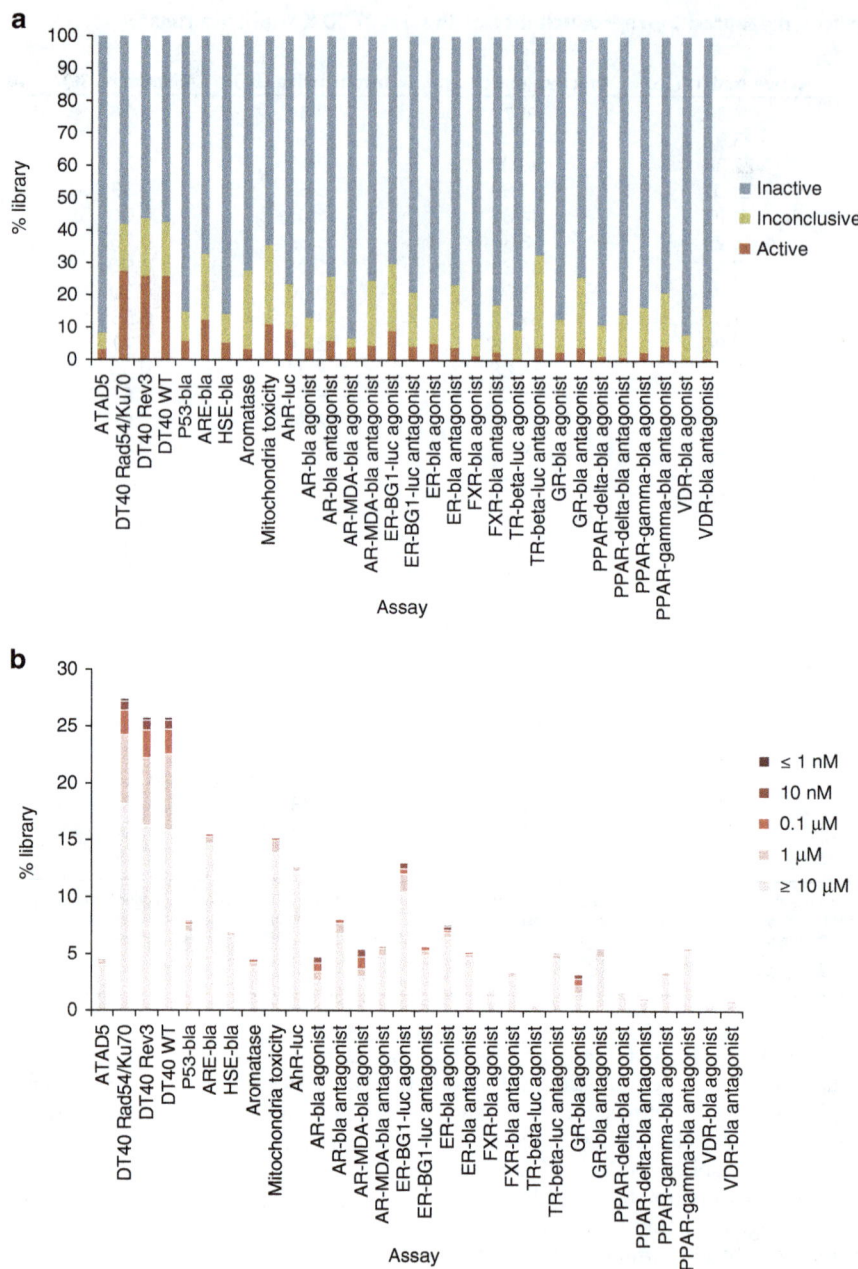

**Figure 2 | Activity distribution of the Tox21 10 K library screened against the 30 assays.** (a) Activity outcome distribution; (b) potency distribution.

for human skin irritation, multiple dose toxicity data (TDLo) through oral exposure from human females, human males and humans (gender not specified), and reproductive toxicity data (TDLo) through oral exposure from human females. The models built for these end points performed significantly better than the models of the 42 mouse/rat toxicity end points and the 7 rabbit toxicity end points comparing the AUC-ROC values (t-test: $P < 0.05$). The average AUC-ROC values were 0.75 for the human toxicity models, 0.65 for the mouse/rat models and 0.59 for the rabbit toxicity models.

The compound structure-based models showed overall better performance than the activity-based models, underlying the ongoing need to further expand the battery of *in vitro* assays. The AUC-ROC values for the 72 toxicity end points ranged from 0.59 (rat multiple dose inhalation) to 0.93 (acute toxicity in dog) with an average of 0.78 and 45 of the 72 end points had good predictive models with average AUC-ROC values $> 0.75$ (Fig. 4 and Supplementary Table 1). However, the models built for

toxicity end points from different species did not show any significant difference in their predictive performance. Compared with the activity-based models, the performance improved significantly for the mouse/rat and rabbit toxicity models with average AUC-ROC values increased to 0.77 and 0.76, respectively, but not as significantly for the human toxicity models (average AUC-ROC = 0.81).

We then attempted to combine the compound structure and assay data in an effort to further improve the models. Models were built for only 67 of the 72 toxicity end points because the more stringent criteria (requiring compounds to co-cluster by both structure and activity) used to form the consensus clusters, which were the basis for the combined models, resulted in smaller clusters such that the compounds with data on the remaining 5 end points all became singletons when split into training and test sets. Nevertheless, the combined models built for the 67 end points showed significantly better performance than the structure-based models with an average AUC-ROC of 0.84

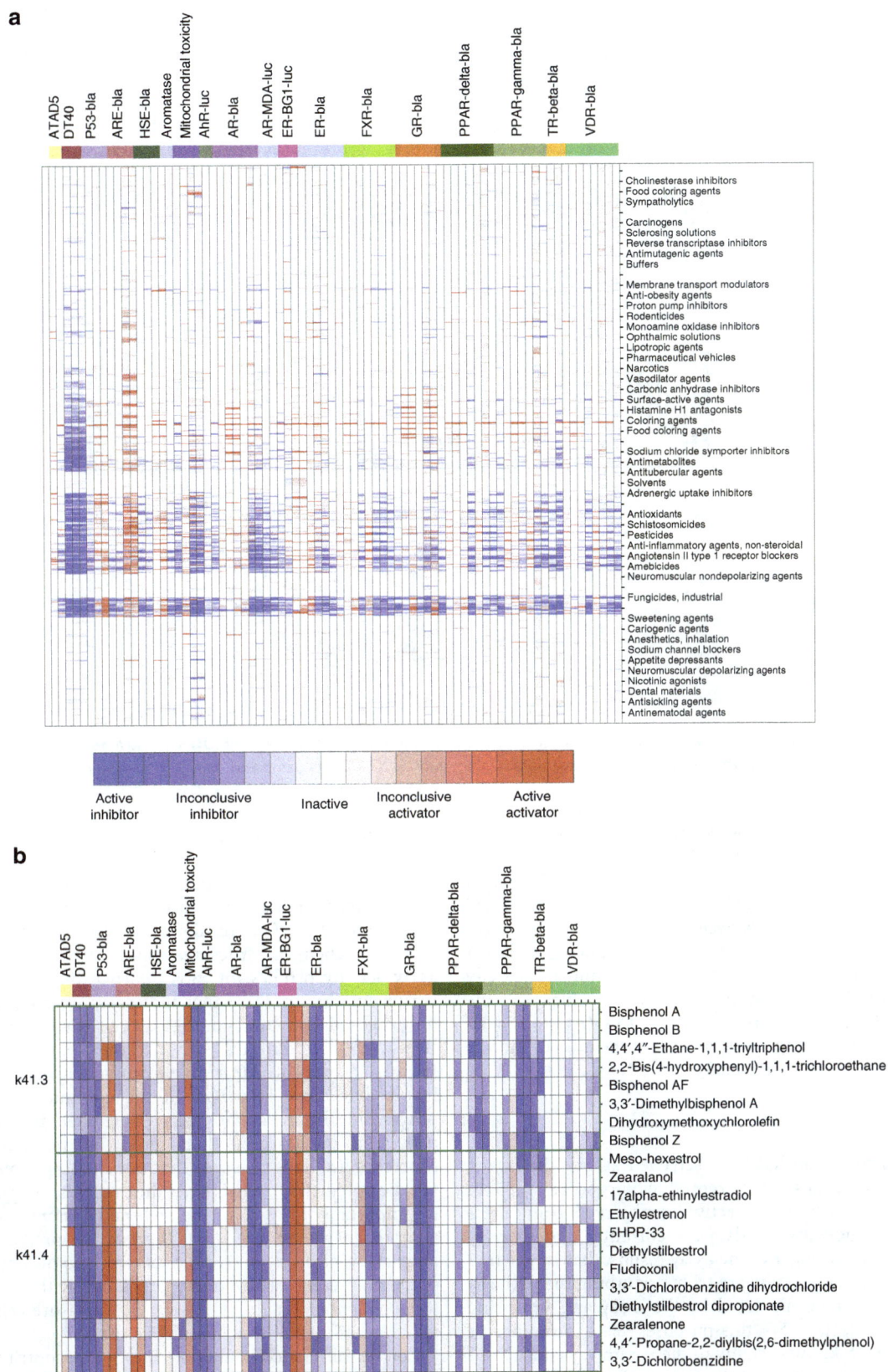

**Figure 3 | Clustered activity profiles of the Tox21 10 K library.** In the heat map, each row is a compound and each column is an assay readout. The heat map is coloured by the compound activity outcome, such that darker red or blue colours indicate more confident activators (red) or inhibitors (blue). Compounds are grouped into clusters of similar activity profiles. Each cluster of compounds is labelled by the most significantly enriched MeSH PA term in that cluster measured by a Fisher's exact test. (**a**) All 10 K compounds; (**b**) example clusters of oestrogenic compounds.

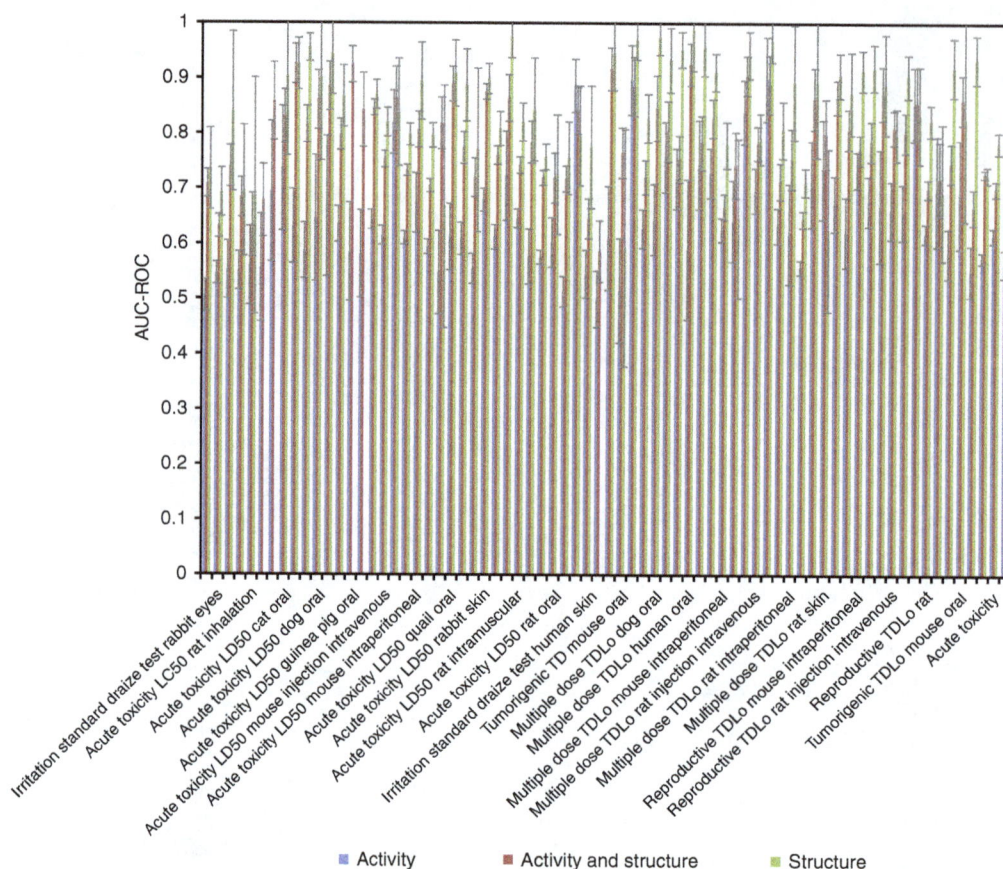

**Figure 4 | AUC-ROC values of predictive models built for 72 *in vivo* toxicity end points using the Tox21 10 K *in vitro* assay activity data, compound structure data and a combination of the assay and structure data.** Columns of AUC-ROC values are shown as mean ± s.d. (n = 100).

(compared to the average AUC-ROC of 0.78 for the structure-based models; *t*-test: $P = 1.7 \times 10^{-7}$), and 55 out of the 67 end points achieved AUC-ROC values > 0.75 (Fig. 4 and Supplementary Table 1). Similar to the structure-based models, the species difference between the model performances disappeared. The animal toxicity models showed a larger improvement in model performance than the human toxicity models, for example, the average AUC-ROC values for the mouse/rat models increased from 0.77 to 0.84 (*t*-test: $P = 3.7 \times 10^{-6}$), whereas the average performance of the human models increased from 0.81 to 0.86 ($P > 0.05$).

## Discussion

The Tox21 10 K collection has been screened against 30 assays, yielding high-quality data sets with reproducibility scores > 85. In this study, we summarized the activities observed from these assays. Further analyses of the data are currently underway to assess the biological relevance of the assay results, that is, whether the actives identified by an assay are truly perturbing the pathway that is purportedly being measured by the assay and not being results of assay artefacts[19]. For this purpose, sets of reference or tool compounds with known activity in these pathways need to be collected to obtain an estimate of the false positive/negative rates of each assay[8-10]. With our current active identification methods[9], we found that the compound activity profiles or signatures generated across the 30 assays are useful for MOA hypotheses generation and chemical prioritization[20]. Compounds with unknown MOA that share similar profiles with compounds with known MOA could be prioritized and tested for that hypothesized MOA.

We tested the applicability of the assay data to building predictive models for *in vivo* toxicity end points in comparison with chemical structure data. The predictive performances of most of these models are reasonable but not ideal, and are end point dependent. Our results show that with the current set of assay data chemical structures appear to be more predictive than assay activity profiles for most *in vivo* toxicity end points. One reason for this could be that the assays we have screened so far only focused on two major areas: nuclear receptor signalling and stress response pathways. Although these pathways are important for toxicity, they are far from encompassing all aspects of biology involved in toxic response. In the continuation of the Tox21 programme, more assays will be included to cover additional pathways and targets that could be relevant for toxicity. Moreover, we have observed that not all assays contribute equally to the predictive power of the models, suggesting that it is important to select the relevant assays and to ensure comprehensive coverage. We checked the predictive capacity of each assay of each *in vivo* toxicity end point, and for each end point only a few assays were predictive with AUC-ROC > 0.7 (Supplementary Table 1). Assays that measure cell viability took up over half of the most predictive assays.

Species difference is another important contributing factor to the less-than-ideal performance of the models based on assay data. All of the screening data we used in this analysis are derived from cell-based assays using human cells or cell lines, whereas most of the *in vivo* data we are trying to model are collected from animals. According to a 2004 Food and Drug Administration report, 92% of new drugs that passed animal testing failed in human clinical trials because of lack of effect or unexpected toxicity[21]. More recent studies show that animal data predicted human outcomes only around half of the time[22]. It is thus not surprising that human *in vitro* cell line data did not show high-predictive power when applied to predict animal toxicity data. To better assess the predictive value of the human *in vitro* assay data, *in vivo* human toxicity data, that is, clinical toxicity data presently not readily available to the public, are required.

Comparison of the models for the few human toxicity end points with the models for animal toxicity showed that the assay data-based models performed markedly better in predicting human toxicity than animal toxicity; in contrast, the structure-based models did not show this species selectivity. Consistent with our findings, a previous small-scale study testing 50 compounds reported similar observations that acute human systemic toxicity was predicted better by human cell lines than animal cell lines[23]. Furthermore, differences in experimental conditions between *in vitro* and *in vivo* studies, such as dosage, timing and metabolic capacity differences, could affect the extrapolation from *in vitro* to *in vivo* results. Under qHTS conditions, most compounds are tested in a fixed concentration range up to 100 µM, and all of the assays used are short-term assays with compound exposure time ranging from a few hours to a day, whereas for *in vivo* studies compounds are tested at much wider dose ranges and time spans up to months and years[24].

Encouragingly, combining structure and activity information significantly improved the model performance for most of the *in vivo* end points. This phenomenon has been observed previously and reviewed recently[25]. Our results further corroborate the value of the *in vitro* assay data when applied to *in vivo* toxicity prediction. However, the applicability domain of the combined models is limited by the availability of *in vivo* data for a number of toxicity end points. This again highlights the importance of having easy access to more high-quality *in vivo* data.

Data quality is another important factor affecting model performance—a prediction could only be as good as the data it is based on. All measurements have errors or variations associated. In this study, we evaluated the quality of the *in vitro* qHTS data in terms of reproducibility (Table 2). We also checked the reproducibility of the *in vivo* data for which we tried to model following a similar approach using compounds with replicate measurements in each *in vivo* toxicity end point. We found that

for those end points with at least 20 compounds that had replicates, there was a significant correlation (Pearson correlation: $r = 0.61$, $P = 1.51 \times 10^{-5}$) between the reproducibility of the replicates and the performance (AUC-ROC) of the model built for that end point, such that models built for more reproducible data showed better predictive power (Fig. 5). This observation suggests that improving data quality would help further improve the performance of *in vivo* toxicity prediction models. The high-quality winning models resulting from the recent Tox21 Data Challenge (https://tripod.nih.gov/tox21/challenge), a crowdsourcing effort that asked participants to build predictive models for the *in vitro* assay data based on chemical structure, provide additional evidence for the importance of data quality.

In summary, the Tox21 10 K chemical library has been screened against a panel of nuclear receptor and stress response pathway assays, producing the largest set of high-quality *in vitro* toxicity data known to date. Although data analysis and interpretation are still underway, the compound activity profiles generated from this study have been shown useful for MOA hypotheses generation and chemical prioritization. Here, we built and assessed predictive models for various *in vivo* toxicity end points using *in vitro* qHTS data and compound structure data. The *in vitro* assay data-based models were distinctly better at predicting human toxicity end points than animal toxicity. More human toxicity data and high-quality *in vivo* data are critical in assessing the true predictive power of *in vitro* data-based models of *in vivo* toxicity. Combing structure and activity data resulted in better models than those built with structure or activity data alone reinforcing the value of *in vitro* assay data in toxicity prediction. The scale and high-resolution nature of the data provided within this screening offer the opportunity for researchers worldwide to derive new insights from this valuable resource in a manner akin to previous crowdsourcing efforts (https://tripod.nih.gov/tox21/challenge/)[26,27]. We have made publicly available all the HTS results (http://www.ncbi.nlm.nih.gov/pcassay?term=tox21) as well as the clustering results used for modelling and the Tox21 compound library information online (http://tripod.nih.gov/tox/filedownload/).

## Methods

**Tox21 chemical library.** The Tox21 10 K library consists of compounds mostly procured from commercial sources by the Environmental Protection Agency (http://www.epa.gov/ncct/dsstox/sdf_tox21s.html), National Toxicology Program and National Institutes of Health Chemical Genomics Center[28], for a total of greater than 10,000 plated compound solutions consisting of 8,599 unique chemical substances including pesticides, industrial chemicals, food additives and drugs. The main criteria for selection of the Tox21 compounds included, but were not limited to, known or perceived environmental hazards or exposure concerns, physicochemical properties indicating suitability for HTS (molecular weight, volatility, solubility, logP), commercial availability and cost. In addition, the Tox21 Chemical Selection Group designated 88 diverse compounds in the Tox21 library to serve as internal controls[9] to assess assay reproducibility and examine positional plate effects: these were included as duplicates in all screening plates[7]. The structures and annotations of the Tox21 10 K library have been deposited into PubChem (http://www.ncbi.nlm.nih.gov/pcsubstance/?term=tox21).

**Assays and qHTS data analysis.** Two areas were the initial focus in the Tox21 phase II screening including nine nuclear receptor targets and seven stress response pathways, selected based on their biological and toxicological relevance, public interest and adaptability to miniaturization and automated screening. The assays were run in different modes (agonist versus antagonist) and/or formats (full length versus partial receptor) as detailed below, totaling 30 assays. Two reporter gene systems, β-lactamase (bla) and luciferase (luc), were used in this study[5]. All cell-based assays were multiplexed with a cell viability assay in the same assay well. Although bla-based assays were multiplexed with a luminescence-based cell viability assay (CellTiter-Glo viability assay, Promega), luc-based assays were multiplexed with a fluorescence-based cell viability assay (Cell Titer-Fluor viability assay, Promega). The nuclear receptor assays, including oestrogen receptor alpha, ligand-binding domain (ER-bla), oestrogen receptor alpha, full length (ER-BG1-luc), androgen receptor, ligand-binding domain (AR-bla), androgen receptor, full length (AR-MDA-luc), glucocorticoid receptor (GR-bla), farnesoid X receptor

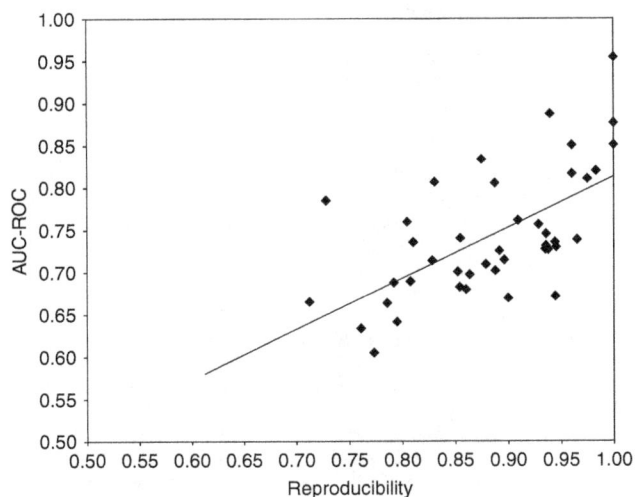

**Figure 5 | Correlation between reproducibility of *in vivo* data and model performance.** A positive correlation (Pearson correlation: $r = 0.61$, $P = 1.51 \times 10^{-5}$) is found between the reproducibility of the compounds tested in replicates and the model performance (AUC-ROC). Models built for more reproducible data showed better predictive power.

(FXR-bla), peroxisome proliferator-activated receptor delta (PPAR-delta-bla), peroxisome proliferator-activated receptor gamma (PPAR-gamma-bla), thyroid hormone receptor (TR-Luc), vitamin D receptor (VDR-bla) and aryl hydrocarbon receptor (AhR-luc) assays, were screened in both agonist and antagonist modes. Aromatase[29,30], mitochondrial toxicity[31] and DNA repair deficient isogenic chicken DT40 cell viability (DT40) assays[32] were screened in antagonist mode. A number of stress response pathway assays, including ATAD5-luc[33], a genotoxicity assay, p53-bla, antioxidant responsive element (ARE-bla)[34] and heat-shock factor response element (HSE-bla) assays, were screened in agonist mode. Detailed assay protocols can be found in PubChem (http://www.ncbi.nlm.nih.gov/pcassay?term=tox21).

In primary screening, all compounds were tested as three independent runs, with each of the three instances of a compound sample residing in a different location on a different compound plate across replicates. Each replicate set of plates was tested on a different day using a different batch of cells. Analysis of compound concentration–response data was performed as previously described[5]. Briefly, raw plate reads for each titration point were first normalized relative to the positive control compound (agonist mode: 100%; antagonist mode: 0%) and dimethylsulphoxide (DMSO)-only wells (agonist mode: 0%; antagonist mode: -100%) as follows: % Activity = $((V_{compound} - V_{DMSO})/(V_{pos} - V_{DMSO})) \times 100$, where $V_{compound}$ denotes the compound well values, $V_{pos}$ denotes the median value of the positive control wells and $V_{DMSO}$ denotes the median values of the DMSO-only wells, and then corrected by applying an in-house pattern correction algorithm using compound-free control plates (that is,, DMSO-only plates) at the beginning and end of the compound plate stack. Concentration–response titration points for each compound were fitted to a four-parameter Hill equation yielding concentrations of half-maximal activity ($AC_{50}$) and maximal response (efficacy) values. Compounds were designated as Class 1–4 according to the type of concentration–response curve observed[5,35]. Curve classes are heuristic measures of data confidence, classifying concentration–responses on the basis of efficacy, the number of data points observed above background activity, and the quality of fit. Each curve class was then converted to a curve rank as previously described[5] such that more potent and efficacious compounds with higher quality curves were assigned a higher rank. Curve ranks should be viewed as qualitative descriptors of the concentration response profile of the compound. Compound reproducibility was assessed by calculating the reproducibility of the curve ranks of each compound generated from the triplicate runs[9]. A reproducibility score was calculated for each assay using the formula: score = 2 × %active match + %inactive match-2 × %mismatch-%inconclusive[5].

**Data sources.** All Tox21 phase II qHTS data are available in PubChem (http://www.ncbi.nlm.nih.gov/pcassay?term=tox21; see Accession Codes section for assay IDs). *In vivo* toxicity data were retrieved from the Registry of Toxic Effects of Chemical Substances database compiled by Leadscope (Leadscope, Inc.). This compilation contains 129 different toxicity end points including acute toxicity, hepatotoxicity, reproductive toxicity, carcinogenicity and skin and eye irritation from various species such as human, rodents, primates and birds on > 10,000 molecules, 6,447 of which overlap with compounds in the Tox21 10 K library. Most of these compounds do not have data available for every toxicity end point. Only the end points that have at least 50 active/toxic calls and 50 inactive/non-toxic calls were kept for further analysis. A total of 68 of the 129 end points met this data availability requirement. In addition, we created three composite end points by aggregating the acute toxicity, reproductive toxicity and the tumorigenic end points, respectively, for a total of 72 end points. For compounds with $LD_{50}$ data available, an $LD_{50}$ of <300 mg kg$^{-1}$ was considered toxic[36]. For other end points, compounds with toxicity measures falling into the top 35 percentile were considered toxic. For composite end points, compounds that are toxic in more than half of the component end points were considered toxic for the composite call. The 72 selected end points and the number of toxic/non-toxic compounds in each end point are listed in Supplementary Table 1. MeSH (http://www.ncbi.nlm.nih.gov/mesh) PA terms were used for compound MOA annotations.

**Clustering and modelling for *in vivo* toxicity.** The 10 K library was clustered based on similarity in its members' activity profiles (measured by curve rank) across the 30 assays using the self-organizing map (SOM) algorithm[37], resulting in 610 clusters. Each cluster was evaluated for enrichment of 363 MeSH PA terms using the Fisher's exact test. The 10 K library was also clustered with the SOM algorithm based on structural similarity using the Leadscope (Leadscope, Inc.) structure fingerprints resulting in 999 clusters. The SOM algorithm clustered the compound activity profiles or structure fingerprints based on the similarity between the profiles measured by pair-wise Euclidean distance, and the analysis was performed using the SOM Toolbox (http://www.cis.hut.fi/projects/somtoolbox/) where detailed documentation of the algorithm can be found. Briefly, the SOM was trained and optimized through 14 phases with 38,000 steps in each phase to minimize the distances between the central data vectors and the compound profiles to form the clusters. The initial learning rate alpha was set to 0.05, which decreased linearly to zero during training. The initial radius of the training area was set to 20 and decreased linearly to one during training. Models were built for the 72 *in vivo* toxicity end points using either the structure

(structure-based models) or assay activity (activity-based models) SOM clusters or both. To build models using both the structure and activity SOM clusters, each compound was reassigned to a 'consensus cluster' such that only compounds that belong to the same structure cluster and the same activity cluster were assigned to the same 'consensus cluster'. The consensus clusters were used to build the structure–activity combined models. For each model, compounds were randomly split into two groups of approximately equal sizes, one used for training and the other used for testing. The randomization was conducted 100 times to generate 100 different training and test sets to evaluate the robustness of the models. For each SOM cluster containing the training compounds, the enrichment of toxic training compounds was determined by a Fisher's exact test. The –log P-value from the Fisher's exact test was used as a measure of the toxic potential (toxicity score) of the compounds in this cluster, and evaluated as a predictor of toxicity for test compounds that fall into the same cluster. More significant P-values (larger –log P-values) indicate a larger probability of toxicity. If a cluster is deficient of toxic compounds, that is, the fraction of toxic compounds in the cluster was smaller than the fraction of toxic compounds in the whole library, the log P-value was used instead. Here we denote the toxicity scores obtained from the activity SOM as p-activity, those from the structure SOM as p-structure, and those using both the activity and structure SOMs as p-both. To test model performance, the corresponding SOM cluster or consensus cluster was located for each test set compound and p-activity, p-structure or p-both obtained from the training set was retrieved and compared with the true toxicity outcome of the test compound to determine whether the test compound should be counted as a true positive (TP: toxic and score > cutoff), false positive (FP: non-toxic and score > cutoff), true negative (TN: non-toxic and score ≤ cutoff) or false negative (FN: toxic and score ≤ cutoff). Model performance was assessed by calculating the AUC-ROC, which is a plot of sensitivity [TP/(TP + FN)] versus (1-specificity [TN/(TN + FP)])[38]. A perfect model would have an AUC-ROC of 1 and an AUC-ROC of 0.5 indicates a random classifier. The random data split and model training and testing were repeated 100 times, and the average AUC-ROC values were calculated for each model.

# References

1. Collins, F. S., Gray, G. M. & Bucher, J. R. Toxicology. Transforming environmental health protection. *Science* **319**, 906–907 (2008).
2. Kavlock, R. J., Austin, C. P. & Tice, R. R. Toxicity testing in the 21st century: implications for human health risk assessment. *Risk Anal.* **29**, 485–487 discussion 492-487 (2009).
3. Tice, R. R., Austin, C. P., Kavlock, R. J. & Bucher, J. R. Improving the human hazard characterization of chemicals: a Tox21 update. *Environ. Health Perspect.* **121**, 756–765 (2013).
4. *NRC Toxicity Testing in the 21st Century: A Vision and a Strategy* (The National Academies Press, 2007).
5. Huang, R. *et al.* Chemical genomics profiling of environmental chemical modulation of human nuclear receptors. *Environ. Health Perspect.* **119**, 1142–1148 (2011).
6. Shukla, S. J., Huang, R., Austin, C. P. & Xia, M. The future of toxicity testing: a focus on *in vitro* methods using a quantitative high throughput screening platform. *Drug Discov. Today* **15**, 997–1007 (2010).
7. Attene-Ramos, M. S. *et al.* The Tox21 robotic platform for the assessment of environmental chemicals - from vision to reality. *Drug Discov. Today* **18**, 716–723 (2013).
8. Hsu, C. W. *et al.* Quantitative high-throughput profiling of environmental chemicals and drugs that modulate farnesoid X receptor. *Sci. Rep* **4**, 6437 (2014).
9. Huang, R. *et al.* Profiling of the Tox21 10 K compound library for agonists and antagonists of the estrogen receptor alpha signaling pathway. *Sci. Rep.* **4**, 5664 (2014).
10. Attene-Ramos, M. S. *et al.* Profiling of the Tox21 chemical collection for mitochondrial function to identify compounds that acutely decrease mitochondrial membrane potential. *Environ. Health Perspect.* **123**, 49–56 (2015).
11. Zhang, J. H., Chung, T. D. & Oldenburg, K. R. A simple statistical parameter for use in evaluation and validation of high throughput screening assays. *J. Biomol. Screen.* **4**, 67–73 (1999).
12. Babula, P., Masarik, M., Adam, V., Provaznik, I. & Kizek, R. From Na + /K + - ATPase and cardiac glycosides to cytotoxicity and cancer treatment. *Anticancer Agents Med. Chem.* **13**, 1069–1087 (2013).
13. Abd, T. T. & Jacobson, T. A. Statin-induced myopathy: a review and update. *Expert Opin. Drug Saf.* **10**, 373–387 (2011).
14. Vidal, B. *et al.* The alkylating carcinogen N-methyl-N'-nitro-N-nitrosoguanidine activates the plasminogen activator inhibitor-1 gene through sequential phosphorylation of p53 by ATM and ATR kinases. *Thromb. Haemost.* **93**, 584–591 (2005).
15. Rogers, J. A., Metz, L. & Yong, V. W. Review: endocrine disrupting chemicals and immune responses: a focus on bisphenol-A and its potential mechanisms. *Mol. Immunol.* **53**, 421–430 (2012).
16. Medjakovic, S. *et al.* Effect of nonpersistent pesticides on estrogen receptor, androgen receptor, and aryl hydrocarbon receptor. *Environ. Toxicol.* **29**, 1201–1216 (2013).

17. Kirpichnikov, D., McFarlane, S. I. & Sowers, J. R. Metformin: an update. *Ann. Intern. Med.* **137**, 25–33 (2002).
18. Carty, T. J. *et al.* Ampiroxicam, an anti-inflammatory agent which is a prodrug of piroxicam. *Agents Actions* **39**, 157–165 (1993).
19. Judson, R. *et al.* Perspectives on validation of high-throughput assays supporting 21st century toxicity testing. *ALTEX* **30**, 51–56 (2013).
20. Huang, R. *et al.* Characterization of diversity in toxicity mechanism using in vitro cytotoxicity assays in quantitative high throughput screening. *Chem. Res. Toxicol.* **21**, 659–667 (2008).
21. FDA. Innovation or Stagnation: Challenge and Opportunity on the Critical Path to New Medical Products, http://www.fda.gov/ScienceResearch/SpecialTopics/CriticalPathInitiative/CriticalPathOpportunitiesReports/ucm077262.htm (2004).
22. Martic-Kehl, M. I., Schibli, R. & Schubiger, P. A. Can animal data predict human outcome? Problems and pitfalls of translational animal research. *Eur. J. Nucl. Med. Mol. Imaging* **39**, 1492–1496 (2012).
23. Walum, E., Nilsson, M., Clemedson, C. & Ekwall, B. The MEIC program and its implications for the prediction of acute human systemic toxicity. *Alternative Methods Toxicol. Life Sci.* **11**, 275–282 (1995).
24. Goldberg, A. M. & Frazier, J. M. Alternatives to animals in toxicity testing. *Sci. Am.* **261**, 24–30 (1989).
25. Low, Y. S., Sedykh, A., Rusyn, I. & Tropsha, A. Integrative approaches for predicting *in vivo* effects of chemicals from their structural descriptors and the results of short-term biological assays. *Curr. Top. Med. Chem.* **14**, 1356–1364 (2014).
26. Abdo, N. *et al.* Population-based hazard and concentration-response assessment of chemicals: The 1000 Genomes High-Throughput Screening Study. *Environ. Health Perspect.* **123**, 458–466 (2015).
27. Eduati, F. *et al.* Prediction of human population responses to toxic compounds by a collaborative competition. *Nat. Biotechnol.* **33**, 933–940 (2015).
28. Huang, R. *et al.* The NCGC pharmaceutical collection: a comprehensive resource of clinically approved drugs enabling repurposing and chemical genomics. *Sci. Transl. Med.* **3**, 80ps16 (2011).
29. Wong, C. & Chen, S. The development, application and limitations of breast cancer cell lines to study tamoxifen and aromatase inhibitor resistance. *J. Steroid Biochem. Mol. Biol.* **131**, 83–92 (2012).
30. Chen, S. *et al.* Cell-based high-throughput screening for aromatase inhibitors in the Tox21 10 K library. *Toxicol. Sci.* **147**, 446–457 (2015).
31. Sakamuru, S. *et al.* Application of a homogenous membrane potential assay to assess mitochondrial function. *Physiol. Genomics* **44**, 495–503 (2012).
32. Yamamoto, K. N. *et al.* Characterization of environmental chemicals with potential for DNA damage using isogenic DNA repair-deficient chicken DT40 cell lines. *Environ. Mol. Mutagen* **52**, 547–561 (2011).
33. Fox, J. T. *et al.* High-throughput genotoxicity assay identifies antioxidants as inducers of DNA damage response and cell death. *Proc. Natl Acad. Sci. USA* **109**, 5423–5428 (2012).
34. Shukla, S. J. *et al.* Profiling environmental chemicals for activity in the antioxidant response element signaling pathway using a high throughput screening approach. *Environ. Health Perspect.* **120**, 1150–1156 (2012).
35. Inglese, J. *et al.* Quantitative high-throughput screening: a titration-based approach that efficiently identifies biological activities in large chemical libraries. *Proc. Natl Acad. Sci. USA* **103**, 11473–11478 (2006).
36. GHS United Nations, Globally Harmonized System of Classification and Labelling of Chemicals (GHS) http://www.unece.org/fileadmin/DAM/trans/danger/publi/ghs/ghs_rev02/English/03e_part3.pdf (2007).
37. Kohonen, T. Self-organizing neural projections. *Neural Netw.* **19**, 723–733 (2006).
38. Zweig, M. H. & Campbell, G. Receiver-operating characteristic (ROC) plots: a fundamental evaluation tool in clinical medicine. *Clin. Chem.* **39**, 561–577 (1993).

## Acknowledgements

This work was supported by the Intramural Research Programs of the National Toxicology Program (Interagency agreement #Y2-ES-7020-01), National Institute of Environmental Health Sciences, the US Environmental Protection Agency (Interagency Agreement #Y3-HG-7026-03) and the National Center for Advancing Translational Sciences, National Institutes of Health. We also thank Nicole Miller, Samuel Michael and Carleen Klumpp-Thomas for assisting with the screens, Paul Shinn, Misha Itkin and Danielle VanLeer for compound management and William Leister for the Tox21 10 K library quality control.

## Author contributions

R.H., M.X., C.P.A and A.S designed the study. S.S., J.Z., S.A.S. and M.A.R. performed the experiments and collected data. R.H. performed statistical analysis of all data. T.Z. aided data analysis and visualization. R.H. and M.X. wrote the manuscript. All authors reviewed the manuscript. The views expressed in this article are those of the authors and do not necessarily reflect the statements, opinions, views, conclusions or policies of the National Center for Advancing Translational Sciences, National Institutes of Health, or the United States government. Mention of trade names or commercial products does not constitute endorsement or recommendation for use.

## Additional information

**Accession codes:** All Tox21 phase II qHTS data have been deposited in PubChem under the following assay IDs: ATAD5: 651632, 720516, 651634; DT40 *Rad54/Ku*: 70743015; DT40 *Rev3*: 743014; DT40 WT: 743015; P53-bla: 651631, 720552, 651633; ARE-bla: 743202, 743219, 743203; HSE-bla: 743210, 743228, 743209; aromatase: 743083, 743139, 743084; mitochondria toxicity: 720635, 720637, 720634; AhR-luc: 743085, 743122, 743086; AR-bla agonist: 743036, 743053; AR-bla antagonist: 743035, 743063, 743033; AR-MDA-luc agonist: 743040; AR-MDA-luc antagonist: 743042, 743054, 743041; ER-BG1-luc agonist: 743079; ER-BG1-luc antagonist: 743080, 743091, 743081; ER-bla agonist: 743075, 743077; ER-bla antagonist: 743069, 743078, 743074; FXR-bla agonist: 743220, 743239, 743218; FXR-bla antagonist: 743217, 743240, 743221; GR-bla agonist: 720691, 720719; GR-bla antagonist: 720692, 720725, 720693; PPAR-delta-bla agonist: 743212, 743227, 743211; PPAR-delta-bla antagonist: 743215, 743226, 743213; PPAR-gamma-bla agonist: 743094, 743140; PPAR-gamma-bla antagonist: 743191, 743199, 743194; TR-beta-luc agonist: 743066; TR-beta-luc antagonist: 743065, 743067, 743064; VDR-bla agonist: 743222, 743241, 743224; VDR-bla antagonist: 743223, 743242, 743225.

# 6

# Stereochemical bias introduced during RNA synthesis modulates the activity of phosphorothioate siRNAs

Hartmut Jahns[1], Martina Roos[1], Jochen Imig[1], Fabienne Baumann[1], Yuluan Wang[1], Ryan Gilmour[2] & Jonathan Hall[1]

An established means of improving the pharmacokinetics properties of oligoribonucleotides (ORNs) is to exchange their phosphodiester linkages for phosphorothioates (PSs). However, this strategy has not been pursued for small interfering RNAs (siRNAs), possibly because of sporadic reports that PS siRNAs show reduced inhibitory activity. The PS group is chiral at phosphorous (Rp/Sp centres), and conventional solid-phase synthesis of PS ORNs produces a population of diastereoisomers. Here we show that the choice of the activating agent for the synthesis of a PS ORN influences the Rp/Sp ratio of PS linkages throughout the strand. Furthermore, PS siRNAs composed of ORNs with a higher fraction of Rp centres show greater resistance to nucleases in serum and are more effective inhibitors in cells than their Sp counterparts. The finding that a stereochemically biased population of ORN diastereoisomers can be synthesized and exploited pharmacologically is important because uniform PS modification of siRNAs may provide a useful compromise of their pharmacokinetics and pharmacodynamics properties in RNAi therapeutics.

[1] Department of Chemistry and Applied Biosciences, ETH Zürich, Vladimir-Prelog-Weg-4, CH-8093 Zürich, Switzerland. [2] Institute for Organic Chemistry, Westfälische Wilhelms-Universität Münster, D-48149 Münster, Germany. Correspondence and requests for materials should be addressed to J.H. (email: jonathan.hall@pharma.ethz.ch).

Chemically modified oligoribonucleotide (ORN) drugs represent an emerging class of pharmaceuticals of which, arguably, the most important is the small interfering RNAs (siRNA). siRNAs are duplexes composed of complementary ORNs of ~21 nucleotides (nt), with 2-nt overhangs. One of the ORNs (the 'guide' strand) directs the RNA-induced silencing complex (RISC) to cleave complementary mRNA targets and thereby represses gene expression[1], whereas the other ORN (passenger strand) is discarded. For most applications *in vivo* the phosphodiester (PO) linkages of double-stranded siRNAs require protection against cleavage by exonucleases[2-4]. However, medicinal chemists have been unable to develop a readily-accessible uniform modification of siRNAs, which resists nuclease attack without sacrificing the reagent's ability to trigger RISC-mediated degradation of its mRNA targets. Hence, still today, siRNAs are rarely used systemically in the absence of complex delivery formulations.

An effective means of protecting single-stranded RNA-based antisense drugs against nucleases is to exchange one of the non-bridging oxygens of each PO linkage with sulfur. Indeed, almost all single-stranded oligonucleotide drugs in clinical trials are phosphorothioates[5]. The phosphorothioate group (PS) is chiral, with either *Rp* or *Sp* configuration at phosphorous[6]. However, an efficient, practicable solid-phase chiral synthesis of PS ORNs using the H-phosphonate method or the phosphoramidite method is not yet available[7,8] (reviewed in ref. 9) and PS ORNs are used as mixtures of $2^n$ diastereoisomers (n, number of PS linkages). PS diastereoisomers have distinct physical and biochemical properties and will interact differently with cellular RNAs and proteins (for example, see refs 10–12). Hence, the net pharmacological properties of PS ORN drugs likely represent the sum effect of the pharmacokinetic (PK) and pharmacodynamic (PD) properties of individual diastereoisomers in an isosequential population.

Full PS modification of siRNAs has been mostly neglected by the field after a few groups reported that PS siRNAs *in vitro* show reduced inhibitory activity compared with unmodified siR-NAs[13,14] and in some[15,16] but not all[17], cases toxicity. In this systematic study of the properties of uniformly modified PS siRNAs, we investigate whether the control of stereochemistry of PS ORN strands in a siRNA can be exploited to improve their pharmacological properties. We show here that the choice of the activating agent during synthesis of a PS RNA shifts the *Rp/Sp* ratio in the PS linkages throughout an ORN and when assembled into duplexes, this bias transforms the PK and PD activities of PS siRNAs. To our knowledge, this account represents the first demonstration of how a stereochemically biased population of PS siRNAs can be prepared and exploited to optimize its pharmacological profile. The findings of the study suggest that the PS modification of siRNA drugs may provide a useful compromise of PK and PD properties for *in vivo* applications. In addition, they encourage continued efforts towards the synthesis of full stereochemically pure PS ORNs.

## Results

**Tetrazole activators bias *Rp/Sp* ratios in PS dinucleotides.** Solid-phase synthesis protocols of ORNs typically use tetrazole reagents to facilitate the attack of the ribose 5′-hydroxyl group on the activated phosphorous centre (Fig. 1)[18]. Tetrazoles act as weakly acidic and nucleophilic catalysts during the displacement of diisopropylamine at the $P^{III}$-stereocentre by the 5′-hydroxyl

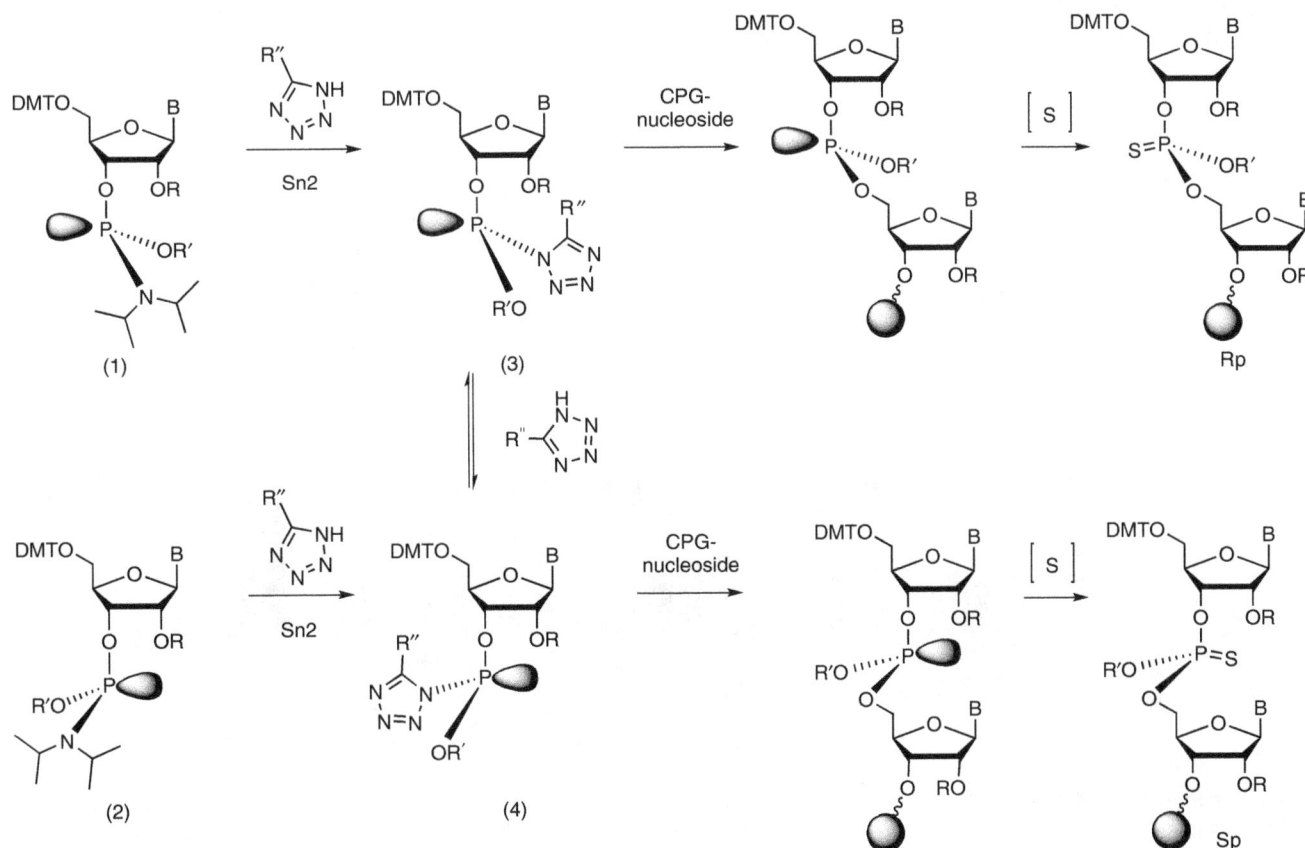

**Figure 1 | Mechanism and stereochemical course of phosphoramidite coupling during solid-phase ORN synthesis.** Epimerization at P occurs at the tetrazolide stage (R, Si-*tert*-BuMe$_2$; R′, CH$_2$CH$_2$CN; R″, H or SCH$_2$Ph; B, nucleobases adenine, guanine, cytosine and uracil; CPG, controlled pore glass solid support).

group of the support-bound ribonucleoside. The final oxidative sulfurization step is stereoretentive, at least in the case of oligodeoxynucleotides (ODNs)[19,20].

We began the investigation with a study of PS diribonucleotides, following the example of a limited investigation using 2'-O-methoxyethyl (MOE) ribonucleotides[21]. We synthesized PS uridine dinucleotide (UsU) using standard 2'-O-tert-butyldimethylsilyl (TBDMS)-protected phosphoramidites and 5-benzylthio-1-H-tetrazole (BTT) as an activator. The diastereoisomeric products were separated using reverse-phase high-performance liquid chromatography (RP-HPLC), which showed an ∼2:1 ratio of diastereoisomers (Fig. 2a). The major isomer was assigned the Rp stereochemistry, as incubation of an equimolar mixture of isomers with nuclease P1 (nP1) attenuated selectively the slower migrating (Sp) compound (Fig. 2b)[22,23].

This product ratio was independent of the configuration of the starting materials as configurationally defined U-phosphoramidites 1 and 2 (Fig. 1; purified by column chromatography: Supplementary Figs 1–4, Supplementary Table 1) yielded the same 2:1 mixture (Fig. 2c), consistent with rapid epimerization at the $P^{(III)}$ stereocentre via repetitive attack of the tetrazole before irreversible coupling of the ribonucleoside (Fig. 1)[24,25]. Analogous observations were made with CsC and AsA dinucleotides that were obtained in diastereoisomeric ratios of ∼5:4 and 10:9, respectively (Fig. 2d). These stereochemical outcomes were mostly reproduced in longer sequences containing UsU, CsC and AsA units. Hence, the incorporation of UsU in the centre of $U_8$, in the centre of a mixed-sequence octanucleotide, and at the 5'- or 3'-termini of octanucleotides yielded ∼2:1 isomer ratios (Fig. 2e), as determined by RP-HPLC (Supplementary Tables 2–4).

We then synthesized all possible 16 PS dinucleotides using two activators, BTT and tetrazole. For comparison, a third electron-deficient activator, dicyanoimidazole was investigated but gave unsatisfactory product yields. We determined the diastereoisomeric ratios of the products using the nP1/RP-HPLC method (Supplementary Fig. 5). For BTT the respective Rp isomer was the dominant product in all cases except GsG (Fig. 3a, Supplementary Table 2), whereas for tetrazole the trend was reversed, that is, the Sp isomers were the major products (Fig. 3b, Supplementary Table 3). The difference in the product profiles may represent different states of the thermodynamic equilibria between tetrazolides 3 and 4 before coupling; however, a kinetic resolution of the rapidly equilibrating 3 and 4 following a Curtin–Hammett principle cannot be discounted (Fig. 1). To further investigate the importance of the 2'-substituent on the Rp/Sp product ratio, we synthesized six UsU-containing PS dinucleotide from DNA and from 2'-F-, 2'-O-Me-, MOE-, 2'-O-[(triisopropylsilyl)oxy]methyl (TOM)- and 2'-O-bis(2-acetoxyethoxy)methyl (ACE)-substituted phosphoramidites, using BTT and tetrazole. In each case we assumed that the stereochemical outcome of the coupling reaction was independent of the Rp/Sp isomeric mixture of the starting phosphoramidites, as shown for 2'-O-TBDMS ribonucleotides (Fig. 2c,d) and for tetrazole-mediated coupling of DNA[26].

The 2'-O-TOM protecting group, which presents a relatively low steric hindrance to phosphoramidite coupling[27], produced a marginal and similar selectivity for the Rp isomer from both the activators (Fig. 3c), suggesting that indeed steric crowding may contribute to reaction diastereoselectivity for the 2'-O-TBDMS derivatives. Results with the 2'-F and DNA phosphoramidites appeared to support this hypothesis (Fig. 3d). The electronic and conformational properties of the 2'-F-deoxyribose resemble those of RNA. However, the small fluoride minimizes steric hindrance for phosphoramidite coupling. Here BTT-mediated reaction produced a slight excess of the faster HPLC-running isomer (presumably Rp), whereas tetrazole was unselective. In the case of DNA, which differs from RNA by its C2'-endo conformation and

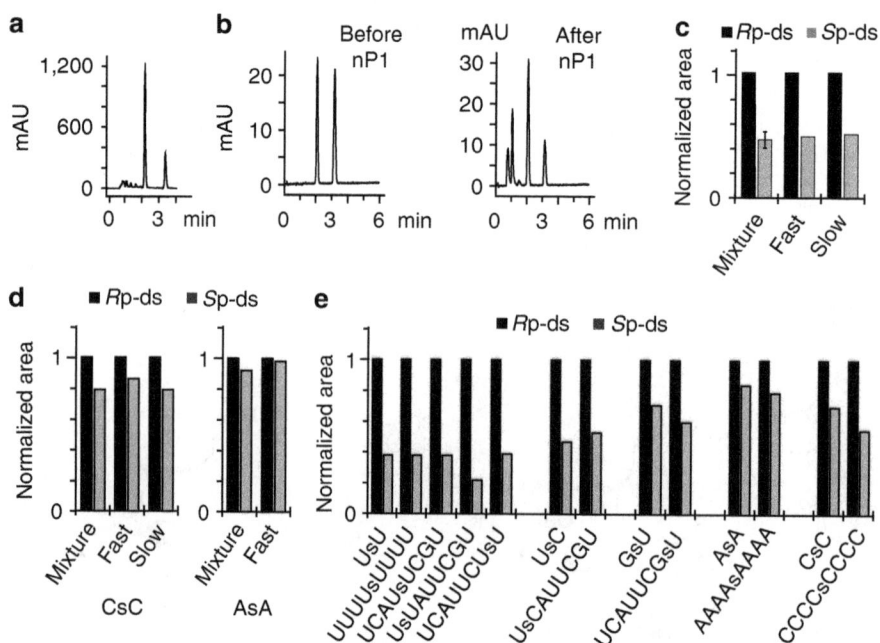

**Figure 2 | Properties of PS diribonucleotides.** (**a**) HPLC chromatogram of BTT-synthesized UsU. (**b**) HPLC chromatogram of an equimolar mixture of UsU diastereoisomers before and after incubation with nP1. (**c**) Diastereoisomer ratios of BTT-synthesized UsU from epimeric (mixture) and from stereopure U-phosphoramidites: fast-migrating phosphoramidite isomer (fast) and slow-migrating isomer (slow). (**d**) Diastereoisomer ratios of BTT-synthesized CsC and AsA starting from epimeric (mixture) and stereopure phosphoramidites (fast-migrating phosphoramidite isomer (fast) and slow-migrating isomer (slow)). (**e**) Diastereoisomer ratios of BTT-synthesized octanucleotides containing single PS dinucleotide units. Error bars are s.d. of three independent experiments. Norm, normal.

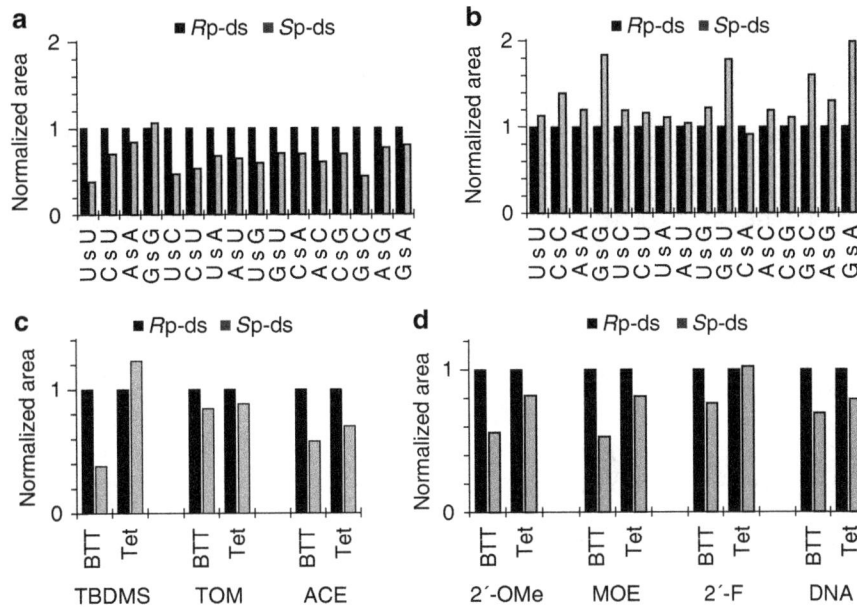

**Figure 3 | Diastereoisomer ratios of 2'-substituted PS dinucleotides synthesized with tetrazole activators.** Ratios of PS diribonucleotides synthesized from TBDMS-protected phosphoramidites using BTT (**a**) and tetrazole (**b**). (**c**) and (**d**) Six UsU-containing PS dinucleotides synthesized from DNA (TsT) and from 2'-F-, 2'-O-methyl (OMe)-, MOE-, 2'-O-[(triisopropylsilyl)oxy]methyl (TOM)- and ACE-substituted phosphoramidites, using BTT and tetrazole (Tet). (Data from five dinucleotides shown in Fig. 2e is reshown in (**a**) and (**c**) for ease of comparison).

its lack of a 2'-substitutent on ribose, BTT and tetrazole produced similarly weak Rp-selectivity (Fig. 3d). For the three 2'-O-alkyl substituents (2'-O-Me, 2'-MOE, 2'-O-ACE) stereoselectivites midway between those of 2'-O-TBDMS and 2'-O-TOM were observed. We also synthesized all 16 dinucleotides of DNA and 2'-O-Me using BTT, as well as four dinucleotides with the both activators to confirm that the product ratios for TsT and 2'-O-Me UsU were not outliers for the dinucleotide series and were highly reproducible (Supplementary Fig. 6, Supplementary Note 1).

For the coupling of a single nucleotide in a PS ORN the difference in the stereochemical ratios arising from the use of the two activators is small. For example, AsA synthesized using BTT has an Rp/Sp ratio of $\sim 1.2$, whereas with tetrazole this ratio is reduced to $\sim 0.8$ (Fig. 3a,b); for tetrazole especially, diastereoselective excesses only exceeded 20% for 6 of the 16 dinucleotides (Fig. 3b). Thus, efforts to model such small differences of this solid-phase reaction would be very difficult. However, in an ORN containing 20 PS linkages these small differences are at least partially cumulative and therefore might affect considerably the bulk biophysical properties of an isosequential population of diastereoisomers. To our knowledge, there are no reports on how the physicochemical and biological properties of PS ORNs may be influenced by a variation in the relative population of individual diastereoisomers.

**Higher Rp content of PS siRNAs improves cellular activity.** To assess the biophysical and biological consequences of the use of different activators during the synthesis of PS ORNs, we prepared 24 PS ORNs and assembled siRNA sequences of various compositions targeting *TP53* (p53), *LIN28B* (Lin28), *TGFB1* (TGFβ1), *TGFB2* (TGFβ2), *RBFOX2* (Fox2) and *Renilla* luciferase (Ren; Table 1, Supplementary Table 5). SiRen was prepared 'blunt-ended'.

Assuming that the stereochemical outcome in each coupling step depends only on the sequence of the newly-formed terminal dinucleotide unit (Fig. 2e) and reflects the diastereomeric ratio

observed for isolated dinucleotide formation, a theoretical total Rp content can be calculated for each PS siRNA by dividing the sum of the individual Rp ratios for all the dinucleotide units (Fig. 3a,b) by the number of nucleotide linkages (Table 1). We then measured the melting temperatures ($T_m$) of the duplexes (Supplementary Figs 7–10). It has been reported that short ODNs[11] and ORNs composed of stereopure Rp strands form more stable duplexes with complementary PO ORNs than their Sp analogues, or even their PO counterparts[7,28]. In our case, the PO siRNAs showed the highest $T_m$s of the three formats (Table 1). The $T_m$s of all siRNAs synthesized using BTT (PS$_{BTT}$ siRNA) were 5–6 °C higher than those of siRNAs synthesized using tetrazole (PS$_{tet}$ siRNA). These data are consistent with a higher Rp content ($\sim 60\%$) of the PS$_{BTT}$ siRNA and support our hypothesis that the stereoselectivity observed during the formation of isolated dinucleotides (Fig. 3a,b) extends to the synthesis of $\sim 21$-nt ORNs.

The two most important pharmacological properties of an siRNA are its stability to nucleases and its potency. We therefore tested the effects of stereochemical bias in PS siRNAs on their stability to nucleases in serum and on their biological activity in cells. Stability of ORNs to serum nucleases is commonly assayed using mammalian sera such as fetal bovine serum (FBS). We incubated BTT-synthesized PS and PO siRNAs targeting Fox2, p53, transforming growth factor β1 (TGFβ1) and TGFβ2 with a solution of 10% FBS in a time-course experiment under conditions to highlight the differences in stabilities to nucleases. We quenched the reactions at regular intervals and analysed the equally loaded mixtures on a non-denaturing gel. Under our conditions the PS siRNAs showed superior stability to their PO counterparts in all the cases (Supplementary Figs 11–14). In two cases (TGFβ2 and p53), we compared the stability of the PS$_{BTT}$ and PS$_{tet}$ siRNAs under identical incubation conditions in a detailed time-course experiment (Fig. 4a,b).

Stability of single-stranded PS ODNs to nucleases in sera is reportedly greater for Sp- than for Rp isomers[29,30]. Also, double-stranded ODNs and ORNs are more resistant than their single-stranded counterparts[17,31]. We assigned the rapid appearance of

**Table 1 | Properties of PO- and PS siRNAs.**

| Target | Sequence of siRNA* | siRNA | Nt link†, (% $R_p$‡) | $T_m$ (°C) |
|---|---|---|---|---|
| p53 | 5′ UUGUUUUCAGGAAGUAGUUUU | si1 | PO | 66.3 |
| | UUAACAAAAGUCCUUCAUCAA | si2 | PS$_{BTT}$ (62) | 58.0 |
| | | si3 | PS$_{tet}$ (45) | 52.7 |
| Lin28 | 5′ AAAUCCUUCCAUGAAUAGUTT | si4 | PO | 67.2 |
| | TTUUUAGGAAGGUACUUAUCA | si5 | PS$_{BTT}$ (56) | 55.5 |
| | | si6 | PS$_{tet}$ (40) | 50.6 |
| TGFβ1 | 5′ CCAACUAUUGCUUCAGCUCUU | si7 | PO | 74.1 |
| | UUGGUUGAUAACGAAGUCGAG | si8 | PS$_{BTT}$ (62) | 67.0 |
| | | si9 | PS$_{tet}$ (44) | 61.7 |
| TGFβ2 | 5′ GGAUUGAGCUAUAUCAGAUUUU | si10 | PO | 70.8 |
| | UUCCUAACUCGAUAUAGUCUAA | si11 | PS$_{BTT}$ (65) | 63.0 |
| | | si12 | PS$_{tet}$ (47) | 57.0 |
| Fox2 | 5′ CCUGGCUAUUGCAAUAUUUUU | si13 | PO | 68.5 |
| | UUGGACCGAUAACGUUAUAAA | si15 | PS$_{BTT}$ (62) | 61.7 |
| | | si14 | PS$_{tet}$ (47) | 55.0 |
| Ren | 5′ GAGCGAAGAGGGCGAGAAAUU | si16 | PO | nd |
| | CUCGCUUCUCCCGCUCUUUAA | si17 | PS$_{BTT}$ (61) | nd |
| | | si18 | PS$_{tet}$ (43) | nd |

BTT, 5-benzylthio-1-H-tetrazole; Nt, nucleotide; PO, phosphodiester; PS, phosphorothioate; siRNA, small interfering RNA; tet, 1H-tetrazole; $T_m$, melting temperature.
*Upper sequence: siRNA guide strand; lower sequence: passenger strand.
†Nucleotide linkages in both strands of the siRNAs: PS$_{BTT}$, PS$_{tet}$: PS ORNs synthesized with BTT and tetrazole, respectively.
‡Calculated $R_p$ content of siRNA.

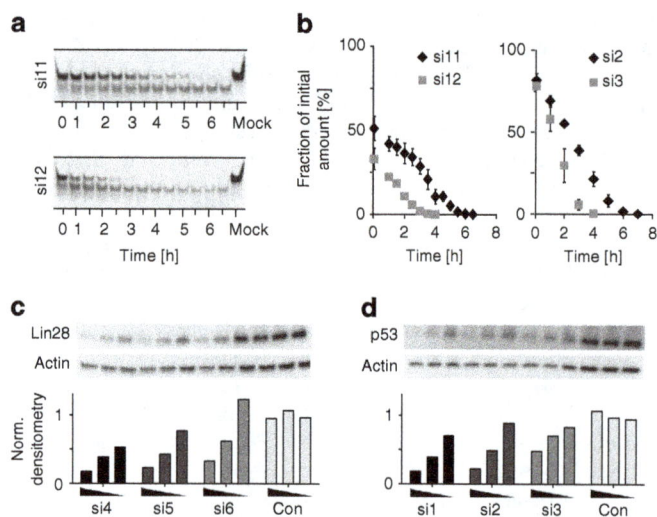

**Figure 4 | Pharmacological properties of PO- and PS- siRNAs.** (**a**) Stability of PS$_{BTT}$-synthesized (si11) and PS$_{tet}$-synthesized (si12) siTGFβ2 siRNAs to nuclease degradation in FBS: 'mock' refers to no FBS treatment. (**b**) Quantification of blots in **a** and Supplementary Fig. 14 using the mock lane for normalization. Sequence-specific inhibition by PO and PS siRNAs of Lin28 protein in Huh7 cells (**c**) (24, 6, 2.5 nM), and p53 protein in HeLa cells (**d**) (6, 2.5, 0.6 nM). Error bars are s.d. of three independent experiments. Con, control; Norm, normalized.

faster migrating (lower) bands for all of the siRNAs to their nuclease-mediated degradation. The PS$_{BTT}$ siRNAs si2 and si11 demonstrated higher stability than si3 and si12, respectively, over the long term. This was perhaps due to the higher thermal stability of the $R_p$-biased duplexes, in agreement with previous suggestions that duplexes of higher stability demonstrate greater nuclease resistance[31].

To compare the potency of the three classes of siRNAs, we tested them at graded concentrations for their inhibition of target proteins

Lin28 and p53 in Huh7 and HeLa cells, respectively (Fig. 4c,d; Supplementary Fig. 15). The negative control (Con), a sequence-unrelated siRNA[32], was inactive. PO siRNAs (si1 and si4) inhibited their targets by 80% at the highest treatment concentrations. Surprisingly, both the PS$_{BTT}$ siRNAs (si2 and si5) appeared to inhibit their targets more efficiently than the PS$_{tet}$ siRNAs (si3 and si6), albeit to a lower extent than their PO counterparts (some toxicity was observed at the highest concentrations). In other words, the higher content of $R_p$ linkages in the PS$_{BTT}$ siRNA was associated with increased potency.

Quantification of bands on western blots cannot be relied upon to determine small differences in the activity. A routine and sensitive method to compare the activity of siRNAs is the luciferase reporter assay, in which a target site for an siRNA can be cloned[32,33]. We employed six luciferase reporter genes, of which five contained a cloned target site for siRNAs of Table 1 (Supplementary Note 2), and one in which an unmodified *Renilla* reporter plasmid was used. SiRNAs were tested at four concentrations. In 5/6 cases the PO siRNAs (si1, si4, si7, si16 and si27) were the most effective, as expected (Fig. 5a–d; Supplementary Fig. 16). Strikingly, however, the PS$_{BTT}$ siRNAs (higher $R_p$ content; si2, si5, si8, si17, si15 and si29) showed roughly twofold higher activity than their PS$_{tet}$ siRNA analogues (higher $S_p$ content).

Many factors affect the potency of siRNAs and include strand selection[34,35], duplex stability[34,36], the presence of nucleotide motifs[37] and chemical modification of guide and/or passenger strands[38]. For unmodified siRNAs, it is known that specific PO cleavage in the centre of the passenger strand by the RNase H-like Argonaute2 (Ago2) is an important step in the activation of the guide strand[10,39,40]. This was elegantly demonstrated employing a PS linkage at the said position: the use of an $R_p$ PS linkage partially attenuated, whereas an $S_p$ PS group greatly attenuated passenger strand cleavage[10]. Our own data are consistent with these observations as the higher $R_p$ PS content in the central position of passenger strands of PS$_{BTT}$ siRNAs correlates with higher potency (Fig. 5a–d).

To shed additional light on the roles of strand modifications, duplex stability and PS stereochemistry in the potency

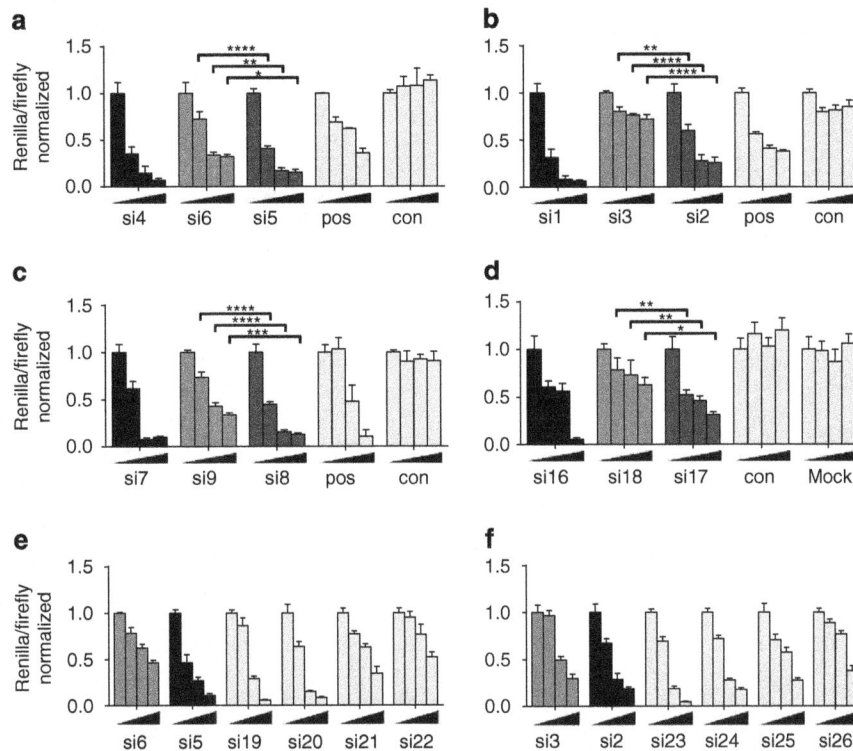

**Figure 5 | Sequence-specific target inhibition of *Renilla*/firefly luciferase reporters in HeLa cells.** Negative and positive controls (con and pos, respectively), PO- and PS- siRNAs (Table 1) targeting (**a**) Lin28, (**b**) p53, (**c**) TGFβ1 and (**d**) Ren were co-transfected with luciferase reporter vectors into HeLa cells at increasing concentrations (0.4, 1.5, 6, 24 nM for (**a,b,d**); 0.3, 1, 4, 16 nM for **c**). *Renilla* luminescence values were normalized against the reference firefly luminescence and then to the value from the lowest treatment concentration. Error bars are s.d. of three transfections. Mixed backbone PO/PS siRNAs (Table 2) were tested against Lin28 (**e**) and p53 (**f**) under identical conditions. (*)$P < 0.05$; (**)$P < 0.01$; (***)$P < 0.001$; (****)$P < 0.0001$.

of PS siRNAs, we assembled Lin28 and p53 siRNAs, combining a PO ORN with either a $PS_{BTT}$ or a $PS_{tet}$ ORN (Table 2). We measured their $T_m$s and tested them in cells in pair-wise comparisons.

For all pairs of siRNAs (si19/si21, si20/si22, si23/si25 and si24/si26), switching the PS backbone from the guide to the passenger strands did not significantly change the $T_m$. However, in each case the potency was clearly highest for siRNAs with a PO guide strand (si19, si20, si23 and si24), regardless of the stereochemistry of the PS passenger strands (Fig. 5e,f). The results suggest that an unmodified guide strand is the dominating factor in the activity of a mixed PO/PS siRNA, consistent with numerous previous accounts on modifications to siRNAs[31,41]. Finally, we compared the properties of two pairs of siRNAs with a PS guide strand and a PO passenger strand (si21/22 and si25/26). The chimeric siRNAs bearing $PS_{BTT}$ guide strands (si21 and si25) showed the higher $T_m$s (3.0–3.3 °C), as well as a superior potency to those comprising $PS_{tet}$ strands (si22 and si26). This is consistent with the properties of their fully phosphorothioated analogues (Table 1, Fig. 5a–d). Furthermore, it demonstrates that a bias for PS linkages with $Rp$ stereochemistry also in the PS guide strand is beneficial for potency, in analogy to the effects described for the RNase H-mediated cleavage of mRNAs by stereopure $Rp$ antisense PS ODNs[11]. Recently, two crystal structures of human Ago2 in complex with RNA guides were published[42,43]. Both the structures show extensive hydrogen bonding contacts between Ago2 residues and the non-bridging oxygens of all PO linkages at the 5'-end and with one PO group at the 3'-end of the guide[42,43]. Given the multitude of these interactions it is to be expected that thiolation of POs in these regions of the guide will affect its interactions with Ago2 and therefore also the properties of PS siRNAs in RISC.

**A high $Rp$ content of PS siRNAs increases loading into RISC.** To add further insight on the interactions of PS siRNAs with RISC, we transfected HeLa cells with PO, $PS_{BTT}$ and $PS_{tet}$ siRNAs (si1, si2 and si3, respectively) and immunoprecipitated the RISC using an Ago2 antibody. We then performed northern blotting for the presence of the individual guide and passenger strands in the complex, as described previously[44]. For the wild-type siRNA (si1), the great majority of the siRNA strands were present in complex with Ago2 compared with the input sample (Fig. 6, Supplementary Fig. 17). In line with the aforementioned study, we assumed that the northern blotting detected individual passenger and guide strands (and not siRNA) of si1 because Ago2 rapidly cleaves the passenger strands of siRNAs[44], and also because the affinity of Ago2 for single-stranded RNA is much higher than for double-stranded siRNA[45]. Consistent with the high potency of si1, the guide strand appeared to occupy a much larger fraction in RISC than the passenger, although the affinity of the two northern probes for the guide and the passenger strands may not be equal.

As expected, bands on the blot were much weaker for the $PS_{BTT}$ and $PS_{tet}$ ORNs, possibly because of an overall lower binding affinity of the radiolabelled probes for PS ORNs. For both the PS siRNAs, loading of the siRNA into RISC was incomplete, as seen by residual bands in the input samples. However, a larger proportion (43%; Fig. 6) of the $PS_{BTT}$ guide strand of si2 was taken into the Ago2 complex from the input sample compared with the $PS_{tet}$ guide strand of si3 (31%; Fig. 6). This is consistent with the higher potency of si2 over si3 against both endogenous p53 protein (Fig. 4d) and the luciferase reporter genes (Fig. 5).

Taken together, our results suggest that an $Rp$ bias in both the PS guide and passenger strands contributes to the increased

**Table 2 | Melting temperatures of mixed backbone PS siRNAs.**

| siRNA | Sequence of siRNA* | Nt link[†] (% $R_p$[‡]) | $T_m$ (°C) |
|---|---|---|---|
| si19 | 5′ AAAUCCUUCCAUGAAUAGUTT TTUUUAGGAAGGUACUUAUCA | PO PS_BTT (58) | 61.4 |
| si20 | 5′ AAAUCCUUCCAUGAAUAGUTT TTUUUAGGAAGGUACUUAUCA | PO PS_tet (40) | 56.9 |
| si21 | 5′ AAAUCCUUCCAUGAAUAGUTT TTUUUAGGAAGGUACUUAUCA | PS_BTT (50) PO | 60.9 |
| si22 | 5′ AAAUCCUUCCAUGAAUAGUTT TTUUUAGGAAGGUACUUAUCA | PS_tet (41) PO | 56.9 |
| si23 | 5′ UUGUUUUCAGGAAGUAGUUUU UUAACAAAAGUCCUUCAUCAA | PO PS_BTT (61) | 62.8 |
| si24 | 5′ UUGUUUUCAGGAAGUAGUUUU UUAACAAAAGUCCUUCAUCAA | PO PS_tet (46) | 60.3 |
| si25 | 5′ UUGUUUUCAGGAAGUAGUUUU UUAACAAAAGUCCUUCAUCAA | PS_BTT (63) PO | 62.3 |
| si26 | 5′ UUGUUUUCAGGAAGUAGUUUU UUAACAAAAGUCCUUCAUCAA | PS_tet (44) PO | 59.0 |

BTT, 5-benzylthio-1-H-tetrazole; Nt, nucleotide; PO, phosphodiester; PS, phosphorothioate; siRNA, small interfering RNA; tet, 1H-tetrazole; $T_m$, melting temperature.
*Sequences of the Lin28 (si19-22) and p53 (si23-26) siRNAs: guide strand (upper); passenger strand (lower).
†Nucleotide linkages in the strands of the siRNAs.
‡Calculated Rp content of the PS ORN, assuming it is composed of independently acting dinucleotides.

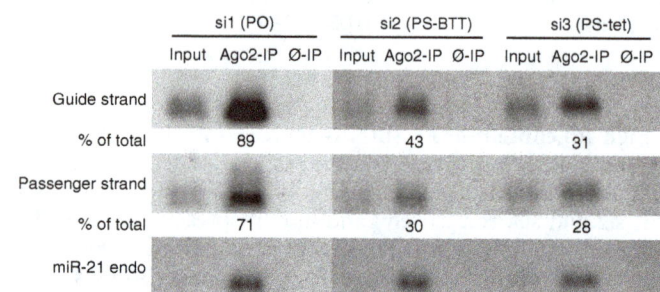

**Figure 6 | Composition of the RISC/Ago2 in HeLa cells transfected with PS- siRNAs.** p53 PO-, PS_BTT- and PS_tet-siRNAs (si1, si2, si3, respectively; 30 nM) were transfected into HeLa cells and worked up under identical conditions after 24 h. Northern blot detection using [32]P-labelled respective counter strands: guide and passenger strands of si1, si2 and si3 (see Supplementary Fig. 17 for original blot). RNA samples loaded were each input RNA (5% of total), Ago2 and control (Ø-IP) immunoprecipitations (47.5% each of total). Endogenously-expressed miR-21 detection served as a loading control. Densitometric quantification of Ago2-IP (in % of total input) is shown. One representative blot of the two independent experiments is shown (see Supplementary Fig. 17).

potency of uniformly modified PS_BTT siRNAs. The origin of the effects is possibly advantageous interactions with Ago2, both during passenger strand cleavage/dissociation, uptake of the guides into RISC and guide-mediated mRNA cleavage. The greater stability to nucleases shown by the PS_BTT siRNAs may also play a contributory role.

## Discussion

Introduction of the PS modification was an enabling advance in the field of oligonucleotide therapeutics. However, the chirality at the new internucleotide linkage was an unwelcome complication

and despite considerable research, a practicable solid-phase synthesis of stereopure PS oligonucleotides is still unavailable. The pharmacological consequences of uniformly modifying siRNAs with PS groups has not been investigated in detail, probably because of sporadic reports that PS siRNAs in vitro are less active than unmodified siRNAs[13,14]. However, medicinal chemists are well aware that small reductions in the potency of a drug can be easily compensated by an improved PK profile.

Here we have confirmed that distinct tetrazole activators produce a stereogenic bias during the solid-phase synthesis of PS ORNs. In a siRNA composed of two ORNs and ∼40 PS linkages, this bias substantially influences the biophysical and biological properties of the duplex. Specifically, compared with tetrazole, ribonucleoside coupling catalysed by BTT yields a higher fraction of Rp PS linkages in an ORN, as evidenced by elevated $T_m$s. Surprisingly, this produces a more potent PS siRNA, albeit with slightly less activity than its PO counterpart. It is likely that other commonly used activating agents, for example, ethylthiotetrazole, may similarly alter these properties and thus, they also deserve to be investigated.

Controlling the configuration of phosphorous centres has been of central importance in various studies charting the stereochemical course of enzymatic processes. Similarly, we envisage that understanding the synergy and additivity of this motif in a supra(bio-)molecular context may be valuable in the field of siRNA therapeutics. The use of PS_BTT siRNAs in animal models of disease is currently under investigation.

## Methods

**General.** Solvents for purification and chromatography were purchased as technical grade. For column chromatography SiO2-60 (230–400-mesh ASTM; Fluka) was used as the stationary phase. Analytical thin layer chromatography was performed on aluminium plates precoated with $SiO_2$-60 F254 (Merck) and visualized with a ultraviolet lamp (254 nm). Concentration in vacuo was performed at ∼10 mbar and 40 °C, drying at ∼$10^{-2}$ mbar and room temperature (rt). NMR (nuclear magnetic resonance) spectra were measured on a Bruker AV400 spectrometer at rt. [31]P NMR spectra are reported as follows: chemical shift in parts per million in $CDCl_3$ and $CD_3CN$.

**Synthesis of ORNs.** Chemicals for oligonucleotide synthesis were purchased from Aldrich and TCI (Sigma-Aldrich Chemie GmbH, D-89555 Steinheim). Phosphoramidites were obtained from Thermo Fisher Scientific (Waltham, MA), Glen Research (TOM-U) (Sterling, VA) and from Dharmacon (ACE-U). The activator 5-benzylthiotetrazole (BTT) was purchased from Biosolve (5555 CE Valkenswaard, the Netherlands). All oligonucleotides used in this work were synthesized with a MM12 synthesizer from Bio Automation Inc. (Plano, TX) on 500 Å UnyLinker CPG from ChemGenes (Wilmington, MA). The coupling time for phosphoramidites was 2 × 90 s. The oligonucleotides were purified on an Agilent 1200 series preparative HPLC fitted with a WatersXBridge OST C-18 column, 10 × 50 mm, 2.5 μm at 60 °C. The RNA phosphoramidites were prepared as 0.08 M solutions in dry acetonitrile (ACN), the activator BTT (Biosolve BV, 5555 CE Valkenswaard, the Netherlands) was prepared as a 0.24 M solution dry ACN, the activator 1H-tetrazole (Tet; Sigma-Aldrich Chemie GmbH, D-89555 Steinheim) was purchased as a 0.45 M solution in dry ACN. Oxidizer was prepared as a 0.02 M $I_2$ solution in THF/Pyridine/$H_2$O (70:20:10, w/v/v/v); the sulfurizing reagent was prepared as a 0.05 M solution of 3-((N,N-dimethylaminomethylidene)amino)-3H-1,2,4-dithiazole-5-thione (DDTT; Sulfurizing Reagent II; Glen Research, Virginia) in dry pyridine/ACN (60:40). Capping reagent A was: THF/lutidine/acetic anhydride (8:1:1) and capping reagent B was: 16% N-methylimidazole/THF. Deblock solution was prepared as a 3% dichloro acetic acid in dichloroethane. The cleavage, the deprotection of the bases and phosphordiesterbackbone was done by the incubation of the CPG-support for 2 h, at 65 °C, 1.8 bar in gaseous methylamine. Deprotection of 2′-O-TBDMS (tert-butyldimethylsilyl-) was performed by incubation of the ORN for 1.5 h, at 70 °C in a mixture of dry 1-N-methyl-2-pyrrolidone/triethylamine/triethylamine x 3HF.

RP-HPLC purification of ORNs. Running buffer for HPLC purification of single-stranded ORNs (up to 23 nt): buffer A (0.1 M triethylammonium acetate), buffer B (methanol); gradient for the DMT-on purification: 20–60% buffer B over 5 min; gradient for the DMT-off purification: 5 –35% buffer B over 5 min. Fractions containing the product were collected and dried in a miVac duo SpeedVac from Genevac. The oligonucleotides were analysed by LC-MS (Agilent 1200/6130 system) on a Waters Acquity OST C-18 column, 2.1 × 50 mm, 1.7 μM, 65 °C. Buffer A: 0.4 M HFIP, 15 mM triethylamine; buffer B: MeOH. Gradient: 7–35% B in 14 min; flow rate: 0.3 ml min$^{-1}$.

**Nuclease degradation assays.** To determine the stereochemistry of PS diastereoisomers, an equal quantity of both the diastereoisomers were mixed and a sample was injected onto HPLC to generate a reference chromatagram. Then mixtures of ZnCl$_2$ (2.5 µl, 5 mM), NaOAc (10 µl, 20 mM), H$_2$O (6.5 µl; Milli-Q) of the RNA-dinucleotide diastereoisomers (1 mM) were prepared. nP1 1 µl (2.4 U µl$^{-1}$) was added and the samples were incubated at 37 and 50 °C (UsC, UsA, CsU, CsA and GsA) for 30 min. The reaction was stopped by adding EtOH/HCl (9:1, 5 µl). To denature the nP1 enzyme the samples were heated to 95 °C for 3 min. To separate the enzyme from the ORNs the samples were extracted with CHCl$_3$ (20 µl). Supernatant (20 µl) was injected and analysed by HPLC.

**Stability of siRNA in 10% FBS.** Solution of the siRNA (25 µl, 50 µM) was diluted by addition of 42.5 µl H$_2$O (Milli-Q) and mixed with 7.5 µl FBS, Sigma-Aldrich, Buchs, CH). The reaction mixture was incubated at 37 °C (controls were treated identically, except FBS solution was replaced by water). At specific time points aliquots of 6 µl were collected, flash frozen in liquid nitrogen and stored at −80 °C. Samples were analysed on a 15% non-denaturing acrylamide gel. Gels were prepared using 30% acrylamide solution diluted with water (Millipore) and 5 × Tris–Borate–EDTA buffer, 10% ammonium persulfate solution to start polymerization. DNA loading buffer II was added to the samples without heating before loading onto the gel. The gel was run for 20 min at 50 V followed by 1 h at 90 V. After electrophoresis the gel was stained with GelRed nucleic acid stain (incubation time: 10 min) and analysed on a Bio-Rad Gel DocTMOR System. All samples were assayed in triplicates (Supplementary Figs 11–14).

**Cell culture and transfection of siRNAs.** HeLa cells (ATCC, no. CCL-2; LGC, Molsheim, FR) and Huh7 cells (Cell Lines Service, no. 300156, Eppelheim, DE) were maintained in Dulbecco's Modified Eagle's medium (Gibco, Invitrogen, Basel, CH) supplemented with 10% FBS (Sigma-Aldrich, Buchs, CH). SiRNA against *Renilla* (siRen: Supplementary Table 5) was from Dharmacon (Chicago, USA) and the control siRNA (Con; no. AM4640) was from Ambion (Austin, USA). RNAs were transfected using Oligofectamine (no. 12252-011, Invitrogen, Basel, CH) according to the manufacturer's instructions.

**Reporter plasmid preparation and sequences.** PsiCHECK2 (no. C8021, Promega, Dübendorf, CH) dual luciferase reporter plasmids were cut with Not1 and Xho1 restriction enzymes, and the inserts were cloned into the 3′-untranslated region of the Renilla gene to yield reporter constructs having fully complementary binding sites for the siRNA. The inserted sequences in the psiCHECK2 vector are reported in Supplementary Note 2. The psiCHECK2 vectors were sequenced and subsequently transfected as described above. For reporting the siRen activity, an empty PSICHECK2 luciferase plasmid was used.

**Luciferase assays.** HeLa cells were seeded in white 96-well plates and the RNAs were transfected after 4 h. There were no overt differences in cell viability between the cells transfected with PO siRNA and the control cells. All transfections were performed in triplicates. After 24 h DNA plasmid (20 ng per well) was transfected using jetPEI (no. 101-10, Polyplus, Illkirch, FR) according to the manufacturer's protocol. After 48 h, the supernatants were removed and the firefly substrate (15 µl; Dual-GloR Luciferase Assay System, Promega, Dubendorf, CH) that was diluted with 15 µl H$_2$O was added. Luminescence was measured on a microtitre plate reader (Mithras LB940, Berthold Technologies, Bad Wildbad, DE). After measurement 15 µl of *Renilla* substrate per well was added and the measurement was repeated. Values were normalized against firefly luciferase and the lowest siRNA concentration transfected. All statistical analyses were performed by ANOVA using Dunnett's post-test. All statistics were run with GraphPad.

**Western blotting.** Cells were lysed with radioimmunoprecipitation assay lysis buffer (R 0278; Sigma) for p53 detection and with lysis buffer (10 mM Hepes, 400 mM NaCl, 3 mM MgCl$_2$, 0.5% Triton-X 100, 1 mM DTT and 10% Glycerol) for LIN28B detection. Protein concentrations were determined using a BCA assay (Thermo Fisher Scientific 23225), the protein (10–20 ng) was mixed with equal quantities of SDS loading buffer (100 mM Tris–HCl, 4% SDS, 20% glycerol and 0.2% bromophenol blue). Samples were heated at 95 °C for 5 min, separated on SDS gels and transferred to polyvinylidene difluoride membranes. Non-specific membrane binding was blocked for 40 min at rt with 5% or milk in phosphate-buffered saline containing 0.05% Tween-20. Membranes were incubated overnight at 4 °C with appropriate primary antibodies p53 (no. sc-126) from Santa Cruz Biotechnology, Lin28B (no. A5316) from Cell Signaling Technology and β-ACTIN (no. A5316) from Sigma Life Science. After washing, the membranes were incubated with horseradish peroxidase-conjugated secondary antibodies for 2 h at rt in blocking buffer and washed again. Signals generated by the chemiluminescent substrate (ECL( + ); Amersham Biosciences) were captured by a cooled CCD (charge-coupled device) camera (Bio-Rad). Protein bands were quantified by densitometry using the analysis software imageJ. Samples were normalized against β-ACTIN protein band and the average of transfection with control siRNA Con.

**RISC affinity assays.** SiRNAs si1, si2 and si3 were reverse-transfected with RNAiMax (Life Technologies) according to the manufacturer's recommendation at 30 nM final concentration into HeLa cells in 15 cm cell culture dishes ($\sim 1.2 \times 10^7$ cells; 60% confluency) and incubated for 24 h. Ago2 immunoprecipitation was performed with modifications as described previously[46]. In brief, PBS-washed cells were collected and lysed in NP40 lysis buffer (50 mM HEPES pH 7.5, 150 mM KCl, 0.5% NP40, 0.5 mM DTT, 2 mM EDTA and 50 U ml$^{-1}$ RNAsin, complete protease inhibitor (Roche)). Each cleared lysate was equally divided for Ago2 (clone 9A11, Ascenion, Munich, Germany) and control IP samples (non-specific rat serum IgG, (Sigma)). Five per cent (of 1st and 2nd replicate) of each input sample was collected for later RNA extraction. Then, 40 µl of Prot G Dynabeads (Life Technologies) per ½ 15 cm dish were washed two times with 1 ml of citrate-phosphate buffer (25 mM citric acid, 66 mM Na2HPO4 and pH 5.0). The Ago2 or rat IgG control serum antibodies (20 µg per 40 µl of Prot G Dynabeads) were immobilized in a total volume of 1,000 µl of citrate-phosphate buffer by gentle rolling for 1 h at 4 °C. Thereafter, the beads were washed three times with each 1 ml of NP40 lysis buffer and blocked for 1 h at 4 °C with 1 ml of NP40 lysis buffer containing BSA (10 µg ml$^{-1}$). The beads were washed three times with 1 ml of NP40 lysis buffer and resuspended in 50 µl therein. Lysates were incubated with the Ago2/control antibody-coupled Prot G beads for 1 h at 4 °C with gentle rolling and then washed five times with 1 ml of IP wash buffer (50 mM HEPES, pH 7.5, 300 mM KCl, 0.05% NP40, 0.5 mM DTT and complete protease inhibitor (Roche)). Finally, bound RNA was released by the addition of 200 µl proteinase K digest buffer (100 mM Tris–HCl pH 7.5, 150 mM NaCl and 12.5 mM EDTA) containing 240–440 µg proteinase K (recombinant PCR grade solution, (Roche)) at 65 °C for 15 min shaking. RNA from Ago2/control IP and input was isolated by chloroform/phenol (Life Technologies) extraction and resuspended in 20 µl PCR grade water and subjected to northern blot analysis as described in ref. 47. One microgram of total input RNA and 10 µl (50%) of IP RNA were denatured in RNA loading buffer for 5 min at 95 °C and separated on a 15% denaturing polyacrylamide/urea gel electrophoresis. RNA transfer onto neutral nylon membrane (Hybond NX, GE Healthcare) and EDC/1-methylimidazole crosslinking was performed as described previously[47]. Detection probes were generated according to ref.48. PO backbone ssRNA sip53 (2.5 µl each; 50 µM stock; guide or passenger strand) and 2′-O-Me anti-miR-21 as normalization control were 5′ phosphorylated using T4 phosphokinase, 3′-phosphatase-free (Roche) according to the providers' protocol with 1 µl γ-$^{32}$P-ATP (6000Ci/mmol Perkin-Elmer) followed by chloroform/phenol purification and used for the detection of the respective counter strand. Membranes were (pre-)-hybridized rotating at 40 °C overnight. Washed membranes were exposed to Phospho-Imager screen (GE Healthcare) for at least 24 h and the signals were detected by Typhoon scanner device FLA-7000 (GE Healthcare). Densitometric signal quantification was done by using ImageQuantTL software (GE Healthcare). Assays were performed in two independent biological replicates.

## References

1. Elbashir, S. M. *et al.* Duplexes of 21-nucleotide RNAs mediate RNA interference in cultured mammalian cells. *Nature* **411**, 494–498 (2001).
2. Soutschek, J. *et al.* Therapeutic silencing of an endogenous gene by systemic administration of modified siRNAs. *Nature* **432**, 173–178 (2004).
3. Morrissey, D. V. *et al.* Potent and persistent in vivo anti-HBV activity of chemically modified siRNAs. *Nat. Biotechnol.* **23**, 1002–1007 (2005).
4. Dorn, G. *et al.* siRNA relieves chronic neuropathic pain. *Nucleic Acids Res.* **32**, e49 (2004).
5. Lightfoot, H. L. & Hall, J. Target mRNA inhibition by oligonucleotide drugs in man. *Nucleic Acids Res.* **40**, 10585–10595 (2012).
6. Eckstein, F. Nucleoside phosphorothioates. *Annu. Rev. Biochem.* **54**, 367–402 (1985).
7. Nukaga, Y., Yamada, K., Ogata, T., Oka, N. & Wada, T. Stereocontrolled solid-phase synthesis of phosphorothioate oligoribonucleotides using 2′-O-(2-cyanoethoxymethyl)-nucleoside 3′-O-oxazaphospholidine monomers. *J. Org. Chem.* **77**, 7913–7922 (2012).
8. Wan, W. B. *et al.* Synthesis, biophysical properties and biological activity of second generation antisense oligonucleotides containing chiral phosphorothioate linkages. *Nucleic Acids Res.* **42**, 13456–13468 (2014).
9. Guga, P. & Koziolkiewicz, M. Phosphorothioate nucleotides and oligonucleotides—recent progress in synthesis and application. *Chem. Biodivers.* **8**, 1642–1681 (2011).
10. Matranga, C., Tomari, Y., Shin, C., Bartel, D. P. & Zamore, P. D. Passenger-strand cleavage facilitates assembly of siRNA into Ago2-containing RNAi enzyme complexes. *Cell* **123**, 607–620 (2005).
11. Koziolkiewicz, M., Krakowiak, A., Kwinkowski, M., Boczkowska, M. & Stec, W. J. Stereodifferentiation—the effect of P chirality of oligo(nucleoside phosphorothioates) on the activity of bacterial RNase H. *Nucleic Acids Res.* **23**, 5000–5005 (1995).
12. Krieg, A. M., Guga, P. & Stec, W. P-chirality-dependent immune activation by phosphorothioate CpG oligodeoxynucleotides. *Oligonucleotides* **13**, 491–499 (2003).
13. Behlke, M. A. Progress towards in vivo use of siRNAs. *Mol. Ther.* **13**, 644–670 (2006).

14. Winkler, J., Stessl, M., Amartey, J. & Noe, C. R. Off-target effects related to the phosphorothioate modification of nucleic acids. *ChemMedChem* **5**, 1344–1352 (2010).
15. Amarzguioui, M., Holen, T., Babaie, E. & Prydz, H. Tolerance for mutations and chemical modifications in a siRNA. *Nucleic Acids Res.* **31**, 589–595 (2003).
16. Harborth, J. *et al.* Sequence, chemical, and structural variation of small interfering RNAs and short hairpin RNAs and the effect on mammalian gene silencing. *Antisense Nucleic Acid Drug Dev.* **13**, 83–105 (2003).
17. Braasch, D. A. *et al.* RNA interference in mammalian cells by chemically-modified RNA. *Biochemistry* **42**, 7967–7975 (2003).
18. Wei, X. Coupling activators for the oligonucleotide synthesis via phosphoramidite approach. *Tetrahedron* **69**, 3615–3637 (2013).
19. Seela, F. & Kretschmer, U. Diastereomerically pure Rp and Sp dinucleoside H-phosphonates: the stereochemical course of their conversion into P-methylphosphonates, phosphorothioates, and [oxygen-18] chiral phosphates. *J. Org. Chem.* **56**, 3861–3869 (1991).
20. Mukhlall, J. A. & Hersh, W. H. Sulfurization of dinucleoside phosphite triesters with chiral disulfides. *Nucleosides Nucleotides Nucleic Acids* **30**, 706–725 (2011).
21. Ravikumar, V. T. & Cole, D. L. Diastereomeric process control in the synthesis of 2′-O-(2-methoxyethyl) oligoribonucleotide phosphorothioates as antisense drugs. *Nucleosides Nucleotides Nucleic Acids* **22**, 1639–1645 (2003).
22. Potter, B. V., Connolly, B. A. & Eckstein, F. Synthesis and configurational analysis of a dinucleoside phosphate isotopically chiral at phosphorus. Stereochemical course of Penicillium citrum nuclease P1 reaction. *Biochemistry* **22**, 1369–1377 (1983).
23. Griffiths, A. D., Potter, B. V. & Eperon, I. C. Stereospecificity of nucleases towards phosphorothioate-substituted RNA: stereochemistry of transcription by T7 RNA polymerase. *Nucleic Acids Res.* **15**, 4145–4162 (1987).
24. Stec, W. J. & Zon, G. Stereochemical studies of the formation of chiral internucleotide linkages by phosphoramidite coupling in the synthesis of oligodeoxyribonucleotides. *Tetrahedron Lett.* **25**, 5279–5282 (1984).
25. Dahl, B. H., Nielsen, J. & Dahl, O. Mechanistic studies on the phosphoramidite coupling reaction in oligonucleotide synthesis. I. Evidence for nucleophilic catalysis by tetrazole and rate variations with the phosphorus substituents. *Nucleic Acids Res.* **15**, 1729–1743 (1987).
26. Cheruvallath, Z. S., Sasmor, H., Cole, D. L. & Ravikumar, V. T. Influence of Diastereomeric Ratios of Deoxyribonucleoside Phosphoramidites on the Synthesis of Phosphorothioate Oligonucleotides. *Nucleosides Nucleotides Nucleic Acids* **19**, 533–543 (2000).
27. Pitsch, S. & Weiss, P. A. Chemical synthesis of RNA sequences with 2′-O-[(triisopropylsilyl)oxy]methyl-protected ribonucleoside phosphoramidites. *Curr. Protoc. Nucleic Acid Chem.* Chapter 3, Unit 3, 8 (2002).
28. Oka, N., Kondo, T., Fujiwara, S., Maizuru, Y. & Wada, T. Stereocontrolled synthesis of oligoribonucleoside phosphorothioates by an oxazaphospholidine approach. *Org. Lett.* **11**, 967–970 (2009).
29. Koziolkiewicz, M. *et al.* Stability of stereoregular oligo(nucleoside phosphorothioate)s in human plasma: diastereoselectivity of plasma 3′-exonuclease. *Antisense Nucleic Acid Drug Dev.* **7**, 43–48 (1997).
30. Wojcik, M., Cieslak, M., Stec, W. J., Goding, J. W. & Koziolkiewicz, M. Nucleotide pyrophosphatase/phosphodiesterase 1 is responsible for degradation of antisense phosphorothioate oligonucleotides. *Oligonucleotides* **17**, 134–145 (2007).
31. Prakash, T. P. *et al.* Positional effect of chemical modifications on short interference RNA activity in mammalian cells. *J. Med. Chem.* **48**, 4247–4253 (2005).
32. Guennewig, B. *et al.* Synthetic pre-microRNAs reveal dual-strand activity of miR-34a on TNF-alpha. *RNA* **20**, 61–75 (2014).
33. Tuschl, T., Zamore, P. D., Lehmann, R., Bartel, D. P. & Sharp, P. A. Targeted mRNA degradation by double-stranded RNA *in vitro*. *Genes Dev.* **13**, 3191–3197 (1999).
34. Khvorova, A., Reynolds, A. & Jayasena, S. D. Functional siRNAs and miRNAs Exhibit Strand Bias. *Cell* **115**, 209–216 (2003).
35. Schwarz, D. S. *et al.* Asymmetry in the assembly of the RNAi enzyme complex. *Cell* **115**, 199–208 (2003).
36. Addepalli, H. *et al.* Modulation of thermal stability can enhance the potency of siRNA. *Nucleic Acids Res.* **38**, 7320–7331 (2010).
37. Huesken, D. *et al.* Design of a genome-wide siRNA library using an artificial neural network. *Nat. Biotechnol.* **23**, 995–1001 (2005).
38. Bramsen, J. B. *et al.* A screen of chemical modifications identifies position-specific modification by UNA to most potently reduce siRNA off-target effects. *Nucleic Acids Res.* **38**, 5761–5773 (2010).
39. Leuschner, P. J., Ameres, S. L., Kueng, S. & Martinez, J. Cleavage of the siRNA passenger strand during RISC assembly in human cells. *EMBO Rep.* **7**, 314–320 (2006).
40. Rand, T. A., Petersen, S., Du, F. & Wang, X. Argonaute2 cleaves the anti-guide strand of siRNA during RISC activation. *Cell* **123**, 621–629 (2005).
41. Parrish, S., Fleenor, J., Xu, S., Mello, C. & Fire, A. Functional anatomy of a dsRNA trigger: differential requirement for the two trigger strands in RNA interference. *Mol. Cell* **6**, 1077–1087 (2000).
42. Schirle, N. T. & MacRae, I. J. The crystal structure of human Argonaute2. *Science* **336**, 1037–1040 (2012).
43. Elkayam, E. *et al.* The structure of human argonaute-2 in complex with miR-20a. *Cell* **150**, 100–110 (2012).
44. Petri, S. *et al.* Increased siRNA duplex stability correlates with reduced off-target and elevated on-target effects. *RNA* **17**, 737–749 (2011).
45. Lima, W. F. *et al.* Binding and cleavage specificities of human Argonaute2. *J. Biol. Chem.* **284**, 26017–26028 (2009).
46. Imig, J. *et al.* miR-CLIP capture of a miRNA targetome uncovers a lincRNA H19-miR-106a interaction. *Nat. Chem. Biol.* **11**, 107–114 (2015).
47. Imig, J. *et al.* microRNA profiling in Epstein-Barr virus-associated B-cell lymphoma. *Nucleic Acids Res.* **39**, 1880–1893 (2011).
48. Beitzinger, M. & Meister, G. Experimental identification of microRNA targets by immunoprecipitation of Argonaute protein complexes. *Methods Mol. Biol.* **732**, 153–167 (2011).

## Acknowledgements

This work was supported in part by grants from the ETH (ETH-01 11-2; ETH-14 09-3) and Krebsliga Schweiz (KFS-2648-08-2010) to J.H. We thank U. Pradère, R. Häner and K.-H. Altmann for discussions on the manuscript.

## Author contributions

J.H. and R.G. conceived the project; H.J., M.R., F.B., J.I. and Y.W. performed the experiments; H.J., R.G. and J.H. wrote the paper.

## Additional information

# 7

# Drug design from the cryptic inhibitor envelope

Chul-Jin Lee[1,*], Xiaofei Liang[2,*], Qinglin Wu[1,*], Javaria Najeeb[1,*], Jinshi Zhao[1], Ramesh Gopalaswamy[2], Marie Titecat[3], Florent Sebbane[3], Nadine Lemaitre[3], Eric J. Toone[1,2] & Pei Zhou[1,2]

Conformational dynamics plays an important role in enzyme catalysis, allosteric regulation of protein functions and assembly of macromolecular complexes. Despite these well-established roles, such information has yet to be exploited for drug design. Here we show by nuclear magnetic resonance spectroscopy that inhibitors of LpxC—an essential enzyme of the lipid A biosynthetic pathway in Gram-negative bacteria and a validated novel antibiotic target—access alternative, minor population states in solution in addition to the ligand conformation observed in crystal structures. These conformations collectively delineate an inhibitor envelope that is invisible to crystallography, but is dynamically accessible by small molecules in solution. Drug design exploiting such a hidden inhibitor envelope has led to the development of potent antibiotics with inhibition constants in the single-digit picomolar range. The principle of the cryptic inhibitor envelope approach may be broadly applicable to other lead optimization campaigns to yield improved therapeutics.

[1] Department of Biochemistry, Duke University Medical Center, Durham, North Carolina 27710, USA. [2] Department of Chemistry, Duke University, Durham, North Carolina 27708, USA. [3] Inserm, Univ. Lille, CHU Lille, Institut Pasteur de Lille, CNRS, U1019-UMR 8204-CIIL-Center for Infection and Immunity of Lille, F-59000 Lille, France. * These authors contributed equally to this work. Correspondence and requests for materials should be addressed to P.Z. (email: peizhou@biochem.duke.edu).

The availability of high-resolution crystal structures of protein-inhibitor complexes has revolutionized the drug development process, enabling structure-aided design of improved therapeutics based on visual inspection of receptor-ligand interactions. However, it is increasingly recognized that high-resolution structures of protein-inhibitor complexes do not necessarily enable a successful lead optimization campaign, as the static structural models often fail to capture the conformational flexibility of receptors or their bound inhibitors[1,2]. In contrast to the largely static view of protein structures provided by crystallography, the discovery of ring flipping events of buried aromatic residues of the basic pancreatic trypsin inhibitor by NMR (ref. 3) has heralded the widespread observation of molecular motions within macromolecules in solution. Conformational dynamics involving side-chain rearrangement, domain reorganization and binding-induced structural remodelling has been shown to play important roles in enzyme catalysis[4-7], allosteric regulation[8] and nucleic acid function[9]. Molecular recognition of small molecules likewise alters protein dynamics[10]. Despite the extensive demonstration of conformational dynamics of macromolecules in solution, the application of such information to drug development has remained an unmet challenge.

In this study, we used solution NMR to investigate the conformational states of small-molecule inhibitors bound to LpxC, an essential metalloamidase that catalyses the deacetylation of UDP-(3-O-acyl)-N-acetylglucosamine during the biosynthesis of lipid A in Gram-negative bacteria[11,12]. We show that these enzyme-bound inhibitors dynamically access alternative, minor conformations in solution in addition to the ligand state observed in the crystal structure. Furthermore, we demonstrate that these ligand conformational states collectively define a cryptic inhibitor envelope that can be exploited for optimization of lead compounds.

## Results

**A cryptic inhibitor envelope invisible in crystal structures.** We chose *Aquifex aeolicus* LpxC (AaLpxC) in the lipid A biosynthetic pathway (Supplementary Fig. 1) for structural and dynamics investigation due to its exceptional thermostability, which has enabled both NMR measurements and crystallographic studies (for example, refs 13–16). *Pseudomonas aeruginosa* LpxC (PaLpxC) was exploited when co-crystal structures of the desired AaLpxC-inhibitor complexes could not be obtained. As a starting point, we investigated the conformations of CHIR-090 and LPC-011 bound to AaLpxC, two inhibitors that share the same threonyl-hydroxamate head group, but differ in their tail groups (Supplementary Fig. 1, Supplementary Table 1). CHIR-090 features a substituted biphenyl acetylene tail group that competes with the acyl chain of the LpxC substrate to occupy the hydrophobic substrate passage of the enzyme[14]. Replacing the tail group of CHIR-090 with a substituted biphenyl diacetylene group generated LPC-011 with improved antibiotic activity due to minimization of vdW clashes with the substrate-binding passage[16,17]. To provide a direct comparison with solution NMR investigations, we determined the crystal structure of AaLpxC in complex with LPC-011 (Fig. 1a, Supplementary Table 2). This structure reveals a single conformation of the threonyl-hydroxamate head group in the active site, with the threonyl Cγ2 methyl group packing against an invariant phenylalanine residue (F180 in AaLpxC) and the Oγ1 hydroxyl group forming a hydrogen bond with the catalytically important lysine residue (K227 in AaLpxC). The threonyl side chain of the inhibitor features a *trans* configuration with a $\chi^1$ angle of 180°, a rotameric state that is less energetically favourable

(7% population of all threonine side chains in proteins) compared with the alternative rotameric states of *gauche-* ($\chi^1 = -60°$) and *gauche+* ($\chi^1 = 60°$) collectively accounting for 92% of the observed threonine side-chain conformations[18]. Since the observed ligand conformation in the crystal structure represents an unfavourable $\chi^1$ rotameric state of the threonyl head group, we investigated whether this group could access alternative ligand conformations in solution.

Database analysis of high-resolution protein structures has indicated that amino acid side-chains adopt specific rotameric conformations[18], and side-chain motions can be approximated as conformational hopping between rotameric states[19]. Such motions occur over a wide range of timescales, from ns movement of surface exposed residues to μs-ms timescale ring flipping in protein cores. To determine rotameric populations of the ligand threonyl side chain over a wide timescale, we synthesized isotopically labelled CHIR-090 and LPC-011 and measured the scalar couplings $^3J_{NC\gamma2}$ and $^3J_{C'C\gamma2}$ that are dependent on the $\chi^1$ angle of the threonyl side chain (Supplementary Fig. 2 and Supplementary Table 3). Specifically, a large $^3J_{NC\gamma2}$ value of ~1.9 Hz is consistent with a *trans*$^{NC\gamma2}$ relationship between the amide nitrogen and the Cγ2 methyl group of the threonyl head group, corresponding to a $\chi^1$ angle of $-60°$ (*gauche-* $\chi^1$), whereas a small value of ~0.2 Hz reflects a *gauche*$^{NC\gamma2}$ relationship (*gauche+*$^{NC\gamma2}$ or *gauche-*$^{NC\gamma2}$), corresponding to $\chi^1$ angles of 180° (*trans* $\chi^1$) or 60° (*gauche+* $\chi^1$), respectively. An intermediate value reflects a population-weighted average between the *trans*$^{NC\gamma2}$ and *gauche*$^{NC\gamma2}$ states[20]. A similar relation is noted for the $^3J_{C'C\gamma2}$ coupling[20]. Thus simultaneous measurements of the $^3J_{NC\gamma2}$ and $^3J_{C'C\gamma2}$ scalar couplings enable the determination of the populations of all three rotameric states of the threonyl side chain[19]. Measurements of LPC-011 yielded a $^3J_{NC\gamma2}$ coupling of 0.58 ± 0.05 Hz and a $^3J_{C'C\gamma2}$ coupling of 0.77 ± 0.04 Hz, corresponding to a predominant *trans* $\chi^1$ configuration with a population of 0.65 ± 0.03 (Fig. 1b, Supplementary Table 3). Such an observation is consistent with the ligand conformation in the AaLpxC/LPC-011 crystal structure (Fig. 1a). However, the measurements also revealed that the threonyl side chain of LPC-011 can readily access alternative, minor conformational states with a population of 0.23 ± 0.03 for the *gauche-* $\chi^1$ conformation and a population of 0.12 ± 0.01 for the *gauche+* $\chi^1$ conformation (Fig. 1b). Measurements of CHIR-090 yielded a similar result, with a $^3J_{NC\gamma2}$ coupling of 0.45 ± 0.07 Hz and a $^3J_{C'C\gamma2}$ coupling of 0.67 ± 0.04 Hz, corresponding to populations of 0.77 ± 0.04 for the *trans* $\chi^1$ configuration, 0.14 ± 0.04 for the *gauche-* $\chi^1$ configuration and 0.09 ± 0.01 for the *gauche+* $\chi^1$ configuration (Fig. 1b, Supplementary Table 3). Modelling of the threonyl side chain in the second-most-populated *gauche-* $\chi^1$ state indicates that the Cγ2 methyl group would experience vdW interactions with the hydrophobic component of the K227 side chain with the Oγ1 hydroxyl group oriented towards solvent, leaving a cavity against the F180 side chain of AaLpxC (Fig. 1c). Although the protein-ligand interactions in the *gauche-* $\chi^1$ rotameric conformation are less favourable than those in the ground state of the *trans* $\chi^1$ rotamer, the lack of optimal interactions is partially compensated by the intrinsic free energy difference of the rotameric states of the threonyl side chain that favours the *gauche-* $\chi^1$ rotamer over the *trans* $\chi^1$ rotamer in the unbound ligand. Taken together, these solution measurement-derived rotamers collectively portray an inhibitor envelope that can accommodate three substitutions at the Cβ position of the threonyl head group (Fig. 1c).

To test this prediction, we merged the two conformations of the threonyl head group and generated Cβ-di-methyl substituted compounds with the third Cβ-substitution containing either a hydroxyl group (LPC-037) or an amino group (LPC-040)

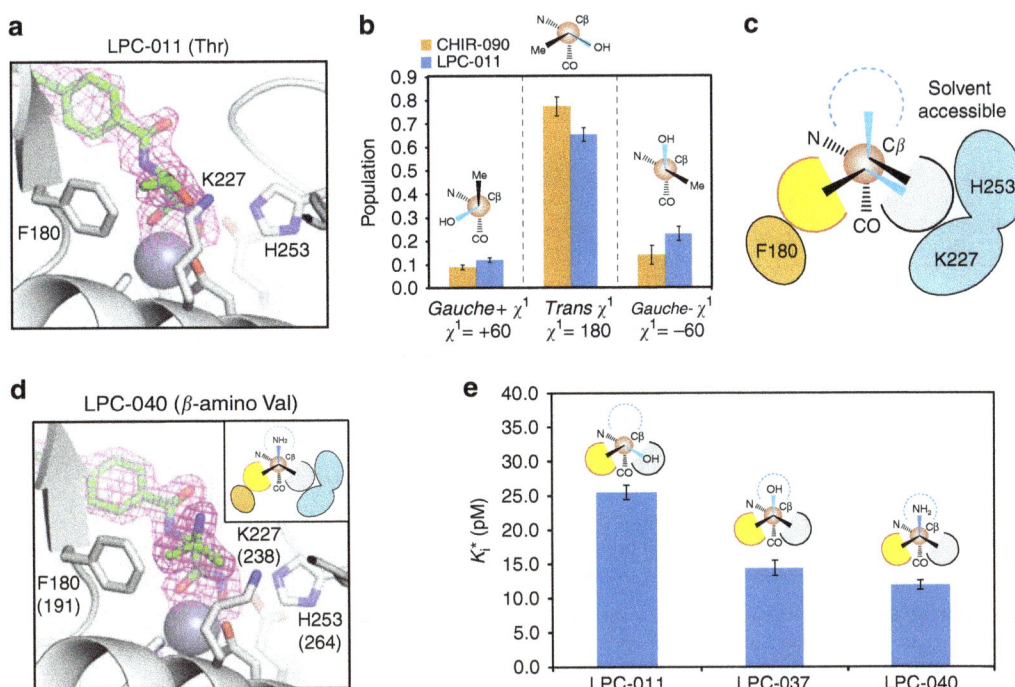

**Figure 1 | Dynamic access of minor conformational states of LpxC inhibitors containing the threonyl head group.** (**a**) Crystal structure of the AaLpxC/ LPC-011 complex, showing a single *trans* $\chi^1$ rotamer of the threonyl side chain of the inhibitor. AaLpxC is shown in the cartoon model and catalytically important residues in the stick model. LPC-011 is shown in the stick model, with the purple mesh representing the inhibitor omit map (2mFo-DFc) contoured at 1.0σ. (**b**) NMR measurements of scalar couplings ($^3J_{NC\gamma2}$ and $^3J_{C'C\gamma2}$) of the threonyl-head-group-containing LpxC inhibitors CHIR-090 (orange) and LPC-011 (blue) reveal a dynamic distribution of all three rotameric $\chi^1$ states. (**c**) Combining the two most-populated ligand states creates a dynamically accessible inhibitor envelope around the Cβ atom of the threonyl head group. The binding pockets near F180 and H253/K227 are coloured in yellow and grey, respectively, and a third binding pocket accessible to solvent is denoted by an open dashed circle in blue. (**d**) The Cβ-triply substituted compound LPC-040 occupies all three pockets within the inhibitor envelope. PaLpxC is shown in the cartoon model, with catalytically important residues shown in the stick model. Residue numbering reflects the corresponding residues in AaLpxC, with PaLpxC residue numbers shown in parentheses. LPC-040 is shown in the stick model, with the purple mesh representing the inhibitor omit map (2mFo-DFc) contoured at 1.0σ. (**e**) Inhibition constants ($K_i^*$) of LpxC inhibitors. Chemical substitutions at the Cβ-position of the inhibitors and their observed (LPC-011 and LPC-040) and predicted (LPC-037) binding modes within the inhibitor envelope are labelled.

(Supplementary Table 1). Structural analysis of LPC-040 in complex with PaLpxC indeed revealed the anticipated ligand conformation (Fig. 1d; Supplementary Table 2) with the two Cβ-substituted methyl groups forming hydrophobic interactions with F180 (F191$^{PaLpxC}$) and the stem of K227 (K238$^{PaLpxC}$) and the amino group directed towards solvent accessible space to form a water-mediated hydrogen bond with the backbone carbonyl of F180 (F191$^{PaLpxC}$).

We next investigated whether these compounds show enhanced LpxC inhibition in enzymatic assays. *Escherichia coli* LpxC inhibition by CHIR-090 and LPC-011 both displayed slow-binding kinetics consistent with the transition from a rapid-forming initial encounter complex (enzyme-inhibitor complex (EI)) to the stable complex (EI*; Supplementary Fig. 3). Therefore, we focused enzymatic assays on the stable EI* complex. CHIR-090 and LPC-011 are potent LpxC inhibitors with $K_i^*$ values of $153 \pm 8$ pM and $26 \pm 1$ pM, respectively. Excitingly, the Cβ-triply substituted compounds LPC-037 and LPC-040 both showed enhanced LpxC inhibition, displaying $K_i^*$ values of $14 \pm 1$ pM and $12 \pm 1$ pM, respectively (Fig. 1e; Supplementary Table 4).

**Drug design from the expanded inhibitor envelope.** Having delineated the hidden inhibitor envelope at the Cβ position of the threonyl head group, we next examined whether the dynamically accessible envelope of LpxC inhibitors can be further expanded at

the γ position. The molecule that fits this purpose is LPC-023 bearing an isoleucine-hydroxamate head group (Supplementary Table 1). Isoleucine shares a basic molecular scaffold with threonine, and its Cγ1-Cδ1 group can be viewed as a substitution of the Oγ1 group of threonine near the conserved lysine (K227$^{AaLpxC}$; K238$^{PaLpxC}$) and histidine (H253$^{AaLpxC}$; H264$^{PaLpxC}$) residues. The isoleucine analogue was crystallized with AaLpxC (Supplementary Table 2), and two copies of the LpxC-inhibitor complexes were found in the asymmetric unit. Among the two protomers of LpxC, the second protomer (chain B) displayed a distorted active site with the catalytic H253 flipped out of the active site in a configuration that has not been observed in any of the previously reported LpxC structures. We reasoned that this would likely reflect a crystallization artifact and consequently focused our analysis on the first LpxC protomer (chain A) in complex with the isoleucine analogue, LPC-023 (Fig. 2a). In this protomer, the isoleucine head group displays a *trans* $\chi^1$ configuration, consistent with the predominant rotameric state observed in the threonyl group of CHIR-090 and LPC-011. The Cδ1 methyl group adopts a *gauche+* $\chi^2$ conformation with regard to the Cα atom. In such a configuration, the Cδ1 methyl group is closest to and potentially forms vdW interactions with the nearby imidazole ring of the catalytic H253. This observation is somewhat surprising as the *gauche+* $\chi^2$ angle is rarely observed for isoleucine in protein structures and contributes to <5% of the observed $\chi^2$ rotamers, indicating that such a rotamer represents a high-energy state of the free ligand.

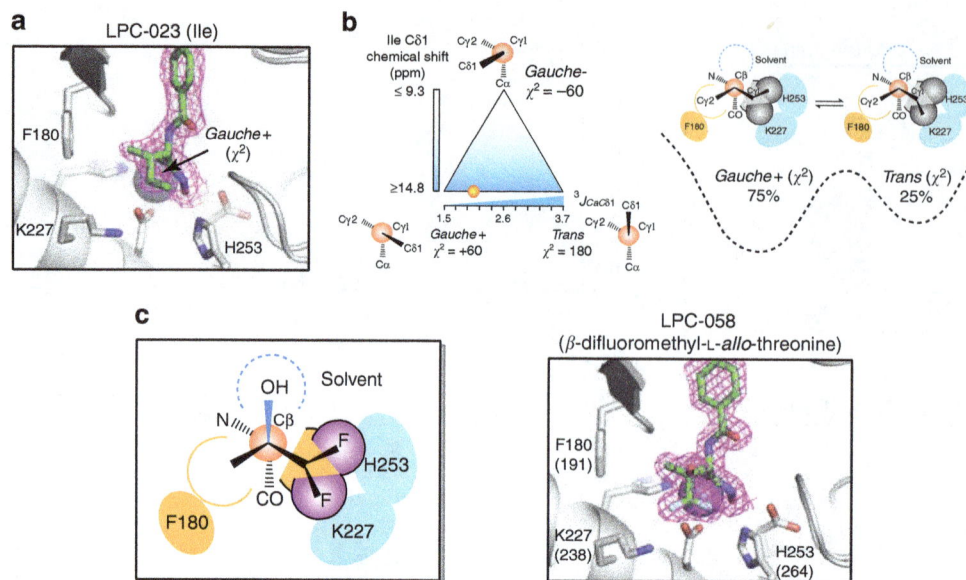

**Figure 2 | Expanded inhibitor envelope enables the design of a potent inhibitor, LPC-058. (a)** Crystal structure of AaLpxC in complex with LPC-023, an isoleucine derivative, reveals a *gauche+* $\chi^2$ rotamer conformation of the inhibitor. AaLpxC is shown in the cartoon model, with the catalytically important residues in the stick model. LPC-023 is shown in the stick model, with the purple mesh representing the inhibitor omit map (2mFo-DFc) contoured at $1.0\sigma$. **(b)** Combined measurements of the $C\delta1$ chemical shift and the $^3J_{C\alpha C\delta1}$ coupling of LPC-023 in the protein-bound complex reveal a dynamic equilibrium between *gauche+* and *trans* $\chi^2$ rotameric states, with the *gauche+* state being the predominant conformation ($\sim75\%$ population) and the *trans* state being the minor conformation ($\sim25\%$). **(c)** Design and structural validation of LPC-058 that optimally occupies the inhibitor envelope. PaLpxC is shown in the cartoon model, with the catalytically important residues in the stick model. Residue numbering reflects the corresponding residues of AaLpxC, with PaLpxC numbers shown in parentheses. LPC-058 is shown in the stick model, with the purple mesh representing the inhibitor omit map (2mFo-DFc) contoured at $1.1\sigma$.

We thus investigated whether the $C\delta1$ methyl group of the isoleucine analogue LPC-023 can access alternative $\chi^2$ rotameric states in solution using the isotopically labelled compound.

The isoleucine $C\delta1$ chemical shift is sensitive to its $\chi^2$ dihedral angle[21]. For the *gauche+* and *trans* $\chi^2$ rotamers, isoleucine $C\delta1$ methyl groups display downfield shifted chemical shifts of $>14.8$ p.p.m., whereas upfield shifted $C\delta1$ chemical shifts of $<9.3$ p.p.m. indicate a *gauche-* $\chi^2$ conformation. The unbound LPC-023 compound has a $C\delta1$ chemical shift of 12.8 p.p.m. (Supplementary Fig. 4), consistent with rotameric averaging between a *gauche-* $\chi^2$ romateric state and the *trans/gauche+* states. In contrast, the LpxC-bound LPC-023 displays a $C\delta1$ chemical shift of 15.2 p.p.m. (Supplementary Fig. 4), indicating that the $\chi^2$ conformation resides entirely in the *trans* or *gauche+* rotameric states or switches between these two states, but has no detectable population in the *gauche-* state (Fig. 2b, Supplementary Table 5). We next measured the $^3J$ coupling between the $C\delta1$ and $C\alpha$ atoms (Supplementary Fig. 4). A *trans* configuration between $C\alpha$ and $C\delta1$ would yield a large scalar coupling of $\sim3.7$ Hz, whereas a *gauche* configuration would yield a small coupling of $\sim1.5$ Hz (ref. 21). Our measurements yielded a $^3J_{C\alpha C\delta1}$ coupling of $2.05\pm0.04$ Hz, corresponding to $75\pm2\%$ population in the *gauche+* $\chi^2$ state with the $C\delta1$ methyl group located adjacent to H253 and $25\pm2\%$ population in the *trans* $\chi^2$ state with the same methyl group oriented towards K227 (Fig. 2b, Supplementary Table 5). Although the predominant *gauche+* $\chi^2$ rotameric state is consistent with the crystallographically observed inhibitor conformation, our NMR measurements support the notion that both the *gauche+* and *trans* states of the $\chi^2$ rotamers are conformationally populated and they collectively expand the inhibitor envelope at the $\gamma$-position, whereas the *gauche-* $\chi^2$ rotamer is energetically occluded and dynamically inaccessible in solution.

The delineation of two additional pockets that can accommodate methyl-sized functional groups to interact with side chains of the catalytically important histidine and lysine residues suggests fluorine as an attractive functional group for substitution. Fluorine has a slightly smaller size compared with the methyl group[22], and the fluorine atom is both strongly electronegative and lipophilic[23]. This renders the fluorine group well-suited for forming hydrophobic interactions with the deprotonated histidine side chain and the stem of the lysine group, or forming electrostatic interactions with a protonated histidine imidazolium and a positively charged lysine terminal ammonium group.

Based on this analysis, we introduced difluoro substitution to the pro-$R$ methyl group of LPC-037 to yield LPC-058. Structural analysis of LPC-058 with PaLpxC indeed revealed the anticipated ligand conformation (Fig. 2c; Supplementary Table 2), with the $\beta$-methyl group occupying the hydrophobic pocket next to F180 (F191[PaLpxC]) for vdW contacts, the $\beta$-hydroxyl group residing in the solvent pocket to form a water-mediated hydrogen bond with the backbone of F180 (F191[PaLpxC]), and finally with the difluoromethyl group oriented towards H253 (H264[PaLpxC]) and K227 (K238[PaLpxC]). One of the fluorine atoms adopts a *gauche+* configuration with respect to $C\alpha$ and forms a hydrogen bond with N$\epsilon1$ atom of the protonated H253 (H264[PaLpxC]), while the second fluorine atom adopts a *trans* configuration with respect to $C\alpha$ and forms an electrostatic interaction with the ammonium group of K227 (K238[PaLpxC]).

Excitingly, LPC-058 is an exceptionally potent inhibitor. It displayed slow-binding kinetics consistent with the rapid formation of an initial encounter complex (EI) followed by slow transition to the stable EI* complex (Supplementary Fig. 5). Accordingly, $k_{obs}$ increased hyperbolically over the inhibitor concentration[24]. Steady-state kinetics analysis of the stable EI*

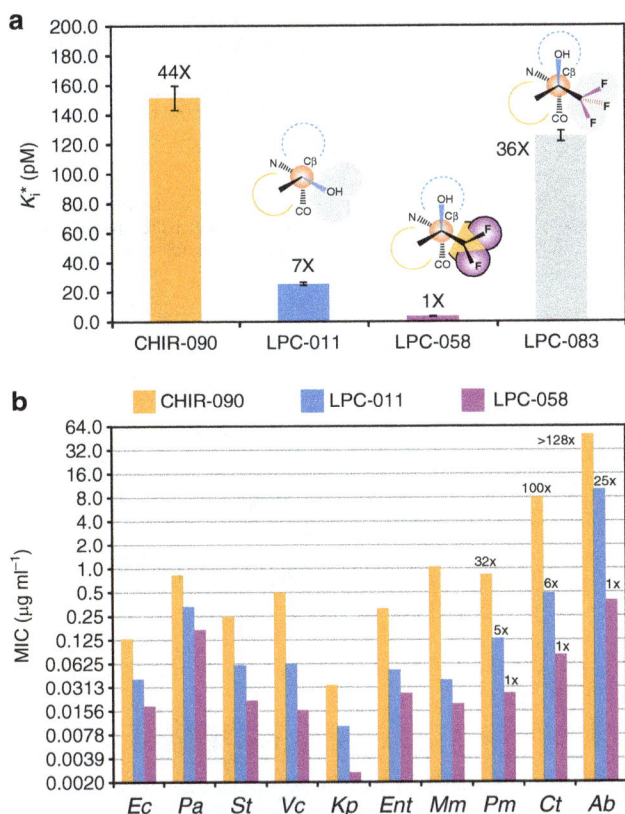

**Figure 3 | LPC-058 is a superior inhibitor compared with the parent compounds LPC-011 and CHIR-090.** (**a**) Inhibition constants of CHIR-090, LPC-011, LPC-058 and LPC-083. The head group of each compound and its conformation within the inhibitor envelope is denoted. (**b**) LPC-058 is a potent antibiotic and displays enhanced activity over LPC-011 and CHIR-090 against a diverse array of Gram-negative pathogens. MIC enhancement of >4-fold over LPC-011 and ≥32-fold over CHIR-090 is labelled. Tested bacterial species include *E. coli* (Ec), *P. aeruginosa* (Pa), *Salmonella typhimurium* (St), *Vibrio cholerae* (Vc), *Klebsiella pneumoniae* (Kp), *Enterobacter cloacae* (Ent), *Morganella morganii* (Mm), *Proteus mirabilis* (Pm), *Chlamydia trachomatis* (Ct) and *Acinetobacter baumannii* (Ab).

complex revealed an inhibition constant ($K_i^*$) of $3.5 \pm 0.2$ pM, a 7-fold enhancement of potency over LPC-011 and a 44-fold enhancement over CHIR-090 (Fig. 3a). Incorporation of the $K_i^*$ value into the analysis of the inhibitor concentration-dependent $k_{obs}$ values enabled accurate determination of the forward rate ($k_5 = 0.39 \pm 0.02$ min$^{-1}$) and reverse rate ($k_6 = 0.0014 \pm 0.0001$ min$^{-1}$) from EI to EI* and the inhibition constant of the initial encounter complex EI ($K_i = 973 \pm 128$ pM).

To examine whether LPC-058 designed from the cryptic inhibitor envelope shows improved antibiotic activity over CHIR-090 and LPC-011, we determined its minimum inhibitory concentration (MIC) values against a range of Gram-negative pathogens. LPC-058 showed uniform improvement over CHIR-090 and LPC-011 against all Gram-negative bacterial strains tested (Fig. 3b, Supplementary Table 6). In general, enhanced antibiotic activities of 2- to 4-fold over LPC-011 and 5- to 55-fold over CHIR-090 were observed for *E. coli*, *P. aeruginosa*, *Salmonella typhimurium*, *Vibrio cholerae*, *Klebsiella pneumoniae*, *Enterobacter cloacae* and *Morganella morganii*. More pronounced improvements (5- to 25-fold over LPC-011 and 32- to >128-folds over CHIR-090) were observed for *Proteus mirabilis*, *Chlamydia trachomatis* and *Acinetobacter baumannii*. Of particular importance is the potent antibiotic activity of LPC-058

against *Acinetobacter baumannii* (MIC = 0.39 μg ml$^{-1}$). To the best of our knowledge, LPC-058 is the first reported LpxC inhibitor with an MIC value below 1 μg ml$^{-1}$ against this clinically important Gram-negative pathogen *in vitro*. The broad-spectrum antibiotic activity of LPC-058 highlights the therapeutic potential of LpxC inhibitors as effective antibiotics against a wide range of Gram-negative infections.

## Discussion

It is widely acknowledged that the dynamic interconversion of multiple conformational states is an intrinsic property of proteins and nucleic acids in solution. In comparison, the conformational dynamics of small molecules in their receptor-bound states has rarely been investigated, let alone utilized for drug design. Here we show that small-molecule inhibitors of LpxC dynamically access alternative, minor-state ligand conformations in addition to the predominant conformational state observed in crystal structures. These minor-state ligand conformations, together with that of the major state, collectively delineate a cryptic inhibitor envelope in solution that is invisible to crystallographic studies. Furthermore, we show that such a cryptic inhibitor envelope provides important molecular insights for the design of high-affinity ligands. In the case of LpxC inhibitors, analysis of the inhibitor envelope has led to the development of a potent antibiotic LPC-058. With inclusion of only three additional heavy atoms, the newly designed compound LPC-058 enhanced the inhibitory effect towards *E. coli* LpxC over its parent compound LPC-011 by 7-fold and improved antibiotic activity by 2- to 25-fold against a wide range of Gram-negative pathogens, rendering it the most potent and the most broad-spectrum LpxC inhibitor *in vitro*. Although some features of the LpxC inhibitor envelope, such as the solvent accessible pocket at the Cβ position of the threonyl head group, might have been envisaged based on structural analysis of the LpxC/CHIR-090 complex[14] and LpxC inhibitors bearing similar head groups to LPC-037 and LPC-040, but different tail groups, have been reported[25,26], the precise definition of two accessory pockets at the Oγ1-position of the threonyl head group could not have been predicted by structural analysis alone. In fact, the most widely employed functional substitution of a pro-*R* methyl of LPC-037 is the trifluoromethyl group (CF$_3$), not the difluoromethyl group (CF$_2$) utilized in LPC-058. However, the trifluoromethyl substituted compound LPC-083 compromised the inhibitory effect ($K_i^* = 125 \pm 4$ pM) over its parent compound LPC-037 ($K_i^* = 14 \pm 1$ pM) by nearly ninefold (Supplementary Table 4). Its inhibition constant is worse than LPC-011 by fivefold (Fig. 3a), and it is a less potent antibiotic against *E. coli* (MIC = 0.1 μg ml$^{-1}$) than LPC-011 (MIC = 0.04 μg ml$^{-1}$), which would have argued away from development of the synthetically more challenging β-difluoromethyl-*allo*-threonyl compound LPC-058 designed from the dynamic inhibitor envelope.

The work presented here departs from the established paradigm of ligand design from the crystallographically visible, static ligand conformation and highlights the potential of drug development from the 'invisible', dynamically accessible inhibitor envelope in solution, which encompasses the receptor-bound ligand conformations from both major and minor states. The framework presented here should be broadly applicable to lead optimization campaigns for small molecules, peptides and peptidomimetics to yield more effective therapeutics.

## Methods

**Chemical synthesis.** Details of chemical synthesis and characterization are described in Supplementary Methods.

**Crystallography structural analysis.** Protein samples of AaLpxC and PaLpxC were prepared as described previously[16]. Before crystallization trials, a fourfold molar excess of each compound, dissolved in DMSO, was mixed with 8 mg ml$^{-1}$ AaLpxC (1–275, C181A) in 100 mM potassium chloride, 2 mM dithiothreitol and 25 mM HEPES (pH 7.0) or 12 mg ml$^{-1}$ PaLpxC (1–299, C40S) in 50 mM sodium chloride, 2 mM tris(2-carboxyethyl)phosphine and 25 mM HEPES (pH 7.0), respectively. For PaLpxC, 10 mM zinc sulfate was added as a crystallization additive. The protein-inhibitor mixture was incubated for 30 min at room temperature to obtain a homogenous sample. All of the LpxC-inhibitor complex crystals were obtained by the sitting-drop vapour diffusion method at 20 °C. Initial crystallization screening yielded microcrystals for the AaLpxC/LPC-011 complex in a reservoir solution containing 0.1 M HEPES (pH 7.0) and 15% PEG 8000 and for the AaLpxC/LPC-023 complex in a reservoir solution containing 0.18 M ammonium chloride, 11.8% PEG3350 and 4% 1,3-propanediol. The microcrystals were used to prepare seeding stocks by the Seed-Bead protocol (Hampton Research, HR2-320). Diffraction quality crystals were obtained by the streak-seeding method. The final crystallization reservoirs contained 0.05 M ammonium acetate and 10% PEG3350 for the AaLpxC/LPC-011 complex and 0.18 M ammonium chloride, 11.8% PEG3350 and 10% 1,3-propanediol for the AaLpxC/LPC-023 complex, respectively. High quality crystals of the PaLpxC/LPC-040 and PaLpxC/LPC-058 complexes were obtained in precipitant solutions containing 0.1 M sodium acetate trihydrate (pH 4.8–5.1) and 2.4–2.6 M ammonium nitrate. Crystals were cryoprotected using the corresponding mother liquor solutions containing 30% 2-methyl-2,4-pentanediol (MPD) for the AaLpxC/LPC-011 complex, 30% ethylene glycol for the AaLpxC/LPC-023 complex and 10% glycerol for the PaLpxC/inhibitor complexes, respectively, before flash-freezing for data collection.

Data sets of the PaLpxC/LPC-040 and PaLpxC/LPC-058 complexes were collected in-house using a Rigaku MicroMax-007 HF rotating anode generator and R-Axis IV + + detector. Data sets of the AaLpxC/LPC-011 and AaLpxC/LPC-023 were collected at the SER-CAT 22-ID beamline at the Advanced Photon Source at Argonne National Laboratory. The collected X-ray diffraction data were processed using HKL2000 (ref. 27) or XDS (ref. 28). The crystal structures of LpxC-inhibitor complexes were solved by molecular replacement with the programme PHASER (ref. 29) using PDB entries 3P3C and 3P3E for the AaLpxC-inhibitor complexes and the PaLpxC-inhibitor complexes, respectively. Restraints of the inhibitors were generated by using ELBOW (ref. 30) and edited manually. Iterative model building and refinement was carried out using COOT (ref. 31) and PHENIX (ref. 32). The 2mFo-DFc omit maps[33] were generated using PHENIX[32].

**Solution NMR measurements.** Deuterated AaLpxC was expressed and purified as described previously[13]. The AaLpxC-inhibitor complexes were prepared by adding individual inhibitors to the purified protein in the presence of 5% deuterated dimethylsulfoxide (DMSO) in a 1:2 protein-inhibitor molar ratio, and incubated initially at room temperature and then at 45 °C to form the complex. Samples were concentrated and exchanged into the NMR buffer containing 25 mM sodium phosphate pH 7.0, 100 mM KCl, 5% deuterated DMSO and 100% D$_2$O. The NMR sample concentration was ~1 mM.

The scalar couplings of $^3J_{C'C\gamma2}$ and $^3J_{NC\gamma2}$ for the AaLpxC/CHIR-090 and AaLpxC/LPC-011 complexes were measured on a Bruker 700 MHz NMR spectrometer at 45 °C, using J-modulated $^1H$–$^{13}C$ constant-time HSQC experiments[34,35]. The reference and scalar coupling-modulated CT-HSQC spectra were recorded in an interleaved manner with a constant-time delay (2T) set to 57.4 ms, and the maximum evolution time for the indirect ($^{13}C$) dimension set 12.1 ms. Data were processed using NMRPIPE (ref. 36) with eightfold zero-filling in the indirect dimension. The peak intensities were measured by SPARKY (ref. 37), and the $^3J_{C'C\gamma2}$ and $^3J_{NC\gamma2}$ couplings were calculated from the ratio of the peak intensities between the reference spectrum (I$_{ref}$) and the J-modulated spectrum (I$_{mod}$) according to equation (1):

$$\frac{I_{mod}}{I_{ref}} = \cos(2\pi J T) \quad (1)$$

Rotameric populations were calculated based on the three-site jump model[19] using values derived from self-consistent parameterization of $^3J$ couplings[20]. The scalar coupling $^3J_{C\alpha C\delta1}$ for the AaLpxC/LPC-023 complex was measured on an Agilent 800 MHz NMR spectrometer at 37 °C using a J-modulated constant-time $^{13}C$ HSQC experiment using selective Ile-C$\alpha$ inversion pulses. The $^3J_{C\alpha C\delta1}$ coupling was calculated from the ratio of the peak intensities between the reference spectrum (I$_{ref}$) and the J-modulated spectrum (I$_{mod}$) according to equation (1). Since the C$\delta1$ chemical shift of 15.2 p.p.m. of LPC-023 excludes the *gauche-* $\chi^2$ rotamer[21], populations of the remaining rotamers were calculated from $^3J_{C\alpha C\delta1}$ based on the two-site jump model between the *gauche +* and *trans* rotameric states[21].

**Enzymatic assays.** The radiolabelled substrate for the LpxC enzymatic assays, [α-$^{32}$P] UDP-3-O-[(R)-3-hydroxymyristoyl]-N-acetylglucosamine, and the unlabelled carrier substrate were prepared as previously described[38]. The assays were performed in a buffer consisting of 25 mM HEPES pH 7.4, 100 mM KCl, 1 mg ml$^{-1}$ BSA, 2 mM dithiothreitol and 5 μM substrate at 30 °C. Serial twofold dilutions of each inhibitor were prepared in DMSO and added to the reaction

mixture with a 10-fold dilution. The assays were initiated by addition of purified LpxC protein into the reaction mix with 1:4 dilution to the final concentration as specified.

The $K_M$ value was determined by varying substrate concentrations from 0.4 to 50 μM with 0.2 nM of LpxC. To study the slow, tight-binding inhibition, LpxC activity was assessed in the presence of varying inhibitor concentrations. The product conversions were determined from 15 s up to 2 h after addition of 0.2 nM enzyme for CHIR-090 and LPC-011 in the presence of 5 μM substrate. Time-dependent inhibition of LPC-058 was assayed in the presence of 30 μM substrate and 0.1 nM enzyme such that $k_{obs}$ can be extracted under the slow, but not tight-binding conditions[39]. The following time-dependent equation was used to fit the data:

$$[P] = v_s t + \frac{v_i - v_s}{k_{obs}}[1 - e^{-k_{obs}t}] + c \quad (2)$$

with $v_s$ representing the steady-state rate, $v_i$ the initial rate, $k_{obs}$ the rate of transition from the initial encounter complex to the final complex and $c$ the baseline.

The $K_i^*$ was determined by analysing the rate of product accumulation after formation of the stable EI* complex. IC$_{50}$ curves for individual compounds were determined in the presence of 20 pM of the enzyme and varying inhibitor concentrations. The Morrison's quadratic equation was used to fit the fractional activity data to determine $K_i^{*app}$:

$$\frac{v_i}{v_0} = 1 - \frac{[E]_T + [I]_T + K_i^{*app} - \sqrt{([E]_T + [I]_T + K_i^{*app})^2 - 4[E]_T[I]_T}}{2[E]_T} \quad (3)$$

where [E]$_T$ and [I]$_T$ represent the total enzyme and inhibitor concentrations, respectively. The inhibition constant $K_i^*$ is converted from $K_i^{*app}$ according to the following relationship:

$$K_i^* = K_i^{*app}/(1 + \frac{[S]}{K_M}) \quad (4)$$

For two-step slow-binding inhibition, kinetic parameters $k_5$, $k_6$ and $K_i$ were extracted from curve fitting of experimental $k_{obs}$ values to inhibitor concentrations based on equations (5 and 6).

$$k_{obs} = k_6 + \frac{k_5[I]}{K_i^{app} + [I]} = k_6 + \frac{k_5[I]}{K_i(1 + [S]/K_M) + [I]} \quad (5)$$

$$K_i = K_i^*(1 + k_5/k_6) \quad (6)$$

**Measurements of the minimum inhibitory concentration (MIC).** The MIC assay protocol was adapted from methods described in National Committee for Clinical Laboratory Standards (NCCLS) to using 96-well plates[40]. Bacteria were grown in the Mueller–Hinton medium at 37 °C in the presence of varying concentrations of inhibitors and 5% DMSO. To obtain more accurate readings of the MICs, three series of twofold dilutions of inhibitors were used. The starting concentrations of the three series are different by factors of 1.33 and 1.67, respectively. MICs were reported as the lowest compound concentration that inhibited bacterial growth.

## References

1. Thanos, C. D., Randal, M. & Wells, J. A. Potent small-molecule binding to a dynamic hot spot on IL-2. *J. Am. Chem. Soc.* **125**, 15280–15281 (2003).
2. Carlson, H. A. & McCammon, J. A. Accommodating protein flexibility in computational drug design. *Mol. Pharmacol.* **57**, 213–218 (2000).
3. Wagner, G. & Wuthrich, K. Dynamic model of globular protein conformations based on NMR studies in solution. *Nature* **275**, 247–248 (1978).
4. Eisenmesser, E. Z. *et al.* Intrinsic dynamics of an enzyme underlies catalysis. *Nature* **438**, 117–121 (2005).
5. Whittier, S. K., Hengge, A. C. & Loria, J. P. Conformational motions regulate phosphoryl transfer in related protein tyrosine phosphatases. *Science* **341**, 899–903 (2013).
6. Bhabha, G. *et al.* A dynamic knockout reveals that conformational fluctuations influence the chemical step of enzyme catalysis. *Science* **332**, 234–238 (2011).
7. Boehr, D. D., McElheny, D., Dyson, H. J. & Wright, P. E. The dynamic energy landscape of dihydrofolate reductase catalysis. *Science* **313**, 1638–1642 (2006).
8. Tzeng, S. R. & Kalodimos, C. G. Dynamic activation of an allosteric regulatory protein. *Nature* **462**, 368–372 (2009).
9. Kimsey, I. J., Petzold, K., Sathyamoorthy, B., Stein, Z. W. & Al-Hashimi, H. M. Visualizing transient Watson-Crick-like mispairs in DNA and RNA duplexes. *Nature* **519**, 315–320 (2015).
10. Frederick, K. K., Marlow, M. S., Valentine, K. G. & Wand, A. J. Conformational entropy in molecular recognition by proteins. *Nature* **448**, 325–329 (2007).
11. Barb, A. W. & Zhou, P. Mechanism and inhibition of LpxC: an essential zinc-dependent deacetylase of bacterial lipid A synthesis. *Curr. Pharm. Biotechnol.* **9**, 9–15 (2008).

12. Raetz, C. R. H. & Whitfield, C. Lipopolysaccharide endotoxins. *Annu. Rev. Biochem.* **71**, 635–700 (2002).

13. Coggins, B. E. *et al.* Structure of the LpxC deacetylase with a bound substrate-analog inhibitor. *Nat. Struct. Biol.* **10**, 645–651 (2003).

14. Barb, A. W., Jiang, L., Raetz, C. R. & Zhou, P. Structure of the deacetylase LpxC bound to the antibiotic CHIR-090: Time-dependent inhibition and specificity in ligand binding. *Proc. Natl Acad. Sci. USA* **104**, 18433–18438 (2007).

15. Whittington, D. A., Rusche, K. M., Shin, H., Fierke, C. A. & Christianson, D. W. Crystal structure of LpxC, a zinc-dependent deacetylase essential for endotoxin biosynthesis. *Proc. Natl Acad. Sci. USA* **100**, 8146–8150 (2003).

16. Lee, C. J. *et al.* Species-specific and inhibitor-dependent conformations of LpxC: implications for antibiotic design. *Chem. Biol.* **18**, 38–47 (2011).

17. Liang, X. *et al.* Syntheses, structures and antibiotic activities of LpxC inhibitors based on the diacetylene scaffold. *Bioorg. Med. Chem.* **19**, 852–860 (2011).

18. Lovell, S. C., Word, J. M., Richardson, J. S. & Richardson, D. C. The penultimate rotamer library. *Proteins* **40**, 389–408 (2000).

19. Hennig, M. *et al.* Side-chain conformations in an unfolded protein: chi1 distributions in denatured hen lysozyme determined by heteronuclear $^{13}$C, $^{15}$N NMR spectroscopy. *J. Mol. Biol.* **288**, 705–723 (1999).

20. Perez, C., Lohr, F., Ruterjans, H. & Schmidt, J. M. Self-consistent Karplus parametrization of $^3$J couplings depending on the polypeptide side-chain torsion chi1. *J. Am. Chem. Soc.* **123**, 7081–7093 (2001).

21. Hansen, D. F., Neudecker, P. & Kay, L. E. Determination of isoleucine side-chain conformations in ground and excited states of proteins from chemical shifts. *J. Am. Chem. Soc.* **132**, 7589–7591 (2010).

22. Meng, H., Clark, G. A. & Kumar, K. in *Fluorine in Medicinal Chemistry and Chemical Biology.* (ed. Ojima, I.) 411–446 (Blackwell Publishing Ltd, 2009).

23. Wang, J. *et al.* Fluorine in pharmaceutical industry: fluorine-containing drugs introduced to the market in the last decade (2001–2011). *Chem. Rev.* **114**, 2432–2506 (2014).

24. Vogt, A. D. & Di Cera, E. Conformational selection or induced fit? A critical appraisal of the kinetic mechanism. *Biochemistry* **51**, 5894–5902 (2012).

25. Mansoor, U. F. *et al.* Design and synthesis of potent Gram-negative specific LpxC inhibitors. *Bioorg. Med. Chem. Lett.* **21**, 1155–1161 (2011).

26. Fei, Z. *et al.* A scalable synthesis of a hydroxamic acid LpxC inhibitor. *Org. Process Res. Dev.* **16**, 1436–1441 (2012).

27. Otwinowski, Z. & Minor, W. Processing of X-ray diffraction data collected in oscillation mode. *Methods Enzymol.* **276**, 307–326 (1997).

28. Kabsch, W. XDS. *Acta Crystallogr. D Biol. Crystallogr.* **66**, 125–132 (2010).

29. McCoy, A. J. *et al.* Phaser crystallographic software. *J. Appl. Crystallogr.* **40**, 658–674 (2007).

30. Moriarty, N. W., Grosse-Kunstleve, R. W. & Adams, P. D. electronic Ligand Builder and Optimization Workbench (eLBOW): a tool for ligand coordinate and restraint generation. *Acta Crystallogr. D Biol. Crystallogr.* **65**, 1074–1080 (2009).

31. Emsley, P. & Cowtan, K. Coot: model-building tools for molecular graphics. *Acta Crystallogr. D Biol. Crystallogr.* **60**, 2126–2132 (2004).

32. Adams, P. D. *et al.* PHENIX: a comprehensive Python-based system for macromolecular structure solution. *Acta Crystallogr. D Biol. Crystallogr.* **66**, 213–221 (2010).

33. Terwilliger, T. C. *et al.* Iterative-build OMIT maps: map improvement by iterative model building and refinement without model bias. *Acta Crystallogr. D Biol. Crystallogr.* **64**, 515–524 (2008).

34. Grzesiek, S., Vuister, G. W. & Bax, A. A simple and sensitive experiment for measurement of JCC couplings between backbone carbonyl and methyl carbons in isotopically enriched proteins. *J. Biomol. NMR* **3**, 487–493 (1993).

35. Vuister, G. W., Wang, A. C. & Bax, A. Measurement of three-bond nitrogen-carbon J couplings in proteins uniformly enriched in $^{15}$N and $^{13}$C. *J. Am. Chem. Soc.* **115**, 5334–5335 (1993).

36. Delaglio, F. *et al.* NMRPipe: a multidimensional spectral processing system based on UNIX pipes. *J. Biomol. NMR* **6**, 277–293 (1995).

37. Goddard, T. D. & Kneller, D. G. *Sparky 3* (Univeristy of California, 2008).

38. Jackman, J. E., Raetz, C. R. H. & Fierke, C. A. Site-directed mutagenesis of the bacterial metalloamidase UDP-(3-O-acyl)-N-acetylglucosamine deacetylase (LpxC). Identification of the zinc binding site. *Biochemistry* **40**, 514–523 (2001).

39. Zhang, R. & Windsor, W. T. In vitro kinetic profiling of hepatitis C virus NS3 protease inhibitors by progress curve analysis. *Methods Mol. Biol.* **1030**, 59–79 (2013).

40. National Committee for Clinical Laboratory Standards. *Approved Standard M7-A1: Methods For Dilution Antimicrobial Susceptibility Test For Bacteria That Grow Aerobically.* Clinical and Laboratory Standards Institute, Wayne, PA, USA (1997).

## Acknowledgements

Diffraction data of LpxC-inhibitor complexes were collected at the Duke Macromolecular X-ray Crystallography Shared Resource and at the Southeast Regional Collaborative Access Team (SER-CAT) 22-ID beamline at the Advanced Photon Source, Argonne National Laboratory, supported by the US Department of Energy, Office of Science and the Office of Basic Energy Sciences under Contract number W-31-109-Eng-38. NMR data were collected at the Duke University NMR Center. This work was supported by National Institutes of Health grants AI055588 and AI094475, and the Duke Bridge Fund awarded to P.Z. The authors would like to acknowledge Drs Bidong D. Nguyen and Raphael Valdivia for providing the MCC (minimal chlamydiacidal concentration) measurements of LpxC inhibitors against *Chlamydia trachomatis*, and Dr Xin Chen for insightful discussions on compound synthesis.

## Author contributions

P.Z. conceived the project. C.-J.L. and J.N. determined crystal structures of LpxC-inhibitor complexes; X.L. and R.G. synthesized LpxC inhibitors under the direction of E.J.T.; Q.W. performed NMR studies; J.Z. carried out enzymatic assays; and J.Z., C.-J. L., M.T., F.S. and N.L. determined the antibiotic activities of LpxC inhibitors. P.Z. wrote the manuscript with critical inputs from all authors.

## Additional information

**Accession codes:** The coordinates for the X-ray structures have been deposited to the Protein Data Bank (PDB) with accession codes of 5DRO, 5DRQ, 5DRP and 5DRR for the AaLpxC/LPC-011, PaLpxC/LPC-040, AaLpxC/LPC-023 and PaLpxC/LPC-058 complexes, respectively.

**Competing financial interests:** P.Z. and E.J.T. declare a competing financial interest. A patent on designed LpxC inhibitors was awarded to P.Z. and E.J.T. The remaining authors declare no competing financial interest.

# 8

# Exclusive photorelease of signalling lipids at the plasma membrane

André Nadler[1,2,*], Dmytro A. Yushchenko[1,3,*], Rainer Müller[1], Frank Stein[1], Suihan Feng[1], Christophe Mulle[4], Mario Carta[4] & Carsten Schultz[1]

Photoactivation of caged biomolecules has become a powerful approach to study cellular signalling events. Here we report a method for anchoring and uncaging biomolecules exclusively at the outer leaflet of the plasma membrane by employing a photocleavable, sulfonated coumarin derivative. The novel caging group allows quantifying the reaction progress and efficiency of uncaging reactions in a live-cell microscopy setup, thereby greatly improving the control of uncaging experiments. We synthesized arachidonic acid derivatives bearing the new negatively charged or a neutral, membrane-permeant coumarin caging group to locally induce signalling either at the plasma membrane or on internal membranes in β-cells and brain slices derived from C57B1/6 mice. Uncaging at the plasma membrane triggers a strong enhancement of calcium oscillations in β-cells and a pronounced potentiation of synaptic transmission while uncaging inside cells blocks calcium oscillations in β-cells and causes a more transient effect on neuronal transmission, respectively. The precise subcellular site of arachidonic acid release is therefore crucial for signalling outcome in two independent systems.

[1] European Molecular Biology Laboratory, Cell Biology and Biophysics Unit, Meyerhofstraße 1, 69117 Heidelberg, Germany. [2] Max Planck Institute of Molecular Cell Biology and Genetics, Pfotenhauerstraße 108, 01307 Dresden, Germany. [3] Institute of Organic Chemistry and Biochemistry, Academy of Sciences of the Czech Republic, Flemingovo náměstí 2, 16610 Prague 6, Czech Republic. [4] Institut Interdisciplinaire de Neurosciences, CNRS UMR 5297 Université Bordeaux 2, 146, rue Léo-Saignat, 33077 Bordeaux, France. * These authors contributed equally to this work. Correspondence and requests for materials should be addressed to C.S. (email: schultz@embl.de).

Cellular signalling networks are crucially dependent on small molecules. The molecular mechanisms involved in the respective signalling events are highly diverse and include direct interactions with target proteins, rapid concentration changes of the respective signalling molecules, intracellular concentration gradients and secondary signalling events because of ongoing metabolism[1,2]. Therefore, it is essential to mimic or modulate such events with high spatial and temporal precision. This is not always easily achieved by traditional approaches such as RNA interference or small-molecule inhibition of key enzymes[3,4]. Successful approaches for manipulating the cellular levels of small signalling molecules within a millisecond to second timeframe usually involve either chemical[5,6] or optogenetic[7,8] protein modulation systems for generating second messengers in situ or photoactivatable (caged) small molecules, which release the active species upon illumination[9,10]. Prominent examples of caged small molecules that have been employed in cell biology include lipids, nucleotides and neurotransmitters[11-17]. One major advantage of utilizing caged compounds is the possibility of stepwise concentration increases of signalling molecules to a fixed level from which they are subsequently metabolized. Only a few other methods enable these intracellular 'relaxation' experiments, such as reversible dimerizer systems introduced by our group[18] and others[19-21] and switchable optogenetic protein modulation systems[22]. These approaches constitute valuable tools, but in most cases there are significant experimental challenges yet to overcome[10]. Uncaging assays for instance are still hindered by the difficulty of measuring the reaction progress of photoreactions at the single-cell level in the midst of live-cell imaging experiments and an almost complete lack of strategies to achieve organelle-specific compound release. To date, two-photon uncaging might be considered as one of the most promising approaches to confine an uncaging event to a distinct organelle[23] and a number of novel caging groups with vastly improved optical properties have been developed over the last years[24-26]. Caged fluorophores offer probably the best approximation of an experimental tool to measure the progress of a photoreaction in living cells[27]. However, quantifying the photo-release of cellular messengers based on this approach is often difficult.

Stringent temporal and spatial control of externally induced signalling events is of particular importance if the investigated process is in part governed by highly dynamic concentration changes of small molecules. Arachidonic acid (AA) signalling provides a particularly striking example for the challenges of investigating small-molecule events. AA is directly involved in various cellular processes, most notably in controlled cell death (that is, through apoptosis), and by regulating vesicle fusion events occurring in neurite and axonal outgrowth, neurotransmitter release and insulin secretion[28-31]. Its direct molecular targets include a number of ion channels, syntaxin, protein kinase C and the G-protein-coupled receptor GPR40 (refs 32–36). Furthermore, AA may serve both as a second messenger in intracellular signalling cascades and as a messenger in intercellular communication[28]. Thus, the experimental possibility to alter AA concentrations at defined locations on a subcellular scale would be greatly beneficial for detailed understanding of its diverse cellular signalling roles. To address this, we developed a sulfonated caging group that allows for (i) uncaging of signalling lipids exclusively at the plasma membrane and (ii) straightforward quantification of uncaging reactions in living cells and the liberation of defined amounts of active compound. Here we demonstrate the applicability of the novel photoactivatable tool in two independent biological systems.

## Results

**Prelocation of caged signaling molecules at the plasma membrane.** A set of caged fatty acid (FA) derivatives (5–10) was synthesized

to study the consequences of rapidly elevated AA concentrations at the plasma membrane versus internal membranes (Supplementary Scheme 1 in the Supplementary Information). To achieve a stable pre-localization at the outer leaflet of the plasma membrane, we functionalized a photocleavable 7-diethylamino coumarin with two sulfonate groups (Fig. 1a). Their negative charges rendered the caged FA membrane-impermeant without significantly affecting the photophysical properties and the kinetics of the photoreaction (Supplementary Figs 1 and 2). Illumination removed the negative charges and enabled FA flip-flop across the plasma membrane, a fairly fast process for hydrophobic signalling lipids[37] (Fig. 1b), and binding to interacting proteins. The sulfonated coumarin alcohol **1** was obtained in a reaction sequence starting from 3-aminophenol (**2**), which was converted into coumarin **3** in two steps. Cleavage of the carbamate protecting group and alkylation of the 7-amino function gave the sulfonated derivative **4**, which was subsequently oxidized to alcohol **1** (Fig. 1a). **1** was coupled to AA to yield the desired caged AA derivative **5**. We further synthesized the neutral variant **6** caged with the standard 7-diethylaminocoumarin caging group and a third compound **7** bearing two additional carboxylates instead of the sulfonate groups. The latter compound enabled us to establish that sulfonate groups are indeed crucial for efficient pre-localization at the plasma membrane (Fig. 1c). Finally, caged butyric acid (**8**) was synthesized as biologically inactive control compound and oleic acid (OA) derivatives **9** and **10** were prepared to distinguish AA signalling properties from other long-chain FAs (Fig. 1a and Supplementary Note).

The cellular localization of the respective caged AA derivatives **5–7** was analysed by confocal fluorescence microscopy in HeLa cells, taking advantage of the intrinsic fluorescence of all three coumarin caging groups. Compound **6** predominantly stained internal membranes in an unspecific manner, whereas the sulfonated derivative **5** was exclusively localized at the plasma membrane (Fig. 1c and Supplementary Information). Staining caused by compound **7** appeared to be fairly unstable and was difficult to reproduce (Fig. 1c and Supplementary Fig. 3b). Therefore, compound **7** was not used in cell experiments. The localization pattern of **5** was stable over a wide concentration range in the loading solution (Supplementary Fig. 4). We established that the photophysical properties of the relevant coumarin esters **5**, **6**, **7**, **8**, **9** and **10** were very similar (Supplementary Fig. 2) thus enabling us to use coumarin fluorescence to quantify lipid loading. However, the efficiency of membrane incorporation varied significantly for the different compounds. This prompted us to adjust the concentrations (15 μM for **5**, 100 μM for **6**) to be able to differentiate the physiological consequences of uncaging a given lipid at the plasma membrane or in internal membranes while excluding simple concentration effects (Supplementary Fig. 5a,b). The final cellular concentration of the caged AA derivatives was estimated to be $81 \pm 13$ μM (**6**) and $67 \pm 17$ μM (**5**), respectively (Supplementary Fig. 5e), well in line with reported AA levels in β-cells[28,38,39]. Next, we addressed the stability of the plasma membrane localization of the caged AA derivative **5**. We found that the amount of vesicular structures bearing the fluorophore was negligible in the first 30 min after exchanging the loading solution for the imaging medium and started to increase slowly afterwards (Supplementary Fig. 6), suggesting a time window of 30–60 min for generating consistent data.

**Quantification of the uncaging reaction in living cells.** We reasoned that a mixed localization of **5** at the plasma membrane and in vesicular structures might be very helpful for developing an assay for quantifying the efficiency of uncaging reactions in living cells. To generate such a mixed localization, we kept the cells at 37 °C for 90–180 min after removal of the loading solution to ensure sufficient vesicle formation by endocytosis. The predominant mechanism for the observed decreases in fluorescence intensity upon uncaging is inherently different for vesicles and the plasma membrane. Photoactivation of **5** at the outer leaflet of the plasma membrane releases the highly hydrophilic coumarin alcohol **1** into the extracellular space where it is readily dispersed by diffusion. The opposite holds true for the intra-vesicular localization. Diffusion of the photo-generated coumarin alcohol **1** is in this case limited by the size of the respective vesicular structure and its fluorescence will therefore continue to contribute to the observed signal. Lower fluorescence intensity after uncaging is in this case only caused by true photobleaching pathways. As expected, 90–180 min after loading to HeLa cells, we observed a pronounced reduction of fluorescence after

uncaging **5** at the plasma membrane as compared with vesicular structures (Fig. 2a and Supplementary Movie M1). We developed a semi-automated image analysis approach to quantify this effect and thus established a general protocol for optimizing the required light intensity for uncaging experiments (Supplementary Methods and Supplementary Figs 7 and 8). The applicability of this protocol was exemplarily shown for two laser lines (375 and 405 nm) on a dual scanner confocal microscope (Olympus Fluoview 1200). Optimal laser settings were determined by systematic variation of the applied laser intensity (Fig. 2b,c and Supplementary Fig. 8). Although the normalized fluorescence intensity detected at the plasma membrane served as readout for judging the completeness of the photoreaction (Fig. 2c), the ratio of the observed decreases in fluorescence intensity at the plasma membrane and in vesicular structures was employed as a measure to determine laser settings for maximal efficiency of the

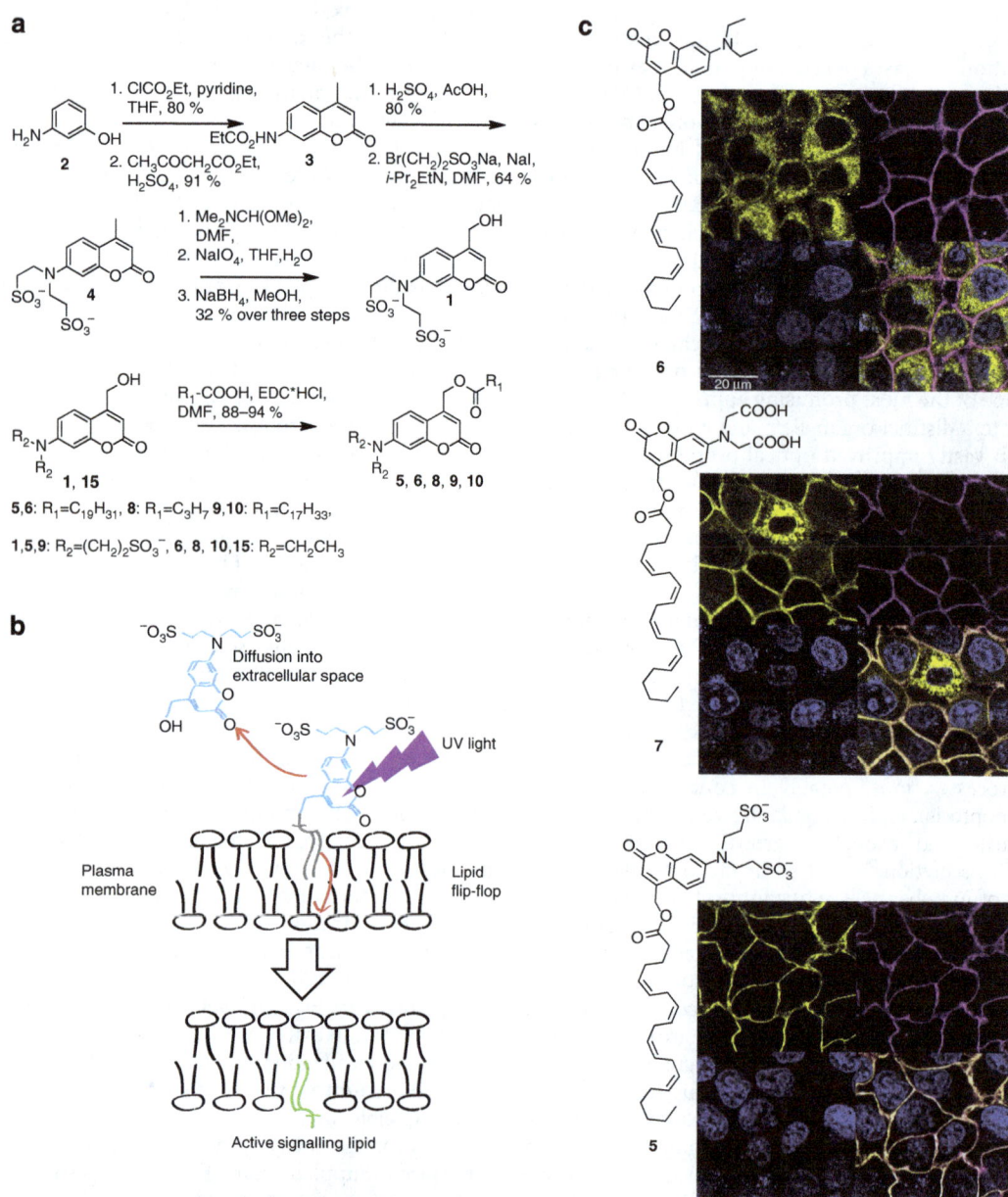

**Figure 1 | Synthesis and cellular localization of caged AA derivatives.** (**a**) Synthesis of the sulfonated coumarin alcohol **1** and caged FAs. (**b**) Schematic representation of plasma membrane-specific photoactivation of caged signalling lipids. (**c**) Cellular localization of AA derivatives **6** (top), **7** (middle) and **5** (bottom) with different coumarin cages (coumarin—yellow, plasma membrane marker—magenta, nuclear marker—blue). Images were acquired using identical acquisition settings directly after removing loading solutions. All images are presented at the same magnification; the scale bar indicates 20 μm.

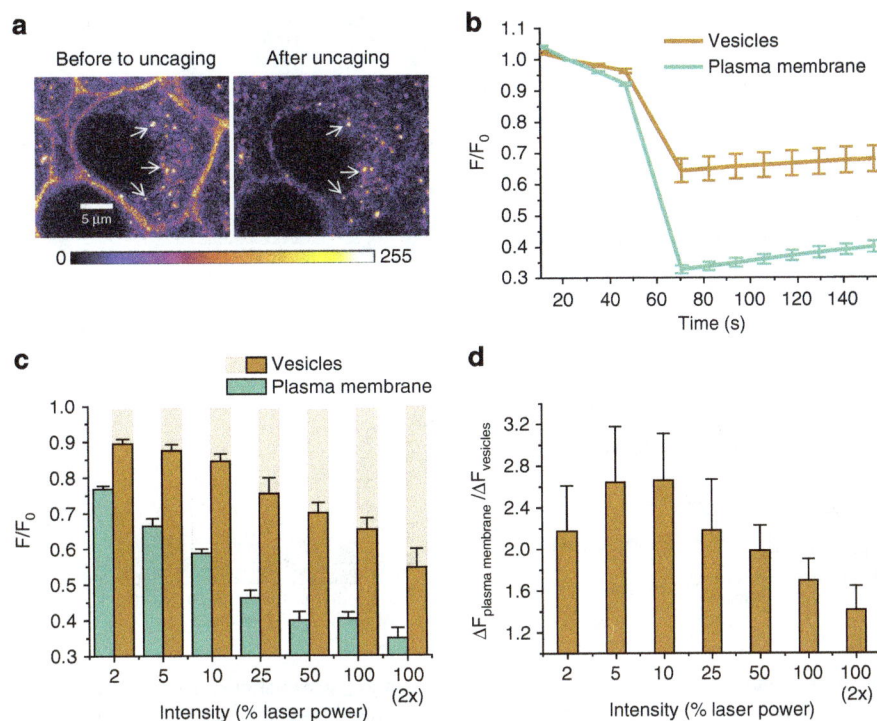

**Figure 2 | Quantifying photoreactions in living cells by uncaging of 5.** Cells were loaded with **5** and kept at 37 °C for 90–180 min after removal of the loading solution to ensure sufficient vesicle formation by endocytosis. (**a**) Distinctly different decreases of fluorescence intensity were observed at the plasma membrane and in vesicles upon photoactivation. The scale bar indicates 5 μm. (**b**) Quantification of normalized fluorescence intensity for the plasma membrane and vesicular structures over time for typical uncaging experiments (photoactivation settings: 1 scan, 50% 405 nm laser intensity, three individual experiments, error bars represent s.d.). (**c**) Remaining fluorescence intensity at the plasma membrane and in vesicular structures upon photoactivation with varied light intensity (error bars represent s.e.m., seven individual experiments). The shaded area represents the detected fluorescence intensity in vesicular structures before photoactivation; the observed difference is a measure for photobleaching. (**d**) Ratio of $\Delta F$ values determined for plasma membrane and vesicular fluorescence as a measure of photoreaction efficiency (errors derived by error propagation using the s.e.m. of the respective $\Delta F$ values, seven individual experiments).

photoreaction (Fig. 2d and Supplementary Fig. 8). In the example illustrated in Fig. 2, complete uncaging is reached between 25 and 50% laser intensity, whereas the highest efficiency is achieved between 5 and 10% laser intensity.

**Modulation of $Ca^{2+}$ oscillations depends on the site of AA photorelease.** We chose to study two processes known to involve AA signalling, namely lipid-induced potentiation of hippocampal mossy fibre (Mf) synaptic transmission and insulin secretion by pancreatic β-cells in order to demonstrate the potential of the method for establishing distinct roles of signalling lipids in different cellular compartments. AA was shown to be present in fairly high concentrations (50–75 μM) in pancreatic β-cells, to modulate calcium release, and as a result to trigger insulin secretion in β-cells[28,38,39]. Similarly, artificially reduced levels of AA led to impaired insulin secretion[40]. However, a conclusive model describing the interplay between AA concentration and insulin secretion has not been proposed so far and our understanding of its molecular targets and their intracellular localization remains incomplete. We chose the mouse insulinoma-derived cell line MIN6 as a model β-cell line to perform live-cell uncaging experiments. These cells exhibit characteristics of glucose metabolism and glucose-stimulated insulin secretion similar to those observed for islets[41]. When exposing MIN6 cells expressing the genetically encoded $Ca^{2+}$ sensor R-GECO[42] to a glucose concentration of 20 mM, we observed characteristic calcium oscillations, which subsequently triggered increased insulin release (Supplementary Fig. 9). Upon addition of free AA, the average calcium concentration decreased

markedly and calcium oscillations were significantly diminished (Supplementary Fig. 10a,b). On the contrary, the addition of compounds **5** and **6** did not have significant long-term effects on calcium oscillations, thereby establishing that both caging groups sufficiently mask the AA moiety to block its cellular signalling activity (Supplementary Fig. 10c,d). As observed for HeLa cells, incubation of MIN6 cells with **5** resulted in a pronounced plasma membrane stain, whereas **6** stained internal membranes in a homogenous manner (Supplementary Fig. 11). To study the effects of elevated AA levels at internal membranes versus the plasma membrane, we performed AA uncaging experiments under optimized conditions (Supplementary Fig. 8) with compounds **5** and **6** in 390 cells while monitoring changes in the intracellular $Ca^{2+}$ concentration using the $Ca^{2+}$ sensor R-GECO. As the analysed data set comprised a large cell population exhibiting widely diverse calcium oscillation patterns with regard to frequency, amplitude and duration of calcium events, we analysed the obtained data set by several independent approaches addressing the distribution of individual calcium events over the time course of all experiments. To this end, we employed a peak detection algorithm ('Package Peaks; R' see Supplementary Information for details) and identified 6,808 individual events, which were included in the analysis. By conducting an internal normalization, we used the relative height of each event with regard to the highest detected peak in each trace as a criterion to group all calcium transients into high- (intensity $\geq 60\%$ of highest peak) and low-intensity (intensity $< 60\%$ of highest peak) events. Uncaging of compounds **5** and **6** resulted in surprisingly different changes in the observed distribution of high-intensity calcium events (Supplementary

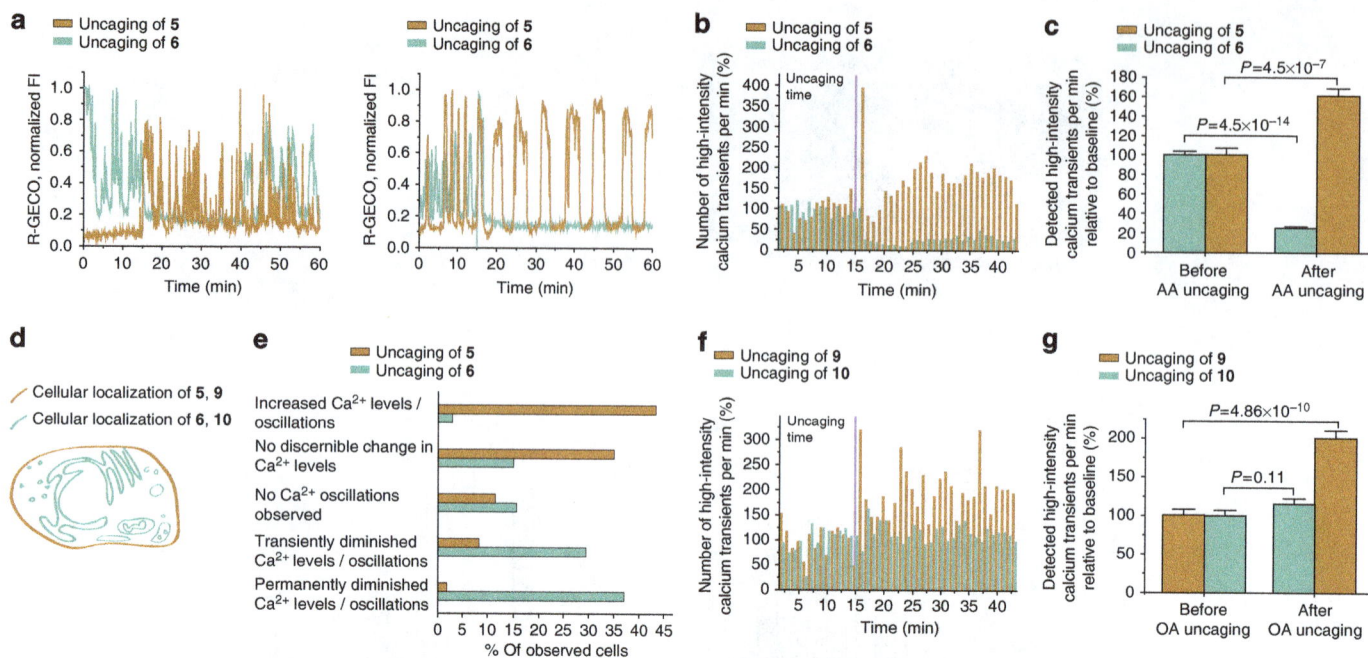

**Figure 3 | Compartment-dependent modulation of intracellular calcium dynamics in glucose-stimulated β-cells by fatty acids. (a)** Exemplary calcium traces after uncaging compounds **5** and **6**, respectively; left: initiated and transiently interrupted $Ca^{2+}$ oscillations; right: potentiated and terminated $Ca^{2+}$ oscillations. **(b)** Number of detected high-intensity calcium events within every 60 s interval before and after uncaging compounds **5** and **6**. **(c)** Distribution of high-intensity calcium events ($\geq 60\%$ of highest event in each trace) in ~390 cells over time before and after AA uncaging from compounds **5** or **6**. Note: a detailed discussion of scope and limitations of the data analysis approach may be found in the Supplementary Fig. 10. *P*-values were obtained by Welch two sample *t*-test. **(d)** Schematic representation of cellular localization of compounds **5** and **6** before uncaging. **(e)** Classification of cellular responses after uncaging AA. **(f)** Number of detected high-intensity calcium events within every 60 s interval before and after uncaging oleic acid (OA) from compounds **9** and **10**. **(g)** Distribution of high-intensity calcium events ($\geq 60\%$ of highest event in each trace) in 304 cells before and after uncaging **9** and **10**. Error bars represent s.e.m. throughout the figure.

Movies M2 and M3). Uncaging AA from **5** at the plasma membrane triggered an immediate fourfold increase in high-intensity calcium events. This acute effect was not observed after uncaging AA from **6** (Fig. 3b and Supplementary Fig. 12). In addition, high-intensity calcium events were detected with a permanently augmented rate after uncaging AA from **5**, whereas uncaging AA from **6** had the opposite effect, that is, blocked or reduced $Ca^{2+}$ oscillations (Fig. 3b,c, Supplementary Fig. 10 in the Supplementary Information). These findings imply that solely higher levels of AA at the plasma membrane as opposed to elevated concentrations at internal membranes are responsible for increased calcium signalling in β-cells.

In an effort to gain further insight into the nature of the observed effects, we classified the cells according to their response pattern and compared the distribution of cellular responses after liberating AA at the plasma membrane or internal membranes. This classification was performed both by manual assignment (blind expert assignment) and an automated approach based on a newly developed algorithm for analysing calcium oscillations (Supplementary Methods) with both methods yielding very similar results (Fig. 3e and Supplementary Fig. 13). After liberating AA at the plasma membrane by uncaging **5**, we observed prolonged increases of the average calcium concentration as well as permanently enhanced amplitudes in a significant fraction of the analysed cells. The duration of individual calcium events was often increased. Furthermore, calcium oscillations were initiated in a number of cells that did not exhibit an oscillatory behaviour before the uncaging event (Fig. 3a,e). On the contrary, AA release at internal membranes caused by uncaging of **6** typically resulted in transiently (typically 4–10 min) or permanently diminished calcium oscillations and lower average calcium levels after occasional initial calcium transients

(Fig. 3a,e). These findings indicate a dual role for AA signalling in β-cells. Although enhanced levels of AA at the plasma membrane trigger and potentiate calcium oscillations and thereby ultimately cause enhanced insulin release, opposing effects are caused by higher AA levels at internal membranes.

We performed a series of control experiments to rule out potential side effects caused by the respective photoreactions or the released coumarin alcohol species. We first established that neither addition nor photoactivation of the sulfonated coumarin alcohol **1** or photoactivation of 7-diethylamino caged butyrate (**8**) as a biologically inactive control compound changed the observed pattern of $Ca^{2+}$ oscillations in MIN6 cells (Supplementary Fig. 14a–c). It can thus be concluded that the observed effects were neither artefacts caused by the photoreactions nor due to release of coumarin alcohol or its further metabolites.

To assess whether the observed dual signalling role is a general response of MIN6 cells to elevated levels of unsaturated FAs at the plasma membrane versus internal membranes, we prepared caged OA derivatives bearing the novel sulfonated (**9**) or the neutral cage (**10**; Fig. 1a) and uncaged them in MIN6 cells. We found that the modulation of $Ca^{2+}$ oscillations triggered by uncaging of OA at the plasma membrane was similar to the effect observed for AA uncaging, namely an initial burst of high-intensity events directly after uncaging followed by long-lasting enhanced calcium signalling (Fig. 3f,g). Importantly, uncaging of OA at internal membranes did not have any noticeable effect on calcium oscillations (Fig. 3f,g).

The effects of higher AA and OA levels on the plasma membrane are well in line with previous studies[35,38,39] and are probably caused by direct interaction of AA and other FAs with a number of signalling effectors, most likely the pancreatic islet GPR40 (refs 32–36), which has been shown to be indiscriminately

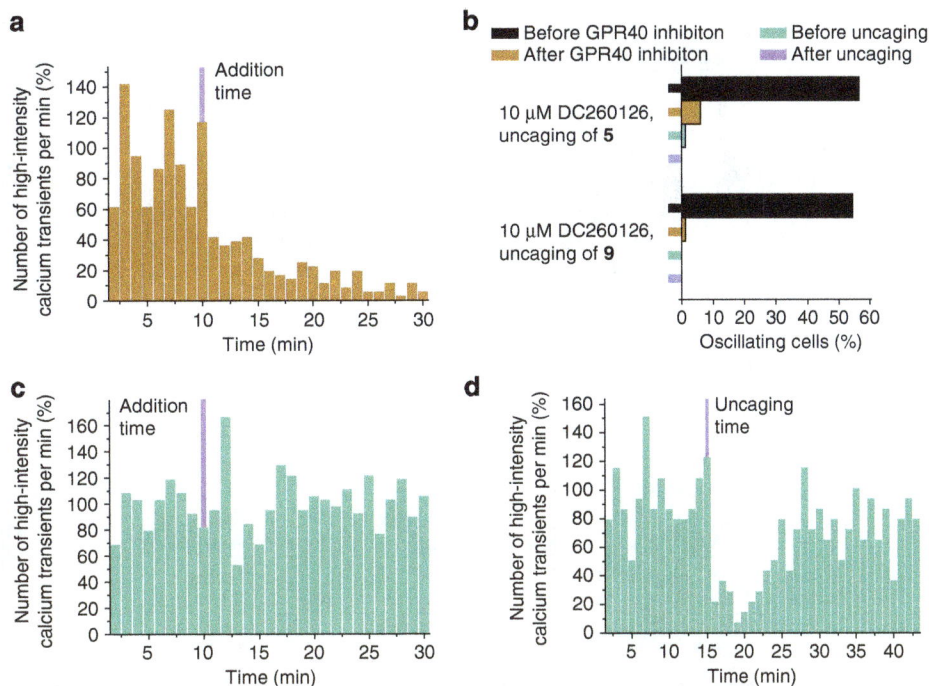

**Figure 4 | Effects of GPR40 and CB1 inhibition on intracellular calcium dynamics in glucose-stimulated β-cells.** (**a**) Number of detected high-intensity calcium events within every 60 s interval before and after addition of the specific GPR40 inhibitor DC260126 (10 μM) in 162 cells. (**b**) Percentage of oscillating cells before and after DC260126 addition as well as before and after subsequent uncaging of OA and AA at the plasma membrane using compounds **5** and **9**, respectively. Two time lapses were recorded in the same field of view with a 15-min delay to ensure a stable baseline after inhibitor treatment and before lipid uncaging. (**c**) Number of detected high-intensity calcium events within every 60 s interval before and after addition of the specific CB1 inhibitor AM251 (100 nM) in 141 cells. (**d**) Number of detected high-intensity calcium events within every interval of 60 s before and after uncaging AA at internal membranes (**6**) in cells pre-treated with 100 nM AM251 in 63 cells.

activated by different FAs[35]. After confirming GPR40 expression in MIN6 cells (Supplementary Fig. 15), we evaluated whether inhibition of GPR40 influenced Ca$^{2+}$ oscillations or the modulation of oscillation patterns caused by AA and OA uncaging at the plasma membrane. When applying the GPR40 inhibitors DC260126 (refs 43,44; 10 μM) or GM1100 (ref. 45; 10 μM), we found a gradual disappearance of Ca$^{2+}$ oscillations (Fig. 4a and Supplementary Fig. 14d). We next performed uncaging experiments of **5** or **9** in MIN6 cells after treatment with 10 μM DC260126. Manual classification of Ca$^{2+}$ traces by blind expert assignment revealed that AA or OA uncaging failed to restore Ca$^{2+}$ oscillations to pre-inhibition levels (Fig. 4b) or even augment oscillatory behaviour relative to the lower post-inhibition baseline. Taken together, these data confirm that GPR40 serves as one of the main plasma membrane receptors for FA-induced Ca$^{2+}$ signalling. In fact, it appears that a basal activation of GPR40 by the autocrine release of FAs is required for maintaining Ca$^{2+}$ oscillations in β-cells as addition of lipid-free bovine serum albumin led to diminished oscillatory behaviour (Supplementary Fig. 14e).

Alternatively, metabolites such as endocannabinoids could be essential players in driving calcium oscillations. We therefore inhibited the CB1 receptor by adding the specific inhibitor AM251 (100 nM), which did not alter oscillation patterns in any discernible way (Fig. 4c). Uncaging AA (**6**) on internal membranes of MIN6 cell in the presence of 100 nM AM251 (refs 46,47) also led to a transient reduction of Ca$^{2+}$ oscillations (Fig. 4d), indicating that acute reduction of high-intensity calcium events is probably not due to endocannabinoid signalling but rather a direct effect of higher AA concentrations. These findings indicate that elevated levels of unsaturated FAs at the plasma membrane constitute a general cue for enhanced, likely GPR40-mediated Ca$^{2+}$ signalling activity. When uncaged at

internal membranes, elevated levels of unsaturated FAs lead to different effects depending on the FA structure. AA caused downregulation of Ca$^{2+}$ oscillations, whereas another long-chain unsaturated FA (OA) had no significant effect on Ca$^{2+}$ oscillations.

**AA release at the plasma membrane potentiates synaptic transmission.** In an attempt to establish whether different roles for AA signalling at individual organelles are a more widespread phenomenon, we compared the efficiency of AA release from compounds **5** and **6** with regard to synaptic transmission in hippocampal brain slices prepared from C57B1/6 mice. Utilizing this experimental model allowed us to demonstrate the applicability of caged signalling lipids equipped with the novel sulfonated caging group in relatively thick (300 μm) tissues. We chose to study mossy fibre to CA3 synapses where the strength of synaptic transmission is tightly controlled by presynaptic Kv channels[48,49], a known sensitive target of AA[32]. We recently established that AA acts as a retrograde messenger at these synapses by broadening presynaptic action potentials mainly due to direct inactivation of presynaptic Kv channels at the plasma membrane[30]. In the current paper, we used compounds **5** and **6** to study the dynamic of this effect in greater detail. We performed whole-cell voltage clamp recordings of excitatory post synaptic currents (EPSCs) from visualized CA3 pyramidal cells by stimulating the granule cell axons (mossy fibres) to release glutamate from synaptic vesicles (Fig. 5).

As previously observed for **6** (ref. 30), uncaging of AA from **5** and **6** at 10 μM loading concentration induced robust and transient potentiation of synaptic transmission. Loading of both compounds was confirmed by fluorescence microscopy (Supplementary Fig. 16). Application of the uncaging protocol without compound loading had no effect on synaptic

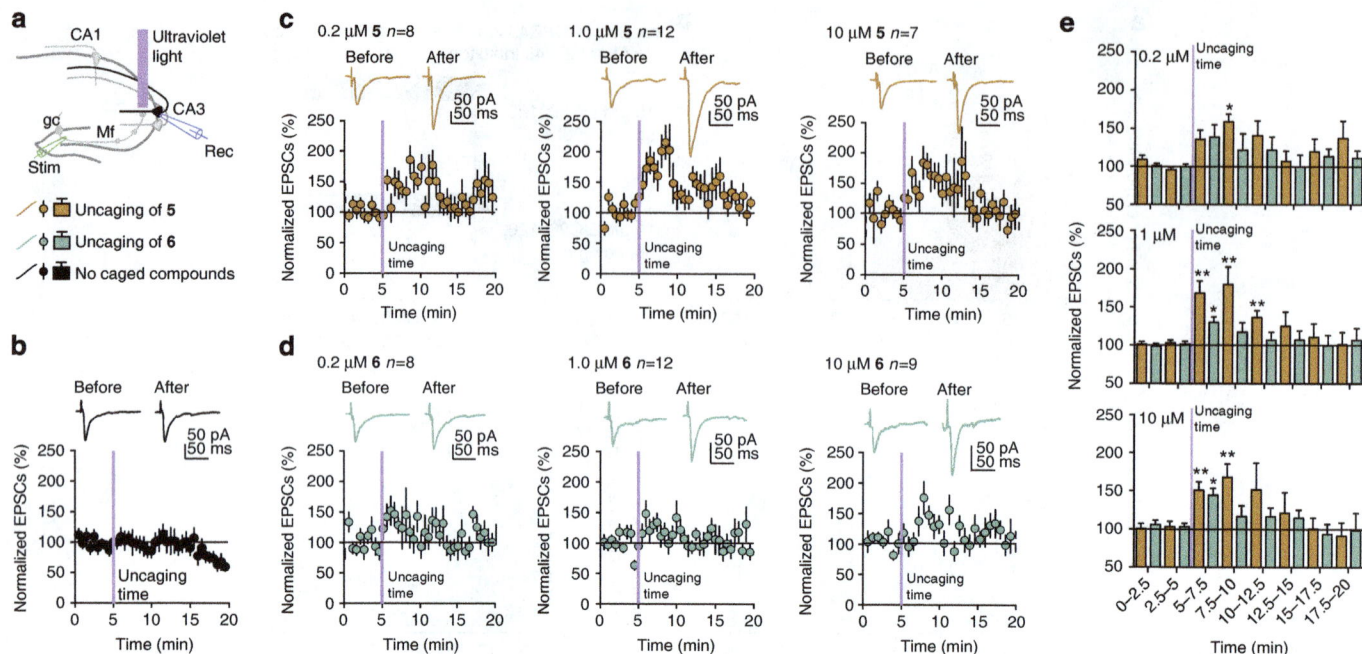

**Figure 5 | Potentiation of mossy fibre (Mf) synaptic transmission following photorelease of AA at the plasma membrane versus internal membranes.** (**a**) Schematic representation of a hippocampal slice with recording (blue, in a CA3 pyramidal cell) and stimulating electrodes (green, inside the dentate gyrus). Slices were perfused with compound **5** or **6** for 10–15 min before flashing ultraviolet light in the *stratum lucidum* near the recorded CA3 pyramidal cell. gc, granule cell; rec, recording (voltage clamp recording of excitatory post synaptic currents); stim, stimulation (stimulation of the granule cell axons). (**b**) Control experiment performed in the absence of caged compounds excluded photoartefacts on synaptic transmission. Upper panel: sample traces (average of 30 sweeps) of mossy fibre (Mf) EPSCs before and after uncaging. Lower panels: summary time course of normalized Mf-EPSCs. (**c**) Modulation of Mf-EPSCs after uncaging compound **5** (0.2, 1.0 or 10 μM). Upper panels: sample traces (average of 30 sweeps) of Mf-EPSCs before and after uncaging. Lower panels: summary time course of normalized Mf-EPSCs. (**d**) Modulation of Mf-EPSCs after uncaging compound **6** (0.2; 1.0 or 10 μM). Upper panels: sample traces (average of 30 sweeps) of Mf-EPSCs before and after uncaging. Lower panels: summary time course of normalized Mf-EPSCs. (**e**) Bar graphs illustrating the different time regimes of the effects illustrated in **c,d**. Error bars represent s.e.m. *P*-values were obtained by Wilcoxon matched pairs test. *$P<0.05$, **$P<0.01$.

transmission, indicating that illumination alone does not alter major electrophysiological properties (Fig. 5b). Importantly, the duration of this effect was markedly dependent on the specific nature of the membranes where AA photorelease took place. Although release of AA at the plasma membrane resulted in prolonged potentiation of synaptic transmission irrespective of the applied concentration (7.5–10 min, Fig. 5e), release at internal membranes only sparked a brief potentiation, which was potentially caused by diffusion of a minute amount of released AA towards its plasma membrane target (Fig. 5c–e). This is in line with the proposed mechanism for the action of AA as a retrograde messenger via Kv channels, located at the plasma membrane.

## Discussion

The number of approaches for organelle-specific localization of photocaged small signalling molecules is extremely limited[50,51] and no method for directly targeting the plasma membrane has been reported so far. We have developed a novel caging group that allows photorelease of hydrophobic signalling molecules exclusively at the outer leaflet of the plasma membrane. To the best of our knowledge, this is the first example of organelle-specific photorelease of cellular signalling lipids achieved by controlling the cellular localization of the caged compound before photoactivation. In addition, this sulfonated caging group is a suitable tool for quantifying the completeness and efficiency of photoreactions inside living cells, thereby greatly facilitating the setup and optimization of uncaging experiments and simultaneously reducing the need for a functional readout. We applied the synthesized caged AA derivatives to study AA

signalling with subcellular precision both in cell culture and tissue. Our results suggest that the initial cellular site of increased AA concentration plays a crucial role in determining signalling outcome in very different cellular settings despite the fact that AA is a readily diffusible small molecule. Our findings underline the intricate nature of the molecular mechanisms governing small-molecule signalling and also the growing need to develop more sophisticated tools for its study. Caged small molecules that can be released in an organelle-specific manner will likely play an important role both as discovery tools and by providing a straightforward approach to analysing metabolism and lifetimes of small-molecule messengers at distinct subcellular sites.

## Methods

**General synthetic procedures.** All chemicals were obtained from commercial sources (Acros, Sigma, Aldrich, Enzo, Lancaster or Merck) and were used without further purification. Solvents for flash chromatography were obtained from VWR and dry solvents were obtained from Sigma. Deuterated solvents were obtained from Deutero GmbH. All reactions were carried out using dry solvents under an inert atmosphere unless otherwise stated in the respective experimental procedure. Thin-layer chromatography was performed on precoated plates of silica gel (Merck, 60 F254) using ultraviolet light (254 or 366 nm) or a solution of phosphomolybdic acid in EtOH (10 g phosphomolybdic acid, in 100 ml EtOH) for analysis. Preparative column chromatography was performed using silica from Merck (silica 60, grain size 0.063–0.200 mm) with a pressure of 1–1.5 bar. HPLC was performed on a Knauer HPLC Smartline Pump 1000 using a Knauer Smartline UV Detector 2,500 instrument. A 250 mm × 10 mm LiChrospher 100 RP-18 column was used for semi-preparative HPLC applications. [1]H- and [13]C-NMR-spectra were obtained on a 400-MHz Bruker UltraShield spectrometer. Chemical shifts of [1]H- and [13]C-NMR-spectra are referenced indirectly to tetramethylsilane. *J* values are given in Hz and chemical shifts in p.p.m. Splitting patterns are designated as follows: s, singlet; d, doublet; t, triplet; q, quartet; m, multiple; b, broad. [13]C-NMR-spectra were broadband hydrogen decoupled. Mass spectra (ESI) were recorded using a Waters Micromass ZQ mass spectrometer or a HP Esquire-LC mass spectrometer.

High-resolution mass spectra (HRMS) were recorded at the University of Heidelberg.

**Brief description of conducted chemical syntheses.** *N*-Ethylcarbamoyl-3-aminophenol **11** was condensed with ethyl acetoacetate in the presence of sulfuric acid to afford coumarin precursor **3**, which after deprotection with 1:1 sulfuric acid/glacial acetic acid mixture was alkylated with sodium 2-bromoethanesulfonate in *N,N*-dimethylformamide. The bis-sulfonated, water-soluble 7-amino-4-methyl-coumarin derivative **4** was purified by reverse phase chromatography employing a triethylammonium bicarbonate buffer system and obtained as its bis-triethyl-ammonium salt. The 4-methyl group was converted to the alcohol using a previously reported protocol[52] in three steps. Esters were prepared from the alcohol in a carbodiimide-mediated reaction utilizing AA or OA to give compounds **5** and **9**, respectively. The target compounds were obtained as readily soluble triethylammonium salts from reverse phase chromatography. Additional neutral or carboxylated coumarin precursors were synthesized according to the literature protocols[53,54]. They were coupled to the respective carboxylic acids using carbodiimide-mediated reactions, which yielded coumarin esters **6**, **8**, **10** and **17**. The synthesis of **6** was published previously by our group[30]. The *t*-Bu protecting groups of **17** were removed by treatment with trifluoroacetic acid. A detailed description of the conducted syntheses as well as NMR and MS data of all new compounds may be found in the Supplementary Information.

**Photophysical characterization of new compounds.** Absorbance spectra were measured using a Cary 60 UV–vis spectrophotometer (Agilent Technologies) in Cary WinUV Scan Application (version 5.0.0.999). The detection range was set to 250–500 nm, the spectral resolution to 0.5 nm and the averaging time to 0.1 s. The path length of the cuvette was 1 cm, baseline correction was carried out by subtraction of the background signal of an EtOH sample and the compound absorbance maxima were below 0.1.

All emission spectra were measured using a Photon Technology International Fluorimeter in FeliX32 Analysis Application (version 1.2). The excitation wavelength was set to 360 nm and emission collection to 370–700 nm with a step size of 3 nm and 0.1 s integration. Spectra were recorded from EtOH solutions and averaged three times. Emission end excitation slits were set to 2 nm (0.5 mm W). Quantum yields (QYs) were calculated as following:

$$QY = QY_{ref} \frac{\eta^2}{\eta_{ref}^2} \frac{I}{A} \frac{A_{ref}}{I_{ref}}$$

where $QY_{ref}$ is the QY of the reference compound (coumarin 1, $QY = 0.73$ (ref. 55)), $\eta$ is the refractive index of the solvent (ethanol for all samples and the reference), $A$ is the absorbance at the excitation wavelength, $I$ is the integrated fluorescence intensity.

**Cell culture and cDNA transfection.** MIN6 cells were kindly provided by the Miyazaki laboratory. HeLa 'Kyoto' cells were obtained from the Kyoto University. The initial characterization of caged lipid performance in cells, loading and localization analyses as well as development and optimization of uncaging protocols and quantification of photoreactions were carried out in HeLa 'Kyoto' cells. Cells were grown in low-glucose DMEM (31885-023, Life Technologies) supplied with 10% fetal bovine serum (F7524, Sigma) and 100 µg ml$^{-1}$ antibiotic Primocin (ant-pm-1, Invitrogen). Cells were seeded in eight-well Lab-Tek microscope dishes 24–48 h (to reach 50–80% confluence) before imaging. The mouse insulinoma-derived cell line MIN6 used in this study as a model β-cell line was initially developed and kindly provided by Miyazaki *et al.*[56]. Cells were grown in high-glucose DMEM (41965-039, Life Technologies) supplied with 15% fetal bovine serum (10270098, Lifetechnologies), 100 U ml$^{-1}$ of antibiotics penicillin-streptomycin (15140122, Life Technologies) and 70 µM β-mercaptoethanol, always added freshly to the cell culture flasks (P07-05100, PAN-Biotech). Cells were seeded in eight-well Lab-Tek microscope dishes 48–64 h (to reach 50–80% confluence) before imaging. For calcium imaging, MIN6 cells were transfected with cDNA coding for the R-GECO Ca$^{2+}$ reporter and cDNA coding for the C1-GFP DAG sensor (as a control) usually 24–48 h after seeding. A transfection cocktail of 200 ng C1-GFP and 200 ng R-GECO in 10 µl of Opti-MEM (31985-070, Life Technologies) and 1.5 µl of Lipofectamine 2000 transfection reagent (11668030, Life Technologies) was added to each well of an eight-well Lab-Tek microscope dish loaded with 200 µl Opti-MEM (37 °C, immediately before the addition) 24–48 h prior imaging.

**Fluorescence microscopy.** Cells were imaged in eight-well Lab-Tek microscope dishes (155411, Thermo Scientific) at 37 °C in imaging buffer containing (mM): 20 HEPES, 115 NaCl, 1.2 CaCl$_2$, 1.2 MgCl$_2$, 1.2 K$_2$HPO$_4$. For HeLa cells, the imaging buffer additionally contained 10 mM glucose. MIN6 cells (if not specified otherwise) were imaged in an imaging buffer containing a stimulatory amount of glucose (20 mM).

Imaging was performed on a dual scanner confocal microscope Olympus Fluoview 1200, with × 20 (air) and/or × 63 (oil) objectives. This microscope houses two independent, fully synchronized laser scanners for simultaneous laser stimulation and confocal observation and permits capturing cellular responses that occur during or immediately following laser stimulation. Microscope settings were adjusted to generate images displaying background fluorescence values slightly larger than zero in order to capture the complete signal stemming from the respective fluorescent dyes or proteins. Coumarin dyes were excited with 405 nm laser and emitted light was collected between 425 and 525 nm. C1-GFP and R-GECO were excited with 488 and 559 nm lasers and emitted light was collected at 500–550 nm and 570–670 nm, respectively.

**Image analysis and data processing.** Images were analysed using Fiji (http://fiji.sc/Fiji) and the previously reported FluoQ macro[57] (for calcium imaging data) or the newly developed ImageJ macro 'PM/background ratio-calculator Macro' (see Supplementary Methods for description of ImageJ macros and the general pipeline used for analysis of coumarin imaging data). Raw images were loaded into Fiji and the respective macro was started to perform all image analysis steps. An R script based on the package 'peaks'[47] was used to automatically detect calcium transients in obtained single-cell traces. A detailed description of the performed image analysis steps, a discussion of the chosen parameters and the source code of all newly developed Fiji macros and R scripts may be found in the Supplementary Information.

**Electrophysiology.** Parasagittal hippocampal slices (320 µm thick) were obtained from 30- to 35-day-old C57Bl/6 mice. Slices were transferred to a recording chamber in which they were continuously superfused with an oxygenated extra-cellular medium (95% O$_2$ and 5% CO$_2$) containing (mM): 125 NaCl, 2.5 KCl, 2.3 CaCl$_2$, 1.3 MgCl$_2$, 1.25 NaH$_2$PO$_4$, 26 NaHCO$_3$, 20 glucose, pH 7.4. Whole-cell recordings were made at ∼25 °C from CA3 pyramidal cells under infrared differential interference contrast imaging using borosilicate glass capillaries, which had resistances between 4 and 8 MΩ. For voltage-clamp recordings from CA3 pyramidal cells, the patch electrodes were filled with a solution containing (mM): 140 CsCH$_3$SO$_3$, 2 MgCl$_2$, 4 NaCl, 5 phospho-creatine, 2 Na$_2$ATP, 0.2 EGTA, 10 HEPES, 0.33 GTP, pH 7.3 adjusted with CsOH. Bicuculline (10 µM) was present in the superfusate of all experiments. A patch pipette (open tip resistance ∼5 MΩ (about 1 µm tip diameter)) was placed in the dentate gyrus to stimulate Mfs. Mf synaptic currents were identified according to the following criteria: robust low-frequency facilitation, low-release probability at 0.1 Hz, rapid rise times of individual EPSCs (∼1 ms) and EPSC decays free of secondary peaks that may indicate the presence of polysynaptic contamination. Fresh aliquots of caged AA derivatives **5** and **6** were used for each experiment and were dissolved in extra-cellular medium. The slices were perfused with extracellular medium containing either **5** or **6** for at least 10–15 min before starting the experiments to ensure homogenous penetration of the caged compound into the slice (Supplementary Fig. 16). During the application of caged AA derivatives **5** and **6**, a total amount of 10–15 ml of extracellular solution containing the caged compound was con-tinuously re-circulated and oxygenated. AA was locally uncaged in the *stratum lucidum* of the patched CA3 pyramidal cell or the patched MfB by an ultraviolet flash photolysis (Xenon flash lamp; Rap OptoElectronic).

Statistical analysis was carried out as follows: Values are presented as mean ± s.e.m. of *n* experiments. Nonparametric tests were used for statistical analysis. Within cell comparisons were made using the Wilcoxon matched pairs test, which was applied to raw non-normalized data of baseline values and values obtained after applying the respective uncaging protocol. The Mann–Whitney test was used for comparison of two groups. Statistical differences were considered as significant at $P < 0.05$. Statistical analysis was performed using GraphPad prism software.

All animals were used according to the guidelines of the University of Bordeaux/CNRS Animal Care and Use Committee.

## References

1. Purvis, J. E. & Lahav, G. Encoding and decoding cellular information through signaling dynamics. *Cell* **152**, 945–956 (2013).
2. Marks, F., Klingmueller, U. & Mueller-Decker, K. *Cellular Signal Processing: an Introduction to the Molecular Mechanisms of Signal Transduction* (Garland Science, 2009).
3. Ouyang, X. & Chen, J. K. Synthetic strategies for studying embryonic development. *ChemBiol.* **17**, 590–606 (2010).
4. Putyrski, M. & Schultz, C. Protein translocation as a tool: The current rapamycin story. *FEBS Lett.* **586**, 2097–2105 (2012).
5. Rutkowska, A. & Schultz, C. Protein tango: the toolbox to capture interacting partners. *Angew. Chem. Int. Ed. Engl.* **51**, 8166–8176 (2012).
6. Fegan, A., White, B., Carlson, J. C. T. & Wagner, C. R. Chemically controlled protein assembly: techniques and applications. *Chem. Rev.* **110**, 3315–3336 (2010).
7. Pastrana, E. Optogenetics: controlling cell function with light. *Nat. Methods* **8**, 24–25 (2011).
8. Kramer, R. H., Mourot, A. & Adesnik, H. Optogenetic pharmacology for control of native neuronal signaling proteins. *Nat. Neurosci.* **16**, 816–823 (2013).
9. Kao, J. P. Caged molecules: principles and practical considerations. *Curr. Protoc. Neurosci.* Chapter 6, Unit 6 20 (2006).

10. Gautier, A. *et al.* How to control proteins with light in living systems. *Nat. Chem. Biol.* **10**, 533–541 (2014).

11. Nadler, A. *et al.* The fatty acid composition of diacylglycerols determines local signaling patterns. *Angew. Chem. Int. Ed. Engl.* **52**, 6330–6334 (2013).

12. Artamonov, M. V. *et al.* Agonist-induced $Ca^{2+}$ sensitization in smooth muscle: redundancy of Rho guanine nucleotide exchange factors (RhoGEFs) and response kinetics, a caged compound study. *J. Biol. Chem.* **288**, 34030–34040 (2013).

13. Ellis-Davies, G. C. A practical guide to the synthesis of dinitroindolinyl-caged neurotransmitters. *Nat. Protoc.* **6**, 314–326 (2011).

14. Hoglinger, D., Nadler, A. & Schultz, C. Caged lipids as tools for investigating cellular signaling. *Biophys. Biochim. Acta* **1841**, 1085–1096 (2014).

15. Huang, X. P., Sreekumar, R., Patel, J. R. & Walker, J. W. Response of cardiac myocytes to a ramp increase of diacylglycerol generated by photolysis of a novel caged diacylglycerol. *Biophys. J.* **70**, 2448–2457 (1996).

16. Thompson, S. M. *et al.* Flashy science: controlling neural function with light. *J. Neurosci.* **25**, 10358–10365 (2005).

17. Neveu, P. *et al.* A caged retinoic acid for one- and two-photon excitation in zebrafish embryos. *Angew. Chem. Int. Ed. Engl.* **47**, 3744–3746 (2008).

18. Feng, S. H. *et al.* A rapidly reversible chemical dimerizer system to study lipid signaling in living cells. *Angew. Chem. Int. Ed. Engl.* **53**, 6720–6723 (2014).

19. Ahmed, S., Xie, J., Horne, D. & Williams, J. C. Photocleavable dimerizer for the rapid reversal of molecular trap antagonists. *J. Biol. Chem.* **289**, 4546–4552 (2014).

20. Zimmermann, M. *et al.* Cell-permeant and photocleavable chemical inducer of dimerization. *Angew. Chem. Int. Ed. Engl.* **53**, 4717–4720 (2014).

21. Liu, P. *et al.* A bioorthogonal small-molecule-switch system for controlling protein function in live cells. *Angew. Chem. Int. Ed. Engl.* **53**, 10049–10055 (2014).

22. Tischer, D. & Weiner, O. D. Illuminating cell signalling with optogenetic tools. *Nat. Rev. Mol. Cell Biol.* **15**, 551–558 (2014).

23. Matsuzaki, M. & Kasai, H. Two-photon uncaging microscopy. *Cold Spring Harb. Protoc.* **2011**, pdb prot5620 (2011).

24. Olson, J. P. *et al.* Optically selective two-photon uncaging of glutamate at 900 nm. *J. Am. Chem. Soc.* **135**, 5954–5957 (2013).

25. Fournier, L. *et al.* Coumarinylmethyl caging groups with redshifted absorption. *Chemistry* **19**, 17494–17507 (2013).

26. Brieke, C., Rohrbach, F., Gottschalk, A., Mayer, G. & Heckel, A. Light-controlled tools. *Angew. Chem. Int. Ed. Engl.* **51**, 8446–8476 (2012).

27. Banala, S., Maurel, D., Manley, S. & Johnsson, K. A caged, localizable rhodamine derivative for superresolution microscopy. *ACS Chem. Biol.* **7**, 288–292 (2012).

28. Brash, A. R. Arachidonic acid as a bioactive molecule. *J. Clin. Invest.* **107**, 1339–1345 (2001).

29. Jacobson, D. A., Weber, C. R., Bao, S. Z., Turk, J. & Philipson, L. H. Modulation of the pancreatic islet beta-cell-delayed rectifier potassium channel Kv2.1 by the polyunsaturated fatty acid arachidonate. *J. Biol. Chem.* **282**, 7442–7449 (2007).

30. Carta, M. *et al.* Membrane lipids tune synaptic transmission by direct modulation of presynaptic potassium channels. *Neuron* **81**, 787–799 (2014).

31. Alexander, L. D. & Hamzeh, M. T. Arachidonic acid-induced apoptosis in human renal proximal tubular epithelial cells. *FASEB J.* **27**, 727.12 (2013).

32. Meves, H. Arachidonic acid and ion channels: an update. *Br. J. Pharmacol.* **155**, 4–16 (2008).

33. Oliver, D. *et al.* Functional conversion between A-type and delayed rectifier $K^{+}$ channels by membrane lipids. *Science* **304**, 265–270 (2004).

34. Connell, E. *et al.* Mechanism of arachidonic acid action on syntaxin-Munc18. *EMBO. Rep.* **8**, 414–419 (2007).

35. Itoh, Y. *et al.* Free fatty acids regulate insulin secretion from pancreatic beta cells through GPR40. *Nature* **422**, 173–176 (2003).

36. O'Flaherty, J. T., Chadwell, B. A., Kearns, M. W., Sergeant, S. & Daniel, L. W. Protein kinases C translocation responses to low concentrations of arachidonic acid. *J. Biol. Chem.* **276**, 24743–24750 (2001).

37. Bennett, W. F. D. & Tieleman, D. P. Molecular simulation of rapid translocation of cholesterol, diacylglycerol, and ceramide in model raft and nonraft membranes. *J. Lipid Res.* **53**, 421–429 (2012).

38. Ramanadham, S., Gross, R. & Turk, J. Arachidonic-acid induces an increase in the cytosolic calcium-concentration in single pancreatic-islet beta-cells. *Biochem. Biophys. Res. Commun.* **184**, 647–653 (1992).

39. Yeung-Yam-Wah, V., Lee, A. K. & Tse, A. Arachidonic acid mobilizes $Ca^{2+}$ from the endoplasmic reticulum and an acidic store in rat pancreatic beta cells. *Cell Calcium* **51**, 140–148 (2012).

40. Ramanadham, S., Gross, R. W., Han, X. L. & Turk, J. Inhibition of arachidonate release by secretagogue-stimulated pancreatic-islets suppresses both insulin-secretion and the rise in beta-cell cytosolic calcium-ion concentration. *Biochemistry* **32**, 337–346 (1993).

41. Ishihara, H. *et al.* Pancreatic beta-cell line min6 exhibits characteristics of glucose-metabolism and glucose-stimulated insulin-secretion similar to those of normal islets. *Diabetologia* **36**, 1139–1145 (1993).

42. Zhao, Y. *et al.* An expanded palette of genetically encoded Ca(2)( + ) indicators. *Science* **333**, 1888–1891 (2011).

43. Hu, H. *et al.* A novel class of antagonists for the FFAs receptor GPR40. *Biochem. Biophys. Res. Commun.* **390**, 557–563 (2009).

44. Wu, J. *et al.* Inhibition of GPR40 protects MIN6 beta cells from palmitate-induced ER stress and apoptosis. *J. Cell. Biochem.* **113**, 1152–1158 (2012).

45. Briscoe, C. P. *et al.* Pharmacological regulation of insulin secretion in MIN6 cells through the fatty acid receptor GPR40: identification of agonist and antagonist small molecules. *Br. J. Pharmacol.* **148**, 619–628 (2006).

46. Lipina, C., Rastedt, W., Irving, A. J. & Hundal, H. S. Endocannabinoids in obesity:brewing up the perfect metabolicstorm? *WIREs Membr. Transp. Signal* **2**, 49–63 (2013).

47. Eckardt, K. *et al.* Cannabinoid type 1 receptors in human skeletal muscle cells participate in the negative crosstalk between fat and muscle. *Diabetologia* **52**, 664–674 (2009).

48. Alle, H., Kubota, H. & Geiger, J. R. P. Sparse but highly efficient K(v)3 outpace BKCa channels in action potential repolarization at hippocampal mossy fiber boutons. *J. Neurosci.* **31**, 8001–8012 (2011).

49. Geiger, J. R. P. & Jonas, P. Dynamic control of presynaptic ca2 + inflow by fast-inactivating $K^{+}$ channels in hippocampal mossy fiber boutons. *Neuron* **28**, 927–939 (2000).

50. Horinouchi, T., Nakagawa, H., Suzuki, T., Fukuhara, K. & Miyata, N. A novel mitochondria-localizing nitrobenzene derivative as a donor for photo-uncaging of nitric oxide. *Bioorg. Med. Chem. Lett.* **21**, 2000–2002 (2011).

51. Leonidova, A. *et al.* Photo-induced uncaging of a specific Re(I) organometallic complex in living cells. *Chem. Sci.* **5**, 4044–4056 (2014).

52. Riesgo, E. C., Jin, X. & Thummel, R. P. Introduction of benzo[h]quinoline and 1,10-phenanthroline subunits by friedlander methodology. *J. Org. Chem.* **61**, 3017–3022 (1996).

53. Ito, K. & Nakajima, K. Selenium dioxide oxidation of alkylcoumarins and related methyl-substituted heteroaromatics. *J. Heterocycl. Chem.* **25**, 511–515 (1988).

54. Hagen, V. *et al.* Coumarinylmethyl esters for ultrafast release of high concentrations of cyclic nucleotides upon one- and two-photon photolysis. *Angew. Chem. Int. Ed. Engl.* **44**, 7887–7891 (2005).

55. Jones, G., Jackson, W. R., Choi, C. & Bergmark, W. R. Solvent effects on emission yield and lifetime for coumarin laser-dyes - requirements for a rotatory decay mechanism. *J. Phys. Chem.* **89**, 294–300 (1985).

56. Miyazaki, J. I. *et al.* Establishment of a pancreatic beta-cell line that retains glucose-inducible insulin-secretion - special reference to expression of glucose transporter isoforms. *Endocrinology* **127**, 126–132 (1990).

57. Stein, F., Kress, M., Reither, S., Piljic, A. & Schultz, C. FluoQ: a tool for rapid analysis of multiparameter fluorescence imaging data applied to oscillatory events. *ACS Chem. Biol.* **8**, 1862–1868 (2013).

## Acknowledgements

We thank the Advanced Light Microscopy Facility (ALMF) of European Molecular Biology Laboratory (EMBL) for expert support in microscopy. We thank the Miyazaki laboratory for kindly providing MIN6 cells and sharing information with regard to MIN6 cell culture. We thank Sébastien Marais at the Bordeaux Imaging Center, part of the FranceBioImaging national infrastructure, for support in two photon microscopy. This work was supported by the DFG (TRR83). D.A.Y. is a fellow of the EMBL Interdisciplinary Postdoc Programme (EIPOD).

## Author contributions

C.S., A.N., D.A.Y. and S.F. conceived the project. C.S., A.N., D.A.Y., M.C. and C.M. designed the experiments. R.M., A.N. and S.F. synthesized caged compounds. D.A.Y. and R.M. examined photophysical properties of new compounds. A.N., D.A.Y., S.F. and M.C. participated in *in vitro* and *ex vivo* experiments. F.S. programmed Fiji macros and R scripts. F.S., A.N. and D.A.Y. analysed imaging data. M.C. and C.M. analysed electrophysiological data. A.N., D.A.Y. and C.S. wrote the manuscript. All authors edited and approved the final manuscript.

## Additional information

# 9

# Lysosome triggered near-infrared fluorescence imaging of cellular trafficking processes in real time

Marco Grossi[1], Marina Morgunova[1], Shane Cheung[1,2], Dimitri Scholz[3], Emer Conroy[3], Marta Terrile[3], Angela Panarella[4], Jeremy C. Simpson[4], William M. Gallagher[3] & Donal F. O'Shea[1,2]

Bioresponsive NIR-fluorophores offer the possibility for continual visualization of dynamic cellular processes with added potential for direct translation to *in vivo* imaging. Here we show the design, synthesis and lysosome-responsive emission properties of a new NIR fluorophore. The NIR fluorescent probe design differs from typical amine functionalized lysosomotropic stains with off/on fluorescence switching controlled by a reversible phenol/phenolate interconversion. Emission from the probe is shown to be highly selective for the lysosomes in co-imaging experiments using a HeLa cell line expressing the lysosomal-associated membrane protein 1 fused to green fluorescent protein. The responsive probe is capable of real-time continuous imaging of fundamental cellular processes such as endocytosis, lysosomal trafficking and efflux in 3D and 4D. The advantage of the NIR emission allows for direct translation to *in vivo* tumour imaging, which is successfully demonstrated using an MDA-MB-231 subcutaneous tumour model. This bioresponsive NIR fluorophore offers significant potential for use in live cellular and *in vivo* imaging, for which currently there is a deficit of suitable molecular fluorescent tools.

[1] Department of Pharmaceutical and Medicinal Chemistry, Royal College of Surgeons in Ireland, 123 St Stephen's Green, Dublin 2, Ireland. [2] School of Chemistry and Chemical Biology, Conway Institute, University College Dublin, Belfield, Dublin 4, Ireland. [3] School of Biomolecular and Biomedical Science, Conway Institute of Biomolecular and Biomedical Research, University College Dublin, Belfield, Dublin 4, Ireland. [4] School of Biology and Environmental Science, Conway Institute of Biomolecular and Biomedical Research, University College Dublin, Belfield, Dublin 4, Ireland. Correspondence and requests for materials should be addressed to D.F.O. (email: donalfoshea@rcsi.ie).

Ehrlich's use of synthetic dyes as a means of staining biological samples can be viewed as one of the foundation stones of modern scientific research. A century later, the use of fluorescence imaging as a technique to visualize specific regions of live cellular[1–4] or whole organisms[5,6] is often central to research programmes, with clinical applications such as fluorescence-guided surgery now emerging[7–11].

The major shortcomings of fluorescence imaging using molecular fluorophores are interference from nonspecific background fluorescence outside the region of interest (ROI), insufficient photostability and cytotoxicity. Poor ROI selectivity necessitates a time delay to allow background fluorophore clearance and/or a washing procedure between fluorophore administration and image acquisition. This can limit imaging to fixed cells or static snapshots, without the possibility of continuous data acquisition throughout the experiment. An innovative approach to enhance target-to-background signal ratio is to exploit a mechanism of selective fluorescence quenching in the background areas, while establishing the emitting potential of the fluorophore only in the ROI[12,13]. Continuous recording of dynamic cellular events in real time may become feasible if the on/off fluorescence switching is reversible.

Developing a responsive fluorophore suitable for real-time live-cell imaging poses a series of challenges. Stringent criteria are required, such as near-perfect response selectivity, exceptional photostability and low dark and light toxicities. Obtaining selective fluorescence responses for intracellular analytes is not trivial, as analyte selectivity observed in a controlled homogeneous environment of a cuvette does not necessarily translate to far more complex in vitro or in vivo settings. Continuous live-cell imaging places very high demands on photostability of the fluorophore, as the same cell(s) are repeatedly imaged over time. Fluorophore dark toxicity must be low, so that cell viability is not compromised and normal cellular processes are unperturbed. To minimize light-induced toxicity, it is preferable to use low-energy wavelengths in the near-infrared (NIR) spectral region ($\lambda = 700$–900 nm). For in vivo imaging, the use of NIR fluorophores is essential. This spectral region is required for effective light transmission through body tissue, as there are reduced levels of absorption and scattering at these longer wavelengths and less intrinsic autofluorescence. In addition, if on/off NIR fluorescence switching could be accomplished in vivo, then similar imaging advantages could be gained as for in-vitro cell imaging.

Currently, there is a small yet growing selection of NIR fluorophore classes but they often suffer from insufficient photostability and lack emission wavelengths above 700 nm[14]. Our recent research focus led to the development of BF$_2$-chelated azadipyrromethene class 1 (Fig. 1)[15–18]. This class is relatively straightforward to synthesize, amenable to structural elaboration and exhibits excellent photophysical properties. For example, the derivative 1 (R = Ph) has an absorption/emission $\lambda_{max}$ at 696 and 727 nm in aqueous solutions, high fluorescence quantum yields (0.3–0.4) and excellent photostability[17]. Yet, in spite of recent progress, a significant need remains for new, more sophisticated intracellularly responsive molecular NIR fluorophores, which can be used to visualize dynamic cellular processes in real time with the potential for in vivo translation.

The goal of our current work was to develop an NIR fluorophore capable of a lysosomal-induced off-to-on fluorescence response, thereby permitting real-time imaging of cellular uptake, trafficking and efflux without perturbing function[19]. Endocytosis, the process through which cells internalize biomolecules, is common to all cells and represents a crucial area of research interest due to the numerous associated biological processes[20,21]. The participating organelles at each stage in the endocytosis pathway maintain a unique intravesicular/localized pH, to provide appropriate conditions for specific biochemical processes. Although the extracellular and cytosolic regions are at pH ~7.2, the lysosomes are significantly more acidic. Along the endocytic pathway, the pH lowers from ~6.3 in early endosomes through ~5.5 in late endosomes, down to ~4.5 in lysosomes (Fig. 2)[22]. As such, a difference of almost three orders of magnitude in proton concentration exists between the lysosome interior and the outside of a cell, which is sufficient to establish a selective trigger for fluorescence switching[23–26]. However, a major additional response selectivity challenge still remains, in that pH-responsive molecular fluorophores can also be responsive to micro-environmental polarity, which can compromise their use in cellular experiments (vide infra).

Our novel lysosomal responsive probe design is illustrated in Fig. 1 in which functionalization of the fluorophore core (orange box) with an ortho-nitro phenolic group was chosen to impart the pH-responsive feature of the probe. It would be expected that the electron withdrawing o-nitro group would result in the ionized phenolate dominating at pH 7.2, resulting in fluorescence quenching due to a non-emissive intramolecular charge transfer excited state (Fig. 1, grey box). Following cellular uptake via endocytosis and compartmentalization into acidic organelles such as lysosomes, protonation would occur giving the neutral phenol species and the NIR emission signal would be established (Fig. 1, red box). This approach is a significant departure from other lysosomal stains, which rely on an amine protonation to form a positively charged ammonium salt to concentrate the fluorophore in the acidic compartments[19,22]. An important additional design feature includes a covalently linked polyethylene glycol (PEG) polymer to provide aqueous solubility and promote cellular uptake via endocytic pathways (Fig. 1, blue box)[27].

## Results

**Synthesis and photophysics.** The starting point of the synthesis was the previously reported BF$_2$-chelated bis-phenol azadipyrromethene 3, accessible in three synthetic steps from 1-(4-hydroxyphenyl)-3-phenylpropenone (Fig. 1)[28]. Monoalkylation of 3 was achieved to produce 4 by reaction with t-butyl bromoacetate and CsF in dimethylsulfoxide (DMSO) at 30 °C. After isolation, compound 4 was then subjected to ortho-nitration of the remaining phenol ring with KHSO$_4$/KNO$_3$ to provide 5. Next, hydrolysis of the t-butyl ester of 5 with trifluoroacetic acid (TFA) gave the carboxylic acid 6, which was converted into its activated ester 7 by reaction with N-hydroxysuccinimide and N-(3-dimethylaminopropyl)-N'-ethylcarbodiimide in DMSO. Formation of the activated ester was monitored by $^1$H NMR via the diagnostic CH$_2$ peaks at 5.48 for 7 and 4.83 p.p.m. for carboxylic acid 6, which showed complete conversion within 2 h (Supplementary Fig. 1). Conjugation of 7 in DMSO with a terminal amine functionalized PEG polymer (average molecular weight of 4,900) was effective, with the final fluorophore 2 obtained in high yield (Fig. 1). Matrix-assisted laser desorption/ionization–time of flight (MALDI–TOF) analysis of 2 showed the expected molecular weight centred at 5,410 Da, indicating that the covalent linkage was effective. Furthermore, $^1$H NMR was consistent with the product structure and analytical high-performance liquid chromatography (HPLC) showed a single peak for 2 with retention time differing from that of both the acid 6 and ester 7 (Supplementary Fig. 2).

Comparative absorption and fluorescence emission spectra were recorded for the organic soluble fluorophore 5 in chloroform and aqueous soluble 2 in phenol red-free imaging DMEM medium adjusted to pH 2 (Fig. 3a). Only small differences were

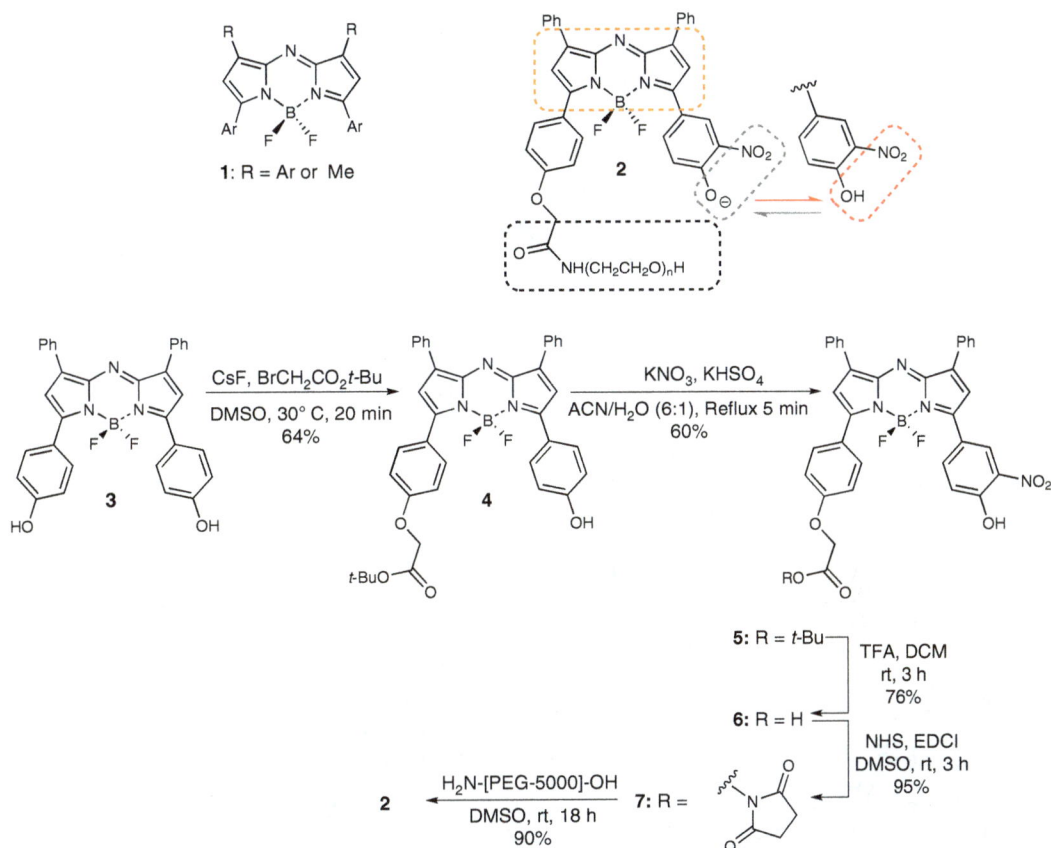

**Figure 1 | BF₂-azadipyrromethene NIR fluorophores.** General structure of BF₂-azadipyrromethenes **1**. Design and synthesis of lysosomal responsive BF₂-azadipyrromethene NIR fluorophore **2**.

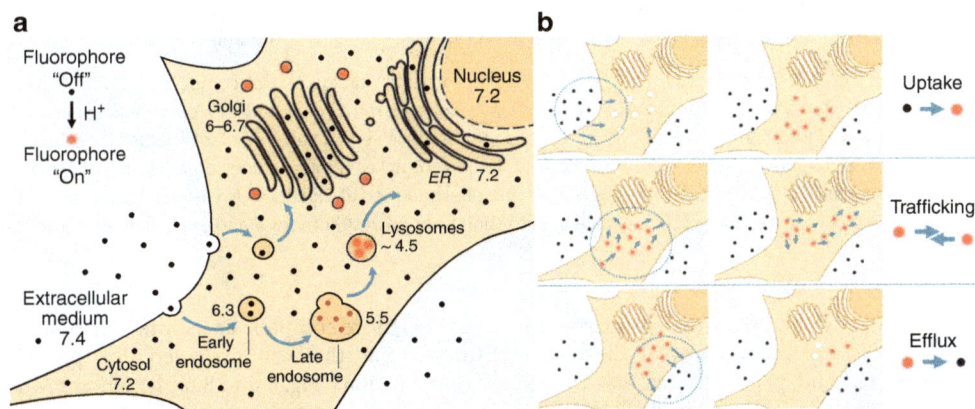

**Figure 2 | Cellular uptake responsive NIR-fluorophore.** (**a**) Simplified endocytosis of a responsive NIR fluorophore. Numbers represent the approximate pH of the corresponding organelles. (**b**) Three observable stages of the path of the pH-responsive fluorophore in the cellular environment: uptake, trafficking and efflux.

observed between the two fluorophores in the differing organic and aqueous media. Encouragingly, probe **2** had fluorescence $\lambda_{max}$ at 707 nm with an absorbance $\lambda_{max}$ at 685 nm. Extinction coefficient and fluorescence quantum yield values for **5** and **2** were similar with polyethylene glycol-substituted **2** having values of 97,000 cm$^{-1}$ M$^{-1}$ and 0.18, respectively (Fig. 3a).

An undesirable feature of some pH-sensitive fluorophores is their strong sensitivity to micro-environmental polarity, which significantly compromises their use in biological settings[29–31]. To test the polarity sensitivity of **2**, its acid/base emission-responsive properties were recorded in toluene, tetrahydrofuran, dimethylformamide and DMSO for both the phenol and phenolate state using 1,8-diazabicyclo[5.4.0]undec-7-ene (DBU)

and TFA to cycle between the two (Fig. 3b). A plot of solvent polarity function $(\Delta f)^{32}$ versus integrated fluorescence intensities in the off states showed highly effective fluorescence quenching as the phenolate irrespective of solvent polarity. A strong fluorescence output was established once protonated to the phenol in all solvents (Fig. 3c). These results predict that the modulation of fluorescence intensity would be selective for pH changes, while remaining unresponsive to differing intracellular micro-environmental polarities, thereby removing the potential for false-positive emissions. An identical study was carried out for fluorophore **5**, giving similar results and indicating that this positive feature is general to the fluorophore class (Supplementary Fig. 3).

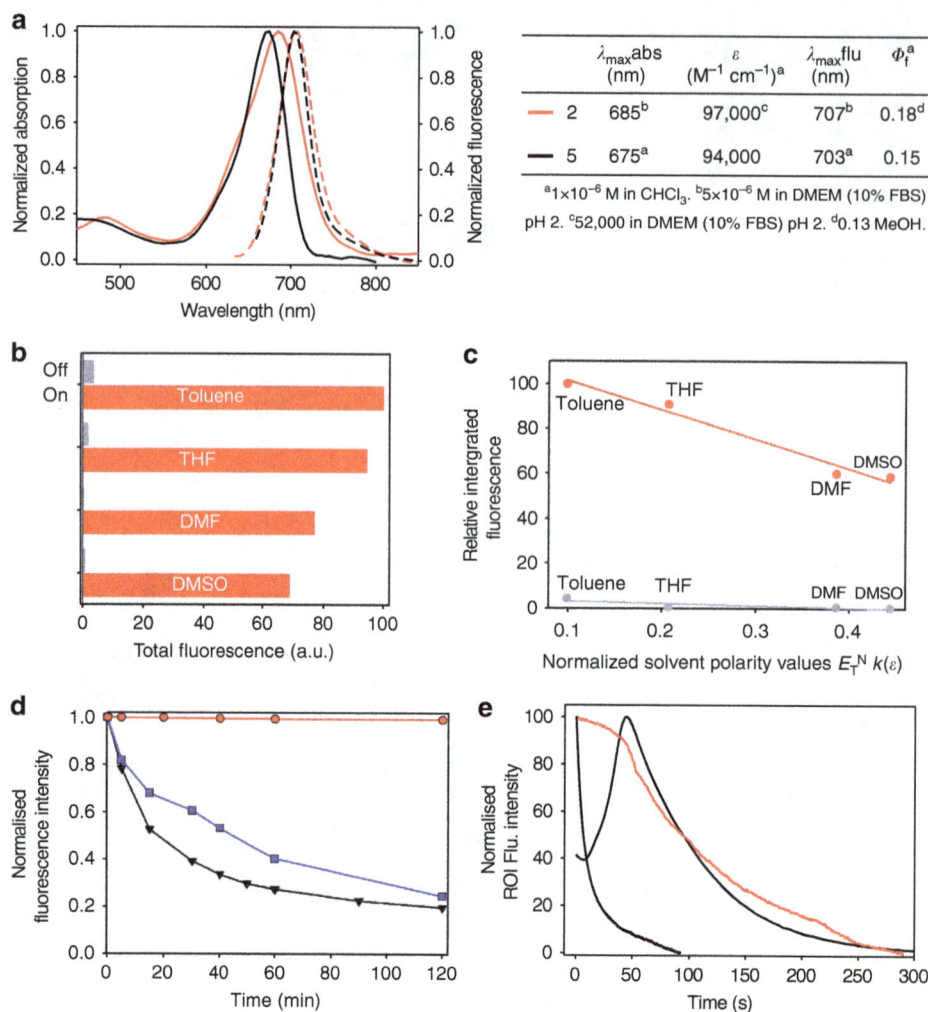

**Figure 3 | Photophysical properties of NIR-fluorophores.** (**a**) Light absorption and emission spectra of compounds **2** and **5**, and their photo-physical parameters. (**b**) Integrated off and on fluorescence states of **2** ($5 \times 10^{-6}$ M) in toluene, tetrahydrofuran (THF), dimethylformamide (DMF) and DMSO with TFA (red bars) and DBU (grey bars). (**c**) Plot of relative off and on integrated fluorescence versus solvent polarity values for toluene, THF, DMF and DMSO. (**d**) Comparative photobleaching of $1 \times 10^{-7}$ M DMEM solutions of **2** (red line), lysotracker red (blue line) and pHrodo red (black line) with 150 W fibre optic delivered light 620(30) nm for **2** and 540(40) nm for lysotracker red and pH-rhodo red at 25 °C. (**e**) In vitro photobleaching of **2** (red), lysotracker red (blue line) and pHrodo red (black line) with maximum LED power using excitation filter 640(14) nm for **2** and excitation filter 563(9) nm for lysotracker red and pH-rhodo red.

As sufficient photostability is an essential property for prolonged live-cell imaging, a comparative study of the photo-degradation of **2**, lysotracker red and pH-rhodo red was carried out. DMEM solutions of the three fluorophores were illuminated with light of 620(30) nm for **2** and 540(40) nm for lysotracker red and pH-rhodo red for 2 h, and their fluorescence intensity monitored. Encouragingly, no photobleaching for **2** was observed, whereas both other fluorophores were ~80% degraded within that time frame (Fig. 3d). Comparison of their stabilities in HeLa Kyoto cells using illumination from a solid-state light emitting diode (LED) light source was also examined. Cells stained with **2**, lysotracker red or pH-rhodo red were constantly illuminated with LED power set to a maximum, to promote a fast rate of photobleaching. The same excitation filters used for imaging (640(14) nm for **2** and 563(9) nm for lysotracker red and pH-rhodo red) were used, allowing images to be acquired at various time intervals. Graphing the average cell fluorescence intensity versus time showed that **2** was the most photostable with 50% loss of signal in 94 s and lysotracker red being the least photostable with 50% of signal loss in just 6 s (Fig. 3e, Supplementary Fig. 4 and Supplementary Movies 1 and 2). The behaviour of pH-rhodo

red was more complex, as its intensity first significantly increased throughout the cell followed by photobleaching (Fig. 3e, Supplementary Fig. 4 and Supplementary Movie 3). This response to irradiation is indicative of a photo-conversion occurring for pH-rhodo red but further studies would be required to fully establish the cause for this. Comparison of these results highlights the distinct advantage of **2** for prolonged live-cell imaging in which fluorophore photostability is an essential parameter.

The pH-responsive properties of **2** were investigated in DMEM containing 10% fetal bovine serum (FBS) before its use in imaging studies. Fluorescence output of **2** was negligible at pH 7.4, but became highly fluorescent at acidic pH with its pKa determined as 4.0 (Fig. 4a). Cy5.5 light filter parameters of 690/50 nm were applied to the emission bands at pH 7.4, 5.5 and 4.5, and the integrated fluorescence intensity differences determined. At pH 5.5, as found in late endosomes, the fluorescence enhancement factor (FEF) was 6-fold, while it reached a remarkable 21-fold at lysosomal pH of 4.5 (Fig. 4b). Taken together, these results predict that at a cellular level **2** would remain non-fluorescent in the extracellular environment and become highly NIR fluorescent on uptake and localization in the lysosomes (Fig. 4c).

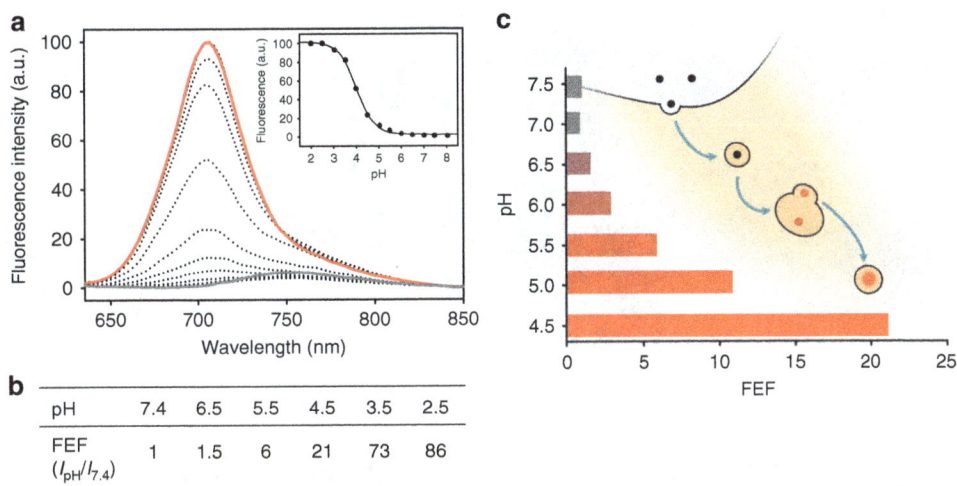

**Figure 4 | Cellular uptake responsive NIR-fluorescence.** (**a**) Emission spectra of **2** ($5 \times 10^{-6}$ M) in DMEM (10% FBS) at pH ranging from 8 (grey) to 2 (red). Exc: 625 nm. Inset: fluorescence intensity at $\lambda_{max} = 707$ nm versus pH; sigmoidal plot fit resulted in apparent $pK_a = 4.0$. (**b**) Corresponding FEF values from differing pH solutions applying Cy5.5 filter parameters. (**c**) Diagram represents the pH changes and increasing fluorescence intensity along endocytic path towards lysosomes.

| pH | 7.4 | 6.5 | 5.5 | 4.5 | 3.5 | 2.5 |
|---|---|---|---|---|---|---|
| FEF ($I_{pH}/I_{7.4}$) | 1 | 1.5 | 6 | 21 | 73 | 86 |

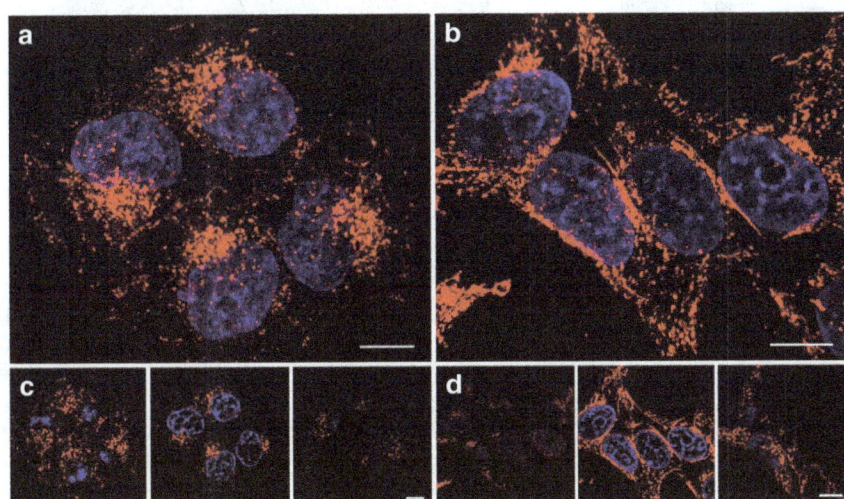

**Figure 5 | Intracellular NIR-emission profile.** CLSM images showing intracellular localization of pH-responsive compound **2** (10 μM, red) and nuclear counterstain Hoechst 33342 (blue) in fixed (**a,c**) HeLa Kyoto and (**b,d**) HEK293 cell lines. Bottom: three corresponding representative slices of the Z-stack for each cell type. Scale bars, 10 μm.

The difference in emission intensity at pH 5.5 (late endosomes) and 4.5 (lysosomes) suggests that the increased activation of **2** in lysosomes may be sufficient to allow differentiation between these organelles.

***In vitro* fixed and real-time live-cell imaging.** Before testing the imaging capabilities of **2**, its cytotoxicity in HeLa Kyoto (cervical cancer) and HEK (human embryonic kidney) cell lines was determined. Following a 24-h incubation of cells with **2**, an MTT (3-(4,5-dimethylthiazol-2-yl)-2,5-diphenyltetrazolium bromide) assay was performed and $EC_{50}$ values of 0.43 and 0.44 mM, respectively, were obtained (Supplementary Fig. 5). These values were used as the basis to select 10 μM as the concentration for its use in imaging experiments. To establish the ability of **2** to internalize in cells, HeLa Kyoto and HEK293 cells were incubated with **2** for 2 h, followed by fixation, nuclei staining (Hoechst 33342) and imaging with confocal laser scanning microscopy (CLSM) (Fig. 5). These images show that **2** was internalized in both cell lines within 2 h (Supplementary Movie 4). The fluorescence signal was predominately localized in the

perinuclear region as would be expected for a lysosomal staining pattern, indicating that the fluorophore had accumulated and become fluorescent in lysosomes[19].

To gain further evidence of selective lysosomal staining, an identical experiment was performed using a HeLa cell line stably expressing the lysosomal-associated membrane protein 1 (LAMP1) fused to green fluorescent protein (GFP)[33]. Following incubation, CLSM-imaged cell images showed very high levels of co-compartmentalization of the red (**2**) and green channel (GFP) emissions to the lysosomes (Fig. 6). Examination of selected focal planes clearly showed a circumferential staining pattern (green) of the LAMP1-GFP in the lysosome membrane and red emission of **2** from within the acidic lumen of the organelles (for Z-stack see Supplementary Movie 5). This ability to resolve the lysosome membrane from the interior is approaching the confocal resolution limit, with ~1 μm being the average diameter of an individual lysosome (for images from an additional independent experiment, see Supplementary Fig. 6). This was achieved due to the high signal-to-noise ratio, as background emission from **2** is not observed, and high red/green contrast with the genetically

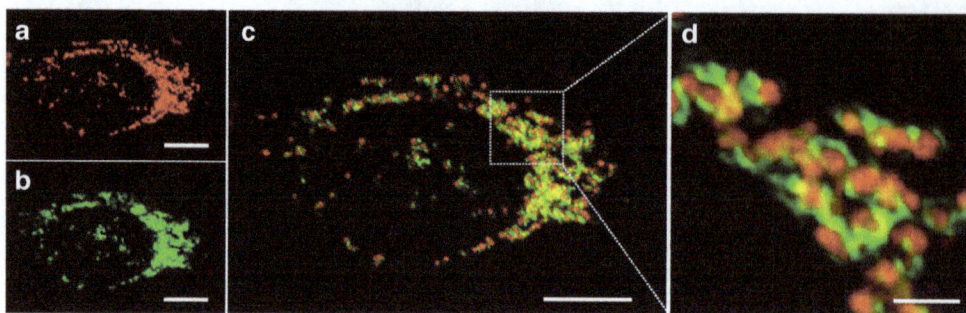

**Figure 6 | Identification of subcellular NIR-fluorescent on switch.** CLSM fluorescent images showing lysosomal localization of the 'on' state of **2** in LAMP1-GFP-expressing HeLa cells (**a**) Cy5.5 channel; (**b**) GFP channel. (**c**) Three-dimensional image of overlaid Cy5.5 and GFP channels. (**d**) Zoom-in of the dashed box. Scale bars, 10 μm (**a–c**) and 2 μm (**d**).

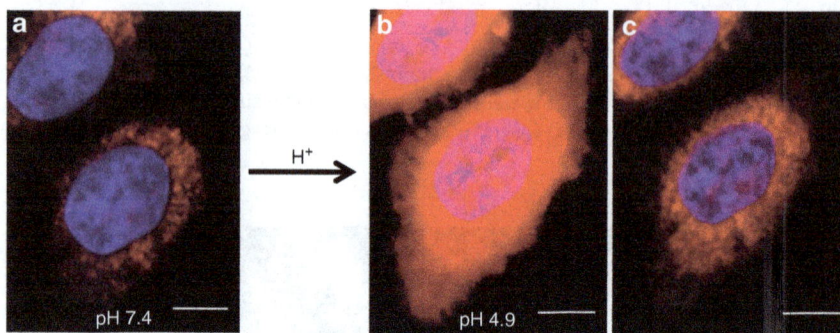

**Figure 7 | Illustration of NIR-fluorescence response selectivity.** (**a**) CLSM imaging of HeLa Kyoto cells following incubation with **2** (10 μM) for 2 h at 37 °C, DAPI nuclei staining and fixing. (**b**) The same set of cells imaged after buffer changed to pH 4.9, keeping the same laser power and PMT voltage. (**c**) The same set of cells after adjustment of microscope laser power and PMT voltage to obtain a non-saturated image. Red: **2**; blue: DAPI stain. Scale bar, 10 μm.

expressed LAMP1–GFP. Further statistical evidence of co-compartmentalization of the NIR and green emissions was provided by the calculated Manders' coefficients of 0.91 (05) for $M_{NIR}$ and 0.95 (05) for $M_{green}$ and a Pearson's coefficient of 0.79 (09)[34]. The lower Pearson's coefficient value when compared with Manders' may be attributable to the fact that the NIR and green emissions are co-compartmentalized to the lysosomes but not fully co-localized as the green is in the outer membrane and the NIR from within the internal lumen. Golgi co-staining of HeLa cells with **2** showed no significant co-localization (Supplementary Fig. 7, Method 1).

Two possible explanations for the excellent co-compartmentalization of red and green emissions could be envisaged. Either **2** is physically located exclusively in the lysosomes or it is present in other organelles along the endocytic pathway and is only emissive from the lysosomes due to its lower pH, with the remaining **2** being fluorescent silent. To visualize all intracellular **2** in its fluorescent on state, HeLa cells were incubated with **2** in media at pH 7.4 for 2 h, nuclei stained with 4,6-diamidino-2-phenylindole (DAPI) for 15 min, fixed and imaged using CLSM (Fig. 7a and Supplementary Movie 6). The media containing the fixed cells was then changed for media at pH 4.9 (adjusted using HCl), which on penetrating the cell forced on the fluorescence of all **2** within cells not localized in a sufficiently low pH environment. Re-imaging of the same cells (30 min after media change) with identical microscope settings showed a significantly increased fluorescence with saturation of the field of view within the cells (Fig. 7b and Supplementary Movie 7). The mean corrected total fluorescence from the cells (calculated using ImageJ) showed an FEF of 9.5 on lowering of the pH (see Supplementary Fig. 8 for images from additional independent experiment). This clearly illustrates that fluorescence from the majority of **2** was not switched on by the cells and only **2** localized in the lysosomes was emissive under normal pH conditions. The same field of view was imaged for the third time using adjusted microscope laser power and photomultiplier tube (PMT) voltage and it became clear that the additional fluorescence of **2** was predominantly from other cellular organelles of higher pH (Fig. 7c and Supplementary Movie 8). Unfortunately, the adjustment to lower pH caused a loss of the LAMP1–GFP signal; thus, a comparison of Manders' coefficients was not possible.

To show that fluorophore **2** was internalized in cells via endocytosis and not passive diffusion, HeLa Kyoto cells were incubated at 4 °C for 30 min with **2**. It is known that endocytosis is inhibited at 4 °C, whereas passive diffusion can still occur[35]. Following incubation, cells were imaged, the buffer adjusted to pH 4.9 and re-imaged as described above. In contrast to the result shown in Fig. 7, no fluorescence from **2** was detected before or after changing the buffer to pH 4.9 (Supplementary Fig. 9). From these experiments, it was concluded that **2** was internalized via an energy requiring endocytosis rather than passive diffusion.

For the results outlined above, it was anticipated that cellular uptake of **2** could be continuously imaged in real-time without the need for washing or manipulating cells. The first live-cell imaging experiment involved imaging cells in a single focal plane over a 1.5-h time period. Once in focus, HeLa Kyoto cells were treated with **2** and imaged with an epi-fluorescence live-cell microscope, under optimal conditions of temperature and atmosphere for the cells to remain fully active (37 °C and 5% $CO_2$ humidified environment). Images were continuously acquired every 30 s for 90 min and then combined to form a movie (Supplementary Movie 9). Representative time-lapse images after 1, 30, 60 and 90 min (in black/white for clarity) are shown below in Fig. 8a with the 90 min time point in red

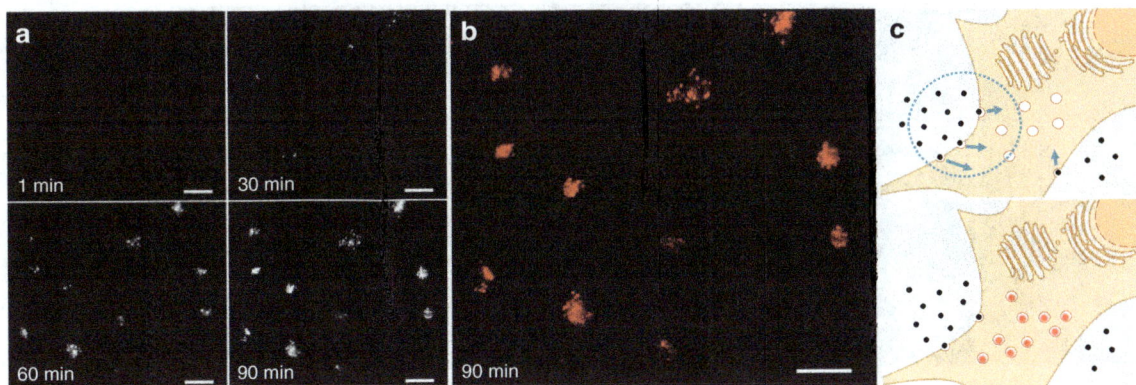

**Figure 8 | Widefield live-cell imaging of the uptake of 2 (10 μM) into HeLa Kyoto cells. (a)** Time-lapse black and white images are shown 1, 30, 60 and 90 min. **(b)** Red-coloured image at 90 min. **(c)** Schematic depiction of the uptake process of responsive fluorophore **(c)**. Scale bars, 20 nm.

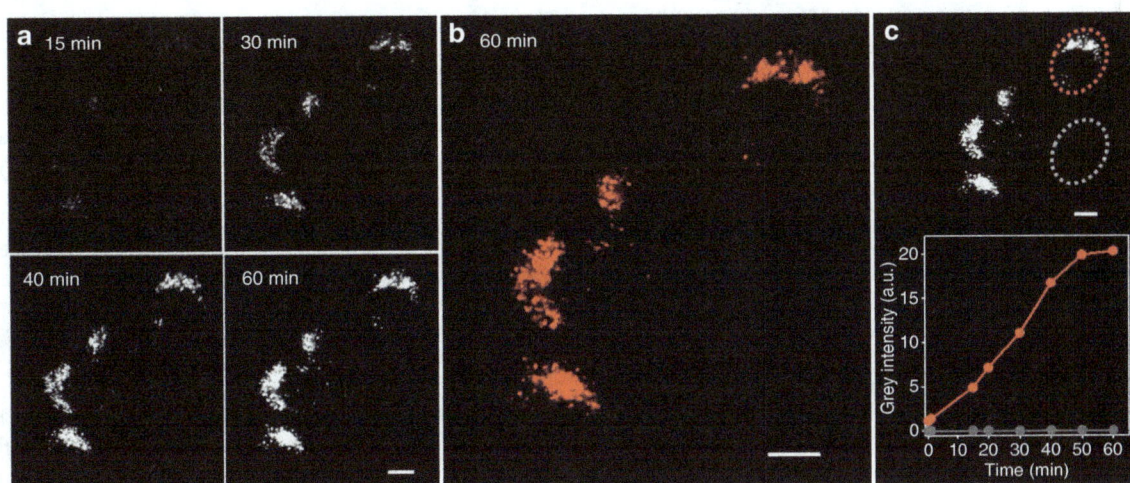

**Figure 9 | Z-axis projections of widefield 4D live-cell imaging of the uptake of 2 (10 μM) from HeLa Kyoto cells.** Images were acquired in 25 focal planes every 1 min for 60 min. **(a)** Time lapse b/w images are shown for 15, 30, 40 and 60 min. **(b)** Red-coloured image at 60 min. **(c)** Fluorescence intensity quantification in two identical volumes around a selected cell (1) and in the extracellular environment (2). Scale bars, 10 μm.

for comparison (Fig. 8b). The first image acquired at 1 min showed no NIR fluorescence, a signal confirming the effective fluorescence quenching of extracellular **2**. However, over the following 90-min time period, a strong signal arose from point-like organelles as a result of cellular uptake of **2** and transport through the endocytic pathway to the lysosomes (Fig. 8: 60- and 90-time points). On close inspection, individual lysosomes can be seen emerging into view over the first 30-min time period, following which they increase in intensity and number from 30 to 90 min (Supplementary Movie 9).

The continuous imaging of live cells in the z-axis provides the most realistic method for following the progress of biological events over time and is of particular relevance when imaging small mobile organelles. To generate a three-dimensional (3D) representation of the cell in real time, a z-stack of 25 focal planes through the cell was acquired every minute[36]. This continual recording of cellular 3D volume over a period of time is termed four-dimensional (4D) imaging as the sample is imaged in the x,y,z and time dimensions, from which a time-lapse video of the 3D cellular volume can be created. Using HeLa cells, a 4D data set of the uptake of **2** over a 60-min time period was acquired followed by deconvolution of the data set to correct for motion of fluorescence objects between focal planes during the 1-min time period required to complete a z-stack of the cell. This experiment showed punctuated regions of fluorescence over the 60-min time period starting from a non-fluorescent background at time zero

with individual lysosome movement clearly observable (Fig. 9a,b and Supplementary Movie 10). Quantification of the fluorescence increase was measured by selecting two identical volumes of the imaged area, one overlapping with a chosen cell and the other on the extracellular environment, and applying an image analysis algorithm (ImageJ) to measure the total fluorescence intensity within the volumes (Fig. 9c). A comparative plot of both intensities over time shows how the background remained non-fluorescent, whereas intracellular fluorescence intensity increased over time, reaching a plateau at 60 min at which point a dynamic equilibrium was established between extracellular and intracellular **2**.

This ability to 4D image with **2** provides a tool for tracking lysosomal movements within the cell. Lysosomal staining of HeLa Kyoto cells was achieved by incubation with **2** for 1 h following which they were imaged in 3D for 35 min (without medium replacement) using a widefield microscope. Image analysis software was used to tag individual lysosomes as white spheres to facilitate visualization and the movement of these lysosomes was tracked over the 35-min time period (Fig. 10). The path the lysosome takes through the cell is illustrated by a lengthening white tail, which extends from the lysosome as the video progresses (Fig. 10c,d and Supplementary Movie 11)[37]. Tracking of all lysosomes within a cell (or field of view) was also possible and is shown for Fig. 10 in the Supplementary Information (Supplementary Movie 12).

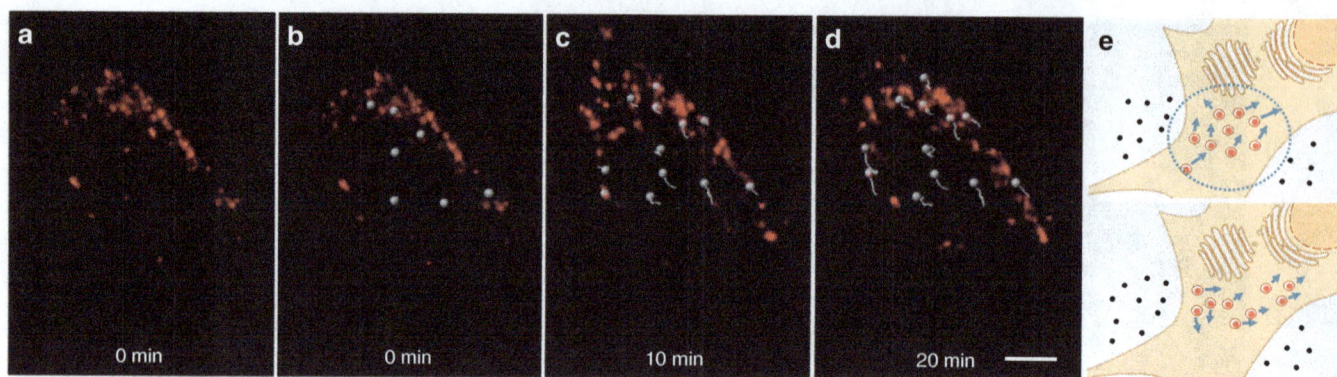

**Figure 10 | Lysosome tracking in living HeLa Kyoto cells post 1 h incubation with 2 (10 μM).** (**a**) Time-lapse representative snapshot of a single cell chosen for image analysis. (**b**) Lysosome selection at 0 min. (**c,d**) Tracking over time. (**e**) Schematic depiction of tracking intracellular vesicular movements with bioresponsive fluorophore. Scale bar, 5 μm.

Our final *in vitro* experiments with pH-responsive **2** were aimed at studying the efflux of the fluorophore from the cell. HeLa Kyoto cells were pre-treated with **2** for 2 h as previously described, then the medium was replaced with fluorophore-free DMEM. Cells were incubated at 37 °C and imaged at time points of 1, 15, 30 and 120 min using confocal microscopy (Fig. 11). The overall NIR fluorescence detected after 2 h incubation with **2** in the intracellular environment showed to be again arising from point-like organelles (Fig. 11a). The quantity of fluorescent vesicles decreased within 30 min after medium change, with significant clearance of **2** by 120 min (Fig. 11b). Using ImageJ analysis, lysosome number per cell (or field of view) was determined for the different time points and were observed to steadily decrease from time 0 to 120 min (Fig. 11d). In addition, after 60 min cells were fixed and acidified with buffer of pH 4.9 and only a small, 1.6-fold increase in total fluorescence was observed.

***In vivo* imaging**. A distinct advantage of NIR fluorophores is their ability to directly transfer from *in vitro* to *in vivo* imaging due to transparency of biological tissue at these longer wavelengths. To test *in vivo* performance of **2**, the luciferase-expressing human breast cell line MDA-MB-231-luc-D3H1 was chosen to grow subcutaneous tumours of size 100–200 mm³, which permitted NIR fluorescence imaging with confirmatory bioluminescence. The ability of PEG polymers to act as a drug delivery vehicle has been well established, with several PEGylated drugs in clinical use for over 20 years. PEGylation is known to influence pharmacokinetic properties resulting in prolonged blood circulation times. As such, it was anticipated that the PEG-conjugated **2** may have passive tumour-targeting properties leading to some preferential uptake into tumour cells, thereby generating a distinguishable fluorescent signal. Following an intravenous tail vein injection of **2** (2 mg kg⁻¹), images were acquired at regular intervals over the course of 24 h. A plot of tumour to background NIR fluorescence showed that the fluorescence signal was low in the beginning in both the background and tumour. Emission from the liver peaked at 1 h and subsided over the following 24 h (Fig. 12d dashed red line). In contrast, the tumour fluorescence intensity reached a maximum at 24 h, allowing for good image discrimination at that time point (Fig. 12d solid red line). Bioluminescence and NIR fluorescence imaging at 24 h confirmed that **2** has indeed been taken up into tumour cells and switched on (Fig. 12a–c). No adverse reactions or animal weight loss were observed during or after imaging. These preliminary *in vivo* results represent a unique example of selective NIR *in vivo* tumour imaging using a pH-responsive fluorochrome.

## Discussion

Imaging with molecular fluorophores is an indispensable tool for all forms of biological and medical research. Although the full spectrum of colours are available from molecular fluorophores, imaging with lowest energy NIR spectral region offers advantage for prolonged live cellular imaging with the possibility of *in vivo* imaging with the same probe[5,37]. Although most fluorescent markers are permanently fluorescent (unless photobleached), huge imaging advantage can be gained if the fluorescence can be modulated from off to on in a reversible bioresponsive manner[14]. With these goals in mind, we have designed a lysosomal-responsive NIR probe with the potential for real-time visualization of their key cellular operations that can be directly translated for use *in vivo* (Figs 1 and 2). NIR probe **2** is synthesized in five synthetic operations from a bisphenolic substituted member of the BF₂-azadipyrromethene fluorophore class (Fig. 1). An *o*-nitro-substituted phenol group on the probe acts as the fluorescent switch with the nitro group tailoring the emission response to the lumen microenvironment of the lysosomes. Photophysical measurements in DMEM shows that **2** remained fluorescent silent at pH 7.2, yet became highly fluorescent in the pH range that corresponds with the acidic micro-environment of the lysosome (Fig. 4). The non-fluorescent state shows little polarity sensitivity, indicating that it could be used in the more complex intracellular environment without resulting in false-positive emissions. Solution and cellular photobleaching experiments indicated high stability, which is an essential feature for imaging over prolonged time periods that is often lacking in both synthetic and genetically expressed probes (Fig. 3).

To illustrate the potential uses of probe **2**, a series of increasingly complex imaging experiments were undertaken in fixed, live cells and *in vivo*. The high fidelity for the switching on of **2** to the lysosomal lumen is observed when imaging with LAMP1-GFP HeLa cells (Fig. 6). We have exploited these responsive emission properties for 3D and 4D real-time live-cell imaging of several fundamental biological events, such as endocytosis, organelle trafficking and efflux. As **2** has extremely low emission in cell media and with fluorescence activated on cellular internalization, a high background-to-noise ratio is achieved, making the continuous acquisition of data as straightforward as just add **2** to cells and image. It could be anticipated that these techniques would be valuable for many types of cellular experiments involving lysosomal response to stimuli.

It is important to note that the mode of action and the use of existing lysosomotropic stains, which are typically amine-functionalized fluorophores to promote retention in acidic

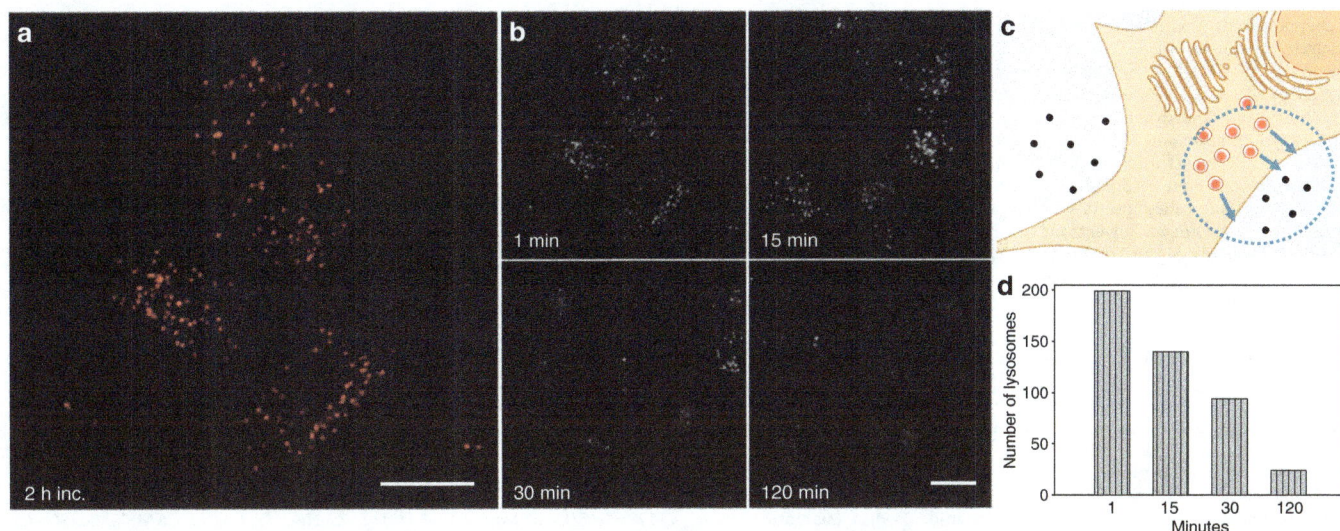

**Figure 11 | Imaging of cellular efflux.** HeLa Kyoto cells were pre-treated with **2** (10 μM) for 2 h, DMEM replaced with fluorophore-free DMEM and cells fixed at various time points. (**a**) Cells imaged after 2 h incubation. (**b**) Cells imaged after 1, 15, 30 and 120 min post media change. (**c**) Schematic depiction of efflux of bioresponsive fluorophore. (**d**) Decrease in number of NIR fluorescent lysosomes from 1 to 120 min. Scale bars, 20 μm.

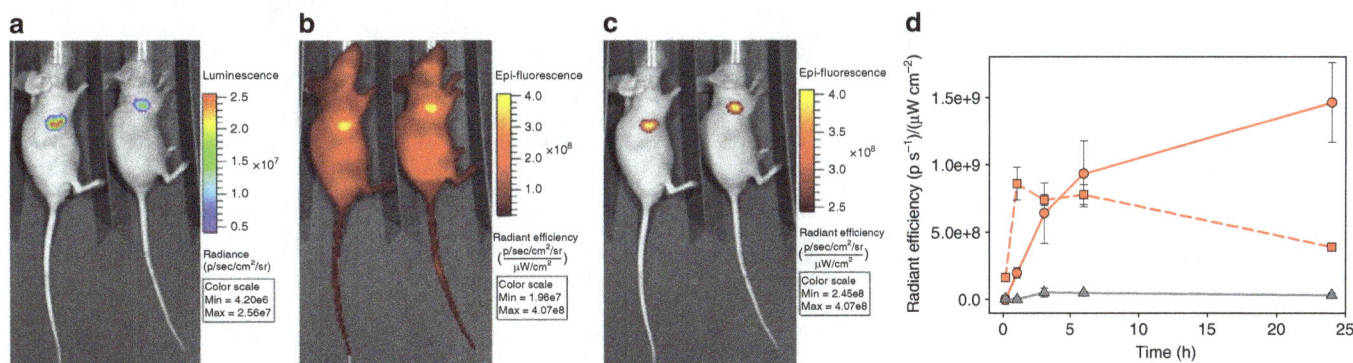

**Figure 12 | *In vivo* imaging of 2 using a MDA-MB-231-luc-D3H1 subcutaneous tumour model in two representative mice.** (**a**) Bioluminescence imaging confirmation of tumour cells. (**b**) NIR fluorescence imaging 24 h post intravenous (i.v.) administration of **2** (excit. 660-690 nm, emis. 710-730 nm). (**c**) NIR fluorescence imaging 24 h post i.v. administration of **2** with intensity scale adjusted (excit. at 675 nm, emiss. at 720 nm). (**d**) Profile of tumour NIR fluorescence (red solid line) and liver (red dashed line) over time following i.v. tail injection of **2**. Non-injected control tumour NIR fluorescence (grey solid line). Values determined from the same sized ROI from background area and tumour averaged for $n = 3$.

lysosomes on protonation, is significantly different from **2** (refs 38,39). These lysosomotropic fluorophores are pre-incubated for a period of time (typically 30 min), followed by cell washing to remove excess nonspecific fluorophore, before imaging can be carried out for up to a maximum of 1 h (ref. 40). In contrast, **2** in its fluorescent on state is uncharged, showing little cytotoxic effect over long incubation times and no background fluorescence due to the highly selective switching 'on' only in the desired lysosomal ROIs. These characteristics have been demonstrated to be particularly advantageous for continuous real-time live-cell imaging and show significant potential for use in a wider range of complex bio-imaging applications. Overall, the cellular imaging performance, ease of utility and the selectivity for lysosomal staining could be judged as excellent in comparison with the most recently developed probes[41].

To complete the imaging portfolio, we wished to illustrate that the bioresponse of **2** within subcellular compartments could also be visualized at the macroscopic scale of a tumour. This translation to *in vivo* tumour imaging is achievable as shown in

Fig. 12 in which tumour can be clearly distinguished, showing a high potential for targeted responsive fluorescence imaging.

In conclusion, the development of **2** as the first phenol/phenolate controlled molecular NIR lysosome-responsive fluorophore has important implications for the study of intracellular transport mechanisms, lysosome-based diseases and *in vivo* targeting. The use of **2** for further 4D real-time studies of more complex dynamic cell mechanisms involving lysosomes are ongoing. Future studies will include the conjugation of active tumour targeting motifs via **7** (instead of PEG) to the fluorophore, to further broaden the possibilities of *in-vivo* imaging targets with potential applications for fluorescence-guided surgery.

## Methods

**General information and materials.** All commercially available solvents and reagents were used as supplied, unless otherwise stated. All reactions were performed under nitrogen or argon atmosphere in oven-dried glassware. Gel chromatography was performed with Davisil 60 silica (230–400 mesh). On the basis of NMR and reverse-phase HPLC, all final compounds were >95% pure. [1]H and [13]C spectra were recorded on 300, 400, 500 or 600 MHz NMR spectrometers and chemical shifts were reported in p.p.m. using solvent residual peak as standard. For spectra of compounds **4**, **5**, **6** and **7**, see Supplementary Figs 10–13, respectively. All $^{19}$F NMR chemical shifts are referenced to CFCl$_3$. All $^{11}$B NMR chemical shifts are referenced to BF$_3$.Et$_2$O/CDCl$_3$. High-resolution mass spectrometry and tandem mass spectrometry experiments were carried on an electrospray ionisation and MALDI–TOF instruments. Infrared spectra were recorded as a KBr pellet using a Fourier Transform infrared spectrometer. Absorbance spectra were recorded with a Varian Cary 50 Scan ultraviolet–visible spectrometer. Fluorescence spectra were recorded with a Varian Cary Eclipse Fluorescence Spectrometer. Solvents for absorbance and fluorescence experiments were of HPLC quality. SigmaPlot, MestreNova, ChemDraw, Zeiss LSM and ImageJ software were used for data analysis. Phenol red-free imaging DMEM medium was used for all experiments.

**Synthesis of compound 4.** Compound (ref. 28) (200 mg, 0.38 mmol) and CsF (288 mg, 1.89 mmol) were dissolved in dry DMSO (6 ml) and stirred at 30 °C under a nitrogen atmosphere for 10 min, during which time the colour changed from dark green to dark purple. *t*Butyl bromoacetate (126 mg, 0.64 mmol) was then added via syringe in one go and the solution was stirred at 30 °C for 20 min. The mixture was partitioned between AcOEt (100 ml) and PBS buffer at pH 7 (100 ml). The organic phase was washed with water (3 × 100 ml), brine (50 ml), dried over Na$_2$SO$_4$, filtered and evaporated to dryness. The crude product was purified by silica gel chromatography, eluting with CH$_2$Cl$_2$:AcOEt (99:1 → 90:10) to yield the product **4** as a red metallic solid (151 mg, 64%). mp: 183–187 °C; $^1$H NMR (400 MHz, DMSO-$d_6$): δ 10.58 (br s, 1H), 8.19–8.10 (m, 8H), 7.65 (s, 1H), 7.57–7.43 (m, 7H), 7.10 (d, J = 8.7 Hz, 2H), 6.95 (d, J = 8.6 Hz, 2H), 4.81 (s, 2H), 1.46 (s, 9H); $^{13}$C NMR (100 MHz, DMSO-$d_6$): δ 167.5, 161.5, 160.0, 158.9, 155.7, 145.0, 143.8, 142.6, 141.0, 132.5, 132.0, 131.7, 131.4, 129.6, 129.3, 129.1, 129.0, 128.7, 128.6, 124.1, 121.3, 120.3, 119.1, 116.0, 114.8, 81.6, 65.0, 27.7; $^{11}$B NMR (128 MHz, DMSO-$d_6$) δ: 1.00 (t, J = 32.6 Hz); $^{19}$F NMR (376 MHz, DMSO-$d_6$): δ – 130.44 (q, J = 32.6 Hz); ultraviolet–visible: $\lambda_{max}$ (CHCl$_3$): 680 nm (ε: 85,000 cm$^{-1}$ M$^{-1}$); emission: $\lambda_{max}$ (CHCl$_3$): 708 nm, Φ$_F$ (CHCl$_3$) = 0.31; high resolution mass spectrometry (HRMS) (m/z): [M-H]$^-$ calcd. for C$_{38}$H$_{31}$BN$_3$O$_4$F$_2$, 642.2376; found, 642.2357.

**Synthesis of compound 5.** A solution of **4** (300 mg, 0.48 mmol) in acetonitrile (6 ml) was heated under reflux for 5 min. A solution of KNO$_3$ (53 mg, 0.53 mmol) and KHSO$_4$ (130 mg, 0.96 mmol) in water (1 ml) was added and the mixture was heated under reflux for 5 min. The suspension was cooled to room temperature (rt) and partitioned between AcOEt (200 ml) and water (100 ml). The organic phase was washed with water (100 ml), brine (2 × 100 ml), dried over Na$_2$SO$_4$, filtered and evaporated to dryness. The crude was purified by silica gel chromatography, eluting with CH$_2$Cl$_2$:AcOEt (99: 1) to yield the product **5** as a dark red metallic solid (198 mg, 60%). mp: 192–194 °C; $^1$H NMR (400 MHz, DMSO-$d_6$): δ 11.88 (br s, 1H), 8.69 (d, J = 2.0 Hz, 1H), 8.27 (dd, J = 8.9, 2.0, 1H), 8.21 (d, J = 9.0 Hz, 2H), 8.19–8.13 (m, 4H), 7.71 (s, 1H), 7.60 (s, 1H), 7.57–7.43 (m, 6H), 7.27 (d, J = 8.9 Hz, 1H), 7.12 (d, J = 9.0 Hz, 2H), 4.84 (s, 2H), 1.45 (s, 9H); $^{13}$C NMR (100 MHz, DMSO-$d_6$): δ 167.4, 160.9, 159.3, 154.1, 153.9, 145.4, 143.9, 143.3, 141.5, 137.3, 135.6, 132.1, 131.8, 131.5, 129.9, 129.5, 129.2, 129.0, 128.7, 128.7, 126.6, 123.3, 121.8, 120.8, 119.4, 119.1, 115.0, 81.7, 65.1, 27.7; $^{11}$B NMR (128 MHz, DMSO-$d_6$): δ 0.92 (t, J = 32.7 Hz); $^{19}$F NMR (376 MHz, DMSO-$d_6$): δ – 130.31 (q, J = 32.7 Hz); ultraviolet–visible: $\lambda_{max}$ (CHCl$_3$) 675 nm (ε: 94,000 cm$^{-1}$ M$^{-1}$); emission: $\lambda_{max}$ (CHCl$_3$) 703 nm, Φ$_F$ (CHCl$_3$) = 0.15; HRMS (m/z): [M-H]$^-$ calcd. for C$_{38}$H$_{30}$BN$_4$O$_6$F$_2$, 687.2226; found, 687.2229.

**Synthesis of compound 6.** TFA (1 ml) was added dropwise to a solution of **5** (175 mg, 0.25 mmol) in dichloromethane (DCM) (9 ml) and the solution was stirred at rt for 3 h. The solvent was removed under vacuo and the residual TFA was removed azeotropically with serial additions of DCM and subsequent removal under vacuo. The solid was suspended in DCM, filtered and washed with DCM, to yield the product as a dark purple solid (136 mg, 84%). The product was pure enough to proceed to the next synthetic step. To remove the last trace of starting material, the solid was partitioned between AcOEt (90 ml) and Na$_2$CO$_3$ sat. (180 ml). The organic layer was discarded and the water layer was extracted with AcOEt (90 ml), separated, carefully acidified with 5 M HCl and extracted again with AcOEt (180 ml). The organic layer was separated, dried over anhydrous Na$_2$SO$_4$, filtered and evaporated to dryness. The product **6** was obtained as a dark purple solid (122 mg, 76%). mp: 214–219 °C; $^1$H NMR (400 MHz, DMSO-$d_6$): δ 13.15 (br s, 1H), 8.68 (d, J = 2.3 Hz, 1H), 8.28 (dd, J = 8.9, 2.3 Hz, 1H), 8.24–8.13 (m, 6H), 7.71 (s, 1H), 7.60 (s, 1H), 7.57–7.44 (m, 6H), 7.28 (d, J = 8.9 Hz, 1H), 7.14 (d, J = 9.0 Hz, 2H), 4.87 (s, 2H); $^{13}$C NMR (100 MHz, DMSO-$d_6$): δ 169.7, 161.0, 159.4, 154.1, 153.8, 145.4, 143.9, 143.4, 141.5, 137.3, 135.5, 132.1, 131.9, 131.5, 129.9, 129.5, 129.2, 129.0, 128.7 (2C), 126.6, 123.2, 121.9, 120.8, 119.5, 119.1, 115.1, 64.6; $^{11}$B NMR (128 MHz, DMSO-$d_6$): δ 0.93 (t, J = 32.8 Hz); $^{19}$F NMR (376 MHz, DMSO-$d_6$): δ – 130.34 (q, J = 32.8 Hz); HRMS (m/z): [M-H]$^-$ calcd. for C$_{34}$H$_{22}$BN$_4$O$_6$F$_2$, 631.1600; found, 631.1603.

**$^1$H NMR monitoring of the formation of activated ester 6.** A mixture of **5** (45 mg, 0.071 mmol), N-(3-dimethylaminopropyl)-N'-ethylcarbodiimide

hydrochloride (27 mg, 0.14 mmol) and N-hydroxysuccinimide (82 mg, 0.71 mmol) was placed in a sealed dry flask. Anhydrous deuterated DMSO-$d_6$ (1.2 ml) was added to the mixture and the solution was stirred at rt under N$_2$ atmosphere. Samples (50 μl) were withdrawn at 15, 30, 60, 120 min and 19 h, diluted with DMSO-$d_6$ in an NMR tube (650 μl) and $^1$H spectra were recorded at a 600-MHz spectrometer.

**Synthesis of compound 7.** A mixture of **6** (40 mg, 0.063 mmol), N-(3-dimethyl-laminopropyl)-N'-ethylcarbodiimide hydrochloride (24 mg, 0.13 mmol) and N-hydroxysuccinimide (73 mg, 0.63 mmol) was dissolved in anhydrous DMSO (1 ml) and stirred at rt for 3 h under N$_2$ atmosphere. The solution was partitioned between DCM (50 ml) and 0.5 M HCl (50 ml). The organic phase was washed with 0.5 M HCl (50 ml), acidic brine (50 ml), dried over Na$_2$SO$_4$, filtered and evaporated to dryness, keeping the temperature of the bath below 35°C. The product **7** was obtained as a purple metallic solid (44 mg, 95%). m.p.: 177–183 °C; $^1$H NMR (400 MHz, DMSO-$d_6$): δ 8.69 (d, J = 2.2 Hz, 1H), 8.29 (dd, J = 8.9, 2.2 Hz, 1H), 8.23 (d, J = 8.9 Hz, 2H), 8.21–8.14 (m, 4H), 7.72 (s, 1H), 7.64 (s, 1H), 7.59–7.45 (m, 6H), 7.28 (d, J = 8.9 Hz, 1H), 7.23 (d, J = 8.9 Hz, 2H), 5.53 (s, 2H), 2.85 (s, 4H); $^{13}$C NMR (100 MHz, DMSO-$d_6$): δ 169.9, 165.2, 159.8, 158.6, 154.5, 145.2, 144.2, 143.1, 142.0, 137.4, 135.5, 132.0, 131.8, 131.6, 129.8, 129.6, 129.2, 129.1, 128.7, 126.8, 124.1, 121.5, 120.6, 119.6, 119.5, 115.2, 63.0, 25.5; $^{11}$B NMR (128 MHz, DMSO-$d_6$): δ 0.93 (t, J = 32.8 Hz); $^{19}$F NMR (376 MHz, DMSO-$d_6$): δ – 130.34 (q, J = 32.8 Hz); HRMS (m/z): [M-H]$^-$ calcd. for C$_{38}$H$_{25}$BF$_2$N$_5$O$_8$, 728.1764; found, 728.1730.

**Synthesis of compound 2.** A mixture of **7** (6.4 mg, 0.0088 mmol) and O-(2-aminoethyl)polyethylene glycol 5000 (CAS 32130–27–1) (40 mg, 0.008 mmol) was dissolved in anhydrous DMSO (0.88 ml) and stirred at rt for 18 h under a N$_2$ atmosphere. The solvent was removed by short-path distillation at rt overnight and the crude was partitioned between DCM (20 ml) and 1 M Na$_2$CO$_3$ (20 ml). The aqueous phase was extracted with DCM (2 × 20 ml). The organic layers were combined, washed with slightly acidic (HCl) water (20 ml), brine (20 ml), dried over anhydrous Na$_2$SO$_4$, filtered and evaporated to dryness. The residue was dissolved in HPLC grade water (8 ml) and the dark solution was passed through a Sep Pak C18 reverse-phase cartridge, then freeze dried. The product **2** was obtained as a dark green solid (40 mg, 90%). mp: 43–45 °C; $^1$H NMR (400 MHz, DMSO-$d_6$): δ 8.74–8.72 (m, 1H), 8.28 (dd, J = 9.0, 2.3 Hz, 1H), 8.24–8.15 (m, 7H), 7.68 (s, 2H), 7.58–7.52 (m, 4H), 7.52–7.45 (m, 2H), 7.25–7.19 (m, 1H), 7.16 (d, J = 9.0 Hz, 2H), 4.66 (s, 2H), 3.70–3.65 (m, 4H), 3.50 (s, 680H); Ultraviolet–visible: $\lambda_{max}$ (CHCl$_3$) 670 nm, (ε: 97,000 cm$^{-1}$ M$^{-1}$); emission: $\lambda_{max}$ (CHCl$_3$) 702 nm, Φ$_F$ (CHCl$_3$) = 0.18; HRMS (m/z): MALDI–TOF distribution maximum centred at 5410.3999 Da.

**Fluorescence quantum yields and extinction coefficients.** The compound of interest (0.005 mmol) was dissolved in CHCl$_3$ (50 ml) to prepare a stock solution (10$^{-4}$ M). The stock was diluted to concentrations 2, 4, 6, 8 and 10 × 10$^{-7}$ M with CHCl$_3$ and each solution was analysed with an ultraviolet–visible spectrometer and a fluorescence spectrometer against CHCl$_3$ background. Excitation = 640 nm; emission range = 660–900 nm; slit width = 5/5 nm; scan rates = 600 nm min$^{-1}$. Plots of abs$_{max}$ versus conc and fluorescence area versus abs (640 nm) allowed the calculation of extinction coefficient and fluorescence quantum yield, respectively. Compound **1** (R = Ph, Ar = pMeOC$_6$H$_4$) was used as standard for fluorescence quantum yields with Φ$_F$ = 0.36 (refs 17,42).

**Fluorescence response of 2 and 5 to addition of DBU/TFA in organic solvents.** Compound **2** or **5** was dissolved in toluene, tetrahydrofuran, dimethylformamide and DMSO (25 ml) to a final concentration of 5 μM. A solution of DBU (29.5 mg in 100 ml of CHCl$_3$) was added (64 μl = 1 eq) gradually, and absorbance and fluorescence spectra were recorded before and after each addition. The addition was stopped once spectra remained unchanged. At this stage, an excess of DBU was added and the spectra were recorded. Subsequently, a higher excess of TFA was added and the spectra were recorded. The area below the last two curves was plotted for off/on histogram (shown in Fig. 3). (Note: the toluene solution of **2** contained 1% CHCl$_3$ for solubility).

**Fluorescence response of 2 to pH variation in DMEM.** Compound **2** (2.8 mg) was dissolved in PBS (500 μl). The stock solution (1 mM) was diluted with DMEM supplemented with 10% FBS to the concentration of 5 μM. The pH of the solution was adjusted with diluted HCl or NaOH, to obtain a range from 8 to 2 at regular intervals, each of which was recorded, and the respective solution analysed by ultraviolet–visible absorption and fluorescence emission. Excitation = 625 nm; emission range = 635–900 nm.

**Comparative solution and cellular photobleaching of 2 and lysotracker red and pHrodo red.** Entire fluorescence cuvettes contain 1 × 10$^{-7}$ M DMEM solutions at pH 4.0 of **2**; lysotracker red and pH-rodo red were continuously irradiated with light of wavelength 620(30) nm for **2** and 540(40) nm for lysotracker red and pH-rhodo red at 25 °C for 2 h. Filtered light from a 150-W light source used with complete cuvette irradiation via a fibre optic with attached light diffuser. Fluorophore fluorescence intensities were recorded every 20 min. The average fluorescence intensity from three independent experiments were normalized and plotted with sigmaplot 8.

Ten thousand HeLa-Kyoyo cells in DMEM were seeded onto chamber slides and incubated with 2 (20 µM) for 60 min or lysotracker red (150 nM) for 30 min, or pH rhodo red (15 µM) for 30 min. DMEM was replaced with fluorophore-free media and cells constantly irradiated with a Lumencor SPECTRA light engine LED used as the light source set to a maximum power for 400 s. Excitation filter 563/9 nm was used for lysotracker red and pH-rhodo red and excitation filter 640(14) nm was used for 2. Cells were imaged with the shutter open, a time intervals of either 0.1 or 1.0 or 5 s with exposure of 10 ms and individual frames complied into movie format. The average cellular ROI fluorescence intensities from three independent experiments were plotted. An Olympus × 60 PLANAPO/1.42 objective and Andor iXon 888 ultra were used for signal detection. Acquisition and analysis performed with MetaMorph v7.8.

**MTT assay of 2.** Compound 2 (4.0 mg) was dissolved in sterile PBS (71 µl) to prepare a stock solution 10 mM. This was serially diluted to prepare samples at 5, 1, 0.5, 0.1 and 0.05 mM. Each of the stock solutions was diluted 1:10 with DMEM medium, which was co-incubated with HeLa or HEK293 cells at 5,000 cells per well on a 96-well plate for 24 h. The solution was removed and substituted with MTT solution (5 mg ml$^{-1}$ in DMEM). The cells were incubated for 3 h. The medium was removed and the wells were treated with DMSO for 10 min. The absorbance of each well was read with a plate reader at 540 nm.

**Production and validation of HeLa Kyoto cell line stably expressing LAMP1-GFP fusion protein.** An expression plasmid encoding the LAMP1-GFP fusion protein was generated via the complete open-reading frame coded by a I.M.A.G.E. Fully Sequenced cDNA Clone (Source BioScience, I.M.A.G.E. ID: 5019745) of the human LAMP1 (GenBank accession number BC021288) was amplified by PCR using primers designed to append an XhoI site upstream of the translation initiation site and to replace the translation termination site by a segment encoding EcoRI site followed by a linker sequence CTCCTC (single-letter nucleotide code). The PCR product was gel purified and cloned in the XhoI–EcoRI sites of a pEGFP-N1 vector (BD Biosciences Clontech). Constructs were verified by DNA sequencing.

For stable transfection, HeLa cells were grown at 37 °C in complete DMEM supplemented with 10% FBS and 1% glutamine, to 30–40% confluency, and subsequently transfected with the LAMP1–GFP-encoding plasmid using FuGENE 6 (Roche) following the manufacturer's instruction. One day later, 0.6 g l$^{-1}$ G418 was added. The medium was changed every day to remove the G418 non-resistant cells and when the cell number looked stabilized the G418 was lowered to 0.5 g l$^{-1}$. Cells displaying resistance to G418 and expressing LAMP1-GFP (as judged by fluorescence microscopy) were cloned by limiting dilution and, sorted on a BD FACSAria flow cytometer (Becton Dickinson). The clones were validated by immunostaining, by western blotting and by two functional assays (lysotracker uptake and dextran uptake).

**Microscopy.** Confocal images (Figs 5–7) were acquired using an Olympus Fluoview FV1000 CLSM and × 60/1.35(oil) UPLSAPO objective with a 635-nm laser at 12%, PMT voltage of ∼750 v, pixel dwell time of 4 µs per pixel, pixel size 0.103 µm and image size 1,024 × 1,024. Nuclear staining was performed using Hoecscht33342 or DAPI. Hoecshct33342 signal was imaged using a 405-nm laser at 10% power and PMT voltage of ∼700 v. GFP signal was imaged using a 488 nm laser line at 5% power and PMT voltage of ∼600v.

Live-cell images (Figs 8–10) were acquired on a Zeiss AxioVert 200 M epi-fluorescent widefield microscope equipped with a Andor iXon 885 EMCCD, CoolLED pE-2 solid-state LEDs capable of excitation at 445, 488 and 635 nm, and Zeiss Plan-Apochromat × 100/1.40 Oil DIC objective. The microscope was surrounded by an incubation chamber that allowed the temperature and CO$_2$ to be maintained at 37 °C and 5%, respectively. Fluorophore 2 channel was recorded using a 649-nm emission long-pass filter, GFP was imaged using a 520/50 emission bandpass filter.

**Fixed cell imaging.** Cells were seeded onto an eight-well chambered glass slide and allowed to attach for 24 h. The media was then replaced with 200 µl of 2 (10 µM) in media and incubated for the appropriate time at 37 °C. Cells were counterstained with Hoechst 33342 or DAPI for 15 min. Cells were then washed once with PBS and fixed in 3.7% paraformaldehyde in PBS solution for 3 min and washed thoroughly with PBS. Images were collected by using an Olympus Fluoview 1,000 CLSM. The fluorescence arising from 2 was detected by a Cy5.5 filter. DAPI and GFP channels were used in parallel when cells were counterstained and/or transfected.

**Fixed cell imaging at different pH.** Cells were seeded onto an eight-well chambered glass slide and allowed to attach for 24 h. The media was then replaced with 200 µl of 2 (10 µM) in media incubated for 2 h at 37 °C. Cells were counterstained with DAPI for 15 min. Cells were then washed once with PBS and fixed in 3.7% paraformaldehyde in PBS solution for 3 min, and washed thoroughly with PBS. A collection of cell were Z-stack imaged using CLSM (PMT voltage = 782v, laser power 12%) and while maintaining focus of the microscope on the same cells the medium was exchanged with medium acidified to pH 4.9 (by addition of HCl (aq)). After allowing 15 min for equilibration the same cells were re-imaged (PMT voltage = 782v) using the same laser power. Following which the same cells were imaged for the third time following the adjustment of the PMT voltage 512v to obtain a non-saturated image. Mean total cell fluorescence was determined from two independent experiments using ImageJ.

**Imaging following 4 °C incubation.** Cells seeded onto an 8-well chambered glass slide and allowed to attach for 24 h. The media was then replaced with 200 µl of 2 (10 µM) in media incubated for 30 min at 4 °C. Cells were counterstained with DAPI for 15 min. Cells were then washed once with PBS and fixed in 3.7% paraformaldehyde in PBS solution for 3 min and washed thoroughly with PBS. A collection of cell were imaged using CLSM, and while maintaining focus of the microscope on the same cells the medium was exchanged with medium acidified to pH 4.9 (by addition of HCl (aq)). After allowing 15 min for equilibration, the same cells were re-imaged using the same exposure times and laser power.

**Real-time live-cell imaging.** HeLa Kyoto cells in Dulbecco's cell growth media containing 10% FBS were seeded onto an eight-well chambered glass slide and incubated for 24 h. The slides were placed on the microscope platform and the microscope was focused on a collection of cells. Next, 2 (final concentration 10 µM) was added and fluorescence images (Cy5.5 filter) were acquired at regular intervals. Images were deconvolved and combined in a video format.

**Time-dependant efflux of 2.** Cells seeded onto an eight8-well chambered glass slide and allowed to attach for 24 h. The media was then replaced with 200 µl of 2 (10 µM) in media incubated for 2 h at 37 °C. Media was replaced with fresh media and the loss of fluorescence monitored over time. Lysosome counting was carried out at 1, 15, 30 and 120 min using ImageJ.

**Image processing.** Deconvolution of widefield data sets was performed using AutQuant X3 deconvolution software with ten iterations of adaptive point spread function calculations. Lysosome detection and tracking were performed using Imaris 7.7.1 software (Bitplane Scientific). Background subtraction was applied to all images before lysosome detection. The Spots module of Imaris was used to detect lysosomes with an estimated diameter of 1.27 µm. Detected spots were filtered using the 'quality' algorithm. Only spots with values higher than the set threshold value (> 91.76) were analysed. Quality is defined as the intensity at the centre of the spot, Gaussian filtered by the spot radius. The success and accuracy of Spot detection was judged by visual inspection. Tracking lysosome movement over the course of the video was performed using an autoregressive motion algorithm. A maximum search distance of 1 µm was defined to disallow connections between a spot and a candidate match if the distance between the predicted future position of the spot and the candidate position exceeded the maximum distance. A gap-closing algorithm was also implemented to link track segment ends to track segment starts, to recover tracks that were interrupted by the temporary disappearance of particles. The maximum permissible gap length was set equal to three frames. Tracking all the lysosomes in the cell were selected by applying filters, which were based on 'Track Length' (> 0.2 µm) and 'Track Duration' (> 60 s).

**Statistical analysis of cell images.** Manders' and Pearson's coefficients used to show co-compartmentalization of LAMP1–GFP and 2 emissions were calculated using the Image J plugin 'Coloc2'. Rolling ball background subtraction (50 pixel diameter) and a Gaussian Filter (1 pixel diameter) were applied to all images before running the 'Coloc2' plugin. The ROI surrounding the cell was selected manually using the freeform drawing tool. Analysis was performed on six cells from two independent experiments.

Corrected total cell fluorescence (CTCF) in Fig. 7 was performed on six cells from two different experiments (Supplementary Fig. 8). Z-stack data acquired on the Olympus FLuoview100 was compressed into a single plane using the 'Sum Slice' function in Image J. Individual cells were selected using the freeform drawing tool to create a ROI (ROI). Selecting the 'Measure' function provided the area, the mean grey value and integrated density of the ROI. The mean background level was obtained by measuring the intensity in three different regions outside the cells and averaging the values obtained. The CTCF for each cell was calculated using the formula: CTCF = Integrated density of cell ROI − (Area of ROI × Mean fluorescence of background). The FEF was calculated by dividing the CTCF value of a cell at pH 7 into the CTCF value of the same cell at pH 4.9.

The number of lysosomes per field of view (Fig. 11) after efflux was counted using Z-stack data acquired on the Olympus Fluoview100, which was compressed into a single plane using the 'Sum Slice' function in Image J. A Max Entropy Threshold 15,000–40,000 was applied to each slice followed by use of the 'Despeckle', 'Erode' and 'Dilate' functions to remove noise. To count the number of lysosomes in each image the 'Analyze Particles' function was used to count objects with a circularity of 0.75–1.00 and size from 0 to 200 pixels.

***In vivo* mouse imaging.** MDA-MB-231-luc-D3H1, a luciferase-expressing human breast adenocarcinoma cell line, was obtained from Caliper Life Sciences. Cells were maintained as a monolayer culture in minimum essential medium containing 10% (v/v) FBS and supplemented with 1% (v/v) L-glutamine, 50 U ml$^{-1}$ penicillin, 50 µl ml$^{-1}$ streptomycin, 1% (v/v) sodium pyruvate and 1% (v/v) non-essential amino acids. All cells were maintained in 5% $CO_2$ (v/v) and 21% $O_2$ (v/v) at 37 °C. Balb/C nu/nu mice (Harlan) were housed in the Biomedical Facility (UCD) in individually ventilated cages in temperature and humidity controlled rooms with a 12-h light–dark cycle. Two to five million MDA-MB-231-luc-D3H2LN cells in 100 µl of a DPBS:Matrigel (50:50) solution were injected subcutaneously behind the fore limb of the 5-week-old mice using a 25-g needle. Tumours reached an average diameter of 6 mm before injection. All animal protocols were approved by University College Dublin's local Animal Research Ethics Committee and under the licence from the Department of Health and Children. Animals were split into two groups ($n = 4$) and **2** dissolved in PBS (200 µl) was administered through the lateral tail vein at a concentration of 2 mg kg$^{-1}$. Optical imaging was performed with an IVIS Spectrum small-animal *in-vivo* imaging system (Caliper LS) with integrated isoflurane anaesthesia. A non-injected control animal was included. Images were acquired at regular intervals post injection of **2** with excitation 675 nm (30 nm band-pass filter) and emission 720 nm (20 nm band-pass filter) narrow band-pass filters and were analysed using Living Image Software v3.0 (Caliper LS).

## References

1. Salipalli, S., Singh, P. K. & Borlak, J. Recent advances in live cell imaging of hepatoma cells. *J. BMC Cell Biol.* **15**, 26 (2014).
2. Correa, Jr I. R. Live-cell reporters for fluorescence imaging. *Curr. Opin. Chem. Biol.* **20**, 36–45 (2014).
3. Dean, K. M. & Palmer, A. E. Advances in fluorescence labelling strategies for dynamic cellular imaging. *Nat. Chem. Biol.* **10**, 512–523 (2014).
4. Baker, M. Cellular imaging: taking a long, hard look. *Nature* **466**, 1137–1140 (2010).
5. de Jong, M., Essers, J. & van Weerden, W. M. Imaging preclinical tumour models: improving translational power. *Nat. Rev. Cancer* **14**, 481–493 (2014).
6. Olivo, M., Ho, C. J. H. & Fu, C. Y. Advances in fluorescence diagnosis to track footprints of cancer progression *in vivo*. *Laser Photon. Rev.* **7**, 646–662 (2013).
7. Vahrmeijer, A. L., Hutteman, M., van der Vorst, J. R., van de Velde, C. J. H. & Frangioni, J. V. Image-guided cancer surgery using near-infrared fluorescence. *Nat. Rev. Clin. Oncol.* **10**, 507–518 (2013).
8. Liu, Y. *et al.* Near-infrared fluorescence goggle system with complementary metal-oxide-semiconductor imaging sensor and see-through display. *Biomed. Opt.* **18** 101303 1–10 (2013).
9. Sevick-Muraca, E. M. Translation of near-infrared fluorescence imaging technologies: emerging clinical applications. *Annu. Rev. Med.* **63**, 217–231 (2012).
10. van Dam, G. M. *et al.* Intraoperative tumor-specific fluorescence imaging in ovarian cancer by folate receptor-alpha targeting: first in-human results. *Nat. Med.* **17**, 1315–1319 (2011).
11. Nguyen, Q. T. *et al.* Surgery with molecular fluorescence imaging using activatable cell-penetrating peptides decreases residual cancer and improves survival. *Proc. Natl Acad. Sci. USA* **107**, 4317–4322 (2010).
12. Li, X., Gao, X., Shi, W. & Ma, H. Design strategies for water-soluble small molecular chromogenic and fluorogenic probes. *Chem. Rev.* **114**, 590–659 (2014).
13. Guo, Z., Park, S., Yoon, J. & Shin, I. Recent progress in the development of near-infrared fluorescent probes for bioimaging applications. *Chem. Soc. Rev.* **43**, 16–29 (2014).
14. Yuan, L., Lin, W., Zheng, K., He, L. & Huang, W. Far-red to near infrared analyte-responsive fluorescent probes based on organic fluorophore platforms for fluorescence imaging. *Chem. Soc. Rev.* **42**, 622–661 (2013).
15. Wu, D. & O'Shea, D. F. Synthesis and properties of BF$_2$-3,3′-dimethyldiaryl azadipyrromethene near-infrared fluorophores. *Org. Lett.* **15**, 3392–3395 (2013).
16. Palma, A. *et al.* Cellular uptake mediated off/on responsive near-infrared nanoparticles. *J. Am. Chem. Soc.* **133**, 19618–19621 (2011).
17. Batat, P. *et al.* BF$_2$-azadipyrromethenes: probing the excited-state dynamics of a NIR fluorophore and photodynamic therapy agent. *J. Phys. Chem. A.* **115**, 14034–14039 (2011).
18. Tasior, M. & O'Shea, D. F. BF$_2$-Chelated tetraarylazadipyrromethenes as NIR fluorochromes. *Bioconjugate Chem.* **21**, 1130–1133 (2010).
19. Kilpatrick, B. S., Eden, E. R., Hockey, L. N., Futter, C. E. & Patel, S. Methods for monitoring lysosomal morphology. *Methods Cell Biol.* **126**, 1–19 (2015).
20. Mayor, S. & Pagano, R. E. Pathways of clathrin-independent endocytosis. *Nat. Rev. Mol. Cell Biol.* **8**, 603–612 (2007).
21. Canton, I. & Battaglia, G. Endocytosis at the nanoscale. *Chem. Soc. Rev.* **41**, 2718–2739 (2012).
22. Casey, J. R., Grinstein, S. & Orlowski, J. Sensors and regulators of intracellular pH. *Nat. Rev. Mol. Cell Biol.* **11**, 50–61 (2010).
23. Lee, H. *et al.* Near-infrared pH-activatable fluorescent probes for imaging primary and metastatic breast tumors. *Bioconjugate Chem.* **22**, 777–784 (2011).
24. Han, J. & Burgess, K. Fluorescent indicators for intracellular pH. *Chem. Rev.* **110**, 2709–2728 (2010).
25. Koide, Y., Urano, Y., Hanaoka, K., Terai, T. & Nagano, T. Evolution of group 14 rhodamines as platforms for near-infrared fluorescence probes utilizing photoinduced electron transfer. *ACS Chem. Biol.* **6**, 600–608 (2011).
26. Urano, Y. *et al.* Selective molecular imaging of viable cancer cells with pH-activatable fluorescence probes. *Nat. Med.* **15**, 104–109 (2009).
27. Knop, K., Hoogenboom, R., Fischer, D. & Schubert, U. S. Poly(ethylene glycol) in drug delivery: pros and cons as well as potential alternatives. *Angew. Chem. Int. Ed.* **49**, 6288–6308 (2010).
28. Murtagh, J., Frimannsson, D. O. & O'Shea, D. F. Azide conjugatable and pH responsive near-infrared fluorescent imaging probes. *Org. Lett.* **11**, 5386–5389 (2009).
29. Zhang, X.-X. *et al.* pH-sensitive fluorescent dyes: are they really pH-sensitive in cells? *Mol. Pharm.* **10**, 1910–1917 (2013).
30. Hall, M. J., Allen, L. T. & O'Shea, D. F. PET modulated fluorescent sensing from the BF$_2$ chelated azadipyrromethene platform. *Org. Biomol. Chem.* **4**, 776–780 (2006).
31. Garcia, M. E. D. & Medel, A. S. Dye-surfactant interactions: a review. *Talanta* **33**, 255–264 (1986).
32. Katritzky, A. R., Fara, D. C., Yang, H. & Tamm, K. Quantitative measures of solvent polarity. *Chem. Rev.* **104**, 175–198 (2004).
33. Falcon-Perez, J. M., Nazarian, R., Sabatti, C. & Dell'Angelica, E. C. Distribution and dynamics of Lamp1-containing endocytic organelles in fibroblasts deficient in BLOC-3. *J. Cell Sci.* **118**, 5243–5255 (2005).
34. Bolte, S. & Cordelieres, F. P. A guided tour into subcellular colocalization analysis in light microscopy. *J. Microsc.* **224**, 213–232 (2006).
35. Firdessa, R., Oelschlaeger, T. A. & Moll, H. Identification of multiple cellular uptake pathways of polystyrene nanoparticles and factors affecting the uptake: relevance for drug delivery systems. *Eur. J. Cell Biol.* **93**, 323–337 (2014).
36. De Mey, J. R. *et al.* Fast 4D Microscopy. *Methods Cell Biol.* **85**, 83–112 (2008).
37. Godley, B. F. *et al.* Blue light induces mitochondrial DNA damage and free radical production in epithelial cells. *J. Biol. Chem.* **280**, 21061–21066 (2005).
38. Freundt, E. C., Czapiga, M. & Lenardo, M. J. Photoconversion of lysotracker red to a green fluorescent molecule. *Cell Res.* **17**, 956–958 (2007).
39. Galindo, F. *et al.* Synthetic macrocyclic peptidomimetics as tunable pH probes for the fluorescence imaging of acidic organelles in live cells. *Angew. Chem. Int. Ed.* **44**, 6504–6508 (2005).
40. Chazotte, B. Labeling lysosomes in live cells with lysotracker. *Cold Spring Harb. Protoc.* **2011**, pdb.prot5570 (2011).
41. Zhang, J. *et al.* Near-infrared fluorescent probes based on piperazine-functionalized BODIPY dyes for sensitive detection of lysosomal pH. *J. Mat. Chem. B* **3**, 2173–2184 (2015).
42. Gorman, A. *et al.* In vitro demonstration of the heavy-atom effect for photodynamic therapy. *J. Am. Chem. Soc.* **126**, 10619–10631 (2004).

## Acknowledgements

D.O.S. gratefully acknowledges Science Foundation Ireland grant number 11/PI/1071(T) for financial support. E.C. and W.M.G. acknowledge the Irish Cancer Society Collaborative Cancer Research Centre BREAST-PREDICT (CCRC13GAL) for financial support. J.C.S. and A.P. acknowledge Science Foundation Ireland grant 09/IN.1/B2604 for financial support.

## Author contributions

M.G. carried out all synthetic chemistry and photophysical measurements. M.M. and S.C. did all fixed and live-cell imaging and imaging analysis. D.S. set up and managed all imaging hardware and software used, and provided expertise advice. E.C. and M.T. conducted the *in vivo* imaging study. A.P. generated LAMP-1 GFP HeLa cells. J.S. provided LAMP-1 GFP HeLa cells and assisted with image and data analysis. W.G. provided expertise on *in vivo* imaging. D.O.S. wrote the manuscript with input from all the co-authors.

## Additional information

**Competing financial interests:** Two authors declare the following competing financial interest. A patent application has been filed on azadipyrromethene based NIR fluorophores (PCT/EP2010/065991) in which both D.O.S. and W.G. have a financial interest.

# 10

# Structural and dynamic insights into the energetics of activation loop rearrangement in FGFR1 kinase

Tobias Klein[1,*,†], Navratna Vajpai[1,*,†], Jonathan J. Phillips[2,*,†], Gareth Davies[1], Geoffrey A. Holdgate[1], Chris Phillips[1], Julie A. Tucker[1,†], Richard A. Norman[1], Andrew D. Scott[1,†], Daniel R. Higazi[2], David Lowe[2], Gary S. Thompson[3] & Alexander L. Breeze[1,3,†]

Protein tyrosine kinases differ widely in their propensity to undergo rearrangements of the N-terminal Asp–Phe–Gly (DFG) motif of the activation loop, with some, including FGFR1 kinase, appearing refractory to this so-called 'DFG flip'. Recent inhibitor-bound structures have unexpectedly revealed FGFR1 for the first time in a 'DFG-out' state. Here we use conformationally selective inhibitors as chemical probes for interrogation of the structural and dynamic features that appear to govern the DFG flip in FGFR1. Our detailed structural and biophysical insights identify contributions from altered dynamics in distal elements, including the αH helix, towards the outstanding stability of the DFG-out complex with the inhibitor ponatinib. We conclude that the αC-β4 loop and 'molecular brake' regions together impose a high energy barrier for this conformational rearrangement, and that this may have significance for maintaining autoinhibition in the non-phosphorylated basal state of FGFR1.

[1] Discovery Sciences, AstraZeneca R&D, Alderley Park, Macclesfield SK10 4TG, UK. [2] MedImmune, Granta Park, Cambridge CB21 6GH, UK. [3] Astbury Centre for Structural Molecular Biology, Faculty of Biological Sciences, University of Leeds, Leeds LS2 9JT, UK. * These authors contributed equally to this work. † Present addresses: Bayer Healthcare, GP Grenzach Produktions GmbH, Postfach 1146, D-79629 Grenzach-Wyhlen, Germany (T.K.); Biological E. Ltd, ICICI Knowledge Park, Shameerpet, Ranga Reddy District, Hyderabad, Telangana 500078, India (N.V.); Department of Chemical Engineering and Biotechnology, University of Cambridge, Cambridge CB2 3RA, UK (J.J.P.); Northern Institute for Cancer Research, Paul O'Gorman Building, Medical School, Newcastle University, Framlington Place, Newcastle upon Tyne NE2 4HH, UK (J.A.T.); Molplex Ltd, BioHub at Alderley Park, Alderley Park, Macclesfield SK10 4TG, UK (A.D.S.); Astbury Centre for Structural Molecular Biology, Faculty of Biological Sciences, University of Leeds, Leeds LS2 9JT, UK (A.L.B.). Correspondence and requests for materials should be addressed to A.L.B. (email: a.l.breeze@leeds.ac.uk).

Receptor tyrosine kinases (RTKs) wield exquisite control over cell differentiation, fate, metabolism and homeostasis. Dysregulation of RTK signalling plays a significant role in the pathogenesis of disease conditions ranging from cancers to inflammatory and neurodegenerative illnesses. Hence, it is not surprising that over the past two decades they have become one of the most important classes of enzyme to be exploited as targets for drug discovery[1]. Conformational plasticity is an essential feature of kinase function and regulation. Inhibitors of kinase domain catalytic activity developed in the course of drug discovery programmes have drawn attention to the importance of mobility in the conserved Asp–Phe–Gly (DFG) tripeptide motif at the proximal end of the activation loop (A-loop). The majority of kinase inhibitors described to date bind competitively with ATP to a presumed basal state conformation (termed 'DFG-in' or the 'type I' binding mode) in which the Phe side chain of the DFG motif resides in a hydrophobic pocket deep within the kinase fold. An early insight into the role of the DFG motif as a conformational switch was provided by the structure of the tyrosine kinase Bcr-Abl complexed with the inhibitor STI-571 (imatinib)[2]. This structure indicated that the DFG motif undergoes a conformational rearrangement whereby the Phe side chain is flipped out of its hydrophobic pocket, vacating space for insertion of an aromatic moiety of the inhibitor. 'DFG-out' conformations have since been observed in many kinases, both inhibitor bound[3] and, occasionally, in the unbound state[4–8]. The DFG-out state is catalytically inactive, since it is sterically incompatible with cofactor and substrate binding, and in some kinases may natively contribute to autoinhibition[8,9]. Indeed, several so-called 'type II' inhibitors, that bind to and stabilize the DFG-out form of a number of kinases, have been described[10]. An intriguing anecdotal observation from drug discovery is that it is relatively easy to identify type II (as opposed to type I) inhibitors against some kinases, but difficult or impossible for others. A plausible explanation for these differences may lie in the chemical space populated by screening libraries, favouring type I binding modes against some kinases and type II inhibitors in others. Alternatively, there could be specific structural or dynamic differences between individual kinases that relatively favour one or other binding mode. This conformational balance has been referred to as the 'DFG-out propensity'[11]. Evidence has been advanced recently that DFG-out propensity and/or the rates of interconversion between DFG-in and DFG-out may be influenced by the side chain properties at, or adjacent to, particular points of the regulatory or catalytic 'spines' of the kinase domain[12].

Members of the FGFR family (FGFR1 to 4) are key mediators of both developmental and disease-associated angiogenesis[13] and are heavily implicated in the pathogenesis of tumour vascularization in a number of different tumour types including breast[14], pancreatic[15], prostate[16] and ovarian[17] carcinomas, as well as being driving oncogenes for malignant transformation in their own right[13,18]. Hence, they have been seen as attractive targets for the development of therapeutic agents aimed at inhibition of tumour growth and metastasis. Despite concerted efforts to develop type II inhibitors of FGFR1 kinase in our own drug discovery programme, we obtained only type I inhibitors as confirmed by X-ray crystallographic analysis of >70 compounds, and none of the >30 FGFR1 kinase structures in the Protein Data Bank adopts the DFG-out conformation. Recently, however, we observed that the Bcr-Abl inhibitor ponatinib (AP24534) also binds potently to FGFR1 kinase, and moreover we and others have now confirmed that it binds to the DFG-out conformation of FGFR kinases[19–21]. Intrigued by this finding, we embarked on an investigation of the factors that underlie the seemingly strong preference for the DFG-in state in FGFR1, using inhibitors that stabilize the respective A-loop conformations as chemical 'free-energy probes'. When compared with well-known type I inhibitors, binding of ponatinib to FGFR1 revealed startling differences in kinetic and thermodynamic behaviour associated with the two binding modes. Our analysis of changes in protein dynamics between the unbound, type I-bound and type II-bound states, using both nuclear magnetic resonance (NMR) and hydrogen–deuterium-exchange mass spectrometry (HDX-MS), shows that both proximal and distal structural elements influence activation loop conformational energetics in FGFR1.

## Results

For our studies of the FGFR1 kinase domain, we have used a construct spanning residues Ala458 to Glu765 of human FGFR1 that contains two mutations (Cys488Ala and Cys584Ser) designed to stabilize the enzyme against covalent aggregation. The protein is non-phosphorylated after co-expression with PTP1B and purification from *Escherichia coli*. An additional mutation of the catalytic aspartate (Asp623Ala) was introduced for NMR studies to increase further the yield of stable isotope-labelled protein. Our previous studies have shown that this mutation does not detectably perturb the structure of the FGFR1 kinase domain[22]. In addition, we have confirmed using surface plasmon resonance (SPR) that binding parameters for a close analogue of the canonical FGFR1 inhibitor PD173074 (ref. 23; henceforth referred to as PDA; Fig. 1a; Supplementary Fig. 1) are unaltered for the Asp623Ala mutant relative to the kinase-active form (data not shown).

**Ponatinib-binding kinetics suggest a low DFG-out propensity.** On binding of ponatinib (Fig. 1a), and in contrast to the binding of PDA, FGFR1 kinase domain was observed to adopt a DFG-out conformation as determined by X-ray crystallography in our laboratory and reported elsewhere[19–21] (Fig. 1b,c). The reason for the apparently refractory behaviour of FGFR1 towards adopting 'DFG-out' has, however, remained elusive to date. We hypothesized that an intrinsically low DFG-out propensity might be the underlying reason, prompting us to compare the binding kinetics of ponatinib (to date the only known high-affinity type II FGFR1 inhibitor) with those of representative type I inhibitors (Supplementary Fig. 1a). A kinetic analysis using SPR highlights the fact that while the type I inhibitors we investigated show fairly uniformly fast association rate constants, the binding of ponatinib is exceptionally slow (Fig. 1d, Supplementary Table 1). With an association rate constant of $2.4 \times 10^4 \, \text{M}^{-1} \text{s}^{-1}$, it is $\sim 70 \times$ slower than that of PDA ($k_{\text{on}} = 1.6 \times 10^6 \, \text{M}^{-1} \text{s}^{-1}$), which has almost identical affinity to ponatinib. Notably, the comparable affinity of ponatinib ($K_D = 7.9 \, \text{nM}$) and PDA ($K_D = 5.7 \, \text{nM}$) for FGFR1 is a result of the outstandingly long lifetime of the ponatinib–FGFR1 complex: the dissociation rate constant for ponatinib ($k_{\text{off}} = 1.9 \times 10^{-4} \, \text{s}^{-1}$, corresponding to a half-life ($t_{1/2}$) for the complex of $\sim 61 \, \text{min}$) is $\sim 50 \times$ slower than that for PDA ($k_{\text{off}} = 9.2 \times 10^{-3} \, \text{s}^{-1}$; $t_{1/2} = 1.3 \, \text{min}$). Slow rate constants have previously been reported for type II inhibitors binding to a number of kinases[24–28]. These observations have been interpreted as consistent with a slow equilibrium between DFG-in and DFG-out states, where the DFG-out conformation is sampled only infrequently, accompanied in some cases by slow interconversion of ligand conformations, as ,for example, in the binding of analogues of BIRB-796 to p38 MAP kinase[29]. Recently, Shan *et al.*[11], using Abl kinase as a model system, suggested that the DFG conformation is controlled by a protonation-dependent energetic switch. According to that analysis, the acidity of the DFG aspartate may be one factor that drives the equilibrium between the DFG-in and DFG-out conformations. To investigate whether protonation of the DFG aspartate also influences the

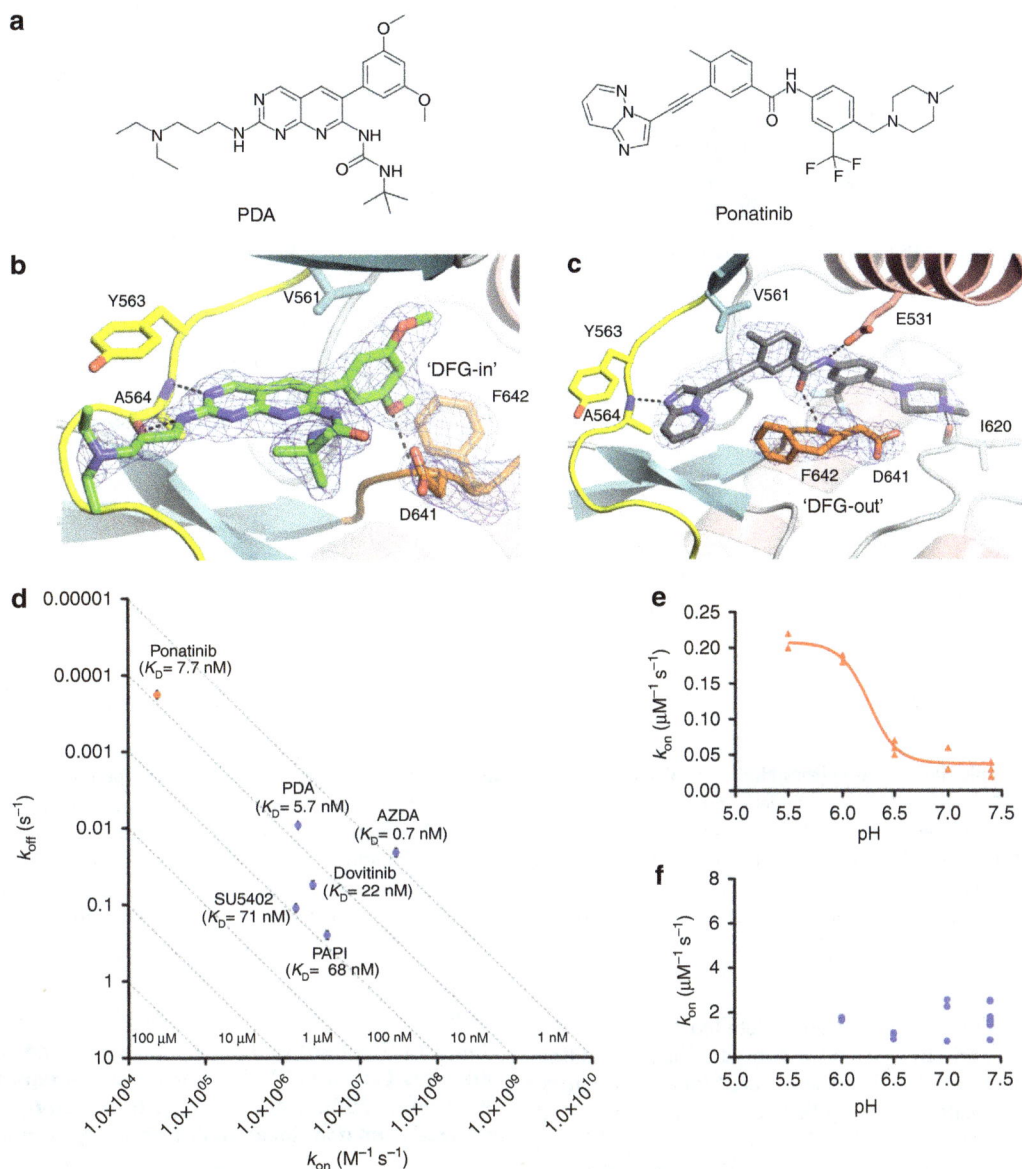

**Figure 1 | Structural and kinetic characteristics of FGFR1 complexes with the type I and type II inhibitors PDA and ponatinib.** (**a**) Chemical structures of PDA and ponatinib. (**b**) Active site of FGFR1 kinase in complex with PDA (green carbons) as determined at 2.09 Å resolution (Supplementary Table 5). The hinge region (yellow) and A-loop (orange) are highlighted. $F_o - F_c$ OMIT electron density for PDA and the DFG motif is represented as a blue mesh contoured at $3.0\sigma$. Polar interactions are indicated as dotted lines. (**c**) Active site of FGFR1 kinase in complex with ponatinib (grey carbons) as determined at 2.33 Å resolution, with $F_o - F_c$ OMIT electron density for ponatinib and the DFG motif represented as a blue mesh contoured at $3.0\sigma$. Colouring of FGFR1 as in **b**. (**d**) Kinetic value plot of association rate constant ($k_{on}$) versus dissociation rate constant ($k_{off}$). Rate constants were determined using SPR at 298 K, pH = 7.4. The affinities ($K_D$) were calculated from the equation $K_D = k_{off}/k_{on}$ and broken lines represent affinity isotherms. Data represent geometric means from at least three independent experiments; standard errors are shown as error bars (values and errors are presented in Supplementary Table 1). (**e**) The association rate constant of ponatinib binding to FGFR1 as a function of pH, as measured by SPR at 298 K. The red line represents the result of the non-linear fitting of the data to the 4 PL model ($R^2 = 0.968$; $pK_a$(Asp641) = 6.25). (**f**) The association rate constant of PDA binding to FGFR1 as a function of pH, as measured by SPR at 298 K.

DFG flip in FGFR1, we derived the association rate constants of the DFG-out inhibitor ponatinib and the DFG-in inhibitor PDA as a function of pH. The rate of ponatinib binding to FGFR1 increased nearly sevenfold as pH decreased from 7.4 to 5.5 (Fig. 1e). Ponatinib is expected to be protonated on its terminal methylpiperazinyl nitrogen across this pH range; thus, the pH dependence very likely reports on the ionization state of Asp641, giving a calculated effective $pK_a$ of 6.25, well above the unperturbed range for aspartate. The observed variation in the on-rate for binding to the DFG-out conformation as a function of Asp641 ionization state lends support to the hypothesis that the

DFG flip is rate-limiting on ponatinib association to FGFR1. In contrast to ponatinib, the binding kinetics of PDA to FGFR1 showed no dependence on pH over a similar range (Fig. 1f), which is consistent with the assumption that the binding of the so-called DFG-in inhibitors is not affected by the conformation of the DFG motif[30].

**Ponatinib binding is accompanied by an enthalpic penalty.** The apparently slow equilibrium between DFG-in and DFG-out conformations in FGFR1 kinase suggests a high free-energy

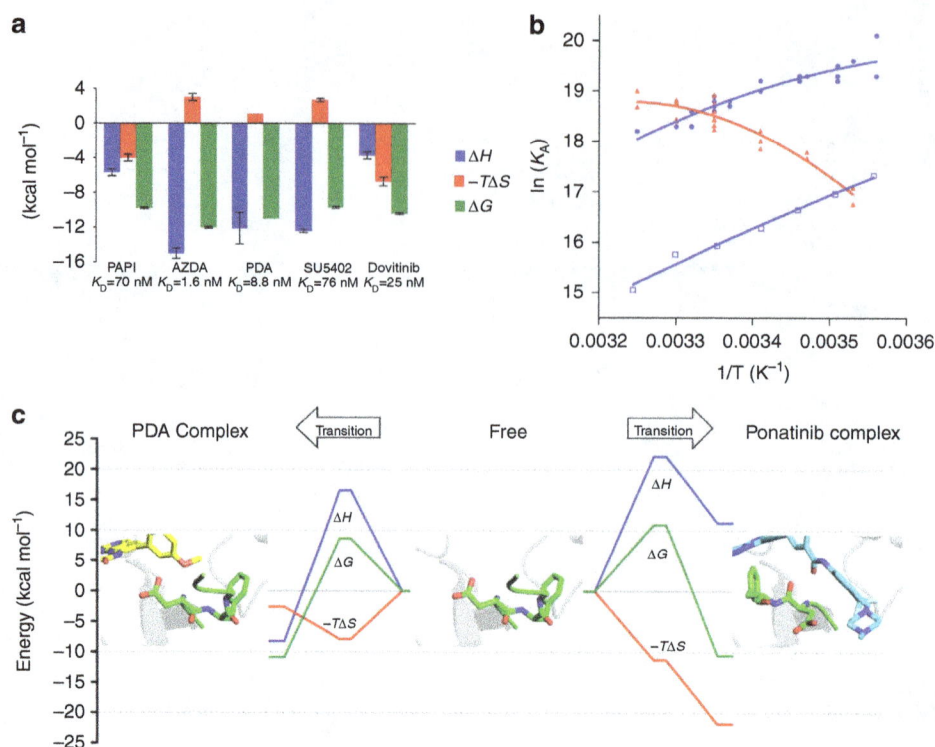

**Figure 2 | Thermodynamic data for inhibitors binding to FGFR1 kinase domain.** (**a**) Thermodynamic signatures for type I inhibitors binding to FGFR1 derived by ITC at 298 K. Data shown are arithmetic mean ± s.d. from at least two independent experiments (values and errors are presented in Supplementary Table 2). (**b**) van't Hoff plot visualization of temperature-dependent FGFR1–ligand interactions measured by SPR for PDA (blue circles), SU5402 (blue open squares) and ponatinib (red triangles). (**c**) Thermodynamic reaction pathway models for FGFR1 interacting with PDA (left) and ponatinib (right). The reaction coordinate depicts the lowest energy continuous pathway between the free (centre of the figure) and bound states (left for PDA complex; right for ponatinib complex) via the transition state, for free energy $\Delta G$ (green), enthalpy $\Delta H$ (blue) and entropy $-T\Delta S$ (red).

barrier for the DFG flip. We carried out a detailed analysis of the changes in enthalpy and entropy that accompany ligand binding to enhance our understanding. For the selected type I inhibitors, isothermal titration calorimetry (ITC) experiments revealed exothermic binding reactions (Fig. 2a, Supplementary Fig. 2, Supplementary Table 2), and the derived binding affinities were largely in agreement with those determined by SPR. In contrast, for ponatinib, which binds to FGFR1 in a DFG-out conformation, the observed titration curve (Supplementary Fig. 2h) was of poor quality and did not allow derivation of thermodynamic parameters. As an alternative to ITC, we analysed kinetic and equilibrium data from SPR as a function of temperature, following the van't Hoff method, to provide independent thermodynamic characterization of binding events. For ponatinib and two selected type I inhibitors (PDA and SU5402), the derived binding enthalpies and entropies revealed another marked difference between the type II inhibitor ponatinib and the selected type I inhibitors (Fig. 2b, Table 1). In the case of PDA and SU5402, van't Hoff analysis confirmed exothermic binding enthalpies (PDA, $\Delta H = -11.5\,\text{kcal mol}^{-1}$; SU5402, $\Delta H = -14.2\,\text{kcal mol}^{-1}$) and the data are in close agreement with the $\Delta H$ values of $-12.1\,\text{kcal mol}^{-1}$ (PDA) and $-12.4\,\text{kcal mol}^{-1}$ (SU5402) determined by ITC. Unexpectedly, the type II inhibitor ponatinib showed an endothermic $\Delta H$ value ($\Delta H = 10.1\,\text{kcal mol}^{-1}$) that indicates enthalpically unfavourable binding. Ponatinib and PDA exhibit comparable van't Hoff free energies of binding (Table 1) that are consistent with their very similar affinities measured directly by SPR; however, breaking this down into enthalpic and entropic components revealed significant differences, as the

binding of PDA and ponatinib were determined to be enthalpy driven and entropy driven, respectively. An endothermic enthalpy, as observed for the equilibrium between free and FGFR1-bound ponatinib, raises the possibility that the conformational rearrangement required to effect the DFG flip in FGFR1 may also be associated with an enthalpic penalty (neglecting net contributions from protein–ligand and protein–solvent interactions of ponatinib). Furthermore, we established that the vascular endothelial growth factor receptor (VEGFR) inhibitor tivozanib (AV-951) also binds to FGFR1 in a DFG-out mode ($K_D = 1.3\,\mu M$ by SPR) and does so endothermically by van't Hoff analysis (Supplementary Fig. 3), lending further support to the notion that this may be a signature of a DFG-out binding mode for FGFR1, rather than a compound-specific characteristic of ponatinib.

**Table 1 | Standard enthalpies, entropies and Gibbs free energies (kcal mol$^{-1}$) for binding of the type I inhibitors, PDA and SU5402, and the type II inhibitor ponatinib, derived from non-linear van't Hoff analysis of data in Fig. 2b.**

| Inhibitor | $\Delta H_0^{\text{van't Hoff}}$ | $-T\Delta S_0^{\text{van't Hoff*}}$ | $\Delta G_0^{\text{van't Hoff*}}$ | $R^{2\dagger}$ |
|---|---|---|---|---|
| PDA | $-11.5 \pm 0.8$ | $0.4 \pm 0.8$ | $-11.1 \pm 0.02$ | 0.862 |
| SU5402 | $-14.2 \pm 0.8$ | $4.8 \pm 0.8$ | $-9.5 \pm 0.02$ | 0.983 |
| Ponatinib | $10.1 \pm 1.6$ | $-21.0 \pm 1.6$ | $-10.9 \pm 0.04$ | 0.879 |

*Standard errors, $T_0 = 298\,\text{K}$.
†From non-linear fitting of data in Fig. 3b.

**A large free-energy barrier for the the DFG flip.** By measuring the temperature dependence of the kinetic association and dissociation rate constants for PDA and ponatinib, we were able to discern the transition state energies for the association and dissociation steps of the binding reaction according to the method of Eyring (for details, see Supplementary Information). From linear Eyring plots (Supplementary Fig. 4), we determined $\Delta H^{\#}$, $-T\Delta S^{\#}$ and $\Delta G^{\#}$ for the association and dissociation steps (Supplementary Table 3), which enabled us to construct detailed thermodynamic reaction pathway models for the binding of PDA and ponatinib to FGFR1 (Fig. 2c). For the binding of the type I inhibitor PDA to FGFR1, we observed a free-energy barrier of 8.7 kcal mol$^{-1}$ associated with the transition state. This energy barrier is dominated by an enthalpic penalty ($\Delta H^{\#}_{ass} = 16.6$ kcal mol$^{-1}$); however, a significant favourable entropy ($-T\Delta S^{\#}_{ass} = -7.9$ kcal mol$^{-1}$) lowers the overall free-energy barrier to reach the transition state. As observed for PDA, the transition state for binding of the type II inhibitor ponatinib to FGFR1 is also associated with an enthalpic penalty, which is partly compensated by a favourable entropic contribution. However, although the transition state entropy for the association of ponatinib ($-T\Delta S^{\#}_{ass} = -11.2$ kcal mol$^{-1}$) is more favourable compared with that of PDA, it is not sufficient to compensate for the extraordinarily unfavourable $\Delta H^{\#}_{ass}$ of 22.2 kcal mol$^{-1}$ for ponatinib binding, resulting in a 2.3 kcal mol$^{-1}$ higher transition state free-energy barrier ($\Delta G^{\#}_{ass} = 11.0$ kcal mol$^{-1}$) associated with the type II binding mode. This difference in transition state free energy is in excellent accord with the $\sim 70$-fold slower association rate constant that we measure for ponatinib binding.

**A partially unfolded intermediate in the DFG-out transition.** Localized unfolding is widely believed to contribute to the crossing of free-energy barriers encountered during protein motion[31], and 'cracking' at the kinase hinge region has been observed to be a key element in the simulated DFG-in/out transition of EGFR kinase[32]. Assuming that protein conformational energetics contribute substantially to the free-energy changes on binding type I and type II inhibitors[33], the thermodynamic signature (unfavourable enthalpy and favourable entropy) that we have observed for the association of ponatinib to FGFR1 (Fig. 2, Table 1) is consistent with the proposed model of partial unfolding, facilitating conformational transitions in proteins. Therefore, our thermodynamic data suggest that the transition state conformations traversed by FGFR1 during the 'in-out' trajectory may be partially, or locally, unfolded. The guanidinium chloride (GdmCl)-induced unfolding transition curve of FGFR1 monitored by the change in far-ultraviolet circular dichroism (CD) shows two folding transitions, the first occurring between 1 and 2 M GdmCl with accumulation of an intermediate at $\sim 2$ M GdmCl. The second transition occurs between 2 and 3.5 M GdmCl, by which point the protein is completely unfolded (Supplementary Fig. 5). Using SPR, we determined the association rate constant of ponatinib in the presence of 1.2 M GdmCl in the phosphate-buffered saline (PBS) running buffer to sample the partly unfolded intermediate. With an association rate constant of $2.3 \times 10^5$ M$^{-1}$ s$^{-1}$, it is almost an order of magnitude faster than that of ponatinib in the absence of GdmCl ($k_{on} = 2.4 \times 10^4$ M$^{-1}$ s$^{-1}$). In contrast, PDA binding is minimally affected by the presence of 1.2 M GdmCl, with a $k_{on} = 6.8 \times 10^5$ M$^{-1}$ s$^{-1}$, consistent with the lack of requirement for a flip of the DFG motif for type I inhibitor binding. The structural loosening induced by intermediate concentrations of denaturant might be expected also to influence dissociation rates, and this was indeed observed for both inhibitors. Interestingly, the $k_{off}$ for PDA was again only

moderately affected ($k_{off} = 1.8 \times 10^{-3}$ s$^{-1}$, against $9.2 \times 10^{-3}$ s$^{-1}$ in the absence of GdmCl; resulting in a $K_D$ $\sim 2$-fold weaker), while that for ponatinib was dramatically increased ($k_{off} = 1.8 \times 10^{-2}$ s$^{-1}$ against $1.9 \times 10^{-4}$ s$^{-1}$ in the absence of GdmCl), leading to an $\sim 10$-fold weaker $K_D$ overall in the presence of 1.2 M GdmCl. The differential effects on association rate constant for DFG-in and DFG-out ligands provide evidence that the partly unfolded FGFR1 intermediate observed from the unfolding curve favours attainment of the DFG-out conformation, and furthermore suggests that it could serve as an intermediate of the DFG flip in FGFR1.

**Ponatinib binds to FGFR1 more slowly than to Abl.** Ponatinib binds to DFG-out conformations of FGFR1 and Abl kinases with an almost identical binding mode and many conserved interactions between inhibitor and protein. In view of these overall similarities, we compared the ponatinib association rates for both kinases, using identical SPR-based protocols, to address the question of whether different underlying DFG-out propensities might play a significant role in type II inhibitor binding in these tyrosine kinases. The binding rate of ponatinib using SPR is over an order of magnitude faster for Abl ($k_{on} = 5.2 \times 10^5$ M$^{-1}$ s$^{-1}$) than for FGFR1 ($k_{on} = 2.4 \times 10^4$ M$^{-1}$ s$^{-1}$). This faster association rate accounts for the greater part of the roughly 10-fold higher affinity of ponatinib for Abl ($K_D = 0.9$ nM) than for FGFR1 ($K_D = 7.9$ nM), as the dissociation rate constants are rather similar for the two kinases (Supplementary Table 4). Our measured association rate for binding of the canonical type II inhibitor, imatinib, to Abl ($k_{on} = 5.5 \times 10^5$ M$^{-1}$ s$^{-1}$) is similar to that for ponatinib, and is in good accord with previously reported data[34]. This suggests that the difference in on-rate constant between the two kinases that we observe for ponatinib may reflect a fundamental difference in conformational energetic balance, with a considerably higher free-energy barrier for adopting the DFG-out conformation in FGFR1 in contrast to Abl kinase.

**Dynamic cross-talk revealed by NMR and mass spectrometry.** To gain insights into the dynamic origins of slow access to the DFG-out state in FGFR1, we employed both NMR spectroscopy and HDX-MS. We have previously reported NMR resonance assignments for FGFR1 kinase domain in the ligand-free state[22]. Titration of either PDA or ponatinib into samples of $^{15}$N-labelled FGFR1 kinase resulted in amide chemical shift perturbations (CSPs) in the slow-exchange regime that were completely saturated at 1:1 molar stoichiometry, typical of high-affinity binding in the nanomolar $K_D$ range (Fig. 3a). Unlike for the unbound[22] and the PDA-complex states of FGFR1, the first six residues (Asp641–Arg646) of the A-loop were observable in the $^1$H-$^{15}$N TROSY-HSQC spectrum of the FGFR1–ponatinib complex, indicative of altered A-loop dynamics in the ponatinib complex compared with the unbound or PDA-bound kinase. Comparison of $^1$H-$^{15}$N TROSY-HSQC spectra of both PDA and ponatinib complexes with unbound FGFR1 showed large amide chemical shift changes for many residues. Mapping of these perturbations onto the crystal structure of FGFR1 (PDB-code: 1FGK) shows that most of them are localized in the catalytically important and structurally conserved regions surrounding the active site (Fig. 3a). Significant CSPs were observed for Ala564 in the hinge region of both complexes, due to direct hydrogen-bond interactions with a ring nitrogen of the inhibitor; that seen in the ponatinib complex is substantially larger and may reflect a stronger hydrogen bond. For the PDA complex, CSPs were detected only for residues in the region of the P-loop, the N-terminus of the αC helix, the hinge region residues, and Ala640, which are all in close promixity to the inhibitor (Fig. 3a,

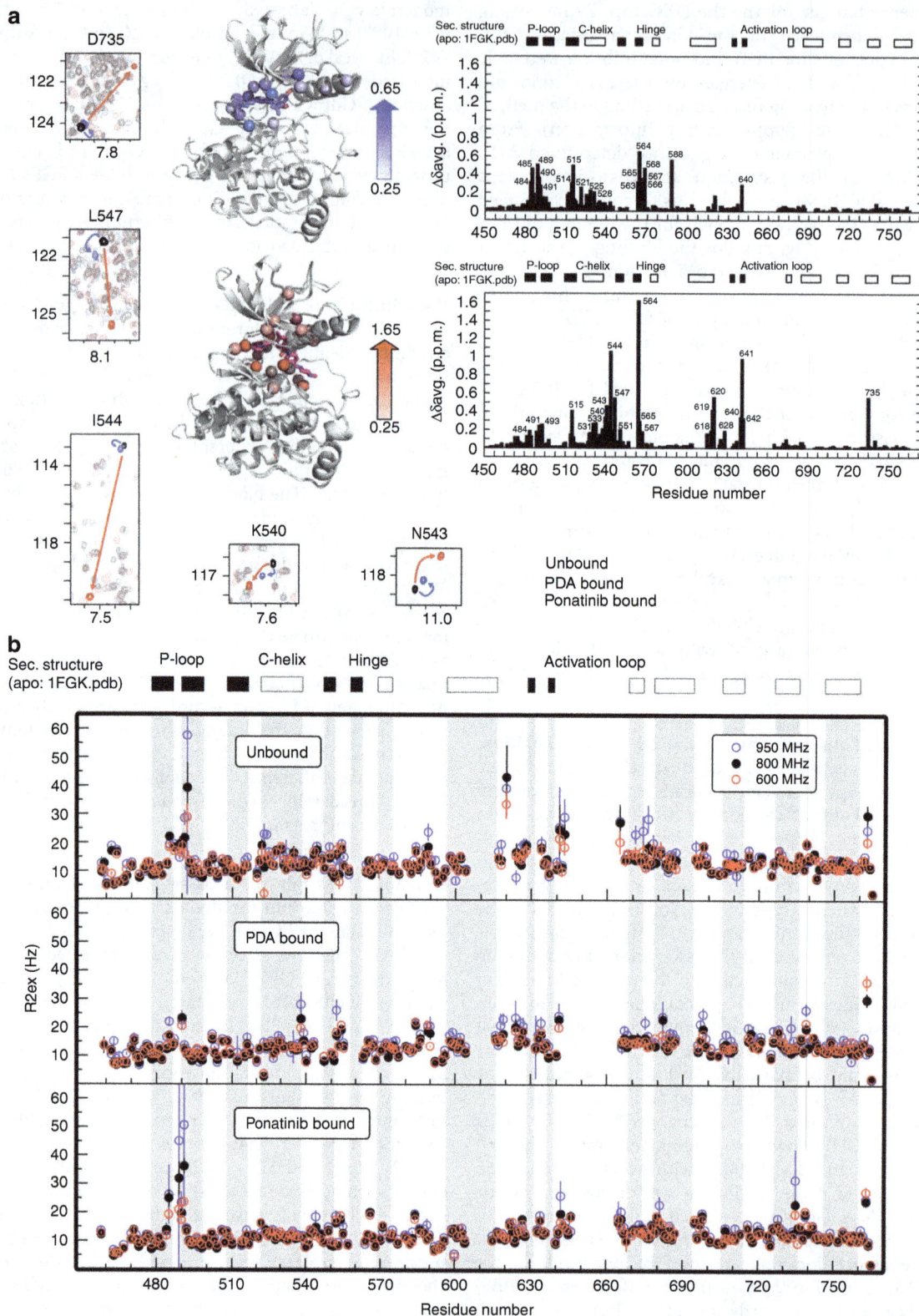

**Figure 3 | NMR analysis of structural and dynamic perturbations to FGFR1 kinase on binding of type I and type II inhibitors.** (a) Backbone amide chemical shift perturbation (CSP) analysis on ligand binding. Weighted CSPs were calculated as $\Delta\delta_{ave} = (\Delta\delta^2(N)/50 + \Delta\delta^2(H)/2)^{1/2}$ between unbound and PDA complex (top right panel), and between unbound and ponatinib complex (bottom right panel). The CSPs $>0.25$ for the two complexes are mapped on the X-ray crystal structure of unbound FGFR1 (PDB-code: 1FGK). Solid bars represent regions of β-strand secondary structure, open bars regions of α-helical secondary structure. Selected regions of overlayed $^{1}$H-$^{15}$N TROSY-HSQC plots of representative amino acids in the αC-β4 loop and D735 in the distal αH helix are shown in small panels (left, bottom). The contour plots are colour coded as follows: unbound (black); PDA bound (blue); ponatinib bound (red). Arrows of the corresponding color connect the same residue in different spectra. (b) Analysis of chemical exchange contributions to transverse relaxation rates ($R_{2,ex}$) measured for ligand-free (top), PDA-bound (middle) and ponatinib-bound (bottom) FGFR1 kinases at static fields of 600 MHz (black), 800 MHz (red) and 950 MHz (blue circles), reflecting motions on time scales $>100$ μs.

upper right panel). Interestingly, ponatinib binding revealed both local and distal changes (Fig. 3a, lower right panel). Local CSPs were observed in the P-loop, αC helix and hinge regions, and for Ile620 in the catalytic loop all of which participate in direct interactions with the inhibitor. The backbone amide nitrogen of Asp641 (of the DFG motif) also engages in a direct hydrogen-bond interaction with the amide carbonyl oxygen of ponatinib, which is likely to dominate the observed CSP for this residue, along with the change in the φ torsion angle associated with the DFG flip (Fig. 3a, lower right panel; Fig. 1c). Notably, substantial CSPs were also observed in the αC-β4 loop around Ile544, and for Asp735 in the αH helix. These are all spatially distant from the active site; thus, the observed chemical shift changes (Fig. 3a, small panels) must be a result of structural or dynamic changes propagated through an interaction network. The CSPs in the αC-β4 loop region are of particular interest, since these amides are likely to be highly sensitive reporters on changes in conformation or dynamics associated with movements of the αC helix[35]. By analogy with other kinases, the hydrophobic spine network[36-38] of FGFR1 is expected to be disrupted on the reorientation of Phe642 in the inactive DFG-out state, which may be reflected in perturbations seen in the chemical shifts of the residues neighbouring His621 in the catalytic loop. Direct contacts with the terminal methylpiperazinyl group of ponatinib from residues including Ile620 and His621 are also likely to contribute to the observed CSPs. Such perturbations are not seen for the PDA-bound state (which is assumed to populate predominantly the DFG-in conformation in solution). The large chemical shift change we observe for Asp735 in the ponatinib-bound complex is surprising, as Asp735 is situated in helix αH, which is rather remote from the active site. The upfield shift of the backbone amide resonance might reflect subtly altered hydrogen bonding and may report on perturbed dynamics in the αH helix as opposed to gross conformational change (vide infra), since the mean structures from X-ray crystallography are essentially superimposable in this region. Further insights into the underlying dynamics of FGFR1 in the three states were obtained from measurements of contributions from chemical exchange effects to the $^{15}N$ transverse relaxation rates of backbone amides, $R_{2,ex}$. Using data acquired at three different magnetic field strengths for unbound, PDA-bound and ponatinib-bound FGFR1, we observe particularly large field-dependent chemical exchange contributions to the $^{15}N$ linewidth (attributable to dynamics on time scales longer than $\sim 100\,\mu s$) for the ponatinib complex in the P-loop, compared with smaller but still significant effects for unbound FGFR1, and a marked suppression of millisecond time-scale P-loop dynamics in the PDA complex (Fig. 3b); this correlates with the additional P-loop protein–ligand contacts that we observe in crystal structures of PDA-bound FGFR1, but also suggests that DFG-out binding of ponatinib is accompanied by loosening of restraining forces on P-loop conformation. However, in contrast to the enhanced P-loop dynamics, slow time-scale motions are markedly suppressed in the αC-helix of the ponatinib complex compared with either ligand-free or PDA-bound states. The $R_{2,ex}$ data further show the presence of significant slow time-scale motion in the αH helix region of the DFG-out ponatinib complex around Asp735, in agreement with CSP data.

Amide protection rates as measures of solvent accessibility determined by HDX-MS[39,40] can provide complementary insights into conformational flexibility. By comparing the deuterium incorporation in the PDA and ponatinib complexes of FGFR1 with the unbound form (Fig. 4), several regions can be seen to exhibit significant relative (de)protection. Both inhibitor complexes are protected relative to ligand-free FGFR1 in the P-loop and inter-lobe hinge, consistent with direct protection

from solvent by the ligand (Fig. 4b). The observed rate of hydrogen exchange in the P-loop follows the order unbound > ponatinib-bound > PDA-bound, reflecting the direct interaction between the t-butyl group of PDA and the P-loop. By comparison with the unbound and PDA-bound forms of FGFR1, the DFG-out ponatinib complex has faster exchange kinetics in the proximal A-loop, including the DFG motif (Fig. 4c middle panel). In contrast, the distal stretch of the A-loop, including the short αEF helical segment, displays the opposite sensitivity to DFG-in or DFG-out binding modes: it is significantly deprotected in the PDA complex (Fig. 4c lower panel), whereas in the ponatinib complex this deprotection is marginal relative to ligand-free FGFR1. This finding may indicate a certain mutual exclusivity in the dynamic perturbations of the proximal and distal sections of the activation loop, which is not obvious from the X-ray crystal structure data in these regions.

Consistent with NMR chemical shift analysis, His621 in the catalytic loop and Phe642 (of the DFG motif) display increased hydrogen exchange in the ponatinib complex. Ponatinib-bound FGFR1 experiences widespread loss of hydrogen-exchange protection factors in peptides throughout the regulatory spine (R spine)[36,37]. Indeed, of the five amino acids in the R spine, four exhibit significant increases in observed hydrogen exchange rate in the ponatinib-bound ensemble (Fig. 4, Supplementary Fig. 9). In contrast, just one amino acid (His621) in the R spine was seen to have been marginally deprotected in the PDA-bound ensemble, while Phe642 was significantly protected. Together, these alterations to FGFR1 solvent accessibility indicate that the hydrophobic spine network is perturbed when the kinase adopts the DFG-out conformation. Significantly, and in agreement with NMR chemical shift data, another region that showed relative deprotection in the ponatinib-bound complex was the C-terminal end of the αC helix and the subsequent αC-β4 loop (Fig. 4c top panel). While the uncomplexed and PDA-bound forms show equivalent hydrogen-exchange profiles, the ponatinib complex displays a markedly greater extent of solvent exposure in this region on average. Again consistent with NMR, the αH helix also exhibits slight deprotection in the ponatinib complex, further supporting the likelihood of a structural loosening of this distal region of the kinase in the DFG-out state that is not evident from X-ray crystal structures.

## Discussion

The flip between active DFG-in and inactive DFG-out states of kinases, besides being exploitable for inhibitor design, has been advanced as a physiologically significant conformational transition that may have a role in modulation of the enzymatic activity of many kinases[11]. This is corroborated by the observation of DFG-out conformations in X-ray crystal structures of the unliganded and/or autoinhibited states of a number of kinases including Abl, c-Kit, FLT3, insulin receptor kinase and B-Raf[4-8]. The influence of the protonation state of the DFG aspartate on the $k_{on}$ for binding of imatinib to the DFG-out state of Abl has been interpreted as a possible factor in the regulation of kinase activity through facilitation of nucleotide binding and release, and as evidence for a physiological role for the DFG flip[11]. Our results support a role for the protonation state of the DFG aspartate in influencing the accessibility of the DFG-out conformation in FGFR1, but the modest difference in p$K_a$ of the DFG aspartate in FGFR1 (6.25) compared with that calculated for Abl (6.6)[11] is insufficient to explain the wide gulf in $k_{on}$ for type II binding to the two kinases. Our kinetic and thermodynamic data strongly suggest that, in contrast to Abl[41], association of type II inhibitors is limited by an exceptionally slow DFG flip in FGFR1, because a particularly high free-energy barrier must be crossed in the

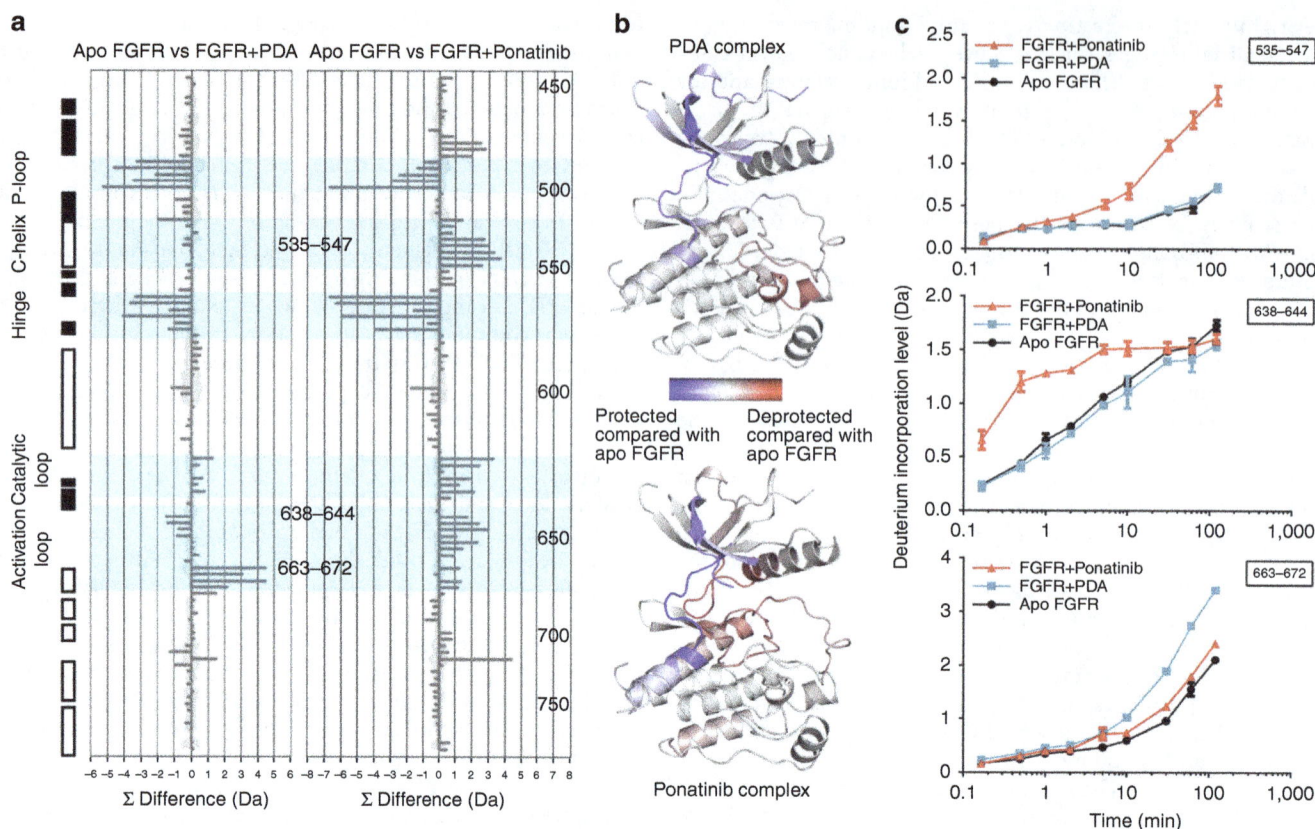

**Figure 4 | Hydrogen/deuterium-exchange changes on ligand binding to FGFR1 kinase domain.** (**a**) Difference in hydrogen exchange relative to unbound FGFR1 for complex with PDA (left) and ponatinib (right). Protection due to ligand binding leads to a reduction in mass relative to the ligand-free form (more negative value); deprotection results in an increase in relative mass (more positive value). Each horizontal bar represents a single peptide from FGFR1. Vertical scale is not linear: peptides are in order of start residue from N (top) to C terminus (bottom). Peptides whose start residue is within a secondary structural element are indicated as filled (β-sheet) or empty (α-helix) bars. Values are the sum of all nine time points sampled (each is a minimum of two experiments and one to seven ions per peptide). Continuous shaded region denotes error at 1 s.d. Peptides from **c** are annotated by residue number. (**b**) Data from **a** as a heat map projected on the unbound FGFR1 structure (PDB-code: 1FGK): complex with PDA (top) and complex with ponatinib (bottom). Only significant changes are shown (>0.4 Da difference from ligand-free form per data point)[43]. Data sets have been normalized to the same scale. (**c**) Deuterium uptake plots for three peptides: residues Met535-Leu547, Lys638-Leu644 and Pro663-Leu672. Data points are the mean of at least two experiments. Error bars indicate 1 s.d.

transition between DFG-in and DFG-out states. Thus, there are likely to be structural and/or dynamic differences between Abl and FGFR1 that influence the accessibility of the DFG-out state. Evidence from Eyring analysis for the elevated free energy associated with the transition state is corroborated by the slow association kinetics for the type II inhibitor ponatinib, and by the differential effects on association and dissociation rate constants for ponatinib and PDA under conditions of partial unfolding or structural loosening in the presence of 1.2 M GdmCl. This calls into question whether it is feasible that such an innately slow DFG flip could play a physiologically relevant role in the catalytic function of FGFR1, as has been postulated for other kinases[4-8].

The thermodynamic signatures for binding to the DFG-in and DFG-out states of FGFR1 appear to be highly distinct, with favourable enthalpy (at relative entropic cost) for PDA binding contrasting with a highly entropically driven interaction for ponatinib. This rather extreme example of enthalpy–entropy compensation[42] between two inhibitors sharing very similar $K_{DS}$ but strikingly different binding modes may point to greater motional freedom as a contributory factor in the energetics of binding of ponatinib to the DFG-out state. Indeed, our HDX-MS data indicate an overall increased exposure of backbone amides to solvent in the DFG-out complex relative to unbound FGFR1, compared with predominantly enhanced protection from

exchange in the DFG-in complex with PDA. This loosening of the structure in the DFG-out conformation is reflected in the region of the αH helix, where we observed a large NMR shift for Asp735, increased chemical exchange contributions to the NMR $R_2$ relaxation rates and enhanced hydrogen exchange rates for surrounding residues in HDX-MS experiments, despite essentially identical mean conformations as judged by X-ray crystallography. We speculate that this effect is mediated through the αF helix, which anchors the hydrophobic spine network. Loss of communication through the spine as a result of the DFG flip may lead to slight destabilization of the αH helix, resulting in increased dynamic freedom in this region. We hypothesize that the increased mobility evident from hydrogen exchange and NMR relaxation data in regions both proximal (P-loop) and distal (αH helix) to the ponatinib-binding site and the DFG motif may contribute enhanced protein conformational entropy towards the markedly favourable gain in global entropy that characterizes the formation of the ponatinib complex[33,43,44]. Conversely, the suppression of slow time-scale motion in the αC helix that we see from NMR $R_2$ relaxation rates in the ponatinib-bound state may contribute to the long residence time that characterizes the DFG-out binding mode.

Crystal structures and molecular dynamics simulations, using Abl kinase as a model system, suggest that displacement of the αC

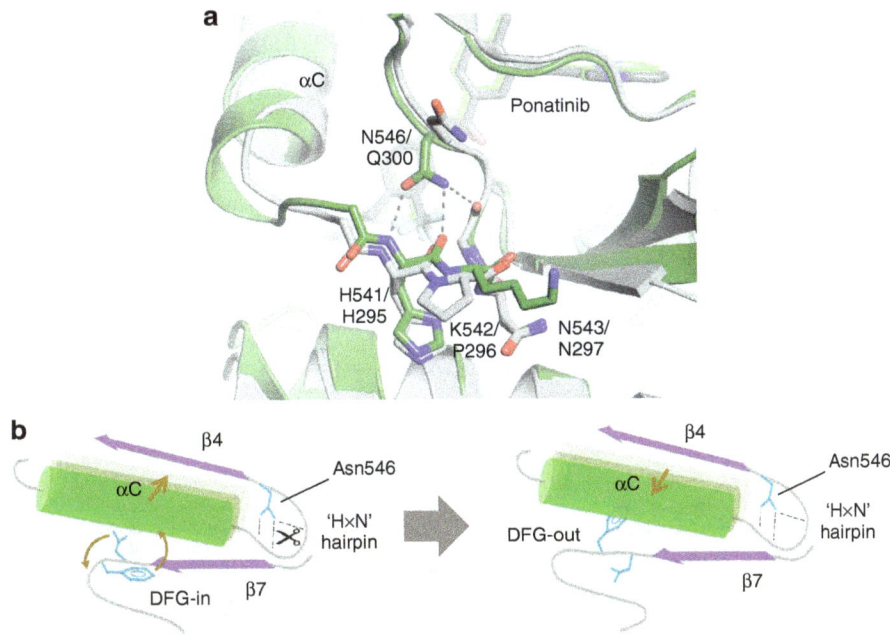

**Figure 5 | The role of the αC-β4 loop and molecular brake regions in the DFG flip of FGFR1.** (a) Comparison of αC-β4 loop and molecular brake regions in FGFR1 and Abl kinase complexes with ponatinib. The structures of the FGFR1/ponatinib complex (PDB ID: 4V01) and the Abl/ponatinib complex (PDB ID: 3OXZ) are displayed in dark green and grey, respectively, with the bound ponatinib inhibitors displayed, respectively, in light green and grey. Relative to Abl, the αC helix of FGFR1 extends approximately one-half turn further, in part due to insertion of a Gly at position 539, and the 'HxN hairpin' contains a Lys rather than a Pro at the middle position. The molecular brake of FGFR1 is engaged via hydrogen bonds (dotted lines) from the side chain of Asn546 and likely inhibits the outward motion of helix αC in FGFR1, whereas Abl, with a Gln at the equivalent position and lacking the Gly insert at position 539, is unable to form the molecular brake interactions. (b) Schematic illustration of the interplay between the DFG flip, outward movement of the αC helix and the proposed role of the Asn546 molecular brake hydrogen bonds in FGFR1. The Asn546 hydrogen bonds (of which Abl lacks an equivalent) may need to be transiently disengaged (scissors) to facilitate the αC-out, and hence DFG-out, movements. View in **b** as if from the left-hand side of **a**.

helix away from the active site facilitates the DFG flip in kinases, with the resulting 'αC-out' conformation being a potential intermediate[11,36]. The αC-β4 loop has been proposed to act as an anchor for the αC helix to the catalytic core, and as a hinge for the αC helix during the transition from active to inactive states of protein kinases[34,45]. The substantial amide CSPs we observed using NMR for residues in the αC-β4 loop in the DFG-out state of FGFR1, coupled with significantly enhanced solvent exchange rates by HDX-MS, indicate that the transition from the active to the inactive state is accompanied by a structural or dynamic perturbation. Compared with Abl, FGFR1 contains an insert (Gly539) C-terminal to the αC helix and a conformationally significant substitution in the relatively conserved 'HxN hairpin'[35,46] that follows (HPN in many kinases including Abl; HKN in FGFRs). The HxN hairpin may function as a pivot for the outward movement of the αC helix that is required to facilitate the excursion of the DFG Phe side chain towards its 'out' configuration[46]. The Gly539 insert results in extension of the C-terminal end of the αC helix of FGFR1 by around half a turn relative to Abl (Fig. 5a), and facilitates the formation of the molecular brake hydrogen-bond network[47] between the side chain of Asn546 and the backbone atoms of His541 of the HxN motif. By contrast, Abl is unable to form these hydrogen bonds to the HxN backbone. Asn546 is a key member of the triad that forms the molecular brake in FGFR isoforms, and is the site of a number of pathogenic gain-of-function mutations that are implicated in developmental disorders and cancers. The hydrogen-bond network involving Asn546 of FGFR1 would be expected to stabilize the αC helix in its 'in' orientation, thereby inhibiting the 'αC-out' movement required to effect the DFG flip (Fig. 5b). Thus, our analysis suggests that a distributed network of individual contributions from several regions of the kinase

structure conspires to hinder the DFG flip in FGFR1, and that the most important of these is likely to reside in the αC-β4 loop region. This is interesting in light of a recent report that the N550K mutation in FGFR2 (equivalent to Asn546 in FGFR1) confers resistance to the type I inhibitors PD173074 and dovitinib, but not to ponatinib, which displays enhanced inhibitory potency against this mutant relative to wild-type in BaF3 cell proliferation assays[48]. Our insights into the structural and dynamic influences on the DFG flip in FGFR1 corroborate the important role of the molecular brake in inhibiting basal kinase activity in unphosphorylated FGFRs, and imply that its function (and its release by pathogenic mutations) may be intimately associated with its ability to suppress the catalytically significant DFG flip[11] by inhibiting the outward movement of the αC helix.

## Methods

**Protein expression and purification.** Human FGFR1 consisting of residues Ala458-Glu765 with an engineered TEV-cleavable N-terminal 6 × His tag and mutations Cys488Ala and Cys584Ser was co-expressed in *Escherichia coli* with protein tyrosine phosphatase 1B (PTP1B) and purified by sequential immobilized metal affinity chromatography (IMAC, QIAGEN NiNTA), ion exchange (ResourceQ) and size exclusion chromatography[49]. The hexa-histidine tag was cleaved from protein by overnight treatment with TEV protease and concomitant dialysis, immediately after the IMAC step. Purified protein in a buffer comprising 20 mM Tris-HCl, pH 8.0, 20 mM NaCl, 2 mM TCEP, was snap frozen in liquid nitrogen and stored at − 80 °C. For NMR studies, an additional mutation (Asp623Ala) was introduced to improve the yield of stable isotope-labelled FGFR1 kinase protein[22]. Uniform isotopic labelling was achieved by growing *E. coli* (DE3) Star cells in $D_2O$-based M9 minimal medium supplemented with $^{15}NH_4Cl$ (Cambridge Isotope Laboratories or Sigma Aldrich) together with U-[$^1$H,$^{13}$C]- or (for fully deuterated samples) U-[$^2$H,$^{13}$C]-glucose (Cambridge Isotope Laboratories or Sigma Aldrich) as sole nitrogen and carbon sources, respectively. Purification was by IMAC and ion exchange chromatography; the 6 × His tag was not cleaved from the protein used for NMR. Human Abl consisting of residues Ser248-Val534 with an engineered TEV-cleavable N-terminal 6 × His tag and mutation Asn355Ser was expressed and purified as described[50] with minor modifications. For biophysical studies, the 6 × His tag was retained intact for both kinases as it was used for immobilization on a nitrilotriacetic acid (NTA) sensor chip.

**Crystallization, crystallographic data collection, structure determination and refinement.** Growth of FGFR1 crystals were grown by the hanging drop vapour diffusion method at 4 °C by mixing equal volumes of purified FGFR1 at 10 mg ml$^{-1}$ with a reservoir solution comprising 18–20% PEG8000 (w/v), 200 mM ammonium sulphate, 100 mM PCTP, pH 6.75 and 20% ethylene glycol (v/v) so as to obtain a 2 μl drop[49]. Crystals were allowed to grow for at least 1 week before harvesting into a soaking solution comprising 22% PEG8000 (w/v), 200 mM ammonium sulphate, 100 mM PCTP, pH 6.75, 20% ethylene glycol (v/v) and 1 mM PDA or 1 mM dovitinib plus 1% DMSO (v/v). Soaks were incubated overnight. All work was carried out at 277 K. Crystals were flash frozen in a stream of nitrogen gas at 100 K directly from the drop. Diffraction data were collected in-house on a Rigaku FRE rotating anode generator ($\lambda = 1.54$ Å) equipped with a Saturn 944 CCD detector or at Diamond Light Source on beamline I04 ($\lambda = 0.92$ Å) using an ADSC Quantum 315 CCD detector. Data were processed with XDS and AIMLESS as implemented within autoPROC[51] and XIA2 (ref. 52), respectively. The FGFR1–PDA and FGFR1–dovitinib crystals belong to the space group C1 2 1 and contain two complexes per asymmetric unit. The structures were solved by molecular replacement using the programme AMORE[53] and an in-house FGFR1 structure as a search model. The structures were completed with iterative rounds of manual building in Coot[54] interspersed with refinement using the programmes REFMAC[55] and autoBUSTER applying NCS restraints and TLS. Quality checks were carried out using the validation tools in Coot and MolProbity[56], while the compound stereochemistry was checked against the Cambridge Structure Database (CSD)[57] using Mogul[58]. Ramachandran analysis revealed 93.6% (favoured), 6.0% (allowed) and 0.4% (generously allowed) for the FGFR1–PDA complex and 91.3% (favoured), 7.3% (allowed), 1.0% (generously allowed) and 0.4% (disallowed) for the FGFR1–dovitinib complex. Crystallographic statistics indicating data and model stereochemical quality are given in Supplementary Table 5. The final structures have been deposited in the PDB with ID code: FGFR1 − PDA complex, 5A4C; FGFR1–dovitinib complex, 5A46. All structural figures were prepared using PyMOL (Schrödinger LLC).

**Surface plasmon resonance.** Non-phosphorylated, histidine-tagged FGFR1 and non-phosphorylated, histidine-tagged Abl were immobilized as the ligand onto NTA sensor chips using a capture coupling method[59]. The NTA surface was first activated with 500 μM NiSO$_4$ in immobilization buffer. The carboxymethyl dextran surface was then activated with a 1:1 ratio of 0.4 M EDC and 0.1 M NHS. Hexa-histidine-tagged protein was diluted into immobilization buffer to a concentration of 30 μg ml$^{-1}$, and immobilized onto the surface with a 7-min injection. Remaining activated groups were blocked with 0.1 M Tris, pH 8.0. Typical immobilization levels ranged from 6,000 to 8,000 resonance units (RU). PBS, pH 7.4, 50 μM EDTA and 0.05% Surfactant P20 (v/v) (for Abl supplemented with 10% glycerol (v/v)) were used as immobilization buffer. Typical immobilization levels ranged from 3,800 to 8,000 RU. SPR experiments were performed using the Biacore 3000, Biacore S51 and Biacore T200 biosensors (GE Healthcare), with NTA and series S NTA sensor chips (GE Healthcare). All FGFR1 binding experiments were done using PBS (pH range 7.0–7.4), 50 μM EDTA, 0.05% Surfactant P20 (v/v) and 1% DMSO (v/v) or 50 mM Bis-Tris (pH range 5.5–6.5), 100 mM NaCl, 50 μM EDTA, 0.05% Surfactant P20 (v/v) and 1% DMSO (v/v), as running buffer. All Abl binding experiments were conducted using PBS, pH 7.4, 50 μM EDTA, 0.05% Surfactant P20 (v/v), 10% glycerol (v/v) and 1% DMSO (v/v) as running buffer. Compounds as DMSO stocks were diluted in DMSO to concentrations 100-fold higher than the final assay concentration. Finally, they were diluted 1:100 (v/v) in running buffer without DMSO to achieve the target concentration resulting in a final DMSO concentration of 1% (v/v).

**SPR kinetic analysis.** To determine the rate constants of association ($k_{on}$) or dissociation ($k_{off}$), either multi-cycle or single-cycle SPR experiments were performed at 298 K. Single-cycle kinetic analysis was done at a constant flow rate of 60 μl min$^{-1}$ in running buffer. The highest compound concentration varied, but for all analytes five sequential injections with constant injection time and a constant dilution factor were done. All analyte concentrations were injected in one cycle, one after the other for 120 s with a short dissociation phase in between injections (~ 60 s) and with a longer dissociation phase at the end of the cycle (1,000 to 20,000 s that varied depending on the expected dissociation rate constant of the analyte). Zero-buffer blank injections were included for referencing. Biacore T200 evaluation software and BIAevaluation 4.1 software, respectively, were used for processing and analysing data. Rate constants were calculated globally from the obtained sensorgram data by fitting to a 1:1 interaction model. Representative sensorgrams are shown in Supplementary Fig. 6. Multi-cycle kinetic analysis was carried out as previously described[59]. Binding affinities ($K_D$) were calculated from the equation $K_D = k_{off}/k_{on}$.

**SPR thermodynamic analysis.** The thermodynamic parameters of ponatinib and PDA binding were determined by performing single-cycle kinetic analysis at different temperatures as described above. SU5402 showed faster association and dissociation rate constants, thereby complicating kinetic analysis at higher temperatures. Binding affinities ($K_D$) were therefore determined from dosage experiments and binding responses at equilibrium were fit to a 1:1 steady-state affinity model available within the Scrubber 2 software (BioLogic Software Ltd., Campbell, Australia). For each analysed ligand, rate constants and/or affinity were determined at a minimum of six different temperatures between 281 and 308 K. Association constants ($K_A$) derived from kinetic or steady-state analysis were plotted as ln ($K_A$) against $1/T$, according to the integrated van't Hoff equation[60,61]—equation (1).

$$\ln(K_A/K_{A0}) = \left[ (\Delta H_0 - T_0 \Delta C_p)(1/T_0 - 1/T) + \Delta C_p \ln(T/T_0) \right]/R \quad (1)$$

where $T_0$ is an arbitrarily selected reference temperature, $K_{A0}$ is the association constant at temperature $T_0$, and $\Delta H_0$ is the van't Hoff enthalpy at temperature $T_0$. $\Delta C_p$ is the temperature-independent heat capacity change (constrained to the experimentally determined values of − 359 and − 172 cal mol$^{-1}$ K$^{-1}$ (Supplementary Fig. 7) for the analysis of PDA and SU5402, respectively) and $R = 1.986$ cal mol$^{-1}$ K$^{-1}$. $\Delta H_0$ was determined by non-linear fitting of equation (1) to the experimental data using Prism 5.1 (GraphPad Software, Inc, La Jolla, USA). Transition state thermodynamic quantities were determined from the kinetic association ($k_{on}$) and dissociation ($k_{off}$) rate constants as previously described[64] by plotting ln ($kh/k_BT$) versus $1/T$ according to the linear Eyring equation (2).

$$\ln(kh/k_BT) = - \Delta H^*/RT + \Delta S^*/R \quad (2)$$

where $h = 6.63 \times 10^{-34}$ J s and $k_B = 1.38 \times 10^{-23}$ J K$^{-1}$ are the Planck and Boltzmann constants, respectively. Here $k$ is either the association rate constant ($k_{on}$) or the dissociation rate constant ($k_{off}$). $\Delta H$ and $\Delta S$ are the changes in free enthalpy and entropy of binding, respectively, while the superscript '*' denotes that these refer to a transition state. $T$ is the absolute temperature, and $R = 1.986$ cal mol$^{-1}$ K$^{-1}$. $\Delta H^*$ and $\Delta S^*$ were determined by linear fitting of equation (2) to the experimental data using Prism 5.1 (Supplementary Fig. 4).

**Isothermal titration calorimetry.** ITC experiments were carried out using an ITC$_{200}$ instrument (Microcal Inc., GE Healthcare)[59]. Final ligand concentrations were achieved by diluting ligand stock solutions in DMSO 1:50 (v/v) in the experimental buffer, resulting in a final DMSO concentration of 2% (v/v). Protein concentration was determined by measuring the absorbance at 280 nm. DMSO concentration in the protein solution was adjusted to 2% (v/v). ITC measurements were routinely performed at 25 °C in 20 mM Tris, pH 7.8, 20 mM NaCl, 2 mM TCEP and 2% DMSO (v/v). The titrations were performed on 10 − 20 μM FGFR1 in the 200 μl sample cell using 2 μl injections of 0.1 − 0.2 mM ligand solution every 120 s. To correct for heats of dilution and mixing, the final baseline consisting of small peaks of identical size at the end of the experiment was subtracted. Representative ITC titrations are shown in Supplementary Fig. 2. To determine the heat capacity $\Delta C_p$ of ligands binding to FGFR1, ITC titrations were performed as described above at 10, 15, 20, 25, 30 and 35 °C. Binding enthalpies derived from ITC experiments were plotted as $\Delta H$ against $T$ and $\Delta C_p$ are given by the slope of the linear regression analysis according to equation (3) (Supplementary Fig. 7):

$$\Delta H(T_2) = \Delta H(T_1) + \Delta C_p(T_2 - T_1) \quad (3)$$

**Equilibrium chemical denaturation of FGFR1 using far-ultraviolet CD Spectroscopy.** Far-ultraviolet CD spectra (190–260 nm) of FGFR1 kinase domain were obtained at different concentrations of GdmHCl. Spectra were measured on a JASCO J-810 spectropolarimeter at 293 K in 10 mM sodium phosphate buffer, pH 7.4; the concentration of FGFR1 used throughout was 2.8 μM. Unfolding experiments on FGFR1 kinase domain were completed by diluting the native FGFR1 protein sample with sequential additions of a second stock solution containing FGFR1 protein unfolded in 5 M GdmHCl, similarly buffered in 10 mM sodium phosphate, pH 7.4. The concentration of GdmHCl was determined using refractive index measurements as described[62]. Appropriate buffer blanks containing the corresponding concentration of denaturant were subtracted from all spectra, to account for the small contribution to the observed signal made by buffer. Molar ellipticity values at 222 nm obtained at varying denaturant concentrations were analysed using non-linear least-squares regression analysis, employing a modified version of the equation described in Morjana et al.[63]:

$$y = ((y_n + m_n[D]) + ((y_i + m_i[D])K_{n \to i}) + ((y_u + m_u[D]K_{n \to i}K_{i \to u}))/(1 + K_{n \to i} + (K_{n \to i}K_{i \to u})) \quad (4)$$

where $y_n$, $y_i$, $y_u$ are the signals of the native (n), intermediate (i) and unfolded (u) states, respectively, at zero denaturant concentration ([D]), $m_n$, $m_i$, $m_u$ represent dy/d[D] or slopes of the native, intermediate and unfolded state signals, respectively, $K_{n \to i} = \exp - (\Delta G_{n \to i} - m_{n \to i}[D])/RT$, $K_{i \to u} = \exp - (\Delta G_{i \to u} - m_{i \to u}[D])/RT$, where $\Delta G$ represents the free-energy change of the indicated transition, $m$ is the slope of the free-energy change versus [D] for the indicated transition, $R$ is the gas constant and $T$ is the experimental temperature.

**NMR spectroscopy.** Uniformly $^{13}C/^{15}N/^2H$-labelled samples of unbound FGFR1 and ligand-bound FGFR1–ponatinib and FGFR1–PDA complexes (1:1) were prepared[22] as 0.35 mM solutions in 450 µl of 95% $H_2O$ and 5% $D_2O$, 50 mM sodium phosphate, 0.1 mM EDTA, 2 mM dithiothreitol and 0.02% sodium azide (pH 7.0). PDA and ponatinib were added from concentrated stock solutions dissolved in DMSO-d6. NMR spectra were recorded at 298 K on Bruker Avance 600 MHz, Avance III 800 and Avance III HD 950 MHz spectrometers equipped with z-axis pulsed-field gradient TCI CryoProbes. TROSY-based detection schemes were used throughout as previously described[22]. Backbone resonance assignments for the FGFR1–PDA and FGFR1–ponatinib complexes followed standard triple-resonance strategies with two- and three-dimensional experiments using TROSY detection[22], and will be reported elsewhere. The presence of backbone amide conformational exchange effects was studied by measuring the relaxation rates of the slowly relaxing $^{15}N$-$\{^1H\}$ TROSY doublet component using a Hahn-echo-based sequence optimized for deuterated proteins as described in Lakomek et al.[64,65]. All NMR data were processed using the NMRPipe suite of programmes[66] and analysed with CARA[67] to obtain assignments. $^{15}N$ relaxation decay curves were fitted using a simplex search minimization and Monte Carlo estimation of errors.

**Hydrogen/deuterium-exchange mass spectrometry.** Hydrogen exchange was performed using an HDX Manager (Waters Corp.) equipped with a CTC PAL sample handling robot (LEAP Technologies). Briefly, FGFR1 kinase domain (52.3 µM) in protonated aqueous buffer (20 mM Tris, 20 mM NaCl, 2 mM TCEP, pH 7.4) was incubated with ligand (100 µM) or DMSO. This gave 99.7 and 99.8% bound FGFR1 following dilution in the labelling solution for ponatinib ($K_D = 7.7$ nM) and PDA ($K_D = 5.7$ nM), respectively. Hydrogen exchange was initiated by dilution of 20-fold into deuterated buffer (20 mM Tris, 20 mM NaCl and 2 mM TCEP, pD 7.4) at 293 K. After incubation between 10 s and 2 h, hydrogen-exchange was quenched by mixing 1:1 with 100 mM potassium phosphate to a final pH of 2.55 at 274 K. Sample was immediately digested by a pepsin–agarose column (Poroszyme) and the resulting peptides separated on a C18 column (1 × 100 mm ACQUITY BEH 1.7 µm, Waters Corp.) with a linear gradient of acetonitrile (3–40%) supplemented with 0.1% formic acid. Peptides were analysed with a Synapt G2 mass spectrometer (Waters Corp.). Peptides were identified by MS$^E$ fragmentation, yielding coverage of 97% of the His-tagged fusion protein construct of FGFR1 kinase domain with a high degree of redundancy (Supplementary Fig. 8). Peptides from Fig. 4c were confirmed by targeted tandem mass spectrometry fragmentation. No correction was made for back-exchange, and all results are reported as relative deuterium level. Deuterium incorporation was measured in DynamX (Waters Corp.) and data normalization was calculated with in-house software written in MatLab (Mathworks) and Python. Structural representations were generated with PyMol and plots in Fig. 4 prepared with Prism. Hydrogen/deuterium-exchange was represented in Fig. 4b by calculating the mean deuteration level per amino acid, according to equation (5).

$$\bar{M}_j = \frac{1}{n}\sum_1^n \frac{1}{q}\sum_0^t (m_i^t - m_i^0) \qquad (5)$$

Where $\bar{M}_j$ is the mean deuteration level at amino acid $j$, $n$ is the number of overlapping peptides, $q$ is the number of exchangeable amides for peptide species $i$, $m_i^t$ is the isotopic weighted midpoint at time $t$ and $m_i^0$ is the midpoint at time 0 (undeuterated).

## References

1. Norman, R. A., Toader, D. & Ferguson, A. D. Structural approaches to obtain kinase selectivity. *Trends Pharmacol. Sci.* **33**, 273–278 (2012).
2. Schindler, T. et al. Structural mechanism for STI-571 inhibition of abelson tyrosine kinase. *Science* **289**, 1938–1942 (2000).
3. Zuccotto, F., Ardini, E., Casale, E. & Angiolini, M. Through the 'gatekeeper door': exploiting the active kinase conformation. *J. Med. Chem.* **53**, 2681–2694 (2010).
4. Hubbard, S. R., Wei, L. & Hendrickson, W. A. Crystal structure of the tyrosine kinase domain of the human insulin receptor. *Nature* **372**, 746–754 (1994).
5. Bollag, G. et al. Vemurafenib: the first drug approved for BRAF-mutant cancer. *Nat. Rev. Drug Discov.* **11**, 873–886 (2012).
6. Nagar, B. et al. Structural basis for the autoinhibition of c-Abl tyrosine kinase. *Cell* **112**, 859–871 (2003).
7. Mol, C. D. et al. Structural basis for the autoinhibition and STI-571 inhibition of c-Kit tyrosine kinase. *J. Biol. Chem.* **279**, 31655–31663 (2004).
8. Griffith, J. et al. The structural basis for autoinhibition of FLT3 by the juxtamembrane domain. *Mol. Cell* **13**, 169–178 (2004).
9. Hubbard, S. R. Autoregulatory mechanisms in protein-tyrosine kinases. *J. Biol. Chem.* **273**, 11987–11990 (1998).
10. Liu, Y. & Gray, N. S. Rational design of inhibitors that bind to inactive kinase conformations. *Nat. Chem. Biol.* **2**, 358–364 (2006).
11. Shan, Y. et al. A conserved protonation-dependent switch controls drug binding in the Abl kinase. *Proc. Natl Acad. Sci. USA* **106**, 139–144 (2009).
12. Hari, S. B., Merritt, E. A. & Maly, D. J. Sequence determinants of a specific inactive protein kinase conformation. *Chem. Biol.* **20**, 806–815 (2013).
13. Turner, N. & Grose, R. Fibroblast growth factor signalling: from development to cancer. *Nat. Rev. Cancer* **10**, 116–129 (2010).
14. Koziczak, M., Holbro, T. & Hynes, N. E. Blocking of FGFR signaling inhibits breast cancer cell proliferation through downregulation of D-type cyclins. *Oncogene* **23**, 3501–3508 (2004).
15. Chen, G. et al. Inhibition of endogenous SPARC enhances pancreatic cancer cell growth: modulation by FGFR1-III isoform expression. *Br. J. Cancer* **102**, 188–195 (2010).
16. Feng, S., Shao, L., Yu, W., Gavine, P. & Ittmann, M. Targeting fibroblast growth factor receptor signaling inhibits prostate cancer progression. *Clin. Cancer Res.* **18**, 3880–3888 (2012).
17. Zhang, Y., Guo, K. J., Shang, H., Wang, Y. J. & Sun, L. G. Expression of aFGF, bFGF, and FGFR1 in ovarian epithelial neoplasm. *Chin. Med. J. (Engl.)* **117**, 601–603 (2004).
18. Knights, V., Cook, S. J. & De-regulated, F. G. F. receptors as therapeutic targets in cancer. *Pharmacol. Ther.* **125**, 105–117 (2010).
19. Tucker, J. A. et al. Structural insights into FGFR kinase isoform selectivity: diverse binding modes of AZD4547 and ponatinib in complex with FGFR1 and FGFR4. *Structure* **22**, 1764–1774 (2014).
20. Huang, Z. et al. DFG-out mode of inhibition by an irreversible type-1 inhibitor capable of overcoming gate-keeper mutations in FGF receptors. *ACS Chem. Biol.* **10**, 299–309 (2014).
21. Lesca, E., Lammens, A., Huber, R. & Augustin, M. Structural analysis of the human fibroblast growth factor receptor 4 kinase. *J. Mol. Biol.* **426**, 3744–3756 (2014).
22. Vajpai, N., Schott, A.-K., Vogtherr, M. & Breeze, A. L. NMR backbone assignments of the tyrosine kinase domain of human fibroblast growth factor receptor 1. *Biomol. NMR Assign.* **8**, 85–88 (2014).
23. Mohammadi, M. et al. Crystal structure of an angiogenesis inhibitor bound to the FGF receptor tyrosine kinase domain. *EMBO J.* **17**, 5896–5904 (1998).
24. Gruenbaum, L. M. et al. Inhibition of pro-inflammatory cytokine production by the dual p38/JNK2 inhibitor BIRB796 correlates with the inhibition of p38 signaling. *Biochem. Pharmacol.* **77**, 422–432 (2009).
25. Iwata, H. et al. Biochemical characterization of a novel type-II VEGFR2 kinase inhibitor: comparison of binding to non-phosphorylated and phosphorylated VEGFR2. *Bioorg. Med. Chem.* **19**, 5342–5351 (2011).
26. Namboodiri, H. V. et al. Analysis of imatinib and sorafenib binding to p38alpha compared with c-Abl and b-Raf provides structural insights for understanding the selectivity of inhibitors targeting the DFG-out form of protein kinases. *Biochemistry* **49**, 3611–3618 (2010).
27. Pargellis, C. et al. Inhibition of p38 MAP kinase by utilizing a novel allosteric binding site. *Nat. Struct. Biol.* **9**, 268–272 (2002).
28. Sullivan, J. E. et al. Prevention of MKK6-dependent activation by binding to p38α MAP kinase. *Biochemistry* **44**, 16475–16490 (2005).
29. Regan, J. et al. The kinetics of binding to p38 MAP kinase by analogueues of BIRB 796. *Bioorg. Med. Chem. Lett.* **13**, 3101–3104 (2003).
30. Vogtherr, M. et al. NMR characterization of kinase p38 dynamics in free and ligand-bound forms. *Angew. Chem. Int. Ed.* **45**, 993–997 (2006).
31. Miyashita, O., Onuchic, J. N. & Wolynes, P. G. Nonlinear elasticity, proteinquakes, and the energy landscapes of functional transitions in proteins. *Proc. Natl Acad. Sci. USA* **100**, 12570–12575 (2003).
32. Shan, Y., Arkhipov, A., Kim, E. T., Pan, A. C. & Shaw, D. E. Transitions to catalytically inactive conformations in EGFR kinase. *Proc. Natl Acad. Sci. USA* **110**, 7270–7275 (2013).
33. Fenley, A. T., Muddana, H. S. & Gilson, M. K. Entropy-enthalpy transduction caused by conformational shifts can obscure the forces driving protein-ligand binding. *Proc. Natl Acad. Sci. USA* **109**, 20006–20011 (2012).
34. Seeliger, M. A. et al. c-Src binds to the cancer drug imatinib with an inactive Abl/c-Kit conformation and a distributed thermodynamic penalty. *Structure* **15**, 299–311 (2007).
35. Kannan, N., Neuwald, A. F. & Taylor, S. S. Analogueous regulatory sites within the αC-β4 loop regions of ZAP-70 tyrosine kinase and AGC kinases. *Biochim. Biophys. Acta* **1784**, 27–32 (2008).
36. Kornev, A. P., Haste, N. M., Taylor, S. S. & Ten Eyck, L. F. Surface comparison of active and inactive protein kinases identifies a conserved activation mechanism. *Proc. Natl Acad. Sci. USA* **103**, 17783–17788 (2006).
37. Kornev, A. P., Taylor, S. S. & Ten Eyck, L. F. A helix scaffold for the assembly of active protein kinases. *Proc. Natl Acad. Sci. USA* **105**, 14377–14382 (2008).
38. Taylor, S. S. & Kornev, A. P. Protein kinases: evolution of dynamic regulatory proteins. *Trends Biochem. Sci.* **36**, 65–77 (2011).
39. Phillips, J. J. et al. Conformational dynamics of the molecular chaperone Hsp90 in complexes with a co-chaperone and anticancer drugs. *J. Mol. Biol.* **372**, 1189–1203 (2007).
40. Wales, T. E. & Engen, J. R. Hydrogen exchange mass spectrometry for the analysis of protein dynamics. *Mass Spectrom. Rev.* **25**, 158–170 (2006).
41. Agafanov, R. V., Wilson, C., Otten, R., Buosi, V. & Kern, D. Energetic dissection of Gleevec's selectivity toward human tyrosine kinases. *Nat. Struct. Mol. Biol.* **21**, 848–853 (2014).

42. Lumry, R. & Rajender, S. Enthalpy-entropy compensation phenomena in water solutions of proteins and small molecules: A ubiquitous property of water. *Biopolymers* **9**, 1125–1227 (1970).

43. Frederick, K. K., Marlow, M. S., Valentine, K. G. & Wand, A. J. Conformational entropy in molecular recognition by proteins. *Nature* **448**, 325–329 (2007).

44. Marlow, M. S., Dogan, J., Frederick, K. K., Valentine, K. G. & Wand, A. J. The role of conformational entropy in molecular recognition by calmodulin. *Nat. Chem. Biol.* **6**, 352–358 (2010).

45. Vajpai, N. *et al.* Backbone NMR resonance assignment of the Abelson kinase domain in complex with imatinib. *Biomol. NMR Assign.* **2**, 41–42 (2008).

46. Kannan, N., Haste, N., Taylor, S. S. & Neuwald, A. F. The hallmark of AGC kinase functional divergence is its C-terminal tail, a cis-acting regulatory module. *Proc. Natl Acad. Sci. USA* **104**, 1272–1277 (2007).

47. Chen, H. *et al.* A molecular brake in the kinase hinge region regulates the activity of receptor tyrosine kinases. *Mol. Cell* **27**, 717–730 (2007).

48. Byron, S. A. *et al.* The N550K/H mutations in FGFR2 confer differential resistance to PD173074, dovitinib, and ponatinib ATP-competitive inhibitors. *Neoplasia* **15**, 975–988 (2013).

49. Norman, R. A. *et al.* Protein-ligand crystal structures can guide the design of selective inhibitors of the FGFR tyrosine kinase. *J. Med. Chem.* **55**, 5003–5012 (2012).

50. Seeliger, M. A. *et al.* High yield bacterial expression of active c-Abl and c-Src tyrosine kinases. *Protein Sci.* **14**, 3135–3139 (2005).

51. Vonrhein, C. *et al.* Data processing and analysis with the autoPROC toolbox. *Acta Crystallogr. Sect. D* **67**, 293–302 (2011).

52. Winter, G., Lobley, C. M. C. & Prince, S. M. Decision making in xia2. *Acta Crystallogr. Sect. D* **69**, 1260–1273 (2013).

53. Navaza, J. AMoRe: an automated package for molecular replacement. *Acta Crystallogr. Sect. A* **50**, 157–163 (1994).

54. Emsley, P., Lohkamp, B., Scott, W. G. & Cowtan, K. Features and development of Coot. *Acta Crystallogr. Sect. D* **66**, 486–501 (2010).

55. Murshudov, G. N. *et al.* REFMAC5 for the refinement of macromolecular crystal structures. *Acta Crystallogr. Sect. D* **67**, 355–367 (2011).

56. Chen, V. B. *et al.* MolProbity: all-atom structure validation for macromolecular crystallography. *Acta Crystallogr. Sect. D* **66**, 12–21 (2010).

57. Allen, F. The Cambridge Structural Database: a quarter of a million crystal structures and rising. *Acta Crystallogr. Sect. B* **58**, 380–388 (2002).

58. Bruno, I. J. *et al.* Retrieval of crystallographically-derived molecular geometry information. *J. Chem. Inf. Comput. Sci.* **44**, 2133–2144 (2004).

59. Klein, T., Tucker, J., Holdgate, G. A., Norman, R. A. & Breeze, A. L. FGFR1 kinase inhibitors: close regioisomers adopt divergent binding modes and display distinct biophysical signatures. *ACS Med. Chem. Lett.* **5**, 166–171 (2013).

60. Naghibi, H., Tamura, A. & Sturtevant, J. M. Significant discrepancies between van't Hoff and calorimetric enthalpies. *Proc. Natl Acad. Sci. USA* **92**, 5597–5599 (1995).

61. Karlsson, R., Nilshans, H. & Persson, A. Thermodynamic analysis of protein interactions with biosensor technology. *J. Mol. Recognit.* **11**, 204–210 (1998).

62. Nozaki, Y. In *Methods Enzymol* Vol. 26 (eds Timasheff, S. N. & Hirs, C. H. W.) pp 43–50 (Academic Press, 1972).

63. Morjana, N. *et al.* Guanidine hydrochloride stabilization of a partially unfolded intermediate during the reversible denaturation of protein disulfide isomerase. *Proc. Natl Acad. Sci. USA* **90**, 2107–2111 (1993).

64. Lakomek, N.-A. *et al.* Internal dynamics of the homotrimeric HIV-1 viral coat protein gp41 on multiple time scales. *Angew. Chem. Int. Ed.* **52**, 3911–3915 (2013).

65. Lakomek, N.-A., Ying, J. & Bax, A. Measurement of 15N relaxation rates in perdeuterated proteins by TROSY-based methods. *J. Biomol. NMR* **53**, 209–221 (2012).

66. Delaglio, F. *et al.* NMRPipe: a multidimensional spectral processing system based on UNIX pipes. *J. Biomol. NMR* **6**, 277–293 (1995).

67. Keller, R. *The Computer Aided Resonance Assignment Tutorial* (CANTINA Verlag, 2011).

## Acknowledgements

We thank J. Griesbach and H.K. Pollard for their assistance in protein purification and the AstraZeneca FGFR chemistry team for synthesis of compounds, M. Vendruscolo and C. Camilloni for helpful discussions and critical reading of the manuscript, and G. Kelly (MRC Biomedical NMR Centre, Francis Crick Institute) for assistance with collection of the 950 MHz NMR data.

## Author contributions

T.K., N.V., J.J.P., G.D., J.A.T., G.A.H., R.A.N., A.D.S., D.R.H., D.L. and A.L.B. designed the research; T.K., N.V., J.J.P., G.D., J.A.T., R.A.N., A.D.S. and G.S.T. performed the research; T.K., N.V., J.J.P., G.A.H., J.A.T., R.A.N., A.D.S., D.R.H., D.L., G.S.T and A.L.B. analysed the data; and T.K., N.V., J.J.P., C.P., J.A.T., D.R.H., D.L. and A.L.B. wrote the paper. All authors have given approval to the final version of the manuscript.

## Additional information

**Accession codes:** coordinates and structure factors have been deposited in the Protein Data Bank with the following accession codes: FGFR1 – PDA complex, 5A4C; FGFR1–dovitinib complex, 5A46.

**Competing financial interests:** This work was funded as part of the AstraZeneca Internal Postdoctoral program. All authors with the exception of G.S.T. are employees (and stockholders) of AstraZeneca UK Ltd or MedImmune LLC, or were at the time that this study was conducted.

# 11

# Crystallographic and spectroscopic snapshots reveal a dehydrogenase in action

Lu Huo[1,2,*,†], Ian Davis[1,2,*], Fange Liu[1,†], Babak Andi[3], Shingo Esaki[1,2], Hiroaki Iwaki[4], Yoshie Hasegawa[4], Allen M. Orville[3,5] & Aimin Liu[1,2]

Aldehydes are ubiquitous intermediates in metabolic pathways and their innate reactivity can often make them quite unstable. There are several aldehydic intermediates in the metabolic pathway for tryptophan degradation that can decay into neuroactive compounds that have been associated with numerous neurological diseases. An enzyme of this pathway, 2-aminomuconate-6-semialdehyde dehydrogenase, is responsible for 'disarming' the final aldehydic intermediate. Here we show the crystal structures of a bacterial analogue enzyme in five catalytically relevant forms: resting state, one binary and two ternary complexes, and a covalent, thioacyl intermediate. We also report the crystal structures of a tetrahedral, thiohemiacetal intermediate, a thioacyl intermediate and an $NAD^+$-bound complex from an active site mutant. These covalent intermediates are characterized by single-crystal and solution-state electronic absorption spectroscopy. The crystal structures reveal that the substrate undergoes an $E/Z$ isomerization at the enzyme active site before an $sp^3$-to-$sp^2$ transition during enzyme-mediated oxidation.

[1] Department of Chemistry, Georgia State University, Atlanta, Georgia 30303, USA. [2] Molecular Basis of Disease Area of Focus Program, Georgia State University, Atlanta, Georgia 30303, USA. [3] Photon Sciences Directorate, Brookhaven National Laboratory, Upton, New York 11973, USA. [4] Department of Life Science and Biotechnology and ORDIST, Kansai University, Suita, Osaka 564-8680, Japan. [5] Biosciences Department, Brookhaven National Laboratory, Upton, New York 11973, USA. * These authors contributed equally to this work. † Present addresses: Department of Pharmaceutical Sciences, University of Connecticut (L.H.); Department of Chemistry, University of Chicago (F.L.). Correspondence and requests for materials should be addressed to A.L. (email: Feradical@gsu.edu).

The dominant route of tryptophan catabolism, the kynurenine pathway, has recently garnered increased attention given its apparent association with numerous inflammatory and neurological conditions, for example, gastrointestinal disorders, depression, Parkinson's disease, Alzheimer's disease, Huntington's disease and AIDS dementia complex[1-6]. Though the precise mechanism by which the kynurenine pathway influences these diseases has not yet been fully elucidated, it has been determined that several metabolites of this pathway are neuroactive. Notably, the concentration of quinolinic acid, a non-enzymatically derived decay product of an intermediate of the kynurenine pathway used for $NAD^+$ biosynthesis, is elevated over 20-fold in patients' cerebrospinal fluid with AIDS dementia complex, aseptic meningitis, opportunistic infections or neoplasms[7], and more than 300-fold in the brain of human immunodeficiency virus-infected patients[8]. This $NAD^+$ precursor has also been shown to be an agonist of N-methyl-D-aspartate receptors, and an increase of its concentration may lead to over-excitation and death of neuronal cells[9,10].

The apparent medical potential of the kynurenine pathway warrants detailed study and characterization of its component enzymes and their regulation. One enzyme in particular, 2-aminomuconate-6-semialdehyde dehydrogenase (AMSDH), is responsible for oxidizing the unstable metabolic intermediate 2-aminomuconate-6-semialdehyde (2-AMS) to 2-aminomuconate (2-AM) (Fig. 1a). On the basis of sequence alignment, AMSDH is a member of the hydroxymuconic-semialdehyde dehydrogenase (HMSDH) family under the aldehyde dehydrogenase (ALDH) superfamily[11]. ALDHs are prevalent in both prokaryotic and eukaryotic organisms and are responsible for oxidizing aldehydes to their corresponding carboxylic acids. They use $NAD(P)^+$ as a hydride acceptor to harvest energy from their primary substrate and generate NAD(P)H, which provides the major reducing power to maintain cellular redox balance[12,13]. In addition to being commonly occurring metabolic intermediates, aldehydes are reactive electrophiles, making many of them toxic. Enzymes of the ALDH superfamily are typically promiscuous with regards to their substrates; however, in recent years, this superfamily has had several new members identified with greater substrate fidelity, especially when the substrate is identified as a semialdehyde[14].

The putative native substrate of AMSDH, 2-AMS, is a proposed metabolic intermediate in both the 2-nitrobenzoic acid degradation pathway of Pseudomonas fluorescens KU-7 (ref. 15) and the kynurenine pathway for L-tryptophan catabolism in mammals[9,10,16]. In the presence of $NAD^+$ and AMSDH, 2-AMS is oxidized to 2-AM (Fig. 1a); however, it can also spontaneously decay to picolinic acid and water with a half-life of 35 s at neutral pH[17]. Due to its instability, 2-AMS has not yet been isolated, leaving its identity as the substrate of AMSDH an inference based on decay products and further metabolic reactions. There are several reasons for the poor understanding of this pathway: it is complex with many branches, some of the intermediates are unstable and difficult to characterize, and several enzymes of the pathway, including AMSDH, are not well understood. Hence, the structure of AMSDH will help to address questions such as what contributes to substrate specificity for the semialdehyde dehydrogenase and how 2-AMS is bound and activated during catalysis.

In the present study, we have cloned AMSDH from Pseudomonas fluorescens, generated an E. coli overexpression system and purified the target protein for molecular study. We also constructed several mutant expression systems to characterize the role of specific active site residues. Enzymatic assays were performed for all forms of the enzyme, and crystal structures were solved for the wild type and one mutant. We were able to capture several catalytic intermediates in crystallo by soaking protein crystals in mother liquor containing either the primary organic substrate or a substrate analogue and discovered that in addition to dehydrogenation, the substrate undergoes isomerization at the active site.

## Results

**Catalytic activity of wild-type AMSDH.** Due to the unstable nature of its substrate, 2-AMS, the activity of AMSDH was detected using a coupled-enzyme assay that employed its upstream partner, α-amino β-carboxymuconate ε-semialdehyde decarboxylase (ACMSD), to generate 2-AMS in situ. ACMSD transforms α-amino β-carboxymuconate ε-semialdehyde (ACMS) ($\lambda_{max}$ at 360 nm) to 2-AMS ($\lambda_{max}$ at 380 nm)[16,17]. As seen in Fig. 1b, in an assay that uses only ACMSD, the absorbance peak of its substrate, ACMS, red-shifts to 380 nm as 2-AMS is formed. The absorbance at 380 nm then quickly decreases as 2-AMS decays to picolinic acid, a compound with no absorbance features above 200 nm. In a coupled-enzyme assay, ACMSD, AMSDH and $NAD^+$ are included in the reaction system. As shown in Fig. 1c, ACMS is still consumed; however, there is no red shift observed because 2-AMS is enzymatically converted to 2-AM ($\lambda_{max}$ at 325 nm) rather than accumulating and decaying to picolinic acid. The production of 2-AM requires that an equimolar amount of $NAD^+$ be reduced to NADH ($\lambda_{max}$ at 339 nm). A stable alternative substrate, 2-hydroxymuconate-6-semialdehyde (2-HMS), was used to pursue kinetic parameters (Fig. 1d), when using saturating $NAD^+$ concentrations ($\geq 1$ mM), the $k_{cat}$ and $K_m$ of AMSDH for 2-HMS were $1.30 \pm 0.01$ s$^{-1}$ and $10.4 \pm 0.2$ μM, respectively (Fig. 1e).

**Structural snapshots of the dehydrogenase catalytic cycle.** We solved five crystal structures of wild-type AMSDH, including the ligand-free (2.20 Å resolution), $NAD^+$-bound binary complex (2.00 Å), ternary complex with $NAD^+$ and substrate 2-AMS (2.00 Å) or 2-HMS (2.20 Å) and a thioacyl intermediate (1.95 Å). All five structures belong to space group $P2_12_12_1$. Data collection and refinement statistics are listed in Supplementary Table 1. The complete AMSDH model includes four polypeptides per asymmetric unit describing one homotetramer (Supplementary Fig. 1a). Each monomer of AMSDH contains three domains: a subunit interaction domain, a catalytic domain and an $NAD^+$ binding domain (Supplementary Fig. 1b). For details of the secondary structure, see Supplementary Discussion.

In the structure of the co-crystallized binary complex, an $NAD^+$ molecule is present in an extended, anti-conformation in the amino-terminal, co-substrate-binding domain of each monomer (Fig. 2a). The electron density map of $NAD^+$ is well defined, and the interactions between the protein and $NAD^+$ are equivalent in all four subunits as shown in Fig. 2e. The $NAD^+$-bound AMSDH structure is similar to the ligand-free structure with an aligned r.m.s.d. of 0.239 Å. Residues that belong to the $NAD^+$-binding pocket are also well aligned with the exception of Cys302, Arg108 and Leu116 (Supplementary Fig. 2). On binding $NAD^+$, the thiol moiety of Cys302 rotates so that the sulfur is 2.3 Å closer to the substrate-binding pocket and away from the nicotinamide head of $NAD^+$.

**Crystal structures of enzyme–substrate ternary complexes.** Structures of AMSDH in ternary complex with co-substrate $NAD^+$ and its primary substrates were obtained by soaking co-crystallized AMSDH-$NAD^+$ crystals with 2-AMS and 2-HMS, respectively. Extra density that fits with the corresponding substrate molecule was observed in the active site of each subunit. The co-substrate $NAD^+$ in the ternary complex structures is

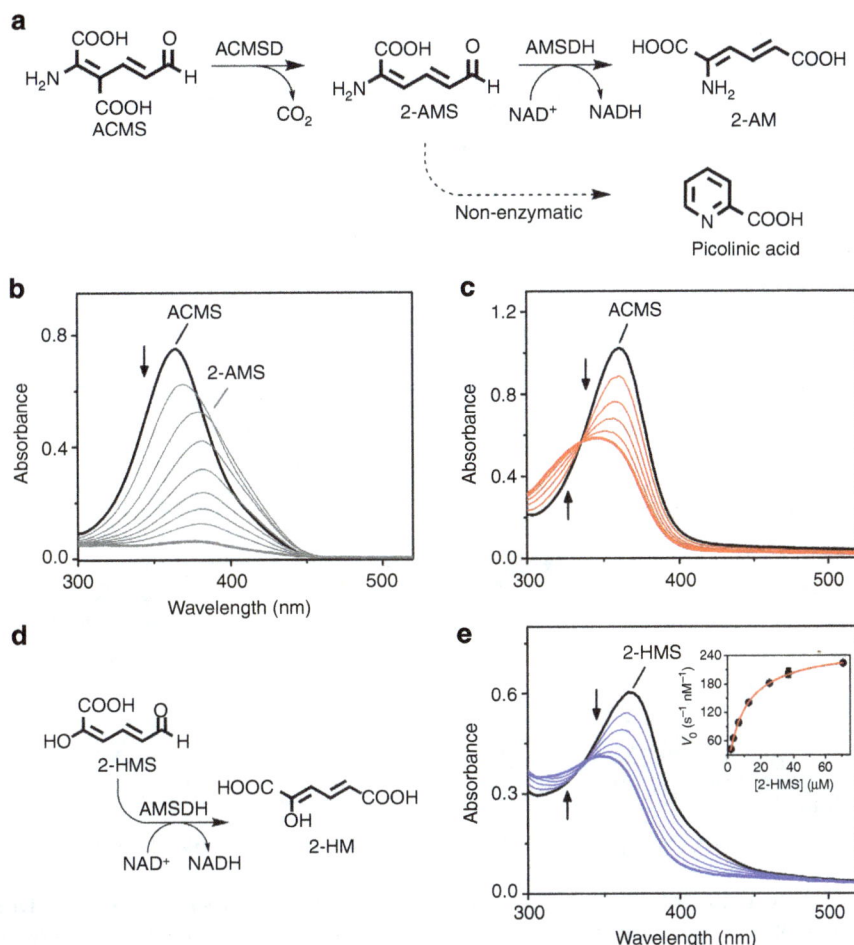

**Figure 1 | Activity of AMSDH.** (**a**) Reaction scheme showing the enzymatic generation of 2-AMS, the reaction catalysed by AMSDH, and the competing non-enzymatic decay of 2-AMS to picolinic acid. (**b**) Representative assay showing the ACMSD (1 μM)-catalysed conversion of ACMS ($\lambda_{max}$ 360 nm) to 2-AMS ($\lambda_{max}$ 380 nm), which decays to picolinic acid (transparent). (**c**) Coupled-enzyme assay in which AMSDH (200 nM) oxidizes 2-AMS, produced *in situ* as shown in **b** in 50 s, to 2-AM ($\lambda_{max}$ 325 nm). (**d**) Reaction scheme showing 2-HMS oxidation by AMSDH. (**e**) Representative assay showing the activity of AMSDH (200 nM) on 2-HMS ($\lambda_{max}$ 375 nm) in 50 s. The inset is a Michaelis–Menten plot.

bound in the same manner as in the binary complex. Binding of the primary substrates introduced minimal change to the protein structure; the r.m.s.d. for the superimposed structures of substrate-free with 2-AMS- and 2-HMS-bound ternary complex structures are 0.170 and 0.276 Å, respectively. These two primary substrates bind to AMSDH in an identical fashion, with two arginine residues, Arg120 and Arg464, playing an important role in stabilizing the substrate by forming two sets of bifurcated hydrogen bonds with one of the carboxyl oxygens and the 2-amino or hydroxyl group of 2-AMS (Fig. 2b) or 2-HMS (Fig. 2c), respectively. The observation of two hydrogen bonds being donated by the active site arginines to the 2-amino group of 2-AMS indicates that in the substrate-bound form, 2-AMS may be in its 2-imine rather than 2-enamine tautomer, as an amino group unlikely to accept two hydrogen bonds. Mutation of Arg120 to alanine causes a moderate decrease of the $k_{cat}$ to $0.7 \pm 0.2\,\text{s}^{-1}$ from $1.30 \pm 0.01\,\text{s}^{-1}$ and a dramatic increase of the $K_m$ with a lower bound of $446.3 \pm 195.9\,\mu\text{M}$ (an accurate determination of the $K_m$ is hindered by insufficient 2-HMS concentrations) compared with $10.4 \pm 0.2\,\mu\text{M}$ in the wild type (Supplementary Fig. 3a). Mutation of Arg464 to alanine decreased the $k_{cat}$ to $\sim 0.3\,\text{s}^{-1}$, and not only increased the $K_m$ to $\sim 170\,\mu\text{M}$, but also leads to a significant substrate inhibition effect with a $K_i$ of $\sim 6\,\mu\text{M}$ (Supplementary Fig. 3b). This substrate inhibition is likely caused by the unproductive binding of a second substrate molecule in the space created by the deletion of

Arg464 or by a failure of the enzyme to properly bind and stabilize the imine form of the substrate.

**Catalytic intermediates trapped after ternary complex formation.** Enzyme–NAD$^+$ binary complex crystals were soaked in mother liquor containing 2-HMS for a range of time points from 5 min to more than 3 h before flash cooling in liquid nitrogen. In a crystal that was soaked for 40 min, an intermediate was trapped and refined to a resolution of 1.95 Å (Fig. 2d). Crystals soaked for longer time points gave a similar intermediate with poorer resolution. In this structure, 2-HMS is observed in the 2*Z*, 4*E* isomer rather than the 2*E*, 4*E* isomer as seen in the substrate-bound ternary structure. Also, the substrate interacts with Arg120 and Arg464 with both of its carboxyl oxygens rather than one carboxyl oxygen and the 2-hydroxy oxygen as shown in the 2-HMS ternary complex structure. Fitting this density with the 2*E*, 4*E* conformation resulted in unsatisfactory $2F_o - F_c$ and $F_o - F_c$ density maps as shown in Supplementary Fig. 4a. Likewise, attempting to fit the 2*Z*, 4*E* isomer to the ternary complex structure did not produce satisfactory results (Supplementary Fig. 4b). On *E* to *Z* isomerization, the carbon chain of the substrate extends, and the distance between its sixth carbon and Cys302's sulfur is now at 1.8 Å, which is within covalent bond distance for a carbon–sulfur bond. Also, the continuous electron density

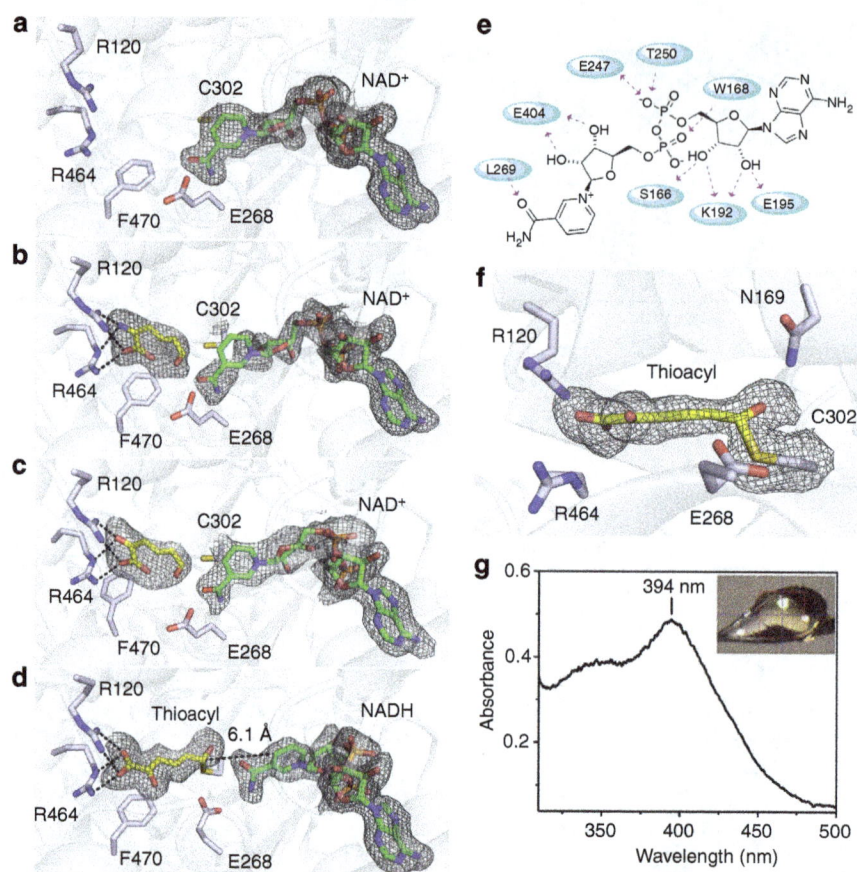

**Figure 2 | Crystal structures of wild-type AMSDH and single-crystal electronic absorption spectrum of a catalytic intermediate.** AMSDH was co-crystallized with NAD$^+$ to give AMSDH-NAD$^+$ binary complex crystals that were used for soaking experiments. (**a**) Active site structure of the binary AMSDH-NAD$^+$ complex, (**b**) the ternary complex of AMSDH-NAD$^+$ crystals soaked with 2-AMS for 5 min before flash cooling, (**c**) the ternary complex of AMSDH-NAD$^+$ soaked with 2-HMS for 10 min before flash cooling, (**d**) the trapped thioacyl, NADH-bound intermediate obtained by soaking AMSDH-NAD$^+$ crystals with 2-HMS for 40 min before flash cooling. (**e**) Two-dimensional interaction diagram for NAD$^+$ binding. (**f**) Close-up of the thioacyl intermediate in **d**. (**g**) Single-crystal electronic absorption spectrum of **d**. Protein backbone and residues are shown as light blue cartoons and sticks, respectively. The substrates and intermediate are shown as yellow sticks, and NAD$^+$ and NADH are shown as green sticks. The omit map for ligands is contoured to 2.0 $\sigma$ and shown as a grey mesh.

between Cys302-SG and 2-HMS-C6 indicates the presence of a covalent bond (Fig. 2f). Another feature of this intermediate is that the nicotinamide ring of NAD$^+$ has moved 4.6 Å away from the active site and adopted a bent conformation (Fig. 2d) compared with the position in the binary or ternary complex structures (Fig. 2a–c). The structural changes of NAD$^+$ associated with reduction has been observed and well documented[18,19]. In the oxidized state, NAD(P)$^+$ lies in the Rossmann fold in an extended conformation, allowing for hydride transfer from the substrate to its nicotinamide carbon during the first half of the reaction. Reduced NAD(P)H then adopts a bent conformation in which the nicotinamide head moves back towards the protein surface. This movement provides more space in the active site for the second half of the reaction, acyl-enzyme adduct hydrolysis, to take place. Thus, the coenzyme in this intermediate structure is likely to have been reduced to NADH and, as such, the structure is assigned as a thioacyl-enzyme–substrate adduct. The single-crystal electronic absorption spectrum of the sample has an absorbance maximum at 394 nm (Fig. 2g). The same absorbance band was observed in crystals soaked with 2-HMS from 30 min to 2 h (Supplementary Fig. 5a). However, this long-lived intermediate in the crystal was not observed in solution with millisecond-to-second time resolution in stopped-flow experiments (Supplementary Fig. 6a). Thus, it is either present in an earlier time domain

(sub-milliseconds), or alternatively, it may not accumulate in solution because NADH can readily dissociate in solution, whereas it may be trapped in the active site when in the crystalline state.

Another notable change in the intermediate structure is the movement of the side chain of Glu268, which rotates 73° towards the active site (Fig. 2c,d). To probe the function of Glu268, we constructed an alanine mutant and found that it exhibited no detectable activity in steady-state kinetic assays. Interestingly, E268A exhibits completely different pre-steady state activity than the wild-type enzyme. As shown in Supplementary Fig. 6b, an absorbance band at 422 nm was formed concomitant with the decay of the 2-HMS peak within 1 s of the reaction. This new species is generated stoichiometrically on titration of 2-HMS with E268A (Fig. 3d). The moiety that gives rise to this new absorbance band is stable for minutes at room temperature and cannot be separated from the protein by membrane filtration-based methods[20], suggesting that it is covalently bound to the protein. The formation of an enzyme–substrate adduct in the E268A mutant was investigated by mass spectrometry. For the as-isolated E268A, the resultant multiply charged states (Supplementary Fig. 7) were deconvoluted to obtain a molecular weight (MW) of 56,252 Da (Fig. 4a). This value is in good agreement with the predicted MW of E268A AMSDH plus an amino-terminal His-tag and linking residues, 56,251 Da. The

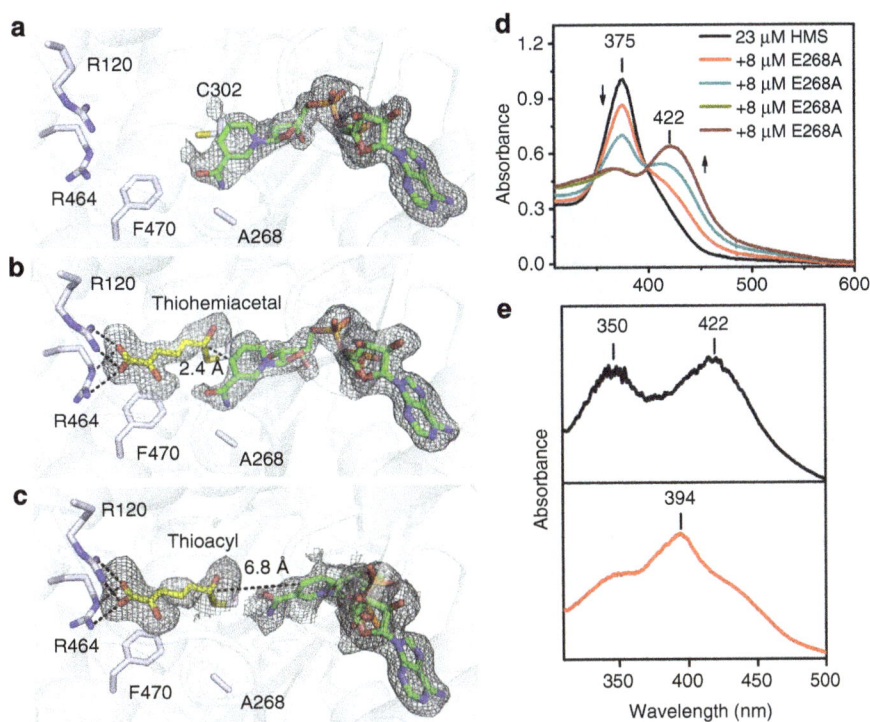

**Figure 3 | Crystal structures of the E268A mutant and its solution and single-crystal electronic absorption spectra.** (**a**) Structure of the active site of the co-crystallized E268A-NAD$^+$ binary complex, (**b**) a thiohemiacetal intermediate obtained by soaking the E268A-NAD$^+$ crystals with 2-HMS for 30 min before flash cooling and (**c**) a thioacyl intermediate obtained by soaking the E268A-NAD$^+$ crystals with 2-HMS for 180 min before flash cooling. (**d**) Solution electronic absorption spectra of a titration of 2-HMS with E268A. (**e**) Single-crystal electronic absorption spectrum of the intermediate in **b** (top panel) and single-crystal electronic absorption spectrum of the intermediate in **c** (bottom panel). Protein backbone and residues are shown as light blue cartoons and sticks, respectively. The substrate and intermediate are shown as yellow sticks, and NAD$^+$ and NADH are shown as green sticks. The omit map for ligands is contoured to 2.0 $\sigma$ and shown as a grey mesh.

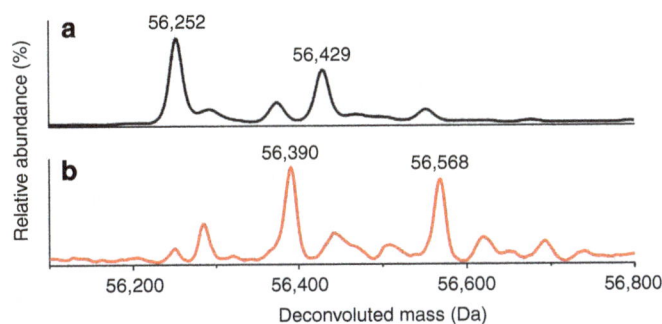

**Figure 4 | Deconvoluted positive-mode electrospray ionization mass spectra of as-isolated E268A (a) and 2-HMS treated-E268A (b).** The two major components are labelled with their respective molecular weights.

second largest peak in the deconvoluted spectrum has a MW 177 Da greater than that of the most abundant signal. This is likely due to post-translational modification of the His-tag; α-N-Gluconoylation of His-tags has been observed in *E. coli*-expressed proteins, which cause 178 Da extra mass[21]. The mutant protein was then treated with the alternate substrate, 2-HMS, and the mass spectrum shows a new major peak at 56,390 Da (Fig. 4b), 138 Da heavier than the as-isolated mutant. Similarly, the second most abundant peak corresponds to a His-tag modified mutant plus 139 Da. In this spectrum, the peaks arising from the as-isolated mutant are substantially reduced, indicating that 2-HMS, 141 Da, is bound to the E268A mutant enzyme.

We determined the crystal structure of E268A co-crystallized with NAD$^+$ and refined it to 2.00 Å resolution (Fig. 3a). The

overall structure aligns very well with the wild-type binary complex structure with an r.m.s.d. of 0.139 Å. The active site of E268A also resembles the native AMSDH structure (Supplementary Fig. 8). The nature of the absorbing species at 422 nm was further investigated by soaking co-crystallized E268A-NAD$^+$ crystals in mother liquor containing 2-HMS. By doing so, two temporally, structurally and spectroscopically distinct intermediates were identified.

When E268A-NAD$^+$ crystals are soaked with 2-HMS for 40 min or less, their single-crystal electronic absorption spectra show an absorbance maximum at 422 nm (Supplementary Fig. 5b), as was observed in the solution-state titration and the stopped-flow assays. An individual electronic absorption spectrum for an E268A-NAD$^+$ crystal soaked with 2-HMS for 15 min can be found in Fig. 3e (top). The structure of E268A-NAD$^+$ soaked with 2-HMS for 30 min before flash cooling was solved and refined to 2.15 Å resolution (Fig. 3b). In this structure, a continuous electron density between Cys302-SG and 2-HMS-C6 is observed, similar to the thioacyl intermediate observed in the wild-type enzyme. However, in contrast to the thioacyl intermediate, the density around C6 is less flat, indicating an $sp^3$- rather than $sp^2$-hybridized carbon (Fig. 5a). The angle between the plane of the carbon backbone of the substrate and the formerly aldehydic oxygen is 55 ± 9°, compared with the angle of the wild-type thioacyl intermediate at 26 ± 4° (Supplementary Table 2). More importantly, the C6 of 2-HMS and the C4N of NAD$^+$ are very close (2.4–2.8 Å), making it unlikely that the hydride has been transferred from the substrate. Taken together, these data allow us to assign this intermediate to a thiohemiacetal enzyme adduct (Fig. 3b). A similar intermediate has only been trapped once previously

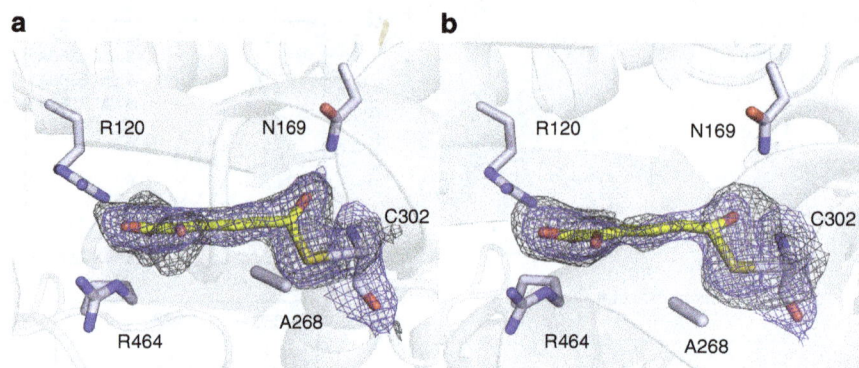

**Figure 5 | Crystal structures of two distinct catalytic intermediates.** (**a**) Electron density map of the thiohemiacetal intermediate obtained from E268A-NAD$^+$ crystal soaked with 2-HMS for 30 min. (**b**) Electron density map of the thioacyl intermediate obtained from E268A-NAD$^+$ crystal soaked with 2-HMS for 180 min. The $2F_o - F_c$ electron density map for ligands and Cys302 is contoured to 1.0 $\sigma$ and shown as a blue mesh. The omit map for ligands and Cys302 is contoured to 2.0 $\sigma$ and shown as a gray mesh.

in a crystal that contains no co-substrate[22]. Hence, this is the first time for this intermediate to be trapped in the presence of NAD$^+$.

If the E268A-NAD$^+$ crystals are soaked with 2-HMS for longer than 1 h, their single-crystal electronic absorption spectra begin to resemble that of the wild-type, thioacyl intermediate with a corresponding absorbance maximum at 394 nm (Supplementary Fig. 5b), as seen in wild-type, thioacyl intermediate crystals. An individual electronic absorption spectrum for an E268A-NAD$^+$ crystal soaked with 2-HMS for 120 min can be found in Fig. 3e (bottom). The structure of an E268A-NAD$^+$ crystal soaked with 2-HMS for 180 min was solved and refined to 2.20 Å (Fig. 3c). The structure of this intermediate is also similar to the wild-type, thioacyl-enzyme adduct with NADH, rather than NAD$^+$ found at the active site. The distance between the C4N of NADH and C6 of 2-HMS is longer than 6.1 Å (Fig. 3c). The electron density around C6 is flatter (Fig. 5b) compared with the thiohemiacetal intermediate and similar to the thioacyl intermediate trapped in the wild-type AMSDH structure (Fig. 2f), and the angle between the plane of the carbon backbone of the substrate and the carbonyl oxygen is 20 ± 5°, which is statistically indistinguishable from that of the wild-type, thioacyl intermediate, 26 ± 4° (Supplementary Table 2). On the basis of the similarities in their absorbance and structures, we conclude that this latter intermediate is equivalent to the wild-type, thioacyl intermediate. It is also worth noting that the strictly conserved asparagine 169 (Fig. 5) is seen to stabilize both the thiohemiacetal and thioacyl intermediates through hydrogen-bonding interactions.

**Investigation of isomerization by computational modelling.** The isomerization of 2-AMS from the 2E to 2Z isomer implied by the solved crystal structures was probed with density functional theory calculations. The free energy profiles obtained were used to help illuminate the nature of 2-AMS and gain insight into how the active site of AMSDH may facilitate the isomerization. The total energies of different isomers and rotamers of 2-AMS in its enamine/aldehyde and imine/eneol tautomers and the rotational barriers about their respective 2–3 bond were compared. For the imine/eneol tautomer, additional computations were performed with the side groups from Arg120 and Arg464 to investigate what effect, if any, they will have on the free energy profile for rotation about the 2–3 bond of 2-AMS.

First, 2-AMS was constructed and optimized in its 2-enamine, 6-aldehyde, 2E isomer with a negatively charged 2-carboxylate group (Fig. 6a). To estimate the energy barrier for an uncatalysed

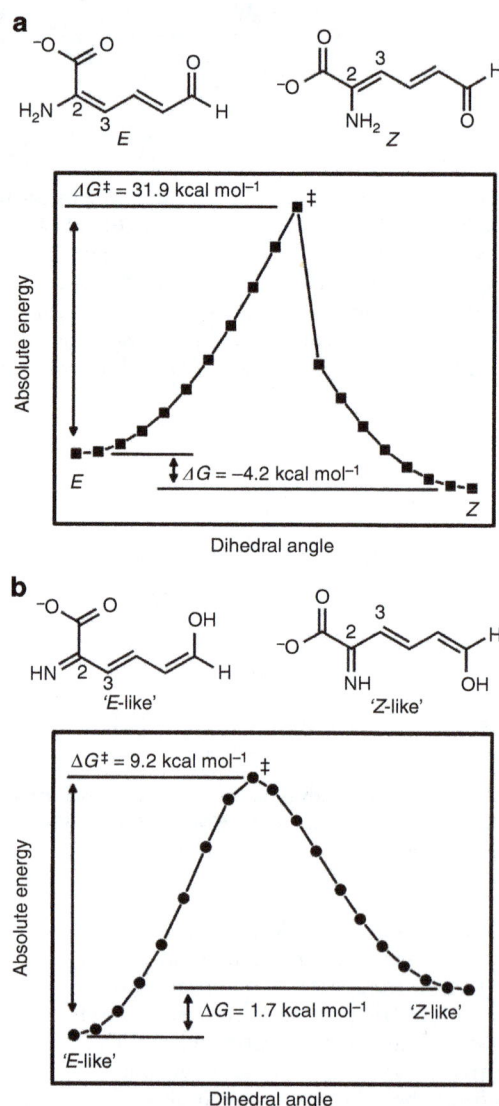

**Figure 6 | Free energy profiles for the rotation about the 2–3 bond of 2-AMS in its (a) enamine and (b) imine form, respectively.** DFT calculations were performed at the B3LYP/6-31G* + level of theory. The dihedral angle about the 2–3 bond was restrained in 10° increments and the structures were optimized at each point.

isomerization from the 2E to the 2Z isomer, the 2–3 double bond was then restrained at 10° intervals from 180 to 0°, and the structure was optimized at each point. On the basis of the free energy profile (Fig. 6a), the uncatalysed isomerization barrier is 31.9 kcal mol$^{-1}$. The profile also shows that the 2Z isomer, as might be expected, is lower in energy than the 2E isomer by 4.2 kcal mol$^{-1}$. Next, the rotational barrier about the 2–3 bond of 2-AMS when in its 2-imine, 6-enol tautomer, as is suggested by the ternary complex structure, was calculated in the same manner. The barrier was found to be 9.2 kcal mol$^{-1}$, and opposite to the enamine tautomer, the '2Z-like' rotamer is higher in energy than the '2E-like' rotamer by 1.7 kcal mol$^{-1}$ (Fig. 6b). Unsurprisingly, the rotational barrier about the 2–3 bond is much lower in the imine tautomer; however, the '2Z-like' rotamer of the imine tautomer is 21.8 kcal mol$^{-1}$ higher in energy than the 2Z isomer of the enamine tautomer.

Possible influences of the two active site arginines on the free energy profile for rotation were also considered. To mimic the conditions of the enzyme active site, similar calculations as those above were performed, which included the guanidinium heads of Arg120 and Arg464. The starting model was built using the active site geometry of the ternary complex crystal structure (PDB entry: 4I25), and on inspection, it is immediately apparent that with two arginines in such close proximity to the substrate, there is insufficient space for two hydrogen atoms on the nitrogen at the 2-position of 2-AMS, and attempts to optimize an enamine tautomer with the hydrogen-bonding pattern of the ternary complex produced structures within which the entire 2-AMS molecule rotates so that only the carboxylate group interacts with the guanidinium moieties. The absolute positions of the guanidinium groups were fixed and the structure of 2-AMS in the imine tautomer was optimized. The dihedral angle of the 2–3 bond of 2-AMS was then increased in 45° increments and the structure optimized while restraining the position of the guanidinium groups and the 2–3 bond to build a rough free energy profile to estimate the rotational barrier. In the presence of the active site arginines, the barrier about the 2–3 bond of 2-AMS is further reduced to 8.5 kcal mol$^{-1}$ (Supplementary Table 3). Another interesting finding is that in the presence of the guanidinium groups, the '2E-like' and '2Z-like' rotamers of 2-AMS are nearly isoenergetic, with a free energy difference of 0.2 kcal mol$^{-1}$ (Supplementary Table 3).

## Discussion

The substrate of AMSDH, 2-AMS, contains an unstable aldehyde in conjugation with an enamine and can decay to picolinic acid and water, presumably through an electrocyclization reaction similar to its metabolic precursor, ACMS[23]. To assay the enzymatic activity, the upstream enzyme was utilized in the reaction mixture to generate substrate, and it was shown that AMSDH is catalytically active. Unfortunately, no kinetic parameters can be reliably determined because the concentration of 2-AMS is not well defined in the coupled-enzyme assay. To circumvent this issue, a previously-identified, stable alternative substrate, 2-HMS[24,25], in which a hydroxyl group replaces the amino group in 2-AMS to prevent cyclization, was used to characterize the activity of AMSDH and to examine the activity of the mutants.

Substrate-bound, ternary complex structures were obtained by soaking co-crystallized protein and NAD$^+$ with 2-AMS or 2-HMS. 2-AMS is an unstable compound which decays with a $t_{1/2}$ of about 9 s at pH 7.5 and 37 °C or 35 s at pH 7.0 and 20 °C. Notably, this is its first reported structure. It appears to be stabilized in the enzyme active site in its imine tautomer by forming two sets of bifurcated hydrogen bonds with Arg120 and

Arg464 so that the electrocyclization reaction cannot occur. Both arginine residues are close to the protein surface and in good positions to serve as gatekeepers, bringing the substrate into the active site. As a residue residing on a loop, Arg464 should be relatively flexible. The electron density for the side chain of Arg120 is partially missing in the binary complex structure but very well resolved in both ternary complex structures. This observation indicates that the presence of substrate can stabilize what may be a flexible residue. It becomes evident from the coordinates that Arg120 and Arg464 play an important role in substrate recognition, stabilization and possibly product release. Two arginine residues are rarely observed in such close proximity, stabilizing one end of the same molecule with multiple hydrogen bonds. With the exception of the hydrogen bonds provided by Arg120 and 464, the substrate-binding pocket is mostly composed of hydrophobic residues. On the basis of sequence alignment (Supplementary Fig. 9), these two arginine residues are strictly conserved throughout the HMSDH family but are not found in other members of the ALDH superfamily. We propose that these dual arginines combined with the size restrictions provided by the hydrophobic pocket endow this enzyme with its specificity towards small α-substituted carboxylic acids with an aldehyde moiety, such as 2-AMS and 2-HMS. Furthermore, our computational work suggests that these arginines are crucial for stabilizing the imine tautomer of 2-AMS to allow for rotation about its 2–3 bond.

Two strictly conserved catalytic residues, Cys302 and Glu268, are present at the interior of the substrate-binding pocket. General features regarding these residues in the ALDH super-family are (1) that the cysteine serves as a catalytic nucleophile, which is anticipated to form a covalent-adduct intermediate with the substrate by a nucleophilic addition during catalysis[26–28] and (2) that the glutamate serves as a base to activate water for hydrolysis of the thioacyl-enzyme adduct[29–32]. Previous studies indicate that the catalytic cysteine can adopt two conformations, resting and attacking[19]. In the ligand-free structure, Cys302 is far from where the carbonyl carbon of the substrate should be and is in the resting state. In the ternary complex structures, Cys302 is located at an ideal position to initiate catalysis, which is the attacking state. It is proposed to attack the aldehydic carbon (C6) of the substrate. In the two ternary complex structures, the distance between the sulfur of Cys302 and the C6 of the substrate is ~3.3 Å. Cys302 and the aldehydic carbon form a covalent bond in both thioacyl and thiohemiacetal intermediates. Mutation of Cys302 to serine led to enzyme with no detectable dehydrogenase activity, further confirming its catalytic significance.

Examining the wild-type AMSDH structures shows that in the NAD$^+$-bound binary complex, Glu268 adopts a 'passive' conformation, pointing away from the substrate-binding pocket, and forms hydrogen bonds with both NE of Trp177 (3.2 Å distance) and the backbone oxygen of Phe470 (3.6 Å) to leave space for the reduction of NAD$^+$. Its electron density is very well resolved and the side chain B-factor is close to average: 28.2 Å$^2$/28.5 Å$^2$. The thiol moiety of Cys302 is 7.14 Å from Glu268 and is unlikely to form interactions. Interestingly, in both substrate-bound structures, Glu268 becomes more flexible and exhibits much weaker electron density and increased side chain B-factors compared with average protein B-factors: 37.8 Å$^2$/28.5 Å$^2$ and 66.37 Å$^2$/39.7 Å$^2$. In the thioacyl intermediate structure, the electron density of Glu268 becomes very well defined again, but its side chain rotates 73° towards the bound substrate and seems to be in an 'active' position to abstract a proton from a deacylating water (Fig. 2d). At this point in the reaction cycle, the NADH molecule needs to leave the active site to make room for the catalytic water molecule. Movement of the nicotinamide ring of

NAD$^+$ coupled with the rotation of an active site glutamate has previously been observed in other ALDHs during catalysis[30–32].

Mutation of Glu268 to alanine led to the accumulation of the thiohemiacetal intermediate in both solution and crystalline states. The strictly conserved glutamate residue in the active site of ALDH enzymes has been proposed to play up to three possible roles during catalysis. It is strictly required to activate the deacylating water that allows for product release, it is in a 'passive' conformation during NAD(P)$^+$ reduction, and in some cases, it may serve to activate cysteine for nucleophilic attack[33]. On the basis of these roles, mutation to alanine would be expected to decrease the rate of hydrolysis of the thioacyl adduct, have no effect on the rate of reduction of NADH and possibly decrease the rate of nucleophilic attack by cysteine. With this understanding, deletion of the active site glutamate should cause an accumulation of the thioacyl intermediate. However, in this work, the E268A mutant is shown to accumulate the preceding thiohemiacetal intermediate both in crystal and in solution. This finding suggests an additional catalytic role for this residue: rotation of Glu268 towards the active site facilitates the hydride transfer from the tetrahedral thiohemiacetal adduct to NAD$^+$. The rapid formation of the intermediate in solution indicates that Glu268 of AMSDH does not play a role in activating cysteine. However, it does appear necessary to complete hydride transfer from the substrate to NAD$^+$, and its removal turns the native, primary substrate into a suicide inhibitor.

On the basis of previous studies of the ALDH mechanism, the eight high-resolution crystal structures solved (Supplementary Table 1) as well as our biochemical and computational studies, we propose a catalytic mechanism for AMSDH. As shown in Fig. 7, NAD$^+$ binds to the enzyme, **1**, to form an NAD$^+$-bound AMSDH complex, **2**. The substrate, 2-AMS, is then recognized by Arg120 and Arg464 through multiple hydrogen-bonding interactions, and its imine tautomer is stabilized in the active site, **3**. At this point, the order of the rotation, tautomerization and nucleophilic attack by C302 on the aldehydic carbon to produce

the tetrahedral, thiohemiacetal intermediate, **4**, is not yet clear. The isomerization and nucleophilic attack drive a translation of the substrate away from Arg120 and Arg464 so that they are only able to interact with the carboxylate group of the substrate. Next, NAD$^+$ is reduced to NADH by abstraction of a hydride from **4**, forming a thioacyl intermediate, **5**, a process which involves an $sp^3$-to-$sp^2$ transition during oxidation of the organic substrate by NAD$^+$. On reduction, the nicotinamide portion of NADH moves away from the substrate as Glu268 rotates into position to activate a water molecule to perform a nucleophilic attack on the same carbon that was previously attacked by Cys302, forming a second tetrahedral intermediate, **6**. Finally, the second tetrahedral intermediate collapses, breaking the C–S bond and releasing the final products, 2-AM and NADH. Species **1–5** are spectroscopically and structurally characterized, while intermediate **6** was not seen to accumulate.

In this work, five catalytically relevant structures of the wild-type AMSDH and three mutant structures yield a comprehensive understanding of the protein's overall structure, co-substrate-binding mode and elucidate the primary residues responsible for substrate specificity among the HMSDH family of the ALDH superfamily. The structural and spectroscopic snapshots capture the crystal structure of an unstable kynurenine metabolite, 2-AMS, and two catalytic intermediates, including stabilizing a tetrahedral intermediate in a mutant protein, which was further verified by mass spectrometry. Capture of a thiohemiacetal intermediate upon deletion of E268 also points to a new role for this well-established active site base in hydride transfer from the substrate to NAD$^+$. Another interesting finding revealed through solving the ternary complex and intermediate crystal structures and supported by computational studies is that an $E$ to $Z$ isomerization of the substrate occurs in this dehydrogenase before hydride transfer. To the best of our knowledge, this is the first piece of structural evidence illustrating an ALDH that proceeds via an $E/Z$ isomerization of its substrate during catalysis.

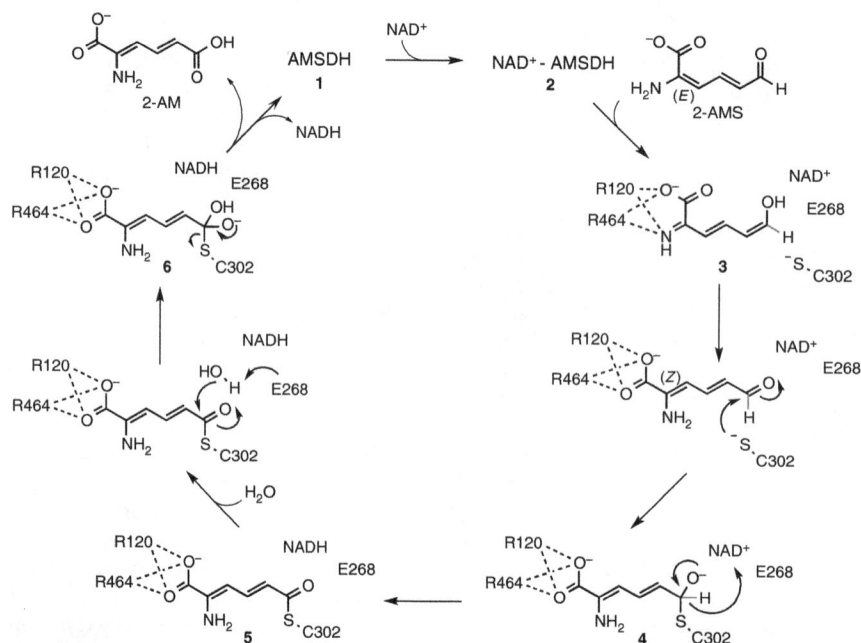

**Figure 7 | Proposed catalytic mechanism for the oxidation of 2-AMS by AMSDH.** The primary substrate (2E, 4E)-2-aminomuconate-semialdehyde binds to the enzyme in its imine tautomer to form the ternary complex (**3**). An isomerization and attack by cysteine on the aldehydic carbon form the (2Z, 4E)-2-aminomuconate-thiohemiacetal adduct (**4**). AMSDH-mediated oxidation of **4** concomitant with reduction of NAD$^+$ to NADH follows, generating a thioacyl-enzyme intermediate (**5**). Both **4** and **5** are the catalytic intermediates covalently attached to the enzyme. Hydrolysis of **5** then allows the release of the products 2-AM and NADH, restoring the ligand-free enzyme for the next catalytic cycle.

## Methods

**General methods.** The cloning, expression, purification and site-directed mutagenesis of AMSDH are described in the Supplementary Methods.

**Preparation of ACMS and 2-HMS.** ACMS was generated by catalysing the insertion of molecular oxygen to 3-hydroxyanthranilic acid by purified, $Fe^{2+}$ reconstituted 3-hydroxyanthranilate 3,4-dioxygenase as described previously[16,20]. 2-HMS is generated non-enzymatically from ACMS following a previously established method[24]. The pH of solutions containing ACMS was adjusted to $\sim 2$ by the addition of hydrochloric acid. 2-HMS formation was monitored on an Agilent 8453 diode-array spectrophotometer at 315 nm. The solutions were then neutralized with sodium hydroxide once the absorbance at 315 nm stopped increasing. 2-HMS at neutral pH has a maximum absorbance at 375 nm (ref. 24).

**Enzyme activity assay using 2-HMS as substrate.** Steady-state kinetics analyses were carried out at room temperature on an Agilent 8453 diode-array spectrophotometer. Reaction buffer contains 25 mM HEPES and 1 mM $NAD^+$, pH 7.5. Consumption of 2-HMS by 200 nM AMSDH was detected by monitoring the decrease of its absorbance at 375 nm with a molar extinction coefficient of $43,000\,M^{-1}cm^{-1}$ (ref. 24) for 15 s with a 0.5 s integration time. For mutants, 700 nM protein and a wavelength of 420 nm, $\varepsilon_{420}$ $11,180\,M^{-1}cm^{-1}$, was used. Absorbance at 375 nm decreased and blue shifted to 295 nm, the maximum ultraviolet absorbance for the product, 2-hydroxymuconic acid. This is consistent with previous reports in which the ending compound was purified and verified as the correct product[24]. The pre-steady state spectra were obtained with an Applied Photophysics Stopped-Flow Spectrometer SX20 (UK) with the mixing unit hosted inside an anaerobic chamber made by Coy Laboratory Products (MI, USA). Pre-steady state activity used the same reaction buffer but with 23 μM AMSDH or E268A and 25 μM 2-HMS and were carried out at 10 °C. The change in absorbance was monitored for 1.0 s.

**X-ray crystallographic data collection and refinement.** Purified AMSDH samples at a final concentration of $10\,mg\,ml^{-1}$ containing no $NAD^+$ or 10 equiv. of $NAD^+$ were used to set up sitting-drop vapour diffusion crystal screening trays in Art Robbins 96-well Intelli-Plates using an ARI Gryphon crystallization robot. The initial crystallization conditions were obtained from PEG-Ion 1/2 (Hampton Research) screening kits at room temperature. The screened conditions were optimized by increasing protein concentration to $40\,mg\,ml^{-1}$ and lowering crystallization temperature to 18 °C. $NAD^+$-bound AMSDH crystals were obtained from drops assembled with 1.5 μl of protein (preincubated for 10 min with 10 equiv. of $NAD^+$) mixed with 1.5 μl of a reservoir solution containing 20% polyethylene glycol 3350 and 0.2 M sodium phosphate dibasic monohydrate, pH 9.1, by hanging drop diffusion in VDX plates (Hampton Research). Pyramid shaped crystals that diffract up to $\sim 1.9$ Å appeared overnight. The reservoir solution for crystallizing the cofactor-free AMSDH crystals contains 12% polyethylene glycol 3350, 0.1 M sodium formate, pH 7.0. Crystals belonging to the same space group formed within 2–3 days with an irregular plate shape and diffracted up to $\sim 2.2$ Å. $NAD^+$-AMSDH crystals were used for substrate-soaking experiments. Crystals were transferred to mother liquor solution containing $\sim 1$ mM 2-HMS and incubated for 10–180 min before flash cooling in liquid nitrogen. Soaking 2-AMS as a substrate is more complicated because of its instability. Crystallization solution containing $\sim 1.5$ mM ACMS were used for soaking. After transferring several crystals to the soaking solution (8 μl), 2 μl of 1 mM purified ACMSD was included to catalyse the conversion of ACMS to 2-AMS. Crystals were flash frozen after a 5 min-incubation. Crystallization solution containing 20% glycerol or ethylene glycol was used as cryoprotectant. X-ray diffraction data were collected on SER-CAT beamline 22-ID or 22-BM of the Advanced Photon Source, Argonne National Laboratory.

**Ligand refinement and molecular modelling.** The first AMSDH structure, the cofactor $NAD^+$ bound structure, was solved by the molecular replacement method with the Advanced Molecular Replacement coupled with Auto Model Building programs from the PHENIX software using 5-carboxymethyl-2-hydroxymuconate semialdehyde dehydrogenase (PDB: 2D4E) as a search model, which shares 39% of amino-acid sequence identity with *P. fluorescens* AMSDH. The ligand-free, mutant and ternary complex structures were solved by molecular replacement using the refined $NAD^+$-AMSDH as the search model. Refinement was conducted using PHENIX software[34]. The program Coot was used for electron density map analysis and model building[35]. $NAD^+$/NADH, substrates 2-AMS and 2-HMS and Cys-substrate covalent-adduct intermediate were well defined and added to the model based on the $2F_o - F_c$ and $F_o - F_c$ electron density maps. Refinement was assessed as complete when the $F_o - F_c$ electron density contained only noise. The structural figures were generated using PyMOL software (http://www.pymol.org/).

**Single-crystal spectroscopy.** Electronic absorption spectra from single crystals held at 100 K were collected at beamline X26-C of the National Synchrotron Light Source (NSLS)[36]. The electronic absorption data were typically obtained between 200 and 1,000 nm with a Hamamatsu (Bridgewater, N.J.) L10290 high-power ultraviolet–visible light source. The lamp was connected to one of several 3-m long

solarization-resistant optical fibres with an internal diameter of 115, 230, 400 or 600 μm (Ocean Optics, Dunedin, FL). The other end was connected to a 40-mm diameter, 35 mm working distance 4 ×, Schwardchild design reflective microscope objective (Optique Peter, Lentilly France). The spectroscopy spot size is a convolution of the optical fibre diameter and the magnification of the objective, which in this case produced 28, 50, 100 or 150 μm diameter spots, respectively. Photons that passed through the crystal were collected with a second, aligned objective that was connected to a similar optical fibre or one with a slightly larger internal diameter. The spectrum was then recorded with either an Ocean Optics USB4000 or QE65000 spectrometer. Anisotropic spectra and an image of the crystal/loop were collected as a function of rotation angle in 5° increments. These were analysed by XREC[37] to determine the flat face and optimum orientation.

**Mass spectrometry.** To prepare samples for ESI mass spectrometry, as-isolated E268A AMSDH was buffer-exchanged to 10 mM Tris (pH 8.0) by running through a desalting column (GE Healthcare). Intermediate bound E268A was obtained by mixing E268A with 3 equiv. of 2-HMS. Excess 2-HMS was removed by desalting chromatography using the same buffer. Desalted proteins were concentrated to a final concentration of 20 μM. Freshly prepared samples were rinsed by acetonitrile and 0.1% formic acid (1:1 ratio) before injection. Mass spectrometry experiments were conducted using a Waters (Milford, MA) Micromass Q-TOF micro (ESI-Q-TOF) instrument operating in positive mode. The capillary voltage was set to 3,500 V, the sample cone voltage to 35 V and the extraction cone voltage to 2 V. The source block temperature and the desolvation temperature were set to 100 and 120 °C, respectively. The samples were introduced into the ion source by direct injection at a flow rate of $5\,\mu l\,min^{-1}$. The raw data containing multiple positively charged protein peaks were deconvoluted and smoothed using MassLynx 4.1.

**Computational studies.** All ground-state density functional theory calculations were performed with Gaussian 03 Revision-E.01 at the B3LYP/6-31G*+ level of theory[38]. The chemical structures were optimized using the ternary complex crystal structure (PDB entry: 4I25) as a starting model. To calculate the isomerization barrier, the dihedral angle about the 2–3 bond was restrained and the rest of the molecule was optimized. For the calculations that included the guanidinium heads of Arg120 and Arg464, the geometry was obtained from the crystal structure, and their positions were fixed while the substrate was optimized.

## References

1. Keszthelyi, D., Troost, F. J. & Masclee, A. A. Understanding the role of tryptophan and serotonin metabolism in gastrointestinal function. *Neurogastroenterol. Motil.* **21**, 1239–1249 (2009).
2. Myint, A. M. *et al.* Kynurenine pathway in major depression: evidence of impaired neuroprotection. *J. Affect. Disord.* **98**, 143–151 (2007).
3. Ogawa, T. *et al.* Kynurenine pathway abnormalities in Parkinson's disease. *Neurology* **42**, 1702–1706 (1992).
4. Guillemin, G. J. *et al.* Quinolinic acid in the pathogenesis of Alzheimer's disease. *Adv. Exp. Med. Biol.* **527**, 167–176 (2003).
5. Guidetti, P. & Schwarcz, R. 3-Hydroxykynurenine and quinolinate: pathogenic synergism in early grade Huntington's disease? *Adv. Exp. Med. Biol.* **527**, 137–145 (2003).
6. Kerr, S. J., Armati, P. J., Guillemin, G. J. & Brew, B. J. Chronic exposure of human neurons to quinolinic acid results in neuronal changes consistent with AIDS dementia complex. *AIDS* **12**, 355–363 (1998).
7. Heyes, M. P. *et al.* Quinolinic acid in cerebrospinal fluid and serum in HIV-1 infection: relationship to clinical and neurological status. *Ann. Neurol.* **29**, 202–209 (1991).
8. Heyes, M. P. *et al.* Sources of the neurotoxin quinolinic acid in the brain of HIV-1-infected patients and retrovirus-infected macaques. *FASEB J.* **12**, 881–896 (1998).
9. Schwarcz, R. The kynurenine pathway of tryptophan degradation as a drug target. *Curr. Opin. Pharmacol.* **4**, 12–17 (2004).
10. Stone, T. W. & Darlington, L. G. Endogenous kynurenines as targets for drug discovery and development. *Nat. Rev. Drug Discov.* **1**, 609–620 (2002).
11. Perozich, J., Nicholas, H., Wang, B. C., Lindahl, R. & Hempel, J. Relationships within the aldehyde dehydrogenase extended family. *Protein Sci.* **8**, 137–146 (1999).
12. Unden, G. & Bongaerts, J. Alternative respiratory pathways of *Escherichia coli*: energetics and transcriptional regulation in response to electron acceptors. *Biochim. Biophys. Acta* **1320**, 217–234 (1997).
13. Nicholls, D. G. Mitochondrial function and dysfunction in the cell: its relevance to aging and aging-related disease. *Int. J. Biochem. Cell Biol.* **34**, 1372–1381 (2002).
14. Hempel, J., Nicholas, H. & Lindahl, R. Aldehyde dehydrogenases: widespread structural and functional diversity within a shared framework. *Protein Sci.* **2**, 1890–1900 (1993).

15. Hasegawa, Y. *et al.* A novel degradative pathway of 2-nitrobenzoate via 3-hydroxyanthranilate in *Pseudomonas fluorescens* strain KU-7. *FEMS Microbiol. Lett.* **190**, 185–190 (2000).

16. Li, T., Walker, A. L., Iwaki, H., Hasegawa, Y. & Liu, A. Kinetic and spectroscopic characterization of ACMSD from *Pseudomonas fluorescens* reveals a pentacoordinate mononuclear metallocofactor. *J. Am. Chem. Soc.* **127**, 12282–12290 (2005).

17. Li, T., Ma, J., Hosler, J. P., Davidson, V. L. & Liu, A. Detection of transient Intermediates in the metal-dependent non-oxidative decarboxylation catalyzed by α-amino-β-carboxymuconate-ε-semialdehyde decarboxylase. *J. Am. Chem. Soc.* **129**, 9278–9279 (2007).

18. Perez-Miller, S. J. & Hurley, T. D. Coenzyme isomerization is integral to catalysis in aldehyde dehydrogenase. *Biochemistry* **42**, 7100–7109 (2003).

19. Muñoz-Clares, R. A., González-Segura, L. & Díaz-Sánchez, A. G. Crystallographic evidence for active-site dynamics in the hydrolytic aldehyde dehydrogenases. Implications for the deacylation step of the catalyzed reaction. *Chem. Biol. Interact.* **191**, 137–146 (2011).

20. Huo, L., Davis, I., Chen, L. & Liu, A. The power of two: arginine 51 and arginine 239* from a neighboring subunit are essential for catalysis in α-amino-β-carboxymuconate-ε-semialdehyde decarboxylase. *J. Biol. Chem.* **288**, 30862–30871 (2013).

21. Geoghegan, K. F. *et al.* Spontaneous α-N-6-phosphogluconoylation of a 'His tag' in *Escherichia coli*: the cause of extra mass of 258 or 178 Da in fusion proteins. *Anal. Biochem.* **267**, 169–184 (1999).

22. Blanco, J., Moore, R. A. & Viola, R. E. Capture of an intermediate in the catalytic cycle of L-aspartate-β-semialdehyde dehydrogenase. *Proc. Natl Acad. Sci. USA* **100**, 12613–12617 (2003).

23. Colabroy, K. L. & Begley, T. P. The pyridine ring of NAD is formed by a nonenzymatic pericyclic reaction. *J. Am. Chem. Soc.* **127**, 840–841 (2005).

24. Ichiyama, A. *et al.* Studies on the metabolism of the benzene ring of tryptophan in mammalian tissues. II. Enzymic formation of α-aminomuconic acid from 3-hydroxyanthranilic acid. *J. Biol. Chem.* **240**, 740–749 (1965).

25. He, Z., Davis, J. K. & Spain, J. C. Purification, characterization, and sequence analysis of 2-aminomuconic 6-semialdehyde dehydrogenase from *Pseudomonas pseudoalcaligenes* JS45. *J. Bacteriol.* **180**, 4591–4595 (1998).

26. Abriola, D. P., Fields, R., Stein, S., MacKerell, Jr. A. D. & Pietruszko, R. Active site of human liver aldehyde dehydrogenase. *Biochemistry* **26**, 5679–5684 (1987).

27. Kitson, T. M., Hill, J. P. & Midwinter, G. G. Identification of a catalytically essential nucleophilic residue in sheep liver cytoplasmic aldehyde dehydrogenase. *Biochem. J.* **275**, 207–210 (1991).

28. Farres, J., Wang, T. T., Cunningham, S. J. & Weiner, H. Investigation of the active site cysteine residue of rat liver mitochondrial aldehyde dehydrogenase by site-directed mutagenesis. *Biochemistry* **34**, 2592–2598 (1995).

29. Steinmetz, C. G., Xie, P., Weiner, H. & Hurley, T. D. Structure of mitochondrial aldehyde dehydrogenase: the genetic component of ethanol aversion. *Structure* **5**, 701–711 (1997).

30. Moore, S. A. *et al.* Sheep liver cytosolic aldehyde dehydrogenase: the structure reveals the basis for the retinal specificity of class 1 aldehyde dehydrogenases. *Structure* **6**, 1541–1551 (1998).

31. D'Ambrosio, K. *et al.* The first crystal structure of a thioacylenzyme intermediate in the ALDH family: new coenzyme conformation and relevance to catalysis. *Biochemistry* **45**, 2978–2986 (2006).

32. Park, J. & Rhee, S. Structural basis for a cofactor-dependent oxidation protection and catalysis of cyanobacterial succinic semialdehyde dehydrogenase. *J. Biol. Chem.* **288**, 15760–15770 (2013).

33. Wang, X. & Weiner, H. Involvement of glutamate 268 in the active site of human liver mitochondrial (class 2) aldehyde dehydrogenase as probed by site-directed mutagenesis. *Biochemistry* **34**, 237–243 (1995).

34. Adams, P. D. *et al.* PHENIX: a comprehensive Python-based system for macromolecular structure solution. *Acta Crystallogr. D Biol. Crystallogr.* **66**, 213–221 (2010).

35. Emsley, P. & Cowtan, K. Coot: model-building tools for molecular graphics. *Acta Crystallogr. D Biol. Crystallogr.* **60**, 2126–2132 (2004).

36. Orville, A. M. *et al.* Correlated single-crystal electronic absorption spectroscopy and X-ray crystallography at NSLS beamline X26-C. *J. Synchrotron Radiat.* **18**, 358–366 (2011).

37. Pothineni, S. B., Strutz, T. & Lamzin, V. S. Automated detection and centring of cryocooled protein crystals. *Acta Crystallogr. D Biol. Crystallogr.* **62**, 1358–1368 (2006).

38. Frisch, M. J. *et al. Gaussian 03, Revision E.01* (Gaussian, Inc., 2004).

## Acknowledgements

This work was supported, in whole or in part, by the National Science Foundation grant CHE-0843537, National Institutes of Health grants GM108988 and GM107529 and Georgia Research Alliance Distinguished Scientist Program (A.L.), Molecular Basis of Disease Area of Focus graduate fellowship (L.H., I.D. and S.E.), Center for Diagnostics and Therapeutics (F.L.), Georgia State University Dissertation Award (L.H.) and funds from Mext Haiteku (Y.H.), Offices of Biological and Environmental Research award FWP BO-70 of the U.S. Department of Energy and NIH grant P41GM103473 (B.A. & A.M.O.). We thank Dr. Siming Wang for assistance with the mass spectrometry analysis and Dr. Donald Hamelberg for valuable discussions. X-ray data were collected at the Southeast Regional Collaborative Access Team (SER-CAT) 22-ID and 22-BM beamlines at the Advanced Photon Source, Argonne National Laboratory. Use of the Advanced Photon Source was supported by the U.S. Department of Energy, Office of Science, Office of Basic Energy Sciences, under Contract No. W-31-109-Eng-38. Single-crystal spectroscopy data were obtained at beamline X26-C of the National Synchrotron Light Source (NSLS), Brookhaven National Laboratory with the support of the U.S. Department of Energy under Contract No. DE-AC02-98CH10886.

## Author contributions

A.L. conceived of and led the study. H.I. and Y.H constructed the initial expression system. F.L. optimized the expression and established protein isolation and activation procedures. I.D. solved the apo-AMSDH structure. L.H. solved all complex and intermediate structures. L.H. and I.D. performed the kinetic assays. L.H. performed the mass spectrometry experiment. S.E. constructed the mutant expression systems. B.A. and A.M.O. collected the single-crystal electronic absorption spectra. I.D. performed the quantum chemical calculations. The manuscript was written by L.H. I.D. and A.L. All authors approved the final submitted manuscript.

## Additional information

**Accession codes:** Coordinates and structure factors for apo-AMSDH, NAD⁺-bound AMSDH, NAD⁺- and 2-AMS-bound AMSDH, NAD⁺- and 2-HMS-bound AMSDH, thioacyl intermediate AMSDH, E268A AMSDH, E268A thiohemiacetal intermediate, and E268A thioacyl intermediate have been deposited in the RCSB Protein Data Bank under accession codes 4I26, 4I1W, 4I25, 4I2R, 4NPI, 4OE2, 4OU2, and 4OUB, respectively.

**Competing financial interests:** The authors declare no competing financial interests.

# 12

# Applying medicinal chemistry strategies to understand odorant discrimination

Erwan Poivet[1,*], Zita Peterlin[2,*], Narmin Tahirova[1], Lu Xu[1], Clara Altomare[1], Anne Paria[1], Dong-Jing Zou[1] & Stuart Firestein[1]

Associating an odorant's chemical structure with its percept is a long-standing challenge. One hindrance may come from the adoption of the organic chemistry scheme of molecular description and classification. Chemists classify molecules according to characteristics that are useful in synthesis or isolation, but which may be of little importance to a biological sensory system. Accordingly, we look to medicinal chemistry, which emphasizes biological function over chemical form, in an attempt to discern which among the many molecular features are most important for odour discrimination. Here we use medicinal chemistry concepts to assemble a panel of molecules to test how heteroaromatic ring substitution of the benzene ring will change the odour percept of acetophenone. This work allows us to describe an extensive rule in odorant detection by mammalian olfactory receptors. Whereas organic chemistry would have predicted the ring size and composition to be key features, our work reveals that the topological polar surface area is the key feature for the discrimination of these odorants.

[1] Department of Biological Sciences, Columbia University, New York, New York 10027, USA. [2] Corporate Research and Development, Firmenich Incorporated, Plainsboro, New Jersey 08536, USA. * These authors contributed equally to this work. Correspondence and requests for materials should be addressed to S.F. (email: sjf24@columbia.edu).

A comprehensive system for classifying odours has been an elusive goal of olfactory inquiry for centuries. The root of the problem, which can be stated most simply as biology versus chemistry, can be seen even in the earliest attempts to bring order to odours. Linnaeus, the master classifier, developed an odour classification scheme using seven primary percepts along a scale of pleasant to unpleasant[1]. Following him, Zwaardemaker[2] proposed the most comprehensive organization of odours, using 9 or 10 perceptual groupings. With the 19th century development of atomic and organic chemistry, numerous researchers attempted to correlate chemical characteristics with odours[3,4]. Perfumers and other fragrance purveyors implemented their own, sometimes less scientific schemes[5]. In the past century more modern attempts generated schemes that encompassed psychophysical descriptors and behavioural responses to complex mixtures[6-8]. However, with the landmark discovery of the unexpectedly large odorant receptor (OR) family of GPCRs by Buck and Axel[9], these efforts largely came to a halt, replaced by the promise of a molecular basis for odour perception. Because typical mammalian odour gene families number over a thousand different receptors, it seemed that the coding problem would soon be solved with high-throughput screening technologies[10].

Mature olfactory sensory neurons (OSNs) are believed to express only one OR gene[11-13]. This property, combined with the unexpectedly large number of receptors, has given rise to the widely accepted proposal that peripheral discrimination works through a reciprocal combinatorial code in which one chemical can be detected by different ORs and one OR can detect a group of different chemicals[14,15]. Additionally, the axons of all OSNs expressing a particular OR project to the same glomerulus in the olfactory bulb, suggesting a labelled line-style 'odour-map' in the brain[16-18]. Taken together, these properties seemed to reduce the odour-coding problem to simply matching particular receptors to their cognate odours. Thus, recent efforts have mainly been directed at identifying ligands for various ORs by screening large sets of supposedly diverse odours[19-21]. However, this programme has run into several obstacles.

First, only a handful of ORs have been successfully de-orphaned, severely limiting the possibility of uncovering hypothesized combinatorial rules. Additional confusion arose when an unexpectedly large repertoire of chemically different molecules were identified as ligands of the single mouse OR, SR1 (ref. 22), complicating the idea of an 'odour-map' and re-opening the question of broadly versus narrowly tuned receptors. Finally, several psychophysical odour paradoxes remain, such as the diversity of compounds that give rise to identical musk percepts. The enormity of the issue was further emphasized by a recent publication claiming that the human olfactory system could discriminate over 1 trillion odours[23]. Absent a systematic understanding of odour detection and discrimination at the periphery, it is difficult to imagine how higher brain centres process the sensory input to develop perceptions and regulate behaviour.

To address these issues from a new perspective, we here take an alternate approach to receptor ligand interactions that is based on medicinal chemistry principles. Medicinal chemistry emphasizes biological function—in this case receptor activation—over chemical form. Similarity between odorants is defined not by strict chemical characteristics but rather by their ability to activate the same receptor or receptors. We use a panel of compounds based on the common odorant acetophenone to investigate the effect of heteroaromatic ring substitution for benzene rings on its odour percept. Using both single cell responses and behavioural tests in mice we find that the classification of the odorants is significantly different from the one expected when classified using classical organic chemical rules. From these results it appears that this approach, based on medicinal chemistry and the related concept of bioisosterism, may reveal a novel strategy for comprehending odour discrimination.

## Results

**Responses of OSNs to aromatic odorants.** Odorants are multi-dimensional stimuli but not all features are necessarily equally weighted by ORs. Here, in a calcium imaging assay, we challenged dissociated mouse OSNs with a panel of related heteroaromatic odorants to investigate whether the ring's sterics (size) or its toplogical polar surface area (TPSA) are better correlated with odorant co-detection.

Panel 1 consisted of acetophenone [1] and five derivatives that replaced the apolar, 6-membered benzene ring with heteroaromatic rings of different sizes and atomic composition (Fig. 1a). Ten per cent of viable OSNs (276/2,750) responded to at least one panel member. Thirty-six distinct patterns were observed when responses were conservatively scored in a binary fashion (Fig. 1c). The analogues varied in their ability to mimic [1] in terms of activation. Of the OSNs detecting [1], 72% also detected 2-acetylthiophene [2], 38% detected 2-acetylpyridine [4], 30% 2-acetylthiazole [5], 25% 2-acetylfuran [3] and 13% 2-acetylpyrazine [6]. [1] and [2] have similar TPSAs but different ring sizes. In contrast, [1], [4] and [6] have the same ring size but different TPSAs. That [1] and [2] are far more frequently co-detected than are [1] and [4] or [6] suggests that TPSA is a more heavily weighted 'epitope' than ring size.

The prioritization of TPSA over ring size appears to be a general trend shaping OSN response patterns. [1], [4] and [6] preserve the same ring size, but as the TPSA increased, the extent of co-detection with [1] decreased. Among OSNs responding to [2], 65% co-detect the similar TPSA but larger-sized ring [1], while only 32% co-detect the higher TPSA but similar-sized ring [5]. Likewise, among OSNs responding to [4], 50% co-detect the similar TPSA but smaller-size ring [5], while only 37% co-detect the higher TPSA but similar-sized ring [6]. This further reinforces that although the geometry of the molecule may be most salient, the TPSA seems to be the driver of these co-recognition patterns.

Strikingly, we found that the diversity of response patterns was constrained by two extensible rules. The first rule is that, at the assay concentration of 30 µM, every OSN that detects both [1] and [3] will also detect [2]. Even when assayed at a higher (150 µM) concentration (Supplementary Fig. 1), this same 'if [1] and [3] then [2]' rule applies, suggesting that there is a conserved biological constraint among OR-binding pockets. At 30 µM, we also note that [5], a five-membered ringed odorant with similar TPSA to [3], can substitute for [3] 95% of the time in this rule, making the relationship 'if [1] and [5] then [2]' a highly predictive one. The second extensible rule is that if an OSN detects [1], [3] and [6] then it will detect all the odorants of Panel 1. Although at first surprising, this rule may be considered to be a fusion of the rule 'If [1] and [3] then [2]' with how OSNs respond to a graded increase in TPSA within a fixed ring size among [1], [4] and [6].

Although the TPSA-based rule was strict for [1], [2] and [3], discrimination based on TPSA partly breaks down when considering [1], [4] and [6]. One might expect that an OSN responding to both [1] and [6] would never reject the intermediate TPSA [4], and yet this occurs 12% of the time. One possibility may be how the appended ketone group interacts with the dual nitrogens of [6]. The benzene ring in [1] has neither a dipole nor a polar constituent. The pyrazine ring in [6] has no dipole (that of the two oppositely situated nitrogens cancelling out), but it is still highly polar (hence its high TPSA). The ketone group, while preferring to lie in plane with the aromatic ring, has

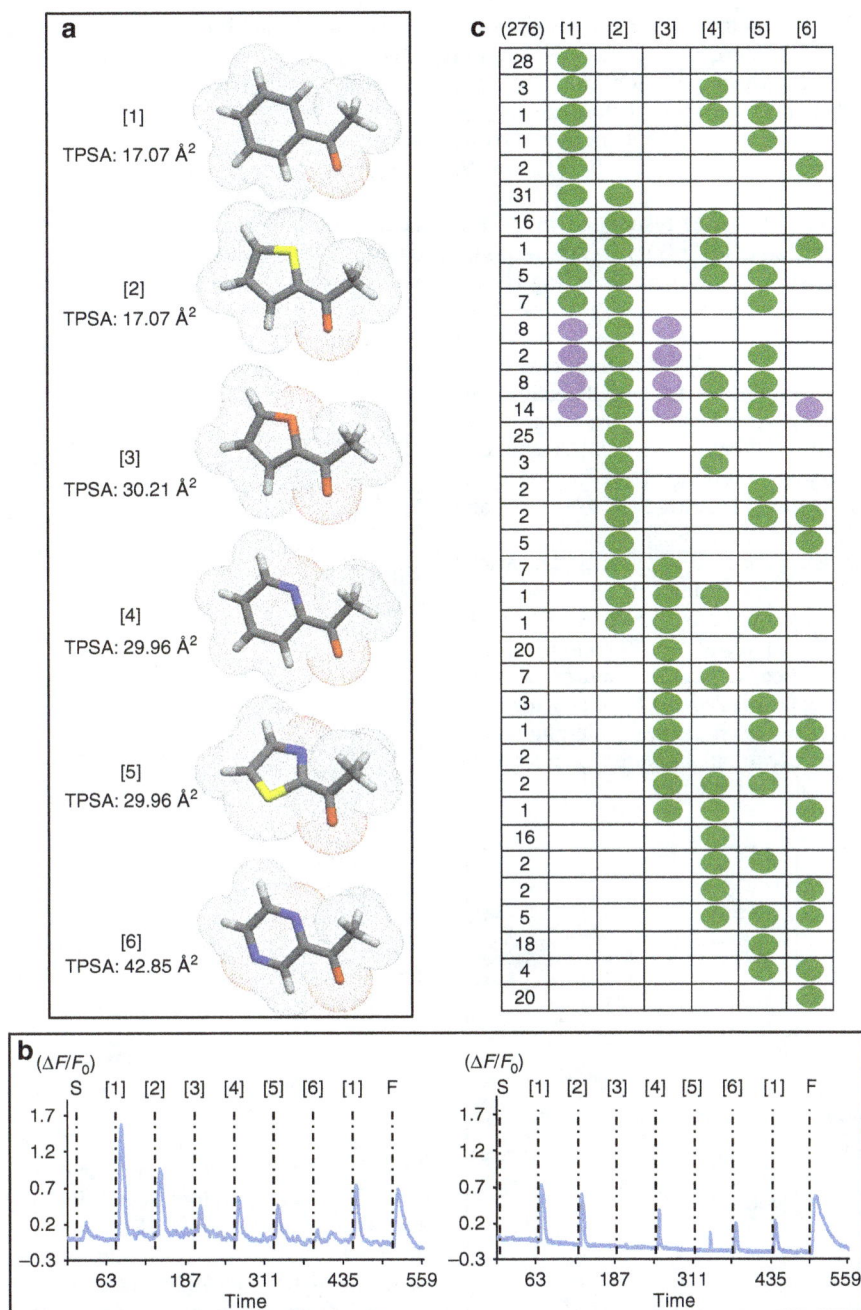

**Figure 1 | Reponses of dissociated OSNs to Panel 1 odorants in calcium imaging.** (**a**) Three-dimensional (3D) representations of Panel 1 odorants. The vertices of the tubes symbolize atoms—grey, carbon; blue, nitrogen; yellow, sulfur; red, oxygen. The dotted red surface around the atoms represents polar regions of the surface area. 3D-representations of these and all odorants used in the study were made using Galaxy 3D Structure Generator free software (www.molinspiration.com.) and TPSA were calculated according to ref. 36. (**b**) Calcium imaging traces of two different OSNs responding to Panel 1 odorants. (**c**) A total of 276 OSNs out of 2,750 viable OSNs responded to at least one Panel 1 odorant, leading to 36 distinct binary response patterns. The numbers indicate how often a particular response pattern was observed. Green dot: activation of the OSN by the corresponding odorant. The OSNs that respond to [1] and [3] always respond to [2] (Purple dot). OSN that respond to [1], [3] and [6] always respond to all the odorants of the panel (Purple dot). S, dimethyl sulfoxide; F, forskolin.

freedom to rotate in [1], but less so in [6], where the two polar nitrogens tend to mutually repulse it. With just one nitrogen to interact with the appended ketone, [4] should then co-activate 50% of ORs detecting [1] and 50% of ORs detecting [6]. This is indeed the response pattern observed in the OSNs: 50% of ORs detecting [1] co-detect [4], and 41% of ORs detecting [4] co-detect [6].

To investigate if the rule of 'If [1] and [3] then [2]' transferred to other contexts, we tested two manipulations. A second panel of odorants (Panel 2) included molecules that were also ketones, but had an extra benzene ring fused to their far end. This manipulation increases the total surface area and affords an extended aromatic system while preserving the TPSA and relationship of the heteroatom to the carbonyl group. Thus, the Panel 2 odorants included 2-acetonaphtone [7] as an analogue to [1], 2-acetylbenzothiophene [8] to [2] and 2-benzofuranyl-methyl-ketone [9] to [3] (Supplementary Fig. 2). Another panel of odorants (Panel 3) replaced the ketone group with a carboxylic acid group. This change allows us to sample a markedly different chemical space as judged by the low frequency of co-recognition of the ketone [1] versus its acid version 2-naphthoic acid [10] (Fig. 1d). Panel 3 includes three acids and their ketone analogues: [10] as the acid analogue to [7], benzo[b]thiophene-2-carboxilic acid [11] to [8] and benzo[b]furan-2-carboxilic acid [12] to [9] (Supplementary Fig. 3).

Twenty-six per cent of OSNs (245/926) responded to at least one Panel 2 member, generating 26 distinct binary OSN response patterns. Consistent with the prior study, OSNs detecting [1] co-detected [2] more frequently than [3] (59% versus 17%, respectively). The benzene-fused analogues showed the same trend; OSNs detecting [7] co-detected [8] more frequently than [9] (83% versus 64%, respectively). The strict co-detection rule that was seen for the single ring [1], [2] and [3] also extended to the benzene-fused [7], [8] and [9]. That is, if an OSN responded to both [7] and [9] it always responded to [8] (Fig. 2; Supplementary Fig. 2). Thus, the 'TPSA rule' is robust among both ketone scaffolds.

Intriguingly, the benzene-fused analogues of Panel 2 activated markedly more OSNs than did their single ring counterparts. Eighteen per cent of OSNs were activated by [7] versus 9% by [1], 18% [8] versus 8% [2] and 13% [9] versus 3% [3] (Supplementary Fig. 2). This suggests that increased surface area and/or extended aromaticity could be a stabilizing factor, perhaps by improving pi–pi stacking with aromatic amino-acid side chains in the binding pocket. This may form the basis of a strategy to rationally design an aromatic odorant to increase the breadth of ORs it targets.

Among the acids of Panel 3, we again observed conservation of the 'TPSA rule' with all OSNs that responded both to [10] and [12] also responding to [11] (Fig. 2c). Twenty-two distinct patterns were observed when responses were conservatively scored in a binary fashion (Supplementary Fig. 3). Of the 308 OSNs recorded, 45% responded to at least one Panel 3 member; for the ketones 26% responded to [7] and 20% to [1]. For the acids only 11% responded to [10], 13% to [11] and 11% to [12]. These results indicate that acids are generally weaker odorants than ketones. Interestingly the OSNs recognizing the acid [10] were mostly distinct from the population responding to either the single-ringed ketone [1] or the double-ringed ketone [7] (Fig. 2d), lending further support to the transferability of the 'TPSA rule'.

**Comparing odorant classifications**. Medicinal chemistry substitutions can be discrepant in form but they nevertheless preserve similar biological functionality across multiple targets. In our panels, several of the heteroaromatic substitutions from the 'lead' odorant [1] were inspired by medicinal chemistry substitutions. We thus compared the classification of the Panel 1 odorants using both traditional chemistry-centric and biology-centric approaches.

For the chemistry-centric approach, we used the e-Dragon software to obtain 1,666 molecular descriptors for each odorant. We generated a dendrogram (Fig. 3a) which revealed two clearly distinguishable branches. The segregation was driven by ring size with the 6-membered ring [1], [4] and [6] forming one cluster, and the 5-membered ring [2], [3] and [5] forming a second cluster. The two families were further fractionated by atomic composition via the presence of nitrogen in the 6-membered ring family and sulfur in the 5-membered ring family. In the 5-membered ring family, [3] and [5] are split apart despite their similar TPSA, leaving [5] to cluster with [2]. This clustering pattern underscores that in the traditional chemistry-centric classification atomic composition is given pre-eminence over TPSA.

For the biology-centric classification the response patterns of the OSNs formed the basis for the hierarchical cluster analysis. The resulting dendrogram has striking differences (Fig. 3b). Notably, in the dendrogram for Panel 1, [1] and [2] are tightly linked, as determined from their biological activity profiles. This branch, which contains the two low TPSA rings, segregates from the odorants with larger TPSA values. TPSA, however, is not the sole determinant of the remaining organization as [5] clusters with the higher TPSA [6] instead of the matched TPSA [4]. When Panel 2 odorants were clustered via their OSN response patterns, the major split was along the lines of total surface area with all the double-ringed odorants segregating from the single-ringed odorants (Fig. 3c). Within each family, however, clustering reflected the division of low TPSA from high TPSA that was seen in the Panel 1 odorants. That is, [1] was tightly linked to [2] and separate from [3], whereas [7] was tightly linked with [8] and separate from [9].

Similarly, biology-centric classification separates Panel 3 odorants according to their TPSA. The three acids segregate from the ketone [7], and [10] was tightly linked to [11] and separated from [12] (Supplementary Fig. 4). A chemistry-centered approach on the other hand separates once again Panel 3 odorants according to their ring size and composition: [7] and [10] group together despite their functional group difference, and separate from [11] and [12].

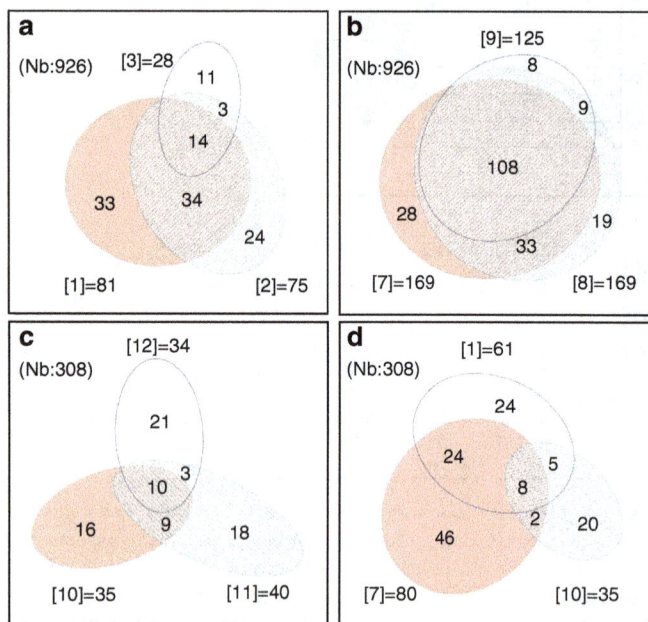

**Figure 2 | Transferability of the intra-ring TPSA rule.** (**a**) Venn diagram of OSNs responding to the single-ring ketones [1], [2] and [3]. If an OSN responds to [1] and [3] it always responds to [2]. (**b**) Venn diagram of OSNs responding to the double-ring ketones [7], [8] and [9]. If an OSN responds to [7] and [9], it always responds to [8]. (**c**) Venn diagram of OSNs responding to the double-ring acids [10], [11] and [12]. If an OSN responds to [10] and [12] it always responds to [11]. (**d**) Venn diagram of OSNs responding to the single-ringed ketone [1], the double-ringed ketone [7] and the double-ringed acid [10]. Note that OSNs co-detecting the acid make up just a small portion of the overall responses. OSNs were counted and converted into surface area for each response combination using the eulerAPE free software. The number of OSNs responding with that pattern is indicated in that sector. All odorants were tested at 30 μM. Nb, total number of viable OSNs screened.

**Behavioural response of mice to the odorants**. Having examined how OSNs parse heteroaromatic odorants, we turned to a habituation–dishabituation test to investigate how readily a

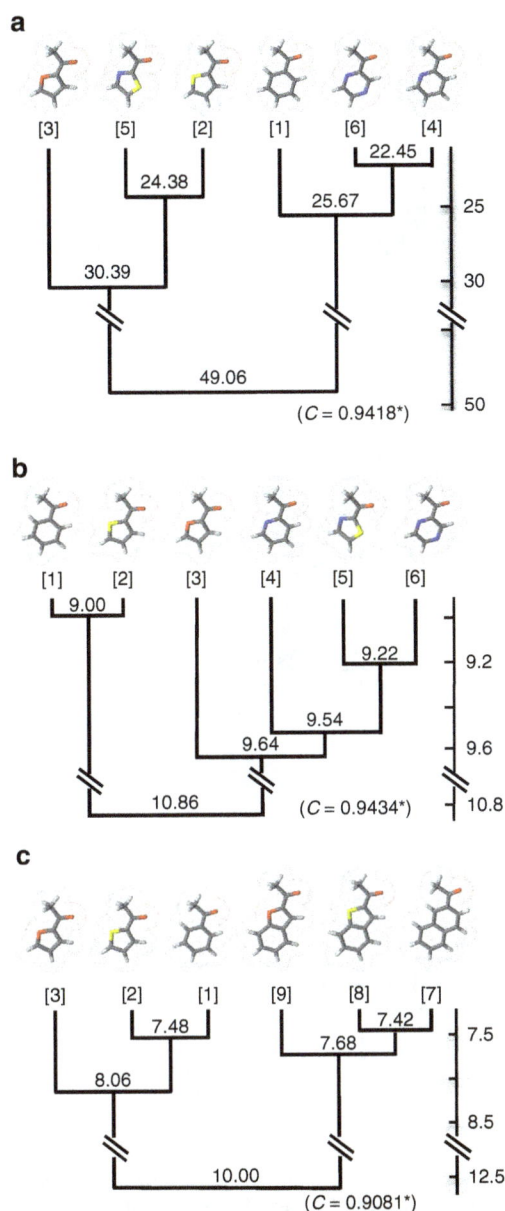

**Figure 3 | Hierarchical clustering analysis of Panel 1 and Panel 2 odorants.** (**a**) Panel 1 odorants clustered according to chemical similarity as evaluated by 1,666 molecular descriptors downloaded through the e-dragon applet. (**b**) Panel 1 odorants clustered according to biological response similarity based on calcium imaging of dissociated OSNs. Note that in the chemical-based clustering the major division is on ring size while in the biological-based clustering the major division is on the TPSA. (**c**) Panel 2 odorants clustered according to their biological responses, as in **b**. Although there is a major division based on the presence of a double-ring scaffold, within each branch further subdivisions follow the TPSA rule as in **b**. *Cophenetic correlation coefficient. All distances in the dendrograms are Euclidian. See online methods for details of dendrogram generation.

robust acceptance for a carbon-to-sulfur swap; mice that habituated to [1] remained habituated to [2] (Fig. 4). This habituation also occurred when odorants were presented in the reverse order (i.e., habituation to [2] then probed with [1]). Reciprocal habituation also occurred when mice were challenged with [7] and [8], the double-ringed analogues of [1] and [2].

Mice also demonstrated reciprocal habituation to [4] and [5] (Table 1). Like [1] and [2], [4] and [5] are related by a carbon-to-sulfur swap but in an overall more polar background. Although [1] and [2] cluster tightly in the OSN response-based dendrogram, [4] and [5] are admittedly more distant (Fig. 3b). Still, they are far closer in the OSN-based response dendrogram than in the molecular descriptor-based dendrogram. This supports that OSN response patterns are indeed a better predictor of olfactory-guided behaviour.

Clear reciprocal dishabituation was noted for certain carbon-to-oxygen and carbon-to-nitrogen swaps. Mice stimulated by [1] failed to generalize to the oxygen-containing [3]. This behaviour may find its basis in that OSNs show a far lower degree of co-detection between [3] and [1] as opposed to [2] and [1]. Dishabituation was also seen when the mouse was stimulated by [7] but probed with [9], the double-ring analogues of [1] and [3], or stimulated by [10] but probed with [12] (their acids analogues) (Fig. 4). For carbon-to-nitrogen swaps, reciprocal dishabituation was observed between [2] followed by [5] and by [1] followed by [4] (Table 1). Within both of these pairings, the ring size is preserved but TPSA changes. This further reinforces the relative pre-eminence of TPSA from a biological standpoint.

Interestingly, not all the habituations were reciprocal. Habituation was observed when the mouse was stimulated with [4] then probed with [3] but not if stimulated with [3] then probed with [4]. The same situation occurred between [5] and [6]. Cases of asymmetrical habituation have been previously reported in the psychophysical literature and are suggestive of non-overlapping sets of receptors that bind the same ligands.

## Discussion

The classification of the vast number and diversity of odorant molecules has been a controversial topic in psychophysics and more recently in molecular physiology and systems biology of the olfactory system[23,24]. Here, our work reveals an extensible rule of odorant detection by OSNs.

In colour perception there is a generally agreed-upon set of rules determining how wavelengths mix to produce millions of hues. In the auditory system the combination of frequencies and amplitudes produces a predictable perception of tonality. No such agreement or scheme is available in olfaction and it remains virtually impossible to predict, from looking at a chemical structure, whether a molecule will have an odour or not, let alone what that quality may be.

One obstacle to gaining this understanding may be that we have adopted a physical and organic chemistry scheme of molecular description and classification. Chemists classify molecules according to characteristics that are useful in synthesis or isolation, features that may be of no importance to a biological sensory system, either at the olfactory receptor level or at higher perceptual levels.

It has been shown that among all molecular features that describe a compound, some are more important than others for odorant perception by receptors[25,26]. Learning how features of odorants are weighted by ORs could clarify the fundamental structure of the stimulus space and help predict similarity of odour quality. Computational models such as the 3D-QSAR can already efficiently identify a few key descriptors common to all the ligands of an OR and then predict and design new ligands for

mouse could discriminate between select pairings of Panel 1 and Panel 2 odorants. Habituation is defined by a progressive decrease in olfactory investigation towards repeated presentation of the same odour stimulus. Dishabituation is defined by reinstatement of olfactory investigation when a novel odour is presented.

Many of the trends seen in the behavioural assay paralleled those seen in the response patterns of OSNs. Notably, there was

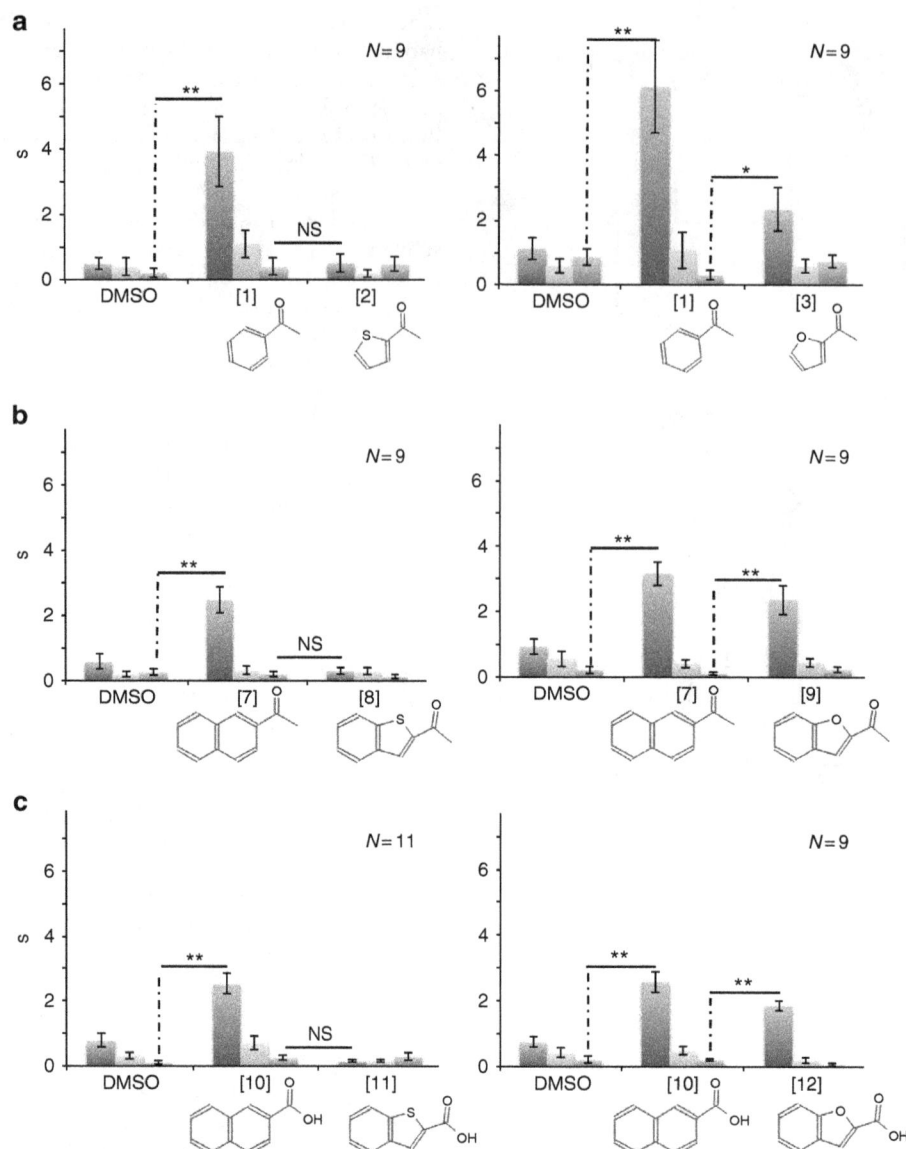

**Figure 4 | Habituation–dishabituation olfactory test.** The average olfactory investigation time (s) by mice during repetitive 2 min exposures to odorant pairs or DMSO (solvent). (**a**) Mice that habituated to [1] remained habituated to [2] but dishabituated to [3]. (**b**) Mice that habituated to [7] remained habituated to [8] but dishabituated to [9]. (**c**) Mice that habituated to [10] remained habituated to [11] but dishabituated to [12]. In all cases, the analogue with the low TPSA thiophene ring was not discriminated from the lead, low TPSA benzene-ringed version, but the analogue with the higher TPSA furan ring was. Behaviour tests were analysed using ANOVA test followed by a *post hoc* paired *t*-test (*$P < 0.05$, **$P < 0.005$ paired *post hoc* t-test). NS, not significant, *$P$-value$< 0.05$, **$P$-value$< 0.005$ paired *t*-test. Error bars: s.e.m. *N*, number of animals.

that OR[26,27]. But this model depends on already partially deorphanized ORs, and these key descriptors appear to be different for every OR. Recently Sobel's lab[28], correctly identified the problem of odour perception as one of quantifying odour characteristics, has taken a mathematical approach to reduce complex odour structures to a small number of vectors. Unfortunately, a reliance on chemical descriptors means elements of the vectors cannot often be identified with any empirical odour structure. (Examples from their Table 1 include, 'the molecular multiple path count number', the 'spectral moment from edge adj. matrix weighted by dipole moments' and 19 other similarly esoteric descriptors).

We have instead taken an approach that is more bio-centric using principles developed in the practice of medicinal chemistry to identify biologically relevant features of an odour stimulus[29]. This is sometimes known as bioisosterism—the practice of exchanging molecular fragments that subtly tweak but largely

preserve chemical structure and performance at a variety of enzyme and receptor targets.

As a proof of principle we assessed one type of bioisosteric exchange, that of heteroaromatic rings for benzene, against the suite of mouse ORs. Starting from acetophenone as the 'lead' odorant, we found that several of the predicted exchanges were, indeed, well tolerated. Acetophenone possesses a benzene ring that can be replaced by alternative ring structures. The most common prediction would be that odour quality varies according to ring steric size and shape or atomic composition. On the contrary, our analysis revealed that the overall TPSA was of greater importance, such that having a ring component with a high TPSA was a generally disfavoured epitope for OSNs responding to acetophenone. These findings at the sensory neuron/receptor level transferred to behavioural testing

A second panel, which included benzene-fused double-ringed versions of the analogues, surprisingly revealed that the

**Table 1 | OSNs discrimination and behaviour response to Panel 1 odorants.**

| Odorant pairs | Changes | Co-activation | Exclusion (%) | Behaviour |
|---|---|---|---|---|
| [1] versus [2] | C->S, polarity and ring size change | 72%/66% | 30 | Habituated |
| [1] versus [3] | C->O, polarity, ring size and TPSA changes | 25%/41% | 69 | Dishabituated** |
| [1] versus [4] | C->N, polarity and ring TPSA change | 38%/55% | 71 | Dishabituated* |
| [1] versus [5] | C->S, C->N, dipolarity, ring size and TPSA changes | 30%/49% | 55 | — |
| [1] versus [6] | C->N (2×), ring TPSA changes | 13%/29% | 74 | Dishabituated* |
| [2] versus [3] | S->O, polarity and ring TPSA changes | 30%/53% | 62 | — |
| [2] versus [4] | Polarity, ring size and TPSA changes | 35%/55% | 57 | — |
| [2] versus [5] | C->N, dipolarity and ring TPSA change | 37%/65% | 53 | Dishabituated** |
| [2] versus [6] | Polarity, ring size and TPSA changes | 16%/37% | 81 | — |
| [3] versus [4] | Polarity, ring size change | 43%/38% | 60 | Dishabituated* |
| [3] versus [5] | O->N, C->S, dipolarity | 40%/40% | 60 | — |
| [3] versus [6] | Ring size and TPSA changes | 23%/30% | 73 | — |
| [4] versus [5] | C->S, dipolarity, ring size change | 42%/47% | 55 | Habituated |
| [4] versus [6] | C->N, polarity, ring size and TPSA changes | 26%/39% | 68 | — |
| [5] versus [6] | Dipolarity, ring size and TPSA changes | 33%/44% | 62 | Habituated |

TPSA, topological polar surface area.
This table recapitulates co-activation among the OSNs and behavioural responses observed with Panel 1 odorant pairs. 'Changes' gives the substitutions and transformations from one odorant to the other. The first co-activation number gives the percentage of OSNs responding to the first odorant of the pair that were co-activated by the second odorant. The second co-activation number gives the percentage of OSNs responding to the second odorant of the pair that were co-activated by the first odorant. 'Exclusion' numbers give the percentage of OSNs that respond exclusively to one odorant of the pair among the total number of OSNs that respond to those two odorants. 'Behaviour' recapitulates the results observed during the habituation/dishabituation olfactory tests with the odorants of the pair. Behaviour results were analysed using ANOVA test followed by a *post hoc* paired *t*-test (*$P < 0.05$, **$P < 0.005$ paired *post hoc* *t*-test).

double-ringed odorants activated far more OSNs than did the single-ringed ones. The added benzene ring not only increased the breadth of activation across the suite of ORs but it also often led to an increased breadth of tuning for a given single OR. This strategy could be exploited to probe binding pocket accommodation.

A third panel used acid analogues of the double-ringed ketones of Panel 2. We observed only minor levels of co-recognition between the single-ring acid [10] and the analogous single-ringed ketone [1] or the double-ring ketone [7], demonstrating that the acids of Panel 3 likely cover distinct sectors of chemical space than do the ketones of Panel 1 and Panel 2. Yet despite this, the 'TPSA rule' translated well, demonstrating its robustness as a predictive tool.

An important caveat to this work is that we used a simple binary accounting for whether an OSN was activated or not, and each panel was conducted at a single concentration (with the exception of a control experiment run at 150 µM, Supplementary Fig. 1). Thus, we did not measure affinity or efficacy as variables. Increasing concentration would likely activate additional receptors and alter the patterns of overlap, although it has been shown that increasing concentration only rarely alters odorant perceptual quality[30]. Although these effects are not uninteresting they would have clouded the main purpose of the present study—to determine the biologically most relevant attribute of related molecules among a group of receptors. In this regard the olfactory system offers a novel forum for evaluating medicinal chemistry strategies because we are not testing various molecules on a single receptor, as is the case in pharmaceutical experiments. In the olfactory system we have a large number (>1,000 in mouse) of G-protein-coupled receptors that are being tested simultaneously with a panel of carefully altered odour compounds. In a sense we are using the receptors simply to 'take a vote,' which is necessarily binary, on the biologically relevant characteristics of a molecule.

Although we can draw no conclusion as to why TPSA should be of special importance there are several interesting speculations. The TPSA is effectively a measure of the solvent accessible surface area presented by a molecule. Given that odorants must pass through both aqueous and lipid environments to access the presumptive binding regions of the receptors, the surface area could raise or lower the entropic cost of accessing that activating region[31]. Access to the receptor, or specific parts of it, may be

more crucial in determining the efficacy of a molecule than the particular fit it may make in a presumptive binding pocket. The popular lock and key model of receptor ligand interactions is too naive to capture the biophysical requirements that play a role in how a molecule may interact with and stabilize an activated conformation of the receptor. Bioisosterism is an empiric method for probing and understanding those functional details.

As a bonus, this approach also revealed an extensible rule—that if an OSN accepted both the low TPSA, 6-membered benzene ring and the high TPSA, 5-membered furan ring, then it will always accept the 'intermediate challenge' of a low TPSA, 5-membered thiophene ring. We witnessed this for ketone odorant sets [1], [3], [2] and [7], [9], [8] and acids [10], [11], [12]. This rule joins the electronegativity rule of 'if an OSN accepts a n-alcohol and the homologous n-acid, it will always accept the electronegative intermediate homologous n-aldehyde'[32], and the backbone continuity rule of 'if an OSN accepts a chain length of N and N + 2 in an n-odorant, then it will always accept a chain length of N + 1' (refs 15,33,34). These three rules show that constraints in detection exist, despite the wide diversity of odorants and receptors.

We anticipate that there are other rules to be discovered through application of this medicinal chemistry strategy, and that these rules may be extended to other, non-olfactory, GPCRs. Notions such as broad versus narrow tuning of receptors could be revisited in terms of sensitivity to molecular features rather than molecular compounds. Indeed it might well be worth revisiting the idea of odour primaries (as in colours or fundamentals in sound) that are recombined in innumerable, but comprehensible, ways to provide a rich odour world.

## Methods
**Chemicals.** Two panels of six ketone odorants (Panel 1 and Panel 2) and a panel of three acid odorants (Panel 3) were designed to test the hypothesis that, among odorants, heteroaromatic rings can substitute for benzene rings with ORs exhibiting a predictable preference between them. All panels are derived around a lead odorant, acetophenone [1]. Panel 1 consisted of acetophenone [1], 2-acetylthiophene [2], 2-acetylfuran [3], 2-acetylpyridine [4], 2-acetylthiazole [5] and acetylpyrazine [6]. Panel 2 consisted of acetophenone [1], 2-acetylthiophene [2], 2-acetylfuran [3], 2-acetonaphthone [7], 2-acetyl-benzothiofene [8] and 2-benzo-furanyl-methyl-ketone [9]. Panel 3 consisted of acetophenone [1], 2-acetonaphthone [7], 2-naphthoic acid [11], benzo[b]thiophene-2-carboxylic acid [11] and benzo[b]furan-2-carboxylic acid [12]. Odorants [1]–[9] were purchased from Sigma-Aldrich (St Louis, MO, USA). Odorants [10]–[12] were purchased from Acrōs Organics (Thermo Fisher Scientific, New Jersey, USA). Odorant stocks were

made in >99% dimethyl sulfoxide (DMSO) (Sigma-Aldrich) and were diluted in freshly prepared Ringer's solution to a final concentration of 30 or 150 μM just before experiments.

**Animals and tissue collection.** All animal procedures conformed to Columbia University guidelines for care and use of animals. *OMP-Cre*-driven *GCaMP3* mice used in this work were generated by crossing the *OMP-Cre* line (*JAX 006668*) with the *Ai38* line (*RCL-GCaMP3, JAX014538*). In these compound mutant mice, the expression of the genetically encoded calcium sensor GCaMP3 is restricted to the mature olfactory sensory neurons. All mice were reared and maintained in the department animal facility.

Olfactory sensory neurons were isolated from 5 to 8-week old *OMP-Cre*-driven *GCaMP3* male mice with a genotype of $OMP\text{-}Cre^{+/-} GCaMP3^{-/-}$. The mice were overdosed with anaesthetics (ketamine 90 mg kg$^{-1}$; xylazine 10 mg kg$^{-1}$, i.p.) and decapitated. The head was cut open sagitally and the septum was removed to expose the medial surface of the olfactory epithelium and turbinates. The olfactory epithelium and turbinates were dissected and collected in divalent-free Ringer's solution (mM: 145 NaCl, 5.6 KCl, 10 Hepes, 10 Glucose, 4 EGTA, pH 7.4). The tissue was incubated at 37 °C for 45 min in 5 ml of divalent-free Ringer's solution containing 0.5 mg ml$^{-1}$ collagenase, 5 mg ml$^{-1}$ bovine serum albumin (Sigma-Aldrich), 8 U ml$^{-1}$ dispase (Roche, Bassel, Switzerland) and 50 μg ml$^{-1}$ deoxyribonuclease II (Sigma). The tissue was then transferred to a clean tube of culture medium and washed. The OSNs were dissociated by tapping the tube containing the tissue. The OSNs (50 μl volume) were split onto four concanavalin-coated glass coverslips (Sigma-Aldrich, 10 mg ml$^{-1}$), placed in 35 mm Petri dishes. After allowing the cells to settle for 20 min, 2 ml of culture medium was added to each dish and the dishes were placed at 37 °C for at least 1 h. Culture medium consisted of DMEM/F12 (Gibco BRL, Grand Island, NY, USA) supplemented with 10% fetal bovine serum, 1 × insulin-transferrin-selenium (Gibco BRL), 100 U ml$^{-1}$ penicillin and 100 μg ml$^{-1}$ streptomycin (Gibco BRL) and 100 μM ascorbic acid (Sigma-Aldrich).

**Calcium imaging recording.** After being washed with fresh Ringer's solution, the coverslips were mounted on a recording chamber. Imaging was carried out at room temperature on an inverted fluorescence microscope (IMT-Olympus, Tokyo, Japan) equipped with a SIT camera (C10600, Hamamatsu Photonics, Hamamatsu, Japan), a Lambda XL light source (Sutter Instrument, Novato, CA, USA), and Lamba-10B optical filter changer (Sutter Instrument). Using a 1260 Infinity HPLC system (Agilent Technologies, Santa Clara, CA, USA) the dissociated OSNs were stimulated with the odorants in random order between two flanking stimulations with the lead odorant, [1]. A final stimulation with a 10 μM Forskolin (Sigma-Aldrich) solution was made to assess the viability of the OSNs. Recordings were made at 490 nm excitation and 520 nm emission. Images were taken every 4 s and there was a 4 min delay between stimulations. The images were then computed using Metamorph Premier software (Molecular Device LLC, Downingtown, PA, USA) and the cells were manually counted.

**Data analysis of calcium imaging recording.** 1,666 molecular descriptors for the P1 and P2 odorants were downloaded through e-dragon free applet (http://www.vcclab.org/)[35]. Normalized descriptors were used for calculating Euclidean distances and for generating dendrograms using Matlab (MathWorks, Boston, MA, USA). Neuron responses to Panel 1 or Panel 2 odorants in calcium imaging were transformed into an $m^*n$ bool matrix where '$m$' is the number of neurons responding to at least one chemical, and '$n$' is the number of chemicals used; '1' means 'response' and '0' means 'no response'. This matrix was used to calculate Euclidean distances and generate dendrograms of the odorants using Matlab. A Coshran's $Q$ test comparison followed by *post hoc* McNemar tests was performed to compare the odorant 'response', 'No response' heatmaps using Statview (SAS institute, Cary, NC, USA).

**Habituation–dishabituation behavioural test.** Similarities in perceptual odour quality among the Panel 1 odorants were evaluated by a habituation–dishabituation olfactory test in the mouse. Thirty minutes before experimentation, 5–8 weeks old $OMP\text{-}Cre^{+/-} GCaMP3^{-/-}$ male mice were placed individually into a hood in an empty mouse cage containing a cotton swab soaked in 1/1,000 DMSO/Ringer's solution. Each animal was then stimulated three consecutive times over 2 min with the DMSO/Ringer's solution soaked cotton swab as a negative control. Then they received three consecutive presentations of a cotton swab soaked in the first odorant solutions at 30 μM. Each presentation lasted 2 min with a 1 min interval between presentations. Following a 1 min rest, animals were then given three presentations of the second odour in a similar manner. Following a final 1-min break, a 30 μM solution of propyl-valerate was given in a 2-min single stimulation as a positive control. The cumulative sniffing time of the cotton swab was recorded using a silent clock. An analysis of variance (ANOVA) statistic comparison, followed by *post hoc* Paired *t*-test, was performed on the results using Statview. Each mouse was used only once with the same odorant. Mice that were unable to detect the first odorant stimulation or that responded to the negative control were removed from further analysis.

## References

1. Linnaeus, C. Odores medicamentorum. *Amoen. Acad.* **3**, 183–201 (1756).
2. Zwaardemaker, H. *Die Physiologie des Geruchs* (Рипол Классик, 1895).
3. Dumas, M. J. Ueber die vegetabilischen Substanzen welche sich dem Kampfer nähern, und über einige ätherische Oele. *Annalen der Pharmacie* **6**, 245–258 (1833).
4. Perkin, W. H. VI. On the artificial production of coumarin and formation of its homologues. *J. Chem. Soc.* **21**, 53–63 (1868).
5. Ohloff, G., Pickenhagen, W. & Kraft, P. Scent and chemistry, the molecular world of odors. *Chem. Listy* **106**, 685–692 (2012).
6. Guillot, M. Physiologie des sensations-anosmies partielles et odeurs fondamentales. *C. R. Acad. Sci.* **226**, 1307–1309 (1948).
7. Amoore, J. E. Specific anosmia: a clue to the olfactory code. *Nature* **214**, 1095–1098 (1967).
8. Wise, P. M., Olsson, M. J. & Cain, W. S. Quantification of odor quality. *Chem. Senses* **25**, 429–443 (2000).
9. Buck, L. & Axel, R. A novel multigene family may encode odorant receptors: a molecular basis for odor recognition. *Cell* **65**, 175–187 (1991).
10. Zhang, X., Zhang, X. & Firestein, S. Comparative genomics of odorant and pheromone receptor genes in rodents. *Genomics* **89**, 441–450 (2007).
11. Chess, A., Simon, I., Cedar, H. & Axel, R. Allelic inactivation regulates olfactory receptor gene expression. *Cell* **78**, 823–834 (1994).
12. Serizawa, S. *et al.* Mutually exclusive expression of odorant receptor transgenes. *Nat. Neurosci.* **3**, 687–693 (2000).
13. Serizawa, S., Miyamichi, K. & Sakano, H. One neuron–one receptor rule in the mouse olfactory system. *Trends Genet.* **20**, 648–653 (2004).
14. Zhao, H. *et al.* Functional expression of a mammalian odorant receptor. *Science* **279**, 237–242 (1998).
15. Malnic, B., Hirono, J., Sato, T. & Buck, L. B. Combinatorial receptor codes for odors. *Cell* **96**, 713–723 (1999).
16. Stewart, W. B., Kauer, J. S. & Shepherd, G. M. Functional organization of rat olfactory bulb analysed by the 2-deoxyglucose method. *J. Comp. Neurol.* **185**, 715–734 (1979).
17. Ressler, K. J., Sullivan, S. L. & Buck, L. B. A zonal organization of odorant receptor gene expression in the olfactory epithelium. *Cell* **73**, 597–609 (1993).
18. Mori, K., Nagao, H. & Yoshihara, Y. The olfactory bulb: coding and processing of odor molecule information. *Science* **286**, 711–715 (1999).
19. Touhara, K. *et al.* Functional identification and reconstitution of an odorant receptor in single olfactory neurons. *Proc. Natl Acad. Sci. USA* **96**, 4040–4045 (1999).
20. Fukuda, N., Yomogida, K., Okabe, M. & Touhara, K. Functional characterization of a mouse testicular olfactory receptor and its role in chemosensing and in regulation of sperm motility. *J. Cell Sci.* **117**, 5835–5845 (2004).
21. Saito, H., Chi, Q., Zhuang, H., Matsunami, H. & Mainland, J. D. Odor coding by a mammalian receptor repertoire. *Sci. Signal.* **2**, ra9–ra9 (2009).
22. Grosmaitre, X. *et al.* SR1, a mouse odorant receptor with an unusually broad response profile. *J. Neurosci.* **29**, 14545 (2009).
23. Bushdid, C., Magnasco, M. O., Vosshall, L. B. & Keller, A. Humans can discriminate more than 1 trillion olfactory stimuli. *Science* **343**, 1370–1372 (2014).
24. Meister, M. On the dimensionality of odor space. *eLife* **4**, 1–12 (2015).
25. Bieri, S., Monastyrskaia, K. & Schilling, B. Olfactory receptor neuron profiling using sandalwood odorants. *Chem. Senses* **29**, 483–487 (2004).
26. Schmuker, M., de Bruyne, M., Hahnel, M. & Schneider, G. Predicting olfactory receptor neuron responses from odorant structure. *Chem. Cent. J.* **1**, 11 (2007).
27. Sanz, G. *et al.* Relationships between molecular structure and perceived odor quality of ligands for a human olfactory receptor. *Chem. Senses* **33**, 639–653 (2008).
28. Snitz, K. *et al.* Predicting odor perceptual similarity from odor structure. *PLoS Comput. Biol.* **9**, e1003184 (2013).
29. Sheridan, R. P. The most common chemical replacements in drug-like compounds. *J. Chem. Inf. Comput. Sci.* **42**, 103–108 (2002).
30. Furudono, Y., Sone, Y., Takizawa, K., Hirono, J. & Sato, T. Relationship between peripheral receptor code and perceived odor quality. *Chem. Senses* **34**, 151–158 (2009).
31. Lee, B. & Richards, F. M. The interpretation of protein structures: estimation of static accessibility. *J. Mol. Biol.* **55**, 379–IN374 (1971).
32. Araneda, R. C., Peterlin, Z., Zhang, X., Chesler, A. & Firestein, S. A pharmacological profile of the aldehyde receptor repertoire in rat olfactory epithelium. *J. Physiol.* **555**, 743–756 (2004).
33. Sato, T., Hirono, J., Tonoike, M. & Takebayashi, M. Tuning specificities to aliphatic odorants in mouse olfactory receptor neurons and their local distribution. *J. Neurophysiol.* **72**, 2980–2989 (1994).
34. Kaluza, J. F. & Breer, H. Responsiveness of olfactory neurons to distinct aliphatic aldehydes. *J. Exp. Biol.* **203**, 927–933 (2000).

35. Tetko, I. *et al.* Virtual computational chemistry laboratory—design and description. *J. Comput. Aided Mol. Des.* **19**, 453–463 (2005).

36. Ertl, P., Rohde, B. & Selzer, P. Fast calculation of molecular polar surface area as a sum of fragment-based contributions and its application to the prediction of drug transport properties. *J. Med. Chem.* **43**, 3714–3717 (2000).

## Acknowledgements

This research was supported by the NIDCD, R01DC013553. The authors want to thank Christian Margot for his comments on the chemistry and Cen Zhang for her help with the animal facility and lab managing.

## Authors contributions

S.F. conceived and supervised the project. E.P. and Z.P. designed the experiments. E.P. conducted the experiments and analysed the results. Z.P. and L.X. helped with chemical classification and analyzes. N.T., A.P. and C.A. helped with behaviour experiment and cell counting. D.-J.Z. developed and maintained mouse line. E.P., Z.P. and S.F. wrote the manuscript. All authors refined the manuscript.

## Additional information

**Competing financial interests:** The authors declare no competing financial interests.

# 13

# Synergistic activation of human pregnane X receptor by binary cocktails of pharmaceutical and environmental compounds

Vanessa Delfosse[1,2,3,*], Béatrice Dendele[3,4,5,6,*], Tiphaine Huet[1,2,3,*], Marina Grimaldi[3,4,5,6], Abdelhay Boulahtouf[3,4,5,6], Sabine Gerbal-Chaloin[3,7], Bertrand Beucher[3,8,9], Dominique Roecklin[10], Christina Muller[10], Roger Rahmani[11], Vincent Cavaillès[3,4,5,6], Martine Daujat-Chavanieu[3,7,12], Valérie Vivat[10], Jean-Marc Pascussi[3,8,9], Patrick Balaguer[3,4,5,6,**] & William Bourguet[1,2,3,**]

Humans are chronically exposed to multiple exogenous substances, including environmental pollutants, drugs and dietary components. Many of these compounds are suspected to impact human health, and their combination in complex mixtures could exacerbate their harmful effects. Here we demonstrate that a pharmaceutical oestrogen and a persistent organochlorine pesticide, both exhibiting low efficacy when studied separately, cooperatively bind to the pregnane X receptor, leading to synergistic activation. Biophysical analysis shows that each ligand enhances the binding affinity of the other, so the binary mixture induces a substantial biological response at doses at which each chemical individually is inactive. High-resolution crystal structures reveal the structural basis for the observed cooperativity. Our results suggest that the formation of 'supramolecular ligands' within the ligand-binding pocket of nuclear receptors contributes to the synergistic toxic effect of chemical mixtures, which may have broad implications for the fields of endocrine disruption, toxicology and chemical risk assessment.

[1] Inserm U1054, Montpellier 34090, France. [2] CNRS UMR5048, Centre de Biochimie Structurale, Montpellier 34090, France. [3] Université de Montpellier, Montpellier 34090, France. [4] IRCM, Institut de Recherche en Cancérologie de Montpellier, Montpellier 34298, France. [5] Inserm, U1194, Montpellier 34298, France. [6] ICM, Institut régional du Cancer de Montpellier, Montpellier 34298, France. [7] Inserm U1040, Montpellier 34295, France. [8] Inserm U661, Montpellier 34094, France. [9] CNRS UMR5203, Institut de Génomique Fonctionnelle, Montpellier 34094, France. [10] NovAliX, Illkirch 67400, France. [11] INRA UMR 1331, TOXALIM, Sophia-Antipolis 06903, France. [12] CHU de Montpellier, Institut de Recherche en Biothérapie, Montpellier 34295, France. * These authors equally contributed to this work. ** These authors jointly supervised this work. Correspondence and requests for materials should be addressed to P.B. (email: patrick.balaguer@inserm.fr) or to W.B. (email: bourguet@cbs.cnrs.fr).

External compounds (xenobiotics) to which humans are continuously exposed include environmental pollutants, drugs or dietary components. Many of them belong to the structurally heterogeneous group of endocrine-disrupting chemicals (EDCs) that trigger adverse health effects by mimicking or antagonizing the action of endogenous signalling molecules[1–3]. More than 20 years of experimental and epidemiological studies have highlighted the pivotal role of nuclear receptors (NRs) in transducing many of the harmful effects of EDCs[4,5]. NRs belong to a large family of evolutionarily related transcription factors that control complex gene networks, resulting in profound physiological changes[6]. They contain a ligand-binding domain (LBD) that responds to a wide variety of endogenous hormonal and metabolic ligands. The endocrine-disrupting action of chemicals relies mostly on their ability to substitute for natural ligands and deregulate NR signalling, causing reproductive, proliferative and metabolic disorders[7–9]. In addition, human exposure to mixtures of xenobiotics can induce unpredictable additive, antagonistic or synergistic adverse effects[10]. Yet, the molecular mechanisms underlying these cocktail effects are largely unknown.

To explore the outcome of combined exposure to chemicals and establish a detailed mechanistic understanding of this emerging paradigm for EDC action, we focused our attention on the xenoreceptor PXR (pregnane X receptor; NR1I2) which has been identified by the US Environmental Protection Agency ToxCast's program as a major front-line target of chemicals. This NR is a key regulator of the body's defense against foreign substances. It forms heterodimers with the retinoid X receptor (RXR) and binds to PXR responsive elements (PXRE) in the regulatory regions of target genes. Upon activation by xenobiotics (for example, bisphenol-A, organophosphate pesticides, alkylphenols, rifampicin), PXR interacts with coactivators, such as the steroid receptor coactivator-1 (SRC-1), and transcriptionally upregulates major detoxification genes such as the phase I cytochrome P450 enzyme CYP3A4 (ref. 11), which metabolizes more than half of all drugs in clinical use. On the other hand, the interaction of PXR with EDCs has been linked to an increased risk of cardiovascular[12] and metabolic[13] diseases.

Here, using compound screening followed by extensive functional analysis, we demonstrate that the combined use of the pesticide *trans*-nonachlor (TNC) and the active component of contraceptive pills, 17α-ethinylestradiol (EE2), produces synergistic effects on PXR activation and expression of its endogenous target gene *CYP3A4*. Biophysical characterization reveals that EE2 and TNC bind cooperatively to PXR and that the binary mixture has considerably improved functional properties over each of the compounds alone. Crystallographic analysis shows that reciprocal stabilization of the compounds in the ligand-binding pocket (LBP) of the receptor accounts for the enhanced efficacy and potency of the chemical mixture. We therefore propose the concept of a 'supramolecular ligand' that defines a molecular assembly consisting of two or more compounds that interact with each other inside the LBP of a receptor, resulting in the creation of a new entity with improved functional characteristics in regard to those of its individual components.

## Results

### Synergistic activation of PXR by EE2 and TNC. Using medium-throughput ligand screening and a mammalian (HeLa) cell-based activation assay (HG5LN GAL4-PXR-LBD reporter cell line)[14,15], we monitored the agonistic potential of 40 chemicals either alone or in binary mixtures (Supplementary Table 1). Most combinations exhibited additive effects, inducing 50–60% of the

transactivation seen with the cholesterol-lowering drug SR12813 (Fig. 1), a potent and well-characterized PXR full agonist[16–18] (EC$_{50}$ in the 100–200 nM range) used as a reference in all our experiments. However, we observed that the combined use of the organochlorine pesticide TNC and the synthetic oestrogen 17α-ethinylestradiol (EE2), the active component of contraceptive pills, produced more than an additive effect with an induction level of 90% (Fig. 1 and Supplementary Fig. 1). This preliminary observation was confirmed by dose–response experiments. The combination of TNC and EE2 led to a shift of the corresponding activation curve by one order of magnitude towards the low concentrations (Fig. 2a). We next examined the effect of TNC, EE2, and their combination in human liver (HepG2) and colon (LS174T) carcinoma cell lines containing the full-length PXR and the *CYP3A4*-XREM luciferase reporter plasmids. Again, we found that co-treatment yielded to much stronger activation of PXR as monitored by the transactivation of the *CYP3A4* reporter gene (Fig. 2b,c). When used simultaneously, the two compounds activated PXR in a synergistic fashion as illustrated by the theoretical activation curve obtained for the additive combination of EE2 and TNC activities (Fig. 2a–c, red dashed lines calculated using the Bliss independence model[19]). Note that synergism was also observed with other steroidal and organochlorine compound combinations. Supplementary Fig. 2 shows two representative examples associating either EE2 and *cis*-chlordane, or TNC with the natural hormone 17β-estradiol (Fig. 1).

**Cocktail effect on CYP3A4 expression and activity.** Consistent with the reporter gene assays, the effectiveness of the individual compounds at increasing the level of endogenous CYP3A4 mRNA was drastically enhanced by the addition of the second compound in LS174T cells expressing PXR but not in control cells (Fig. 2d). We then compared the ability of EE2 and TNC, alone or in combination, to increase *CYP3A4* gene expression in freshly isolated primary human hepatocytes (PHHs) in culture, the most biologically relevant model regarding PXR function. As shown in Fig. 2e, the CYP3A4 mRNA expression was considerably augmented when both EE2 and TNC were used. Accordingly, the enhanced induction of the CYP3A4 protein by the binary mixture (Fig. 2f, upper panel and Supplementary Fig. 3) closely correlated with higher CYP3A4 enzymatic activity (Fig. 2f, lower panel).

As a whole, cell-based assays clearly show that EE2 and TNC act as poor PXR agonists when used separately whereas their combination triggers PXR activation nearly as efficiently as the reference agonist SR12813. Notably, synergism could be observed in various cellular contexts and with different compound combinations, including the natural hormone 17β-estradiol.

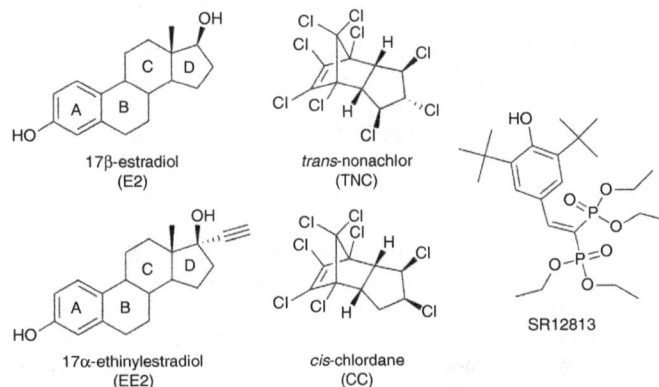

**Figure 1 | Chemical structures of compounds used in this study.**

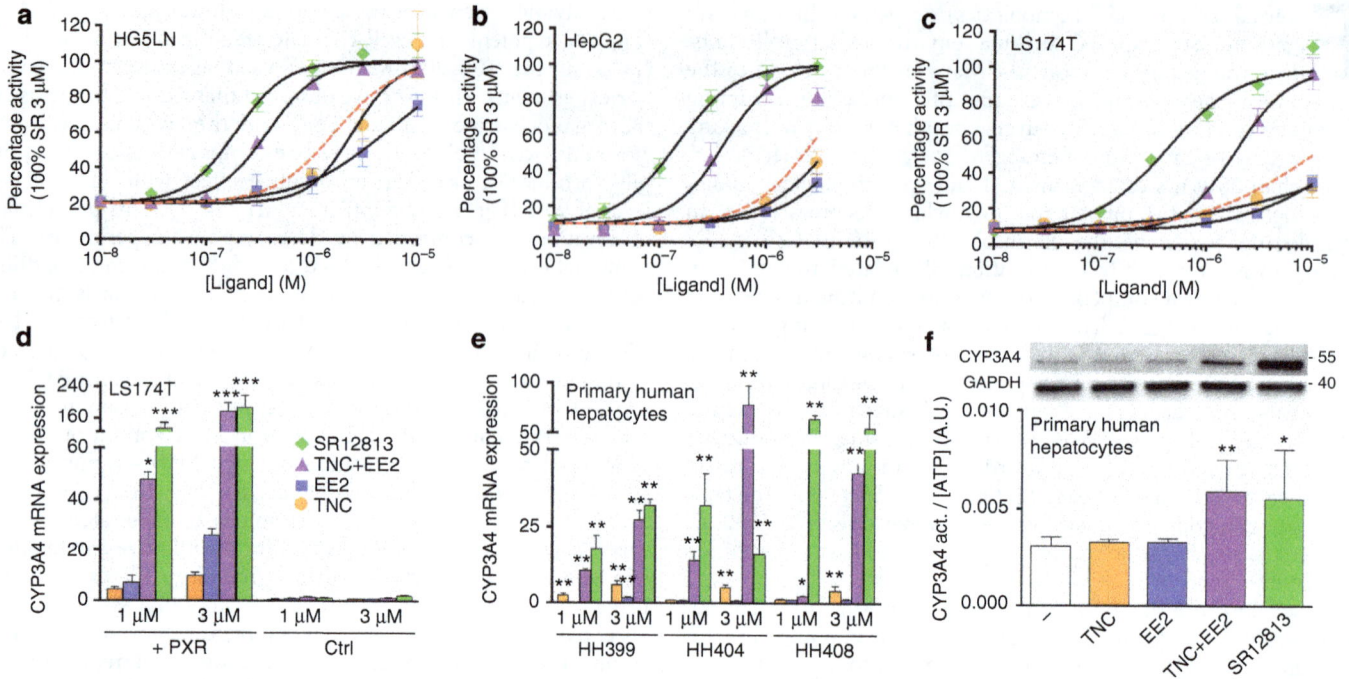

**Figure 2 | EE2 and TNC activate PXR in a synergistic fashion.** HG5LN GAL4-PXR-LBD (**a**), HepG2-PXR 3A4-Luciferase (**b**), and LS174T-PXR 3A4-Luciferase (**c**) cells were exposed to different concentrations of SR12813, TNC and EE2 either alone or in combination. Assays were performed in quadruplicate in at least three independent experiments and data are expressed as mean ($\pm$ s.e.m.). Red dashed lines represent the theoretical activation curves obtained for the additive combination of EE2 and TNC activities calculated using the Bliss independence model[19]. (**d**) RT-qPCR analysis of CYP3A4 mRNA expression in control (Ctrl) or PXR overexpressing LS174T cells treated 48 h by solvent (0.1% DMSO) or the indicated concentration of ligand. Results were obtained from three separate experiments performed in duplicates. Data are expressed as mean ($\pm$ s.e.m.) compared with DMSO treated cells, ***$P < 0.001$ **$P < 0.01$ *$P < 0.05$ (Student's $t$-test) compared to LS-CTRL cells. (**e**) RT-qPCR analysis of CYP3A4 mRNA expression in primary cultures of human hepatocytes (three independent donors: HH399, HH404 and HH408) treated 48 h by solvent (0.1% DMSO) or the indicated concentration of ligand. Results were obtained from experiments performed in triplicates. Data are expressed as mean ($\pm$ s.d.) compared to DMSO treated cells, *$P < 0.05$ (Student's $t$-test). (**f**) Quantifications of CYP3A4 and GAPDH protein expression by western-blot (single experiment, upper panel) and CYP3A4 enzymatic activity (lower panel) in primary culture of human hepatocytes (HH408) treated 72 h by solvent (0.1% DMSO) or 1 μM ligand. Results for enzymatic activity were obtained from one experiment performed in triplicates. Data are expressed as CYP3A4/CellTiter Glow activities ratio as mean ($\pm$ s.d.), **$P < 0.005$ (Student's $t$-test) compared with DMSO treated cells.

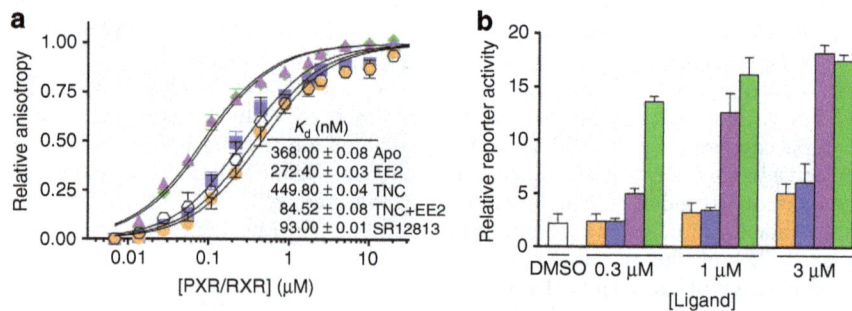

**Figure 3 | Coactivator recruitment by PXR upon co-treatment.** (**a**) Fluorescence anisotropy data showing the relative affinity of the fluorescein-labelled SRC-1 NID for PXR/RXR LBDs heterodimer (20 μM) in the presence of saturating concentrations (60 μM) of TNC, EE2, alone or in mixture, or the PXR agonist SR12813. Assays were performed in three independent experiments and data are expressed as mean ($\pm$ s.e.m.). (**b**) Mammalian two-hybrid experiment. Gal4-hPXR and VP16-TIF2 interaction monitored in U2OS cells in presence of DMSO, SR12813, EE2, TNC or EE2 + TNC. Assays were performed in duplicate in three independent experiments and data are expressed as mean ($\pm$ s.e.m.).

**Coactivator recruitment by PXR upon co-treatment.** In order to decipher the molecular mechanism involved in the synergistic activation of PXR-mediated transcription by EE2 and TNC, we characterized their impact on coactivator recruitment. For this purpose, we used fluorescence anisotropy assays with the purified PXR/RXR LBD heterodimers and the fluorescein-labelled NR interaction domain (NID) of SRC-1. We found that, as expected,

the PXR agonist SR12813 efficiently enhanced SRC-1 recruitment (Fig. 3a). Interestingly, EE2 and TNC had modest effects on their own but their combination produced a strong increase in the coactivator recruitment, similar to that observed with SR12813. Comparable results were obtained from mammalian two-hybrid experiments (Fig. 3b), suggesting that the active form of the receptor is highly stabilized in presence of both ligands.

**Simultaneous binding of EE2 and TNC to PXR.** To characterize further the interaction of PXR with EE2 and TNC, we used electrospray ionization mass spectrometry (ESI-MS) under native conditions. Analysed separately, TNC and EE2 were shown to interact with PXR with affinities in the low micromolar range; about 80 and 70% complex were detected when PXR (10 μM) was incubated with 2 molar equivalent excess of TNC or EE2, respectively (Fig. 4a-c). TNC bound to PXR in a 1:1 molar ratio, while EE2 was found to interact with the receptor with 1:1 and 1:2 binding stoichiometries. Binding specificity of the two compounds for PXR ligand-binding pocket was assessed using competition experiments against SR12813. As shown in Supplementary Fig. 4, SR12813 was able to compete efficiently for EE2 and TNC. No non-specific binding was detected with TNC, whereas some residual EE2 was observed in presence of SR12813, suggesting that part of the secondary EE2 binding sites could reside outside the PXR LBP. Analysis of PXR after incubation with the binary cocktail showed a large predominance of a ternary complex corresponding to PXR interacting with EE2 and TNC in a 1:1:1 molar ratio; the global range of affinity of the two compounds binding to PXR was sub-micromolar (Fig. 4d). Together, these data showed that PXR can accommodate EE2 and TNC simultaneously and suggested greater binding affinity of the binary mixture compared to individual compounds.

**EE2 and TNC bind cooperatively to PXR.** We next assessed the binding characteristics of EE2 and TNC, alone or in mixture, to PXR. Competitive binding assays using time resolved fluorescence resonance energy transfer between a fluorescent PXR ligand and purified human PXR-LBD (LanthaScreen TR-FRET PXR Competitive Binding Assay) showed that the binary mixture binds 10- to 30-fold more avidly to PXR than TNC and EE2 alone (Fig. 5a). To define the receptor–ligand interactions by a direct method, we next measured the binding affinity constants of various EE2 and TNC combinations by isothermal titration calorimetry (ITC). Before ITC experiments, dose-dependent aggregation of the compounds in working buffer was assessed using dynamic light scattering. Critical aggregation concentration (CAC) was estimated to be in the 50-100 μM range when compounds were diluted serially from 500 μM down to 3 μM in Tris-HCl 20 mM, pH 8.5, NaCl 200 mM, TCEP 1 mM, Tween20 0.05%, DMSO 5%. Accordingly, we cannot rule out that when used at 200-300 μM in some experimental conditions, a fraction of the compounds undergo aggregation despite the dilution process occurring during titration experiments (see Methods). Nevertheless, ITC analyses clearly converged towards similar conclusion of cooperative binding of EE2 and TNC to PXR resulting in increased global binding affinity. ITC data showed much better affinities of EE2 and TNC for PXR when the receptor was pre-incubated with TNC or EE2, respectively ($K_d$ values in

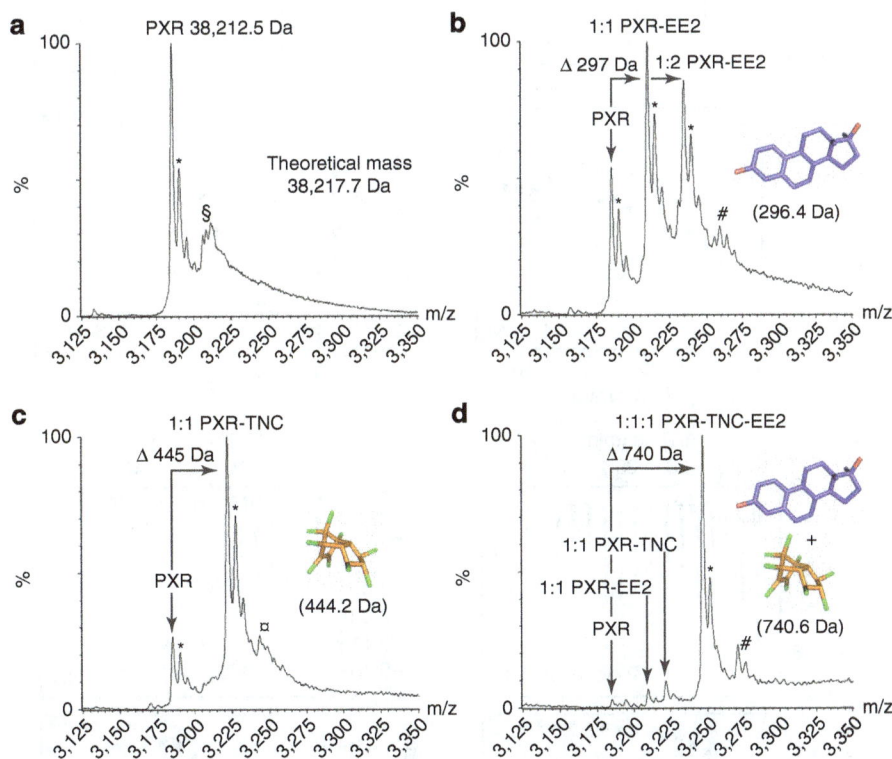

**Figure 4 | Simultaneous binding of EE2 and TNC to PXR.** Mass spectrometry analysis. Non-denaturing ESI-MS was used to characterize PXR LBD (10 μM) in (**a**) the unliganded form or in the presence of (**b**) EE2 (20 μM), (**c**) TNC (20 μM), or (**d**) a mixture of EE2 (20 μM) and TNC (20 μM). *, acetate adducts; $, fortuitous binders 254-324 Da; ☐, fortuitous binder 254 Da; #, non-specific EE2 adducts.

**Figure 5 | EE2 and TNC bind cooperatively to PXR.** (**a**) Inhibition of FRET between fluorescein-labelled PXR ligand and recombinant GST-PXR by SR12813, TNC and EE2, alone or in combination. Results are expressed as the signal from the fluorescein emission divided by the terbium signal to provide a TR-FRET emission ratio. Data are the mean (± s.e.m.) from three separate cocktails. (**b-f**) Isothermal titration calorimetry (ITC) characterization of PXR interaction with EE2 and TNC. Ligands were titrated either independently (**b,c**), after pre-incubating the receptor with EE2 (**d**) or TNC (**e**), or simultaneously (**f**). In **b-f**, representative thermograms (upper row) and corresponding binding isotherms (lower row) are shown. $K_d$ values are expressed as the mean of two independent experiments.

**a**

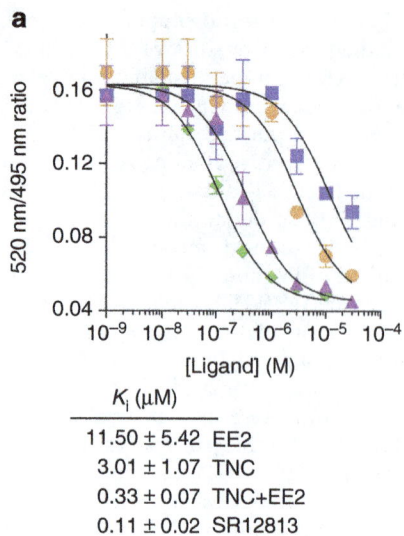

$K_i$ ($\mu$M)

| | |
|---|---|
| 11.50 $\pm$ 5.42 | EE2 |
| 3.01 $\pm$ 1.07 | TNC |
| 0.33 $\pm$ 0.07 | TNC+EE2 |
| 0.11 $\pm$ 0.02 | SR12813 |

**b**

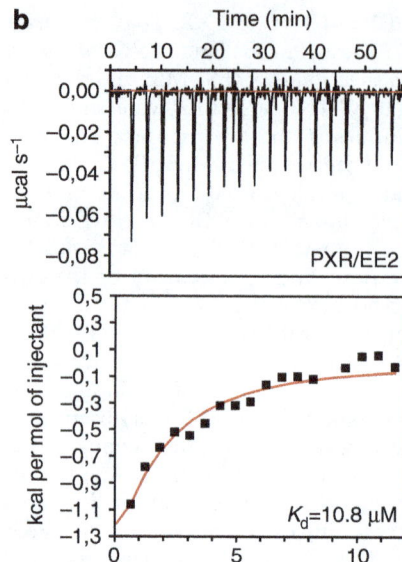

PXR/EE2

$K_d$=10.8 $\mu$M

**c**

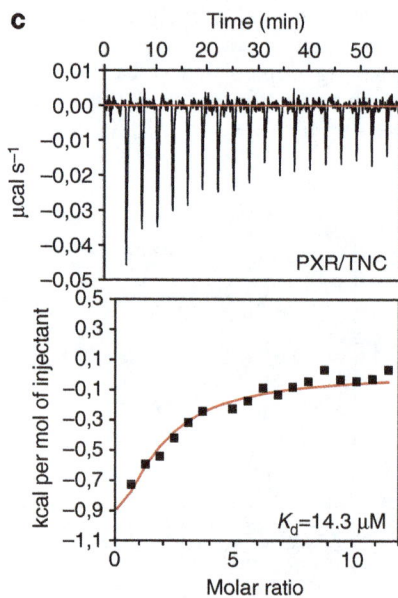

PXR/TNC

$K_d$=14.3 $\mu$M

**d**

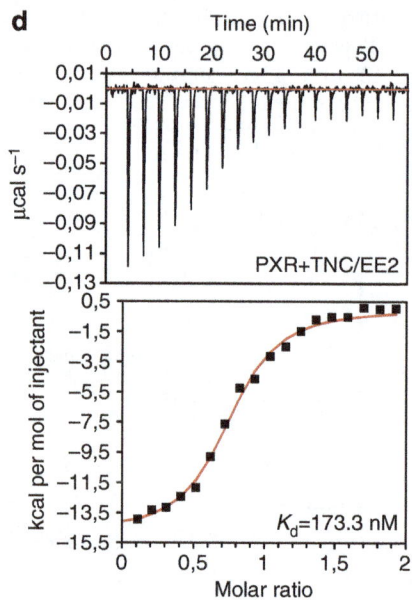

PXR+TNC/EE2

$K_d$=173.3 nM

**e**

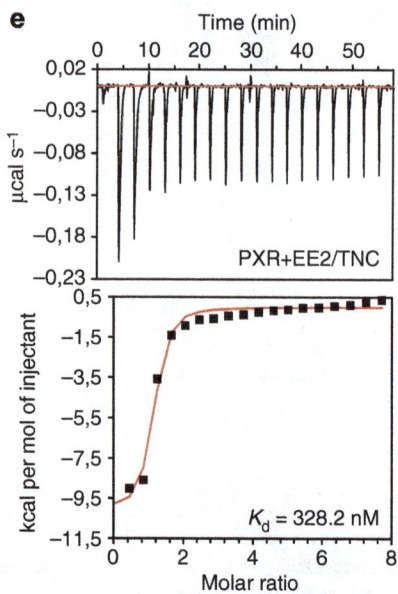

PXR+EE2/TNC

$K_d$ = 328.2 nM

**f**

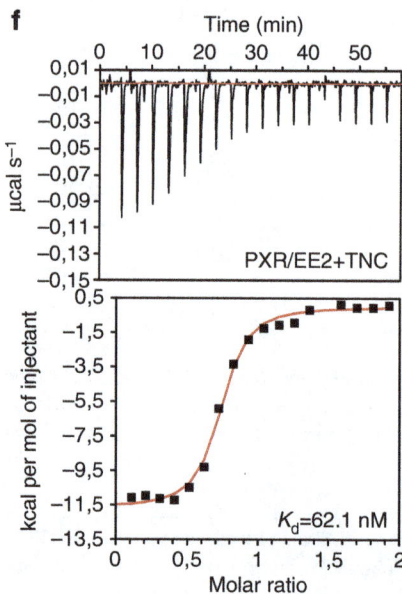

PXR/EE2+TNC

$K_d$=62.1 nM

**Table 1 | Data collection and refinement statistics.**

|  | EE2 (4X1F) | TNC (4XAO) | EE2 + TNC (4X1G) |
|---|---|---|---|
| **Data collection** |  |  |  |
| Space group | $P\,4_3\,2_1\,2$ | $P\,4_3\,2_1\,2$ | $P\,4_3\,2_1\,2$ |
| Cell dimensions |  |  |  |
| $a, b, c$ (Å) | 91.34, 91.34, 85.35 | 92.30, 92.30, 86.30 | 91.34, 91.34, 85.49 |
| $\alpha, \beta, \gamma$ (°) | 90.00, 90.00, 90.00 | 90.00, 90.00, 90.00 | 90.00, 90.00, 90.00 |
| Resolution (Å) | 45.67-2.00 (2.11-2.00)* | 41.28-2.58 (2.69-2.58)* | 40.85-2.25 (2.38-2.25)* |
| $R_{sym}$ | 0.080 (0.489) | 0.110 (0.472) | 0.091 (0.453) |
| $I\,/\,\sigma I$ | 20.1 (4.7) | 14.7 (4.1) | 12.4 (3.1) |
| Completeness (%) | 100.0 (100.0) | 96.8 (78.3) | 97.7 (99.0) |
| Redundancy | 10.4 (10.4) | 7.8 (7.3) | 5.4 (5.2) |
| **Refinement** |  |  |  |
| Resolution (Å) | 45.67-2.00 | 41.28-2.58 | 38.71-2.25 |
| No. reflections | 24,998 | 12,258 | 17,316 |
| $R_{work}/R_{free}$ | 0.182/0.209 | 0.189/0.239 | 0.174/0.218 |
| No. atoms |  |  |  |
| Protein | 2,158 | 2,171 | 2,208 |
| Ligand/ion | 38 | 12 | 53 |
| Water | 170 | 67 | 122 |
| B-factors |  |  |  |
| Protein | 30.19 | 39.48 | 34.80 |
| Ligand/ion | 40.29 | 57.14 | 48.99 |
| Water | 37.31 | 40.64 | 39.36 |
| R.m.s.deviations |  |  |  |
| Bond lengths (Å) | 0.009 | 0.002 | 0.008 |
| Bond angles (°) | 0.996 | 0.569 | 1.067 |

R.m.s, root-mean square; TNC, *trans*-nonachlor. *Values in parentheses are for highest-resolution shell.

the high-nanomolar range compared to the mid-micromolar range interaction of the compounds tested independently; Fig. 5b-e). Simultaneous titration of EE2 and TNC against PXR was associated with a binding affinity of the binary mixture that was substantially higher than those of the compounds tested alone (Fig. 5f), corroborating the observations derived from ESI-MS and LanthaScreen experiments.

Taken altogether, these *in vitro* data performed with purified material strongly support the notion that both the cooperative binding and synergism observed with EE2 and TNC rely on direct interactions with PXR and not on other cellular mechanisms such as cellular influx/efflux, metabolism, or binding to other cellular targets.

**Structural basis for supramolecular ligand activity.** To gain structure-based insight into the binding mode of EE2 and TNC to PXR, we solved the crystal structures of PXR-LBD in complex with EE2, TNC, or both at 2.00 Å, 2.55 Å, and 2.25 Å resolution, respectively (Table 1 and Supplementary Fig. 5). In all cases, the PXR-LBD adopts the canonical active conformation, with the C-terminal activation helix H12 capping the LBP (Fig. 6a and Supplementary Fig. 6a,b). Whereas EE2 could be precisely placed in the electron densities obtained for both the PXR–EE2 and PXR–TNC–EE2 complexes, TNC could be positioned unambiguously in the ternary complex only (Fig. 6b and Supplementary Fig. 6c,d). The poorly defined electron density of TNC in the PXR–TNC complex likely reflects high mobility of this ligand in the pocket. In contrast to EE2 which is engaged in several hydrogen bonds maintaining the ligand in a defined location, TNC has no polar group and likely adopts an ensemble of different positions and/or orientations in the LBP.

EE2 binds closely adjacent to H12 with a binding mode that is reminiscent of that seen for 17β-estradiol[20] (Supplementary Fig. 7a), both in presence or absence of TNC. The 3-hydroxyl group on the A-ring of EE2 forms an hydrogen bond with S247

(helix H3), whereas the 17β-hydroxyl group on the D-ring is hydrogen bonded with R410 in helix H11 and the main chain oxygen atom of D205 from helix H2' (Fig. 6c). The position of R410 is further stabilized by a network of hydrogen bonds involving S208 (helix H2') and E321 from the loop preceding H7. The remaining contacts between EE2 and the protein involve van der Waals interactions with several hydrophobic residues. The particular position of EE2 in a restricted region of PXR LBP leaves a significant portion of the pocket unoccupied and available for additional interactions (Supplementary Fig. 6a,c). As seen in the ternary complex, this empty region can accommodate one molecule of TNC which is stabilized in a well-defined position *via* a number of interactions with both EE2 and the protein. As illustrated in Fig. 6c, eight van der Waals contacts of 3.7 to 4.5 Å in length could be measured between EE2 and TNC. These inter-ligand contacts generate a mutual stabilization of the compounds in the LBP and account for the enhanced binding affinity of the binary mixture (Fig. 5). On the protein side, TNC forms essentially nonpolar interactions with ten residues, including F281, F288, W299, Y306, M323 (Fig. 6c) and two weak halogen bonds with Q285 and C284 from helix H5.

In keeping with the robust agonistic activity of the binary mixture, superposition of the supramolecular ligand-bound structure with that of PXR in complex with SR12813 (ref. 21) or rifampicin[22] reveals that the binding sites of EE2 and TNC overlap those of the two PXR agonists (Supplementary Fig. 7b,c). Hence, the molecular assembly of EE2 and TNC into the LBP of PXR can be regarded as a supramolecular ligand whose functional properties rely on intermolecular interactions and differ from those of the individual components.

## Discussion

Most current knowledge of EDCs action is derived from data sets that use single molecule exposure, with few studies taking into account the more realistic situation where humans are

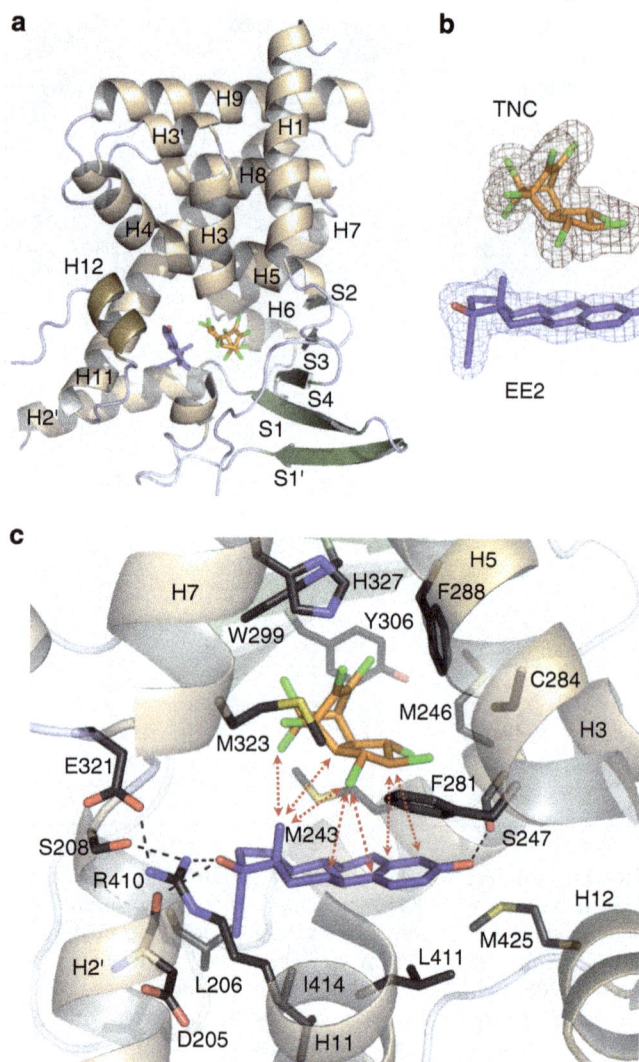

**Figure 6 | Structural basis for supramolecular ligand activity. (a)** Overall structure of PXR-LBD in complex with EE2 (blue) and TNC (orange); the structure shows the LBP bordered by helix H12 (light brown) on one side and the β-sheet (green) on the other side; for clarity helix H3 is partially cut. **(b)** Electron density of ligands ($2F_o$-$F_c$ map contoured at 1σ). **(c)** Interaction network of ligands with residues of the LBP (grey); red dashed lines highlight the interactions between EE2 and TNC resulting in a mutual stabilization of the compounds in the LBP. Colour code: red, oxygen; blue, nitrogen; yellow, sulphur; green, chlorine; black dashed lines, hydrogen bonds.

Reports describing the simultaneous binding of two or more compounds to a common protein binding site are very few in number. One of the rare examples is that of the *Staphylococcus aureus* multidrug-binding transcription repressor QacR which was shown to bind concomitantly to ethidium and proflavin. However no cooperative binding mechanism was observed in this case[23]. Other examples include the cytochrome CYP3A4 bound to two molecules of ketoconazole[24] or the peroxisome proliferator activated receptor gamma (PPARγ) which can accommodate two copies of FMOC-L-Leucine[25], or endogenous fatty acids[26]. However, in both cases the bound molecules were of the same type and the possibility of synergism between the two ligands was not addressed. Besides the interaction of several compounds with a unique binding site, two studies reported on the oestrogen antagonist 4-hydroxytamoxifen and a synthetic ligand binding to two alternate sites of the oestrogen receptor β and PPARγ LBDs, respectively[27,28]. Again, no cooperativity was reported between these dual binding sites. Our study shows that individually, EE2 and TNC are too small to make all the necessary interactions ensuring high binding affinity and effective stabilization of the active conformation of the receptor. In contrast, when associated in a binary mixture, EE2 and TNC form a supramolecular ligand that fills a larger fraction of the PXR LBP, and displays apparent functional properties (*e.g.* activity and binding affinity) comparable to those of the full PXR agonists, SR12813 and rifampicin.

Cloning of PXR orthologues from human, rabbit, rat and mouse has shown that the ligand-binding domain has diverged considerably between the different species, leading to specific ligand-binding and activation profiles[29,30]. This species-specific induction pattern of PXR is possibly an adaptive response to the environment and a need to adjust toxicological responses to endogenously produced substances. Therefore, additional studies should be performed to test the potential EE2/TNC synergy *in vivo* in these species. Notably future studies will be completed in mice, along with the use of PXR-knock-out models and long term exposures with these compounds, to confirm the role of PXR and the physiological relevance of the EE2/TNC synergism on drug and bile acid metabolism, hepatic steatosis, or liver regeneration for instance.

Large LBPs such as those of PXR or PPARγ are obviously predisposed to accommodate several molecules at the same time. However, in contrast to the perception that most NRs possess a well-defined pocket to account for the specific binding of a unique endogenous ligand, several structural studies have revealed that their LBPs exhibit a greater conformational flexibility than previously thought. Importantly, this structural adaptability allows the oestrogen (ERα)[31], thyroid (TRβ)[32], or glucocorticoid (GR)[33] receptors to expand their binding pocket and accommodate much larger ligands[34]. Another interesting example is the oestrogen-related receptor gamma (ERRγ) whose LBP appears too small to bind any ligand, yet can expand it to accommodate rather large compounds[35]. Considering the huge chemical and size diversity of xenobiotics and the high structural plasticity of NR LBPs, the mechanism defined for PXR is likely to occur also with other members of the NR superfamily, with broad reaching implications in the fields of endocrine disruption, chemicals risk assessment and toxicology. The development of novel supramolecular chemistry-based therapeutic options could also benefit from the discovery reported here.

simultaneously and chronically exposed to low doses of multiple EDCs. Indeed, a growing number of studies indicate that human risk assessment approaches based on single molecule exposure underestimate the risk for adverse effects of chemicals[10]. The evaluation of mixture effects by regulatory bodies has been mainly hampered by the huge numbers of pollutants and potential combinations, but also by the lack of knowledge of the molecular pathways involved. This study provides both the first detailed mechanistic explanation and a proof of concept for the synergistic action of two compounds *via* their simultaneous interaction with the LBP of a NR. Our results provide not only new insight as to how low doses of EDCs or drugs may affect physiology and homeostasis, but also suggest that the concomitant binding of chemicals stabilizing each other in NR LBPs likely corresponds to one of the possible mechanisms accounting for the cocktail effect by which compounds toxicity is exacerbated.

## Methods

**Ligands.** TNC, EE2, E2 and CC were purchased from Sigma-Aldrich. SR12813 was purchased from Tocris Bioscience. The provenance of compounds used for medium-throughput screening is described in Supplementary Table 1. All compound stock solutions were prepared at 10 mM in DMSO.

**Plasmids.** The [ − 7.8/ − 7.2] XREM-CYP3A4 (XREM-CYP3A4 pGL3b) luciferase plasmid has been described previously[36]. The 17Mx5-Glob-LUC containing five GAL4 binding sites upstream of the luciferase reporter gene and the VP16 (activation domain)-TIF2 (amino acids 624-869 containing three LXXLL motifs) are gifts from Hinrich Gronemeyer (IGBMC, Illkirch, France). The pM-hPXR expression vector was generated by inserting a PCR fragment corresponding to the full-length hPXR in the pM vector (CLONTECH).

**Cell lines.** LS174T stable human PXR transfectant (LS-PXR2) and the corresponding control cells (LS-CTRL), HEPG2-PXR and HG5LN GAL4-PXR reporter cell lines were previously described[17,37,38]. Briefly, LS-PXR2 was obtained after stable transfection of the LS174T cells with the pcDNA3.1-hPXR (residues 1-434)-neomycin expressing plasmid. HEPG2-PXR was obtained after stable transfection of the same PXR expressing plasmid and the XREM-CYP3A4 pGL3b reporter plasmid[38]. HG5LN GAL4-PXR reporter cell line was established by stable transfection of pSG5-GAL4DBD (residues 1-147)-hPXR LBD (residues 107-434)-puromycin and GAL4-RE5-βGlobin-luciferase-neomycin plasmids[17]. Finally, LS174T-PXR reporter cell line was established by stable transfection of the CYP3A4-luciferase-hygromycine plasmid in the LS-PXR2 cell line. All cell lines were grown in DMEM medium (Invitrogen) supplemented with 10% fetal calf serum, L-glutamine, and antibiotics (Invitrogen).

**Medium-throughput screening.** HG5LN GAL4-PXR reporter cell lines (100 µl) were seeded at a density of 25,000 cells per well in 96-well white opaque tissue culture plates (Greiner CellStar). 24 h later, negative (DMSO 0,1%) and positive (SR12813 3 µM) controls, and the 40 tested compounds (50 µl) were added into the wells as indicated in Supplementary Table 1. Then the ligand to be tested in combination with the different ligands was added (50 µl). Cells were incubated at 37 °C for 16 h. At the end of the incubation period, culture medium was replaced with medium containing $3.10^{-4}$ M luciferin. Luciferase activity was measured for 2 s in intact living cells using a plate reader (PerkinElmer Luminometer).

**Transactivation assays.** HG5LN GAL4-PXR, HEPG2- and LS174T-PXR reporter cell lines were seeded at a density of 25,000 cells per well in 96-well white opaque tissue culture plates (Greiner CellStar). Compounds to be tested were added 24 h later, and cells were incubated at 37 °C for 16 h. At the end of the incubation period, culture medium was replaced with medium containing $3.10^{-4}$ M luciferin. Luciferase activity was measured for 2 s in intact living cells using a plate reader (PerkinElmer Luminometer). EC$_{50}$ values were measured using GraphPad Prism (GraphPad Software Inc.).

**Preparation of primary human hepatocytes.** Liver samples were obtained from liver resections performed in adult patients for medical reasons unrelated to our research program or from donors when the liver was considered unsuitable for organ transplantation. The use of human specimens for scientific purposes was approved by the French National Ethics Committee. Written or oral informed consent was obtained from each patient or family prior to surgery. The clinical characteristics of the liver donors are presented in Supplementary Table 2. Hepatocytes were isolated by using a two-step perfusion protocol and cultured as described previously[39]. Briefly, several veins apparent on the cut edge of the lobectomy were used for sequential perfusions with a washing buffer (10 mM HEPES, 136 mM NaCl, 5 mM KCl, 0.5% glucose, pH 7.6), with a calcium chelating buffer (washing buffer complemented with 0.5 mM EGTA), with the washing buffer and then with a collagenase IV solution (washing buffer supplemented with collagenase IV, 200 U ml$^{-1}$, Sigma) for 20 min. After gentle disruption of the tissue and filtration through a 250 µm mesh, the post-collagenase homogenate was centrifuged at low speed (50 g for 5 min) to pellet the hepatocytes. Hepatocytes were seeded in collagen type-I coated dishes (Becton Dickinson) at 1.7 10$^5$ cells per cm$^2$ in a hormonally and chemically defined medium for short term culture consisting of a mixture of William's E and Ham's F-12 (1:1 in volume) and additives as previously described[40] supplemented with 2% heat inactivated fetal calf serum (FCS). After overnight attachment, the medium and unattached cells were removed and fresh medium without FCS was added. Hepatocytes were treated with the molecules at day 2 post-seeding for 24 or 48 h.

**Reverse transcription-PCR and real-time quantitative PCR.** Total RNA was extracted from 200,000 cells using the RNeasy mini kit (Qiagen) or Trizol reagent (Invitrogen) and treated with DNAse-1. The first strand cDNA was synthetized using Superscript II (Invitrogen) or MMLV (Invitrogen) and random hexamers. Primer sequences are provided in Supplementary Table 3. Gene expression was normalized with GAPDH and the level of expression was compared with the mean level of the corresponding gene expression in untreated cells and expressed as n-fold ratio. The relative amount of RNA was calculated with the $2^{\Delta\Delta CT}$ method.

**Western-immunoblotting analysis.** Total protein extracts were prepared from 500,000 cells with RIPA buffer (Tebu-Bio) in presence of antiproteases (Roche). The protein concentration was determined by the bicinchoninic acid method,

according to the manufacturer's instructions (Pierce Chemical Co.). Bovine serum albumin (Pierce Chemical Co.) was used as standard. Cell lysates were resolved on SDS-PAGE and transferred to a Hybond-ECL membrane (GE Healthcare). Membranes were incubated with rabbit anti-CYP3A4 (5316, 1/10,000, Epitomics) or mouse anti-GAPDH (sc#32233, 1/5,000, Santa Cruz) monoclonal antibodies. Immunocomplexes were detected with horseradish peroxidase-conjugated rabbit (A0545, 1/10,000, Sigma) or mouse (A9044, 1/10,000, Sigma) secondary antibodies followed by enhanced chemiluminescence reaction (Millipore). Chemiluminescence was monitored using a ChemiDoc-XRS$^+$ apparatus (Bio-Rad Laboratories).

**Measurement of CYP3A4 activity.** Primary human hepatocytes were seeded in 96-well plates and treated with the test compounds or vehicle for 48 h. CYP3A4 activity was detected using the P450-Glo CYP3A4 luciferin-IPA Enzyme Activity Kit (Promega) according to the manufacturer's instructions. Cell number was normalized using CellTiter-Glo Luminescent Cell Viability Assay (Promega).

**Statistical analyses.** For the analysis of the correlation between parametric data, Stutent's $t$-test was used, while the Mann–Whitney $U$-test was used for nonparametric data. Differences were considered statistically significant when $P$-values were ***$P < 0.001$ **$P < 0.01$ *$P < 0.05$.

**Preparation of PXR/RXR for fluorescence anisotropy assays.** The human PXR-LBD (130-434; pDB-His-MBP vector) and the human RXR-LBD (223-462; pET-3a vector) were overproduced in *Escherichia coli* BL21(DE3). Cells were grown at 37 °C in LB medium supplemented with the appropriate antibiotic until OD$_{600}$ reached 0.6. Expression of T7 polymerase was induced by addition of iso-propyl-β-D-thiogalactoside to a final concentration of 0.5 mM. After 16 h of incubation at 20 °C, cell cultures were harvested by centrifugation at 6,000g for 15 min. Cell pellets from 3 L of His$_6$-MBP-PXR-LBD culture and from 1 l of non-tagged RXR-LBD culture were resuspended in 150 ml buffer A (20 mM HEPES pH 7.5, 250 mM NaCl, 10 mM imidazole, 5% (v/v) glycerol) supplemented with a protease inhibitor cocktail tablet (cOmplete, EDTA-free, Roche). The suspension was then lysed by sonication and centrifuged at 18,000g at 4 °C for 30 min. The supernatant was loaded onto a Ni$^{2+}$-affinity column (HisTrap 5 ml; GE Healthcare) equilibrated with buffer A, using the ÄKTA pure system (GE Healthcare). The column was washed with 20 volumes of buffer A and 20 volumes of buffer A containing 50 mM imidazole. The His$_6$-MBP-PXR/RXR heterodimer was eluted with buffer A containing 100 mM imidazole. The fractions containing the heterodimer were pooled and incubated for 48 h at 4 °C with the tobacco etch virus protease to cleave the His$_6$-MBP tag, and then further reloaded onto the Ni$^{2+}$-affinity column to eliminate the tag. The flow-through containing the PXR/RXR heterodimer was then purified using a gel filtration column (Superdex 200 16/60; GE Healthcare) equilibrated with buffer B (10 mM Tris-HCl pH 7.5, 250 mM NaCl, 1 mM DTT and 5% (v/v) glycerol). The fluorescein-labelled SRC-1 fragment (residues 570–780) was prepared using the protocol previously described[41]. Briefly, the SRC-1fragment was expressed in *E. coli* BL21(DE3) as a fusion with an inducible self-splicing intein (Sce VMA) and a chitin-binding domain using the vector pTYB1 (New England Biolabs). The cells were grown at 37 °C to 0.6 at OD$_{600}$ and then at 17 °C overnight. The fusion protein was purified using chitin resin (New England Biolabs). Intein cleavage was induced using 2 mM cys-fluor with 50 mM MESNA, releasing C-terminally labelled SRC-1. Excess cys-fluor was removed using a phenyl sepharose resin (Amersham). SRC-1 was further purified using size-exclusion chromatography. Thin-layer chromatography was used to confirm that there was no free fluorescein label in the purified samples.

**Steady-state fluorescence anisotropy.** Measure of the binding affinities of the coactivator fragment for the PXR/RXR heterodimer in the absence and presence of various ligands was performed using a Safire2 microplate reader (TECAN). The excitation wavelength was set at 470 nm and emission measured at 530 nm for the fluorescein-tagged fragment. Assays were carried out in the following buffer solution 20 mM Tris-HCl, pH 7.5, 150 mM NaCl, 1 mM TCEP and 5% (v/v) glycerol. We initiated the measurements at the highest concentration of protein (20 µM) and diluted the protein sample twofold successively with the buffer solution. For each point of the titration curve, the protein sample was mixed with 5 nM of fluorescent fragment and a 3 molar excess of ligand (60 µM final concentrations). Binding data were fitted using a sigmoidal dose–response model using GraphPad Prism (GraphPad Software Inc.).

**Mammalian two-hybrid experiments.** Gal4-hPXR and VP16-TIF2 interaction was monitored on 17Mx5-Glob-LUC reporter construct. Transient transfections assays were performed in U2OS cells using Jet-PEI (Ozyme) according to manufacturer's instructions. Luciferase assays were performed with the Promega dual-reporter kit, according to the manufacturer's instructions. *Renilla* luciferase encoded by the normalization vector phRLTK (Promega) was used as internal control for firefly luciferase normalization.

**Mass spectrometry.** Mass spectrometry experiments were carried out on an electrospray time-of-flight mass spectrometer (LCT, Waters) equipped with an automated chip-based nanoESI device (Triversa Nanomate, Advion Biosciences). External calibration was done in the positive ion mode over the mass range $m/z$ 500–5,000 using the multiply charged ions produced by 0.5 μM horse heart myoglobin solution diluted in water/acetonitrile 50/50 mixture acidified with 0.5% (v/v) formic acid. Purified PXR(130-434)-SRC1 was buffer exchanged against 50 mM ammonium acetate (NH4Ac), pH 8.0 using NAP5 desalting columns (illustra NAP-5 Columns, GE Healthcare Life Sciences). Protein concentration was determined spectrophotometrically ($\varepsilon_{280\,nm} = 26,210\,l\,mol^{-1}\,cm^{-1}$). Analysis of EE2 and TNC binding to PXR(130-434)-SRC1 was achieved in 50 mM NH4Ac pH 8.0 keeping a constant 5% amount of isopropanol (v/v). Protein concentration was set to 10 μM and different compound concentrations ranging from 20 to 80 μM were tested. Incubations lasted 5 min at 18 °C. Mass spectra were recorded using low cone voltage ($V_c$, 20 V) and elevated interface pressure (Pi, 5 mbar).

**Lanthascreen TR-FRET PXR competitive binding assay.** GST-hPXR-LBD (10 nM) was incubated with different concentrations (10–30 μM) of TNC, EE2, TNC and EE2, and SR12813 in the presence of Fluormone PXR ligand (40 nM) and Lanthascreen terbium-anti-GST antibody (10 nM). To read a LanthaScreen TR-FRET assay, the fluorimeter (PHERAstar FS; BMG LABTECH) is configured to excite the terbium donor around 340 nm, and to separately read the terbium emission peak that is centred at ~490 nm, and the fluorescein emission that is centred at ~520 nm. Results are expressed as the signal from the fluorescein emission divided by the terbium signal to provide a TR-FRET emission ratio. Fluorescence ratio data were fitted using a sigmoidal dose–response model using GraphPad Prism (GraphPad Software Inc.).

**Isothermal titration calorimetry.** Purified PXR(130-434)-SRC-1 was dialysed overnight against Tris-HCl 20 mM, pH 8.5, NaCl 200 mM, TCEP 1 mM using 10 kDa molecular weight cut-off dialysis cassettes (Slide-A-Lyzer 0.5 ml 10 K MWCO, Thermo Scientific). Protein concentration was determined spectrophotometrically ($\varepsilon_{280\,nm} = 26,210\,l\,mol^{-1}\,cm^{-1}$). Duplicate experiments were performed on Microcal ITC200 (Malvern) operating at 25 °C. Titrations were carried out in Tris-HCl 20 mM, pH 8.5, NaCl 200 mM, TCEP 1 mM supplemented with 0.05% Tween 20 and 5% DMSO (syringe, sample and reference cells). PXR (5 μM) was disposed in 200 μl cell and compounds were delivered from 40 μl syringe. Compound solutions were set to 300 μM when tested individually (Fig. 5b,c), 50 μM each when used simultaneously (Fig. 5f) and 50 μM (EE2, Fig. 5d) or 200 μM (TNC, Fig. 5e) when tested after pre-incubation of PXR with 50 μM TNC or EE2, respectively. Heat exchanges were monitored throughout titrations consisting of 19 injections (one time 0.5 μl followed by 18 times 2 μl) of compound solutions into the cell containing PXR solution. Data analysis and thermodynamic parameter fitting used Microcal Origin software (Malvern).

**Protein production and purification for structural studies.** The human PXR-LBD (130-434) was co-produced with a fragment of the steroid receptor coactivator-1 (SRC-1, 623-710) to enhance PXR stability. The PXR-LBD gene was cloned into pRSET-A with a His6 tag at the N-terminus (gift from Matthew Redinbo, University of North Carolina at Chapel Hill, USA), and the SRC-1 fragment have been inserted into the pACYC184 vector. Proteins were overproduced in *E. coli* BL21(DE3) cells overnight at 20 °C in LB medium without any ligand. After culture cells were harvested by centrifugation and the pellets resuspended in lysis buffer (20 mM Tris pH 7.5, 250 mM NaCl, 5% (v/v) glycerol) supplemented with lysozyme (1 μg ml$^{-1}$) and a protease inhibitor cocktail tablet (cOmplete, EDTA-free, Roche), and then subjected to sonication. The clarified cell lysate was applied onto a Ni$^{2+}$-affinity column (HisTrap 5 ml; GE Healthcare) equilibrated with lysis buffer supplemented with 10 mM imidazole. The eluted PXR-LBD was then applied onto a gel filtration column (Superdex 75 26/60; GE Healthcare) equilibrated with a buffer containing 20 mM Tris pH 7.8, 250 mM NaCl, 5% (v/v) glycerol, 5 mM DTT, 1 mM EDTA. The PXR-LBD was concentrated and stored at −40 °C.

**Crystallization.** Prior to crystallization assays the purified PXR-LBD (2.4 mg ml$^{-1}$) was mixed with EE2 (2.5 molar equivalents), TNC (2.5 molar equivalents), or EE2 + TNC (2.5 molar equivalents each). Co-crystals with EE2 were obtained in 100 mM NaCl, 100 mM imidazole pH 7.1, 10% (v/v) isopropanol, in 100 mM imidazole pH 7.1, 10% (v/v) isopropanol with TNC alone, and in 50 mM NaCl, 50 mM LiCl, 100 mM imidazole pH 7.1, 10% (v/v) isopropanol with EE2 + TNC mixture.

**Data collection and structure determination.** For all complexes, native data were collected from one crystal cryoprotected with 20% (v/v) MPD on the ID23-2 beamline at the European Synchrotron Radiation Facilities ($\lambda = 0.8726$ Å, 100 K), Grenoble, France. Data were processed and scaled with XDS and XSCALE[42]. Crystals belong to space group $P\,4_32_12$. The X-ray structures were solved and refined using Phenix (phenix.refine)[43] and COOT[44]. The percentages of residues located in the favoured Ramachandran plot region are 98.1, 98.2 and 97.9% for the PXR – EE2, PXR – TNC and PXR – EE2 – TNC complex structures respectively

(calculated with MolProbity[45]). Data collection and refinement statistics are summarized in Table 1. Figures were prepared with PyMOL (http://pymol.org/).

## References

1. Diamanti-Kandarakis, E. *et al.* Endocrine-disrupting chemicals: an Endocrine Society scientific statement. *Endocr. Rev.* **30,** 293–342 (2009).
2. De Coster, S. & van Larebeke, N. Endocrine-disrupting chemicals: associated disorders and mechanisms of action. *J. Environ. Public Health* **2012,** 713696 (2012).
3. Schug, T. T., Janesick, A., Blumberg, B. & Heindel, J. J. Endocrine disrupting chemicals and disease susceptibility. *J. Steroid Biochem. Mol. Biol.* **127,** 204–215 (2011).
4. Janosek, J., Hilscherova, K., Blaha, L. & Holoubek, I. Environmental xenobiotics and nuclear receptors–interactions, effects and *in vitro* assessment. *Toxicol. In Vitro* **20,** 18–37 (2006).
5. Swedenborg, E., Ruegg, J., Makela, S. & Pongratz, I. Endocrine disruptive chemicals: mechanisms of action and involvement in metabolic disorders. *J. Mol. Endocrinol.* **43,** 1–10 (2009).
6. Germain, P., Staels, B., Dacquet, C., Spedding, M. & Laudet, V. Overview of nomenclature of nuclear receptors. *Pharmacol. Rev.* **58,** 685–704 (2006).
7. Grun, F. & Blumberg, B. Perturbed nuclear receptor signaling by environmental obesogens as emerging factors in the obesity crisis. *Rev. Endocr. Metab. Disord.* **8,** 161–171 (2007).
8. Newbold, R. R., Jefferson, W. N. & Padilla-Banks, E. Prenatal exposure to bisphenol a at environmentally relevant doses adversely affects the murine female reproductive tract later in life. *Environ. Health Perspect.* **117,** 879–885 (2009).
9. Rubin, B. S. & Soto, A. M. Bisphenol A: perinatal exposure and body weight. *Mol. Cell. Endocrinol.* **304,** 55–62 (2009).
10. Kortenkamp, A. Low dose mixture effects of endocrine disrupters and their implications for regulatory thresholds in chemical risk assessment. *Curr. Opin. Pharmacol.* **19C,** 105–111 (2014).
11. Lehmann, J. M. *et al.* The human orphan nuclear receptor PXR is activated by compounds that regulate CYP3A4 gene expression and cause drug interactions. *J. Clin. Invest.* **102,** 1016–1023 (1998).
12. Sui, Y. *et al.* Bisphenol A increases atherosclerosis in pregnane X receptor-humanized ApoE deficient mice. *J. Am. Heart Assoc.* **3,** e000492 (2014).
13. Chaturvedi, N. K., Kumar, S., Negi, S. & Tyagi, R. K. Endocrine disruptors provoke differential modulatory responses on androgen receptor and pregnane and xenobiotic receptor: potential implications in metabolic disorders. *Mol. Cell. Biochem.* **345,** 291–308 (2010).
14. Lemaire, G. *et al.* Identification of new human pregnane X receptor ligands among pesticides using a stable reporter cell system. *Toxicol. Sci.* **91,** 501–509 (2006).
15. Mnif, W. *et al.* Estrogens and antiestrogens activate hPXR. *Toxicol. Lett.* **170,** 19–29 (2007).
16. Jones, S. A. *et al.* The pregnane X receptor: a promiscuous xenobiotic receptor that has diverged during evolution. *Mol. Endocrinol.* **14,** 27–39 (2000).
17. Lemaire, G. *et al.* Discovery of a highly active ligand of human pregnane x receptor: a case study from pharmacophore modeling and virtual screening to "in vivo" biological activity. *Mol. Pharmacol.* **72,** 572–581 (2007).
18. Watkins, R. E. *et al.* The human nuclear xenobiotic receptor PXR: structural determinants of directed promiscuity. *Science* **292,** 2329–2333 (2001).
19. Zhao, W. *et al.* A new bliss independence model to analyze drug combination data. *J. Biomol. Screen.* **19,** 817–821 (2014).
20. Xue, Y. *et al.* Crystal structure of the pregnane X receptor-estradiol complex provides insights into endobiotic recognition. *Mol. Endocrinol.* **21,** 1028–1038 (2007).
21. Watkins, R. E., Davis-Searles, P. R., Lambert, M. H. & Redinbo, M. R. Coactivator binding promotes the specific interaction between ligand and the pregnane X receptor. *J. Mol. Biol.* **331,** 815–828 (2003).
22. Chrencik, J. E. *et al.* Structural disorder in the complex of human pregnane X receptor and the macrolide antibiotic rifampicin. *Mol. Endocrinol.* **19,** 1125–1134 (2005).
23. Schumacher, M. A., Miller, M. C. & Brennan, R. G. Structural mechanism of the simultaneous binding of two drugs to a multidrug-binding protein. *EMBO J.* **23,** 2923–2930 (2004).
24. Ekroos, M. & Sjogren, T. Structural basis for ligand promiscuity in cytochrome P450 3A4. *Proc. Natl Acad. Sci. USA* **103,** 13682–13687 (2006).
25. Rocchi, S. *et al.* A unique PPARgamma ligand with potent insulin-sensitizing yet weak adipogenic activity. *Mol. Cell.* **8,** 737–747 (2001).
26. Itoh, T. *et al.* Structural basis for the activation of PPARgamma by oxidized fatty acids. *Nat. Struct. Mol. Biol.* **15,** 924–931 (2008).
27. Wang, Y. *et al.* A second binding site for hydroxytamoxifen within the coactivator-binding groove of estrogen receptor beta. *Proc. Natl Acad. Sci. USA* **103,** 9908–9911 (2006).
28. Hughes, T. S. *et al.* An alternate binding site for PPARgamma ligands. *Nat. Commun.* **5,** 3571 (2014).

29. Iyer, M., Reschly, E. J. & Krasowski, M. D. Functional evolution of the pregnane X receptor. *Expert Opin. Drug Metab. Toxicol.* **2**, 381–397 (2006).

30. Ekins, S., Reschly, E. J., Hagey, L. R. & Krasowski, M. D. Evolution of pharmacologic specificity in the pregnane X receptor. *BMC Evol. Biol.* **8**, 103 (2008).

31. Nettles, K. W. *et al.* Structural plasticity in the oestrogen receptor ligand-binding domain. *EMBO Rep.* **8**, 563–568 (2007).

32. Borngraeber, S. *et al.* Ligand selectivity by seeking hydrophobicity in thyroid hormone receptor. *Proc. Natl Acad. Sci. USA* **100**, 15358–15363 (2003).

33. Suino-Powell, K. *et al.* Doubling the size of the glucocorticoid receptor ligand binding pocket by deacylcortivazol. *Mol. Cell. Biol.* **28**, 1915–1923 (2008).

34. Togashi, M. *et al.* Conformational adaptation of nuclear receptor ligand binding domains to agonists: potential for novel approaches to ligand design. *J. Steroid Biochem. Mol. Biol.* **93**, 127–137 (2005).

35. Greschik, H., Flaig, R., Renaud, J. P. & Moras, D. Structural basis for the deactivation of the estrogen-related receptor gamma by diethylstilbestrol or 4-hydroxytamoxifen and determinants of selectivity. *J. Biol. Chem.* **279**, 33639–33646 (2004).

36. Drocourt, L., Ourlin, J. C., Pascussi, J. M., Maurel, P. & Vilarem, M. J. Expression of CYP3A4, CYP2B6, and CYP2C9 is regulated by the vitamin D receptor pathway in primary human hepatocytes. *J. Biol. Chem.* **277**, 25125–25132 (2002).

37. Raynal, C. *et al.* Pregnane X Receptor (PXR) expression in colorectal cancer cells restricts irinotecan chemosensitivity through enhanced SN-38 glucuronidation. *Mol. Cancer* **9**, 46 (2010).

38. Lemaire, G., de Sousa, G. & Rahmani, R. A PXR reporter gene assay in a stable cell culture system: CYP3A4 and CYP2B6 induction by pesticides. *Biochem. Pharmacol.* **68**, 2347–2358 (2004).

39. Pichard, L. *et al.* Human hepatocyte culture. *Methods. Mol. Biol.* **320**, 283–293 (2006).

40. Ferrini, J. B., Pichard, L., Domergue, J. & Maurel, P. Long-term primary cultures of adult human hepatocytes. *Chem. Biol. Interact* **107**, 31–45 (1997).

41. Nahoum, V. *et al.* Modulators of the structural dynamics of the retinoid X receptor to reveal receptor function. *Proc. Natl Acad. Sci. USA* **104**, 17323–17328 (2007).

42. Kabsch, W. Xds. *Acta. Crystallogr. D Biol. Crystallogr.* **66**, 125–132 (2010).

43. Adams, P. D. *et al.* PHENIX: a comprehensive Python-based system for macromolecular structure solution. *Acta. Crystallogr. D Biol. Crystallogr.* **66**, 213–221 (2010).

44. Emsley, P. & Cowtan, K. Coot: model-building tools for molecular graphics. *Acta. Crystallogr. D Biol. Crystallogr.* **60**, 2126–2132 (2004).

45. Chen, V. B. *et al.* MolProbity: all-atom structure validation for macromolecular crystallography. *Acta. Crystallogr. D Biol. Crystallogr.* **66**, 12–21 (2010).

## Acknowledgements

We thank Matthew Redinbo for providing the expression plasmid used for crystallographic studies. We acknowledge the experimental assistance from the staff of the European Synchrotron Radiation Facility (ESRF, Grenoble, France) during data collection. We thank Diego Tosi for calculation of the theoretical additive activity curves and Catherine Teyssier for preparation of fluorescein-labelled SRC-1 NID. We acknowledge the financial support from the Agence Nationale de la Recherche, project TOXSYN ANR-13-CESA-0017-03 (P.B. and W.B.), and Plan Cancer Inserm, project SYNERPXR A12156FS (J.M.P., P.B. and W.B.).

## Author contributions

J.M.P., P.B. and W.B. conceived the project. V.D. and T.H. performed the crystallographic and fluorescence anisotropy studies. B.D., M.G., A.B., B.B. and P.B. performed screening and cell-based assays. S.G.C., M.D.C. and J.M.P performed CYP3A4 analysis. V.C. analysed data. R.R. provided biological tools. D.R., C.M. and V.V. performed mass spectrometry and ITC measurements. J.M.P., P.B. and W.B. wrote the manuscript. All authors contributed to editing of the manuscript.

## Additional information

**Accession codes:** The atomic coordinates and structure factors have been deposited in the Protein Data Bank under accession codes 4X1F (EE2), 4XAO (TNC) and 4X1G (EE2 + TNC).

# 14

# Non-canonical active site architecture of the radical SAM thiamin pyrimidine synthase

Michael K. Fenwick[1], Angad P. Mehta[2], Yang Zhang[1], Sameh H. Abdelwahed[2], Tadhg P. Begley[2] & Steven E. Ealick[1]

Radical S-adenosylmethionine (SAM) enzymes use a [4Fe-4S] cluster to generate a 5'-deoxyadenosyl radical. Canonical radical SAM enzymes are characterized by a β-barrel-like fold and SAM anchors to the differentiated iron of the cluster, which is located near the amino terminus and within the β-barrel, through its amino and carboxylate groups. Here we show that ThiC, the thiamin pyrimidine synthase in plants and bacteria, contains a tethered cluster-binding domain at its carboxy terminus that moves in and out of the active site during catalysis. In contrast to canonical radical SAM enzymes, we predict that SAM anchors to an additional active site metal through its amino and carboxylate groups. Superimposition of the catalytic domains of ThiC and glutamate mutase shows that these two enzymes share similar active site architectures, thus providing strong evidence for an evolutionary link between the radical SAM and adenosylcobalamin-dependent enzyme superfamilies.

[1] Department of Chemistry and Chemical Biology, Cornell University, 120 Baker Lab, Ithaca, New York 14853, USA. [2] Department of Chemistry, Texas A&M University, College Station, Texas 77843, USA. Correspondence and requests for materials should be addressed to S.E.E. (email: see3@cornell.edu).

Evidence that S-adenosylmethionine (SAM) could be the source of 5′-deoxyadenosyl radicals in enzymes was first discovered through studies of lysine 2,3-aminomutase[1,2]. Frey and colleagues[3-6] showed that radical formation occurred through homolytic reductive cleavage of a SAM C5′–S bond by using a [4Fe-4S] cluster as the source of an electron. The radical then typically abstracts a substrate hydrogen atom to initiate downstream chemistry. Shortly thereafter, Sofia et al.[7] used bioinformatics to identify over 600 enzymes that constituted the radical SAM superfamily. Today, radical SAM enzymes are known to catalyse a wide range of chemical reactions, including posttranslational modifications of proteins and nucleic acids[8], with tens of thousands of family members predicted to exist[9]; however, only a few of these have been biochemically and structurally characterized. The radical SAM superfamily is characterized by a $(\beta/\alpha)_8$-barrel or modified β-barrel fold in which the [4Fe-4S] cluster inserts near the C-terminal ends of the β-barrel strands[10]. The $CX_3CX_2C$ cluster-binding motif is usually located near the N terminus of the sequence. The three cysteine residues ligate three irons of the cluster and SAM is anchored through its amino and carboxylate groups to the fourth iron[11]. This canonical mode of SAM binding positions SAM for cleavage of the C5′–S bond and generation of the 5′-deoxyadenosyl radical and L-Met (Fig. 1a). 5′-deoxyadenosine (5′-dAdo) is often a product of the reaction, although in some cases SAM serves as a cofactor and is regenerated at the end of the reaction cycle[12].

Recently, examples of radical SAM enzymes have been identified that differ significantly from the canonical radical SAM superfamily. Variation in the cluster-binding motif, or its location in the sequence, prevents bioinformatics from identifying these as radical SAM enzymes. The radical SAM enzyme PhnJ catalyses cleavage of the C–P bond of phosphonates and has a $CX_2CX_{21}CX_5C$ motif near the C terminus of the sequence; however, no structure is available and only recently have the three cysteine residues important for cluster binding been identified[13,14]. HmdB, involved in the biosynthesis of hydrogenase cofactor, contains a $CX_5CX_2C$ cluster-binding motif near the N terminus of the sequence[15]. No structure of the enzyme

is available. QueE catalyses the synthesis of 7-carboxy-7-deazaguanine and contains a $CX_{14}CX_2C$ cluster-binding motif. A recent structure shows that QueE contains a $(\beta_6/\alpha_3)$ core and a canonical cluster-binding motif with one extended loop[16].

ThiC, the enzyme discussed here, catalyses the complex rearrangement of aminoimidazole ribonucleotide (AIR) to 4-amino-5-hydroxymethyl-2-methylpyrimidine phosphate (HMP-P) in the thiamin biosynthetic pathways of bacteria and plants (Fig. 1b)[17] and is an example of a non-canonical radical SAM enzyme[18]. The $CX_2CX_4C$ motif of ThiC varies from the canonical radical SAM motif and is located near the C terminus of the sequence. A mechanism, consistent with detailed labelling studies and the observation of formate and carbon monoxide as additional products, has been proposed[19]. In this mechanism, the 5′-deoxyadenosyl radical initiates the reaction by abstraction of a C5′ hydrogen atom of AIR. The homodimeric structure of Caulobacter crescentus ThiC (CcThiC) showed a $(\beta/\alpha)_8$-barrel fold and suggested that the cluster-binding domain, which was disordered, resides near the C terminus and inserts into the active site of an adjacent protomer through domain swapping[18]. The high-resolution structure of Arabidopsis thaliana ThiC (AtThiC)[20] confirmed the earlier results; however, neither ThiC structure contained a [4Fe-4S] cluster nor an ordered cluster-binding motif. We have now used AtThiC and CcThiC to determine a series of structures containing the [4Fe-4S] cluster and various combinations of substrates, products and analogues. These structures define the fold of the ThiC cluster-binding domain and map out the details of the active site.

## Results

**Structure of ThiC with a [4Fe-4S] cluster.** ThiC containing the [4Fe-4S] cluster was crystallized anaerobically under various conditions (Table 1). We prepared two different crystal forms of AtThiC, each containing the substrate analogue imidazole ribonucleotide (IRN) and the SAM analogue S-adenosylhomocysteine (SAH), which is also a competitive inhibitor of ThiC[21].

The ThiC structures revealed the details of the [4Fe-4S] cluster and the cluster-binding domain (Fig. 2a,b and Supplementary

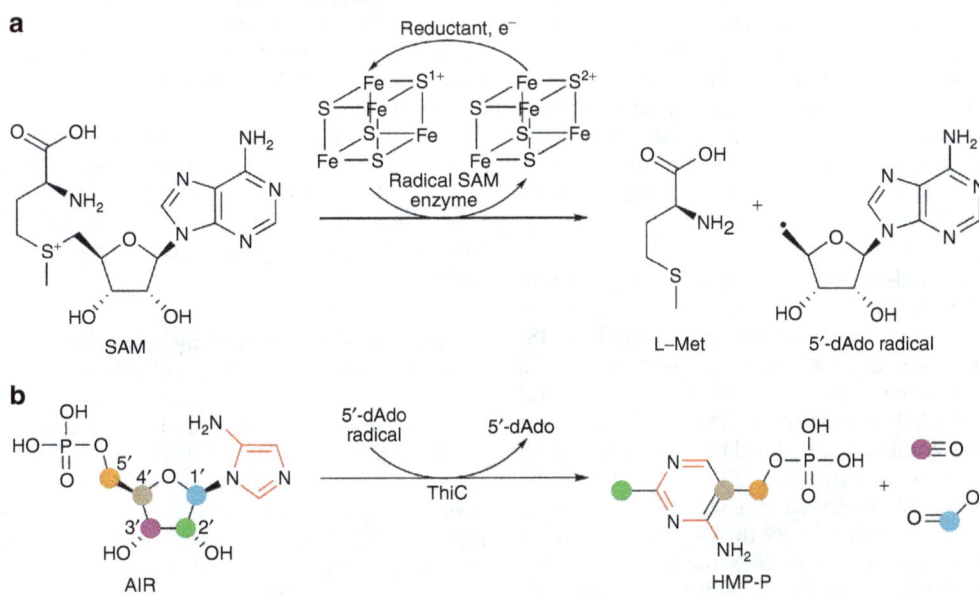

**Figure 1 | Radical SAM reactions and ThiC. (a)** Canonical radical SAM reaction scheme. Canonical radical SAM enzymes use a [4Fe-4S] cluster to convert SAM into L-Met and a 5′-deoxyadenosyl radical. The reduced [4Fe-4S] cluster delivers an electron to SAM, resulting in homolytic cleavage of the C5′-S bond. **(b)** Complex rearrangement reaction catalysed by ThiC. ThiC converts AIR to HMP-P, formate and carbon monoxide. The coloured circles indicate the origin of carbon atoms. In a recently proposed mechanism, the reaction commences when the 5′-deoxyadenosyl radical abstracts a C5′ hydrogen atom of AIR[19].

**Table 1 | Data collection and refinement statistics.**

| | AtThiC Fe4S4/Zn/ IRN/SAH | AtThiC Fe4S4/Zn/ IRN/SAH | AtThiC Fe4S4/Fe/ AIR/5′-dAdo/Met | AtThiC Fe4S4/Fe/ AIR/SAH | AtThiC Fe/IRN | CcThiC Fe4S4 |
|---|---|---|---|---|---|---|
| *Data collection* | | | | | | |
| Space group | $P3_221$ | C2 | $P3_221$ | $P3_221$ | $P3_221$ | C2 |
| *Cell dimensions* | | | | | | |
| $a, b, c$ (Å) | 107.5, 107.5, 87.6 | 175.9, 95.6, 71.4 | 106.9, 106.9, 87.6 | 107.1, 107.1, 87.7 | 107.3, 107.3, 87.8 | 116.7, 68.9, 97.6 |
| $\alpha, \beta, \gamma$ (°) | 90, 90, 120 | 90, 104.2, 90 | 90, 90, 120 | 90, 90, 120 | 90, 90, 120 | 90, 120.3, 90 |
| Resolution (Å) | 46.54–1.45 | 48.95–1.84 | 33.89–1.27 | 40.99–1.25 | 46.46–1.38 | 45.55–2.93 |
| | (1.50–1.45)* | (1.92–1.84) | (1.31–1.27) | (1.30–1.25) | (1.43–1.38) | (3.03–2.93) |
| $R_{sym}$ | 0.06 (0.43) | 0.07 (0.46) | 0.04 (0.37) | 0.05 (0.33) | 0.05 (0.32) | 0.15 (0.52) |
| $I/\sigma I$ | 13.2 (2.7) | 9.9 (1.7) | 15.5 (3.0) | 14.5 (3.2) | 16.5 (3.5) | 8.0 (2.7) |
| Completeness (%) | 97.2 (99.2) | 96.9 (97.1) | 99.9 (99.8) | 99.0 (96.5) | 98.8 (93.7) | 94.2 (97.3) |
| Redundancy | 4.8 (4.6) | 2.8 (2.5) | 5.3 (5.0) | 5.3 (4.5) | 5.4 (5.1) | 2.4 (2.4) |
| | | | | | | |
| *Refinement* | | | | | | |
| Resolution (Å) | 1.45 | 1.84 | 1.27 | 1.25 | 1.38 | 2.93 |
| No. of reflections | 100712 | 94814 | 152447 | 158556 | 118249 | 13613 |
| $R_{work}/R_{free}$ | 0.142/0.163 | 0.166/0.197 | 0.119/0.142 | 0.116/0.138 | 0.118/0.144 | 0.194/0.246 |
| *No. of atoms* | | | | | | |
| Protein | 4288 | 8087 | 4315 | 4336 | 3943 | 4527 |
| Ligand/ion | 66 | 108 | 68 | 93 | 43 | 13 |
| Water | 520 | 543 | 503 | 546 | 545 | 0 |
| *B-factors* | | | | | | |
| Protein | 12.9 | 24.6 | 15.8 | 14.0 | 13.7 | 27.0 |
| Ligand/ion | 15.0 | 24.8 | 17.2 | 15.3 | 12.5 | 28.6 |
| Water | 25.7 | 29.1 | 33.8 | 31.5 | 32.6 | — |
| *Root mean squared deviations* | | | | | | |
| Bond lengths (Å) | 0.007 | 0.008 | 0.006 | 0.006 | 0.005 | 0.002 |
| Bond angles (°) | 1.4 | 1.2 | 1.2 | 1.3 | 1.1 | 1.0 |

AIR, aminoimidazole ribonucleotide; AtThiC, *A. thaliana* ThiC; CcThiC, *C. crescentus* ThiC; IRN, imidazole ribonucleotide; SAH, *S*-adenosylhomocysteine; 5′-dAdo, 5′-deoxyadenosine.
*Highest-resolution shell is shown in parenthesis.

Fig. 1a), and the binding sites for SAH (Fig. 2c and Supplementary Fig. 1b) and IRN (Fig. 2d and Supplementary Fig. 1c). The cluster-binding domain inserts into the active site of an adjacent catalytic domain through domain swapping as previously predicted[18]. The cluster-binding loop (CSMCGPKFC) is preceded by a 12-residue tether, which connects to the adjacent catalytic domain through a three-helix bundle. The three-helix bundle is located at the dimer interface and is the C-terminal feature of the previously reported ThiC structures[18,20]. The cluster-binding loop is followed by a 10-residue α-helix; however, the final 55 residues are not visible in the electron density, even at very high-resolution. Crystals of CcThiC with bound cluster, IRN, SAH and Zn diffracted to lower resolution. The structure is consistent with the AtThiC cluster and cluster-binding domain (Supplementary Fig. 1d), but was not refined.

**SAH-binding site.** The mode of SAH binding was unexpected. In the canonical radical SAM enzymes, the SAM amino and carboxylate groups anchor to the differentiated iron in the [4Fe-4S] cluster (Fig. 2e)[22]; however, in ThiC the fourth iron binds to a chloride ion (Supplementary Fig. 2), and the amino and carboxylate groups of SAH anchor to an additional transition metal site (Fig. 2f). The conformation of SAH is also stabilized by van der Waals interactions and through a salt bridge between the carboxylate and absolutely conserved Arg386 (Fig. 2c). SAH forms two hydrogen bonds with Glu489 through its 2′- and 3′-hydroxyl groups. The adenine base is sandwiched between Leu259 and Met572, the N7 atom forms hydrogen bonds with the guanidinium of Arg343 and the amino group forms a hydrogen bond with the backbone carbonyl of Gly230.

**AIR- and IRN-binding sites.** The three crystal structures determined for AtThiC with bound IRN show that IRN adopts the same binding mode observed in the previously determined structure of CcThiC that lacked a [4Fe-4S] cluster (Fig. 2d)[18]. Its dominant interaction with ThiC occurs through its phosphate group, which forms hydrogen bonds with the guanidinium group of Arg386, the hydroxyl group of Tyr286, the hydroxyl group of Ser342, the imidazole ring of His322 and the backbone amide groups of Arg343 and Gly344. The ribose hydroxyl groups form hydrogen bonds with the side chains of Asn228 and Glu422, and the N3 atom of the imidazole ring hydrogen bonds with the side chain of Asp383. IRN does not make any direct interactions with SAH.

The high-resolution crystal structures of AtThiC with bound AIR show that AIR makes the same interactions with ThiC as IRN. In addition, the structures reveal the orientation of the imidazole ring of AIR (Fig. 3a) and show that its C5 amino group stacks against the aromatic face of Tyr449.

**5′-dAdo- and L-Met-binding sites.** The high-resolution crystal structure of AtThiC with bound [4Fe-4S] cluster, 5′-dAdo, L-Met and AIR shows that the binding configurations of the SAM cleavage products 5′-dAdo and L-Met are similar to the binding configurations of the 5′-dAdo and homocysteine portions, respectively, of SAH (Fig. 3a and Supplementary Fig. 1e,f). The adenine ring of 5′-dAdo packs between the side chains of Leu259 and Met572. The N7 atom forms hydrogen bonds with the guanidinium of Arg343 and the amino group forms a hydrogen bond with the backbone carbonyl of Gly230. The ribose O2′ and O3′ hydroxyl groups form hydrogen bonds with the side chain of Glu489. The only difference between the conformations of 5′-dAdo and the 5′-dAdo group of SAH is a change in ring pucker of the ribose group, which results in a 3-Å displacement of the C5′ atom.

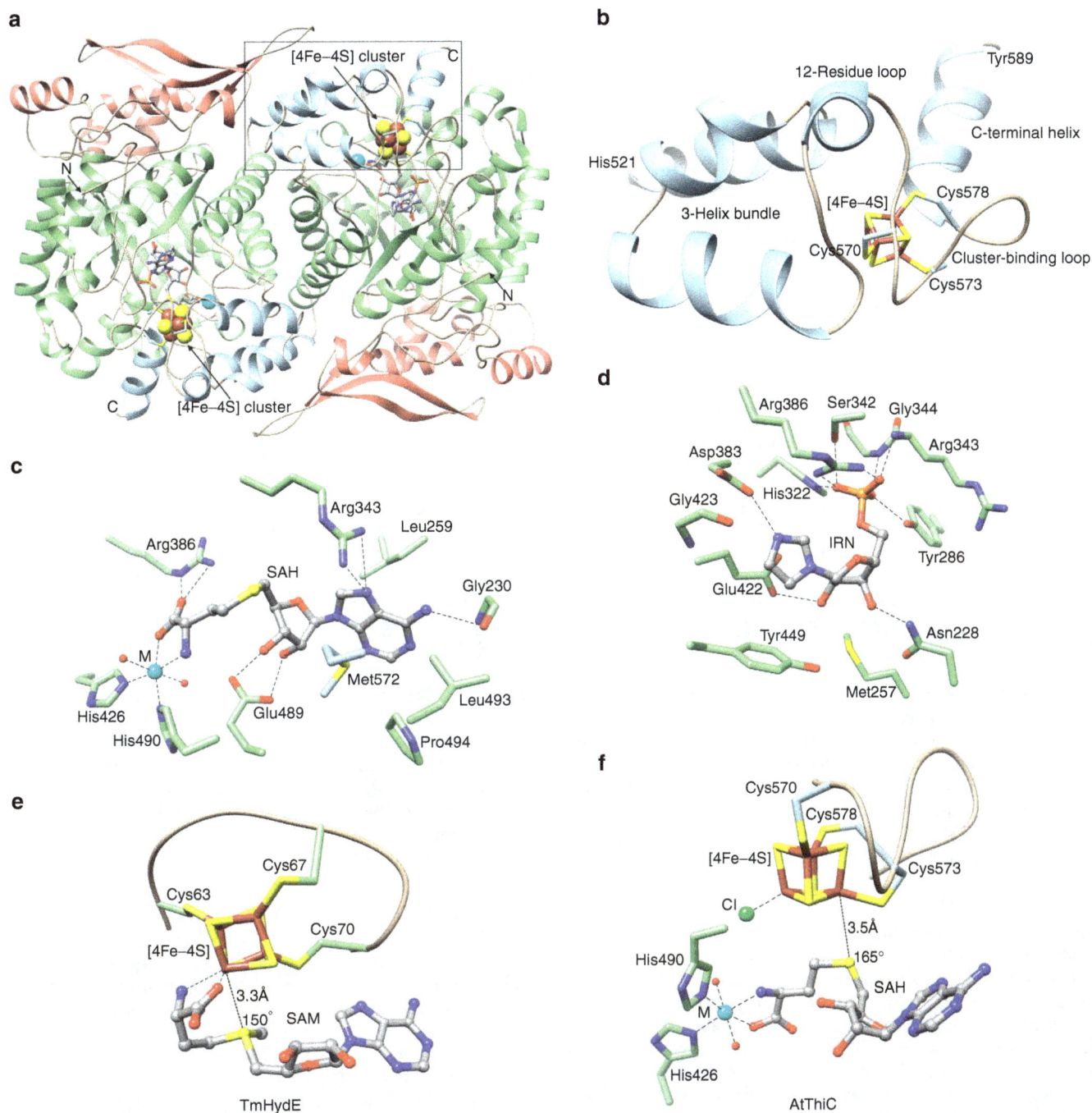

**Figure 2 | Structure of the [4Fe-4S] cluster-binding domain and active site in AtThiC.** (**a**) Structure of AtThiC with [4Fe-4S] cluster. AtThiC is a homodimer with each monomer having an N-terminal domain (salmon), a $(\beta/\alpha)_8$ core domain (green) and a C-terminal cluster-binding domain that inserts into the active site of a twofold related monomer (blue). (**b**) AtThiC cluster-binding domain. The [4Fe-4S] cluster-binding domain is anchored to the adjacent catalytic domain through a three helix bundle. The $CX_2CX_4C$ cluster-binding motif is preceded by a 12-residue loop and followed by a 10-residue α-helix. (**c**) SAH-binding site from the ThiC/SAH/IRN complex. (**d**) IRN-binding site from the ThiC/SAH/IRN complex. Hydrogen bonds for **c** and **d** are shown as dashed lines. (**e**) Mode of SAM binding in canonical radical SAM enzymes. SAM anchors to the differentiated iron of the [4Fe-4S] cluster via its amino and carboxylate groups. The C5′-S bond is positioned such that the SAM sulfonium ion is typically 3.1–3.6 Å from the differentiated iron of the [4Fe-4S] cluster with a nearly linear arrangement of Fe...S-C5′ (typically about 150°). (**f**) Mode of SAH binding in ThiC. SAH does not anchor to the [4Fe-4S] cluster, but instead anchors to a secondary metal via its amino and carboxylate groups. The differentiated iron binds to chloride in our structures. The SAH sulfur atom is 3.5 Å from iron and the Fe...S-C5′ angle is 165°.

L-Met chelates to the additional metal site through its amino and carboxylate groups similar to the way the homocysteine portion of SAH chelates to the metal site. The side chain of L-Met is also oriented similarly to the side chain of the homocysteine portion of SAH. The distance between the sulfur atom of L-Met and the Cys573-ligated Fe of the cluster is 3.0 Å.

**EXAFS of AtThiC crystals**. In addition to the amino and carboxylate groups of SAH, the metal is coordinated to two absolutely conserved histidine residues (His426 and His490 in AtThiC and His417 and His481 in CcThiC) and two water molecules (Fig. 2f). We used X-ray fluorescence spectroscopy of single crystals (Supplementary Fig. 3) to identify

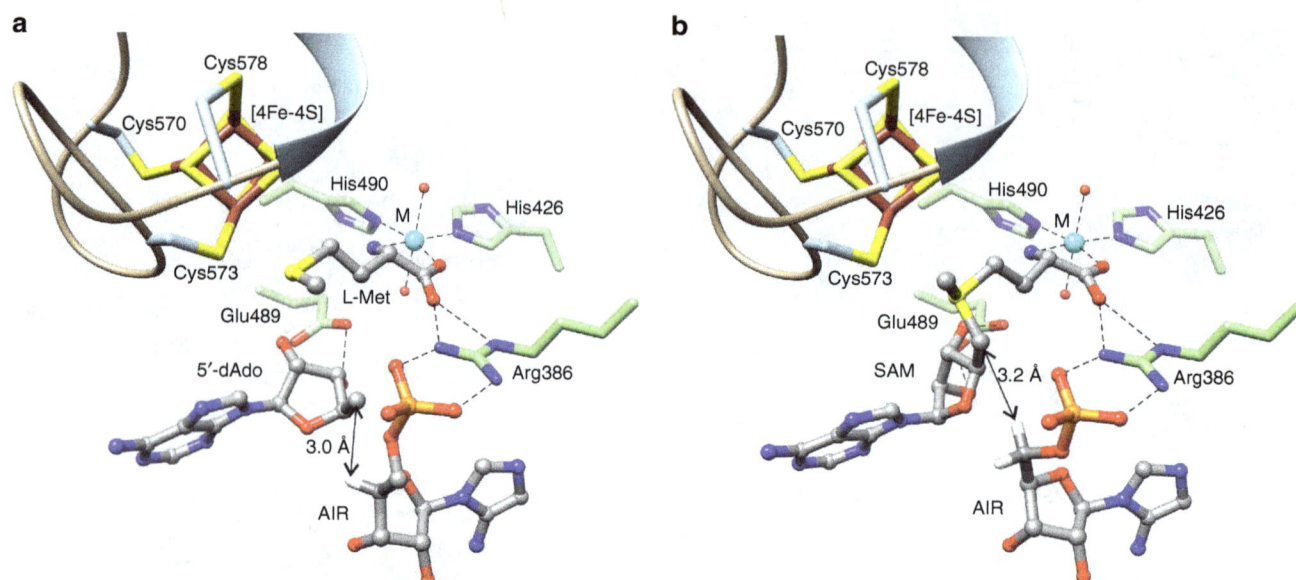

**Figure 3 | Structures of the AtThiC product complex and model of the ThiC substrate complex.** (**a**) X-ray structure of AtThiC with AIR and bound products 5'-dAdo and L-Met. The structures of 5'-dAdo and L-Met are consistent with cleavage of SAM having its carboxylate and amino groups anchored to the additional metal. After cleavage, the carboxylate and amino groups of L-Met remain anchored to the metal. (**b**) Model for the complex of AIR and SAM bound to the AtThiC active site. The model was generated by adding a methyl group to SAH, to give the *S*-enantiomer of SAM (note: the inactive *R*-enantiomer could not be generated without creating steric clashes). In this model, the distance between the C5' atom of SAM and the proS hydrogen atom of AIR is 3.2 Å. After S–C5' bond cleavage, the orientation of the 5'-deoxyadenosyl radical would be favourable for abstraction of the AIR C5'-proS hydrogen atom.

the metal at this site. For AtThiC crystals grown without addition of $ZnSO_4$, a strong fluorescence emission peak was observed for Fe and peaks were observed at trace levels for Zn and Ni (Supplementary Fig. 3a). For the crystals grown with addition of $ZnSO_4$, strong fluorescence emission peaks were observed for both Fe and Zn, and a trace-level peak was observed for Ni (Supplementary Fig. 3b). The higher level of Zn suggested a bound Zn with the probable site being the metal bound by His426 and His490.

**Multiwavelength anomalous difference Fourier analyses.** To provide further evidence for the identity of the additional metal, a multiple wavelength anomalous diffraction experiment was performed. The resulting anomalous difference Fourier maps were very clean and showed strong anomalous scattering signals (Supplementary Fig. 4). Peak heights for a metal at its absorption edge were 60–80 times the root mean square value of the map. The results of the multiwavelength anomalous diffraction experiments using the cluster iron atoms as a reference are shown in Supplementary Table 1 and indicate that in the absence of added $ZnSO_4$ the metal is iron, and that when $ZnSO_4$ is added during crystallization the iron is displaced by zinc.

**Activity studies of mutant proteins.** The prediction that the additional metal site anchors SAM through its amino and carboxylate groups was tested by mutating His417 and His481 in CcThiC individually and together. CcThiC activity was reduced 5-fold for each individual mutation and 15-fold for the double mutation (Supplementary Fig. 5).

**Activity of AtThiC under crystallization conditions.** To confirm that the crystallization buffer does not lead to an inhibited form of the enzyme and the failure to properly coordinate SAM due to the presence of chloride, the activity of AtThiC was tested under starting and intermediate crystallization conditions. AtThiC mostly precipitated under final equilibrium crystallization

conditions. Both the production of HMP-P and 5'-dAdo were monitored and the results are shown in Supplementary Table 2. Under initial crystallization conditions, AtThiC showed 75% HMP-P formation and 88% 5'-dAdo formation relative to control conditions, and 80% HMP-P formation and 100% 5'-dAdo formation halfway between initial and equilibrium conditions, suggesting that the conformation observed in the crystal form represents a *bona fide* catalytic intermediate.

**Exploration of alternate SAM conformations.** A methyl group was added to SAH in the *S*-configuration to create a model of SAM. Although the SAM methionyl group can be pointed generally in the direction of the differentiated iron, it is not long enough to reach the iron and the sulfonium ion is nearly 5 Å from the differentiated iron. In addition, clashes between SAM and the protein atoms occur when the SAM methionyl group is pointed towards the differentiated iron. To more completely explore the alternate SAM conformations, we carried out a Monte Carlo/energy minimization simulation in which the torsion angles of the methionyl portion of SAM were allowed to vary. The lowest energy structure corresponded to the crystal structure. None of the additional low-energy structures corresponded to a conformation in which both the SAM amino and carboxylate groups were anchored to the differentiated iron.

**Discussion**

The high-resolution structure of AtThiC co-crystallized with SAH or with 5'-dAdo and L-Met showed a consistent arrangement with atoms of SAH closely aligned with the corresponding atoms in 5'-dAdo and L-Met (Figs 2f and 3a). The L-Met amino and carboxylate groups chelate to the additional metal ion and the adenosyl moiety superimposes closely with that of SAH. We also determined the structure of AtThiC co-crystallized with AIR and SAH (Supplementary Fig. 1g,h), and showed that AIR and IRN superimpose closely. The ensemble of five high-resolution AtThiC structures containing various combinations of AIR,

IRN, SAH, L-Met and 5′-dAdo show an unexpected, but self-consistent, pattern of substrate, product and analogue-binding geometry.

Although the identity of the additional metal *in vivo* is not known, it was first observed in our original structure of CcThiC and determined to be zinc by EXAFS[18]. The metal was modelled as cobalt in the original structure of AtThiC, because CoCl$_2$ was present in the crystallization conditions; however, the identity of the metal was not confirmed[20]. Formation of the [4Fe-4S] cluster in ThiC requires elevated concentrations of iron in the culture medium, and for crystals prepared this way the metal was assigned as iron. Zinc when added to AtThiC at a 1:1 molar ratio during crystallization largely displaces iron at the additional metal site.

A comparison of the [4Fe-4S] cluster with bound SAM from canonical radical SAM enzymes and AtThiC led to another unexpected observation (Fig. 2e,f). In the canonical radical SAM enzymes, the conformation of SAM places the SAM sulfur atom near the differentiated iron with an approximately linear Fe...S-C5′ arrangement as required for cleavage of the C5′–S bond and formation of the 5′-deoxyadenosyl radical[11,23]. Using deposited high- or very-high-resolution structures of radical SAM structures with bound SAM (PDB IDs 1OLT, 2A5H, 2FB2, 3CB8, 3IIZ, 3RFA, 3T7V, 4FHD, 4K39 and 4M7T), the Fe...S distance ranges from 3.1 to 3.6 Å and the Fe...S-C angle ranges from 139° to 161°. In ThiC, the fourth iron of the cluster bonds to chloride. In the high-resolution crystal structures of AtThiC containing SAH, the average Fe...S distance from the iron covalently bound to Cys573 is 3.5 Å (range 3.3–3.7 Å) with an average Fe...S-C angle of 165° (range 162°–169°). In the structures containing L-Met, the Fe...S distance is ∼3.0 Å from this iron.

The structure of ThiC with bound AIR and SAH, and the demonstration of consistent binding geometries among our collection of structures allowed us to readily generate a model of the ThiC/AIR/SAM complex by adding a methyl group in the S-configuration to the sulfur atom of SAH (Fig. 3b). Adding the methyl group in the R-configuration resulted in steric clashes. In the model of ThiC/AIR/SAM, all SAH atoms were kept fixed and no ThiC conformational changes were required. The model shows that SAM is poised for C5′–S bond cleavage after which the C5′ position of the 5′-deoxyadenosyl radical would be ideally positioned to abstract the C5′-proS hydrogen atom of AIR (C5′...proS H distance 3.2 Å and C5′...proR H distance 4.2 Å).

The unexpected active site geometry of SAM, inferred from SAH or 5′-dAdo and L-Met, raises the possibility that our structures represent inhibitory states of the enzyme or crystallization artefacts, and that in the active form of ThiC SAM would be anchored by the differentiated iron just as it is in the canonical radical SAM superfamily members. To test this possibility, we superimposed a representative [4Fe-4S] cluster and its attached SAM molecule (taken from HydE[22]) onto the ThiC cluster in the three possible orientations (corresponding to 120° rotations about the cluster diagonal containing the differentiated iron) that would allow the SAM amino and carboxylate groups to anchor to the differentiated iron of the ThiC cluster. In each case, the superimposition showed severe clashes, with SAM interpenetrating protein side chains and no possibility of alleviating the clashes (Supplementary Fig. 6). Furthermore, none of the possibilities would place the 5′-deoxyadenosyl radical close enough to the substrate AIR for abstraction of the 5′-hydrogen atom as required by the proposed mechanism[19]. Attempts to reposition the cluster through conformational changes in the tethered region also failed to achieve the canonical radical SAM active site architecture. Finally, studies using conformational searching showed that the lowest energy

conformation for SAM bound to the AtThiC active site corresponds to the crystal structure with SAH. All of these observations support a new, non-canonical active site architecture for AtThiC.

The original structure of CcThiC showed that its most closely related structural homologues were adenosylcobalamin (AdoCbl)-dependent enzymes[18]. This observation is consistent with the proposal based on functional and structural considerations that radical SAM enzymes and AdoCbl-dependent enzymes might have an evolutionary relationship[24,25]. Drennan and colleagues[10] proposed a model for 5′-dAdo binding in ThiC, using a superimposition of the structures of ThiC lacking the [4Fe-4S] cluster[18] and *Clostridium cochlearium* glutamate mutase complexed with AdoCbl and L-glutamate[26]. ThiC and glutamate mutase have structurally homologous catalytic domains containing the substrate (AIR or L-glutamate) binding site. ThiC contains a tethered [4Fe-4S] cluster-binding domain, while glutamate mutase contains a separate chain for binding the AdoCbl cofactor (Fig. 4a). The model predicted that a conserved glutamate side chain (Glu489 in AtThiC) would hydrogen bond to the 2′- and 3′-hydroxyl groups of 5′-dAdo. The model also proposed that a hydrophobic residue (Leu259 in AtThiC) would pack against the adenine ring. Superimposition of the ThiC structures from our crystallographic studies with glutamate mutase (PDB ID 1I9C) confirmed these predictions (Fig. 4b,c). In addition to Glu489 and Leu259, Gly230, Leu493 and Pro494 are also structurally and functionally conserved in glutamate mutase. Furthermore, the ThiC AIR-binding site and the glutamate mutase L-Glu-binding site overlap spatially, and the cobalamin cobalt is near the iron in ThiC that is predicted to interact with the sulfonium ion of SAM (Fig. 4d). This comparison not only supports the non-canonical active site architecture of ThiC but also provides strong evidence for an evolutionary link between the radical SAM and AdoCbl-dependent enzyme superfamilies.

One structure of CcThiC and one structure of AtThiC stand apart and provide insight into conformational changes occurring in ThiC. In the holo CcThiC structure, the entire C-terminal cluster-binding domain (Supplementary Fig. 7a) and [4Fe-4S] cluster (Supplementary Fig. 7b) are clearly defined; however, the cluster-binding domain extends away from the active site and the cluster itself is ∼25 Å from its active site location (Fig. 5a,b). The structure of AtThiC co-crystallized with only IRN shows well-defined density for the catalytic domain and clear density for IRN (Supplementary Fig. 7c); however, the density beyond Glu557, which includes the cluster-binding domain, is absent, indicating that the cluster-binding domain is disordered. In contrast, every structure containing SAH or L-Met and 5′-dAdo shows a well-ordered cluster-binding domain. In addition to the disordered cluster-binding domain, the AtThiC/IRN structure shows conformational changes for residues 227–247, 257–277 and 487–500 (Fig. 6). These protein regions are involved in contacts between the catalytic domain and the cluster-binding domain, and contain residues Leu259, Glu489 and Leu493, which make contacts with SAM in the ThiC/AIR/SAM model.

The ensemble of X-ray structures and the ThiC/substrate model suggest the substrate AIR binds first (similar to IRN in Fig. 2d) and is oriented through hydrogen bonds between the phosphate and the side chains of Tyr286, His322, Ser342 and Arg386, and backbone amides of Arg343 and Gly344. The phosphate group is located at the N-terminal end of an α-helix (343–356) and further stabilized by the helix dipole. Hydrogen bonds also form between the AIR ribosyl hydroxyl groups and Asn228 and Glu422. The imidazole N3 atom is within hydrogen-bonding distance from the Asp383 side chain. The large number of interactions with phosphate and the similarity of the phosphate

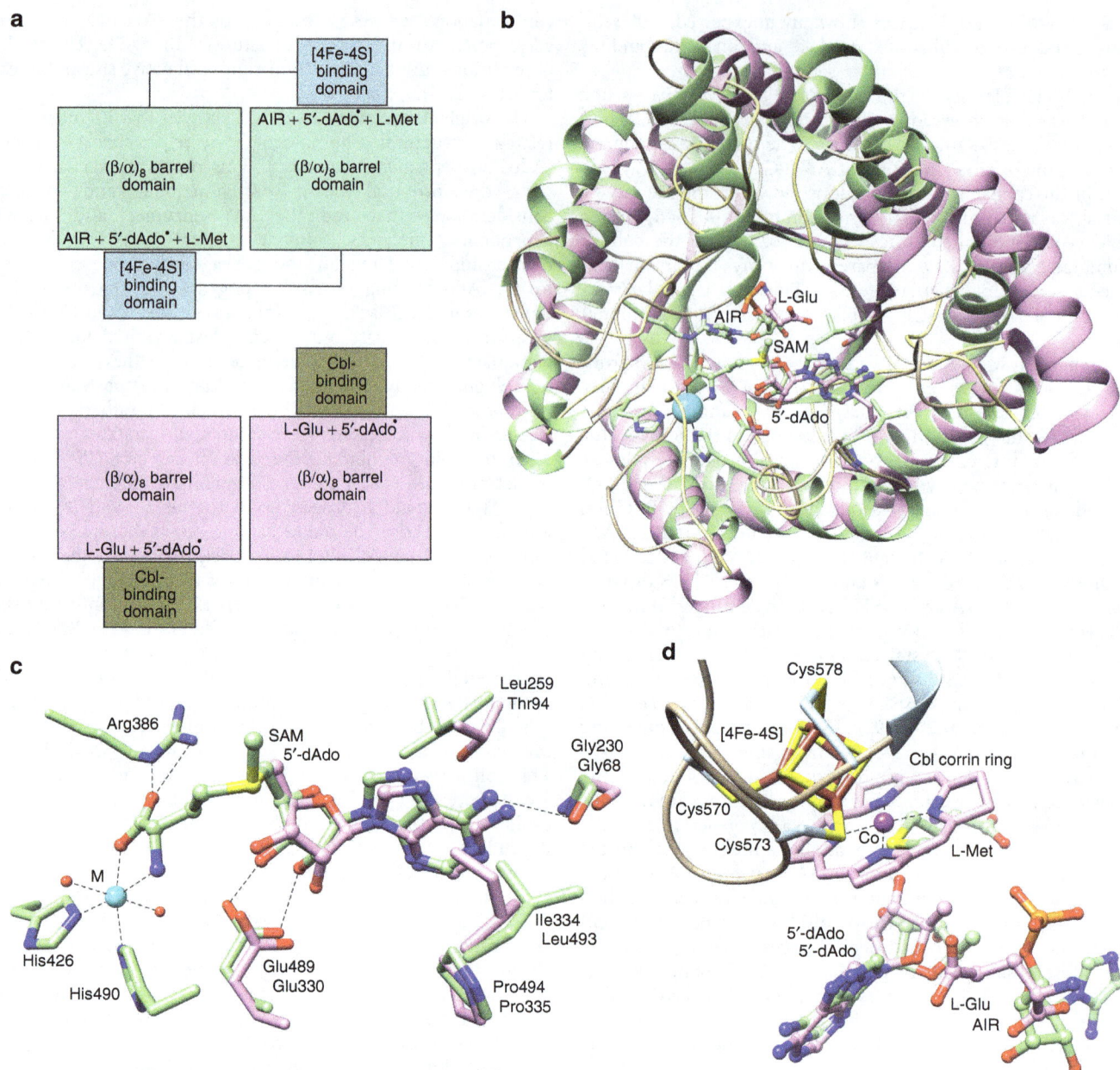

**Figure 4 | Comparison of ThiC- and AdoCbl-dependent glutamate mutase. (a)** Organization of ThiC and glutamate mutase dimers. The enzymes have closely related $(\beta/\alpha)_8$ folds and dimer interfaces. ThiC contains a tethered cluster-binding domain that inserts into the twofold related catalytic domain through domain swapping. The AdoCbl-binding domain is a separate chain in glutamate mutase. **(b)** Superimposition of the catalytic domains of AtThiC and glutamate mutase showing the locations of the substrates (AIR or L-glutamate) and SAM (modelled from SAH), and the 5′-deoxyadenosyl moiety of AdoCbl. The ThiC cluster-binding domain and the glutamate mutase AdoCbl-binding domain, which have very different folds, were omitted for clarity. **(c)** Close-up view of the conserved binding site for the 5′-deoxyadenosyl moieties of ThiC and glutamate mutase. Functionally conserved residues that contact the 5′-deoxyadenosyl moieties are shown. **(d)** Location of the metal centres, 5′-dAdo and substrates after superimposition of the ThiC and glutamate mutase catalytic domains. The Cys573-ligated iron that binds the sulfonium atom of SAM and the cobalt that binds the 5′-deoxyadenosyl moiety of glutamate mutase are ∼2 Å apart, consistent with the insertion of the SAM sulfonium ion.

site to that of HMP-P in our previously reported structure[18] suggest that it acts as an anchor throughout the catalytic cycle. SAM binds second (similar to SAH in Fig. 2c) with its amino and carboxylate groups anchored to the additional metal ion, its ribosyl hydroxyl groups hydrogen bonded to Glu489 and the adenine ring sandwiched between Leu259 and Met572, which comes from the cluster-binding loop. This geometry positions the SAM 5′-carbon atom 3.2 Å from the AIR 5′-proS hydrogen atom; however, no other direct contacts with AIR are observed. This is in contrast to canonical radical SAM enzymes, where interactions between SAM and the substrate are often extensive and orient

reactants for initiation of radical chemistry. SAM binding induces three conformational changes, which facilitate docking of the cluster-binding domain to the active site (Fig. 6). Docking of the catalytic and cluster-binding domains positions the iron bonded to Cys573 near the SAM sulfonium ion. Subsequent C5′–S bond cleavage results in 5′-deoxyadenosyl radical formation and abstraction of the AIR 5′-proS hydrogen atom. This order of binding, where AIR binds first and SAM stabilizes interactions with the cluster-binding domain, controls initiation of radical chemistry and most probably prevents uncoupled turnover of SAM *in vivo*.

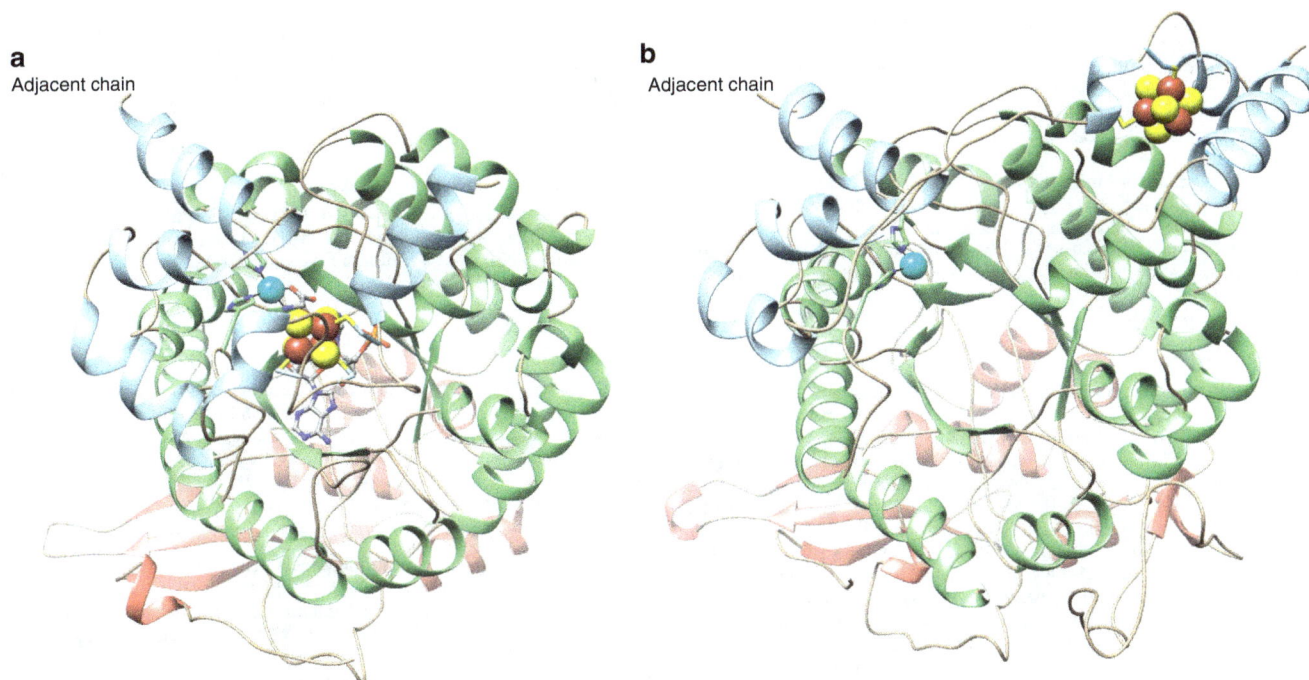

**Figure 5 | Movement of the ThiC cluster and cluster-binding domain.** (**a**) AtThiC protomer structure with bound [4Fe-4S] cluster, SAH, AIR and Fe. (**b**) Holo CcThiC protomer structure. In the absence of substrates, the [4Fe-4S] cluster is located at a remote site in the twofold related protomer, 25 Å away from the active site location of the cluster. The metal (cyan) and conserved histidine residues are included for reference.

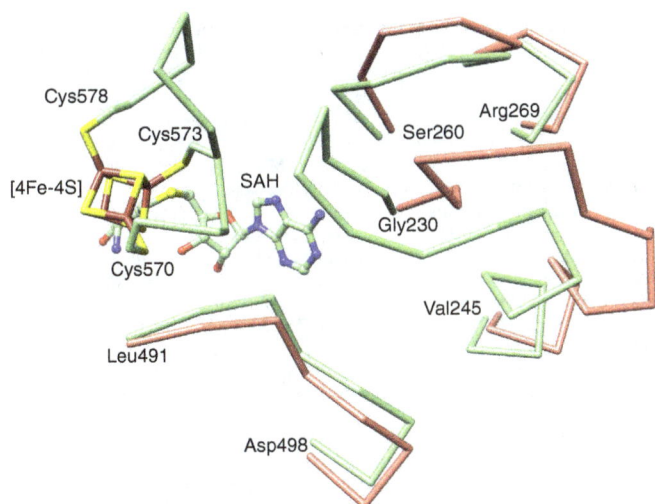

**Figure 6 | Comparison of the structure of AtThiC with bound IRN to the structure of AtThiC with bound SAH and AIR.** The presence of SAH (or 5'-dAdo and L-Met) results in conformational changes for three loops near the SAM-binding site. These loops interact with the cluster-binding domain and promote docking to the active site. The largest change occurs in loop 230–245; this loop is disordered in the holo CcThiC structure. Smaller changes occur in loops 257–277 and 487–500. Ser236, which has clear electron density and forms a strong hydrogen bond with Glu241, is an outlier in the Ramachandran plot when SAH is bound, but not for the AtThiC/IRN structure.

In summary, the ensemble of ThiC structures reported here demonstrates several unprecedented features of ThiC *vis a vis* the radical SAM enzyme superfamily. (i) The binding mode of SAM (inferred from structures with SAH or 5'-dAdo and L-Met) is unique. (ii) The amino and carboxylate groups of SAH are not anchored by the [4Fe-4S] cluster, but instead are anchored by an additional metal ion. (iii) The sulfur atom of SAH is in close proximity to an iron of the [4Fe-4S] cluster, but unlike canonical radical SAM enzymes it is not the differentiated iron but is the iron ligated to Cys573. (iv) Comparison of the structures of ThiC with a [4Fe-4S] cluster and glutamate mutase with AdoCbl provides further evidence for an evolutionary relationship between the two superfamilies. (v) The positioning of the [4Fe-4S] cluster in the active site, along with two structures lacking SAH or L-Met and 5'-dAdo, suggests that the tethered ThiC cluster-binding domain moves into the ThiC active site to initiate catalysis and out of the active site to release products.

## Methods

**Anaerobic production and crystallization of AtThiC.** Previous studies reported that overexpression of full-length AtThiC containing 644 residues results in insoluble protein, whereas overexpression of ΔN71-AtThiC with residues 1–71 removed results in soluble protein with high yield[20,27]. In the present study, *ΔN71-AtThiC* was cloned into a modified pET-28 plasmid that encodes the following hexahistidine-tagged protein product: NH₂- MGSDKIHHHHHHSSG ENLYFQGHMK₇₂...K₆₄₄-COOH.

This plasmid along with plasmid pSuf containing the *Escherichia coli suf* operon were transformed into *E. coli* NiCo21(DE3) cells (New England Biolabs)[28–30]. Large-scale cultures were grown in baffled shaker flasks containing 1.5 l of Luria–Bertani (LB) medium, 0.051 g chloramphenicol and 0.06 g kanamycin. The flasks were shaken at 180 r.p.m. and 37 °C until an optical density at 600 nm (OD₆₀₀) of 0.55–0.6 was reached and then moved to a 4 °C cold room. About 2.5 h later, 200 mg Fe(NH₄)₂(SO₄)₂·6H₂O, 200 mg L-Cys and 36 mg isopropyl β-D-1-thiogalactopyranoside (IPTG) were added per flask and the flasks were shaken at 90 r.p.m. and 15 °C for about 20 h. The flasks were then moved to the cold room for about 3 h before being pelleted by centrifugation.

Cell pellets were brought into a PVC anaerobic chamber (Coy Laboratory Products) and re-suspended in lysis buffer (100 mM Tris, pH 7.6, 5 mM dithiothreitol (DTT), 0.4 mg ml⁻¹ lysozyme, 1.9 kU benzonase). The cells were further lysed via sonication and the lysate was moved out of the glove box for centrifugation at 60,000*g* and 4 °C for 20 min. The spun lysate was returned to the glove box and the supernatant was subjected to immobilized nickel affinity chromatography employing wash (100 mM Tris, 300 mM NaCl, 20 mM imidazole and 2 mM DTT, pH 7.6) and elution (100 mM Tris, 300 mM NaCl, 250 mM imidazole, pH 7.7) buffers. Eluted protein was buffer exchanged into 25 mM Tris and 150 mM NaCl, using a Bio-Rad Econo-Pac 10DG desalting column, and incubated for 5–14 h in the glove box with tobacco-etch virus protease for cleavage of the hexa-His tag. Subtractive immobilized nickel affinity chromatography was then performed, after which time the isolated protein was buffer exchanged into

5 mM HEPES buffer and 22 mM NaCl, pH 7.0. The protein used for crystallization was previously shown to be active[20].

Crystals were grown in the anaerobic chamber at room temperature using hanging-drop vapour diffusion, with drops containing a 1:1 ratio of protein to reservoir solution. The concentration of AtThiC was $\sim 17$ mg ml$^{-1}$. The final concentrations of ligands co-crystallized with AtThiC were as follows: AIR, 3 mM; IRN, 3 mM; L-Met, 7 mM; 5′-dAdo, 4 mM; SAH (in dimethylsulfoxide), 5 mM; SAM (Sigma or Cayman), 5–7 mM; and ZnSO$_4$, 0.23 mM. For the AtThiC crystals having space group P3$_2$21, typical reservoir solutions were 100 mM Na acetate pH 5.3–5.6 and 1–9% (v/v) 1,4 butanediol. For the AtThiC crystals having space group C2, the reservoir solution was 100 mM imidazole pH 6.6, 200 mM NaCl and 20% (w/v) polyethylene glycol (PEG) 8000. Crystals were harvested and cryocooled in liquid N$_2$ using 32% (v/v) 1,4-butanediol and 38% (w/v) PEG4000 as cryoprotectants, respectively.

**Anaerobic production and crystallization of CcThiC.** Full-length *CcThiC* was cloned into the aforementioned modified pET28 vector to give the expressed product: NH$_2$- MGSDKIHHHHHHSSGENLYFQGHM$_1$...E$_{612}$-COOH.

This plasmid and plasmid pDB1282 (ref. 31) were transformed into *E. coli* B834 (DE3) cells. Large-scale cultures were grown in 1.8 l of minimal medium (21 g M9 salts, 7.5 g glucose, 0.45 g MgSO$_4$ and 28 mg CaCl$_2 \cdot$ 2H$_2$O) per flask supplemented with 0.21 g L-Met, 0.19 g ampicillin and 0.075 g kanamycin. The flasks were shaken at 180 r.p.m. and 37 °C until an OD$_{600}$ of 0.20–0.25 was reached, at which point $\sim 140$ mg Fe(NH$_4$)$_2$(SO$_4$)$_2 \cdot$ 6H$_2$O, 140 mg L-Cys and 5 g L-(+)-arabinose were added per flask. The flasks were then shaken at 50 r.p.m. and 37 °C until the OD$_{600}$ reached about 0.65 and then cultures were moved to a cold room set at 4 °C. After 4–5 h, 6.4 mg of IPTG were added per flask and the flasks were shaken at 50 r.p.m. and 15 °C for about 20 h. Afterwards, the cultures were moved to the cold room for about four hours prior to centrifugation.

The procedures for anaerobic preparation of CcThiC including cleavage of the hexa-His tag and crystallization were identical to those followed for AtThiC. To crystallize holo CcThiC with the [4Fe-4S] cluster remotely docked at His347, protein samples of $\sim 22$ mg ml$^{-1}$ CcThiC were combined with typical reservoir solutions containing 140 mM HEPES pH 9.2 and 7.5% (w/v) PEG8000; the cryoprotectant was 40% (w/v) PEG4000. For co-crystallization of CcThiC, IRN, SAH and Zn, protein samples containing $\sim 12$ mg ml$^{-1}$ CcThiC were supplemented with 5 mM IRN (prepared previously[18] in NaH$_2$PO$_4$ and NaCl), 20 mM HEPES pH 7.8, 5 mM SAH, 1.7% (v/v) dimethylsulfoxide, 0.18 mM ZnSO$_4$ and 0.9 mM HEPES, pH 6.9. Typical reservoir solutions consisted of 0.79–0.81 M sodium citrate and between 100 and 200 mM imidazole pH 7.7–8.6; 1.6 M sodium citrate was used for cryoprotection.

**X-ray diffraction data for CcThiC and AtThiC.** All X-ray diffraction data were measured at 100 K at beamline 24-ID-C of the Advanced Photon Source, Argonne National Laboratory, using a PILATUS 6M-F detector and a wavelength of $\lambda = 0.979$ Å. Diffraction images were recorded in shutterless mode at a rate of 1 degree per second. Typical rotation ranges were 80°–120° for space group P3$_2$21 and 120°–155° for space group C2. Data were integrated and scaled by using HKL2000 (ref. 32).

**Structure determination and refinement of CcThiC and AtThiC.** The structures of CcThiC and AtThiC complexed with various ligands were determined by either difference Fourier synthesis or molecular replacement using the published structures PDB ID 4N7Q (AtThiC)[20] or PDB ID 3EPN (CcThiC)[18] as the starting models. The structures were refined iteratively using programmes Phenix[33] and COOT[34], and validated using MolProbity[35]. The final refinement statistics are summarized in Table 1. Supplementary Fig. 1, showing the quality of the electron density maps and confirming the assignments of the ligands, was prepared using PyMOL[36].

During the refinement of the very high-resolution structures, the occupancy for the [4Fe-4S] cluster was found to be $\sim 70\%$. Electron density near the differentiated iron of the cluster was much too large to be explained by a water molecule (average peak height of 20 times the root mean square value of the map). Assigning this peak as chloride resulted in an average B-factor after refinement of $\sim 18$ Å$^2$ (average protein atom B-factor around 15 Å$^2$). The occupancies of the cluster atoms and chloride ion were constrained to the same value during refinement. The 1.25-Å resolution structure of AtThiC with SAH and AIR showed two conformations for SAH. Although atoms of the adenine ring and the aminocarboxypropyl group were the same, the ribosyl group pucker was different (C2′-*exo*, C3′-*endo* for one conformation and C1′-*exo* for the other) causing changes mainly in the positions of O2′, C3′, O3′, C4′, C5′, Cγ and S. In both conformations the SAH amino and carboxyl groups chelate to the additional iron.

A Ramachandran analysis shows that for all structures all torsion angles are in the allowed or favoured regions, except for Ser236, which has clear electron density. The Ser236 side chain forms a strong hydrogen bond with the side chain of Glu241 (2.5 Å) and is located at the end of a loop that interacts with the cluster-binding domain.

**X-ray fluorescence spectra for AtThiC crystals.** Crystal structures of AtThiC with a [4Fe-4S] cluster showed a metal bound by His426 and His490. This metal presumably originated from iron in the culture medium, which included LB medium and Fe(NH$_4$)$_2$(SO$_4$)$_2 \cdot$ 6H$_2$O during protein induction, or from ZnSO$_4$ added during crystallization. To determine the identity of the metal, X-ray fluorescence spectra were recorded at beamline 24-ID-C of the Advanced Photon Source (Argonne National Lab), which is equipped with an Amptek XR-100SDD silicon drift detector. Spectra were obtained for AtThiC crystals and associated cryo-solutions for the energy range 500–16,000 eV. Spectra were recorded for co-crystals of AtThiC, AIR and SAM, and co-crystals of AtThiC, AIR, SAM and Zn. For the latter, ZnSO$_4$ was added to a final concentration of 0.22 mM before crystallization (corresponding to a 1:1 molar ratio with AtThiC); Zn was not included in the cryoprotectant. Spectra were also recorded for co-crystals of AtThiC, IRN and SAH, with or without added ZnSO$_4$.

**Multiwavelength anomalous difference Fourier analyses.** Four data sets were collected for each of two crystals, one without and one with added ZnSO$_4$. Wavelengths were chosen at (7,132 eV) and below (7,100 eV) the Fe absorption edge and at (9,673 eV) and below (9,650 eV) the Zn absorption edge. Anomalous diffraction data sets were processed using HKL2000 (ref. 32) and analysed using COOT to determine relative anomalous difference Fourier peak heights[34].

**DNA primers for cloning and mutagenesis.** The primers used for cloning AtThiC are 5′-aaaacatatgaaacacaccattgatcctt-3′ (forward) and 5′-aaaactcgagt-tatttctgagcagctttt-3′ (reverse); the recognition sequences for NdeI and XhoI are underlined. The primers used for cloning CcThiC are 5′-gggtagcatatgaatatcca-gagcaccatcaagg-3′ (forward) and 5′-cccctaggatcctcactcggtcttcagatagatc-3′ (reverse); the recognition sequences for NdeI and BamHI are underlined. The primers used for making the mutant CcThiC H417A are 5′-gatgatcgaagggccgggcGCGgtggccat gcacaagatcaagg-3′ (forward) and 5′-ccttgatcttgtgcatggccacCGCgcccggcccttcgatc atc-3′ (reverse); capitalized nucleotides mark the site of mutation. The primers used for making CcThiC H481A are 5′-ctacgtcacgcccaaggagGCGctgggcctgccggaccgcg-3′ (forward) and 5′-cgccggtccggcaggcccagCGCctccttgggcgtgacgtag-3′ (reverse).

**Production of CcThiC mutant proteins.** The mutant plasmids were co-transformed with pDB1282 in *E. coli* B834 (DE3). ThiC variants were co-expressed in the presence of a pDB1282 plasmid encoding the *Isc* operon for *in vivo* biosynthesis of [4Fe-4S] in *E. coli* B834 (DE3). An overnight 15-ml culture was grown in LB media in the presence of kanamycin (40 mg l$^{-1}$) and ampicillin (100 mg l$^{-1}$). This was then added to 1.8 l minimal media (containing M9 minimal salts 1 ×, 27 ml 50% (w/v) glucose, 7 ml 1 M MgSO$_4$, 200 µl 1 M CaCl$_2$, 200 mg L-Met, 72 µg kanamycin and 180 µg ampicillin). The cultures were incubated at 37 °C with shaking (180 r.p.m.) until the OD$_{600}$ reached 0.2 to 0.25. Next, 5 g of L-(+)-arabinose, 120 mg of Fe(NH$_4$)$_2$(SO$_4$)$_2 \cdot$ 6H$_2$O and 120 mg of L-Cys were added. The cultures were incubated at 37 °C with shaking (50 r.p.m.) until the OD$_{600}$ reached 0.65 to 0.7. The cultures were then incubated at 4 °C without shaking for 3–4 h. This was followed by induction of the culture with 15 µM IPTG and incubation at 15 °C with shaking (50 r.p.m.) for 18–20 h. The cultures were then incubated at 4 °C for 3 h without shaking. The cells were then harvested and stored in liquid nitrogen overnight before enzyme purification. For enzyme purification, the cell pellets were thawed at room temperature in an anaerobic chamber and suspended in lysis buffer (100 mM Tris-HCl, pH 7.5) in the presence of 2 mM DTT, lysozyme (0.2 mg ml$^{-1}$) and benzonase (100 units). This mixture was then cooled in an ice-bath for 2 h. The suspension of cells was sonicated and centrifuged, to give the cell-free extract. The enzyme was purified using standard Ni-NTA chromatography. The column was first incubated with the lysis buffer. The cell-free extract was passed over the column, which was then washed with 8–9 column volumes of wash buffer (100 mM Tris-HCl, 300 mM NaCl, 20 mM imidazole, 2 mM DTT, pH 7.5). The enzyme was eluted using 100 mM Tris-HCl, 300 mM NaCl, 250 mM imidazole and 2 mM DTT, pH 7.5. The purified enzyme was buffer exchanged into 100 mM potassium phosphate, 30% (v/v) glycerol, 2 mM DTT, pH 7.5, using an Econo-Pac 10DG desalting column (Bio-Rad) and the purified enzyme was stored in liquid nitrogen. The proteins as isolated bind to a [4Fe-4S] cluster.

**Activity studies of mutant proteins.** The CcThiC mutant (200 µM) was incubated with 10 mM dithionite, 5 mM AIR and 7 mM SAM at room temperature for 90 min. The protein was filtered out using 10-kDa cutoff filters and the small molecule pool was analysed by HPLC. An Agilent 1260 HPLC was used for detection of the products by ultraviolet absorption at 254 nm. An SPLC-18 column (3.0 × 150 mm, 3 µm, Supelcosil, 25 cm × 10 mm, 5 µm) was used for the HPLC analysis. These experiments were performed twice for each CcThiC mutant.

Chromatography conditions were as follows: (A) water, (B) 100 mM potassium phosphate, pH 6.6, (C) methanol. Flow rate: 2 ml min$^{-1}$. The following gradients were used: 0 min—100% B; 7 min—10% A:90% B; 12 min—25% A:60% B: 15%C; 17 min—25% A: 10% B: 65% C; 19 min—100% C; 29 min—100%A.

**Activity of AtThiC under crystallization conditions.** Activity assays were performed at the crystallization conditions to test whether the crystallization solution, in particular chloride, inhibits the AtThiC-catalysed conversion of AIR and SAM to 5′-dAdo and HMP-P. Co-crystals of AtThiC and SAH were grown via vapour diffusion from drops formed from a 1:1 ratio of protein solution (230 μM AtThiC, 5 mM SAH, 5 mM HEPES, 22 mM NaCl, pH 7.0) and reservoir solution (1–9% (v/v) 1,4-butanediol and 100 mM Na acetate, pH 5.3–5.6). Drop equilibration causes protein precipitation and crystallization, and precipitation occurs more readily at the lower end of the pH range. To reduce the amount of precipitation for the activity assays, reduced protein concentrations (100 μM) were required and assays were performed for initial and midway-equilibrated drop solutions at pH 5.6.

Activity was measured for the following three solutions: (i) the initial crystallization drop solution (2.5 mM HEPES, 11 mM NaCl, 1% (v/v) 1,4-butanediol and 50 mM Na acetate, pH 5.6). This corresponds to a 1:1 mixture of the protein and reservoir solutions. The chloride concentration is 17.3 mM, which derives from 11 mM NaCl plus the amount of HCl required to titrate 50 mM Na acetate to pH 5.6 (6.3 mM). (ii) A midway-equilibrated drop solution (3.75 mM HEPES, 16.5 mM NaCl, 1.5% (v/v) 1,4-butanediol and 75 mM Na acetate, pH 5.6). The chloride concentration is 26 mM. (iii) The standard solution used for activity assays (100 mM $K_2HPO_4$ pH 7.5 and 30% (v/v) glycerol).

Activity assays were performed within a Coy anaerobic chamber. Frozen aliquots of AtThiC in 100 mM $K_2HPO_4$ pH 7.5 and 30% (v/v) glycerol were thawed and desalted into each of the above solutions, using BioRad spin desalting columns. AIR (5 mM), SAM (7 mM) and dithionite (20 mM) were then added and reactions were incubated at room temperature for 90 min. The reactions were then quenched by heating followed by centrifugation and the supernatant was passed through a 10-kDa cutoff filter. This was analysed by HPLC for 5′-dAdo and HMP-P formation. Amounts of 5′-dAdo and HMP-P were normalized to protein concentrations. These experiments were performed once for each condition.

**Exploration of alternate SAM conformations.** A conformational search was performed to explore the possibility of alternate conformations of SAM in which SAM is bound to the differentiated iron of the [4Fe-4S] cluster through both its amino and carboxylate groups. The calculations were carried out using the programme Macromodel (Schrödinger, Inc.) with the OPLS 2005 force field[37]. SAM was modelled by adding a methyl group to SAH in the crystal structure of AtThiC with bound AIR, SAH and Fe. The Monte Carlo Mulitple Minima programme was used to randomly vary the torsion angles of the methionyl portion of SAM, while the ribose and adenine portions were held fixed. Full residues and ligands having at least one atom within 5 Å of the methionyl moiety of SAM were included in the model. A total of 100,000 conformations were sampled and an energy range of 50 kcal mol$^{-1}$ was used to save an extensive number of candidate structures above the lowest energy structure, which corresponded to the crystal structure. Conformations were energy minimized with the energy gradient threshold set to 1 kJ mol$^{-1}$ Å$^{-1}$.

**Figure preparation.** Figure 1 was made using ChemDRAW Pro 14.0 (Perkin-Elmer, Inc.). Figs 2–6 and Supplementary Fig. 6 were made using Chimera[38]. Supplementary Figures showing electron density were made using PyMOL[36].

# References

1. Baraniak, J., Moss, M. L. & Frey, P. A. Lysine 2,3-aminomutase. Support for a mechanism of hydrogen transfer involving S-adenosylmethionine. *J. Biol. Chem.* **264,** 1357–1360 (1989).
2. Moss, M. & Frey, P. A. The role of S-adenosylmethionine in the lysine 2,3-aminomutase reaction. *J. Biol. Chem.* **262,** 14859–14862 (1987).
3. Ballinger, M. D., Reed, G. H. & Frey, P. A. An organic radical in the lysine 2,3-aminomutase reaction. *Biochemistry* **31,** 949–953 (1992).
4. Broderick, J. B. *et al.* Pyruvate formate-lyase-activating enzyme: strictly anaerobic isolation yields active enzyme containing a [3Fe-4S]( + ) cluster. *Biochem. Biophys. Res. Commun.* **269,** 451–456 (2000).
5. Frey, P. A. & Booker, S. J. Radical mechanisms of S-adenosylmethionine-dependent enzymes. *Adv. Protein Chem.* **58,** 1–45 (2001).
6. Lieder, K. W. *et al.* S-adenosylmethionine-dependent reduction of lysine 2,3-aminomutase and observation of the catalytically functional iron-sulfur centers by electron paramagnetic resonance. *Biochemistry* **37,** 2578–2585 (1998).
7. Sofia, H. J., Chen, G., Hetzler, B. G., Reyes-Spindola, J. F. & Miller, N. E. Radical SAM, a novel protein superfamily linking unresolved steps in familiar biosynthetic pathways with radical mechanisms: functional characterization using new analysis and information visualization methods. *Nucleic Acids Res.* **29,** 1097–1106 (2001).
8. Atta, M. *et al.* S-adenosylmethionine-dependent radical-based modification of biological macromolecules. *Curr. Opin. Struct. Biol.* **20,** 684–692 (2010).
9. Shisler, K. A. & Broderick, J. B. Emerging themes in radical SAM chemistry. *Curr. Opin. Struct. Biol.* **22,** 701–710 (2012).
10. Dowling, D. P., Vey, J. L., Croft, A. K. & Drennan, C. L. Structural diversity in the AdoMet radical enzyme superfamily. *Biochim. Biophys. Acta* **1824,** 1178–1195 (2012).
11. Walsby, C. J., Ortillo, D., Broderick, W. E., Broderick, J. B. & Hoffman, B. M. An anchoring role for FeS clusters: chelation of the amino acid moiety of S-adenosylmethionine to the unique iron site of the [4Fe-4S] cluster of pyruvate formate-lyase activating enzyme. *J. Am. Chem. Soc.* **124,** 11270–11271 (2002).
12. Frey, P. A., Hegeman, A. D. & Ruzicka, F. J. The radical SAM superfamily. *Crit. Rev. Biochem. Mol. Biol.* **43,** 63–88 (2008).
13. Kamat, S. S., Williams, H. J. & Raushel, F. M. Intermediates in the transformation of phosphonates to phosphate by bacteria. *Nature* **480,** 570–573 (2011).
14. Kamat, S. S., Williams, H. J., Dangott, L. J., Chakrabarti, M. & Raushel, F. M. The catalytic mechanism for aerobic formation of methane by bacteria. *Nature* **497,** 132–136 (2013).
15. McGlynn, S. E. *et al.* Identification and characterization of a novel member of the radical AdoMet enzyme superfamily and implications for the biosynthesis of the Hmd hydrogenase active site cofactor. *J. Bacteriol.* **192,** 595–598 (2010).
16. Dowling, D. P. *et al.* Radical SAM enzyme QueE defines a new minimal core fold and metal-dependent mechanism. *Nat. Chem. Biol.* **10,** 106–112 (2014).
17. Jurgenson, C. T., Begley, T. P. & Ealick, S. E. The structural and biochemical foundations of thiamin biosynthesis. *Annu. Rev. Biochem.* **78,** 569–603 (2009).
18. Chatterjee, A. *et al.* Reconstitution of ThiC in thiamine pyrimidine biosynthesis expands the radical SAM superfamily. *Nat. Chem. Biol.* **4,** 758–765 (2008).
19. Chatterjee, A., Hazra, A. B., Abdelwahed, S., Hilmey, D. G. & Begley, T. P. A 'radical dance' in thiamin biosynthesis: mechanistic analysis of the bacterial hydroxymethylpyrimidine phosphate synthase. *Angew. Chem. Int. Ed. Engl.* **49,** 8653–8656 (2010).
20. Coquille, S. *et al.* High-resolution crystal structure of the eukaryotic HMP-P synthase (THIC) from *Arabidopsis thaliana. J. Struct. Biol.* **184,** 438–444 (2013).
21. Palmer, L. D. & Downs, D. M. The thiamine biosynthetic enzyme ThiC catalyzes multiple turnovers and is inhibited by S-adenosylmethionine (AdoMet) metabolites. *J. Biol. Chem.* **288,** 30693–30699 (2013).
22. Nicolet, Y. *et al.* X-ray structure of the [FeFe]-hydrogenase maturase HydE from *Thermotoga maritima. J. Biol. Chem.* **283,** 18861–18872 (2008).
23. Kampmeier, J. A. Regioselectivity in the homolytic cleavage of S-adenosylmethionine. *Biochemistry* **49,** 10770–10772 (2010).
24. Berkovitch, F., Nicolet, Y., Wan, J. T., Jarrett, J. T. & Drennan, C. L. Crystal structure of biotin synthase, an S-adenosylmethionine-dependent radical enzyme. *Science* **303,** 76–79 (2004).
25. Frey, P. A. Radical mechanisms of enzymatic catalysis. *Annu. Rev. Biochem.* **70,** 121–148 (2001).
26. Gruber, K., Reitzer, R. & Kratky, C. Radical shuttling in a protein: ribose pseudorotation controls alkyl-radical transfer in the coenzyme B(12) dependent enzyme glutamate mutase. *Angew. Chem. Int. Ed. Engl.* **40,** 3377–3380 (2001).
27. Raschke, M. *et al.* Vitamin B1 biosynthesis in plants requires the essential iron sulfur cluster protein, THIC. *Proc. Natl Acad. Sci. USA* **104,** 19637–19642 (2007).
28. Hanzelmann, P. *et al.* Characterization of MOCS1A, an oxygen-sensitive iron-sulfur protein involved in human molybdenum cofactor biosynthesis. *J. Biol. Chem.* **279,** 34721–34732 (2004).
29. Mehta, A. P. *et al.* Catalysis of a new ribose carbon-insertion reaction by the molybdenum cofactor biosynthetic enzyme MoaA. *Biochemistry* **52,** 1134–1136 (2013).
30. Takahashi, Y. & Tokumoto, U. A third bacterial system for the assembly of iron-sulfur clusters with homologs in archaea and plastids. *J. Biol. Chem.* **277,** 28380–28383 (2002).
31. Zheng, L., Cash, V. L., Flint, D. H. & Dean, D. R. Assembly of iron-sulfur clusters. Identification of an iscSUA-hscBA-fdx gene cluster from *Azotobacter vinelandii. J. Biol. Chem.* **273,** 13264–13272 (1998).
32. Otwinowski, Z. & Minor, W. Processing of X-ray diffraction data collected in oscillation mode. *Methods Enzymol.* **276,** 307–326 (1997).
33. Adams, P. D. *et al.* The Phenix software for automated determination of macromolecular structures. *Methods* **55,** 94–106 (2011).
34. Emsley, P., Lohkamp, B., Scott, W. G. & Cowtan, K. Features and development of Coot. *Acta Crystallogr. D Biol. Crystallogr.* **66,** 486–501 (2010).
35. Chen, V. B. *et al.* MolProbity: all-atom structure validation for macromolecular crystallography. *Acta Crystallogr. D Biol. Crystallogr.* **66,** 12–21 (2010).
36. *The PyMOL Molecular Graphics System* (DeLano Scientific, San Carlos, CA, 2002).
37. Banks, J. L. *et al.* Integrated modeling program, applied chemical theory (IMPACT). *J. Comput. Chem.* **26,** 1752–1780 (2005).
38. Pettersen, E. F. *et al.* UCSF Chimera--a visualization system for exploratory research and analysis. *J. Comput. Chem.* **25,** 1605–1612 (2004).

# Acknowledgements

We thank C. Kinsland for cloning CcThiC, H. Xu for cloning AtThiC, and I. Kourinov, F. Murphy, J. Schuermann and K. Rajashankar for assistance with the EXAFS experiments. We thank L. Kinsland for assistance in the preparation of this manuscript. This work was supported by the National Institutes of Health (DK67081 and DK44083) and by the Robert A. Welch Foundation (A-0034). This work is based on research conducted at the Advanced Photon Source on the Northeastern Collaborative Access Team beamlines, which are supported by grant from the National Institute of General Medical Sciences (GM103403) of the National Institutes of Health. The use of the Advanced Photon Source, an Office of Science User Facility operated for the U.S. Department of Energy (DOE) Office of Science by Argonne National Laboratory, was supported by the U.S. DOE under contract number DE-AC02-06CH11357.

## Author contributions

M.K.F. crystallized CcThiC and AtThiC, collected X-ray diffraction data and solved the initial structures. A.P.M. worked out conditions for preparing CcThiC and AtThiC with the [4Fe-4S] cluster, and carried out initial purifications, enzyme assays and mutant characterization. Y.Z. refined the CcThiC and AtThiC structures. S.H.A. synthesized AIR and IRN. T.P.B. directed the biochemical studies and assisted in writing the manuscript. S.E.E. directed the structural studies and wrote the manuscript, with contributions from all authors.

## Additional information

**Accession codes**: Atomic coordinates and structure factors for the reported crystal structures have been deposited with the Protein Data Bank (http://www.pdb.org/) under accession codes 4S25 (AtThiC/[4Fe-4S]/IRN/SAH/zinc trigonal crystal form), 4S26 (AtThiC/[4Fe-4S]/IRN/SAH/zinc monoclinic crystal form), 4S27 (AtThiC/[4Fe-4S]/Fe/ 5′-dAdo/Met/AIR), 4S28 (AtThiC/[4Fe-4S]/Fe/AIR/SAH), 4S29 (AtThiC/Fe/IRN) and 4S2A (CcThiC/[4Fe-4S]).

**Competing financial interests:** The authors declare no competing financial interests.

# 15

# *In vivo* covalent cross-linking of photon-converted rare-earth nanostructures for tumour localization and theranostics

Xiangzhao Ai[1], Chris Jun Hui Ho[2], Junxin Aw[1], Amalina Binte Ebrahim Attia[2], Jing Mu[1], Yu Wang[3], Xiaoyong Wang[4], Yong Wang[5], Xiaogang Liu[3,6], Huabing Chen[5], Mingyuan Gao[5], Xiaoyuan Chen[4], Edwin K.L. Yeow[1], Gang Liu[4], Malini Olivo[2] & Bengang Xing[1,6]

The development of precision nanomedicines to direct nanostructure-based reagents into tumour-targeted areas remains a critical challenge in clinics. Chemical reaction-mediated localization in response to tumour environmental perturbations offers promising opportunities for rational design of effective nano-theranostics. Here, we present a unique microenvironment-sensitive strategy for localization of peptide-premodified upconversion nanocrystals (UCNs) within tumour areas. Upon tumour-specific cathepsin protease reactions, the cleavage of peptides induces covalent cross-linking between the exposed cysteine and 2-cyanobenzothiazole on neighbouring particles, thus triggering the accumulation of UCNs into tumour site. Such enzyme-triggered cross-linking of UCNs leads to enhanced upconversion emission upon 808 nm laser irradiation, and in turn amplifies the singlet oxygen generation from the photosensitizers attached on UCNs. Importantly, this design enables remarkable tumour inhibition through either intratumoral UCNs injection or intravenous injection of nanoparticles modified with the targeting ligand. Our strategy may provide a multimodality solution for effective molecular sensing and site-specific tumour treatment.

[1] Division of Chemistry and Biological Chemistry, School of Physical and Mathematical Sciences, Nanyang Technological University, 637371 Singapore, Singapore. [2] Singapore Bioimaging Consortium, Agency for Science Technology and Research (A*STAR), 138667 Singapore, Singapore. [3] Department of Chemistry, National University of Singapore, 117543 Singapore, Singapore. [4] State Key Laboratory of Molecular Vaccinology and Molecular Diagnostics, Center for Molecular Imaging and Translational Medicine, School of Public Health, Xiamen University, 361102 Xiamen, China. [5] School of Radiation Medicine and Protection, Soochow University, 215123 Suzhou, China. [6] Institute of Materials Research and Engineering, A*STAR (Agency for Science, Technology and Research), 117602 Singapore, Singapore. Correspondence and requests for materials should be addressed to G.L. (email: gangliu.cmitm@xmu.edu.cn) or to M.O. (email: malini_olivo@sbic.a-star.edu.sg) or to B.X. (email: Bengang@ntu.edu.sg).

Currently, therapeutic and diagnostic techniques based on supramolecular assemblies and functional nanomaterials have been extensively recognized as promising nano-medicine platforms for the battle against many urgent health concerns including cancer, cardiovascular and neurodegenerative diseases as well as other life-threatening illnesses[1-3]. The remarkable biomedical application of nanomaterials could be mainly attributed to their unique photo-physical properties, high surface area and multivalent binding ability[4,5]. Despite the leap forward in the continuous breakthroughs in biomedical research, critical challenge still remains in designing targeted nanoplatforms that are capable of selectively localizing at the specific diseases—in particular—tumour sites for early-stage diagnosis and effective treatment[6-8]. One emerging strategy to achieve high targeting selectivity is to conjugate the nanomaterials with affinity ligands including small organic moieties or bioactive molecules that can bind to receptors in the tumour cells[9-12]. However, varying expression levels of the receptors, complex and dynamic physiological cell environments may potentially pose the issue of nonspecific recognition for this ligand-mediated tumour affinity. Therefore, more specific targeting approaches are demanded that do not solely rely on receptors to differentiate tumour and normal cells[11,12]. Indeed, some bioorthogonal reactions provide feasibility to locate functional nanostructures into tumour cells mostly through their electrostatic or covalent binding to biomolecules in living system[13-18]. Nevertheless, the effective bioorthogonal functionalities that can selectively respond to the dynamic processes of native environment are still ongoing challenges for *in vivo* applications[18-20]. Hence, different approaches that enable sensitive recognition of dynamic tumour microenvironment, and more importantly, can further trigger the tumour-specific localization of theranostic nanomaterials *in vivo* are highly desirable, and extensive studies still need to be further investigated.

Recently, rare-earth doped upconversion nanocrystals (UCNs) have been widely demonstrated for use in biomedical applications. In general, UCN particles offer deep tissue penetration capability for enhanced bioimaging and better tumour treatment arising from their unique non-linear photon upconverting process upon light irradiation at near-infrared (NIR) window[21-29]. As with the majority of nanomaterials for theranostic tumour studies, the effective targeting of upconversion materials mainly relies on receptor-mediated interactions, and the specific cellular localization of UCN nanostructures at the tumour site upon the sensitive response to microenvironment stimulation have not been fully solved[30-34]. Moreover, despite the great potential of UCNs in meeting biomedical demands *in vitro* and *in vivo*, most of the conventionally used UCNs employ long-wavelength laser irradiation at 980 nm. This often leads to an inevitable overheating effect because of significant water absorption[35,36]. To this end, the development of new strategies to selectively direct UCNs into targeted tumour areas, while notably minimizing undesired side effects is highly essential. Unfortunately, such relevant studies are still limited so far.

In this work, a tumour microenvironment-sensitive UCNs platform has been designed and prepared to perform *in vivo* covalent localization of particles at the tumour site. Different from the process involving nonspecific tumour targeting, such unique platform can respond to tumour-specific enzyme and undergo cross-linking reaction, which thus enables the selective tumour accumulation. More significantly, compared with the particles that cannot undergo cross-linking reaction, the enzyme-triggered covalent cross-linking of UCNs possess an enhanced light upconverting emission when illuminated at 808 nm. Such enhancement can effectively amplify the production of reactive singlet oxygen (for example, $^1O_2$) from the photosensitizers

loaded on the particle surface, which therefore represents promising nanomedicine for improved photodynamic tumour treatment (PDT) as well as non-invasive fluorescence and photoacoustic imaging *in vitro* and *in vivo*.

## Results

### Enzyme-triggered covalent cross-linking of UCN *in vitro*.

Figure 1 illustrates the rational design of our enzyme-responsive cross-linking of rare-earth UCNs (CRUN) for tumour localization and targeted therapeutics. Generally, the core-shell UCNs were synthesized according to an established thermal decomposition method[26]. In order to minimize undesired overheating effect typically associated with 980 nm laser irradiation, we constructed lanthanide platform doped with $Nd^{3+}$ ions for improved pumping efficiency at 808 nm. These $Nd^{3+}$-doped nanocrystals were further coated with polyacrylic acid (PAA) and polyethylenimine (PEI) to improve biocompatibility and to facilitate subsequent surface modification (Supplementary Fig. 1). Transmission electron microscopy and spectroscopic analysis revealed a cubic morphology of the particles with a narrow size distribution of $40 \pm 2$ nm. Both core-shell UCNs and polymer-coated UCNs exhibited similar upconverted luminescences with visible emission at 545 and 655 nm upon excitation at 808 nm (Supplementary Fig. 2).

To ensure the theranostic efficacy, an effective photosensitizer, chlorin-e6 (Ce6), with favourable optical properties was chosen and coupled to the PEI/PAA@UCNs. In order to achieve selective tumour localization, an enzyme-responsive peptide, Ac-FKC(StBu)AC(SH)-CBT (Supplementary Fig. 3) containing a side-protected cysteine and 2-cyanobenzothiazole (CBT) was designed for specific reaction with cathepsin B (CtsB), one important lysosomal cysteine protease overexpressed in various malignant tumours to process intracellular protein degradation and regulate cancer pathology[37]. The selected sequence was further connected to Ce6-PEI/PAA@UCNs through the reaction of free thiol in terminal cysteine and short maleimide oligo (ethyl glycol; dPEG$_2$) linker on the particle surface. The successful formation of peptide and Ce6-modified UCNs (CRUN) conjugate was confirmed by spectroscopic characterization (Supplementary Fig. 4). The optimal amount of peptide on UCNs was finally determined to be $11.4 \, nM \, mg^{-1}$, and there was 2.1% Ce6 found on the UCN surface (Supplementary Fig. 5). In principle, cellular uptake of the peptide and Ce6-conjugated UCNs may first reduce the disulfide bond of protected cysteine due to the reductive cellular environment. The subsequent lysosomal enzyme processing will cleave the peptide and expose a free 1,2-aminothiol group in cysteine, which easily undergoes condensation reaction with –CN moiety in CBT structure[13,18], and thus triggers the localization of cross-linked UCNs at the tumour region (Supplementary Fig. 6). During the reaction, great care needs to be taken to avoid nonspecific cross-linking by providing optimal spacing in the enzyme-responsive peptide and the short PEG linker to control the distance between the peptide and particle surface.

First, we examined the covalent cross-linking reactions of CRUN upon enzyme treatment in buffer through particle morphology and spectroscopic analysis. After incubation of CRUN ($1.5 \, mg \, ml^{-1}$) with CtsB (55 nM), transmission electron microscopy and dynamic light scattering analysis indicated significant particle aggregation in solutions (Fig. 2a,b and Supplementary Fig. 7), while CRUN alone or the enzyme-treated CRUN together with CtsB inhibitor (antipain), remained unchanged over time. Furthermore, the absorption spectrum showed a new peak at 489 nm during the enzyme reaction, indicating the formation of firefly luciferin structure in solutions[13]. These results clearly suggest that the particles

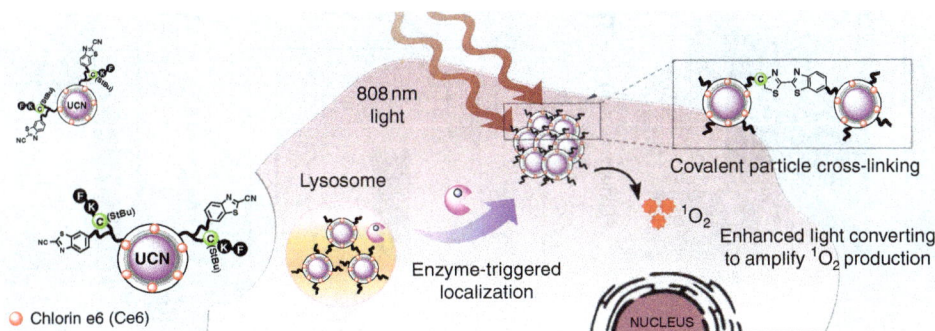

**Figure 1 | Illustration of the microenvironment-sensitive strategy for covalent cross-linking of peptide-premodified UCNs in tumour areas.**
Upon tumour-specific cathepsin B (CtsB) enzyme reactions, the peptide cleavage on the particle surface induces covalent cross-linking between the exposed free cysteine and 2-cyanobenzothiazole on neighbouring particles, which triggers the accumulation of UCNs into the tumour site. Such enzyme-triggered covalent cross-linking of UCNs leads to an enhanced upconversion emission after 808 nm laser irradiation, and in turn amplifies the singlet oxygen generation from the photosensitizers (for example, Ce6) attached on UCNs for enhanced PDT treatment in vitro and in vivo.

aggregation is mainly due to the specific CtsB enzyme-triggered cross-linking reaction (Supplementary Fig. 8). Compared with CRUN without CtsB treatment, the covalent cross-linking of CRUN could lead to an enhanced upconversion luminescence. There was up to 2.2-fold luminescence emission intensity enhancement observed at 655 nm after 4 h reaction (Fig. 2c, d and Supplementary Fig. 9). Similar enzyme-triggered CRUN cross-linking was also displayed by the luminescence colour change in the absence and presence of CtsB inhibitor (Supplementary Fig. 10), further confirming that it is the specific enzyme reaction that induces the enhanced luminescence from CRUN particles.

Considering deep tissue penetration capacity and great advantages of the photoacoustic imaging technique that integrates both functions of ultrasound and optical imaging, we assessed the difference in photoacoustic signal during the specific enzyme cross-linking reactions. Contrary to the increased luminescence observed, there was an obvious photoacoustic phantom signal decrease ($\sim 40\%$ in intensity) detected in solution after 4 h reaction (Fig. 2d). Generally, laser illuminations on enzyme-responsive UCN conjugates would excite the internal energy levels of lanthanide dopants. The excited states can relax to the ground state through both radiative and nonradiative processes. The enhanced radiative emission relaxation may correspondingly lead to a deactivation of the nonradiative relaxation in photoacoustic, therefore resulting in the decreased photoacoustic signals as observed in the experiments[38–41]. These results demonstrate that such inverse trend observed in photoacoustic may also be attributed to the specific enzyme-triggered cross-linking reaction.

Furthermore, we also investigated the underlying reason of the enzyme-responsive luminescence enhancement of CRUN by monitoring the decays of the luminescence lifetime. As shown in Fig. 2e, the presence of Ce6 on CRUN particles would lead to a sharp decay (from 500 to 205 µs) in lifetime at 655 nm, indicating the strong absorption of Ce6 at around 660 nm (ref. 35). Moreover, upon enzyme treatment, there was no significant lifetime change of lanthanide dopants in CRUN (Fig. 2f and Supplementary Fig. 11). In addition, we also examined the resonance light scattering properties of CRUN before and after CstB reactions (Supplementary Fig. 12). There was obvious scattering increment observed upon 808 nm laser excitation, mostly due to the covalent cross-linking of UCN conjugates triggered by enzyme reaction. The increase in scattering efficiency enhanced the amount of light absorption among the cross-linked particles, which could be the possible factor contributing to the increased upconversion luminescence[42–43].

**Enzyme-triggered cross-linking of UCN in living cells.** The intracellular uptake and enzyme-triggered cross-linking of UCNs in living cells were further investigated by fluorescence and photoacoustic imaging techniques. Typically, CtsB-over-expressing HT-29 tumour cell and CtsB-deficient NIH/3T3 cells were chosen as models to incubate with enzyme-responsive CRUN ($50 \mu g\,ml^{-1}$) at 37 °C for 4 h. The different expression levels of CtsB were verified by western blotting and quantitative human CtsB ELISA kit in HT-29 ($40.1 \pm 4.4\,ng\,mg^{-1}$ protein) and NIH/3T3 ($2.2 \pm 1.4\,ng\,mg^{-1}$ protein) cells, respectively (Supplementary Fig. 13)[44]. As shown in Fig. 2g, the obvious yellow fluorescence contributed from light-upconversion emissions at 545 and 655 nm was clearly observed in the HT-29 cells, indicating the effective cellular internalization of CRUN. Moreover, compared with the cells incubated with CtsB inhibitor or control NCRUN conjugates (the control particles with surface modified by non-cross-linking Ac-FKC(StBu)AC sequence), HT-29 cells treated with enzyme-responsive CRUN exhibited an increased luminescence. In addition, there was no obvious fluorescence enhancement detected in CtsB-deficient NIH/3T3 cells after incubation (Supplementary Figs 14,15). These results clearly suggested that the overexpressed CtsB enzyme in HT-29 cell could induce covalent cross-linking of CRUN and such cross-linking leads to the desired particle localization in the tumour microenvironment.

Similar cellular studies based on photoacoustic imaging were also carried out to monitor CRUN localization in living cells. As shown in Fig. 2h, the enzyme-responsive CRUN, when incubated with HT-29 cells, exhibited less photoacoustic signal ($\sim 56\%$) than that incubated with NIH/3T3 cells (CtsB-deficient). Moreover, processing of CRUN in HT-29 cells upon blocking the CtsB activity with inhibitors prevented the decrease in photoacoustic signal, corresponding to the specific enzyme reactivity. Moreover, incubation of CRUN with NIH/3T3 cells would not induce obvious photoacoustic signal change, further confirming that the enzyme processing can cause the localization of CRUN particles within HT-29 cells and the enhanced upconverting emissions resulted in the photoacoustic signal decrease. This was in agreement with the observation tested in buffer.

We also evaluated the capability of CRUN to generate cytotoxic singlet oxygen ($^1O_2$) through the standard singlet oxygen sensor green[45]. Basically, 808 nm laser irradiation of CRUN conjugates resulted in the converted emission 655 nm, corresponding to the absorption of photosensitizer Ce6 labelled on UCNs surface. Compared with CRUN alone or enzyme-treated CRUN with inhibitor, the CtsB enzyme-induced cross-linking can cause

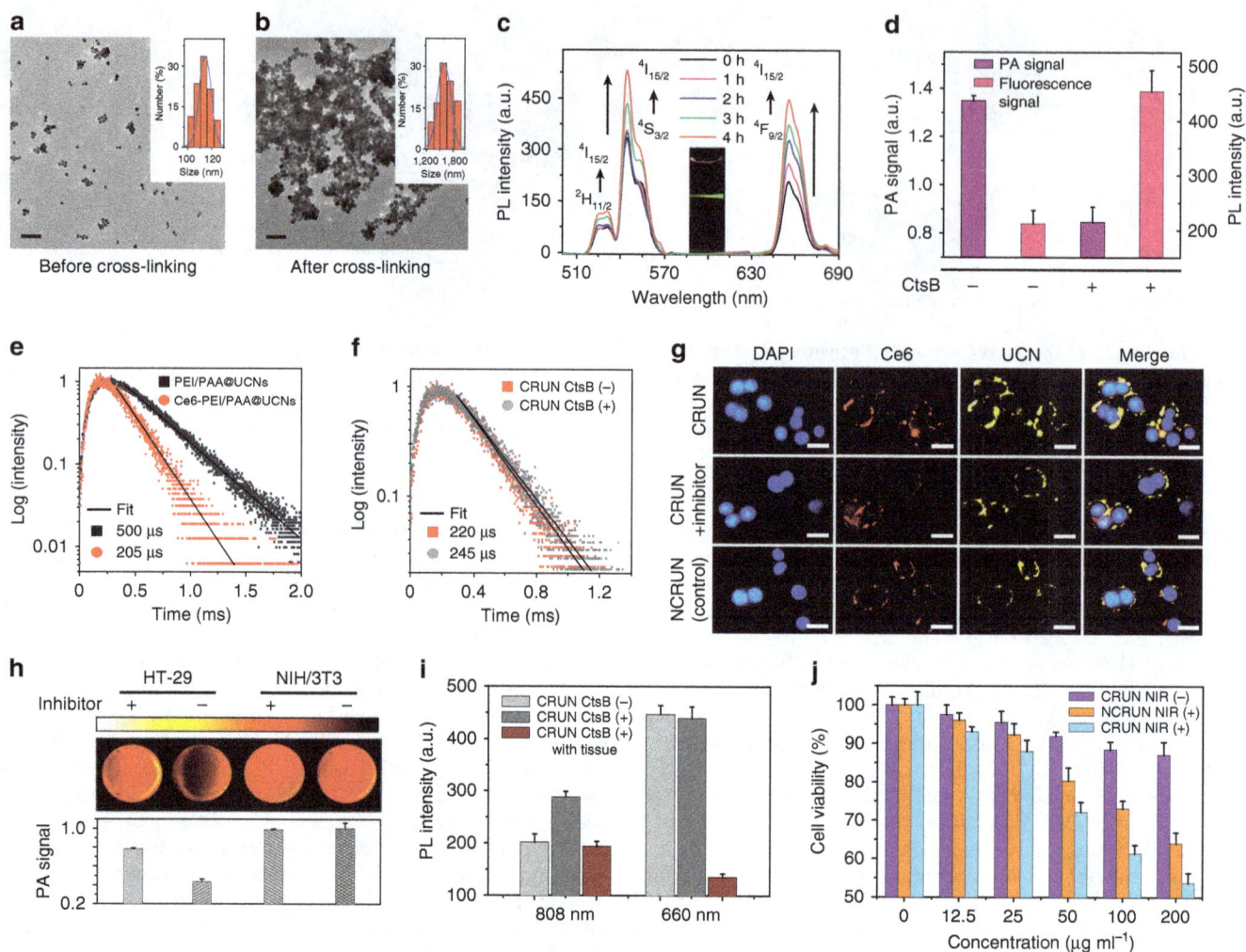

**Figure 2 | *In vitro* studies of tumour-specific enzyme (CtsB) triggered cross-linking of CRUN upon 808 nm illuminations.** TEM and DLS analysis of covalent cross-linking of CRUN (1.5 mg ml$^{-1}$) without (**a**) and with (**b**) CtsB (55 nM) at 4 h. The hydrodynamic diameters in DLS are about 110 nm (**a**) and 1,580 nm (**b**). Scale bar, 200 nm. (**c**) UCL of CRUN at different time intervals with CtsB treatment. The inset shows the luminescence photograph of CRUN incubate with CtsB for 4 h. (**d**) Photoacoustic (PA) imaging and UCL signal after 4 h incubation of CRUN with CtsB. (**e**) Luminescence decay curves of the emission at 655 nm of UCNs in the absence and presence of Ce6 excited by 808 nm laser. (**f**) Lifetime of CRUN in the absence and presence of CtsB treatment. (**g**) Confocal imaging of HT-29 cells incubated with CRUN, CRUN and inhibitor, and control NCRUN, respectively. Blue: DAPI ($\lambda_{ex} = 350/50$ nm, $\lambda_{em} = 460/50$ nm), red: Ce6 ($\lambda_{ex} = 545/25$ nm, $\lambda_{em} = 610/75$ nm), yellow: UCN ($\lambda_{ex} = 980$ nm, $\lambda_{em} = 350$–690 nm). Scale bar, 20 μm. (**h**) PA signals of HT-29 and NIH/3T3 living cells without (left) and with (right) CtsB inhibitor. The PA signals were shown with a laser excitation at 680 nm. (**i**) Quantification of singlet oxygen generation before and after enzyme-triggered CRUN or covered by 2-mm thick pork tissue when exposed to 808 or 660 nm laser irradiation for 1 h (0.4 W cm$^{-2}$). (**j**) Cell viability of HT-29 at different concentrations of CRUN and NCRUN after NIR light treatment for 1 h. Data are mean ± s.d. ($n = 3$ technical replicates).

enhanced light converting emission, which thus amplifies the production of $^1O_2$ from the activated photosensitizers after 1 h illumination (Fig. 2i and Supplementary Fig. 16). Importantly, a 2-mm thick pork tissue (adipose tissue) was also used to mimic clinical skin for further investigation of the different penetration of 808 nm (NIR) and 660 nm light irradiation. As shown in Fig. 2i, the presence of pork tissue significantly decreased the efficiency of $^1O_2$ generation (2.9-fold decrease) with 660 nm irradiation, while there was only 0.6-fold decrease in $^1O_2$ generation when illuminated at 808 nm. The higher singlet oxygen production with 808 nm excitation in the presence of pork tissue clearly indicated a better penetration depth as compared with irradiation at 660 nm, which makes UCN-based PDT suitable for effective tumour treatment in living system.

The potential photodynamic cytotoxicity of CRUN was further evaluated through an *in vitro* toxicology assay (TOX8). As shown in Fig. 2j, there was negligible cytotoxicity in the

CRUN-incubated HT-29 cell without laser exposure at 808 nm, suggesting minimum toxicity caused by the particle itself. However, upon 808 nm laser illumination of CRUN (200 μg ml$^{-1}$) incubated HT-29 cells, there was higher cytotoxicity (~53% cell viability) observed in HT-29 cells than that of cells treated with the control NCRUN (~64% cell viability). In addition, similar photodynamic inactivation was also studied in CRUN-incubated HT-29 cells but with CtsB inhibitor or in the negative control (CtsB-deficient NIH/3T3 cells). Compared with the laser-excited HT-29 cells with CRUN, there was no significant enhancement of cytotoxicity detected in these control experiments. Meanwhile, 808 nm laser irradiation alone did not induce obvious cytotoxicity in both HT-29 and NIH/3T3 cells (Supplementary Fig. 17), indicating that the covalent cross-linking of CRUN triggered by specific enzyme would increase the singlet oxygen generation, and thus induce the enhanced cell death in the targeted tumour cells with overexpression of the

CstB enzyme. Moreover, the cytotoxicity of HT-29 cells incubated with CRUN was also evaluated upon 660 or 808 nm light irradiation. Although the photodynamic cytotoxicity was found to be higher at direct 660 nm excitation ($\sim$86% in cytotoxcity) than that at 808 nm irradiation ($\sim$39% in cytotoxcity) (Supplementary Fig. 18), the presence of pork tissue can significantly lower the cytotoxicity at 660 nm illumination ($\sim$13%, $\sim$6.6-fold decrease), while only less cytotoxicity decreases at 808 nm irradiation ($\sim$28%, $\sim$1.4-fold decrease). Such cytotoxicity studies were consistent with the trends of $^1O_2$ generation as above, clearly demonstrating the great advantage of deep-tissue penetration of 808 nm (NIR) light for living cell PDT treatment.

### In vivo cross-linking of CRUN in tumour-bearing mice.
Inspired by the promising imaging and PDT results in living cells, we examined the in vivo theranostic efficacy via intratumoural injection of CRUN into living female Balb/c nude mice. We implanted HT-29 cells (CtsB overexpressing) into the left and right flanks of nude mice. At 2–3 weeks post-implantation, the tumours grew to 3–5 mm in diameter and the living mice underwent photoacoustic imaging and PDT studies. Typically, one group of xenograft mice bearing two tumours were subjected to CRUN and controlled NCRUN conjugates, respectively (3 mg in 100 µl saline), followed by photoacoustic imaging. As compared with the photoacoustic intensity before injection, the mice exhibited a significant increase in photoacoustic signals at 4 h post-injections for both conjugates. Moreover, the photoacoustic signals in the tumour injected with CRUN were found lower ($\sim$30% decrease in intensity) than that subjected to NCRUN (Fig. 3a and Supplementary Fig. 19). These results suggested the feasibility of the enzyme processing to trigger the localization of CRUN at the tumour site and it was the covalent cross-linking that led to the decreased photoacoustic signals.

Furthermore, we also investigated the therapeutic effect of CRUN as a PDT agent for tumour therapy in living mice. As shown in Fig. 3b, in group 1 ($n = 8$), CRUN was injected into

the implanted tumour on the left flank, and followed by an 808 nm laser irradiation at the tumours after 4 h injection. As contrast, group 2 ($n = 8$) mice were intratumourally injected with the control NCRUN on the right flank and subsequent light exposure under identical conditions. Tumour-bearing mice with direct injection of CRUN but no NIR light irradiation were assigned as group 3 ($n = 8$). The last group (group 4, $n = 8$) consisted of mice injected with saline only. The tumour sizes in the different groups were measured to assess the PDT efficacy over a period of 2 weeks. After CRUN-mediated PDT treatment ($P = 0.038$), the tumour growth in group 1 had been significantly inhibited as compared with that observed in group 2 with control NCRUN ($P = 0.033$, Fig. 3c), indicating that the enzyme-triggered cross-linking of UCN particles at the tumour region had an important role in enhancing therapeutic outcomes. On the other hand, in group 3 ($P = 0.021$) and 4, the tumour sizes grew exponentially over the period of time, suggesting that the treatment with 808 nm laser only or CRUN without light irradiation induced minimum inhibition of tumour growth. To further evaluate the therapeutic efficacy, we also performed TUNEL staining on the tissue sections from different groups after 11 days of PDT treatment (Fig. 3d and Supplementary Fig. 20). As expected, the treated tumours in group 1 mice displayed significant damage than those in group 2 mice. The mice treated with NIR light alone (group 4) and probe without light (group 3) displayed no detectable damage. This is in good accordance with the trends observed in tumour inhibition after NIR light-mediated PDT treatment. More importantly, there were no significant burnt scars on the tumour skins in similar light-mediated treatment by conventionally used 980 nm laser illuminations, clearly implying that the novel enzyme-responsive CRUN conjugates with laser irradiation at 808 nm could serve as a reliable platform for effective in vivo antitumour treatment.

### In vivo PDT treatment with affinity ligand-modified CRUN.
To further improve the tumour uptake and PDT efficiency, we conjugate an affinity ligand, folic acid (FA), to the surface of

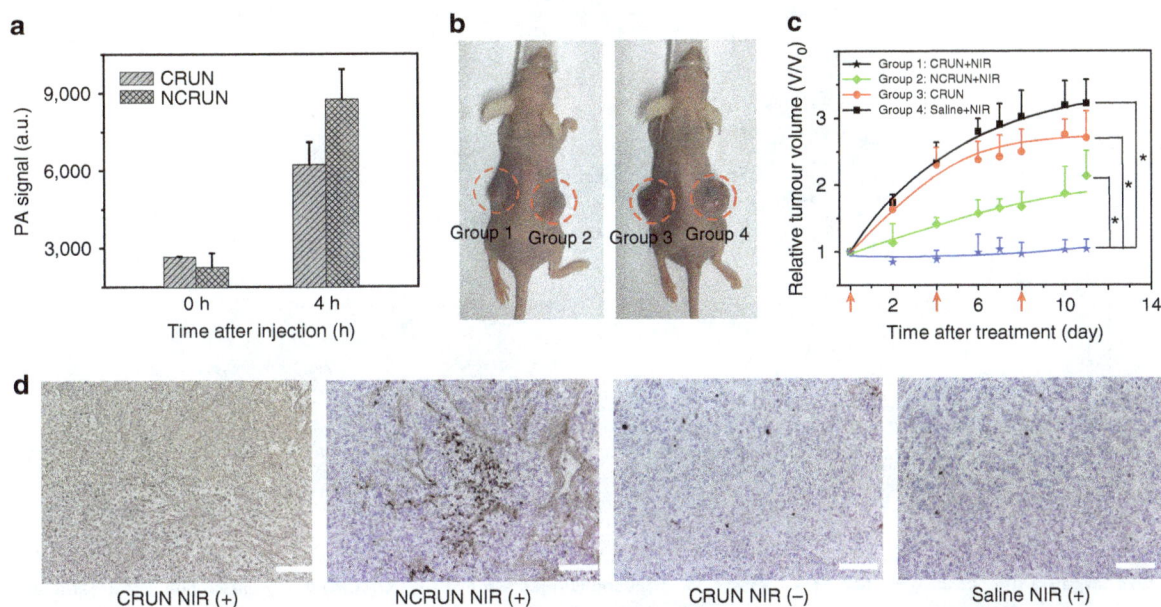

**Figure 3 | In vivo photoacoustic imaging and PDT efficiency by intratumoural injection of CRUN.** (a) PA signals in tumour region after injection of CRUN (3 mg in 100 µl saline). (b) Photographs of tumour-bearing mice after 11 days treatment with CRUN (group 1), NCRUN (group 2), CRUN without NIR light irradiation (group 3) and saline (group 4). (c) Relative tumour volumes in the treated groups after PDT therapy under 808 nm laser irradiation. The treatment after 4 h of intratumoural injection (3 mg in 100 µl saline) was started on day 0 and repeated twice at day 4 and day 8 (arrows) followed by 808 nm laser irradiation. (d) TUNEL histology of tumour tissues after NIR light-mediated tumour treatment. Scale bar, 100 µm. Date are means ± s.d. ($n = 8$ mice per group). Statistical significance is assessed by a Student's t-test (heteroscedastic, two-sided). *$P < 0.05$.

CRUN for their recognition to folate receptors that overexpress in most tumours. Meanwhile, considering the fact to enhance the biocompatibility of CRUN in living system, the FA was first coupled with polyethene glycol (PEG$_{5000}$) spacer containing amino and thiol group at each terminus. Such FA-PEG$_{5000}$-SH linkers were then coupled to enzyme-responsive CRUN conjugate through NHS-dPEG$_2$-Mal as a linkage moiety to afford final product of FA-PEG@CRUN (Fig. 4a and Supplementary Fig. 21). Upon reaction with CtsB enzyme, FA-PEG@CRUN indicated similar luminescence enhancement as compared with previous CRUN platform without FA modification, suggesting that introduction of FA and PEG linker into CRUN platform would not influence its activity for enzyme recognition (Supplementary Fig. 22).

To study whether FA conjugation can aid tumour uptake in living mice, we used fluorescence imaging to real-time monitor the targeting effect *in vivo*. The female Balb/c nude mice bearing HT-29 tumours were divided into three groups. Group 1 consisted of the mice with intravenous injection of saline. The mice subjected to PEG-modified CRUN particles but no FA ligands were assigned to group 2. In group 3, the tumour-bearing mice were injected with FA-PEG@CRUN via the tail vein. The fluorescent images were recorded at different time intervals after the first injection. Relative to control studies, a significant increase in fluorescence was observed in living mice at 4 h post-injection of probes (Fig. 4b and Supplementary Fig. 23). In contrast, the tumours in group 3 with injection of FA-PEG@CRUN displayed

stronger fluorescence than that in group 2 subjected to PEG@CRUN injection. Both fluorescence signals gradually decreased 24 h post-administration through the process of circulation. In addition, similar tumour targeting was also validated by photoacoustic imaging in living mice. After administration of CRUN conjugates at 1 h, the mice showed obvious photoacoustic signal increase in tumours. Moreover, the observed photoacoustic signals in the tumours of group 3 were higher than that in group 2 (Fig. 4c). The higher fluorescence and photoacoustic signals present in tumours confirmed the possible efficacy of the FA ligand in improving the targeting of CRUN conjugates at the tumour region.

Encouraged by the imaging data that revealed the effective tumour uptake of CRUN particles, we continued the light-mediated PDT therapy via intravenous injection of surface-modified CRUN conjugates or saline into six different groups of mice, and *in vivo* tumour inhibition effect was evaluated upon 808 or 660 nm light illumination. As shown in Fig. 4d, after 808 nm laser illumination, the tumour sizes in group 2 ($n = 8$, $P = 0.050$, with injection of PEG@CRUN) and 3 ($n = 8$, $P = 0.038$, injected with FA-PEG@CRUN) were found to be reduced significantly over the therapeutic period, whereas the tumour sizes in control group 1 ($n = 8$) with injection of saline revealed a fast growth rate. In addition, during the treatment, the tumours in group 3 displayed more effective inhibition than those in group 2. These data suggested that the facilitated tumour uptake by FA ligands and effective tumour reduction was mainly

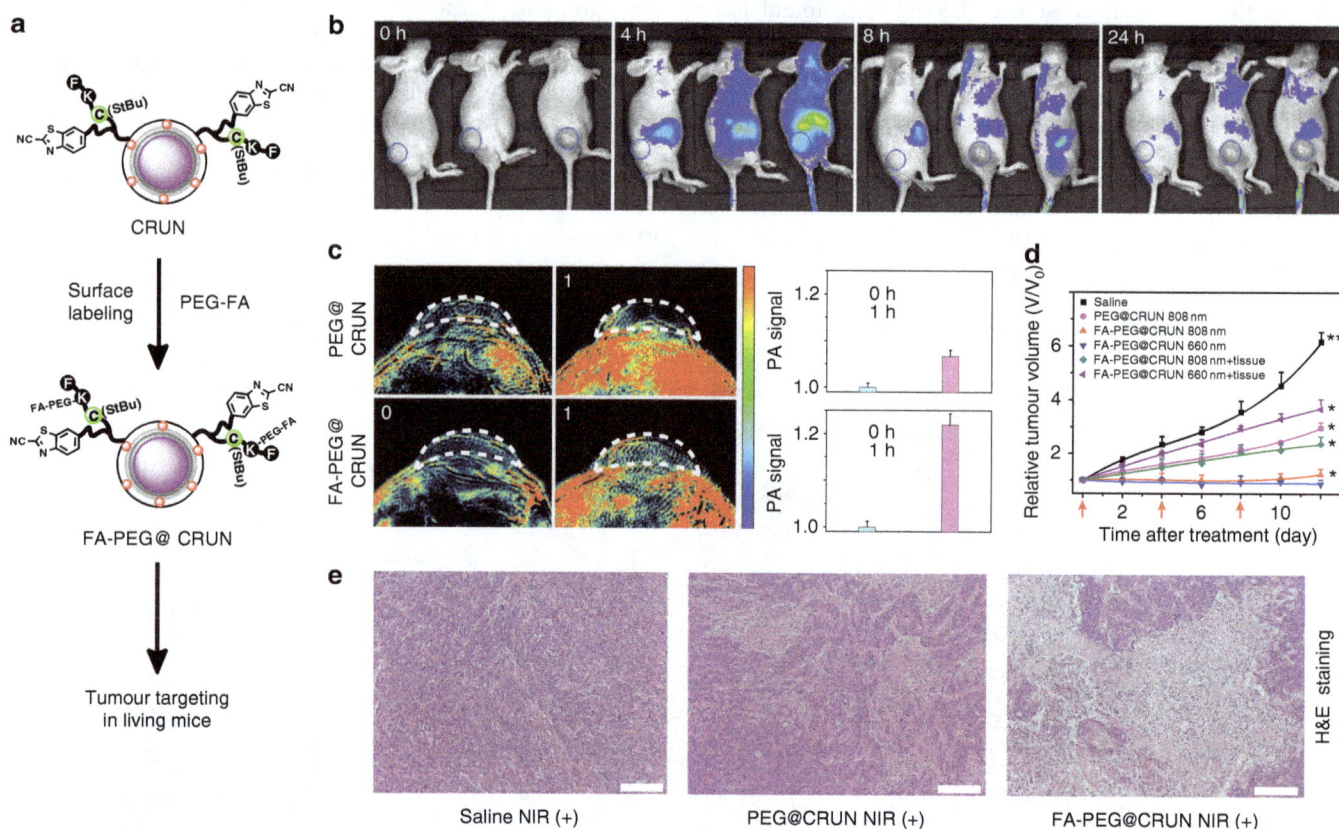

**Figure 4 | *In vivo* targeted tumour imaging and PDT therapy by intravenous injection of FA-PEG@CRUN.** (**a**) Illustration of the targeting strategy of FA-PEG@CRUN through the conjugation of FA and PEG$_{5000}$ on CRUN. (**b**) Fluorescence imaging of tumours (blue circle) in living mice at different time intervals after injection (from left to right: Saline, PEG@CRUN, FA-PEG@CRUN, respectively; $\lambda_{ex} = 640$ nm, $\lambda_{em} = 670$ nm). (**c**) PA imaging signals in the tumour region at different time intervals after intravenous injection (top: PEG@CRUN, bottom: FA-PEG@CRUN, 3 mg in 100 μl saline). (**d**) Tumour volumes change as a function of time in treated groups to evaluate the effectiveness of PDT treatment *in vivo*. The PDT treatment was carried out after 4 h drug-light interval of intravenous injection (3 mg in 100 μl saline) and periodically repeated at indicated time points (arrows) following by 808 nm laser irradiation every two days. (**e**) H&E histology of tumour tissues after 808 nm mediated tumour treatment by intravenous injection. Scale bar, 200 μm. Date are means ± s.d. ($n = 8$ mice per group). Statistical significance is assessed by a Student's *t*-test (heteroscedastic, two-sided). *$P < 0.05$, **$P < 0.01$.

from the production of $^1O_2$ from Ce6 anchored on the CRUN surface.

Notably, direct 808 nm (group 3) and 660 nm (group 5, $P = 0.008$) laser irradiation of living animals with intravenous injection of FA-PEG@CRUN resulted in comparable PDT efficiency, in which tumour growth has been greatly suppressed (Fig. 4d). However, considering the different light penetration, similar *in vivo* PDT treatments were also performed when the tumours were covered with a 2-mm thick pork tissue. Compared with mice with FA-PEG@CRUN injection and laser illumination at 808 nm (group 4, $P = 0.019$), the presence of pork tissue could significantly compromise tumour therapy for the mice treated with FA-PEG@CRUN and 660 nm laser excitation (group 6, $P = 0.029$), further demonstrating the benefits of deep tissue penetration in 808 nm laser-mediated tumour treatment.

Besides the promising antitumour therapy *in vivo*, the histological tumour tissues analysis also revealed the efficacy of PDT treatment. After 12 days of light-mediated PDT at 808 nm, hematoxylin and eosin and TUNEL staining of tumour tissues displayed more significant damage in the mice with intravenous injection of FA-PEG@CRUN (group 3) than those tumours treatment in control groups (Fig. 4e and Supplementary Fig. 24). Moreover, there was no obvious damage observed in other organs during the treatment (Supplementary Fig. 25). And our inductively coupled plasma spectral analysis also validated more $Y^{3+}$ ion distribution in the tumour region (Supplementary Fig. 26). These results clearly indicated that CRUN conjugates modified with active targeting agents can significantly improve the therapeutic efficacy of *in vivo* tumour treatment with a desirable PDT outcome.

## Discussion

Currently, nanomedicine based on lanthanide-doped UCNs towards cancer imaging and therapy has received considerable attention due to their intrinsic photon-converting nature upon exposure to long-wavelength light irradiation[21-29]. Despite such exceptional optical capabilities in the NIR ranges offered by UCN nanoparticles, the ability to provide specific targeting strategy for effective localization of UCNs within pathological areas, especially at the tumour site, remains a critical challenge. So far, the ligand-mediated recognition strategies based on the conjugation of affinity moieties to the particle surface have been mostly applied to achieve tumour targeting specificity[30-34]. Considering the complex and dynamic physiological cell conditions, the results along with the stochastic nature of ligand–receptor interactions between normal and tumour cells are not so satisfactory. Alternative to develop 'smart' conjugating nanoplatforms that are sensitive to local tumour environment conditions (for example, a particular pH, temperature or enzyme level and so on), and then facilitate specific localization of UCN particles at tumour sites would be thus a highly desirable option[10,11].

In this study, by exploiting the advantages of intrinsic enzymatic stimulus to specifically target the tumour areas, we developed such a 'smart' upconversion nanoconjugate that can selectively react with protease enzyme CtsB, which has been considered as one important biomarker in many types and stages of cancers[37,44]. Upon tumour-specific enzymatic CtsB reaction, the CtsB-sensitive peptide on the UCN particle surface can be cleaved, which exposes free cysteine and induces its condensation reaction with CBT groups on the surface of neighbouring particles, therefore triggering the covalent cross-linking of UCNs particles. Our extensive *in vitro* and *in vivo* fluorescence and photoacoustic imaging studies have validated the successful enzyme-triggered cross-linking reaction, which easily recognized

the enhanced light converting emission and unique targeting of enzyme-sensitive UCNs at tumour sites.

So far, photodynamic antitumour therapy has been regarded as a minimally invasive treatment modality, and has been widely applied in clinics for various types of cancer treatment[28,30,32]. Compared with conventional chemotherapy, light-mediated PDT exhibits some unique advantages, in which cytotoxic photosensitizers can be selectively activated by manipulating the location of light exposure. Currently, the most clinically used photosensitizers are excited at a shorter wavelength range from 630 to 700 nm, which unfortunately has limited penetration depth in biological tissues. One promising strategy to avoid this limitation is to shift the excitation wavelength to the NIR region (700–1,000 nm), in which biological tissues have maximum light transparency, and thus ideal for *in vivo* optical imaging and phototherapy[35,36]. Actually, latest advances in rare-earth doped UCNs have witnessed unique upconverting optical properties, which can transduce low-energy NIR light illumination into high-energy emissions from ultraviolet to visible range. Such exceptional light-converting capabilities could greatly match the demands and thus achieve deeper tissue penetration to benefit biomedical applications including photodynamic therapy, biological imaging and remotely controlled release of therapeutic agents *in vitro* and *in vivo*[5,27-30,32].

Current progress of UCN-based PDT system mostly utilizes the NIR light excitation wavelength at 980 nm, at which, however, the absorption of water is significant[27-29]. Such 980 nm light-mediated PDT is facing tremendous hurdles to travel into deeper tissue and bypass the inevitable heat effect[30-34]. Therefore, efforts have been intensively made to adjust the excitation range of UCNs to a shorter wavelength to meet the demands of the medical spectral window (that is, ∼700–900 nm) (refs 35,36). Very recently, $Nd^{3+}$-doped upconversion processes based on a new excitation wavelength around 808 nm have been investigated to manipulate optical properties of lanthanide UCNs to minimize this photothermal effect[26,36]. By taking this promising advantage, in this study, we have carried out a series of UCN-based PDT experiments, and systematically compared the differences of photodynamic antitumour effect *in vitro* and *in vivo* through 808 and 660 nm laser irradiation. Our data clearly demonstrated that the PDT treatment efficacy upon 808 nm light illumination did not show significant decrease when the pork tissue (for example, 2 mm thickness) was used to increase the tissue depth; however, the dramatic drop of therapeutic effect was observed in the similar treatment under 660 nm laser excitation (Figs 2i and 4d, and Supplementary Fig. 18), suggesting that 808 nm light-mediated UCNs exhibit deeper tissue penetration and better antitumour effect than those in the conventional PDT treatment system. Meanwhile, during the process of effective UCN-based PDT treatment, the prolonged 808 nm laser irradiation would not induce significant overheating and tissue damage, thus providing promising prospect for their safer applications in the future.

More significantly, in response to specific enzyme stimulation within tumours, our unique UCNs platform could undergo covalent cross-linking reactions and enable the selective accumulation of particles at the tumour region. The *in situ* enzyme-triggered interparticle cross-linking exhibited enhanced light converting emission and could amplify the singlet oxygen generation. Both our *in vitro* and *in vivo* theranostic studies successfully demonstrated that such unique enzyme-responsive localization of light-converted rare-earth nanostructures could significantly inhibit the tumour growth in process of PDT treatment as compared with those UCNs that could not undergo cross-linking reaction.

In summary, a novel targeting strategy based on specific enzymatic reaction was presented to effectively localize light converting UCN nanocarriers at tumour area. Such tumour microenvironment-responsive enzyme reactions would trigger the covalent cross-linking of single UCN particles, and thus display unique particle localization to facilitate promising multi-modality imaging and amplified PDT therapy of malignant tumours *in vitro* and *in vivo* upon NIR light illumination at 808 nm. This well-defined rational design could be easily extended into other unique functional materials and open new doors for precision nanomedicine in the future clinical applications.

## Methods

**General.** Synthetic procedures and chemical characterizations of all the probes are described in the Supplementary Methods.

*In vitro* **fluorescence imaging in living cells.** The human colorectal adeno-carcinoma cell line (HT-29, cat. no. HTB-38) and mouse embryonic fibroblast cell line (NIH/3T3, cat. no. CRL-1658) were provided from American type culture collection and checked for mycoplasma contamination. These cells are not listed by ICLAC as misidentified cell lines (3 October 2014). CtsB-overexpressing HT-29 cells and CtsB-deficient NIH/3T3 cells were seeded with a cell density of $1 \times 10^5$ in an ibidi confocal μ-dish (35 mm, plastic bottom) in 1 ml DMEM media which was incubated for 24 h at 37 °C. The seeded cells were subsequently treated with CRUN ($50 \mu g\, ml^{-1}$) and incubated for 4 h. The cell nuclei were stained by DAPI ($1 \mu g\, ml^{-1}$) for 20 min and then the cells were washed with PBS (pH 7.4) for three times. Control experiments were setup by treating the two cell lines with CRUN in the presence of CtsB inhibitor and controlled NCRUN conjugates. For the cell-inhibition assay, the cells were first pretreated with CtsB inhibitor (antipain hydrochloride, 100 μM in DMEM) for 2 h at 37 °C followed by the addition of CRUN and incubated for further 4 h. Fluorescence imaging of the samples were then carried out under confocal microscopy system using a continuous-wave 980 nm laser as excitation source (EINST Technology). The quantification of CtsB enzyme expression in HT-29 and NIH/3T3 cells were performed by Human CtsB ELISA kit according to the standard procedure described by the manufacturer (CUSABIO, USA).

*In vivo* **animal studies.** All animal experimental procedures were performed in accordance with the protocol #120774 approved by the Institutional Animal Care and Use Committee (IACUC). Female Balb/c nude mice ($\sim$6–8-weeks old) were purchased from Charles River Laboratories (Shanghai, China). Xenograft mice models were established by injecting 0.1 ml of tumour cells suspension (PSB/matrigel, BD biosciences, 1:1 v/v) containing $3 \times 10^6$ HT-29 cells into both flanks or only right side of mouse. When the tumour volume reached a palpable size, the mouse was used for further studies.

*In vivo* **cross-linking of CRUN in tumour-bearing mice.** Xenograft mice models were treated when the tumour volumes approached 3–5 mm in diameter. For *in vivo* PDT treatment, the tumour-bearing mice were randomized into four groups ($n = 8$, each group) and treated by intratumoural injection of particle samples. In group 1, CRUN (3 mg in 100 μl of saline) was directly injected into implanted tumours, followed by an 808 nm laser irradiation after 4 h. In group 2, same amounts of controlled NCRUN conjugates were injected with subsequent laser exposure. Control experiments with injection of CRUN but no NIR light irradiation (group 3) and saline with laser treatment (group 4) were used, respectively. The effective exposure time for each mouse was 45 min (5 min interval after 5 min irradiation with total treatment time of 90 min) with a power density at $0.4\,W\, cm^{-2}$ (fluence $1,080\,J\, cm^{-2}$). A second- and third-dose injection followed by PDT treatment described above was repeated at 4 days and 8 days after the first-dose injection, respectively. Tumour size was measured three times a week using a vernier caliper upon PDT treatment. The tumour volume was calculated using the following equation: tumour volume $(V) = length \times width^2/2$. Relative tumour volume was calculated as $V/V_0$ ($V_0$ was the initial tumour volume before PDT treatment).

*In vivo* **targeted PDT treatment with intravenous injection.** Xenograft HT-29 tumour model on the right side of female Balb/c mice were first developed as described above. For *in vivo* tumour targeting PDT treatment, the tumour-bearing mice had been randomly arranged into six different groups ($n = 8$, each group) to perform a series of intravenous injections of saline (group 1), PEG@CRUN (group 2) and FA-PEG@CRUN (group 3–6), respectively (3 mg in 100 μl saline). After 4 h of intravenous injection, laser treatment was performed on groups 1–6 by irradiating the tumour region with a continuous 808 or 660 nm laser. The effective exposure time for each mouse was 45 min (5 min interval after 5 min irradiation with a total treatment time of 90 min) with a power density at $0.4\,W\, cm^{-2}$ (fluence $1,080\,J\, cm^{-2}$). The tumours in group 4 and 6 were covered by 2-mm thick pork tissue and then irradiated with 808 or 660 nm lasers under the same condition.

The PDT treatment with laser irradiation was repeated every 2 days. A second- and third-dose injection of the PDT treatment described above was repeated at 4 days and 8 days after the first-dose injection, respectively. Tumour size was measured and calculated as described above.

## References

1. Gao, Y., Shi, J., Yuan, D. & Xu, B. Imaging enzyme-triggered self-assembly of small molecules inside live cells. *Nat. Commun.* **3**, 1033–1040 (2012).
2. Williams, R. J. *et al.* Enzyme assisted self-assembly under thermodynamic control. *Nat. Nanotechnol.* **4**, 19–24 (2009).
3. Lovell, J. F. *et al.* Porphysomenanoesicles generated by porphyrin bilayers for use as multimodal biophotonic contrast agents. *Nat. Mater.* **10**, 324–332 (2011).
4. Petros, R. A. & Desimone, M. J. Strategies in the design of nanoparticles for therapeutic applications. *Nat. Rev. Drug Discov.* **9**, 615–627 (2010).
5. Mitragotri, S., Burke, P. A. & Langer, R. Overcoming the challenges in administrating biopharmaceuticals: formulation and delivery strategies. *Nat. Rev. Drug Discov.* **13**, 655–672 (2014).
6. Cheng, C. J., Tietjen, G. T., Saucier-sawyer, J. K. & Saltzman, W. M. A holistic approach to targeting disease with polymeric nanoparticles. *Nat. Rev. Drug Discov.* **14**, 239–247 (2015).
7. Davis, M. E., Chen, Z. G. & Shin, D. M. Nanoparticle therapeutics: an emerging treatment modality for cancer. *Nat. Rev. Drug Discov.* **7**, 771–882 (2008).
8. Giljohann, D. A. & Mirkin, C. A. Drivers of biodiagnostics development. *Nature* **580**, 461–464 (2009).
9. Koo, H. *et al. In vivo* targeted delivery of nanoparticles for theranosis. *Acc. Chem. Res.* **44**, 1018–1028 (2011).
10. Yu, M. K., Park, J. & Jon, S. Y. Targeting strategies for multifunctional nanoparticles in cancer imaging and therapy. *Theranostics* **2**, 3–44 (2012).
11. Thaker, A. S. & Gambhir, S. S. Nanooncology: the future of cancer diagnosis and therapy. *Cancer J. Clin.* **63**, 395–418 (2013).
12. Mura, S., Nicolas, J. & Couvreur, P. Stimuli-responsive nanocarriers for drug delivery. *Nat. Mater.* **12**, 991–1003 (2013).
13. Liang, G., Ren, H. & Rao, J. A biocompatible condensation reaction for controlled assembly of nanostructures in living cells. *Nat. Chem.* **2**, 54–60 (2010).
14. Haun, J. B., Devaraj, N. K., Hilderbrand, S. A., Lee, H. & Weissleder, R. Bioorthogonal Chemistry amplifies nanoparticles binding and enhances the sensitivity of cell detection. *Nat. Nanotechnol.* **5**, 660–665 (2010).
15. Tsukiji, S., Miyagawa, M., Takaoka, Y., Tamura, T. & Hamachi, I. Ligand-directed tosyl chemistry for protein labeling *in vivo. Nat. Chem. Biol.* **5**, 341–343 (2009).
16. Laughlin, S. T., Baskin, J. M., Amacher, S. L. & Bertozzi, C. R. *In vivo* imaging of membrane-associated glycans in developing zebrafish. *Science* **320**, 664–667 (2008).
17. Prescher, J. A., Dube, D. H. & Bertozzi, C. R. Chemical remodelling of cell surfaces in living animals. *Nature* **430**, 873–877 (2004).
18. Ye, D. *et al.* Bioorthongonal cyclization-mediated *in situ* self-assembly of small-molecule probes for imaging caspase activity *in vivo. Nat. Chem.* **6**, 519–526 (2014).
19. Lang, K. & Chin, J. W. Cellular incorporation of unnatural amino acids and bioorthogonal labeling of proteins. *Chem. Rev.* **114**, 4764–4806 (2014).
20. Grammel, M. & Hang, H. C. Chemical reporters for biological discovery. *Nat. Chem. Biol.* **9**, 475–484 (2013).
21. Deng, R. *et al.* Temporal full-colour tuning through non-steady-state upconversion. *Nat. Nanotechnol.* **10**, 237–242 (2015).
22. Feng, W., Zhu, X. & Li, F. Recent advances in the optimization and functionalization of upconversion nanomaterials for *in vivo* bioapplications. *NPG Asia Mater.* **5**, e75 (2013).
23. Yang, Y. *et al. In vitro* and *in vivo* uncaging and bioluminescent imaging through photocaged upconversion nanoparticles. *Angew. Chem. Intl. Ed.* **51**, 3125–3127 (2012).
24. Idris, N. M. *et al. In vivo* photodynamic therapy using upconversion nanoparticles as remote-controlled nanotransducers. *Nat. Med.* **18**, 1580–1585 (2012).
25. Yang, D. *et al.* Current advances in lanthanide ion ($Ln^{3+}$)-based upconversion nanomaterials for drug delivery. *Chem. Soc. Rev.* **44**, 1416–1448 (2015).
26. Xie, X. J. *et al.* Mechanistic investigation of photon upconversion in $Nd^{3+}$-sensitized core-shell nanoparticles. *J. Am. Chem. Soc.* **135**, 12608–12611 (2013).
27. Kachynski, A. V. *et al.* Photodynamic therapy by *in situ* nonlinear photon conversion. *Nat. Photon.* **8**, 455–461 (2014).
28. Xu, C. T. *et al.* Upconverting nanoparticles for pre-clinical diffuse optical imaging, microscopy and sensing: current trends and future challenges. *Laser Photon. Rev.* **7**, 663–697 (2013).
29. Zou, W., Visser, C., Maduro, J. A., Pshenichnikov, M. S. & Hummelen, J. C. Broadband dye-sensitized upconversion of near-infrared light. *Nat. Photon.* **6**, 560–564 (2012).
30. Wang, C., Cheng, L. & Liu, Z. Drug delivery with upconversion nanoparticles for multi-functional targeted cancer cell imaging and therapy. *Biomaterials* **32**, 1110–1120 (2011).

31. Zhou, L. *et al.* Single-band upconversion nanoprobes for multiplexed simultaneous *in situ* molecular mapping of cancer biomarkers. *Nat. Commun.* **6,** 6938 (2015).
32. Lu, S. *et al.* Multifunctional nano-bioprobes based on rattle-structured upconverting luminescent nanoparticles. *Angew. Chem. Int. Ed.* **54,** 7915–7919 (2015).
33. Xiao, Q. *et al.* A core/satellite multifunctional nanotheranostic for *in vivo* imaging and tumor eradication by radiation/photothermal synergistic therapy. *J. Am. Chem. Soc.* **135,** 13041–13048 (2013).
34. Li, L. *et al.* Biomimetic surface engineering of lanthanide-doped upconversion nanoparticles as versatile bioprobes. *Angew. Chem. Int. Ed.* **51,** 6121–6125 (2012).
35. Ai, F. *et al.* A core-shell-shell nanoplatform upconverting near-infrared light at 808 nm for luminescence imaging and photodynamic therapy of cancer. *Sci. Rep.* **5,** 10785 (2015).
36. Shen, J. *et al.* Engineering the upconversion nanoparticle excitation wavelength: cascade sensitization of tri-doped upconversion colloidal nanoparticles at 800 nm. *Adv. Opt. Mater.* **1,** 644–650 (2013).
37. Lecaille, F., Kaleta, J. & Bromme, D. Human and parasitic papain-like cysteine proteases: their role in physiology and pathology and recent developments in inhibitor design. *Chem. Rev.* **102,** 4459–4488 (2002).
38. Wang, L. V. & Hu, S. Photoacoustic tomography: *in vivo* imaging from organelles to organs. *Science* **335,** 1458–1462 (2012).
39. Nie, L. & Chen, X. Structural and functional photoacoustic molecular tomography aided by emerging contrast agents. *Chem. Soc. Rev.* **43,** 7132–7170 (2014).
40. Sheng, Y., Liao, L. D., Thakor, N. & Tan, M. C. Rare-earth doped particles as dual-modality contrast agent for minimally-invasive luminescence and dual-wavelength photoacoustic imaging. *Sci. Rep.* **4,** 6562 (2014).
41. Chris, H. J. H. *et al.* Multifunctional photosensitizer-based contrast agents for photoacoustic imaging. *Sci. Rep.* **4,** 5342 (2014).
42. Atwater, H. A. & Polman, A. Plasmonics for improved photovoltaic devices. *Nat. Mater.* **9,** 205–213 (2010).
43. Zeng, Z., Mizukami, S., Fujitab, K. & Kikuchi, K. An enzyme-responsive metal-enhanced nearinfrared fluorescence sensor based on functionalized gold nanoparticles. *Chem. Sci.* **6,** 4934–4939 (2015).
44. Talieria, M. *et al.* Cathepsin B and cathepsin D expression in the progression of colorectal adenoma to carcinoma. *Cancer Lett.* **205,** 97–106 (2004).
45. Lovell, J. F., Liu, T. W. B., Chen, J. & Zheng, G. Activatable photosensitizers for imaging and therapy. *Chem. Rev.* **110,** 2839–2857 (2010).

## Acknowledgements

We thank Professor Lei Lu and Dr Meng Shi for the Western blot assay. We also thank Xiangyang Wu, Fang Liu, Yang Zhang and France Widjaja for the helpful suggestions and discussions provided in data analysis. We acknowledge the financial support from Nanyang Technological University, Singapore, through a Start-Up Grant (SUG), Tier 1 RG (11/13), RG (64/10) and RG (35/15). The Major State Basic Research Development Program of China (973 Program) (Grant Nos. 2014CB744503 and 2013CB733802), the National Natural Science Foundation of China (NSFC) (Grant Nos. 81422023, 81371596 and 51273165), the Fundamental Research Funds for the Central Universities, China (Grant No. 2013121039), and the Program for New Century Excellent Talents in University (NCET-13-0502).

## Author contributions

X.A., J.A., J.M., and B.X. conceived the projects and are responsible for the design, experiments and preparation of the manuscript. E.K.L.Y. assisted with spectroscopic measurement. X.A., J.A., C.J.H.H., A.B.E.A. and M.O. performed PA imaging and intratumoural theranostic study. Y.W. and X.L. contributed to lifetime and quantum yield test. X.A., X.W., Y.W., H.C., M.G., X.C. and G.L. carried out *in vivo* imaging and intravenous PDT treatment analysis. All authors reviewed the manuscript.

## Additional information

**Competing financial interests:** The authors declare no competing financial interests.

# 16

# Targeting bacteria via iminoboronate chemistry of amine-presenting lipids

Anupam Bandyopadhyay[1], Kelly A. McCarthy[1], Michael A. Kelly[1] & Jianmin Gao[1]

Synthetic molecules that target specific lipids serve as powerful tools for understanding membrane biology and may also enable new applications in biotechnology and medicine. For example, selective recognition of bacterial lipids may give rise to novel antibiotics, as well as diagnostic methods for bacterial infection. Currently known lipid-binding molecules primarily rely on noncovalent interactions to achieve lipid selectivity. Here we show that targeted recognition of lipids can be realized by selectively modifying the lipid of interest via covalent bond formation. Specifically, we report an unnatural amino acid that preferentially labels amine-presenting lipids via iminoboronate formation under physiological conditions. By targeting phosphatidylethanolamine and lysylphosphatidylglycerol, the two lipids enriched on bacterial cell surfaces, the iminoboronate chemistry allows potent labelling of Gram-positive bacteria even in the presence of 10% serum, while bypassing mammalian cells and Gram-negative bacteria. The covalent strategy for lipid recognition should be extendable to other important membrane lipids.

---

[1] Department of Chemistry, Merkert Chemistry Center, Boston College, 2609 Beacon Street, Chestnut Hill, Massachuetts 02467, USA. Correspondence and requests for materials should be addressed to J.G. (email: jianmin.gao@bc.edu).

I t is increasingly clear that membrane lipids do not merely provide a physical barrier for a cell; instead they play active roles in regulating numerous processes in cell physiology and disease[1]. To support the diverse functions of a membrane, the composite lipids, while maintaining the common feature of amphiphilicity, do vary in their chemical structures to give a complex lipidome (Fig. 1a)[2,3]. The lipid composition of a membrane has significant ramifications in biology. For example, it is well known that the plasma membranes of bacterial and mammalian cells display distinct compositions of lipids: while a mammalian cell membrane primarily consists of phosphatidylcholine (PC) and sphingomyelin (SM), bacterial cells display highly enriched phosphatidylethanolamine (PE) and phosphatidylglycerol (PG)[4,5]. In addition, some bacterial species present a lysine-modified PG (Lys-PG, Fig. 1a) in high percentages as a resistance mechanism to cationic antibiotics[6].

Synthetic molecules that specifically target bacterial lipids may give rise to new imaging methods of bacterial infection, as well as novel solutions to the antibiotic-resistance problem. The critical importance of lipids also manifests in the subcellular distribution of certain lipids in mammalian cells, a change of which may alter the homeostasis of important signalling proteins[7,8]. To further elucidate the diverse roles of membrane lipids, it is highly desirable to have molecular probes that specifically target a lipid of interest as well. Currently known lipid-targeting agents, which are primarily lipid-binding proteins and their synthetic mimetics, achieve lipid recognition by employing networks of *noncovalent* interactions, such as hydrogen bonds and salt bridges[9,10]. It remains to be seen whether membrane lipids can be selectively recognized by *covalently* targeting their unique chemical structure and reactivity with synthetic molecules.

In this contribution, we report the design and synthesis of an unnatural amino acid that selectively conjugates with amine-presenting lipids via formation of iminoboronates. By targeting the membrane lipids enriched in bacterial cells, namely PE and Lys-PG, the iminoboronate chemistry allows highly selective labelling of bacteria over mammalian cells.

## Results

**Design and synthesis of AB1.** The two major bacterial lipids, PE and Lys-PG, differ from their mammalian counterparts (PC and SM) by the presence of primary amino groups. We postulated that these nucleophilic amines could be captured by a 2-acetylphenylboronic acid (2-APBA) motif to form an iminoboronate (Fig. 1b). Although theoretically possible, amines in biology milieu only forms a Schiff base with simple ketones at high concentrations[11]. For example, the association constant of acetone and glycine was reported to be $3.3 \times 10^{-3} M^{-1}$. Usually, the imine formation is trapped with a reduction step for biological applications[12]. With the *ortho* boronic acid group serving as an electron trap, the 2-APBA motif conjugates with an amine much more readily to give an iminoboronate[13–17]. Importantly, the reaction proceeds under physiological conditions and in a reversible manner. Furthermore, an iminoboronate conjugate can exchange with other amines to allow for thermodynamic control of the final iminoboronate formation (Supplementary Fig. 1)[15]. These features make the iminoboronate chemistry particularly suitable for facilitating molecular recognition in biological systems.

To test our hypothesis, we have designed and synthesized a novel unnatural amino acid (AB1, Fig. 2) that presents a 2-APBA motif as its side chain. We envisioned that the amino-acid scaffold should allow the 2-APBA motif to be readily conjugated to fluorescent labels or other functional peptides. The synthetic route of AB1 is summarized in Fig. 2. Briefly, with 2′,4′-dihydroxy acetophenone 1 as the starting material, regioselective alkylation of the 4′-OH followed by triflate protection of the 2′-OH yielded 3 with an overall 81% yield. By taking advantage of the powerful thiol-ene chemistry[18], compound 3 was conjugated to two cysteine derivatives, respectively, to give the protected amino acids 4 and 7 in high yields. The key transformation of our synthesis is the Miyaura borylation[19], which converts the triflate to the Bpin moiety. In our hands, rigorous control of temperature was critical to the success of the borylation step: the reaction did not initiate below 95 °C and prolonged heating at higher temperatures caused the complete loss of the Bpin moiety to give the protodeboronated product, a protected AB2 (ref. 20). With optimized conditions, the Bpin moiety was introduced with 70–80% yield. Fortuitously, with the boronic acid moiety eliminated, AB2 served as a perfect negative control for AB1 in the following membrane-binding studies.

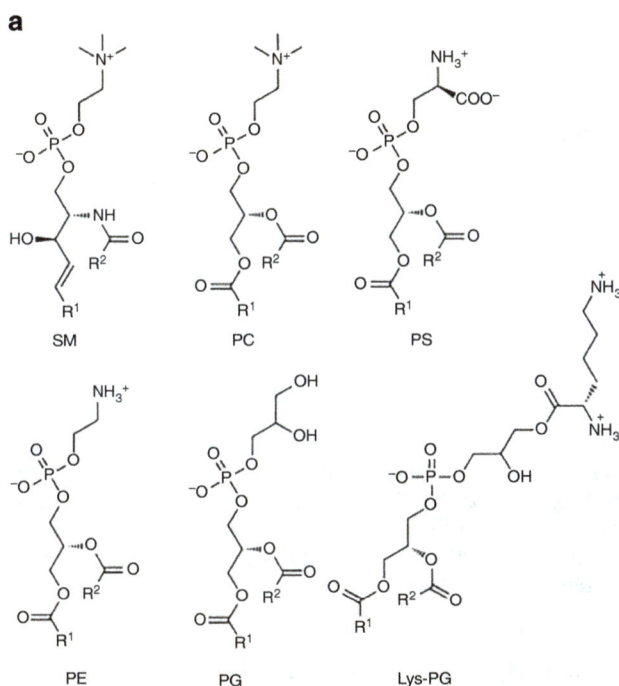

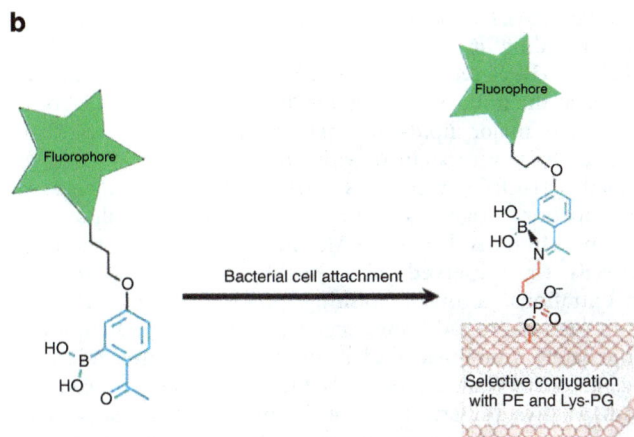

**Figure 1 | Covalent recognition of membrane lipids. (a)** Structures of the major membrane lipids from mamallian (sphingomyelin (SM) and phosphatidylcholine (PC)) and bacterial (phosphatidylethanolamine (PE), phosphatidylglycerol (PG), lysylphosphatidylglycerol (Lys-PG)) cells. PE and phosphatidylserine (PS) exist in mammalian cells as minority lipids. **(b)** Illustration of the iminoboronate chemistry for targeting PE on bacterial cell surfaces.

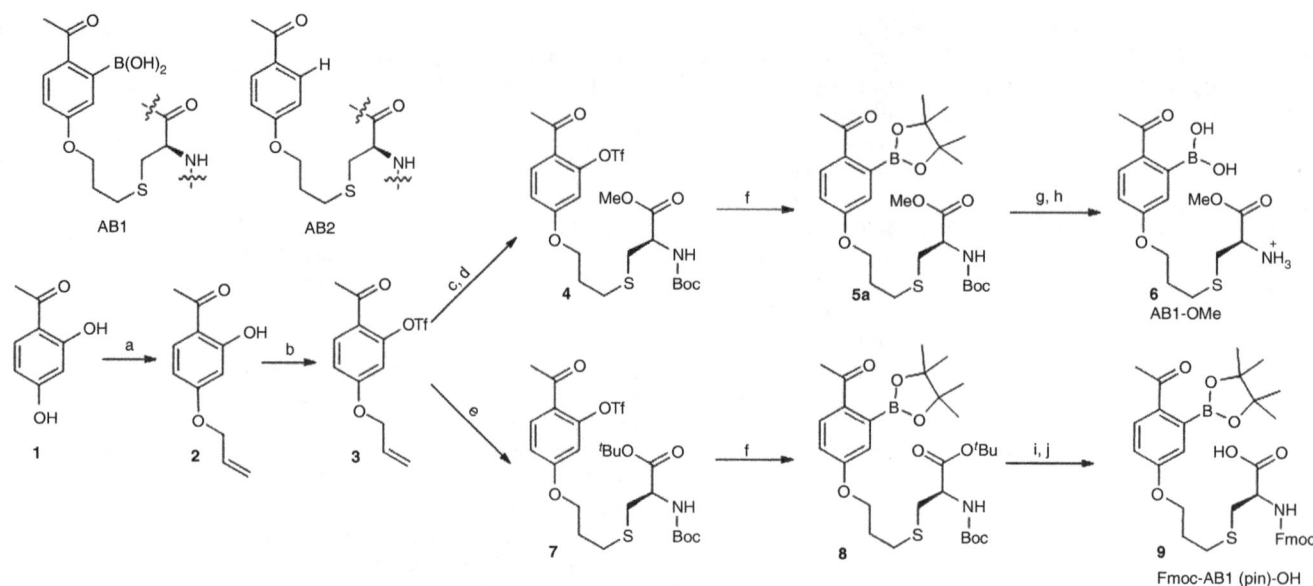

**Figure 2 | Synthesis of AB1 and its derivatives.** (a) Allyl bromide, $K_2CO_3$, NaI, acetone, 81%. (b) $(CF_3SO_2)_2O$, $Et_3N$, DCM, 95%. (c) Cys-OMe, DMPA, MeOH, ~365 nm ultraviolet irradiation. (d) Boc anhydride, $Na_2CO_3$, $THF/H_2O$, 80% over two steps. (e) Boc-Cys-OtBu, DMPA, MeOH, ~365 nm ultraviolet irradiation, 75%. (f) Pd(dppf)$Cl_2$/dppf, $B_2Pin_2$, KOAc, dioxane, ~70–80%. (g) 40% TFA in DCM. (h) diethanolamine, 1N HCl, 74% over two steps. (i) 60% TFA in DCM. (j) Fmoc-OSu, $Na_2CO_3$, $THF/H_2O$, 81% over two steps. DCM, dichloromethane; DMPA, 2,2-dimethoxy-2-phenylacetophenone; THF, tetrahydrofuran; TFA, trifluoroacetic acid.

**AB1 selectively conjugates with PE and Lys-PG.** The use of cysteine methyl ester (Cys-OMe) in the thiol-ene coupling step yielded the AB1 methyl ester (AB1-OMe, Fig. 2), which can be readily labelled with amine-reactive fluorophores. To assess the binding propensity towards different lipids, a fluorescein isothiocyanate-labelled AB1 methyl ester (Fl-AB1-OMe) was tested against lipid vesicles of varied composition. Specifically, 100nm-sized vesicles were prepared with PC alone or with 40% guest lipids including PE, PS, PG and Lys-PG. The fluorescence anisotropy values of Fl-AB1-OMe were recorded with increasing concentrations of lipids and the data are summarized in Fig. 3a. Interestingly, significant anisotropy increases were observed only with vesicles that present PE and Lys-PG, with other vesicle compositions eliciting marginal changes of anisotropy. Specifically, the presence of PG or PS did not induce more AB1 binding than PC-alone, showcasing the unique reactivity of PE and Lys-PG towards AB1. The lack of PS labelling by AB1 is perhaps surprising given that PS does display an amino group. This is presumably because the amino group of PS, in comparison to that of PE, is sterically more challenging for iminoboronate formation. This observation is consistent with a recent report, in which 2-APBA was found to preferentially react with lysine side chains over the main chain amino group[17]. Importantly, in contrast to Fl-AB1-OMe, Fl-AB2-OMe did not show significant association with the PC/PE or PC/Lys-PG vesicles (Fig. 3b), highlighting the importance of the boronic acid moiety in AB1 binding into vesicles.

To further validate the binding mechanism, we directly characterized the postulated iminoboronate conjugate of Lys-PG. Briefly, the PC/Lys-PG vesicles were treated with 2-APBA, the 'warhead' structure of AB1. Then the mixture was lyophilized, redissolved in $CDCl_3/CD_3OD$ (2:1) and subjected to $^{11}B$-NMR and mass spectrometry analysis. The $^{11}B$-NMR spectrum of the treated lipids displays a peak around 13 p.p.m. as expected for iminoboronate structures (Fig. 3c). The mass-spec data clearly present the molecular ions that correspond to the 2-APBA adduct of Lys-PG (Fig. 3d). Further, mass-spec analysis also reveals the iminoboronate conjugate of Lys-PG and an AB1-presenting peptide (Supplementary Fig. 2). These data consistently support

the iminoboronate mechanism for the association of AB1 with lipid membranes.

The iminoboronate mechanism predicts that the iminoboronate formation between AB1 and lipids can be inhibited by the presence lysine and lysine-presenting proteins. Indeed, lysine and bovine serum albumin (BSA) were found to disrupt the association of AB1 and PC/PE vesicle with an $IC_{50}$ of ~0.3 mM and 5 μM, respectively (Supplementary Fig. 3). It is interesting to note that BSA at 5 μM gives ~0.3 mM in lysine concentration given that BSA has a total of 59 lysine residues. In a later section, we will present strategies that minimize the protein interference of AB1-labelling lipids.

**AB1 selectively labels Gram-positive bacteria.** Encouraged by the model membrane studies, we sought to investigate the potential of AB1 in staining bacterial cells. Three strains of bacteria, including B. subtilis (American Type Culture Collection (ATCC) 663), S. aureus (ATCC 6538) and E. coli (BL 21), were selected as the initial set, which are known to have PE and/or Lys-PG as the major lipids of their plasma membranes[4-6]. The bacterial cells were stained with an Alexa Fluor 488 (AF488)-labelled AB1-OMe, which was chosen for cell studies because of the superior brightness and stability of the fluorophore. At concentrations below 1 μM, little fluorescence staining of the cells was observed with AF488-AB1-OMe. With higher concentrations, a quick washing procedure was included to minimize background fluorescence, after which the samples were immediately examined under an epi-fluorescence microscope (Fig. 4a). With wash, AF488-AB1-OMe effectively stained the two Gram-positive bacteria (B. subtilis and S. aureus) at ≥100 μM concentrations. In sharp contrast, the Gram-negative E. coli showed no fluorescence staining at all. As a negative control, AF488-AB2-OMe failed to stain any of the bacterial strains under the same conditions (Supplementary Fig. 4), showcasing the critical importance of the boronic acid moiety for bacteria labelling by AB1. The labelling can be inhibited by the addition of lysine (Supplementary Fig. 5) or BSA (Supplementary Fig. 6), lending further support to the iminoboronate mechanism of

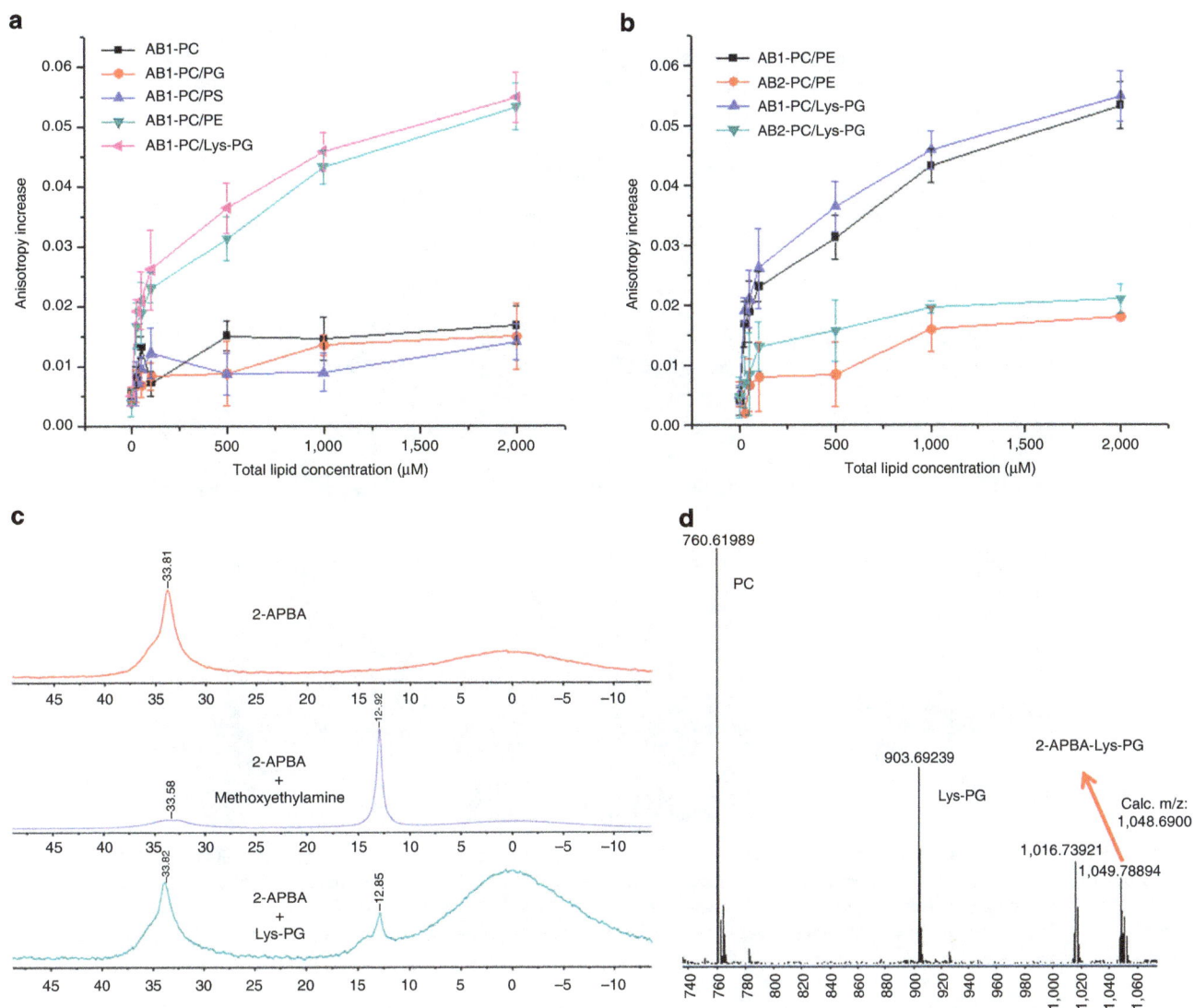

**Figure 3 | Iminoboronate formation on synthetic vesicles.** (**a**) Binding curves of Fl-AB1-OMe to lipid vesicles highlighting its selectivity for PE and Lys-PG. (**b**) Comparison of Fl-AB1-OMe and Fl-AB2-OMe for lipid binding showcasing the critical importance of the boronic acid moiety in vesicle association. All data points were measured with triplicate samples, from which error bars (s.e.m.) were generated. (**c**) $^{11}$B-NMR spectra of 2-APBA (2-acetylphenylboronic acid) and its conjugates with methoxyethylamine and Lys-PG. The peaks around 13 p.p.m. correspond to the iminoboronates and the broad peaks around 0 p.p.m. originate from the NMR tube. (**d**) Mass-spec analysis of the iminoboronate conjugation of 2-APBA to PC/Lys-PG vesicles. The specific lipids used are POPC (1-palmitoyl-2-oleoylphosphatidylcholine) and Lys-DOPG (lysyl 1,2-dioleoylphosphatidylglycerol). 2-APBA-Lys-PG denotes the iminoboronate conjugate of 2-APBA and Lys-DOPG.

conjugation. To gain more mechanistic insights, we analyzed the lipid extract of the *S. aureus* cells treated with 2-APBA: the $^{11}$B-NMR spectrum clearly revealed the characteristic peak ($\sim$13 p.p.m.) for iminoboronates (Supplementary Fig. 7), although our trials with mass-spec failed to identify the expected conjugates directly.

The failure of AB1 to stain *E. coli* is consistent with the fact that the outer membrane of *E. coli* does not have PE or Lys-PG. It further indicates that AF488-AB1-OMe is unable to permeate through the outer membrane to reach the plasma membrane, where PE does exist. Consistent with the membrane impermeability of AF488-AB1-OMe, we found that the fluorescence staining of *S. aureus* could be rapidly and completely washed away with a pH 5.0 buffer (Supplementary Fig. 8). The failure of *E. coli* staining suggests that AB1 does not label cell surface proteins under the experimental conditions, the exact mechanism of which remains to be further investigated. One possible explanation is that certain features of the membrane, such as

local membrane curvature[21], create kinetic traps for AB1. Supporting this hypothesis, we found that, after the initial washing step, the cell-bound AB1 molecules dissociate from the cell very slowly at neutral pH (Supplementary Fig. 8). Another possibility is that the number of surface proteins might be significantly smaller than that of lipids; consequently labelled surface proteins afford negligible fluorescence in comparison to labelled lipids.

We further assessed the selectivity of AB1 for bacteria over mammalian cells. Excitingly, when a co-culture of Jurkat lymphocytes and *S. aureus* cells was treated with AF488-AB1-OMe and analysed under a confocal microscope, strong fluorescence staining was observed for *S. aureus* cells, whereas the Jurkat cells were minimally labelled (Fig. 4b). Similar to the result of *E. coli* staining, the lack of Jurkat cell staining also indicates that AF488-AB1-OMe does not conjugate with cell surface proteins under our experimental conditions. Further, as we learned from the bacterial staining experiments, AF488-AB1-

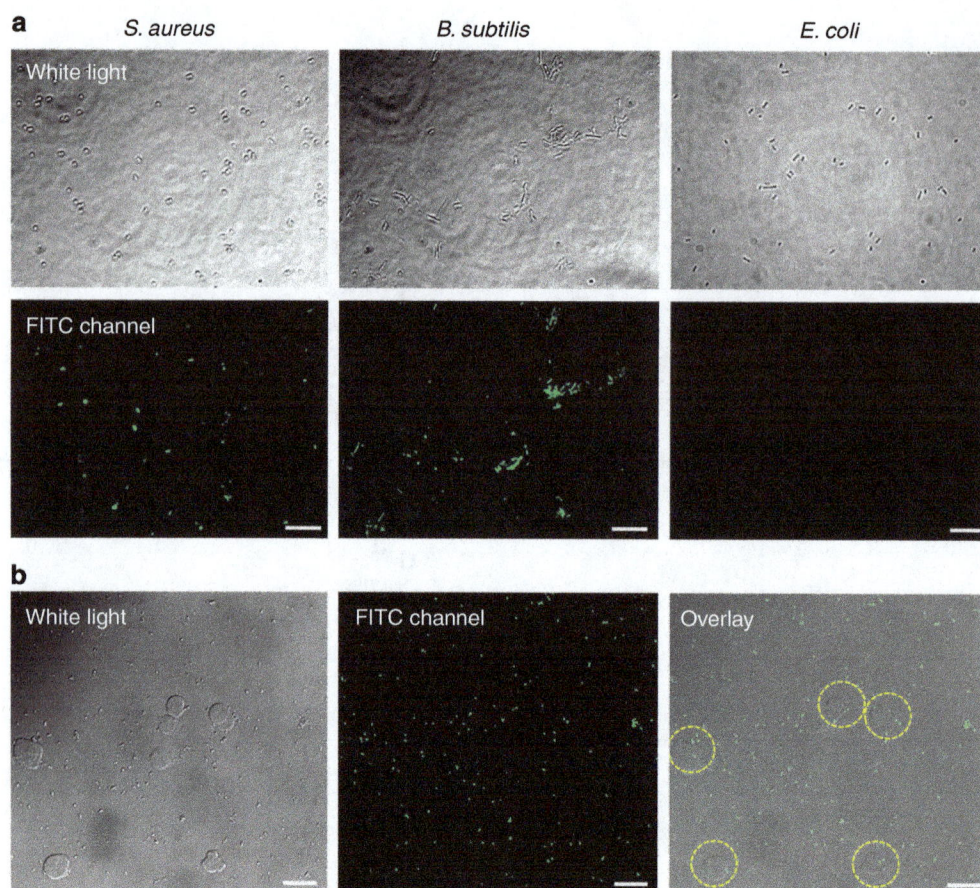

**Figure 4 | Assessing the selectivity of AB1 for various cell types.** (**a**) Microscopic images of three bacterial strains stained with 200 μM AF488-AB1-OMe (scale bar, 10 μm), showing that the AB1 derivative readily labels Gram-positive bacteria, but not Gram-negatives. (**b**) Confocal microscopic images of a mixed cell culture consisting of *S. aureus* and Jurkat lymphocytes stained with 100 μM AF488-AB1-OMe (scale bar, 25 μm). The Jurkat cells are highlighted by the yellow circles on the overlay image. These results collectively demonstrate the superb selectivity of AB1 for Gram-positive bacteria. FITC, fluorescein isothiocyanate.

OMe is membrane impermeable, which precludes labelling of intracellular targets. Finally and importantly, there are few AB1-reactive lipids on the outer surface of Jurkat cells: a mammalian cell does not unusually produce Lys-PG. Although PE can account for up to 20% of the total lipids of a mammalian cell[1], it is primarily confined to the cytosolic leaflet and therefore not available at the cell surface either[7].

**AB1 synergizes with cationic peptides for potent bacteria labelling.** Despite the remarkable selectivity for Gram-positive bacteria, simple AB1 derivatives like AF488-AB1-OMe suffer from the high concentrations needed to achieve effective bacteria labelling. Furthermore, dictated by the mechanism of iminoboronate formation, AB1 derivatives are expected to react with lysine and lysine residues of various proteins[17], which in turn inhibit the association of AB1 with membranes. For example, BSA was found to inhibit the bacterial cell labelling by AF488-AB1-OMe with an apparent $IC_{50}$ of $\sim 1.5 \, \text{mg ml}^{-1}$ ($\sim 22 \, \mu\text{M}$). At $10 \, \text{mg ml}^{-1}$ concentration, BSA resulted in $\sim 90\%$ reduction of the fluorescence staining of the AF488-AB1-OMe-treated *S. aureus* cells (Supplementary Fig. 6). We surmised that these problems could be resolved by conjugating AB1 to a directing functionality to bacterial cells. Towards this end, we have synthesized AB1 in its properly protected form (Fmoc-AB1(pin)-OH, Fig. 2) for solid-phase peptide synthesis, which should allow facile conjugation of AB1 to a variety of peptides or peptidomimetics that can serve as bacteria-directing motifs. Given that bacterial

cells are known to be enriched with negatively charged lipids, such as PG and cardiolipin, we thought to employ cationic peptides to direct AB1 to bacterial cell surfaces[22,23]. A small group of peptides were synthesized to incorporate AB1 as the C-terminal residue (Fig. 5a). The bacteria-targeting elements we tested include single cationic residues Lys and Arg, as well as a polycationic peptide Hlys[24,25] with the sequence of RYWVAWRNR. Hlys was reported to give a minimal inhibitory concentration of $24 \, \mu\text{M}$ against *S. aureus*, yet minimal haemolytic activity[24]. In addition, we chose Hlys because of its small size and the absence of lysine residues, which could in principle form an intramolecular iminoboronate with AB1. Nevertheless, we did not see intramolecular iminoboronate formation with the peptide K-AB1 (Supplementary Fig. 9). This lack of intramolecular conjugation is possibly due to the potential steric constraint that results from the fact that lysine and AB1 are contiguous in sequence. An analogous observation was recently reported for a cysteine-mediated macrocyclization with adjacent residues[26]. A control peptide (G-AB1) was also synthesized that incorporates a glycine instead of cationic motifs. All peptides were synthesized with an N-terminal cysteine so that they can be easily labelled with AF488-C5-maleimide (Fig. 5a).

The fluorescently labelled peptides were first assessed via flow cytometry analysis of the *S. aureus* cells stained with peptides at varied concentrations (Fig. 5b and Supplementary Fig. 10). The control peptide G-AB1 was only able to give a small fluorescence increase even at concentrations up to $100 \, \mu\text{M}$. This is consistent

**a**

| Name | Sequence |
|------|----------|
| G-AB1 | Ac-C*G(AB1)-NH$_2$ |
| K-AB1 | Ac-C*K(AB1)-NH$_2$ |
| R-AB1 | Ac-C*R(AB1)-NH$_2$ |
| Hlys-AB1 | Ac-C*RYWVAWRNRG(AB1)-NH$_2$ |
| Hlys | Ac-C*RYWVAWRNR-NH$_2$ |

**b**

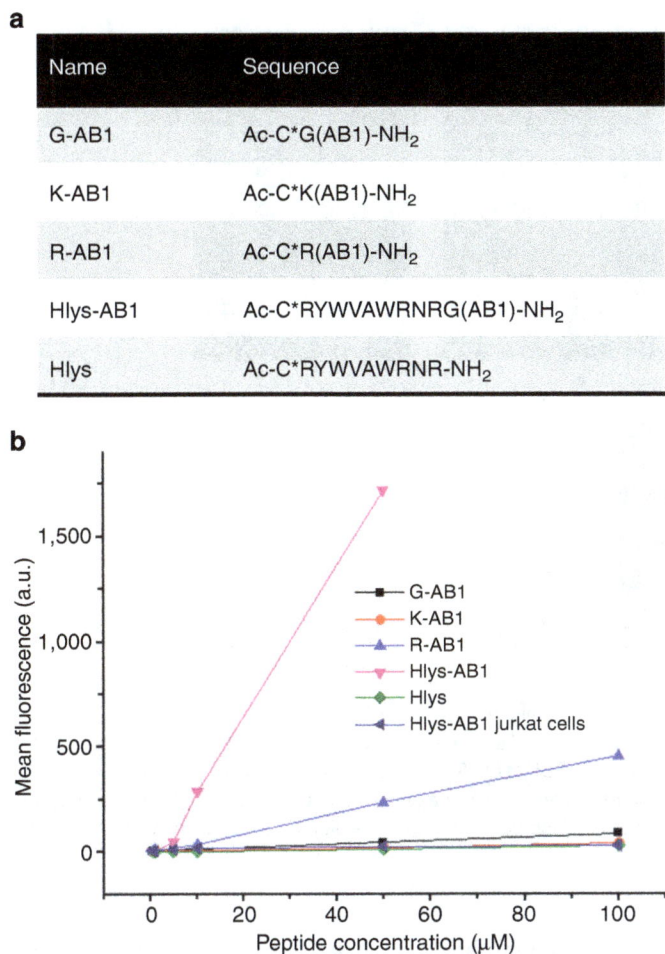

**Figure 5 | Synergizing covalent and noncovalent interactions for bacteria targeting. (a)** Sequences of AB1-presenting peptides, where C* represents a cysteine labelled with AF488-C5-maleimide. **(b)** Concentration profiles of *S. aureus* cell staining by the AB1-presenting peptides. All samples were prepared with the washing step right before analysis. The data for Jurkat cell staining with Hlys-AB1 were included to show its superb bacterial selectivity. The flow cytometry experiments were performed twice, which gave consistent results. One set of the data is presented herein.

appearance of dying or dead bacterial cells, which may stain with AB1 differently.

With much improved potency, we assessed Hlys-AB1 for bacteria labelling at nanomolar concentrations, with which the washing procedure is no longer necessary. The microscopic images show that, without wash, Hlys-AB1 effectively stained *S. aureus* cells at concentrations of 100 nM or higher (Fig. 6 and Supplementary Fig. 12). The confocal images revealed the cell envelope localization of the Hlys-AB1 (Supplementary Fig. 13), as expected for its membrane-targeting mechanism. Including the washing step in sample preparation resulted in approximately sevenfold reduction of the fluorescence staining of the cells (Supplementary Fig. 14). This is perhaps not surprising considering the reversible nature of the iminoboronate chemistry. Importantly, Hlys alone did not label the *S. aureus* cells under the same conditions (Fig. 6a), highlighting the critical importance of the AB1 moiety for the Hlys-AB1 staining of bacterial cells. Excitingly, Hlys-AB1 remained highly selective for Gram-positive bacteria under the no-wash conditions: the peptide failed to afford any fluorescence staining for the Gram-negative *E. coli* (Fig. 6b), as well as the Jurkat lymphocytes (Fig. 6e).

With the design of Hlys-AB1, the antimicrobial peptide Hlys is expected to selectively bind bacterial cell membranes and direct AB1 to covalently label PE or Lys-PG on bacterial cells. We hypothesized that this synergistic mechanism would not only improve the potency for bacteria labelling, but also minimize the protein interference of the iminoboronate chemistry. To prove this hypothesis, we assessed the inhibitory effect of fetal bovine serum (FBS) on the bacterial staining of Hlys-AB1. *S. aureus* cells were treated with Hlys-AB1 in presence of FBS at varied concentrations and the samples were analyzed with fluorescence microscopy and flow cytometry (Fig. 6d,f). The results show that, even with 10% FBS, submicromolar concentrations of Hlys-AB1 readily allowed the visualization of *S. aureus* cells under a fluorescence microscope (Fig. 6d), although reduced brightness was observed in comparison to the cells treated without FBS. Flow cytometry analysis yielded consistent results with microscopy: the presence of 10% FBS elicited ~30% reduction of the median fluorescence of the stained *S. aureus* cells (Fig. 6f). Again we attribute the much reduced protein interference of Hlys-AB1 to the synergy of covalent (AB1) and noncovalent (cationic peptide) mechanisms for bacterial cell targeting. The high potency and bacterial selectivity makes Hlys-AB1 potentially useful for targeting bacteria in blood serum or further in living organisms.

## Discussion

To summarize, we have demonstrated that the two major membrane lipids of bacterial cells, namely PE and Lys-PG, can be selectively targeted by synthetic molecules that induce formation of iminoboronate structures. Specifically, we have synthesized an unnatural amino acid (dubbed AB1) that displays a 2-APBA motif and can therefore conjugate with primary amines to form iminoboronates. By targeting the differential abundance and accessibility of Lys-PG and PE on cell surfaces, a fluorophore-labelled AB1 effectively stains Gram-positive bacteria (*B. subtilis* and *S. aureus*), bypassing Gram-negative bacteria and mammalian cells. Conjugating AB1 to cationic peptides greatly enhances its potency for bacteria labelling and importantly minimizes the interference of serum proteins to its bacterial association. Specifically, a hybrid peptide Hlys-AB1 was found to label *S. aureus* cells at nanomolar concentrations even in the presence of 10% FBS.

Nature primarily employs noncovalent mechanisms, such as hydrogen bonding, to achieve specific molecular recognition. Covalent chemistry has been largely avoided in targeting

with the fact that high concentrations of AF488-AB1-OMe are needed to achieve effective staining of bacterial cells. Conjugating AB1 to a lysine (K-AB1) did not improve, and perhaps even compromised AB1's association with *S. aureus*. In contrast, conjugation to an arginine (R-AB1) significantly enhanced the cell labelling. This contrasting results for K-AB1 and R-AB1 can be rationalized by the fact that an arginine side chain can afford stronger interaction with phospholipids than a lysine[27]. The AB1 conjugate with the polycationic peptide Hlys (Hlys-AB1) afforded a dramatic improvement of its potency for bacterial cell staining. For example, at 50 μM, Hlys-AB1 afforded a mean fluorescence intensity 8 times higher than that of R-AB1 and over 40 times better than G-AB1. In contrast, Hlys alone did not afford any fluorescence staining of *S. aureus* cells (Fig. 5b), again highlighting the critical importance of the AB1 moiety that covalently conjugates with the membrane lipids of bacterial cells. Importantly, Jurkat cells were not stained by Hlys-AB1 even with the highest concentration tested (Fig. 5b). Finally, we note that under our experimental conditions Hlys-AB1 caused marginal reduction of the viability of the bacterial cells (Supplementary Fig. 11), indicating the Hlys-enhanced staining is not due to the

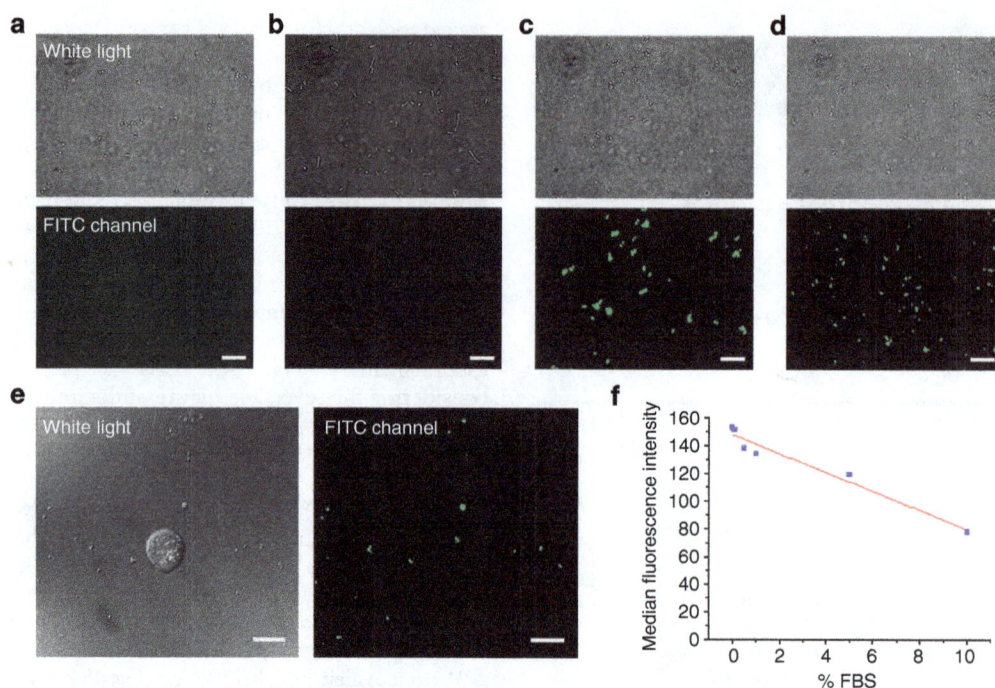

**Figure 6 | Bacterial labelling with submicromolar concentrations of Hlys-AB1.** (**a**) *S. aureus* cells treated with 0.5 μM Hlys as a negative control. (**b**) *E. coli* treated with 0.5 μM Hlys-AB1. (**c**) *S. aureus* cells stained with 0.5 μM Hlys-AB1. (**d**) *S. aureus* cells stained with 0.5 μM Hlys-AB1 in the presence of 10% FBS. (**e**) Confocal images of a *S. aureus* and Jurkat cell mixture stained with 0.5 μM Hlys-AB1. (**f**) FBS inhibition of *S. aureus* cell staining by Hlys-AB1 analysed via flow cytometry. 0.2 μM Hlys-AB1 was used for this experiment. The median fluorescence (*y*-axis) appears to give a linear relationship against the percentage of FBS (*x*-axis). The flow cytometry experiments were performed twice, which gave consistent results. One set of the data is presented. FITC, fluorescein isothiocyanate.

biomolecules of interest as irreversibility could result in modification of unintended targets and consequently toxicity[28,29]. However, reversible covalent chemistry circumvents this problem and should be able to complement the noncovalent mechanisms for molecular recognition. Among the limited number of examples, Wells and co-workers reported a strategy for protein ligand discovery that utilizes reversible disulfide chemistry to target reactive cysteines[30]. More recently, a group of nitrile modified acrylamides was reported by the Taunton group that reacts with cysteines of a protein kinase in a rapidly reversible manner[31]. In addition to targeting thiols, various boronic acid-presenting structures have been developed to target certain carbohydrates via reversible boronic ester formation[32,33]. Our work presented here, together with some recent publications by others[16,17], expands the reversible covalent chemistry toolbox for targeting biological amines. Although the iminoboronate chemistry was previously shown to label purified proteins[17] and aminosugars[16], the selectivity over other biomolecules has not been addressed. In comparison, our work here clearly demonstrates the applicability of the iminoboronate chemistry in complex biological systems (for example, bacteria labelling in the presence of blood serum).

It is highly desirable, yet challenging to differentiate various membrane lipids. A recent report[34] describes the covalent modification of mammalian aminophospholipids (PE and PS) on cell surfaces with an amine reactive reagent named sulfo-NHS-biotin, which allows the capture and quantification of externalized aminophospholipids. The nonselective reactivity of this reagent towards amines precludes its use in complex biological milieu. With the goal of better understanding lipid biology, a number of chemically modified lipids have been developed to display bioorthogonal reacting groups[35–38]. Once incorporated into a membrane, lipids as such can be selectively

labelled to reveal their subcellular distribution and homeostasis behaviour. However, these synthetic lipid probes are not known to afford specificity for bacterial cells. A recent report by Dumont *et al.* describes the metabolic incorporation of an azide-modified sugar into lipopolysaccharide[39], which enables fluorescence labelling of Gram-negative bacteria without genetic modification. Our work differs from these previous reports because the AB1 derivatives selectively target natural endogenous lipids of Gram-positive bacteria. For the purpose of bacterial detection, this contribution complements the elegant work by Dumont and co-workers, which is limited to selected Gram-negative bacteria.

The results of *in vitro* characterization presented here clearly demonstrate the superb bacteria selectivity, as well as the minimal serum interference, of AB1 and derivatives. Ongoing research in our lab seeks to further improve the potency of the AB1 derivatives for fast and more efficient labelling of bacterial cells, as well as to improve their stability towards proteolytic degradation. Our future research will evaluate the potential of optimized AB1 derivatives for biomedical applications, such as detecting bacteria in blood samples or imaging bacterial infection in animal models[40]. Finally, we submit that the covalent strategy for molecular recognition should be extendable to other important lipids of biological membranes. Research towards this end is also currently underway.

## Methods

**Materials and instrumentation.** Chemical reagents for small molecule and peptide synthesis were purchased from various vendors and used as received. The phospholipids were purchased from Avanti Polar Lipids (Alabaster, Al). PBS buffer, DMEM/high-glucose media, RPMI 1640 media and penicillin/streptomycin were purchased from Thermal Scientific. The Gram-positive bacteria (*B. subtilis* (ATCC 663) and *S. aureus* (ATCC 6538)) were purchased from Microbiologics as lyophilized cell pellet. *E. coli* (BL 21) was a gift from the lab of Professor Mary F.

Roberts at Boston College. NMR data of the small molecules were collected on a VNMRS 500 MHz NMR spectrometer. Mass spectrometry data were generated by using an Agilent 6230 LC TOF mass spectrometer. Peptide synthesis was carried out on a Tribute peptide synthesizer from Protein Technologies. The fluorescence anisotropy experiments were performed by using a SpectraMax M5 plate reader. Fluorescence images were taken on a Zeiss Axio Observer A1 inverted microscope. Confocal images were taken on the Leica SP5 confocal fluorescence microscope housed in the Biology Department of Boston College. Flow cytometry analyses were carried out on a BD FACSAria cell sorter also housed in the Biology Department of Boston College.

**Synthesis.** Details of the amino acid and peptide synthesis are provided as Supplementary Methods. Also presented in Supplementary Figs 15–25 are the NMR spectra of novel compounds, exemplary high-performance liquid chromatography traces of the fluorophore-labelled AB1 derivatives, and in Supplementary Table 1 are mass-spec data of the fluorophore-labelled amino acids and peptides.

**Binding assays with lipid vesicles.** Liposomes were prepared by dissolving and mixing the desired phospholipids in chloroform. After evaporating chloroform, the residue was suspended in 50 mM phosphate buffer, pH 7.4. The lipid suspensions were treated through 10 cycles of freeze-and-thaw process, and extruded 11 times through a membrane with pore size of 100 nm. The concentrations of liposome stocks were characterized via the Stewart Assay[41]. The size distribution of each vesicle sample was characterized with a dynamic light scattering instrument (DynaproTM NanoStar, Wyatt Technology Corp.). The diameter of all vesicles were found to fall into the narrow range of 100–120 nm. Lipid vesicles at varied concentrations (25, 50, 100, 500, 1,000, 2,000 $\mu$M total lipids) were incubated with 0.5 $\mu$M of Fl-AB1/2-OMe for 40 min in a phosphate buffer (50 mM Na•Pi, pH = 7.4). Then the fluorescence anisotropy values of each sample were recorded. To correct for the interference of light scattering, the lipid binding data of fluorescein isothiocyanate-alaninamide were used for blank subtraction. All samples were measured in triplicates and the data were averaged to generate the binding curves.

**NMR and mass spectrometry characterization of iminoboronates.** The PC/Lys-PG (3:2) vesicles (200 $\mu$l, 2 mM total lipids) were incubated with 2-APBA (200 $\mu$l, 10 mM) and Ac-R-AB1-amide (200 $\mu$l, 2 mM), respectively, for 40 min. Then the mixtures were lyophilized and dissolved in CDCl$_3$:CD$_3$OD (2:1; 600 $\mu$l). The iminoboronate formation was confirmed by $^{11}$B NMR and mass spectrometry. All the $^{11}$B-NMR experiments were carried out with BF$_3$ as an external standard, the chemical shift of which was set at 0 p.p.m. BF$_3$ was not used as an internal standard because of its acidic nature, which might disrupt the iminoboronate conjugates.

**Bacterial cell culture and staining.** Bacterial staining experiments were performed against three strains: B. subtilis (ATCC 663), S. aureus (ATCC 6538) and E. coli (BL21). For each strain, bacterial cells from a single colony were grown overnight in LB broth at 37 °C with agitation. An aliquot was taken and diluted (1:50 for E. coli, 1:20 for B. subtilis, 1:200 for S. aureus) in fresh broth and cultured for another ~3 h until the cells reached the mid-logarithmic phase (OD$_{600}$ ~0.5). Then the bacterial cell culture was diluted ten times and used immediately for small molecule labelling. For a typical labelling experiment, 100 $\mu$l of the diluted bacterial cell culture was spun down at 7,000 r.p.m. in a centrifuge tube (1.5 ml). The cells were washed once with 100 $\mu$l phosphate buffer (50 mM Na•Pi, pH = 7.4), and then mixed with 100 $\mu$l solution of an AB1 derivative at desired concentrations. After 40 min incubation, the samples with low AB1 concentrations ($\leq 1\,\mu$M) were directly analyzed. The samples with higher AB1 concentrations were subjected to a washing procedure: the cells were spun down at 7,000 r.p.m. and the supernatant was discarded. Then the cells were washed twice with the phosphate buffer (100 $\mu$l, 2 min incubation), after which the spun-down cells were re-suspended in 50 $\mu$l of the phosphate buffer for analysis.

**Mammalian cell culture and staining.** Jurkat cells were grown and maintained in RPMI 1640 media with 10% FBS and 1% penicillin/streptomycin at 37 °C, 5% CO$_2$ and passed for less than 50 generations. The cell viability and density was checked and counted daily by using 0.2 $\mu$M trypan blue as a viability testing dye on a haemocytometer. Before staining with a small molecule, the cells were cultured to a density of 1.5–2.0 × 10$^6$ cells per ml in a Corning cell culture flask (with vent cap). Small-molecule staining was carried out by a similar protocol as used for the bacterial cells except the speed of centrifugation (Jurkat cells were spun down at 200 r.c.f. (relative centrifugal force)). Samples with high AB1 concentrations (>1 $\mu$M) were washed right before analysis.

**Co-culture preparation.** For labelling with AF488-AB1-OMe, 500 $\mu$l of Jurkat (1.5~2.0 × 10$^6$ cells per ml) cells and 100 $\mu$l of S. aureus (2-3 × 10$^8$ cells per ml) were separately stained with 100 $\mu$M AF488-AB1-OMe in centrifuge tubes for 40 min. Then the Jurkat and S. aureus cells were spun down at 200 r.c.f. and

7,000 r.p.m., respectively. The supernatants were discarded and the cells were further washed twice with 100 $\mu$l of the phosphate buffer. Finally, the Jurkat and S. aureus cells were re-suspended in 50 $\mu$l of phosphate buffer for imaging study. For the labelling experiment with Hlys-AB1, 500 $\mu$l of Jurkat (1.5~2.0 × 10$^6$ cells per ml) cells and 100 $\mu$l of S. aureus (2-3 × 10$^8$ cells per ml) were mixed. The mixture was incubated with 0.5 $\mu$M Hlys-AB1 for 40 min and then immediately subjected to microscopy analysis.

**Microscopic analysis of AB1-stained cells.** For epi-fluorescence microscopy, 5 $\mu$l of the bacterial cell suspension was dropped on a glass slide (Fisherfinest premium, 75 × 25 × 1 mm$^3$). A coverslip (Fisherbrand, 22 × 22 × 0.15 mm$^3$) was pressed down on the cell droplet to give a single layer of cells on the glass slides. White light and fluorescence images were taken on a Zeiss Axio Observer A1 inverted microscope equipped with a filter cube (488 nm excitation, 515–520 nm emission) suitable for detection of AF488 fluorescence. A Plan-NeoFluar × 100 oil objective from Zeiss was used to visualize the bacterial cells. All images were captured with the exposure time of 300 ms for AF488-labelled AB1 derivatives. All fluorescence images were processed following a fixed protocol with the software Fiji ImageJ[42]. For confocal analysis, 5 $\mu$l of cells were placed on a glass slide and a 22 × 22 × 1.5 Fisherbrand microscope cover glass was placed on top. Images were taken on a Leica SP5 confocal fluorescence microscope with filters that allowed detection of AF488 (488 nm excitation, 496–564 nm emission). A × 63 oil objective was used with an Argon laser at 10% laser power. Gain was adjusted to between 900 HV and 1,100 HV with an offset of − 0.5%. The images were captured with the software LAS 2.6 and then processed with Fiji ImageJ[42].

**Flow cytometry analysis of AB1 stained cells.** The samples were prepared and stained following the same protocol described for microscopy. The cells stained with sub-micromolar concentrations of Hlys-AB1 were analyzed without wash, while all other samples were subjected to the wash procedure right before analysis. The samples were analyzed on a BD FACSAria cell sorter (BD Biosciences). Data analysis was performed with FlowJo (Tree Star, Inc.), from which the median fluorescence intensities of the stained cells were extracted and plotted against AB1 concentration. For the protein inhibition experiments, the cell samples were prepared in the presence of BSA or FBS at desired concentrations before the addition of the AB1 compounds. The median fluorescence intensity of these cell samples was extracted and plotted against BSA or FBS concentration.

## References

1. van Meer, G., Voelker, D. R. & Feigenson, G. W. Membrane lipids: where they are and how they behave. *Nat. Rev. Mol. Cell Biol.* 9, 112–124 (2008).
2. Shevchenko, A. & Simons, K. Lipidomics: coming to grips with lipid diversity. *Nat. Rev. Mol. Cell Biol.* 11, 593–598 (2010).
3. Wenk, M. R. Lipidomics: new tools and applications. *Cell* 143, 888–895 (2010).
4. Epand, R. F., Schmitt, M. A., Gellman, S. H. & Epand, R. M. Role of membrane lipids in the mechanism of bacterial species selective toxicity by two alpha/beta-antimicrobial peptides. *Biochim. Biophys. Acta* 1758, 1343–1350 (2006).
5. Epand, R. F., Savage, P. B. & Epand, R. M. Bacterial lipid composition and the antimicrobial efficacy of cationic steroid compounds (Ceragenins). *Biochim. Biophys. Acta* 1768, 2500–2509 (2007).
6. Roy, H. Tuning the properties of the bacterial membrane with aminoacylated phosphatidylglycerol. *IUBMB Life* 61, 940–953 (2009).
7. Balasubramanian, K. & Schroit, A. J. Aminophospholipid asymmetry: a matter of life and death. *Annu. Rev. Physiol.* 65, 701–734 (2003).
8. Yeung, T. et al. Membrane phosphatidylserine regulates surface charge and protein localization. *Science* 319, 210–213 (2008).
9. Lemmon, M. A. Membrane recognition by phospholipid-binding domains. *Nat. Rev. Mol. Cell Biol.* 9, 99–111 (2008).
10. Gao, J. & Zheng, H. Illuminating the lipidome to advance biomedical research: peptide-based probes of membrane lipids. *Future Med. Chem.* 5, 947–959 (2013).
11. Crugeiras, J., Rios, A., Riveiros, E., Amyes, T. L. & Richard, J. P. Glycine enolates: the effect of formation of iminium ions to simple ketones on alpha-amino carbon acidity and a comparison with pyridoxal iminium ions. *J. Am. Chem. Soc.* 130, 2041–2050 (2008).
12. McFarland, J. M. & Francis, M. B. Reductive alkylation of proteins using iridium catalyzed transfer hydrogenation. *J. Am. Chem. Soc.* 127, 13490–13491 (2005).
13. Arnal-Herault, C. et al. Functional G-quartet macroscopic membrane films. *Angew. Chem. Int. Ed.* 46, 8409–8413 (2007).
14. Hutin, M., Bernardinelli, G. & Nitschke, J. R. An iminoboronate construction set for subcomponent self-assembly. *Chemistry* 14, 4585–4593 (2008).
15. Galbraith, E. et al. Dynamic covalent self-assembled macrocycles prepared from 2-formyl-aryl-boronic acids and 1,2-amino alcohols. *N. J. Chem.* 33, 181–185 (2009).
16. Gutierrez-Moreno, N. J., Medrano, F. & Yatsimirsky, A. K. Schiff base formation and recognition of amino sugars, aminoglycosides and biological

polyamines by 2-formyl phenylboronic acid in aqueous solution. *Org. Biomol. Chem.* **10**, 6960–6972 (2012).

17. Cal, P. M. *et al.* Iminoboronates: a new strategy for reversible protein modification. *J. Am. Chem. Soc.* **134**, 10299–10305 (2012).

18. Hoyle, C. E. & Bowman, C. N. Thiol-ene click chemistry. *Angew. Chem. Int. Ed.* **49**, 1540–1573 (2010).

19. Ishiyama, T., Itoh, Y., Kitano, T. & Miyaura, N. Synthesis of arylboronates via the palladium(0)-catalyzed cross-coupling reaction of tetra(alkoxo)diborons with aryl triflates. *Tetrahedron Lett.* **38**, 3447–3450 (1997).

20. Lozada, J., Liu, Z. & Perrin, D. M. Base-promoted protodeboronation of 2,6-disubstituted arylboronic acids. *J. Org. Chem.* **79**, 5365–5368 (2014).

21. McMahon, H. T. & Gallop, J. L. Membrane curvature and mechanisms of dynamic cell membrane remodelling. *Nature* **438**, 590–596 (2005).

22. Hancock, R. E. & Diamond, G. The role of cationic antimicrobial peptides in innate host defences. *Trends Microbiol.* **8**, 402–410 (2000).

23. Tew, G. N., Scott, R. W., Klein, M. L. & Degrado, W. F. *De novo* design of antimicrobial polymers, foldamers, and small molecules: from discovery to practical applications. *Acc. Chem. Res.* **43**, 30–39 (2010).

24. Gonzalez, R., Albericio, F., Cascone, O. & Iannucci, N. B. Improved antimicrobial activity of h-lysozyme (107-115) by rational Ala substitution. *J. Pept. Sci.* **16**, 424–429 (2010).

25. Iannucci, N. B., Curto, L. M., Albericio, F., Cascone, O. & Delfino, J. M. Structure-activity relationship analysis of a novel antimicrobial peptide derived from the 107-115 h-lysozyme fragment. *Biopolymers* **100**, 279–279 (2013).

26. Bionda, N., Cryan, A. L. & Fasan, R. Bioinspired strategy for the ribosomal synthesis of thioether-bridged macrocyclic peptides in bacteria. *ACS Chem. Biol.* **9**, 2008–2013 (2014).

27. Tang, M., Waring, A. J., Lehrer, R. I. & Hong, M. Effects of guanidinium-phosphate hydrogen bonding on the membrane-bound structure and activity of an arginine-rich membrane peptide from solid-state NMR spectroscopy. *Angew. Chem. Int. Ed.* **47**, 3202–3205 (2008).

28. Singh, J., Petter, R. C., Baillie, T. A. & Whitty, A. The resurgence of covalent drugs. *Nat. Rev. Drug. Discov.* **10**, 307–317 (2011).

29. Johnson, D. S., Weerapana, E. & Cravatt, B. F. Strategies for discovering and derisking covalent, irreversible enzyme inhibitors. *Future Med. Chem* **2**, 949–964 (2010).

30. Erlanson, D. A. *et al.* Site-directed ligand discovery. *Proc. Natl Acad. Sci. USA* **97**, 9367–9372 (2000).

31. Serafimova, I. M. *et al.* Reversible targeting of noncatalytic cysteines with chemically tuned electrophiles. *Nat. Chem. Biol.* **8**, 471–476 (2012).

32. James, T. D., Sandanayake, K. R. A. S. & Shinkai, S. Saccharide sensing with molecular receptors based on boronic acid. *Angew. Chem. Int. Ed.* **35**, 1910–1922 (1996).

33. Dai, C. F. *et al.* Carbohydrate biomarker recognition using synthetic lectin mimics. *Pure Appl. Chem.* **84**, 2479–2498 (2012).

34. Thomas, C. P. *et al.* Identification and quantification of aminophospholipid molecular species on the surface of apoptotic and activated cells. *Nat. Protoc.* **9**, 51–63 (2014).

35. Neef, A. B. & Schultz, C. Selective fluorescence labeling of lipids in living cells. *Angew. Chem. Int. Ed.* **48**, 1498–1500 (2009).

36. Best, M. D., Rowland, M. M. & Bostic, H. E. Exploiting bioorthogonal chemistry to elucidate protein-lipid binding interactions and other biological roles of phospholipids. *Acc. Chem. Res.* **44**, 686–698 (2011).

37. Yang, J., Seckute, J., Cole, C. M. & Devaraj, N. K. Live-cell imaging of cyclopropene tags with fluorogenic tetrazine cycloadditions. *Angew. Chem. Int. Ed.* **51**, 7476–7479 (2012).

38. Erdmann, R. S. *et al.* Super-resolution imaging of the Golgi in live cells with a bioorthogonal ceramide probe. *Angew. Chem. Int. Ed.* **53**, 10242–10246 (2014).

39. Dumont, A., Malleron, A., Awwad, M., Dukan, S. & Vauzeilles, B. Click-mediated labeling of bacterial membranes through metabolic modification of the lipopolysaccharide inner core. *Angew. Chem. Int. Ed.* **51**, 3143–3146 (2012).

40. Panizzi, P. *et al. In vivo* detection of Staphylococcus aureus endocarditis by targeting pathogen-specific prothrombin activation. *Nature Med.* **17**, 1142–1153 (2011).

41. Stewart, J. C. M. Colorimetric Determination of Phospholipids with Ammonium Ferrothiocyanate. *Anal. Biochem.* **104**, 10–14 (1980).

42. Schindelin, J. *et al.* Fiji: an open-source platform for biological-image analysis. *Nat. Methods* **9**, 676–682 (2012).

## Acknowledgements

We gratefully acknowledge the financial support provided by the Boston College, the National Science Foundation (CHE1112188), and the National Institute of General Medical Sciences (R01GM102735). We also thank Dr Bret Judson for his help on fluorescence microscopy and Dr Patrick Autissier for his help on the flow cytometry experiments.

## Author Contributions

J.G. and A.B. conceived the project, analyzed the data and wrote the manuscript; A.B. performed the majority of the experiments; K.A.M. assisted in the synthesis of the AB1 derivatives; M.A.K. performed the confocal microscopy work with Jurkat cells.

## Additional Information

# 17

# Super-resolution microscopy reveals structural diversity in molecular exchange among peptide amphiphile nanofibres

Ricardo M.P. da Silva[1,2,3], Daan van der Zwaag[2], Lorenzo Albertazzi[2,4], Sungsoo S. Lee[5], E.W. Meijer[2] & Samuel I. Stupp[1,5,6,7,8]

The dynamic behaviour of supramolecular systems is an important dimension of their potential functions. Here, we report on the use of stochastic optical reconstruction microscopy to study the molecular exchange of peptide amphiphile nanofibres, supramolecular systems known to have important biomedical functions. Solutions of nanofibres labelled with different dyes (Cy3 and Cy5) were mixed, and the distribution of dyes inserting into initially single-colour nanofibres was quantified using correlative image analysis. Our observations are consistent with an exchange mechanism involving monomers or small clusters of molecules inserting randomly into a fibre. Different exchange rates are observed within the same fibre, suggesting that local cohesive structures exist on the basis of β-sheet discontinuous domains. The results reported here show that peptide amphiphile supramolecular systems can be dynamic and that their intermolecular interactions affect exchange patterns. This information can be used to generate useful aggregate morphologies for improved biomedical function.

[1] Simpson Querrey Institute for BioNanotechnology (SQI), Northwestern University, Chicago, Illinois 60611, USA. [2] Laboratory of Macromolecular and Organic Chemistry and Institute for Complex Molecular Systems, Eindhoven University of Technology, Eindhoven MB 5600, The Netherlands. [3] Craniofacial Development & Stem Cell Biology, King's College London, London, SE1 9RT, UK. [4] Nanoscopy for Nanomedicine Group, Institute for Bioengineering of Catalonia (IBEC), Barcelona 08028, Spain. [5] Department of Materials Science and Engineering, Northwestern University, Evanston, Illinois 60208, USA. [6] Department of Chemistry, Northwestern University, Evanston, Illinois 60208, USA. [7] Department of Medicine, Northwestern University, Chicago, Illinois 60611, USA. [8] Department of Biomedical Engineering, Northwestern University, Evanston, Illinois 60208, USA. Correspondence and requests for materials should be addressed to E.W.M. (email: e.w.meijer@tue.nl) or to S.I.S. (email: s-stupp@northwestern.edu).

Reversible supramolecular interactions are ubiquitous in nature, controlling the self-assembly of ordered functional structures that need to be dynamic to perform their biological functions. One-dimensional cytoskeletal filaments such as actin and tubulin are typical examples of structures that use dynamics to mediate the adaptive behaviour of cells, resulting in cell motility, shape change, cell division, signalling and muscular contraction at larger length scales[1–6]. Artificial supramolecular materials could offer this bio-inspired dynamic behaviour, thus allowing enhanced interaction with natural systems and increased biomedical functionality. Since many natural processes are carefully regulated, optimization of an artificial supramolecular material requires a detailed understanding of its dynamic properties, for example its exchange kinetics.

Molecular mixing experiments typically assess exchange kinetics by utilizing Förster resonance energy transfer (FRET) between a pair of donor and acceptor fluorophores[7,8], and alternatively, radio-labelled molecules[9] or time-resolved small-angle neutron scattering[10,11]. Although these ensemble experiments can provide the timescale of the processes, they cannot distinguish different mechanisms. Moreover, they fail to detect the structural diversity among fibres, or within an individual fibre, for example, the occurrence of segregated domains. Local variations of molecular composition can have important biological implications, as they can greatly influence the signalling potency through multivalency effects[12–14], and thus understanding the exchange heterogeneity is important to design materials in which function is connected with dynamics for adaptive or responsive behaviour[15].

Super-resolution techniques are powerful tools to reveal the spatial distribution of molecules at the nanoscale, but these techniques have thus far been mainly applied to imaging fine details of cellular structures[16,17]. For instance, a resolution of $\sim 20$ nm can be achieved using stochastic optical reconstruction microscopy (STORM)[18], which is an order of magnitude below the diffraction limit and near the molecular scale. The enhanced resolution is achieved through the accurate localization of single fluorescent molecules; to identify individual fluorophores only a sparse subset of labels should be active at any given time. This sparse population of fluorophores is obtained using probes that can be photo-switched to a temporary non-fluorescent 'off' state by light. By repeatedly activating different subsets and overlaying the resulting localizations, an image can be reconstructed. We have previously reported how to apply STORM to probe the dynamics of supramolecular fibres[19]. Using two-colour STORM and quantitative image analysis, we were able to resolve the monomer distribution along the fibre backbone. By following the monomer distribution during the molecular exchange process, we were able to infer the exchange mechanism.

Peptide amphiphiles (PAs) that self-assemble into high aspect ratio objects offer exciting opportunities for regenerative medicine and other therapeutic applications[20–24]. As illustrated in Fig. 1a, this class of molecules is composed of an unbranched alkyl chain linked to a peptide segment, which can be further subdivided in several domains. A segment with propensity to form β-sheets is conjugated to the alkyl tail, followed by a charged segment for solubility. Additional domains can be conjugated to the canonical structure to introduce biofunctionality in the nanofibres. Whereas the hydrophobic collapse of the aliphatic tails induces self-assembly, experimental[25,26] and theoretical[27] evidence suggests that the formation of directional hydrogen bonds within the β-sheet domain is an additional important component of the driving force for assembly of the molecules into one-dimensional filamentous shapes. The facile incorporation of multiple bioactive signals at controlled concentrations[14,28], together with their structural resemblance with extracellular matrix fibres makes PA assemblies useful as bioactive artificial extracellular matrix components for cell signalling. Furthermore, they are also intrinsically biocompatible and biodegradable and can therefore disappear easily after fulfilling their biological functions. PAs have been extensively studied as a platform for applications that include bone, cartilage, enamel, neuronal regeneration, angiogenesis for ischaemic disease, targeted drug delivery and cancer therapeutics[20–24,29].

Ensemble measurements of PA dynamics revealed partial self-healing using rheological techniques[30] or the existence of a critical aggregation concentration[31], while kinetic spectroscopy showed pathway selection of PAs into different morphologies[32,33]. However, these approaches do not consider structural heterogeneity and its effect on dynamic molecular exchange at the level of individual filaments. In this paper, we use STORM to image individual PA nanofibres, addressing the distribution of molecules along the fibre during exchange and therefore analysing diversity among supramolecular nanofibres.

## Results

**Nanofibre design and preparation.** The PA molecule studied in this work is shown in Fig. 1a. It consists of a palmitic acid tail, six alanines in the β-sheet-forming region, followed by three glutamic acids as charged solubilizing moieties. This PA molecule self-assembles in water, forming nanofibres around 7 nm in diameter and lengths in the range of micrometres as observed by cryogenic transmission electron microscopy (cryoTEM; Fig. 1e). Amino acids with different propensities to form β-sheets affect considerably the internal order of PA assemblies, as well as nanofibre stiffness[26]. Since alanine has a weaker tendency to form β-sheets than valine[34], placing this amino acid in the β-sheet-forming segment was expected to yield relatively dynamic PA fibres. Circular dichroism spectroscopy of this PA is consistent with the typical β-sheet conformation found for other PAs[26], at physiological pH and ionic strength (Fig. 1c). To perform imaging by fluorescence microscopy, PA molecules were labelled with cyanine dyes, namely Cy3 and Cy5 (Fig. 1b). Water soluble sulfonated dyes with a net charge of $-1$ were used to preserve the amphiphilic asymmetry of the PA molecule and, since sulfonate groups are fully ionized in the vicinity of physiological pH, their behaviour is insensitive to pH in the range of interest. The Cy3 and Cy5 dyes have been chosen for their suitable photophysical properties for STORM imaging; moreover, they constitute a good FRET pair with a Förster radius of 50 Å (ref. 35). Single-colour labelled nanofibres were created by mixing a stock solution of either Cy3-PA or Cy5-PA with a stock solution of unlabelled PA, lyophilization, brief co-dissolution in trifluoroacetic acid (TFA) and immediate TFA removal, lyophilization and final reconstitution in the aqueous working solution to form labelled supramolecular aggregates. The degree of labelling was accurately controlled by premixing the different PAs at the desired ratios. Fluorescently labelled nanofibres revealed a morphology that was indistinguishable from their non-labelled counterparts, as shown by cryoTEM (Fig. 1f).

**Ensemble molecular exchange kinetics.** The timescale of molecular exchange in PA-based nanofibres was measured using FRET kinetic experiments. Two sets of PA nanofibres were separately pre-assembled from a mixture of non-labelled PAs with either Cy3-PA (0.5%) or Cy5-PA (0.5%), as illustrated in Fig. 2a. Next, the two solutions were mixed and the FRET ratio, defined as the ratio between the fluorescence intensities of Cy5 acceptor and Cy3 donor, was monitored over time. As shown in Fig. 2b, the FRET ratio increases with time, reaching a plateau after several hours. This means that PA molecules are able to migrate between

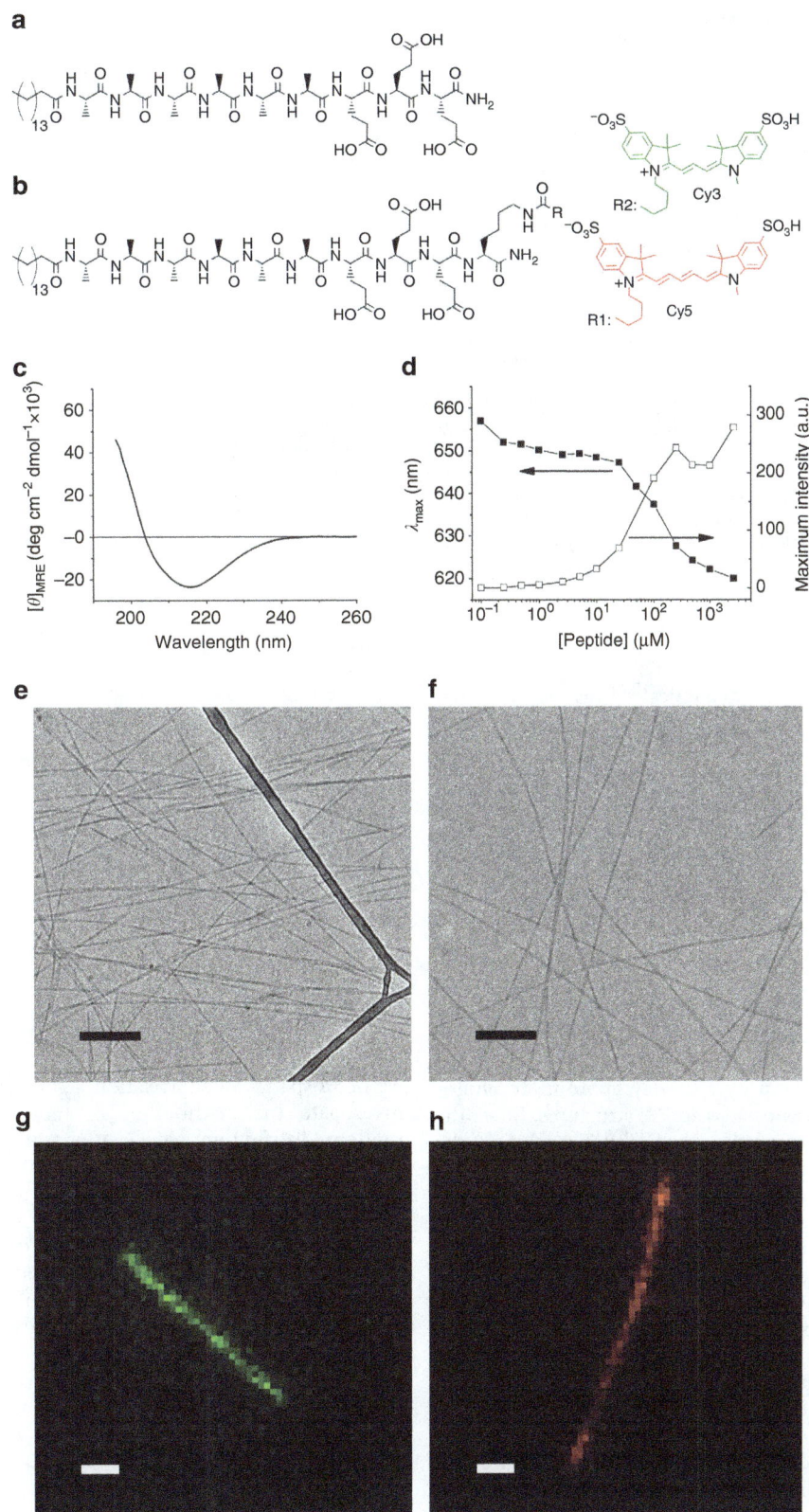

**Figure 1 | PA self-assembly.** Molecular structure of (**a**) non-labelled PA and (**b**) PA molecules labelled with photo-switchable sulfonated cyanine dyes, namely Cy3 (green) and Cy5 (red). (**c**) Circular dichroism spectrum and (**d**) Nile Red assay of non-labelled PA. CryoTEM images of nanofibres self-assembled at pH 7.5 and NaCl 150 mM (**e**) from non-labelled PA alone and (**f**) from a molecular mixture of non-labelled and Cy5-labelled PAs (scale bar, 200 nm). Diffraction-limited fluorescence microscopy images of Cy3- (**g**) and Cy5-labelled (**h**) PA nanofibres (scale bar, 1 μm).

nanofibres, resulting in mixed fluorophore fibres. Figure 2b depicts kinetic experiments at different temperatures, showing that the exchange rate is remarkably faster at 37 °C than at 20 °C. On the other hand, we observed a limited effect of concentration on the exchange rate (Fig. 2c), proving that the exchange process is not diffusion limited in this concentration range. These results resemble observations on self-assembled polymeric micelles, in which unimer expulsion and insertion is thought to be the

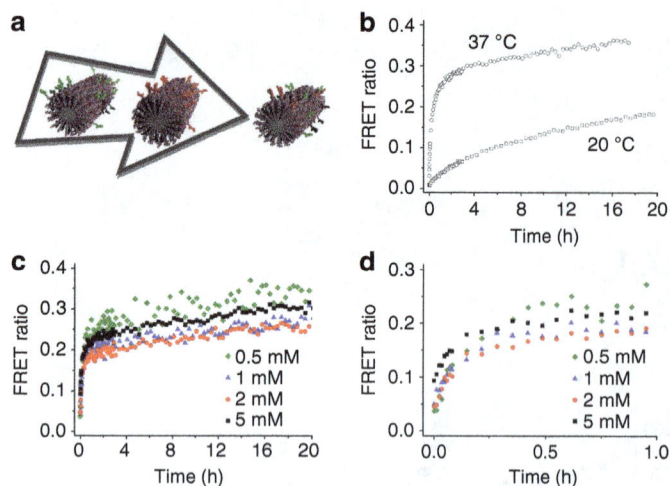

**Figure 2 | Molecular exchange by FRET. (a)** Schematic representation of a molecular exchange kinetic measurement. The molecular exchange progress over time is estimated by means of FRET ratio (dividing Cy5 by Cy3 fluorescence intensities), either at (**b**) a constant PA concentration (0.5 mM) or at (**c,d**) a constant temperature (37 °C), where **d** shows the FRET ratio for short timescale.

rate-determining step of system dynamics[10,11], as well as other synthetic supramolecular polymers[7]. So, while FRET measurements provide useful information about the timescale of the exchange, FRET is an ensemble technique, and it does not provide the spatially resolved information required to elucidate the mechanism of exchange.

**Imaging PA nanofibres.** Due to its high spatial resolution and multicolour imaging ability, STORM microscopy can be used to investigate the spatial details of the exchange process in PA nanofibres. Cy3- or Cy5-labelled PA nanofibres have been immobilized on a glass slide by physisorption and subsequently have been imaged using STORM (see the 'Methods' Section for details). The resulting images show well-reconstructed aggregates (Fig. 3a,b), indicating that both dyes display appropriate photo-switching behaviour when associated to PA structures. Since the actual width of the nanofibre is below the STORM resolution, the apparent thickness of the fibre can be used to provide an estimate for the experimental resolution of these measurements, approximately equal to 50 nm (Supplementary Fig. 1). The fibre characteristics observed by STORM, for example, rigidity and fibre length, match those reported by the cryoTEM images in Fig. 1e,f.

To analyse the monomer distribution inside PA nanofibres, which is crucial to understand the exchange mechanism, we utilized an image analysis routine previously developed in the Meijer laboratory[19]. This method removes background localizations, identifies the contour of the fibre backbone to study its mechanical properties (Supplementary Fig. 2) and computes the localization density along the polymer. Figure 3c,d display the localization density plots for Cy5- and Cy3-labelled PA, respectively. These profiles contain information about the distribution of the dye-functionalized PA molecules in a nanofibre. As can be clearly observed, the number of localizations shows fluctuations along the fibre. These fluctuations can be attributed to two causes: (i) a heterogeneous distribution of monomers or (ii) the stochastic processes taking place during STORM image acquisition, for example, fluorophore blinking[36] and fluorophore bleaching. Therefore, variations in the density of localizations cannot be unequivocally attributed to changes in local monomer concentration. To address this issue,

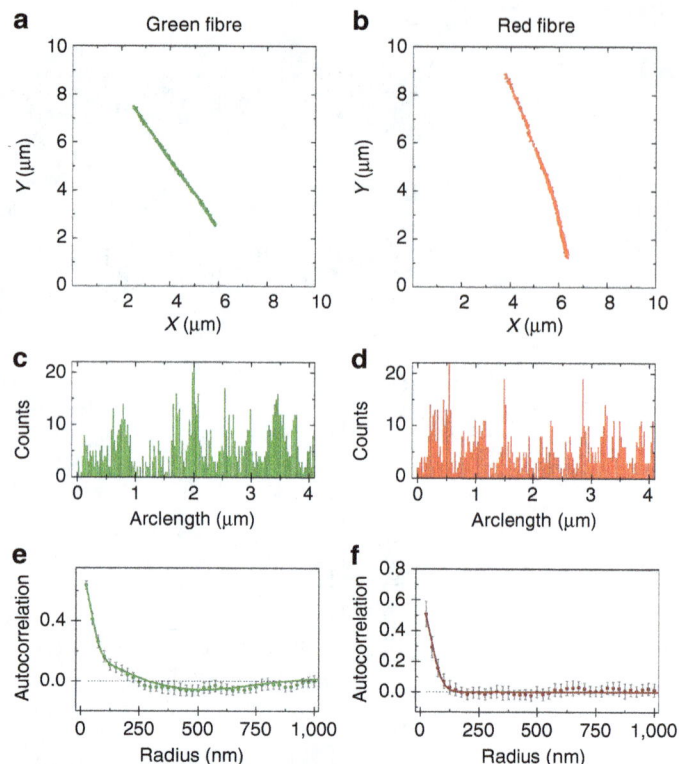

**Figure 3 | STORM analysis of single-colour nanofibres.** PA nanofibres were labelled with 5 mol% of either Cy3-PA (**a,c,e**) or Cy5-PA (**b,d,f**) before the molecular exchange experiment. Localization maps after applying a clustering algorithm, background cleaning and backbone finding (**a,b**). Each fibre localization distribution profile was determined along the fibre backbone (arc length) using a bin size of 25 nm (**c,d**). Averaged autocorrelation (**e,f**) was computed from distribution profiles of a large set of fibre images ($n \geq 19$). Solid lines correspond to model fittings (equation (1) for red channel and equation (2) for green channel) and error bars represent 95% confidence level.

we use a spatial autocorrelation algorithm, a powerful method to investigate the distribution of these localizations. In purely random distributions, localizations at distance $r$ apart are not correlated with each other, by definition giving an autocorrelation $g(r)$ with a value of zero. Positive values of $g(r)$ represent an increased probability that a second localization is found at a certain distance ($r$) from a first given localization, while negative values indicate a decreased probability. The autocorrelation curves of fibres in single-colour samples are shown in Fig. 3e,f. The autocorrelation of Cy5-labelled nanofibres (Fig. 3f) is strongly positive in the short range (< 100 nm) and zero for longer distances. This short-ranged contribution is the typical signature for multiple localizations of the same dye (overcounting) and is present in all STORM images. As previously described, the overcounting contribution in the spatial autocorrelation traces can be modelled using equation (1) (ref. 37),

$$g(r) = A_1 exp(-r^2/4\sigma^2) \qquad (1)$$

where the standard deviation ($\sigma$) provides an estimate of the resolution achieved. Equation (1) provides a good fit to the autocorrelations of Cy5-labelled single-colour nanofibres (Fig. 3f). Therefore, the localization density fluctuations visible in the STORM images for these PA nanofibres (Fig. 3d) are due to stochastic processes inherent to the technique, while the monomer distribution is homogeneous. On the other hand, the autocorrelation function for the Cy3-channel (Fig. 3e) is not well-described by Equation (1), since the Cy3-functionalized fibres display a consistent anti-correlation ($g(r) < 0$) at intermediate distances ($\sim 500$ nm). The autocorrelation behaviour of these fibres was correctly described using a micro-emulsion model, that accounts for the existence of 'microdomains' richer in the fluorescent monomer, according to equation (2) (ref. 37):

$$g(r) = A_1 exp(-r^2/4\sigma^2) + A_2 cos(\pi r/2r_0)exp(-r/\alpha) \quad (2)$$

The first term of this equation is equal to equation (1), since the overcounting phenomenon is still present. In the second term, the parameter $r_0$ is the average size of the 'microdomains' and $\alpha$ is the coherence length of the domains. As defined by equation (2), 'microdomain' corresponds to an extended fibre region that is enriched in fluorescently labelled molecules compared with the fibre average, thus increasing the likelihood of finding those molecules in that particular region. In this context, 'microdomain' does not mean that molecules are completely clustered and confined to those regions.

An improvement in the goodness of the fit was not observed for the Cy5-labelled nanofibres, therefore it is reasonable to adopt the simpler model with less parameters, which excludes the existence of 'microdomains'. On the other hand, in the case of Cy3-labelled nanofibres a considerable improvement on the $\chi^2$-test ($> 3.5$ times lower) is obtained for the model with the 'microdomain' component, compared with the model that only considers the contribution of the overcounting. A 'microdomain' size of $\sim 300$ nm and coherence length of $\sim 500$ nm were obtained from the fit (Fig. 3e), showing a regular fluctuation of localization density in the Cy3-fibres at the submicron length scale. The difference observed between the distributions of Cy3 and Cy5-PAs is surprising, because of the great chemical and structural resemblance of these dyes. These results show that a small change in the PA structure may have noticeable effects on the self-assembly process, and that the STORM-based autocorrelation analysis can be used to investigate structure at the single-aggregate level. In addition, it shows that Cy5-functionalized PA is suitable for use in quantitative exchange experiments, since the Cy5 dye does not affect the molecular distribution of the PA monomers.

**Fibre bundling and multicolour imaging.** Since the final morphology in supramolecular polymers is highly sensitive to not only the molecular structure, but also the polymerization conditions, it is very important to design a correct sample preparation protocol to obtain the desired structure. A typical problem that may occur is the aggregation of fibres into bundles, a phenomenon that is hard to detect using traditional techniques. Avoiding fibre bundling is crucial when investigating molecular mixing, because bundles of nanofibres with different colours confound the observation of exchanging monomers. In this work we show that STORM-based analysis can be applied to detect PA fibre bundling. When nanofibre preparation was performed at higher PA concentration ($[PA]_{final} > 1$ μM), extremely long and curved fibrillar structures were observed in diffraction-limited fluorescence microscopy (Supplementary Fig. 3). However, the increased resolution attained using STORM imaging revealed finer details of the adsorbed structures, showing extensive fibre bundling. What appeared to be curvature at low resolution was actually the intersection of smaller stiff fibres with different orientations. Two-colour STORM has been performed on PA samples of different concentrations and ionic strengths to unambiguously prove the presence or absence of bundled fibres. Two aqueous solutions of PA nanofibres labelled with either Cy3 or Cy5 were mixed together and immediately adsorbed onto a glass coverslip at room temperature, freezing all kinetics. For this short mixing timescale ($< 1$ min), molecular exchange is negligible according to the FRET measurements shown in Fig. 2b. Therefore, single fibres should be either fully green or fully red. If fibres are adsorbed at high ionic strength and low concentration ($< 0.1$ μM), this behaviour was indeed observed (Fig. 4a). On the contrary, as shown in Fig. 4b, fibres adsorbed at concentrations higher than 1 μM can be observed in both channels simultaneously, indicating bundles of PA nanofibres. It is also possible to observe that some of the fibres visible in the green channel are not perfectly aligned with the overlapping fibres visible in the red channel (Fig. 4b). This provides further evidence that at this concentration bundles are observed instead of fully mixed fibres. Therefore, STORM allows us to verify the absence of fibre bundling and therefore select the optimal sample preparation to perform the molecular exchange experiments.

**Molecular exchange kinetics and mechanisms.** The ability of STORM to image individual nanofibres with high resolution and to study the distribution of different molecular species inside aggregates (*vide supra*), makes it the perfect tool to study exchange in PA samples. We prepared samples for these experiments in similar manner to the FRET measurements. First, two sets of single-colour PA nanofibres with either Cy3-PA (5%) or Cy5-PA (5%) were separately preassembled in aqueous buffer solution, followed by a 16-h aging period. Subsequently the two aged solutions were brought to 37 °C and mixed, allowing molecular exchange. Aliquots of the solution were withdrawn over the course of 48 h and nanofibres were adsorbed onto a glass

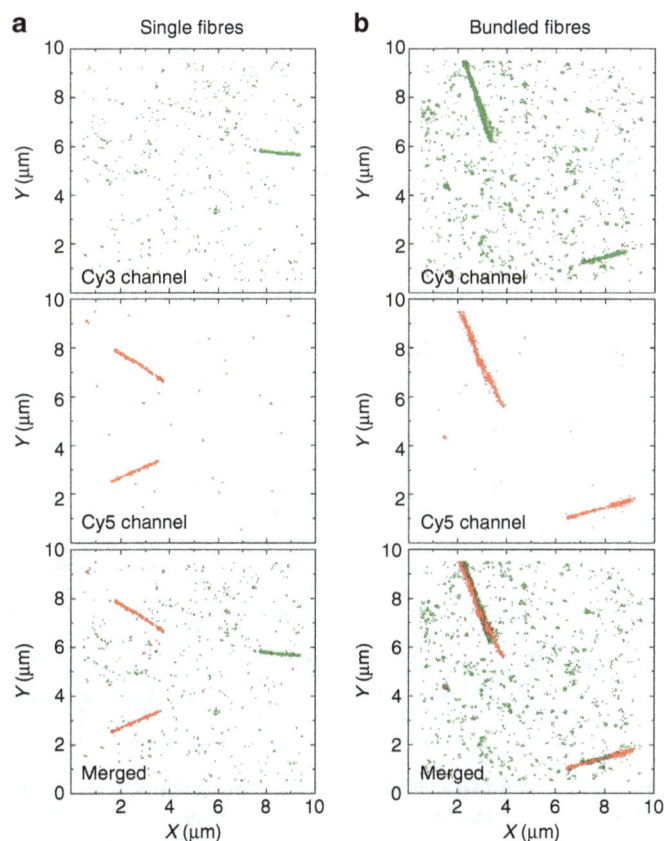

**Figure 4 | Single fibre attachment.** STORM images (with background) of PA nanofibres attached at a final PA concentration of either (**a**) 0.1 μM or (**b**) 1 μM, showing single fibres or bundled fibres, respectively. Attachment was performed immediately on a glass coverslip after dilution (NaCl 1 M).

coverslip at room temperature to freeze the exchange. Two-colour STORM images of assemblies were then acquired at the different time points. Fig. 5 shows representative PA nanofibres at $t = 1$ min and after 48 h of mixing; as clearly shown in the images and in the corresponding localization density plot (Fig. 5b) the fibres initially containing a single label fully mix during the course of the experiment.

We verified that the exchange process is bidirectional by following green fibres incorporating red monomers (Supplementary Fig. 4) and vice versa (Supplementary Fig. 5). The intermediate time points provide insight into the exchange mechanism. Visual inspection of STORM images (Supplementary Fig. 6) showed insertion of labelled monomers over the entire length of the nanofibres throughout the exchange experiment. This finding suggests an exchange mechanism on the basis of ·expulsion and reinclusion of PA molecules. Polymerization–depolymerization at the nanofibre ends would have resulted in preferential exchange in those regions, while fragmentation–recombination would have

resulted in a block-like structure; neither of these mechanisms is consistent with the acquired STORM images.

To confirm this observation and further quantify the exchange, we studied the change in distribution of both dyes with time in initially green-labelled nanofibres, thus tracking the insertion of Cy5-PA. Figure 6 shows the auto- and cross-correlation plots for the green and red channels at the different time points. The green channel displays the previously observed clustered distribution described by equation (2) rendering it less amenable for quantitative analysis. Therefore, we monitored the autocorrelation of the red channel in time to analyse monomer exchange. The red channel autocorrelation, shown in Fig. 6b, can be fit with equation (1) for all time points, suggesting a random exchange of monomers. However, for later time points (that is, after incubation longer than 6 h) the quality of the fit deteriorates, indicating some non-random variations of the Cy5-PA concentration in the nanofibres other than the single decay originated from overcounting. However, the formation of regular 'microdomains' does not seem plausible, because equation (2) did not improve the quality of the fit or yield reasonable fitting parameter values. Since clustering is not observed in single-colour Cy5-PA nanofibres (Fig. 3f), this observation points to a heterogeneous exchange and indicates the presence of structural variations along the fibre or between fibres. This heterogeneity can also be perceived visually by observing single fibre images (Supplementary Fig. 6). The cross-correlation between both channels was computed to assess how the distribution of one labelled PA would affect the distribution of the other, as shown in Fig. 6c. A cross-correlation of zero was found for the entire range, indicating that the distribution of Cy5-PA is not influenced by the presence of Cy3-PA, and vice versa. In other words, the heterogeneous exchange of Cy5-PA is not due to the pre-existing clusters of Cy3-PA, but rather a consequence of structural variation in the PA nanofibre.

The observation of this heterogeneity in exchange, not detected by the FRET experiment, was possible due to the ability of STORM to evaluate the progress of molecular exchange for each individual supramolecular nanofibre. The variability of the exchange progress between different aggregates could be measured, thus allowing us to further probe the heterogeneity in the system during and after molecular exchange. The linear density of localizations, defined as the number of localizations per nanometre arc length, has been computed and could be used as a rough estimate for the ensemble concentration of labelled molecules[19]. The density of Cy5-PA in originally single-colour green fibres was measured as a function of time, showing an increase in the average value as molecular exchange proceeded (Fig. 7a). However, the standard deviation also increased markedly, another indication for the presence of intrinsic structural diversity. In Fig. 7b, it is possible to observe how the distribution of the Cy5-PA linear density gets progressively wider as molecular exchange evolves in time. This can be confirmed visually by inspecting STORM images acquired after 48 h mixing time (Fig. 7c-f), which displayed a range of different behaviours. It was possible to observe a large subset of fibres that had a Cy5-PA density consistent with full mixing (Fig. 7c), as well as fibres comprised of domains of Cy3-PA only (Fig. 7d,e), and both subsets co-existed with a smaller subpopulation of nanofibres that had undergone very little exchange all over their length (Fig. 7f). The existence of regions displaying minimal or no exchange after 48 h implies that full equilibration of these regions will take weeks to months, making them persistent for most practical purposes. This analysis suggests that the population of fibres is considerably more heterogeneous after 48 h than would be expected on the basis of the FRET ensemble measurements.

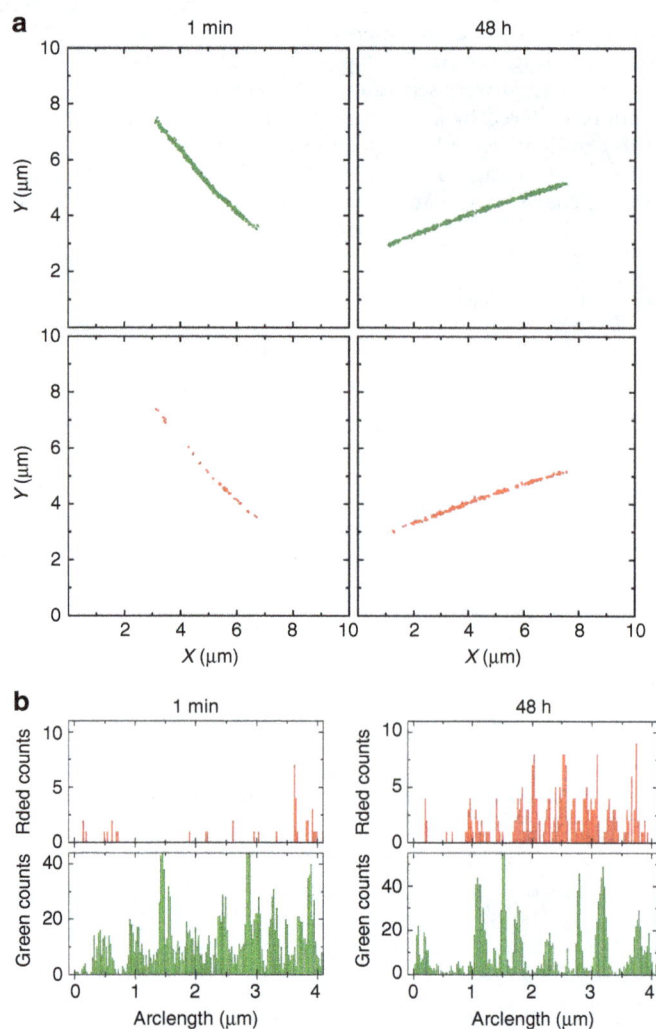

**Figure 5 | Quantitative analysis of STORM imaging during a molecular exchange experiment.** (**a**) Localization maps of PA nanofibres immobilized on a glass coverslip at different time points, after applying a clustering algorithm, background cleaning and backbone finding. The better reconstruction of the fibres in the green channel reflects the selection of initially Cy3-labelled fibres for this experiment. (**b**) Histograms depict the localization density profiles along the nanofibre backbone (bin size 25 nm).

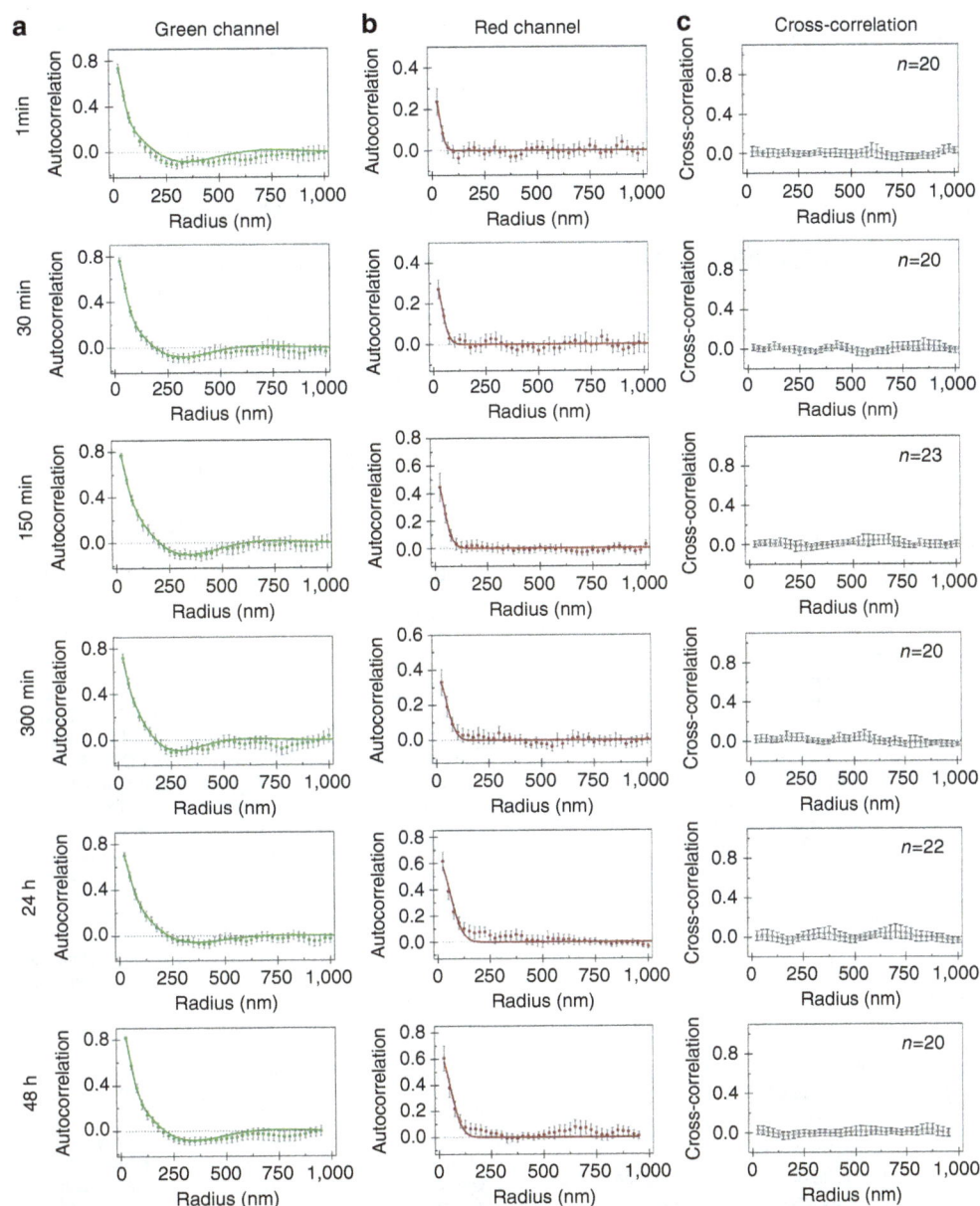

**Figure 6 | Correlation of cyanine dyes along PA nanofibres initially only labelled with Cy3 during molecular exchange.** Averaged autocorrelation plots for (**a**) Cy3 and (**b**) Cy5-labelled PAs and (**c**) averaged cross-correlation between both dyes. Solid lines correspond to model fittings, namely equation (2) for (**a**) green channel and equation (1) for (**b**) red channel (error bars represent 95% confidence level, $n \geq 20$).

Previous molecular dynamics simulations of a similar PA provide a theoretical framework for this structural diversity[27]. In these simulations, a broad distribution of secondary structures was found in the equilibrated fibre. The heterogeneous molecular exchange pattern observed in our study might stem from this conformational diversity. The key feature that led to the discovery of PA supramolecular nanofibres in the Stupp laboratory was the use of a β-sheet peptide domain to drive one-dimensional self-assembly. Using electron paramagnetic resonance, the β-sheet domain has been recently shown to give rise to locally solid-like behaviour in the interior of nanofibre PA assemblies[32]. On the other hand, the surface moieties in these nanofibres have been identified by the electron paramagnetic resonance experiments as regions where liquid-like dynamics prevail. β-Sheets within the nanofibres are highly cohesive assemblies and therefore the dynamic exchange among supramolecular nanofibres should give rise to a large diversity of supramolecular environments. This is in contrast to supramolecular systems with weaker internal cohesion

that could exchange molecules to produce completely homogeneous environments[19]. The formulation of supramolecular systems with high levels of internal order and cohesion such as the PA nanofibres will therefore open new avenues to generate structural diversity for functional purposes. One example will be their ability to adapt structurally to bind important bioactive targets.

## Discussion

We have studied the dynamics of PA nanofibres using FRET and super-resolution localization microscopy. While ensemble FRET measurements prove that PAs exchange between different fibres, two-colour STORM imaging reveals a mechanism on the basis of the transfer of monomers and small clusters. Remarkably, the coexistence of fully dynamic and kinetically inactive areas in the aggregate architecture was observed, demonstrating the existence of structural diversity in PA nanofibres. This intriguing dynamic

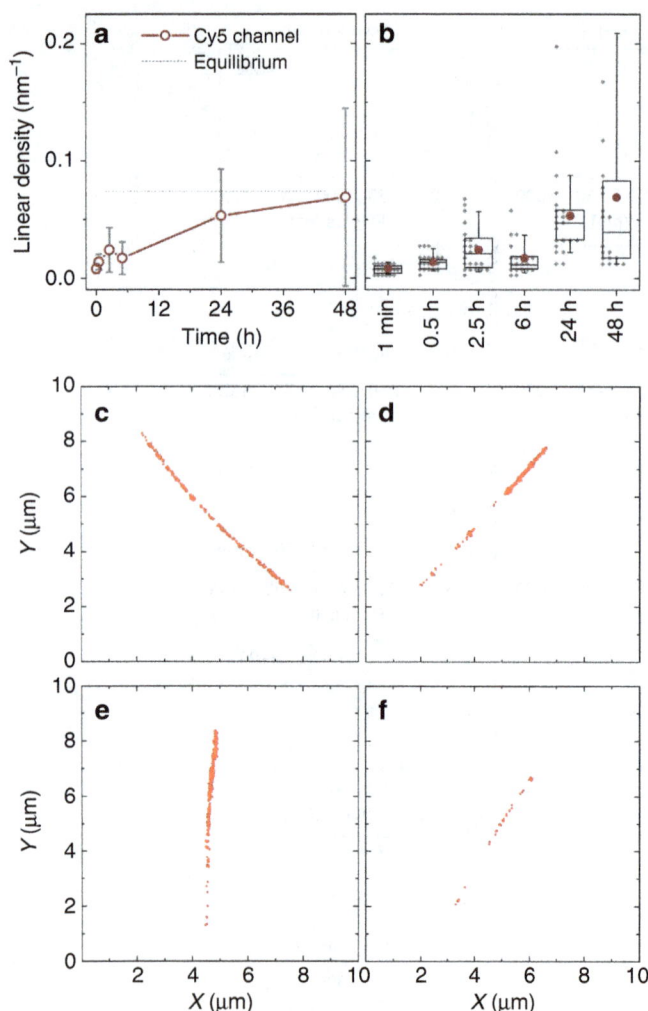

**Figure 7 | PA molecular exchange kinetics estimated using the linear density of localizations calculated from STORM data.** (**a**) Time course-averaged density of red dye (Cy5) on fibres that were initially only labelled with green dyes (Cy3) ($n \geq 20$, error bars correspond to standard deviation). The equilibrium guideline is estimated from the localization density of original single-colour red fibres. (**b**) Box chart of data represented on **a** depicting average (circle), median (box middle line), quartiles (box edges), 5 and 95% percentiles (error bars) and full data set points (diamond symbols, bin size $5 \, \mu m^{-1}$); (**c–f**) Examples of localization plots (after background cleaning) of individual fibres at 48 h.

behaviour is foreseen to have important implications in the biological performance of supramolecular systems.

## Methods

**Peptide synthesis and purification.** PAs were synthesized using standard Fmoc solid-phase peptide synthesis. MBHA rink amide resin, Fmoc-protected amino acids and other solid-phase peptide synthesis reagents were purchased from Novabiochem (USA). Cyanine dyes were supplied by Cyandye (USA). ACS-grade solvents were used for synthesis. Water was purified on an EMD Milipore Milli-Q Integral Water Purification System. For each amino acid coupling, 4.1 equiv of Fmoc-protected amino acid was activated for 1 min with 4 equiv of HBTU (O-benzotriazole-$N,N,N',N'$-tetramethyluroniumhexafluorophosphate) and 6 equiv DIPEA ($N,N$-diisopropylethylamine) in 20 ml DMF. The coupling cocktail was then added to 0.5 mmol of MBHA rink resin and reacted for 1 h. Fmoc removal was performed with 30% piperidine in DMF. Fmoc-Glu(OtBu)-OH ($N$-$\alpha$-Fmoc-L-glutamic acid $\gamma$-t-butyl ester) (3 ×), and Fmoc-Ala-OH ($N$-$\alpha$-Fmoc-L-alanine) (6 ×) were successively coupled to the resin. Finally, the palmitic acid tail (8.1 equiv) was coupled using 8 equiv of HBTU and 12 equiv DIPEA in DMF/DCM (50:50). PAs were cleaved from the resin using a cleavage solution of 95% TFA, 2.5% TIPS (triisopropylsilane) and 2.5% $H_2O$. Cleavage solution was concentrated

by rotary evaporation and the PA precipitated in cold diethyl ether. The crude solid was dissolved in a diluted $NH_4OH$ aqueous solution ($\sim 10 \, mg \, ml^{-1}$) and purified by preparative-scale reversed-phase HPLC on a Varian Prostar 210 HPLC system, using acetonitrile/water gradient containing 0.1% $NH_4OH$. All eluents and additives were HPLC grade. Separation was achieved using a Phenomenex C18 Gemini NX column (150 × 30 mm) with 5 µm particle size and 110 Å pore size. Product-containing fractions were confirmed by ESI mass spectrometry (Agilent 6510 Q-TOF LC/MS) and combined. ACN was removed by rotary evaporation, lyophilized to yield a white solid product and stored at $-30 \, °C$.

**Cyanine dye coupling.** PAs labelled with cyanine dyes were synthesized first coupling Fmoc-Lys(Mtt)-OH ($N$-$\alpha$-Fmoc-$N$-$\epsilon$-4-methyltrityl-L-lysine) to the resin. The rest of the synthetic procedure proceeded as described above. After coupling the palmitic acid tail, Lys(Mtt) was selectively deprotected with 3% TFA, 5% TIPS and 92% DCM for 5 min (3 ×), to expose the side chain amine moiety. Disulfo-cyanine carboxylic acid dyes (Cy3 and Cy5) were dissolved in DMF and stock solutions were stored at -30 °C. (Benzotriazol-1-yloxy)tripyrrolidinophosphonium hexafluorophosphate (PyBOP) was also dissolved in DMF. Cyanine dyes (1 equiv) were activated with PyBOP (1 equiv) and DIPEA (2 equiv) for 1 min and coupled to the lysine-free amine. Cy3- and Cy5-labelled PAs were cleaved and purified as described above. Lyophilization yielded a red powder for Cy3 and blue for Cy5.

**Purity.** Purity was determined by reversed-phase analytical HPLC (see Supplementary Figs 7–9) using a Phenomenex Gemini C18 column (100 × 4.6 mm), with 5 µm particle size and 110 Å pore size, connected to a HPLC system equipped with an autosampler (Shimadzu SIL-20A XR), degasser (Shimadzu DGU-20A3) and a high-pressure gradient system comprising two LC-20AD XR pumps (Shimadzu). Separation was performed at a flow of 1 ml min$^{-1}$ and using a 20-60% acetonitrile linear gradient in water containing 0.1% $NH_4OH$. All eluents and additives were HPLC grade. Peptides, cyanine dye-coupled peptides and eventual contaminants were detected using a PDA (Shimadzu SPD-M20A), acquiring a full ultraviolet-visible spectrum between 200 and 750 nm at any time point. Main chromatogram peaks were collected and product mass confirmed by ESI mass spectrometry using a LCQ Deca XP Max (Thermo Finnigan) ion-trap mass spectrometer. For that, samples were manually injected bypassing the column at a flow rate of 0.20 ml min$^{-1}$ and positive ion mass spectra were acquired in standard enhanced mode.

**Sample preparation.** Stock solutions of non-labelled PA (10 mM), Cy3-labelled PA (1 mM) and Cy5-labelled PA (1 mM) were prepared in Milli-Q water, aliquoted and stored at $-30 \, °C$. Similar to other anionic PA systems, these formed hydrogels in the presence of $CaCl_2$. The degree of labelling of the fibres was accurately controlled by simply mixing PA stock solutions at the desired ratio. Solutions of non-labelled PA were mixed with aqueous solutions of either one or both labels at certain mole ratios, flash-frozen in liquid nitrogen and lyophilized. The obtained PA mixture was molecularly dissolved in TFA and readily evaporated under vacuum. The obtained material was redissolved in $NH_4OH$ (aq.), flash-frozen and further lyophilized. These two last steps resembled the cleavage and purification conditions, respectively. These steps were undertaken to provide molecular mixing, as well as to reset PA self-assembly history. Non-labelled PA was treated using the same method before measurements where the fluorescence-labelled molecules were absent. The obtained powders were reconstituted in appropriate buffers before measurements. After the harsh acidic conditions (dissolution in TFA) used to create a homogenous molecular mixture, PAs were checked by analytical HPLC to rule out potential degradation of the original molecules. Analytical HPLC was performed as described above. Supplementary Fig. 10 shows that the PA does not degrade during the harsh procedure of molecular mixing in TFA. Fluorescence measurements showed pronounced FRET for nanofibres labelled with both Cy3 and Cy5. The FRET intensity increased with concentration of acceptor. This indicates that the sample preparation is suitable to assure co-assembly of the three PAs in a single supramolecular object.

**Circular dichroism spectroscopy.** Circular dichroism spectra were recorded using Jasco Circular Dichroism Spectrometer (model J-715). PA was dissolved in water at 10 mM. Before each measurement, PA stock solution was diluted in appropriate buffers to yield a final PA concentration of 0.1 mM. A quartz cuvette of 1 mm path length was used for the measurements. Each trace represents the average of five scans.

**Cryogenic transmission electron microscopy.** PA nanofibres were separately labelled with either PA-Cy3 (5%) or PA-Cy5 (5%). Both labelled and non-labelled PA nanofibres were dissolved at 1 mg ml$^{-1}$ in 10 mM HEPES buffer (pH 7.5, NaCl 150 mM). CryoTEM was performed using a JEOL 1230 TEM at an accelerating voltage of 100 kV. Using a Vitrobot Mark IV (FEI) vitrification instrument at 25 °C with 100% humidity, a 6.5-µl drop of the sample was deposited on a 300-mesh copper grid with lacey carbon support (Electron Microscopy Sciences, EMS), blotted twice, plunge-frozen in liquid ethane, then stored in liquid $N_2$. For imaging, the sample was transferred to a Gatan 626 cryo-holder under liquid $N_2$, and images were obtained with a Gatan 831 CCD camera.

**Critical micelle concentration by Nile Red assay.** Nile Red is a hydrophobic solvatochromic dye (water soluble at low concentration) with high affinity to the hydrophobic pocket of micelles. In aqueous environment, the molecule is weakly fluorescent. In hydrophobic environments, fluorescence intensity is around 300 times higher and a pronounced blue shift is observed. Therefore, Nile Red is commonly used to probe the existence of hydrophobic pockets formed by surfactants in aqueous environment[38]. Briefly, a Nile Red 2 mM stock solution was prepared in MeOH. Non-labelled PA was dissolved at 10 mM in 10 mM HEPES buffer (pH 7.5, NaCl 150 mM). PA solution was diluted serially to obtain concentrations in the range 200 nM-10 mM. Nile Red stock solution was diluted in the same buffer to obtain a final concentration of 1 μM. Equal volumes of Nile Red aqueous diluted solution and PA solutions were mixed for each PA concentration. Solutions were aged for 24 h to assure full nanofibre disassembly at concentrations below the critical micelle concentration. Fluorescence-emission spectra (excitation 550 nm) were recorded for an emission range between 570 and 700 nm. The maximum intensity and respective wavelength at maximum intensity were both represented as a function of the logarithm of the PA concentration. At concentrations close to critical micelle concentration, it should be observed a sharp increase in fluorescence intensity and a hypsochromic effect.

**Förster resonance energy transfer.** Single-colour PA nanofibres labelled with either Cy3 or Cy5 at 1 mol% were assembled in 10 mM HEPES buffer (pH 7.5, NaCl 150 mM). Fluorescence spectra were recorded in a Varian Cary Eclipse Fluorimeter from Agilent Technologies. Emission spectra were acquired using excitation wavelengths of 520 or 615 nm for Cy3 or Cy5, respectively. Excitation spectra were collected using emission wavelengths of 600 or 710 nm for Cy3 or Cy5, respectively. Temperature was kept at 20 °C using the in-built peltier system. Supplementary Fig. 11a depicts the spectra of the individually labelled PA nanofibres, showing significant overlap between Cy3 emission (donor) and Cy5 excitation (acceptor). On an independent measurement, PA nanofibres labelled with both Cy3 (1 mol%) and Cy5 (0, 0.1, 0.2 and 0.5 mol%) were assembled in 10 mM HEPES buffer (pH 7.5, NaCl 150 mM) at a final PA concentration of 0.5 mM. Fluorescence-emission spectra of dual-labelled PAs were recorded at the excitation wavelength of Cy3 520 nm, at which direct excitation of Cy5 is negligible. Supplementary Fig. 11b shows considerable energy transfer by FRET, which increases monotonically with acceptor concentration, providing evidence for co-assembly of both Cy3- and Cy5-labelled PAs in the same nanofibres.

**Ensemble molecular exchange kinetics.** For molecular exchange kinetic experiments, two sets of PA nanofibres were pre-assembled from a mixture of non-labelled PAs and either Cy3 (0.5%) or Cy5 (0.5%). These independently labelled PA solutions were mixed at 1:1 ratio in a 50 μl quartz cuvette and fluorescence was followed over around 24 h (lag time ∼ 30 s). FRET ratio was determined by dividing Cy5 emission at 668 nm by Cy3 emission at 565 nm, at the Cy3 excitation wavelength (520 nm). The contribution of Cy3 emission at 668 nm was first subtracted from the Cy5 emission. At the beginning of the kinetics experiment, Cy3-PA and Cy5-PA are distributed in different nanofibres, being physically separated by distances much higher than the expected FRET radius[35]. Therefore, energy transfer at time zero should be negligible and, consequently, FRET ratio should be zero. A total PA concentration range of 0.5-5 mM was used. Kinetics was recorded at two different temperatures (20 or 37 °C).

**Sample preparation for microscopy.** Glass microscope coverslips were cleaned by successively immersing in acetone, isopropanol and Milli-Q water. Bath sonication was performed for 10 min with each solvent, followed by drying under $N_2$ flow. The glass coverslips were then etched with a fresh Piranha solution (3:1 v/v $H_2SO_4$ (98%):$H_2O_2$ (30%)) for 30 min. To finish the cleaning procedure, the slides were washed thoroughly with Milli-Q water and rinsed with acetone before drying under $N_2$ flow. To determine the labelled PA distribution along the nanofibres length, single fibres should be immobilized on a glass surface, and remain still during STORM acquisition. A flow chamber was assembled using a glass slide and a clean coverslip separated by double-sided tape. PA nanofibres were immobilized by adsorption onto the surface of the clean coverslip by flushing diluted PA solutions at high ionic strength (NaCl 1 M). After incubation for 1 min, the unbound nanofibres were washed out of the chamber by flushing with HEPES buffer (2 ×) followed by STORM buffer (2 ×). STORM buffer is composed by 50 mM HEPES buffer (pH = 8.5), 1 M NaCl, an oxygen-scavenging system (0.5 mg ml$^{-1}$ glucose oxidase, 40 μg ml$^{-1}$ catalase, 5 wt% glucose) and 200 mM 2-aminoethanethiol. The high ionic strength conditions were used to screen the electrostatic repulsion between highly charged PA nanofibres and the glass surface. However, repulsion between PA nanofibres is also reduced, causing fibre bundling. To avoid bundling, PA solutions were rapidly diluted at low ionic strength (1 μM). After a final (10 ×) dilution step in NaCl 1M, nanofibres were immediately attached on the glass coverslip. At this concentration isolated nanofibres are mainly observed in the microscopic mounting.

**Stochastic optical reconstruction microscopy.** STORM images were acquired using a Nikon N-STORM system configured for total internal reflection fluorescence imaging. Excitation inclination was tuned to maximize the signal-to-noise ratio of the glass-absorbed fibres. Cy3-labelled samples were illuminated by 561 and 647 nm laser lines. No activation ultraviolet light was used. Fluorescence was collected by means of a Nikon 100 ×, 1.4NA oil immersion objective and passed through a quad-band pass dichroic filter (97335 Nikon). All time-lapses were recorded onto a 64 × 64 pixel region (pixel size 0.17 μm) of an EMCCD camera (ixon3, Andor). For each channel, 20,000 frames were sequentially acquired. STORM movies were analysed with the STORM module of the NIS Elements Nikon software.

**Molecular imaging of fibres during exchange.** PA nanofibres were separately labelled with either PA-Cy3 (5%) or PA-Cy5 (5%) at 0.5 mM in 10 mM HEPES buffer (pH 7.5, NaCl 150 mM). Subsequently, 250 μl-aliquots of each solution were mixed in a microcentrifuge tube and gently shaken at 37 °C for different amounts of time. At predefined time points, 2 μl of the solution was withdrawn, diluted and adsorbed on a glass surface as described above. We verified that once adsorbed on the slide, fibres are not able anymore to considerably exchange monomers with the solution in the time frame of the experiment. STORM imaging of Cy5 and Cy3 channels was performed sequentially. Fibres that are initially labelled only with Cy3 and incorporate Cy5 monomers over time were selected observing the fluorescence intensity in the low-resolution images. Original single-colour nanofibres were also observed before mixing as a control.

**Image analysis.** During STORM imaging, the Nikon software generates a list of localizations by 2D Gaussian fitting of blinking chromophores in the acquired movie of conventional microscopic images. This localizations list was subsequently analysed with our custom-made Matlab scripts, as described in detail elsewhere[19]. Briefly, a first script uses a density-based clustering algorithm to automatically identify the fibres in the image and to remove the background. Examples of output localization maps are shown in Figs 3a,b and 5a. A second script is run on the well-reconstructed fibres obtained in the Cy3-channel to find the fibre backbone. The lower localization density obtained in the Cy5-channel for shorter time periods in the molecular exchange experiment is not enough to fully reconstruct the nanofibres. The polymer backbone coordinates are used to obtain structural information. Next, histograms of localization density profiles along the backbone are generated for both channels, as shown in Figs 3c,d and 5b. The spatial autocorrelation (that is, the average correlation between the localization density at one backbone point in the fibre and the density at another backbone point, as a function of the distance between these points) was also computed using the Matlab-function xcov with unbiased normalization. The spatial cross-correlation of the two channels was calculated in a similar manner. Averaging over multiple fibres is required to obtain accurate correlation decay graphs, such as depicted in Figs 3e,f and 6. Investigation of the average correlation decays yields the distribution of dyes in a fibre, without interference from the stochastic fluctuations inherent to the technique.

## References

1. Dominguez, R. & Holmes, K. C. Actin structure and function. *Annu. Rev. Biophys.* **40**, 169–186 (2011).
2. Janke, C. & Kneussel, M. Tubulin post-translational modifications: encoding functions on the neuronal microtubule cytoskeleton. *Trends Neurosci.* **33**, 362–372 (2010).
3. Pollard, T. D. & Borisy, G. G. Cellular motility driven by assembly and disassembly of actin filaments. *Cell* **112**, 453–465 (2003).
4. Pollard, T. D., Blanchoin, L. & Mullins, R. D. Molecular mechanisms controlling actin filament dynamics in nonmuscle cells. *Annu. Rev. Biophys. Biomol. Struct.* **29**, 545–576 (2000).
5. Stournaras, C., Gravanis, A., Margioris, A. N. & Lang, F. The actin cytoskeleton in rapid steroid hormone actions. *Cytoskeleton* **71**, 285–293 (2014).
6. Symons, M. & Rusk, N. Control of vesicular trafficking by Rho GTPases. *Curr. Biol.* **13**, R409–R418 (2003).
7. Albertazzi, L. *et al.* Spatiotemporal control and superselectivity in supramolecular polymers using multivalency. *Proc. Natl Acad. Sci. USA* **110**, 12203–12208 (2013).
8. Marchi-Artzner, V. *et al.* Selective adhesion, lipid exchange and membrane-fusion processes between vesicles of various sizes bearing complementary molecular recognition groups. *Chemphyschem Eur. J. Chem. Phys. Phys. Chem.* **2**, 367–376 (2001).
9. Ferrell, J. E., Lee, K. J. & Huestis, W. H. Lipid transfer between phosphatidylcholine vesicles and human-erythrocytes–exponential decrease in rate with increasing acyl chain-length. *Biochemistry* **24**, 2857–2864 (1985).
10. Lund, R., Willner, L., Richter, D. & Dormidontova, E. E. Equilibrium chain exchange kinetics of diblock copolymer micelles: tuning and logarithmic relaxation. *Macromolecules* **39**, 4566–4575 (2006).
11. Zinn, T., Willner, L., Lund, R., Pipich, V. & Richter, D. Equilibrium exchange kinetics in n-alkyl–PEO polymeric micelles: single exponential relaxation and chain length dependence. *Soft Matter* **8**, 623–626 (2011).

12. Fasting, C. *et al.* Multivalency as a chemical organization and action principle. *Angew. Chem. Int. Ed.* **51**, 10472–10498 (2012).

13. Helms, B. A. *et al.* High-affinity peptide-based collagen targeting using synthetic phage mimics: from phage display to dendrimer display. *J. Am. Chem. Soc.* **131**, 11683 (2009).

14. Silva, G. A. *et al.* Selective differentiation of neural progenitor cells by high-epitope density nanofibers. *Science* **303**, 1352–1355 (2004).

15. Baker, M. B. *et al.* Consequences of chirality on the dynamics of a water-soluble supramolecular polymer. *Nat. Commun.* **6**, 6234 (2015).

16. Bates, M., Huang, B., Dempsey, G. T. & Zhuang, X. Multicolor super-resolution imaging with photo-switchable fluorescent probes. *Science* **317**, 1749–1753 (2007).

17. Shim, S.-H. *et al.* Super-resolution fluorescence imaging of organelles in live cells with photoswitchable membrane probes. *Proc. Natl Acad. Sci. USA* **109**, 13978–13983 (2012).

18. Huang, B., Bates, M. & Zhuang, X. Super-resolution fluorescence microscopy. *Annu. Rev. Biochem.* **78**, 993–1016 (2009).

19. Albertazzi, L. *et al.* Probing exchange pathways in one-dimensional aggregates with super-resolution microscopy. *Science* **344**, 491–495 (2014).

20. Webber, M. J., Berns, E. J. & Stupp, S. I. Supramolecular nanofibers of peptide amphiphiles for medicine. *Isr. J. Chem.* **53**, 530–554 (2013).

21. Boekhoven, J. & Stupp, S. I. 25th anniversary article: supramolecular materials for regenerative medicine. *Adv. Mater.* **26**, 1642–1659 (2014).

22. Stupp, S. I. & Palmer, L. C. Supramolecular chemistry and self-assembly in organic materials design. *Chem. Mater.* **26**, 507–518 (2014).

23. Soukasene, S. *et al.* Antitumor activity of peptide amphiphile nanofiber-encapsulated camptothecin. *ACS Nano* **5**, 9113–9121 (2011).

24. Moyer, T. J. *et al.* Shape-dependent targeting of injured blood vessels by peptide amphiphile supramolecular nanostructures. *Small* **11**, 2750–2755 (2015).

25. Paramonov, S. E., Jun, H.-W. & Hartgerink, J. D. Self-assembly of peptide – amphiphile nanofibers: the roles of hydrogen bonding and amphiphilic packing. *J. Am. Chem. Soc.* **128**, 7291–7298 (2006).

26. Pashuck, E. T., Cui, H. G. & Stupp, S. I. Tuning supramolecular rigidity of peptide fibers through molecular structure. *J. Am. Chem. Soc.* **132**, 6041–6046 (2010).

27. Lee, O.-S., Stupp, S. I. & Schatz, G. C. Atomistic molecular dynamics simulations of peptide amphiphile self-assembly into cylindrical nanofibers. *J. Am. Chem. Soc.* **133**, 3677–3683 (2011).

28. Storrie, H. *et al.* Supramolecular crafting of cell adhesion. *Biomaterials* **28**, 4608–4618 (2007).

29. Stephanopoulos, N., Ortony, J. H. & Stupp, S. I. Self-assembly for the synthesis of functional biomaterials. *Acta Mater.* **61**, 912–930 (2013).

30. Greenfield, M. A., Hoffman, J. R., de la Cruz, M. O. & Stupp, S. I. Tunable mechanics of peptide nanofiber gels. *Langmuir* **26**, 3641–3647 (2010).

31. Newcomb, C. J. *et al.* Cell death versus cell survival instructed by supramolecular cohesion of nanostructuresl. *Nat. Commun.* **5**, 3321 (2014).

32. Korevaar, P. A., Newcomb, C. J., Meijer, E. W. & Stupp, S. I. Pathway selection in peptide amphiphile assembly. *J. Am. Chem. Soc.* **136**, 8540–8543 (2014).

33. Tantakitti, F. *et al.* Energy landscapes and functions of supramolecular systems. *Nat. Mater.* **15**, 469–476 (2016).

34. Levitt, M. Conformational preferences of amino acids in globular proteins. *Biochemistry* **17**, 4277–4285 (1978).

35. Bastiaens, P. I. & Jovin, T. M. Microspectroscopic imaging tracks the intracellular processing of a signal transduction protein: fluorescent-labelled protein kinase C beta I. *Proc. Natl Acad. Sci. USA* **93**, 8407–8412 (1996).

36. Rust, M. J., Bates, M. & Zhuang, X. Sub-diffraction-limit imaging by stochastic optical reconstruction microscopy (STORM). *Nat. Methods* **3**, 793–796 (2006).

37. Veatch, S. L. *et al.* Correlation functions quantify super-resolution images and estimate apparent clustering due to over-counting. *PLoS ONE* **7**, e31457 (2012).

38. Stuart, M. C. A., van de Pas, J. C. & Engberts, J. B. F. N. The use of Nile Red to monitor the aggregation behavior in ternary surfactant–water–organic solvent systems. *J. Phys. Org. Chem.* **18**, 929–934 (2005).

## Acknowledgements

Molecular synthesis, purification and characterization were supported by the U.S. Department of Energy, Office of Science, Basic Energy Sciences, under Award # DE-FG02-00ER45810. The work in Eindhoven was supported by was financed by the Dutch Ministry of Education, Culture and Science (Gravity program 024.001.035) and the European Research Council (FP7/2007-2013, ERC Grant Agreement 246829). R.M.P.d.S. and his research activities were funded by the Marie Curie FP7-PEOPLE-2010-IOF program (ref. no. 273295). Compound preparation, purification and characterization were partially performed at the Peptide Synthesis Core at Simpson Querrey Institute. The Biological Imaging Facility (BIF) and Keck Biophysics facilities at Northwestern provided further instrumentation used in this work. Mark Seniw is acknowledged for the preparation of graphics.

## Author contributions

R.M.P.d.S., L.A., D.v.d.Z., S.I.S. and E.W.M. conceived the project and contributed to the concept of the manuscript. R.M.P.d.S., L.A., D.v.d.Z. designed the experiments and analysed the results. R.M.P.d.S. performed peptide synthesis, spectroscopic and STORM experiments. S.S.L. performed cryoTEM imaging. R.M.P.d.S., L.A., D.v.d.Z., S.I.S. and E.W.M. wrote the manuscript and contributed to discussions on the project.

## Additional information

**Competing financial interests:** The authors declare no competing financial interests.

# 18

# A general strategy for expanding polymerase function by droplet microfluidics

Andrew C. Larsen[1], Matthew R. Dunn[1,2], Andrew Hatch[3], Sujay P. Sau[1], Cody Youngbull[3] & John C. Chaput[1,4,5]

Polymerases that synthesize artificial genetic polymers hold great promise for advancing future applications in synthetic biology. However, engineering natural polymerases to replicate unnatural genetic polymers is a challenging problem. Here we present droplet-based optical polymerase sorting (DrOPS) as a general strategy for expanding polymerase function that employs an optical sensor to monitor polymerase activity inside the microenvironment of a uniform synthetic compartment generated by microfluidics. We validated this approach by performing a complete cycle of encapsulation, sorting and recovery on a doped library and observed an enrichment of $\sim$1,200-fold for a model engineered polymerase. We then applied our method to evolve a manganese-independent $\alpha$-L-threofuranosyl nucleic acid (TNA) polymerase that functions with $>99\%$ template-copying fidelity. Based on our findings, we suggest that DrOPS is a versatile tool that could be used to evolve any polymerase function, where optical detection can be achieved by Watson–Crick base pairing.

[1] The Biodesign Institute, Arizona State University, Tempe, Arizona 85287-5301, USA. [2] School of Life Sciences, Arizona State University, Tempe, Arizona 85287-5301, USA. [3] School of Earth and Space Exploration, Arizona State University, Tempe, Arizona 85287-5301, USA. [4] Department of Chemistry and Biochemistry, Arizona State University, Tempe, Arizona 85287-5301, USA. [5] Department of Pharmaceutical Sciences, University of California, 147 Bison Modular, Building 515, Irvine, California 92697. Correspondence and requests for materials should be addressed to J.C.C. (email: jchaput@uci.edu).

Recent advances in polymerase engineering have made it possible to synthesize nucleic acid polymers with a wide range of chemical modifications, including xeno-nucleic acid polymers (XNAs) with backbone structures that are not found in nature[1–3]. While this technological advance has generated significant interest in XNA as a synthetic polymer for future applications in molecular medicine, nanotechnology and materials science[4–7], the current generation of XNA polymerases function with markedly lower activity than their natural counterparts[8,9]. The prospect of developing synthetic polymerases with improved activity and more diverse functions has driven a desire to apply molecular evolution as a strategy for altering the catalytic properties of natural polymerases[10,11]. Compartmentalized self-replication (CSR) and compartmentalized self-tagging (CST) are examples of technologies that have been developed to evolve polymerases with expanded substrate specificity[1,12]. However, these methods use the parent plasmid as template for the primer-extension reaction, which limits the range of polymerase functions to enzymes that promote DNA-templated synthesis.

Evolving enzymes with new or improved function requires iterative rounds of *in vitro* selection and amplification[13]. The outcome of a selection depends on the number of variants that can be screened and the quality of the separation technique used to partition functional members away from the non-functional pool. The miniaturization of directed evolution experiments into artificial compartments with cell-like dimensions provides access to larger enzyme libraries by reducing sample volumes to the picolitre-scale[14,15]. The simplest approach to water-in-oil (w/o) droplet formation involves the bulk mixing of aqueous and organic phases with vigorous stirring, but this method produces polydisperse droplets with large volumetric differences[14,15]. Given the cubic dependence of volume on diameter, polydisperse droplets cannot be partitioned by optical sorting due to massive differences in enzyme–substrate concentration[16].

To overcome this problem, microfluidic devices have been developed that can generate monodisperse populations of w/o droplets by manipulating fluids at the microscale[17,18]. While this approach has been used to change the specificity of several natural enzymes[19–21], this technique has not yet been applied to problems in polymerase engineering due to the challenges of generating a fluorescent signal with a signal-to-noise ratio (SNR) that is high enough to distinguish droplets containing functional polymerases from those that are empty or contain non-functional enzymes. Here we describe a microfluidics-based polymerase engineering strategy that combines droplet microfluidics with optical cell sorting. Using droplet-based optical polymerase sorting (DrOPS), a library of polymerase variants is expressed in *Escherichia coli* and single cells are encapsulated in microfluidic droplets containing a fluorescent substrate that is responsive to polymerase activity. On lysis, the polymerase is released into the droplet and challenged to extend a primer–template complex with XNA. Polymerases that successfully copy a template strand into full-length product produce a fluorescent signal by disrupting a donor–quencher pair. Although we originally developed the DrOPS method to evolve a manganese-independent TNA polymerase, the generality of this technique suggests that it could be used to evolve any polymerase function where optical detection can be achieved by Watson–Crick base pairing.

## Results

### Fluorescence-based PAA.
Molecular beacons previously developed to monitor polymerase function suffer from a low SNR that precludes their use in w/o droplets[22,23]. We therefore set out to design a polymerase activity assay (PAA) that would produce a strong optical signal when a primer–template complex is extended to full-length product, but remain dim when the primer goes unextended (Fig. 1a). With this goal in mind, a DNA-quencher probe was designed to dissociate from the primer–template complex at elevated temperatures where thermophilic polymerases function with optimal activity and re-anneal at room temperature when the sample is assayed for function (Fig. 1b). By coupling polymerase activity to fluorescence, genes encoding functional polymerases are identified by the optical signal of their droplet, while variants that fail to extend the primer remain dim and are removed from the pool during cell sorting.

Recent advances in the chemistry of dark quenchers caused us to speculate that a donor–quencher pair could be identified with improved spectral properties[24]. By surveying a small number of fluorescent dyes, we found that Cy3 produces an optical signal that is 200-fold higher than its quenched state with Iowa Black FQ (Fig. 1c), which is substantially higher than previous donor–quencher pairs developed to monitor polymerase activity[22].

To test the Cy3-Iowa Black FQ donor–quencher pair in a PAA, the RNA synthesis activity of an engineered DNA polymerase was compared with its wild-type (wt) DNA polymerase counterpart. For this experiment, we used 9n-GLK, which is an engineered version of a DNA polymerase isolated from *Thermococcus* sp. 9° N that carries the mutations Y409G, A485L and E664K (ref. 25). Exonuclease-deficient versions of 9n-GLK and wt 9n were challenged to extend a DNA primer–template complex with deoxyribonucleoside triphosphates (dNTP) and ribonucleoside triphosphates (NTP). Analysis of the primer-extension reactions by denaturing PAGE and fluorescence confirmed that full-length product is obtained in all cases except when the wt polymerase is incubated with NTPs (Fig. 1d). This result is consistent with the strong steric gate activity of natural DNA polymerases[26]. More importantly, however, the strong concordance observed between the PAGE and fluorescence-analysed data (Fig. 1d) demonstrates that the Cy3-Iowa Black FQ donor–quencher produces an optical signal suitable for monitoring polymerases activity in a bulk aqueous environment.

**Miniaturizing the PAA.** Next, we sought to miniaturize the PAA by encapsulating the primer–template complex in uniform w/o droplets. We began by making w/o droplets in a flow-focusing, fluorocarbon-coated microfluidic device (Supplementary Fig. 1). In this system, droplet formation occurs at the flow-focusing junction where the aqueous phase meets a fluorous oil carrier phase. The droplets are stabilized by surfactants in the oil that prevent coalescence at elevated temperatures (as high as 90 °C) and allow for long-term storage at room temperature.

To demonstrate that the PAA functions within the environment of a w/o droplet, strains of *E. coli* expressing the wt and 9n-GLK mutant polymerases were encapsulated with the reagents needed for RNA synthesis on a DNA primer–template complex. Droplets were formed following a Poisson distribution ($\mu = 0.1$) to ensure that 99% of the occupied droplets contain at most a single *E. coli* cell. This prediction was empirically validated using cells expressing the green fluorescence protein (GFP; Supplementary Fig. 2). Once formed, the droplets were heated to promote *E. coli* lysis and incubated for 3 h at 55 °C to facilitate primer extension. Fluorescence and bright-field images were taken to assess polymerase activity in a population of w/o droplets. As shown in Fig. 1e, droplets containing the 9n-GLK *E. coli* strain produce a highly fluorescent signal due to the strong RNA synthesis activity of 9n-GLK, while empty droplets or droplets that contain the wild-type 9n *E. coli* strain

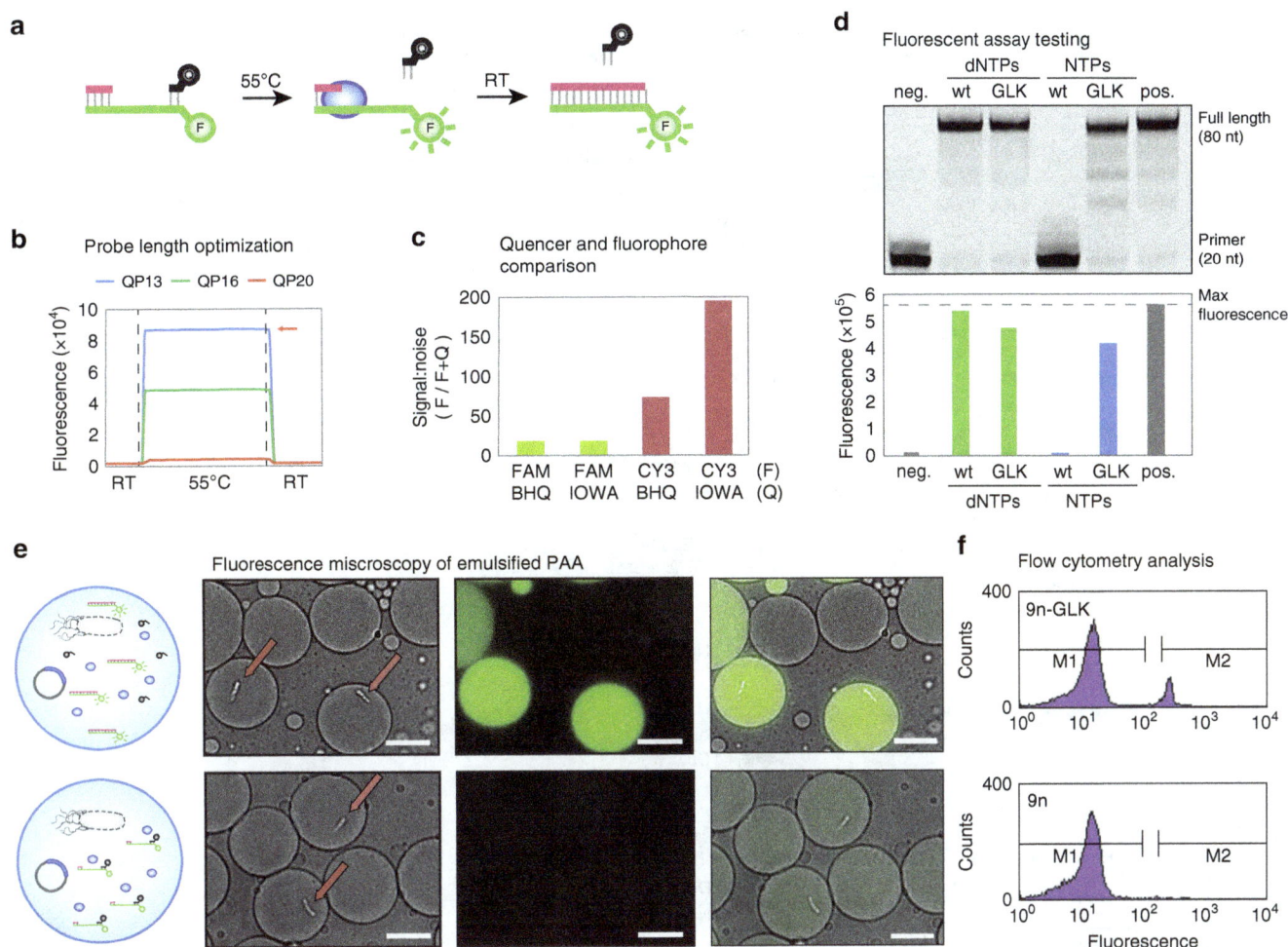

**Figure 1 | Droplet-based optical polymerase sorting.** (**a**) We have developed a fluorescent reporter system that produces an optical signal when a primer–template complex is extended to full-length product. The reporter consists of a primer–template complex (pink and green) containing a downstream fluorophore that is quenched when a DNA-quencher (black) anneals to the unextended region. (**b**) The assay was designed with a metastable probe to allow dissociation at elevated temperatures, where thermophilic polymerases function with optimal activity. Red arrow marks the maximum fluorescence observed in the absence of the quencher probe. (**c**) Flourophore (F)/quencher (Q) pairs were screened to identify a dye pair with the maximum signal-to-noise ratio. (**d**) Primer-extension analysis by denaturing PAGE (top) and fluorescence (bottom) for 9n and 9n-GLK polymerases using dNTP and NTP substrates. Negative control: no NTPs. Positive control: dNTPs or no DNA-quencher probe. (**e**) Single-emulsion droplets containing a functional 9n-GLK polymerase that extends a primer–template complex with RNA (top) and non-functional (bottom) wild-type 9n polymerase. The panel shows a cartoon depiction of the droplet, a bright-field micrograph of encapsulated *E. coli* (arrow), a fluorescence micrograph of the same field of view and an overlay of the two images. Scale bars, 10 μm. (**f**) Flow cytometry analysis of 9n and 9n-GLK polymerases following NTP extension in water-in-oil-in-water (w/o/w) droplets.

remain dim. Taken together, these images demonstrate that the PAA functions with high activity in uniform w/o compartments, which is a necessary criterion for developing a microfluidics-based method for polymerase evolution.

**Formation of double-emulsion droplets.** While w/o droplets provide a physical barrier for maintaining the genotype-phenotype linkage of functional enzymes, the organic carrier phase poses an obstacle for isolating fluorescent droplets using a commercial fluorescence-activated cell sorter (FACS). This problem can be overcome by performing a second compartmentalization step in which w/o droplets are emulsified in water-in-oil-in-water (w/o/w) double-emulsion droplets that have an aqueous carrier phase[27]. We therefore prepared a set of double-emulsion compartments using a hydrophilic microfluidics device that combines the w/o droplets with an aqueous carrier phase at the flow-focusing junction (Supplementary Fig. 1). Two populations of single-emulsion droplets containing either

9n-GLK or wt 9n DNA polymerase were converted to w/o/w droplets and analysed by flow cytometry (Fig. 1f). The population generated with *E. coli* cells expressing the wt 9n polymerase display uniformly low fluorescence, while droplets generated with *E. coli* cells expressing 9n-GLK have a bimodal distribution with low and high fluorescence. The fraction of highly fluorescent droplets correlates with the expected bacterial occupancy of ~10% as predicted by statistical analysis and the GFP encapsulation assay (Supplementary Fig. 2). Moreover, the difference in average fluorescence intensity between the two populations is >10-fold, which is sufficient to separate the two populations by FACS.

**Enrichment efficiency.** To test the ability of the PAA to support a complete round of *in vitro* selection (Fig. 2a, Supplementary Fig. 3), we performed a mock selection to measure the amount of enrichment that occurs per round of selection using the DrOPS method. *E. coli* cells expressing the 9n wt polymerase were

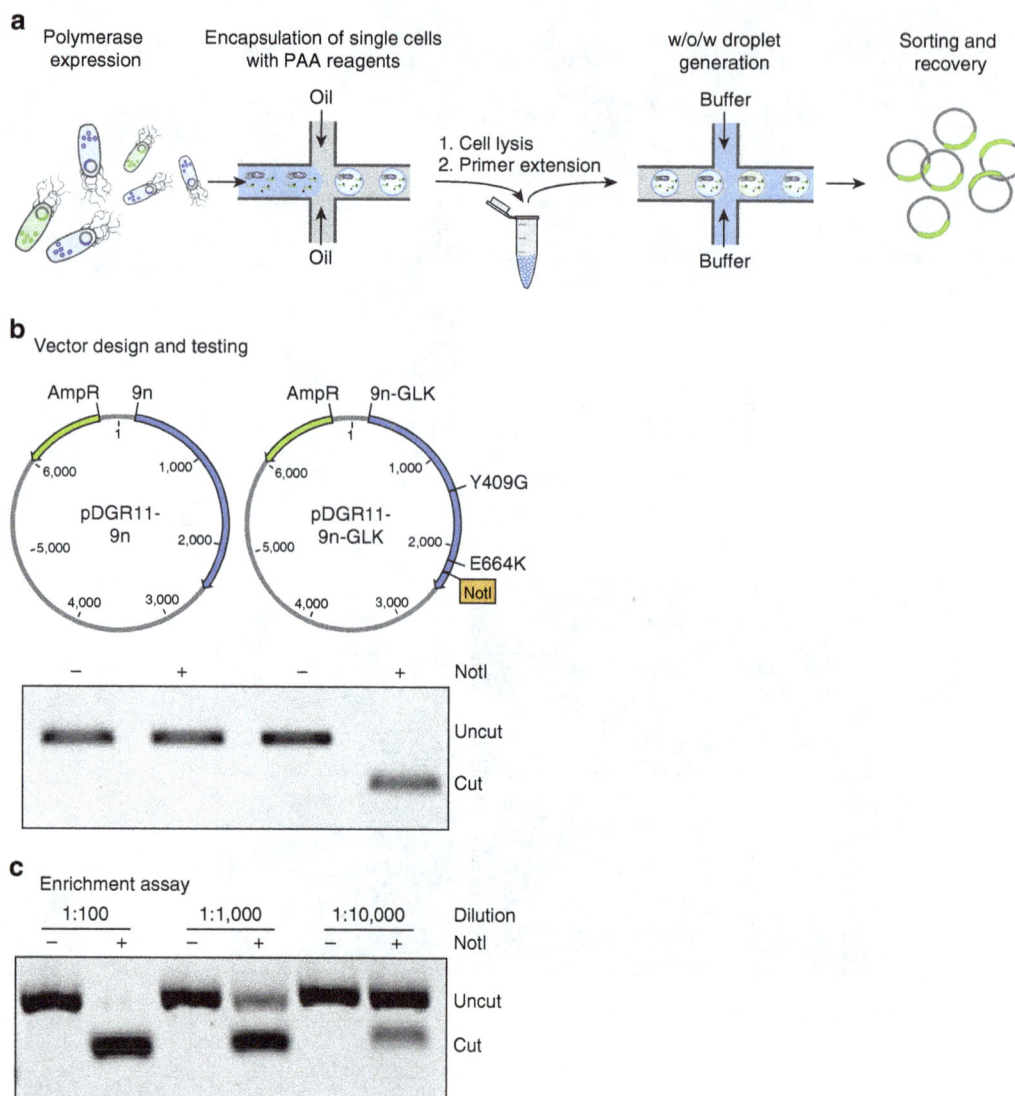

**Figure 2 | Model selection of an engineered polymerase with RNA synthesis activity.** (**a**) Overview of the microfluidic polymerase enrichment strategy. A pool of polymerase genes containing functional (green) and non-functional (blue) members are expressed in *E. coli* and encapsulated in w/o droplets generated in a microfluidics device. Polymerases are liberated from their bacteria by heat lysis and incubated at 55 °C to allow for primer extension. Using a second microfluidics device, droplets are emulsified into a bulk aqueous phase to generate water-in-oil-in-water compartments (w/o/w). Fluorescent w/o/ws are FACS sorted and the vectors encoding functional polymerases are recovered. (**b**) Vector design. The 9n-GLK vector was engineered to contain a unique NotI restriction site. Control digestion showing that NotI only cuts PCR-amplified DNA from the 9n-GLK vector. (**c**) Following a complete cycle of selection and amplification (see Supplementary Fig. 1) PCR-amplified DNA was digested with NotI to measure the enrichment of 9n-GLK from libraries that were doped at levels of 1:100, 1:1,000 and 1:10,000 (9n-GLK to 9n). NotI digestion of the PCR-amplified DNA reveals an enrichment of ~1,200-fold per round of microfluidics selection.

combined with 1/100th, 1/1,000th and 1/10,000th of one equivalent of *E. coli* cells expressing 9n-GLK as a positive control for RNA synthesis activity. The 9n-GLK plasmids were engineered to contain a unique NotI restriction site to distinguish 9n-GLK from wt 9n in a restriction enzyme digestion (Fig. 2b). Accordingly, the three populations of *E. coli* were encapsulated in w/o droplets at an occupancy level of ~10%, which ensured that 99% of the occupied droplets contained no more than one *E. coli* per compartment. Following cell lysis and primer extension, the samples were passed through a second microfluidics device to generate three populations of w/o/w droplets that were each sorted by FACS (Supplementary Fig. 1). Plasmid DNA recovered from the different populations was amplified by PCR and digested with NotI restriction enzyme. Comparison of the digested DNA before and after sorting revealed an enrichment of ~1,200-fold of 9n-GLK (Fig. 2b,c),

which is consistent with previous literature results where model libraries have been sorted in w/o/w double-emulsion droplets[27].

**Evolving a manganese-independent TNA polymerase.** To demonstrate the DrOPS technology in a practical application, we sought to evolve a polymerase that could synthesize an artificial genetic polymer with a backbone structure unrelated to natural DNA and RNA. For this experiment, we chose α-L-threofuranosyl nucleic acid (TNA)—an unnatural genetic polymer composed of repeating units of α-L-threofuranosyl sugars linked by 2′,3′-phosphodiester bonds (Fig. 3a)[28]. TNA is an attractive candidate for therapeutic and diagnostic applications due to its stability against nuclease degradation and ability to undergo Darwinian evolution[3]. However, the current generation of TNA

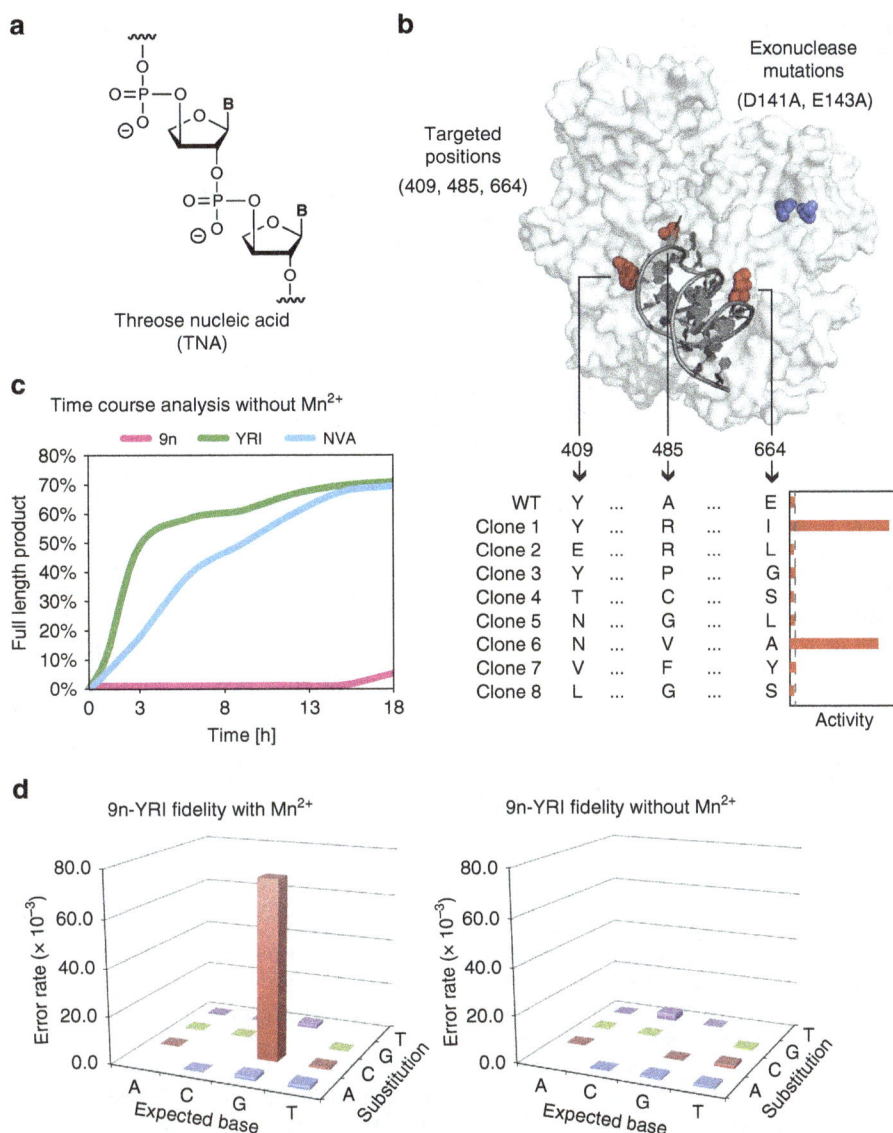

**Figure 3 | Selection of a $Mn^{2+}$-independent TNA polymerase from a focused library.** (**a**) Constitutional structure for the linearized backbone of threose nucleic acid (TNA). (**b**) Positions 409, 485 and 664 mapped onto the structure of 9n DNA polymerase (PDB: 4K8X). Polymerases isolated after one round of selection were analysed for TNA synthesis activity in the absence of $Mn^{2+}$. Activity is defined as the amount of full-length product generated in 18 h. Basal activity of wild-type 9n polymerase (dashed grey line). (**c**) Time course of TNA synthesis for 9n-YRI and 9n-NVA polymerases compared with wild-type 9n. (**d**) Fidelity analysis of 9n-YRI polymerase in the presence and absence of manganese ions yields a mutational profile of 8 errors per 100 bases and 2 errors per 1,000 bases, respectively.

polymerases suffers from low fidelity due to a propensity for G–G mispairing in the enzyme active site[29].

We hypothesized that the low fidelity of TNA synthesis was due to the presence of manganese ions ($Mn^{2+}$), which are used to relax the substrate specificity of natural polymerases[30]. We therefore designed an *in vitro* selection strategy to evolve a $Mn^{2+}$-independent TNA polymerase in hopes of generating an enzyme that functions with higher fidelity. A polymerase library was constructed in which positions 409, 485 and 664 in the 9n DNA polymerase scaffold were fully saturated with all possible amino acid mutations. These positions were chosen based on their known propensity to alter the substrate specificity of natural polymerases[31]. The 8,000 member library was assembled from commercial gene blocks (Supplementary Fig. 4), cloned into *E. coli* and sequence verified. Because the sequencing results revealed a number of random mutations in the gene-coding region, including unwanted stop codons, a single round of

selection was performed under standard DNA synthesis conditions to increase the proportion of active clones. SDS–PAGE analysis of randomly selected clones before and after active polymerase enrichment revealed a dramatic increase in the number of full-length enzymes, indicating that neutral selection removed the truncated non-functional polymerases from the pool (Supplementary Fig. 5).

Next, the plasmid library was taken through a complete round of *in vitro* selection and amplification (Supplementary Fig. 3). Following w/o droplet formation and *E. coli* lysis, the polymerases were challenged to extend a DNA primer–template complex with chemically synthesized TNA triphosphates (tNTPs) in manganese-deficient reaction buffer for 18 h at 55 °C (refs 32,33). The w/o droplets were converted to double emulsions and sorted by FACS. Plasmid DNA was extracted, transformed into a new population of *E. coli* and library members were cloned and sequenced.

**Characterizing selected TNA polymerases**. Eight polymerase variants were chosen for functional analysis (Fig. 3b). Each polymerase was purified by affinity chromatography, quantified and assayed for the ability to extend a DNA primer–template complex with chemically synthesized tNTPs. Control experiments performed in the presence and absence of dNTP substrate confirmed that each polymerase was functional and free of cellular contaminants that could lead to a false positive result in the PAA (Supplementary Fig. 6). Of the eight polymerases tested, two variants showed a significant propensity for TNA synthesis in the absence of manganese ions (Fig. 3b).

Clone 1 (9n-YRI) carries the mutations A485R and E664I and retains the wt tyrosine residue (Y) at position 409. Clone 6 (9n-NVA) carries the mutations Y409N, A485V and E664A, as well as two additional point mutations (D432G and V636A). A time course analysis comparing 9n-YRI and 9n-NVA with wt 9n (Fig. 3c) indicates that both engineered polymerases function as strong $Mn^{2+}$-independent TNA polymerases, generating ~50% full-length product in 3 h and 9 h, respectively. By contrast, wt 9n shows very little full-length product after 18 h of incubation under identical conditions (Fig. 3c, Supplementary Fig. 6), indicating that the selected mutations enable 9n DNA polymerase to synthesize TNA in the absence of manganese ions.

**TNA replication fidelity**. The strong TNA synthesis efficiency of 9n-YRI provided an opportunity to compare the effect of manganese ions on the fidelity of TNA synthesis. We therefore measured the fidelity of TNA synthesis by sequencing >2,000 nucleotide positions isolated from the complementary DNA (cDNA) product generated after a complete cycle of TNA replication (DNA→TNA→DNA) (Supplementary Fig. 7). Unlike kinetic fidelity assays which examine a single-nucleotide insertion event[34], DNA sequencing provides a more complete view of the replication cycle by identifying insertions, deletions and mutations that occur when genetic information is converted from DNA into TNA and then in a separate reaction from TNA back into DNA[1].

A series of controls was used to ensure that the sequencing data reflected the accuracy of TNA 'transcription' and 'reverse transcription' in the primer-extension reactions. The first control was a PCR assay that tested for DNA contaminants in the TNA product isolated by PAGE purification (Supplementary Fig. 8). In no cases did we observe a PCR product that amplified with the same number of cycles as the cDNA strand isolated from the reverse transcription of a TNA template into DNA. The second control involved checking the sequencing product to ensure that a T to A mutation occurred in the primer-binding site. Our TNA synthesis reaction was performed with a primer that contained a single-nucleotide mismatch that would lead to a T to A mutation when the TNA strand was reverse transcribed into DNA but lacking in sequences that were amplified from DNA contaminants[3] (Supplementary Table 2).

Analysis of the sequencing results indicates that a TNA replication cycle performed with 9n-YRI as the TNA polymerase and SuperScript II as the reverse transcriptase produces ~2 mistakes out of 1,000 nucleotide incorporations when manganese ions are absent from the TNA synthesis reaction. By contrast, the mutation rate is ~50-fold higher when the same reaction was performed in the presence of manganese ions (Fig. 3d). This striking result confirms the hypothesis that manganese ions lower the fidelity of TNA synthesis and provide a viable strategy for faithful TNA synthesis under conditions that more closely approximate natural DNA synthesis. In this regard, 9n-YRI and 9n-NVA, represent the first demonstration of TNA polymerases

that functions in the absence $Mn^{2+}$ (Fig. 3c and Supplementary Fig. 6b).

## Discussion

Synthetic genetics aims to develop artificial genetic polymers that can replicate in vitro and eventually in model cellular organisms[4]. Achieving this ambitious goal will require major advances in chemical synthesis and polymerase engineering, as both fields of science are needed to develop the tools necessary for copying information back and forth between DNA and XNA and eventually between XNA polymers themselves. Recognizing that some of the most interesting XNAs can only be obtained by chemical synthesis[7], researchers are facing a pressing need for new synthetic protocols that can be used to generate XNA monomers on the gram scale. Coupled with this effort is the equally challenging demand for new XNA polymerases that can synthesize kilobases of information with no mistakes. While this later goal may seem modest in comparison to natural polymerases, which can faithfully copy a megabase of DNA, the applications envisioned for XNA are less demanding than the biological requirements imposed by cellular organisms[35].

In the current study, we present DrOPS as a new strategy for engineering polymerases with non-natural functions. Our method relies on single and double-emulsion droplets that are produced using commercially available microfluidic chips and reagents. The two-chip design simplifies the procedure for generating monodisperse droplets and provides flexibility for controlling such parameters as droplet size and oil-layer thickness[27]. With this system, droplets can be produced and screened in a matter of hours, which allows a round of selection to take place in 3–4 days. For example, we screened a library of 36 million double-emulsion droplets in 2 h (at 5 kHz) by fluorescence-activated cell sorting. Based on this rate of sorting, we suggest that it should be possible to screen $>10^8$ droplets per day, which may be necessary for some polymerase functions that require greater library diversity.

The DrOPS method has several advantages over existing polymerase engineering technologies. Relative to screening procedures that assay variants in microlitre-scale reactions, miniaturization of the PAA into microdroplets reduces the assay volume to the picolitre scale, which is a $\sim 10^6$-fold reduction in reaction volume per polymerase assay.

This improvement in assay volume size coupled with the ability to sort $>10^8$ droplets per day leads to enormous cost savings for chemically synthesized substrates like tNTPs that require more than 12 synthetic steps to produce[32]. In the case of our 8,000 member library, we performed 1 round of DrOPS using 200 μl of tNTP containing reaction buffer, which is equivalent to 20 primer-extension reactions performed under standard bulk-phase conditions. By comparison, traditional screening of the same library with 98% coverage would require 32,000 PAAs and consume >320 ml of reaction buffer. This striking difference leads to an economy of scale that benefits microfluidics-based reactions by reducing the consumption of chemically synthesized substrates, which is critical to realizing the long-term of goals of synthetic genetics[4].

DrOPS also compares favourably with other polymerase technologies, like CSR and CST, that use w/o emulsions generated by bulk mixing[1,12]. While CSR and CST have been used to evolve polymerases with enhanced activity and expanded substrate recognition, both methods use the parent plasmid as template for the primer-extension reaction, which limits the range of polymerase functions to enzymes that promote DNA-templated synthesis. In addition, CST requires affinity purification on a solid-support matrix, which lowers the partitioning efficiency of functional members due to non-specific DNA binding to the resin. By contrast, DrOPS uses an optical sensor that is amenable to any nucleic acid polymer that is capable of Watson–Crick base pairing and relies on solution-based separation methods, like FACS to separate functional droplets from the non-functional pool. In addition, the ability to specify the sequence composition and length of the template provides enormous control over the stringency of the selection. Together, these properties of template control and solution-based separation make DrOPS a versatile tool that could be applied to a wide range of problems in polymerase engineering.

Although this study examined a specific problem in TNA polymerase engineering, namely, the ability to synthesize TNA in the absence of manganese ions, the DrOPS technology is unique in the sense that it could be applied to other more challenging problems in polymerase engineering. For example, the quantitative aspect of DrOPS could be used to identify new XNA polymerases with superior activity, while the template control aspect provides an avenue for discovering future polymerases that can copy XNA into DNA or possible even XNA into XNA, thereby demonstrating direct XNA replication.

In summary, we have developed a microfluidics-based method for evolving novel polymerase functions *in vitro*. The strategy functions with high partitioning efficiency, using an optical sensor that could be engineered for other substrate–template combinations. While further advances in optical detection and droplet separation are possible[36], the ability to use commercial chips and reagents provides a technology that is readily available to most laboratories.

## Methods

**General information.** DNA oligonucleotides (Supplementary Table 1) were purchased from Integrated DNA Technologies (Coralville, IA), purified by denaturing polyacrylamide gel electrophoresis, electroeluted, ethanol precipitated and quantified by ultraviolet absorbance using a NanoDrop spectrophotometer. NTPs and dNTPs were purchased from Sigma (St Louis, MO). tNTPs were obtained by chemical synthesis as previously described[1,2]. Accuprime DNA Polymerase was obtained from Invitrogen (Grand Island, NY). Hen egg lysozyme was purchased from Sigma. Fluorinated oil HFE-7500 was purchased from 3M Novec (St Paul, MN) and microfluidic chips were purchased from Dolomite (UK). The 9n gene was kindly provided by Andreas Marx in a pGDR11 expression vector. DNA sequencing was performed at the ASU Core Facility. Full-length gels to main text figures are provided in Supplementary Fig. 9.

**Generating emulsion droplets.** The formation of w/o single emulsions was performed using a quartz glass microfluidic device with a single inlet flow-focusing junction geometry of $14 \times 17\,\mu m$ with a hydrophobic/fluorophilic coating (Cat. C000525G, Dolomite). The device was connected by FEP tubing through a top interface linear connector (Cat. 3000109, Dolomite) to syringes (100 μl, 500 μl SGE glass syringes, 2500 μl Hamilton Gastight syringe or 3 ml plastic syringe (Becton-Dickinson, Madrid, Spain)), which were driven by either NE1002× syringe infusion pumps (New Era Pump Systems Inc., USA) or a pump manifold of neMESYS low pressure syringe pumps (Cetoni Gmbh, Germany) with accompanying control software. Carrier fluid was filtered using a 0.2 μm inline syringe filter, while the aqueous phase was filtered using an inline 10 μm frit filter. Droplet generation was monitored using a Nikon eclipse TS100 microscope with 20× ELWD Nikon objective and captured using a QIclick 12 bit monochrome CCD camera (QImaging, BC Canada). Flow rates were adjusted based on visual inspection with an average rate of $5\,\mu l\,min^{-1}$ for the aqueous phase and $12\,\mu l\,min^{-1}$ for the carrier oil. These flow rates yielded droplets with an average diameter of 14 μm (~1 pl volume). A low-viscosity fluorinated oil (HFE-7500) containing 1% (w/w) Pico-Surf surfactant (Dolomite) was used as the carrier fluid.

The formation of w/o/w double emulsions was performed using a quartz glass microfluidic device with a single inlet flow-focusing junction geometry of $14 \times 17\,\mu m$ (Cat. 3200136, Dolomite). The w/o emulsion and aqueous carrier phase were delivered to the device using syringes connected in the same manner as described above for single-emulsion formation. The w/o emulsion was slowly drawn into a 250 μl SGE glass syringe, mounted into an infusion pump in a vertical position and left to settle for at least 30 min prior to delivery. Carrier fluid (25 mM NaCl, 1% Tween-80) was filtered using a 0.2 μm inline syringe filter, while the w/o emulsion was filtered using an inline 10 μm frit filter. Flow rates were adjusted based on visual inspection with an average rate of $1\,\mu l\,min^{-1}$ for the single emulsion and $8\,\mu l\,min^{-1}$ for the carrier aqueous phase.

**Cell compartmentalization in droplets.** Cell populations were grown and polymerase variants were expressed as described above. After expression, an aliquot (2 ml) of cell culture was centrifuged for 5 min (2,000 r.c.f.) and the supernatant discarded. The cells were washed three times with $1 \times$ ThermoPol buffer (20 mM Tris-HCl, 10 mM $(NH_4)_2SO_4$, 10 mM KCl, 2 mM $MgSO_4$, 0.1% Triton X-100, pH 8.8). After each wash, the cells were centrifuged for 5 min (2,000 r.c.f.) and the supernatant discarded. The rinsed bacterial pellet was re-suspended in 500 μl $1 \times$ ThermoPol buffer and the absorbance was measured at 600 nm. Cells were diluted to enable encapsulation at occupancies of 0.1 cells per droplet, according to the assumption that 1 ml of E. coli suspension at an $A_{600}$ value of 1.0 contains $5 \times 10^8$ cells. Just prior to emulsification, the cells were mixed with the fluorescence-based PAA (see section below). The w/o emulsion was collected under a layer of mineral oil in an Eppendorf tube. Following emulsification, the reactions were incubated for 5 min at 90 °C to lyse cells, followed by incubation at 55 °C for the indicated amount of time.

**Microscopy.** Images were collected using a bright-field microscope (Eclipse TE300, Nikon) equipped with a Hamamatsu Orca 3CCD camera using a $60 \times$, 1.32 numerical aperture (NA), oil-immersion objective lens and Immersion Oil Type DF (Cargille Laboratories) imaging medium. QED InVivo 3.2 (Media Cybernetics) was used to collect images, which were processed with Photoshop CS4 (Adobe) or ImageJ (NIH) software. Microfluidic droplet generation was monitored using a Nikon eclipse TS100 inverted microscope with either a $10 \times$, 0.3 NA Plan fluor or $20 \times$, 0.45 NA ELWD S Plan Fluor, Nikon objectives and captured using a QIclick 12 bit monochrome CCD camera.

**Flow cytometric analysis of double-emulsion droplets.** W/o/w double-emulsion droplets were diluted into 150 mM NaCl and subjected to flow cytometric analysis (FACSCalibur, BD Biosciences). The sample was excited with a 488 nm argon laser and the emission was detected using a $530 \pm 15$ nm band-pass filter. Double-emulsion populations were gated on logFSC/logSSC. Fluorescent readout was obtained from $>15,000$ droplets for each measurement and analysed using Cytometer software (Cell Quest, BD Biosciences).

**Polymerase library generation.** The focused 9n DNA polymerase library was generated by replacing the region coding for the finger, thumb and palm domains with a DNA cassette containing unbiased, random codons (NNN) at amino acid positions 409, 485 and 664, respectively. The DNA cassette was generated from three gBlock fragments that were combined by overlapping PCR using AccuPrime DNA polymerase (Supplementary Fig. 3). The second fragment contains a 5′ region that is conserved with the 3′ end of the first fragment and a 3′ region that is conserved with the 5′ end of the third fragment. Each fragment was individually amplified using three sets of unique primers (P1.For, P1.Rev, P2.For, P2.Rev, P3.For, P3.Rev) with an optimized number of PCR cycles determined by quantitative PCR analysis to prevent over-amplification. The full-length cassette was then assembled by combining 15 ng of each fragment and DNA primers P1.For and P3.Rev into a single PCR reaction. The PCR-amplified cassette was digested with AscI and BglII restriction enzymes, ligated into the pGDR11 expression vector and transformed into electrocompetent 10-beta E. coli (New England Biolabs Inc., Massachusetts, USA).

**Polymerase selections.** Polymerase variants were grown as a population of E. coli carrying the pGDR11 plasmid encoding the polymerase of interest in Luria–Bertani (LB) broth supplemented with ampicillin ($100\,\mu g\,ml^{-1}$). Cultures were grown at 37 °C with shaking at 240 r.p.m. and protein expression was induced by adding IPTG to a final concentration of 1 mM at an OD-600 of 0.6. Induced cultures were grown for an additional 3 h at 37 °C with shaking. Prior to emulsion formation, the cells were washed three times with $1 \times$ ThermoPol buffer (NEB, USA) and then diluted to enable encapsulation at occupancies of 0.1 cells per droplet. Just prior to emulsification, the cells were mixed with the fluorescence-based PAA. The w/o emulsion was collected under a layer of mineral oil in an Eppendorf tube. Following emulsification, the reactions were incubated for 5 min at 90 °C to lyse cells, followed by incubation at 55 °C for the indicated amount of time. Single emulsions were then converted to double emulsions as described above. Prior to sorting droplets using a FACS, the aqueous carrier phase (1% w/w Tween-80 in 25 mM NaCl) was exchanged for a solution of 25 mM NaCl to reduce the presence of surfactant in the aqueous phase. Samples were sorted in a BD FACS Aria (BD Biosciences) using PBS as a sheath fluid. A set-up with a 70 μm nozzle was chosen to give an average sort rate of 5,000–8,000 events per second. The threshold trigger was set on side scatter. The sample was excited with a 488 nm argon laser and the emission was detected using a $530 \pm 15$ nm band-pass filter. The double-emulsion population was gated from other populations in the sample on logFSC/logSSC. DNA samples were recovered from sorted emulsions by extraction with ~2 vol of Pico-Break 1 (Dolomite) to disperse the emulsions. The extracted aqueous phase was concentrated using a spin column (Zymo Research) and used to transform electrocompetent E. coli cells (β-10, NEB). Plasmid recovery efficiency was determined by comparing the number of sorted droplets to the number of colonies obtained after transformation and plating.

**Fluorescence-activated droplet sorting.** Prior to sorting droplets using a FACS, the aqueous carrier phase (1% w/w Tween-80 in 25 mM NaCl) was exchanged for a solution of 25 mM NaCl to reduce the presence of surfactant in the aqueous phase. Samples were sorted in a BD FACS Aria using PBS as a sheath fluid. A set-up with a 70 μm nozzle was chosen to give an average sort rate of 5,000–8,000 events per second. The threshold trigger was set on side scatter. The sample was excited with a 488 nm argon laser and the emission was detected using a $530 \pm 15$ nm band-pass filter. The double-emulsion population was gated from other populations in the sample on logFSC/logSSC.

**DNA recovery and transformation.** DNA samples were recovered from sorted emulsions by extraction with ~2 vol of Pico-Break 1 (Dolomite), which contains 1H,1H,2H,2H-perfluorooctanol (PFO). After addition of Pico-Break 1, the samples were vortexed, followed by centrifugation (15 s, 2,000 r.c.f.) to attain phase separation. The top, aqueous layer containing the plasmid DNA was recovered. The bottom layer was extracted second time with 1 vol of molecular grade water to improve recovery yields. The combined aqueous layers containing the plasmid DNA were concentrated using a spin column (DNA Clean & Concentrator-5, Zymo Research) and eluted with molecular biology grade water (10 μl). The DNA Clean & Concentrator-5 also facilitates removal of protein from the sample. Electrocompetent *E. coli* cells (50 μl, β-10 *E. coli* cells, NEB) were transformed with 5 μl of purified DNA by applying one electric pulse of 1.80 kV (using an *E. coli* Pulser cuvette, 0.1 cm electrode; Bio-Rad MicroPulser). Sterile S.O.C. Medium (500 μl, Invitrogen) was added immediately after pulsing and the sample was grown for 30 min at 37 °C with shaking at 240 r.p.m. before plating on LB agar containing ampicillin (100 μg ml$^{-1}$) followed by incubation at 37 °C overnight. Plasmid recovery efficiency was determined by comparing the number of sorted droplets with the number of colonies obtained after transformation and plating. In some cases, dilution plating was used to estimate the number of successful transformants.

**Polymerase expression.** Individual polymerase variants were tested by growing a clonal population of XL-1 blue *E. coli* carrying the pGDR11 plasmid encoding the polymerase of interest in LB broth supplemented with ampicillin (100 μg ml$^{-1}$). Cultures were grown at 37 °C with shaking at 240 r.p.m. and protein expression was induced by adding IPTG to a final concentration of 1 mM at an OD-600 of 0.6. Induced cultures were grown for an additional 3 h at 37 °C with shaking. The cells were then pelleted and re-suspended in nickel-binding buffer (50 mM phosphate, 250 mM sodium chloride, 10% glycerol, pH 8) with 0.1 mg ml$^{-1}$ hen egg lysozyme, and incubated for 15 min at 37 °C. Following lysozyme treatment, the samples were heated for 15 min at 75 °C. Aggregated cellular debris was removed by centrifugation for 15 min at 3,200 r.c.f. Polymerases were purified from the lysate based on an N-terminal 6× His-tag by binding to a nickel-affinity resin. After binding, the resin was washed three times with nickel-binding buffer followed by elution with nickel-binding buffer supplemented with 75 mM imidazole. Protein expression was confirmed by SDS–PAGE analysis with coomassie blue staining. Polymerases were exchanged into storage buffer (10 mM Tris-HCl, 100 mM KCl, 1 mM DTT, 0.1 mM EDTA, pH 7.4) using a Microcon-30 kDa column (Millipore, USA) and stored at 4 °C.

**Polymerase activity assays.** Polymerase activity was evaluated as the ability to extend a DNA primer–template complex with natural, non-cognate and unnatural nucleotide triphosphates. Primer-extension reactions were analysed by denaturing PAGE or fluorescence. The primer–template complex was annealed in ThermoPol buffer (1×: 20 mM Tris-HCl, 10 mM (NH4)2SO4, 10 mM KCl, 2 mM MgSO4, 0.1% Triton X-100, pH 8.8; New England Biolabs) by heating for 5 min at 95 °C and cooling for 5 min at 4 °C. Nucleotide triphosphates (100 μM final) and polymerase were added to the reaction after primer annealing and the reaction was incubated at 55 °C for the indicated amount of time. Fluorescence-based PAAs were performed using an unlabelled DNA primer, a template with a fluorophore label at the 5′ end and a quencher probe labelled with a quencher dye at the 3′ end. The concentration of primer, template and quencher strands were 2 μM, 1 μM and 3 μM, respectively. Fluorescence was measured using a 2014 EnVision multilabel plate reader (PerkinElmer). For PAGE assays, the DNA primer carried an IR800 fluorophore label at the 5′ end and an unlabelled DNA template strand. The concentration of primer and template were 0.5 μM and 1 μM, respectively, and no quencher strand was added. Reactions were quenched by adding 10 equiv. of stop buffer (1× Tris-boric acid buffer, 20 mM EDTA, 7 M urea, pH 8). Samples were denatured for 5 min at 90 °C prior to separation by denaturing PAGE and visualization of the IR800 dye using a LICOR Oddysey CLx imager.

For the polymerase time courses, the reaction volume was increased to 25 μl. At each desired time point, 1 μl of the reaction was removed and added to 30 μl of stop buffer. Samples were then denatured for 5 min at 90 °C prior to separation by denaturing PAGE and visualization of the IR800 dye using a LICOR Oddysey CLx imager. The amount of full-length and truncated products were quantified using the Image Studio software version 4.0. All time course assays were completed with the PBS2-IR800 DNA primer and ST.1G DNA template.

**Fidelity analysis.** Fidelity reactions were performed by sequencing the cDNA strand following a complete cycle of transcription and reverse transcription. The primer-template complex was extended in a 100 μl reaction volume containing 100 pmol of fidelity.temp and 100 pmol of PBS2.mismatch primer. The primer and template were annealed in 1× ThermoPol buffer by heating for 5 min at 95 °C and cooling for 10 min at 4 °C. The 9n-YRI polymerase (10 μl) was added to the reaction mixture. For TNA extensions in the presence of $Mn^{2+}$, the polymerase was pretreated with 1 mM MnCl₂. The reactions were initiated by addition of the TNA nucleotide triphosphates (100 μM). Following a 4-h incubation with $Mn^{2+}$ or an 18-h incubation without $Mn^{2+}$ at 55 °C, the reactions were quenched in stop buffer and denatured at 90 °C for 5 min. Elongated primers were purified by denaturing PAGE, electroeluted and concentrated using a YM-30 concentrator device.

The purified transcripts were reverse transcribed in a final volume of 100 μl. PBS1 primer (100 pmol) was annealed to the template in 1× First Strand Buffer (50 mM Tris-HCl, 75 mM KCl, 3 mM MgCl₂, pH 8.3) by heating for 5 min at 90 °C and cooling for 10 min at 4 °C. Next, 500 μM dNTPs and 10 mM DTT were added and the reaction was allowed to incubate for 2 min at 42 °C. Finally, 3 mM MgCl₂, 1.5 mM MnCl₂ and 10 U μl$^{-1}$ SuperScript II reverse transcriptase were added and the reaction was allowed to incubate for 1 h at 42 °C.

After reverse transcription, the PCR-amplified DNA (1 pmol) was ligated into a pJET vector following manufacturer's protocol. The ligated product was transformed into XL-1 blue *E. coli*, grown in liquid media and individual colonies were isolated, cloned and sequenced (ASU Core Facility). Sequencing results were analysed using CLC Main Workbench. Sequences lacking the T to A watermark were discarded as they were generated from the starting DNA template rather than replicated material. The error rate for each of the nine possible substitution (for example, T→C, T→G or T→A) was determined as follows: $\mu_{exp \to obs}$ = (#observed/#expected) × 1,000. The total error rate was determined by summing the error rate for each substitution.

## References

1. Pinheiro, V. B. *et al.* Synthetic genetic polymers capable of heredity and evolution. *Science* **336**, 341–344 (2012).
2. Yu, H., Zhang, S. & Chaput, J. C. Darwinian evolution of an alternative genetic system provides support for TNA as an RNA progenitor. *Nat. Chem.* **4**, 183–187 (2012).
3. Yu, H., Zhang, S., Dunn, M. & Chaput, J. C. An efficient and faithful in vitro replication system for threose nucleic acid. *J. Am. Chem. Soc.* **135**, 3583–3591 (2013).
4. Chaput, J. C., Yu, H. & Zhang, S. The emerging world of synthetic genetics. *Chem. Biol.* **19**, 1360–1371 (2012).
5. Pinheiro, V. B., Loakes, D. & Holliger, P. Synthetic polymers and their potential as genetic materials. *Bioessays* **35**, 113–122 (2012).
6. Joyce, G. F. Toward an alternative biology. *Science* **336**, 307–308 (2012).
7. Anosova, I. *et al.* The structural diversity of artificial genetic polymers. *Nucleic Acids Res.* **44**, 1007–1021 (2016).
8. Horhota, A. *et al.* Kinetic analysis of an efficient DNA-dependent TNA polymerase. *J. Am. Chem. Soc.* **127**, 7427–7434 (2005).
9. Kempeneers, V., Vastmans, K., Rozenski, J. & Herdewijn, P. Recognition of threosyl nucleotides by DNA and RNA polymerases. *Nucleic Acids Res.* **31**, 6221–6226 (2003).
10. Loakes, D. & Holliger, P. Polymerase engineering: towards the encoded synthesis of unnatural polymers. *Chem. Commun.* 4619–4631 (2009).
11. Chen, T. & Romesberg, F. E. Directed polymerase evolution. *FEBS Lett.* **588**, 219–229 (2014).
12. Ghadessy, F. J., Ong, J. L. & Holliger, P. Directed evolution of polymerase function by compartmentalized self-replication. *Proc. Natl Acad. Sci. USA* **98**, 4552–4557 (2001).
13. Turner, N. J. Directed evolution drives the next generation of biocatalysts. *Nat. Chem.. Biol.* **5**, 567–573 (2009).
14. Griffiths, A. D. & Tawfik, D. S. Man-made enzymes—from design to in vitro compartmentalization. *Curr. Opin. Biotechnol.* **11**, 338–353 (2000).
15. Tawfik, D. S. & Griffiths, A. D. Man-made cell-like compartments for molecular evolution. *Nat. Biotechnol.* **16**, 652–656 (1998).
16. Kaltenbach, M., Devenish, S. R. & Hollfelder, F. A simple method to evaluate the biochemical compatibility of oil/surfactant mixtures for experiments in microdroplets. *Lab Chip* **12**, 4185–4192 (2012).
17. Anna, S. L., Bontoux, N. & Stone, H. A. Formation of dispersions using 'flow focusing' in microchannels. *Appl. Phys. Lett.* **82**, 364–366 (2003).
18. Umbanhowar, P. B., Prasad, V. & Weitz, D. A. Monodisperse emulsion generation via drop break off in a coflowing stream. *Langmuir* **16**, 347–351 (2000).
19. Agresti, J. J. *et al.* Ultrahigh-throughput screening in drop-based microfluidics for directed evolution. *Proc. Natl Acad. Sci. USA* **107**, 4004–4009 (2010).
20. Sjostrom, S. L. *et al.* High-throughput screening for industrial enzyme production hosts by droplet microfluidics. *Lab Chip* **14**, 806–813 (2014).
21. Fischlechner, M. *et al.* Evolution of enzyme catalysts caged in biomimetic gel-shell beads. *Nat. Chem.* **6**, 791–796 (2014).

22. Summerer, D. & Marx, A. A molecular beacon for quantitative monitoring of the DNA polymerase reaction in real-time. *Angew. Chem. Int. Ed.* **31**, 3620–3622 (2002).

23. Dorjsuren, D. *et al.* A real-time fluorescence method for enzymatic characterization of specialized humn DNA polymerases. *Nucleic Acids Res.* **37**, e128 (2009).

24. Marras, S. Selection of fluorophore and quencher pairs for fluorescent nucleic acid hybridization probes. *Methods Mol. Biol.* **335**, 3–16 (2006).

25. Dunn, M. R., Otto, C., Fenton, K. E. & Chaput, J. C. Improving polymerase activity with unnatural substrates by sampling mutations in homologous protein architectures. *ACS Chem. Biol.* doi:10.1021/acshembio.5b00949 (2016).

26. Brown, J. A. & Suo, Z. Unlocking the sugar 'steric gate' of DNA polymerases. *Biochemistry* **50**, 1135–1142 (2011).

27. Zinchenko, A. *et al.* One in a million: flow cytometric sorting of single cell-lysate assays in monodisperse picoliter double emulsion droplets for directed evolution. *Anal. Chem.* **86**, 2526–2533 (2014).

28. Schoning, K. U. *et al.* Chemical etiology of nucleic acid structure: the alpha-threofuranosyl-(3'-->2') oligonucleotide system. *Science* **290**, 1347–1351 (2000).

29. Dunn, M. R. *et al.* Therminator-mediated synthesis of unbiased TNA polymers requires 7-deazaguanine to suppress G-G mispairing during TNA transcription. *J. Am. Chem. Soc.* **137**, 4014–4017 (2015).

30. Tabor, S. & Richardson, C. C. Effect of manganese ions on the incorporation of dideoxynucleotides by bacteriophage T7 DNA polymerase and *Escherichia coli* DNA polymerase I. *Proc. Natl Acad. Sci. USA* **86**, 4076–4080 (1989).

31. Cozens, C., Pinheiro, V. B., Vaisman, A., Woodgate, R. & Holliger, P. A short adaptive path from DNA to RNA polymerases. *Proc. Natl Acad. Sci. USA* **109**, 8067–8072 (2012).

32. Sau, S. P., Fahmi, N. E., Liao, J.-Y., Bala, S. & Chaput, J. C. A scalable synthesis of α-L-threose nucleic acid monomers. *J. Org. Chem* **81**, 2302–2307 (2016).

33. Zhang, S., Yu, H. & Chaput, J. C. Synthesis of threose nucleic acid (TNA) triphosphates and oligonucleotides by polymerase-mediated primer extension. *Curr. Protoc. Nucleic Acid Chem.* **52** 4.54 (2013).

34. Goodman, M. F., Creighton, S., Bloom, L. B. & Petruska, J. Biochemical basis of DNA replication fidelity. *Crit. Rev. Biochem. Mol. Biol.* **28**, 83–126 (1993).

35. Steitz, T. A. DNA polymerases: structural diversity and common mechanisms. *J. Biol. Chem.* **274**, 17395–17398 (1999).

36. Romero, P. A., Tran, T. M. & Abate, A. R. Dissecting enzyme function with microfluidic-based deep mutational scanning. *Proc. Natl Acad. Sci. USA* **112**, 7159–7164 (2015).

## Acknowledgements

We would like to thank members of the laboratory of J.C.C. for helpful discussions and critical reading of the manuscript. This work was supported by the DARPA Folded Non-Natural Polymers with Biological Function (Fold F(x)) Program under award number N66001-14-2-4054. Any opinions, findings and conclusions or recommendations expressed in this publication are those of the author(s) and do not necessarily reflect the views of DARPA.

## Author contributions

A.H. and C.Y. designed the microfluidic droplet generation strategy. A.C.L. and A.H. performed microfluidic droplet generation. A.C.L. and M.R.D. generated polymerase variants and performed activity assays. A.C.L. performed *in vitro* selection rounds. S.P.S. synthesized TNA triphosphates. A.C.L., M.R.D. and J.C.C. designed the study and wrote the paper. All authors discussed the results and commented on the manuscript.

## Additional information

# 19

# Structural basis for the targeting of complement anaphylatoxin C5a using a mixed L-RNA/L-DNA aptamer

Laure Yatime[1], Christian Maasch[2], Kai Hoehlig[2], Sven Klussmann[2], Gregers R. Andersen[1] & Axel Vater[2]

L-Oligonucleotide aptamers (Spiegelmers) consist of non-natural L-configured nucleotides and are of particular therapeutic interest due to their high resistance to plasma nucleases. The anaphylatoxin C5a, a potent inflammatory mediator generated during complement activation that has been implicated with organ damage, can be efficiently targeted by Spiegelmers. Here, we present the first crystallographic structures of an active Spiegelmer, NOX-D20, bound to its physiological targets, mouse C5a and C5a-desArg. The structures reveal a complex 3D architecture for the L-aptamer that wraps around C5a, including an intramolecular G-quadruplex stabilized by a central $Ca^{2+}$ ion. Functional validation of the observed L-aptamer:C5a binding mode through mutational studies also rationalizes the specificity of NOX-D20 for mouse and human C5a against macaque and rat C5a. Finally, our structural model provides the molecular basis for the Spiegelmer affinity improvement through positional L-ribonucleotide to L-deoxyribonucleotide exchanges and for its inhibition of the C5a:C5aR interaction.

[1] Department of Molecular Biology and Genetics, Aarhus University, Gustav Wieds Vej 10C, DK-8000 Aarhus, Denmark. [2] NOXXON Pharma AG, Max-Dohrn-Strasse 8-10, 10589 Berlin, Germany. Correspondence and requests for materials should be addressed to L.Y. (email: lay@inano.au.dk) or to A.V. (email: avater@noxxon.com).

Complement is a central component of innate immunity that provides a first line of defence against invading pathogens and participates in immune surveillance[1-3]. It also bridges the innate and adaptive immunity, initiates the inflammatory response and helps maintaining homeostasis[2-5]. Detection of foreign microorganisms or damaged host cells triggers complement activation, leading to opsonization and assembly of the lytic membrane attack complex (MAC)[1,2]. In addition, the anaphylatoxins C3a and C5a are released and thereafter function as signalling molecules through their cognate G-protein-coupled receptors present on the host cells[6-8]. By targeting its receptors C5aR1 and C5aR2 on a wide range of immune cells[6,9] as well as endothelial cells[10], C5a promotes inflammation, vascular permeability and coagulation[6-9,11]. Although complement activation and inflammation are essential for the host defence and the healing process following tissue damage and infection, unduly elevated levels of C5a may promote and/or exacerbate pathological conditions underlying various acute and chronic inflammatory disorders such as acute lung injury, ischaemia-reperfusion injuries, sepsis, transplant rejection, rheumatoid arthritis, allergy and asthma[9,11-14].

Over the last decades, intensive efforts have been made to design potent inhibitors applicable for the treatment of complement-mediated diseases[15-17]. In particular, specific targeting of C5a and its receptors allows for tuning down anaphylatoxin-mediated inflammation while maintaining opsonization and MAC-mediated bacteriolysis[18,19]. This strategy is likely to be of particular value in patients at increased risk of infection such as critically ill and immunosuppressed patients. In this line of efforts, a series of plasma-stable C5a-inhibiting aptamers composed of non-natural, mirror-image L-nucleotides has been generated. Mirror-image aptamers, also referred to as Spiegelmers (from German Spiegel = mirror), possess the high target specificity of conventional aptamers and are in addition plasma nuclease resistant, therefore having high therapeutic and diagnostic potential[20]. Spiegelmers are generated by the SELEX process (Systematic Evolution of Ligands by EXponential enrichment)[21] from oligonucleotide libraries in the natural D-configuration but with the addition of two chiral inversion steps. In the first step, the enantiomer of the target molecule is synthesized (here murine D-C5a). Next, aptamers from a D-oligoribonucleotide library with $10^{15}$ different sequences are identified using SELEX and ranked for binding to the selection target. In a second chiral inversion step, the best aptamer sequences are truncated to the minimal size without loss of affinity and are then synthesized in their mirror-image L-configuration, thereby resulting in Spiegelmers binding to the natural murine L-C5a (Fig. 1a). Finally, through post-SELEX optimization, the initial C5a-binding, 44 nucleotides-long, L-RNA Spiegelmer NOX-D19 was modified by six positional ribo-to-deoxyribonucleotide exchanges, leading not only to improved affinity but also allowing a further truncation by four nucleotides, thus yielding the 40 nucleotide-long Spiegelmer NOX-D20 (ref. 22). This L-aptamer binds both human and murine C5a with picomolar affinities and has shown efficacy in a rodent model of polymicrobial sepsis induced by cecal ligation and puncture[22]. NOX-D20 is currently under consideration for preclinical and clinical development.

Here we report the crystal structures of NOX-D20 in complex with either murine C5a (mC5a) or its C-terminally desarginylated version mC5a-desArg, at 1.8 and 2.0 Å resolution, respectively. The structures reveal a complex 3D architecture for the Spiegelmer, including a left-turning double-helix and an intramolecular G-quadruplex stabilized by a $Ca^{2+}$ ion. These features allow the L-aptamer to wrap around the C5a molecule, forming a complex with tight shape complementarity. The NOX-D20:C5a binding mode observed in the structure is further validated by mutational studies using surface plasmon resonance (SPR) and allows to explain NOX-D20 specificity for human and mouse C5a species as opposed to macaque and rat C5a[22], as well as the increased affinity obtained following ribo-to-deoxyribonucleotide exchange. Finally, these data provide a molecular basis for the inhibitory properties of NOX-D20 towards the C5a:C5aR interaction.

## Results

### Structure of the NOX-D20:mC5a/mC5a-desArg complexes.
The NOX-D20:C5a complexes were formed by mixing the Spiegelmer with murine recombinant C5a/C5a-desArg[23] in a 1:1 molar ratio, in the presence of monovalent ($K^+$ and $Na^+$) and bivalent ($Ca^{2+}$ and $Mg^{2+}$) cations[22]. Crystals for the NOX-D20:mC5a and NOX-D20:mC5a-desArg complexes both displayed a $P2_12_12$ symmetry and diffracted to a maximal resolution of 1.8 and 2.0 Å, respectively (Table 1). The structure of the NOX-D20:mC5a complex was determined by single-wavelength anomalous diffraction (SAD) phasing using crystals derivatized with $Os(NH_3)_6$ (Table 1 and Supplementary Figs 1 and 2). The resulting model was then used to solve the NOX-D20:mC5a-desArg structure by molecular replacement (MR). Unexpectedly, the asymmetric units of both NOX-D20:mC5a and NOX-D20:mC5a-desArg crystals contained three molecules of mC5a and only two molecules of NOX-D20 (Supplementary Fig. 3a). One molecule of mC5a (red molecule in Supplementary Fig. 3a) has, however, only very few contacts with the Spiegelmer molecules and its role seems to be restricted to crystal packing stabilization. The two other mC5a molecules (beige and purple molecules in Supplementary Fig. 3a) each interact with one Spiegelmer molecule in a comparable manner. The root-mean-square deviation (r.m.s.d.) on all atoms between the two complexes present in the asymmetric unit (complexes A and B in Supplementary Fig. 3a) is 1.29 Å. The major difference between these two complexes is a repositioning of the loop connecting helices H2 and H3 in mC5a (residues Arg708 to Glu713; Supplementary Fig. 3b), which results from the rotation of the uracil base of dU30 by approximately 65° towards the neighbouring mC5a molecule, forcing the side chains of Val709, Asn710, Phe711 and Tyr712 to reorient (Supplementary Fig. 3c). Apart from this region, the two complexes are almost completely equivalent. As a more extended model could be traced for mC5a in complex A, we will describe this in the following section.

The NOX-D20:mC5a and NOX-D20:mC5a-desArg complexes are represented in Fig. 1 and Supplementary Fig. 3d, respectively. The final models encompass residues Asn679 to Pro750 for mC5a, Asn679 to Lys744 for mC5a-desArg and all 40 nucleotides for the Spiegelmer. The two structures are equivalent, with an r.m.s.d. on all atoms of 0.25 Å between the two complexes, in agreement with the fact that NOX-D20 binds both anaphylatoxins equally well[22]. mC5a adopts the canonical four-helix bundle conformation which superimposes well with the structure of isolated mC5a (PDB 4P3A[23]) with an r.m.s.d. on C-alpha atoms of 0.43 Å. The mC5a C-terminus extends away from the four-helix bundle core in the prolongation of helix H4 (Fig. 1c), reinforcing the idea that this region is highly flexible. NOX-D20 folds into a compact V-shape with the concave face of the V wrapping around the C5a molecule along the H1-H2-H3 side (Fig. 1c,d). The two binding partners display a strong shape complementarity (Fig. 1d), resulting in a total surface area buried at the NOX-D20:mC5a interface of 1,791 $Å^2$, as estimated with AREAIMOL[24].

### The anti-C5a Spiegelmer adopts a complex 3D architecture.
NOX-D20 adopts a complex three-dimensional (3D)

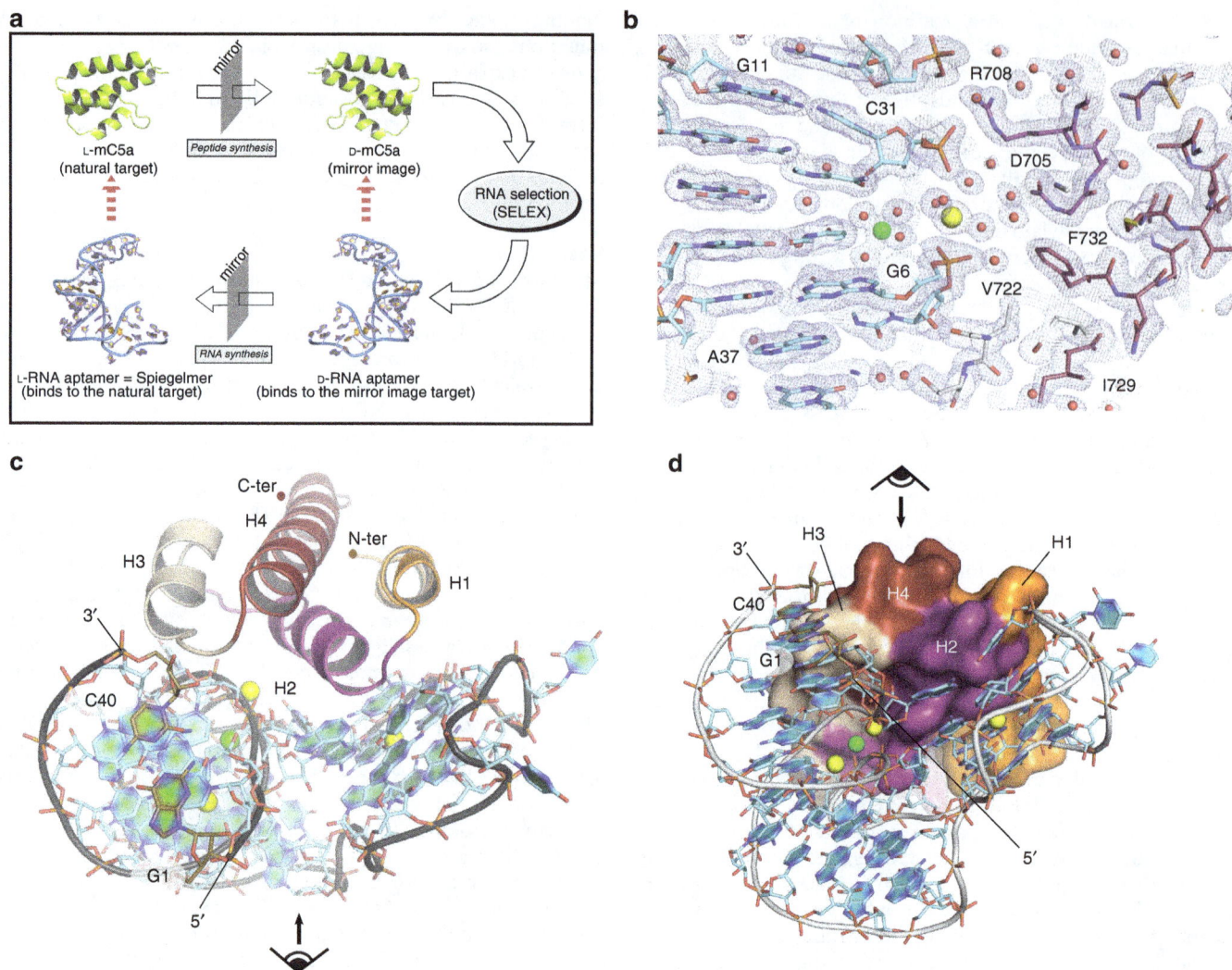

**Figure 1 | The NOX-D20:mC5a complex. (a)** Schematic representation of the selection principle to generate mC5a-binding Spiegelmers. **(b)** Final electron density maps and final model centered on the NOX-D20:mC5a interface. The 2mF$_o$-DF$_c$ map is shown as grey mesh and contoured at 1$\sigma$. Red spheres indicate water molecules. **(c)** Structure of the NOX-D20:mC5a complex at 1.8 Å resolution. The four mC5a helices (H1–H4) are highlighted in distinct colours. The Spiegelmer is represented in grey and cyan and the 5'- and 3'-terminal bases are shown in brown. Divalent cations are indicated as spheres (yellow for Ca$^{2+}$ and green for Mg$^{2+}$). The corresponding direction of view in **d** is indicated by an eye and arrow. **(d)** Same as in **c** but the view is rotated by 90° and mC5a is shown as surface representation to illustrate the strong shape complementarity with the L-aptamer fold. The corresponding direction of view in **c** is indicated by an eye and arrow.

organization in its active, target-bound form (Figs 1 and 2). The position of the six deoxyribonucleotides (dU7, dG14, dA15, dU28, dU30 and dC38) could be unambiguously assigned in the electron density maps and almost all nucleosides are in the *anti* N-glycosidic conformation with the exception of G8, dG14, G17 and G25, which are in *syn*. Furthermore, the solvent-exposed U23 alternates between the *anti* and *syn* conformations in the two molecules of NOX-D20 contained in the asymmetric unit. The majority of the riboses adopt the energetically favourable 3'-endo and 2'-endo conformations except dU7 (4'-exo), G9 (2'-exo), dG14 (4'-exo), G25 (4'-exo) and G26 (2'-exo).

The Spiegelmer molecule can be subdivided into two distinct structural domains (Fig. 2a,b), which are strongly interconnected, thereby preserving the NOX-D20 overall architecture and maintaining the integrity of the mC5a-binding pocket. The first domain, corresponding to the first leg of the V-shape, is built around a double-stranded helical stem that connects the first third of the molecule (nucleotides 1–15) to the last third (nucleotides 29–40) in an antiparallel manner. As expected, due to the L-configuration of all the riboses, this double-helix is left-handed. The domain is shaped by an extensive network of base pairing interactions between the two strands, including eight Watson–Crick base pairs, two non-Watson–Crick base pairs (dU7-G35 and dG14-G29, dG14 being in the *syn* N-glycosidic conformation to permit the interaction), and a non-canonical interaction between the Watson–Crick face of dA15 and the ribose face of G11. Finally, G8, G11, G12, U13 and dA15 help maintaining the domain overall fold through hydrogen bonds and stacking interactions.

The second domain, that is, the second leg of the V-shape, corresponds to a large loop encompassing nucleotides 16–28 and is articulated around an intramolecular G-quadruplex consisting of two G-tetrads of same polarity that stack in a 'partial 5/6 ring' geometry[25]. The bases of the two G-tetrads are twisted around a common perpendicular axis and the G-quadruplex is stabilized by a central Ca$^{2+}$ ion (see later) coordinated by the O6 atoms from all eight guanosine bases (Fig. 2c,d). The packing of the two G-tetrads is further maintained by π-stacking interactions with A16 and U24 on one side of the G-quadruplex, and with A4 and U21 on the other side (Fig. 2c). These nucleotides are themselves

**Table 1 | Data collection and refinement statistics for the native and anomalous data sets.**

| | NOX-D20:mC5a (native) | Os(NH$_3$)$_6$ data set | RbCl data set | High wavelength data set | NOX-D20:mC5a-desArg (native) |
|---|---|---|---|---|---|
| **Data collection** | | | | | |
| Space group | P2$_1$2$_1$2 | P2$_1$2$_1$2 | P2$_1$2$_1$2 | P2$_1$2$_1$2 | P2$_1$2$_1$2 |
| Cell dimensions | | | | | |
| $a$, $b$, $c$ (Å) | 46.47, 282.25, 45.90 | 45.69, 283.24, 45.92 | 45.76, 282.15, 45.78 | 45.78, 282.88, 45.85 | 45.84, 282.63, 45.76 |
| $\alpha$, $\beta$, $\gamma$ (°) | 90, 90, 90 | 90, 90, 90 | 90, 90, 90 | 90, 90, 90 | 90, 90, 90 |
| Wavelength (Å) | 0.91 | 1.1403 | 0.81 | 2.4984 | 1.0 |
| Resolution (Å)*,† | 50-1.8 (1.9-1.8) | 50-1.8 (1.9-1.8) | 50-2.5 (2.6-2.5) | 50-2.5 (2.6-2.5) | 50-2.0 (2.1-2.0) |
| $R_{meas}$ | 6.5 (93.3) | 9.0 (72.0) | 16.8 (85.0) | 8.5 (38.1) | 9.8 (59.8) |
| $I/\sigma I$ | 21.62 (2.98) | 17.65 (3.0) | 9.23 (1.84) | 18.52 (6.10) | 12.01 (2.72) |
| Completeness (%) | 99.9 (100) | 100 (100) | 99.9 (99.9) | 86.7 (46.4) | 99.8 (100) |
| Redundancy | 9.8 (8.2) | 8.3 (8.4) | 3.7 (3.7) | 6.8 (5.9) | 4.7 (4.8) |
| | | | | | |
| **Refinement** | | | | | |
| Resolution (Å) | 35-1.8 | | | | 35-2.0 |
| No. reflections | 56,027 | | | | 40,265 |
| $R_{work}/R_{free}$ | 16.70/19.26 | | | | 16.26/19.94 |
| No. of atoms | | | | | |
| Protein | 1,755 | | | | 1,670 |
| RNA/DNA | 1,716 | | | | 1,716 |
| Ligand/ion | 32 | | | | 20 |
| Water | 508 | | | | 373 |
| B-factors | | | | | |
| Protein | 36 | | | | 38 |
| RNA/DNA | 32 | | | | 37 |
| Ligand/ion | 47 | | | | 45 |
| Water | 39 | | | | 36 |
| R.m.s. deviations | | | | | |
| Bond lengths (Å) | 0.006 | | | | 0.007 |
| Bond angles (°) | 1.366 | | | | 1.469 |

*One crystal for each structure was used for data collection and structure determination.
†Values for the highest resolution shell are shown in parentheses.

held in position by either Watson–crick base pairing (A16-U24) or by single hydrogen bonding (A4-U21). The first G-tetrad is formed by nucleotides G17, G19, G25 and G27, whereas the second G-tetrad encompasses G18, G22, G26 and G32 (Fig. 2b–d). To allow for classical pairing of the guanine bases on both the Watson–Crick and the Hoogsteen faces, G17 and G25 from the first tetrad are in the *syn*-glycosidic conformation. The topology of the NOX-D20 G-quadruplex, with a mixed parallel–antiparallel strand orientation, two connecting loops and two connecting phosphodiester bonds, is quite unusual, in particular due to the fact that one of the G-nucleotides, G32, is provided by a distant and separate structural domain (Fig. 2b). To our knowledge, substitution of a single G from a distant site on the primary sequence has never been encountered before in G-quadruplex structures.

**The NOX-D20 fold is stabilized by divalent cations.** Divalent cations are important stabilizers of RNA folds[26]. The *in vitro* selection of NOX-D20 was performed in the presence of physiological concentrations of MgCl$_2$/CaCl$_2$, and the NOX-D20:C5a complexes were also crystallized in a buffer containing Mg$^{2+}$ and Ca$^{2+}$. To assess the influence of these ions on the NOX-D20:C5a interaction, binding of NOX-D20 to mC5a was followed by SPR measurements in the presence of increasing concentrations of EDTA (Fig. 3a). EDTA inhibited the NOX-D20:mC5a interaction in a dose-dependent manner, with a complete loss of binding at concentrations above 2 mM. Furthermore, titration of MgCl$_2$ or CaCl$_2$ induced a dose-dependent increase in the association rate constant $k_a$ of NOX-D20 binding to mC5a (Fig. 3b–f). The addition of either

CaCl$_2$ or MgCl$_2$ above the physiological concentration of 1 mM further increased target binding by NOX-D20. Ca$^{2+}$ showed a more pronounced effect, suggesting a much stronger contribution to C5a binding, whereas Mg$^{2+}$ only showed a minor influence on $k_a$ (Fig. 3c,f). Furthermore, Ca$^{2+}$ is essential for the stability of the NOX-D20:mC5a complex. When Ca$^{2+}$ is removed from the buffer, the NOX-D20:mC5a complex quickly dissociates (Fig. 3d). In contrast, a stable complex is maintained when a physiological calcium concentration (1 mM) is present during the dissociation phase, even if Ca$^{2+}$ was lower during the association phase (Fig. 3e).

In agreement with these findings, the structure of the NOX-D20:mC5a complex revealed the presence of four ions in each Spiegelmer molecule, including one ion at the centre of the G-quadruplex (Fig. 2d,e). Anomalous difference maps calculated from diffraction data collected at high wavelength ($\lambda = 2.498$ Å) revealed the presence of three strong peaks in each Spiegelmer molecule (Supplementary Fig. 4a). Both Ca$^{2+}$ and K$^+$ display comparable anomalous signal at this wavelength and could therefore be candidate for these positions. To identify the precise nature of the four ions, and knowing that G-quadruplexes tend to favour monovalent cations[27], Rb$^+$ or Mn$^{2+}$ were introduced in the NOX-D20:mC5a crystallization buffer as congeners for K$^+$ and Mg$^{2+}$, respectively, that give a strong anomalous signal at specific wavelengths. No crystals could be obtained in the presence of MnCl$_2$, suggesting that Mg$^{2+}$ plays a role in NOX-D20 stabilization and/or interaction with mC5a that cannot be fulfilled by Mn$^{2+}$. Anomalous data sets for Rb$^+$-derivatized crystals in the presence of Ca$^{2+}$ were collected at $\lambda = 0.81$ Å, but the resulting anomalous difference maps did not reveal the presence of Rb$^+$ (Supplementary Fig. 4b), suggesting that either

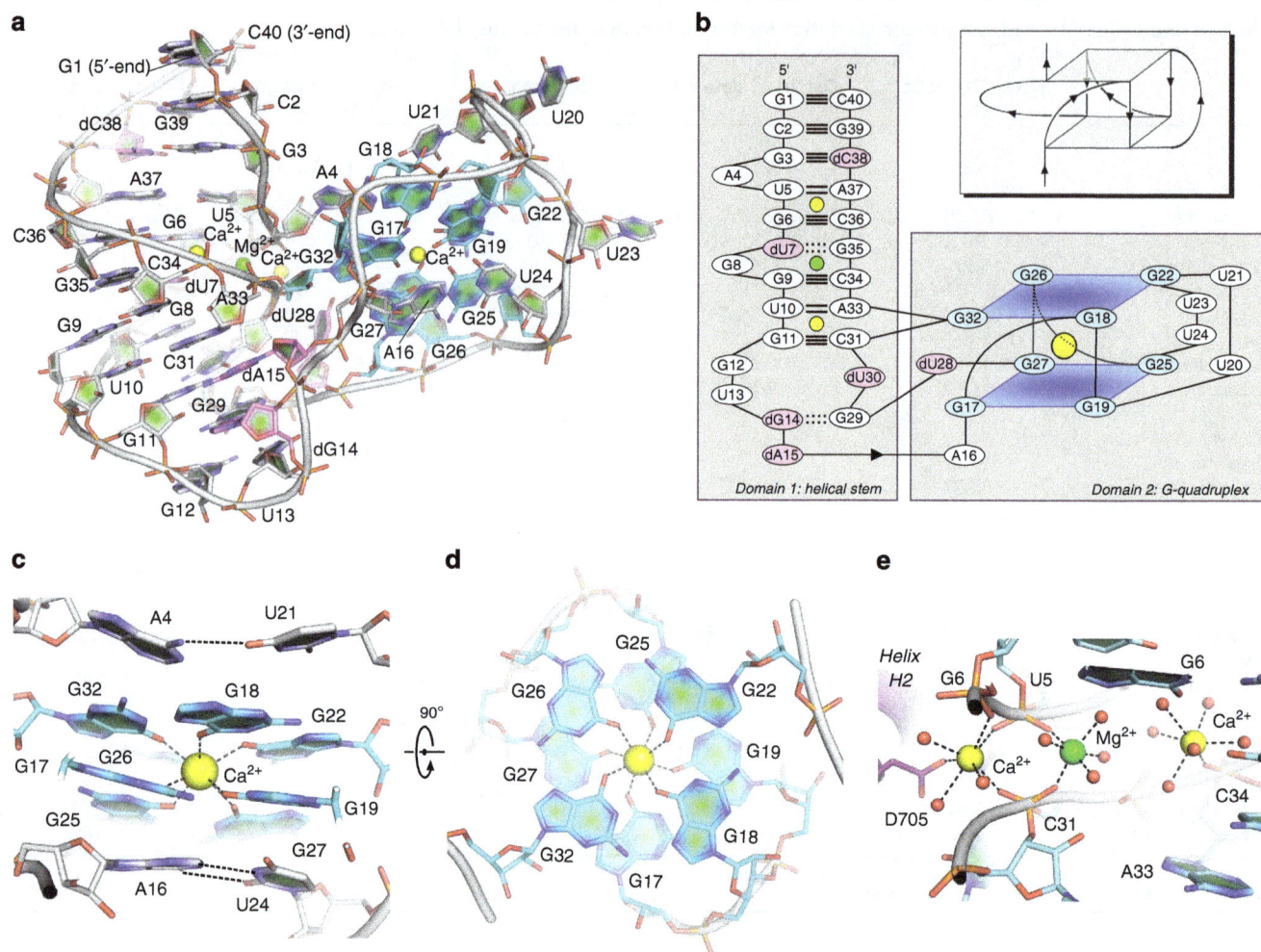

**Figure 2 | NOX-D20 adopts a complex 3D architecture. (a)** Overview of the NOX-D20 structural organization. The deoxyribonucleotides are indicated in purple and the G-nucleotides forming the G-quadruplex are highlighted in cyan. **(b)** Secondary structure representation of the NOX-D20 Spiegelmer reflecting the presence of two strongly interconnected structural domains. The ribose-phosphate backbone is indicated by a single plain line. Watson–Crick and non-Watson–Crick base pairs in the stem domain are indicated by double (A-U) or triple (G-C) plain and double dotted lines, respectively. The topology of the G-quadruplex is indicated in the insert above the scheme. **(c)** Zoom-in on the G-quadruplex domain showing a lateral view of the two G-tetrads and the additional stacking interactions that stabilize their packing on both sides of the quadruplex. **(d)** Same as in **c** but viewed from the top of the G-quadruplex. The stabilizing $Ca^{2+}$ ion lies in the centre of the ion channel in an almost perfect square antiprismatic coordination geometry. **(e)** Zoom-in on the minor groove of the helical stem domain, which is stabilized by three divalent cations arranged along a linear path.

$K^+$ does not play a role in NOX-D20 folding or that in the absence of $K^+$, another ion is stabilizing the Spiegelmer, in particular the G-quadruplex. Furthermore, NOX-D20 binds its target equally well in the absence of $K^+$, even when $Li^+$ is added (Fig. 3g–i). $Li^+$ is known to alter G-quadruplex structures when replacing $K^+$ (refs 28,29). These data add another line of evidence that $K^+$ does not play a role in NOX-D20 folding. Analysis of the ion–oxygen distances observed for these three ions in the final model revealed values ranging from 2.43 to 2.52 Å (Supplementary Fig. 4e), in agreement with values expected for Ca–O distances ($2.43 \pm 0.11$ Å (ref. 30)) and significantly lower than the values expected for K–O distances ($2.81 \pm 0.10$ Å (ref. 30)). To ensure that these distances were not model-biased by the presence of calcium, the $Ca^{2+}$ ions were removed and refinement with simulated annealing was performed to remove model bias. The distances calculated between the coordinating oxygens and the centres of the peaks obtained in the resulting $mF_o$-$DF_c$ map were again much closer to the expected Ca–O distances than the K–O ones (Supplementary Fig. 4e). Finally, adding back either $Ca^{2+}$ or $K^+$ at these positions and

performing refinement with tight restraints on the Ca–O or K–O distances led to a model superimposable to the final one when $Ca^{2+}$ was present, whereas in the presence of $K^+$, the coordinating oxygen ligands (from water molecules, from the RNA phosphate backbone or from the guanine bases of the G-quadruplex) were pushed away to the periphery of the electron density to match more closely the ideal K–O distance values implemented through the restraints, which are also observed in G-quadruplexes stabilized by potassium (Supplementary Fig. 4c,d). Taken together, these observations allowed us to unambiguously assign the three ions present in NOX-D20 structure giving strong anomalous signal at high wavelength as $Ca^{2+}$. As SPR data suggested a slightly enhancing effect of $Mg^{2+}$ on NOX-D20:mC5a binding and as all the other possible ions were ruled out, the last ion present in the structure was assigned as $Mg^{2+}$, which also agrees with the hexacoordination and the average ion-oxygen distance of $2.08 \pm 0.06$ Å observed for this ion in our structure (Fig. 2e).

Three of the four ions present in NOX-D20 structure align along a straight line in the middle of the helical stem in the first

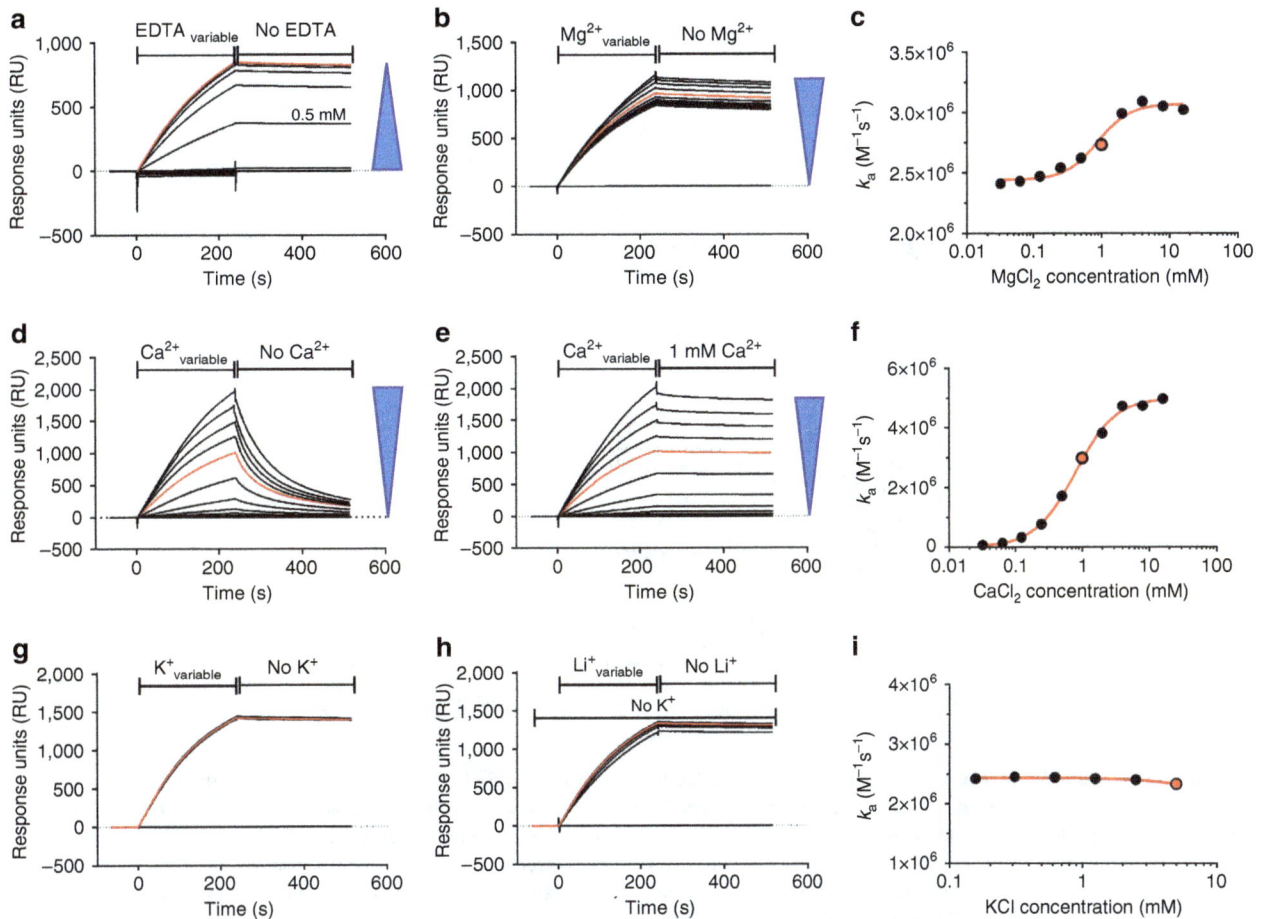

**Figure 3 | Influence of monovalent and divalent cations on mC5a recognition by NOX-D20 measured by SPR.** The blue arrows indicate increasing concentrations and the red line/dot corresponds to the measurement under physiological conditions (0 mM EDTA/1 mM CaCl$_2$/1 mM MgCl$_2$/5 mM KCl/5 mM LiCl). (**a**) Titration of EDTA. (**b**) Titration of MgCl$_2$. (**c**) Plot of the MgCl$_2$ effect on the association rate constant $k_a$. (**d**) Titration of CaCl$_2$. (**e**) Titration of CaCl$_2$ during the association phase and 1 mM CaCl$_2$ during the dissociation phase. (**f**) Plot of the CaCl$_2$ effect on the association rate constant $k_a$. (**g**) Titration of KCl. (**h**) Titration of LiCl in the absence of KCl. (**i**) Plot of the KCl effect on the association rate constant $k_a$.

domain (Fig. 2a,e). Their position right within the minor groove of the double-helix leads to the tightening of the packing between the two strands, thereby bringing them to a distance of 4.5 Å in the narrowest part, corresponding to half the width of the minor groove in standard A-form RNA. The central Mg$^{2+}$ is in an almost perfect octahedral coordination sphere formed by four water molecules and two phosphate oxygens from U5 and C31. The two Ca$^{2+}$ surrounding the Mg$^{2+}$ ion are heptacoordinated by water molecules, oxygens from the phosphate groups of U5, G6, C31 and C34 and, for the most peripheral calcium ion, by the side chain of Asp705 protruding from mC5a helix H2 (Fig. 2e). Finally, our anomalous difference maps conclusively show that the last Ca$^{2+}$ is located in the second structural module of NOX-D20, at the centre of the G-quadruplex, and is coordinated by all eight guanine bases (Fig. 2c,d).

**A complex built on strong shape complementarity.** Within the complex, mC5a inserts into the V-shaped groove formed at the interface between the two structural domains of NOX-D20 and the Spiegelmer's binding pocket almost perfectly complements the surface shape of mC5a helices H1, H2 and H3 (Fig. 1c,d). H2 runs along the bottom of the binding cavity and forms the most extended network of interactions with the aptamer molecule at the junction between NOX-D20 domains. On the sides of the cavity, mC5a helix H1 interacts with the core region of the

G-quadruplex domain, whereas helix H3 is recognized by the helical stem domain. The presence of mC5a therefore locks NOX-D20 into a rigid conformation by tightening the packing between its two structural domains. Contacts between NOX-D20 and mC5a H2 are mainly mediated by polar residues (Fig. 4a). Lys701 and Asp705, located in the middle of H2, form hydrogen bonds with U5 and G32 phosphate groups, along the minor groove of the NOX-D20 stem domain, and with the base of dU28. At the C-terminus of H2, Arg708 makes electrostatic interactions with the phosphate groups of G29 and C31. The H2-H3 loop and helix H3 further interact with the NOX-D20 helical stem domain, through long-range interactions between the side chains of Asn710 and Glu713 and the O4 atom of dU30, as well as between Arg721 and the dU7 phosphate group (Fig. 4b). These interactions are, however, loosened in the other complex of the asymmetric unit because of the alternative conformation of dU30 (Supplementary Fig. 3c). The main-chain carbonyl of Arg721 also directly engages the 2′-OH of U5 ribose in a hydrogen bond. At the end of H3, Thr723 connects the 2′-OH ribose groups from U5 and G39, and makes water-mediated hydrogen bonds with the bases from G3, U5 and dC38. On the other side of H2, residues from H1 and the H1–H2 loop make direct contacts with the G-quadruplex domain, with His696 and Tyr704 making stabilizing stacking interactions with U21 and G26, respectively, whereas the main-chain carbonyl of Lys695 forms a hydrogen bond with the 2′-OH of U21 ribose (Fig. 4c). In addition, Glu688

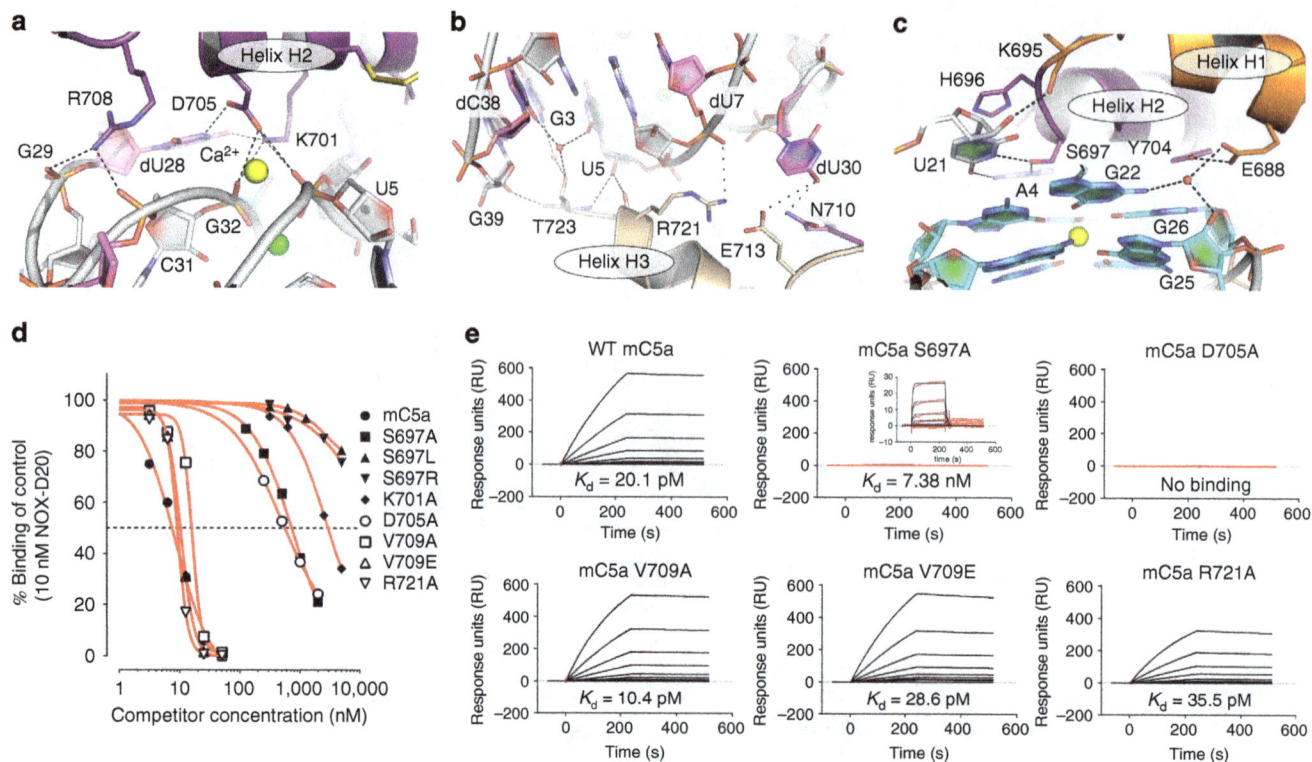

**Figure 4 | The NOX-D20:mC5a interface.** (**a**) Zoom-in on the detailed interactions of mC5a helix H2 with NOX-D20 at the bottom of the Spiegelmer's binding cavity. (**b**) Detailed interactions between mC5a helix H3 and the Spiegelmer stem domain. (**c**) Detailed interactions between mC5a helices H1–H2 and the Spiegelmer G-quadruplex. (**d**) NOX-D20 binding of mC5a WT and mutants analysed by competitive SPR measurement with immobilized mC5a on the sensor chip surface, a fixed NOX-D20 concentration and increasing competitor concentrations. (**e**) Direct SPR measurement of the binding affinities of NOX-D20 for relevant mC5a mutants compared with WT mC5a using immobilized mC5a and increasing NOX-D20 concentrations. The weak interaction with mC5a S697A is shown in the magnified inset.

engages in water-mediated hydrogen bonds with G22 and G25 and holds in place Tyr704. Finally, Ser697, at the tip of the H1-H2 loop, inserts its side chain between A4 and U21 and bridges the two bases through hydrogen bonding, thereby creating a closed surface parallel to the second G-tetrad.

To confirm the NOX-D20:mC5a binding mode observed in the structure (Fig. 5a for summary), mutants of mC5a were generated and tested for their ability to compete the binding of NOX-D20 to immobilized mC5a using SPR measurements (Supplementary Fig. 5a). Mutants Ser697Leu, Ser697Arg and Lys701Ala showed the weakest competition with wild-type (WT) mC5a for binding to NOX-D20 (Fig. 4d), suggesting that Ser697 and Lys701 are key residues for the Spiegelmer:anaphylatoxin interaction. In addition, mutants Ser697Ala and Asp705Ala showed substantially reduced competition, whereas the competition of mutants Val709Ala, Val709Glu and Arg721Ala was comparable to that of WT mC5a. Detailed measurement of the binding affinity of NOX-D20 for these mutants revealed that the Ser697Ala mutant could still bind but with a 400- to 1,000-fold lower affinity, whereas no binding could be detected for the Asp705Ala mutant (Fig. 4e). In contrast, replacement of Arg721 by an alanine had only a minor effect on NOX-D20 binding despite its apparent contacts with the dU7 phosphate. Such replacement would, however, not disrupt the main-chain carbonyl Arg721:U5 interaction. All the mC5a mutants could efficiently induce chemotaxis of hC5aR1-expressing BA/F3 cells, suggesting that their defect in NOX-D20 binding was not due to improper folding (Supplementary Fig. 5b). The Asp705Ala mutant had, however, slightly lower chemotactic activity. In any case, mC5a mutational analysis confirmed the binding mode observed in the structure.

**A rationale for species selectivity and C5a inhibition.** A similar binding mode to NOX-D20 is expected for human and murine anaphylatoxins, with a slightly higher affinity of NOX-D20 for mC5a[22]. Indeed, as shown in a previous study, characterization of NOX-D20-binding properties using commercial recombinant proteins displayed an almost tenfold higher affinity for mC5a compared with hC5a[22]. Here, we used a recombinant hC5a protein bearing an engineered Cys704Arg mutation to avoid nonspecific disulfide crosslinking via the free cysteine[23]. Surprisingly, this mutation resulted in a tenfold increase in the affinity of NOX-D20 for hC5a, as determined by SPR, yielding a $K_d$ value comparable to that for mC5a (Supplementary Fig. 6a). Furthermore, the potency of NOX-D20 to inhibit the hC5a-Cys704Arg-induced chemotaxis of BA/F3 cells increased (Supplementary Fig. 6b). In agreement with our structure, these functional data therefore suggest that Arg708 in mC5a (hC5a Arg704) does not only contribute to the binding but is responsible for the increased affinity of NOX-D20 towards the murine anaphylatoxin as compared with WT hC5a.

To further confirm that mC5a and hC5a have a similar binding mode to NOX-D20, mutational analysis of our hC5a Cys704Arg recombinant protein was undertaken. The hC5a mutants Ser693Ala and Asp701Ala showed a 150- to 300-fold reduction in their affinity for the Spiegelmer and mutants Ser693Leu and Lys697Ala showed complete loss of binding to NOX-D20 (Supplementary Fig. 6c). Again, all hC5a mutants could efficiently trigger chemotaxis in BA/F3 cells, suggesting that their loss of activity towards NOX-D20 is solely mediated by a binding defect (Supplementary Fig. 6d). Interestingly, the Asp701Ala mutant showed full chemotactic activity, whereas the activity of its murine counterpart Asp705Ala was impaired. This residue

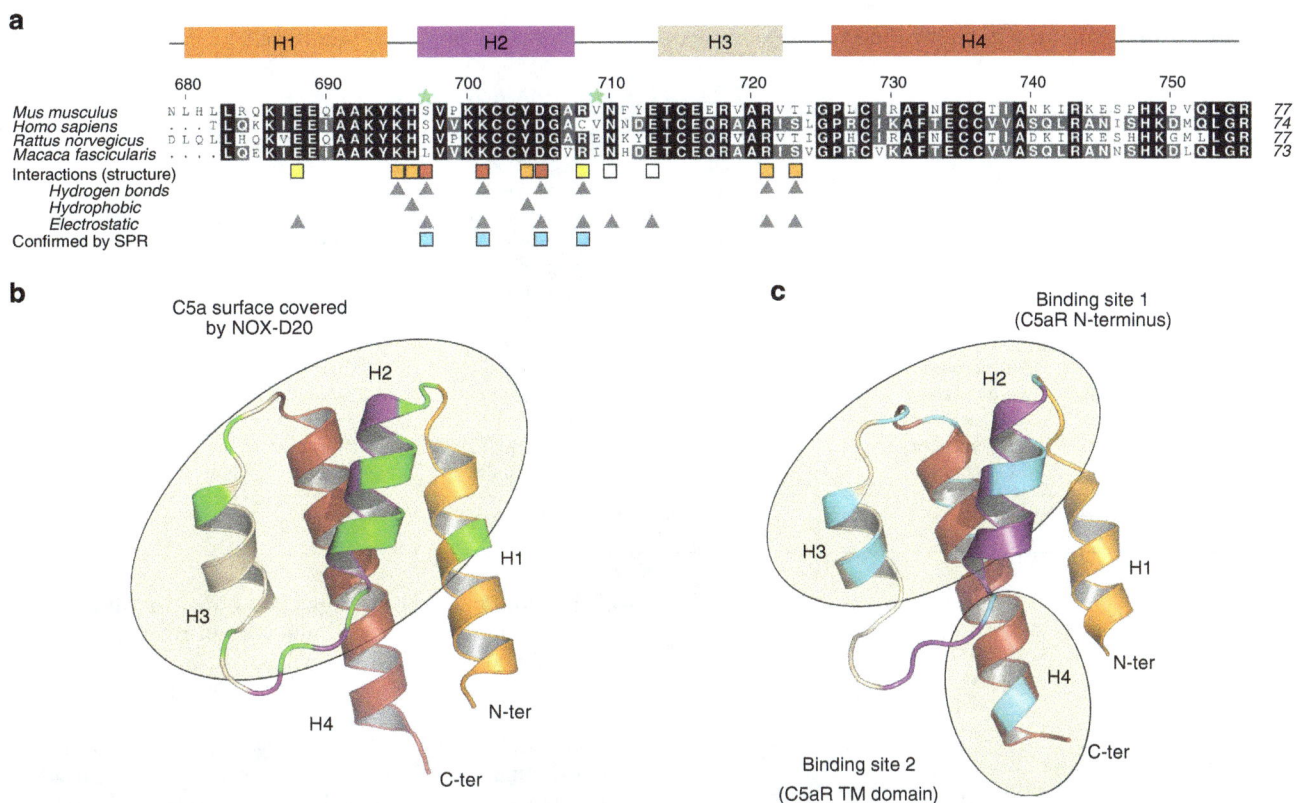

**Figure 5 | C5a alignment showing conserved residues involved in NOX-D20 binding and comparison of NOX-D20 and C5aR footprints onto C5a. (a)** Sequence alignment for murine, human, rat and macaque C5a proteins. The secondary structure elements in mC5a are indicated above the alignment. Residues that are potentially involved in NOX-D20 selectivity towards human and murine C5a are indicated by green stars. C5a residues involved in NOX-D20 binding are marked with coloured squares below the alignment, with a gradient scale from red (strong) to white (weak) to indicate the strength of the interaction. The types of interactions involved for each residue are summarized below with grey triangles. Residues for which the interaction was confirmed by SPR are indicated with blue squares. **(b)** Footprint of NOX-D20 on mC5a based on the NOX-D20:mC5a structure. The mC5a residues directly involved in NOX-D20 binding are highlighted in green and the surface of mC5a covered by the Spiegelmer is indicated by an ellipse. **(c)** Putative footprint of C5aR onto hC5a (four-helix bundle model derived from the hC5a moiety of intact C5 (ref. 60)). The hC5a residues possibly involved in C5aR recognition are forming two distinct binding sites: positively charged residues from the C5a core presumably interact with the C5aR N-terminus, whereas the C5a C-terminus would insert in the receptor transmembrane domain[32-37].

therefore potentially contributes to the differential C5aR activation properties observed for human and murine C5a anaphylatoxins[23]. Nevertheless, our SPR data show that interaction with NOX-D20 involves the same residues on hC5a and mC5a.

NOX-D20 has a strong affinity for hC5a and mC5a but no reactivity towards rat or macaque C5a[22]. Sequence comparison between these different species highlighted two residues potentially involved in the Spiegelmer selectivity, Ser697 and Val709 (mC5a numbering), which are conserved in human and murine C5a but differ in other species[22] (Fig. 5a). In accordance with the determined binding mode of NOX-D20 to mC5a/hC5a, our mutational analysis revealed that Ser697 (hC5a Ser693) indeed plays a crucial role in NOX-D20 binding, which is in agreement with its position on top of the G-quadruplex domain where it holds the stacked layers together (Fig. 4c). Introduction of a hydrophobic residue (such as a leucine in monkey C5a) or a large polar residue (such as an arginine in rat C5a) would break the stabilizing interaction between nucleotides A4 and U21, on the upper side of the G-quadruplex, and pull away the entire domain, thus strongly destabilizing the NOX-D20:C5a interface. In contrast to Ser697, none of the Val709 mutants (hC5a Val705) showed reduced affinity for NOX-D20, suggesting that this residue is not important for the Spiegelmer recognition (Fig. 4d,e, Supplementary Fig. 6). Indeed, Val709 is located in the H2–H3 loop and points towards the solvent on the opposite side of mC5a

as compared with the NOX-D20:mC5a interface. Thus, Ser697 is solely responsible for NOX-D20 selectivity towards human and murine C5a.

Our structural and biochemical data also suggest that hC5a and hC5a-desArg most likely bind NOX-D20 as four-helix bundles, although hC5a-desArg and the shorter C5aR-antagonist hC5a-A8 have been shown to crystallize as three-helix bundles[23,31]. In agreement with this idea, superimposition of the different three-helix bundle conformations observed for hC5a-desArg[31] onto the NOX-D20:mC5a complex reveals that helix H1 from these conformations would extensively clash with the G-quadruplex domain (Supplementary Fig. 6e). As a control of the specific binding of NOX-D20 to the ligands hC5a and hC5 in solution, a competitive SPR binding assay showed no binding to hC3a (Supplementary Fig. 6f). Finally, as the C-terminus of C5a extends away from the Spiegelmer molecule and does not participate in the interaction, glycosylations borne by native hC5a on Asn741 (equivalent to Glu745 in mC5a) most probably do not interfere with hC5a binding to NOX-D20.

The capacity of NOX-D20 to inhibit chemotaxis of C5aR1-expressing cells and to antagonize C5a-induced activation of primary human polymorphonuclear leukocytes suggests that the Spiegelmer directly competes with C5aR binding to the anaphylatoxin[22]. Extensive mutational studies led to propose a two-sites binding mode for the C5a:C5aR interaction[32], according to which positively charged residues located in helices H1–H3

**Table 2 | Effect of positional L-ribonucleotide to L-deoxyribonucleotide exchanges on C5a recognition.**

| NOX-D19 | $1.37 \pm 0.22$ nM (44 nt all-RNA NOX-D20 predecessor) | | | |
|---|---|---|---|---|
| Modified nucleotide | X-fold effect on C5a affinity | Increase/decrease by | Structural basis for the observed effect upon –OH to –H shift | |
| A4 | **7.2x Decrease** | Both | Target binding | Destabilization of $H_2O$ network around the target |
| U5 | **6.1x Decrease** | Off-rate | Target binding | Hydrogen bond to Thr723 main chain broken |
| U7 | **2.0x Increase** | On-rate | Spiegelmer fold | Shift to 4′-exo ribose conformation |
| G11 | **5.1x Decrease** | Both | Spiegelmer fold | –OH interaction with N7 of dG14 disrupted |
| G14 | **1.4x Increase** | On-rate | Spiegelmer fold | Shift to 4′-exo ribose conformation |
| A15 | **1.6x Increase** | On-rate | Spiegelmer fold | Stabilization of the dG14 *syn* conformation |
| G22 | **12.5x Decrease** | On-rate | Spiegelmer fold | Destabilization of G-quadruplex loop region |
| G26 | **1.7x Decrease** | Off-rate | Target binding | Destabilization of $H_2O$ network around the target |
| U28 | **1.9x Increase** | Both | Both | Stabilization of loop region around R708-N710 |
| U30 | **1.6x Increase** | Both | Both | Stabilization of loop region around R708-N710 |
| C31 | **12.0x Decrease** | Off-rate | Spiegelmer fold | Destabilization of $H_2O$ network around $Mg^{2+}$ |
| G32 | **10.2x Decrease** | Off-rate | Spiegelmer fold | Destabilization of the Domains 1-2 interface |
| C38 | **1.6x Increase** | Off-rate | Target binding | Stabilization of $H_2O$ network around the target |

Influence of the removal of the ribose 2′-OH group on the Spiegelmer affinity for C5a as evaluated by SPR[22] and detailed analysis of the influence of backbone modifications on the structural properties of NOX-D20.

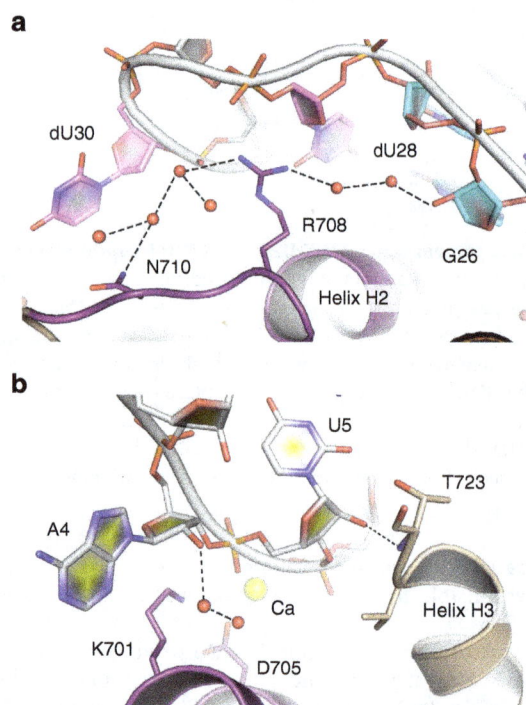

**Figure 6 | A rationale for NOX-D20 affinity improvement through backbone modifications.** (**a**) Introduction of DNA nucleotides at positions 28 and 30 stabilizes the stacking of mC5a Arg708 against the dU28 ribose ring, whereas removal of the 2′-OH group in G26 would destabilize the water network around this interface. (**b**) Modification of A4 and U5 nucleotides directly interferes with mC5a recognition by breaking the U5:Thr723 interaction and by disturbing the water network around Lys701 and Asp705.

interact with acidic residues and sulfo-tyrosines contained in the C5aR N-terminus, whereas the C5a C-terminus inserts into the C5aR transmembrane domain[33–37]. Comparison of the footprint of NOX-D20 onto mC5a with the position of the C5a residues presumably involved in C5aR binding reveals that NOX-D20 strongly overlaps with the first C5aR-binding site (Fig. 5b,c). Although the C5a C-terminus is not masked by the Spiegelmer, blocking C5a access to its primary docking site on C5aR N-terminus appears sufficient to prevent the C5a:C5aR interaction, therefore providing a structural basis for NOX-D20 inhibitory properties.

**Affinity modulation through backbone modifications.** The more affine NOX-D20 Spiegelmer was generated from NOX-D19, which only contains L-RNA nucleotides, by introducing L-DNA nucleotides at specific positions[22]. This affinity increase can arise from a fine tuning of the target recognition mode and/or a stronger stabilization of the Spiegelmer active conformation. Detailed analysis of the influence of sugar modifications on NOX-D20 structure reveals that both effects play a role for the modulation of NOX-D20 affinity towards C5a (Table 2). Structure-stabilizing modifications are restricted to the helical stem domain, in regions of strong constriction where the presence of a 2′-OH group on the ribose would loosen the backbone conformation. For example, the absence of 2′-OH groups on the riboses of dU28 and dU30 allows tight packing with the anti-sense strand in the minor groove of the double-helix, thereby shaping the binding interface around the mC5a H2-H3 loop and thus improving target interaction (Fig. 6a). In addition, ribose to deoxyribose modifications leading to decreased target affinity can easily be explained by the NOX-D20:mC5a structure. Such changes can either disrupt direct or water-mediated NOX-D20:mC5a contacts, as observed for A4, U5 or G26 modifications (Fig. 6a,b), or destabilize the base pair stacking and/or the structural domains of the Spiegelmer (G11, G22, C31 and G32 modifications). Thus, the NOX-D20:mC5a structure allows to rationalize the increased affinity of NOX-D20 for mC5a as compared with its all L-RNA predecessor NOX-D19.

## Discussion

The NOX-D20:mC5a and NOX-D20:mC5a-desArg structures reveal for the first time the active conformation adopted by an L-nucleic acid aptamer, in particular of a mixed L-RNA/L-DNA aptamer, in complex with its physiological target. To date, only structures of a short, non-functional, double-stranded L-RNA and of a L/D-RNA racemate were available[38,39]. Interestingly, NOX-D20 adopts a complex 3D architecture built on the one hand by a helical stem expanding over 30 Å, and on the other hand, by an intramolecular G-quadruplex stabilized by a central $Ca^{2+}$ ion. This allows the L-aptamer to wrap around its target with tight shape complementarity, thereby providing an affinity for C5a in the picomolar range. The binding mode observed in the structures and confirmed by mutational analysis shows that the Spiegelmer recognizes a large epitope also required for C5aR1 binding and thus provides a rationale for NOX-D20 inhibitory properties. Furthermore, NOX-D20 specificity for human and murine C5a over macaque and rat C5a can be explained by the sole presence of a serine residue interacting directly with the G-quadruplex domain, and a single Cys to Arg mutation is responsible for the increased affinity of NOX-D20 for mC5a as compared with its human counterpart. Finally, our data provide a structural basis for the improved affinity of NOX-D20 that was generated from NOX-D19 by the introduction of six ribo-to-deoxyribonucleotide exchanges[22]. Some of these modifications enhanced target recognition directly, whereas the majority led to intramolecular stabilization, also allowing for a four nucleotides truncation in the terminal helix.

G-quadruplexes are often encountered in nucleic acid aptamers as they provide a much more versatile scaffolding element than simple DNA/RNA duplexes for target recognition, because of the huge diversity of structural motifs that can be obtained by varying the length and connectivity of the loops that bridge the G-tetrads[40]. Well-known structures containing G-quadruplexes are thrombin-binding DNA aptamers[41]. G-quadruplexes are generally stabilized by monovalent cations, mostly $Na^+$ and $K^+$, which stack either in between two G-tetrad layers ($K^+$) or in the plane of the G-tetrads ($Na^+$)[27] (Supplementary Fig. 7). Divalent cations seem to have a more complex role with respect to G-quadruplexes, as they have been reported to destabilize G-quadruplex formation or induce topological transitions in the connectivity between the various G-strands, leading for example to a switch from antiparallel to parallel topology[27,42,43]. Despite these observations, still only limited structural information is available to study the influence of divalent cations on G-quadruplex structures and to our knowledge, only one structure of a calcium-containing G-quadruplex has been reported so far, the structure of the bimolecular telomeric G-quadruplex d($TG_4T$) from *Oxytricha nova* in a mixed $Ca^{2+}/Na^+$ environment[43]. The NOX-D20 structure therefore provides the first example of a $Ca^{2+}$-stabilized unimolecular G-quadruplex within an aptamer molecule. The crystal structure also clearly shows that the two G-tetrad layers are twisted by 45°. This different spatial arrangement of the planes may therefore be more compatible with a central $Ca^{2+}$ ion than with a monovalent ion.

Our data suggest that the high affinity of NOX-D20 for C5a results from a bimodular binding mode to which both the G-quadruplex and the stem region contribute. In particular, the helical stem is kept in place by providing one of the guanine nucleotides of the G-quartet, thereby forming a double-pseudoknot-like structure that stabilizes the overall V-shape of the Spiegelmer. Similarly but with less inter-domain interactions, second-generation thrombin-binding aptamers incorporate an additional stem domain to allow for increased affinity and specificity towards their target[44]. Another example of a bimodular aptamer is the recently described IL-6 SOMAmer, which contains DNA nucleotides with modified bases mimicking aromatic side chains[45]. Other RNA aptamers forming duplex-quadruplex junctions and for which the presence of the G-quadruplex is essential for target recognition have also been reported. These include the *in vitro* selected, guanine-rich *sc1* RNA aptamer, which recognizes the RGG peptide of human fragile X mental retardation protein[46]. The RGG peptide binds the aptamer in a pocket at the interface between the stem domain and the G-quadruplex, through tight shape complementarity[46], similarly to what we observe in our NOX-D20:C5a complex. Another good example is the Spinach aptamer, which binds to and thereby activates the fluorescence of 3,5-difluoro-4-hydroxybenzylidene imidazolinone, a mimic of the intrinsic fluorophore of green fluorescent protein[47,48]. In this case, the 3,5-difluoro-4-hydroxybenzylidene imidazolinone ligand stacks in a plan parallel to the first G-tetrad layer, right at the duplex-quadruplex junction[47,48]. The present structures reveal that the

Spiegelmer NOX-D20 adopts an as complex architectural fold as standard aptamers, thereby achieving a highly efficient target recognition thanks to a strong interconnectivity between two geometrically distinct structural modules.

In summary, we describe the first target-engaged mixed L-RNA/L-DNA Spiegelmer structure which features, besides an expected left-turning helix, an unusual G-quadruplex with a central $Ca^{2+}$ ion, and gives a rationale for affinity improvement by positional ribo-to-deoxyribonucleotide exchanges. From the structure, it becomes clear that binding to a discontinuous epitope on C5a leads to the inhibition of the anaphylatoxin-induced receptor signalling thus supporting the published results from cell-based assays and preclinical research. Further work is warranted to show the exact interaction of the Spiegelmer with C5.

## Methods

**L-Aptamers.** The L-aptamers (Spiegelmers) used in this study were manufactured at NOXXON Pharma AG by solid phase synthesis on controlled pore glass support using *tert*-butyl-dimethylsilyl-protected phosphoramidites of L-nucleosides. In previous publications, NOX-D19 and NOX-D20 referred to the oligonucleotides with a 40-kDa Y-shaped methoxy-PEG attached to their 5′-ends via an aminohexyl linker[22,49]. In this study, unPEGylated variants were used since PEG may sterically inhibit the formation of crystals. Consequently, affinity measurements and cell-based assays presented here were also done with unPEGylated Spiegelmers. The sequence for NOX-D20 oligonucleotide is L-RNA/L-DNA (40 nt): 5′-GCGAUG (dU)GGUGGU(dG)(dA)AGGGUUGUUGGG(dU)G(dU)CGACGCA(dC)GC-3′. The sequence of the NOX-D19 oligonucleotide is all L-RNA (44 nt): 5′-GCC UGAUGUGGUGGUGAAGGGUUGUUGGGGUGUCGACGCACAGGC-3′.

**Protein expression and purification.** Recombinant hC5a (mutant Cys704Arg), mC5a and mC5a-desArg were expressed recombinantly in bacteria[23]. All proteins were expressed as fusions with an N-terminal thioredoxin tag (Trx-A) followed by a hexahistidine tag and a TEV protease cleavage site (Trx-His₆-TEV-C5a), in Shuffle T7 Express *E. coli* cells (New England Biolabs). The proteins were purified using a two-step Ni-column affinity chromatography, including removal of the affinity-tag by overnight incubation with TEV protease, followed by a cation exchange chromatography on a Source 15S column (GE Healthcare Life Sciences). The final protein buffer was adjusted to 20 mM HEPES, pH 7.5, 150 mM NaCl before flash freezing in liquid nitrogen and storage at − 80 °C until use. Mutants of hC5a and mC5a were generated using the Quick-Change Lightning Site Directed Mutagenesis Kit from Agilent Technologies and all mutants were expressed and purified using the same protocol as for native proteins. Human C3a was expressed recombinantly following the same protocol as for the C5a proteins[50].

**Crystallization and data collection.** Before crystallization, the mC5a and mC5a-desArg samples were concentrated to 20–25 mg ml⁻¹ and NOX-D20 was dissolved in water at a final concentration of 2 mM. The mC5a proteins and the L-aptamer were mixed in a 1:1 molar ratio at a final concentration of 1 mM for each component in a final buffer adjusted to 10 mM HEPES, pH 7.5, 137 mM NaCl, 3 mM KCl, 3 mM MgCl₂, 3 mM CaCl₂. The mix was allowed to incubate for 2 h at room temperature before setting up crystallization plates. Initial crystallization experiments were carried out in 96-well sitting-drop plates using a MOSQUITO robot (TTP LabTech) and commercial screens from Hampton Research and Molecular Dimensions Ltd. Crystals for both complexes appeared after a few days at 19 °C over a reservoir containing 0.2 M NaCl, 0.1 M Na cacodylate, pH 6.0, 8% (w/v) PEG 8000, for the NOX-D20:mC5a complex, and a reservoir with 0.2 M ammonium acetate, 0.1 M Na acetate pH 4.0, 15% (w/v) PEG 4000, for the NOX-D20:mC5a-desArg complex. In both cases, crystals grew to maximal size within a few weeks. For data collection, the crystals were cryoprotected by soaking into the reservoir solution supplemented with 25–30% glycerol followed by flash freezing in liquid nitrogen. To determine the ion composition in the complex structure, mC5a and NOX-D20 were also mixed as described above in a buffer where KCl had been replaced by 3 mM RbCl and in a buffer where MgCl₂ had been replaced by 3 mM MnCl₂. Crystals similar to those observed in the standard buffer grew in the presence of RbCl, whereas no crystals could be obtained when MnCl₂ was present. For phase determination, 0.3–0.5 µl of osmium hexamine (dissolved in water at 20 mM) were added in the drop containing the crystals of the NOX-D20:mC5a complex (final concentration in the drop between 1.5 and 2.5 mM). Crystals were allowed to incubate with osmium hexamine for 6 h over the reservoir solution before cryoprotection and flash freezing in liquid nitrogen as described above. Native data sets for both complexes, as well as the data sets for the NOX-D20:mC5a crystals derivatized with osmium hexamine or grown with RbCl were collected at 100 K on beamline 911-3 at MAX-lab (Lund, Sweden). The data sets at higher wavelength to identify the ion composition of the asymmetric unit were collected on beamline P13 at PETRA III. All data sets were processed with XDS (X-ray Detection Software)[51].

**Structure determination.** The crystals of the NOX-D20:mC5a complex displayed a $P2_12_12$ symmetry and diffracted to a maximal resolution of 1.8 Å (Table 1). Initial attempts to solve the structure were made by performing MR in PHASER[52] using the structure of mC5a in a four-helix bundle conformation (PDB 4P3A[23]). As hC5a is also able to form a three-helix bundle in its shorter versions hC5a-desArg[31] and hC5a-A8 (ref. 23), models of mC5a in the different three-helix bundle conformations observed for hC5a-derived proteins were also constructed and used as MR search models (Supplementary Fig. 8). None of these models gave a solution, most probably due to the fact that one molecule of mC5a only represents a small fraction of the asymmetric unit content. Three-dimensional models for the L-aptamer were also constructed in a two-step procedure based on secondary structure predictions (Supplementary Fig. 9). First, a standard D-RNA/D-DNA aptamer was constructed using RNAComposer[53]. In a second step, the mirror-image of this model was obtained by inverting the sign of the Z-coordinate for all atoms of the model. All these models failed to give a MR solution. Despite the presence of 78 phosphate groups arising from the two Spiegelmer molecules expected in the asymmetric unit, experimental phasing using P-SAD from data sets collected at higher wavelength (Table 1) was not successful in providing initial phases either. The NOX-D20:mC5a crystals were therefore derivatized by soaking with various compounds containing heavy atoms, including osmium hexamine, $Os(NH_3)_6$, which mimics almost perfectly a hexahydrated $Mg^{2+}$ and has therefore been extensively used to solve large RNA 3D structures[54]. Anomalous data sets extending to 1.8 Å resolution were collected for the $Os(NH_3)_6$-derivatized crystals and the structure was solved by SAD phasing in PHENIX.AUTOSOLVE[55] (Table 1, Supplementary Figs 1 and 2). Sixteen refined $Os(NH_3)_6$ sites were identified. The overall figures of merit after SAD phasing and after density modification using RESOLVE were 0.48 and 0.78, respectively (35–2.5 Å resolution cutoff).

The quality of the initial electron density maps allowed to unambiguously identify the elements composing the asymmetric unit, that is, three molecules of mC5a and two molecules of NOX-D20 (Supplementary Fig. 3). In particular, it was readily visible that mC5a adopts a four-helix bundle conformation and the three mC5a molecules were placed manually in the electron density map using the structural model available for mC5a[23]. The density patterns left accounted for two molecules of Spiegelmer arranging in an upside-down manner without apparent discontinuity in the base-pair stacking between their helical stems. Careful inspection of the density allowed for positioning of the 5' and 3' ends of both Spiegelmer molecules. As the Spiegelmer is built from non-conventional L-nucleotides, geometry files for each of the eight L-RNA/L-DNA nucleotides were parameterized manually to allow rebuilding and refinement of the Spiegelmer molecules using standard programmes. The 40 nucleotides composing each NOX-D20 molecule were then placed manually one by one in the density and refinement of the model was carried out using conventional procedures, after having defined the geometrical parameters for the phosphate bonds linking each pair of nucleotides in PHENIX.REFINE[55]. After an initial round of refinement, the obtained model was used to perform MR in the native data set, which displayed slightly better statistics (Table 1). The model was then further improved by manual rebuilding in COOT[56] and energy minimization in PHENIX.REFINE[55] using individual isotropic Atomic Displacement Parameters (ADP) and Translation–Libration–Screw (TLS) parameterization. Non-Crystallographic Symmetry restraints were not imposed as they were observed to increase the R-factor values during refinement. The final model yielded $R_{work}$ and $R_{free}$ values of 16.70% and 19.26%, respectively, and was used as a MR search model to solve the structure of the NOX-D20:mC5a-desArg complex. Refinement of the NOX-D20:mC5a-desArg model was carried out by alternating between cycles of manual rebuilding in COOT[56] and cycles of energy minimization with PHENIX.REFINE[55] using individual isotropic ADP, TLS refinement, and the same geometry restraints as for the NOX-D20:mC5a complex, again without imposing any Non-Crystallographic Symmetry restraints. The final model yielded $R_{work}$ and $R_{free}$ values of 16.26% and 19.94%, respectively. The quality of both models was assessed with MOLPROBITY[57]. The Ramachandran statistics for the NOX-D20:mC5a and NOX-D20:mC5a-desArg complexes are, respectively: outliers, 0.5%/0%; allowed, 2.3%/2.5%; favoured 97.2%/97.5%. Simulated annealing refinement was performed in PHENIX.REFINE[55] after removal of the $Ca^{2+}$ ions in the final model using a start temperature of 1,500 K and cooling down to 300 K. Refinement with tight metal-O geometry restraints was performed by adding back either $Ca^{2+}$ or $K^+$ ions in the model obtained after simulated annealing refinement and by doing energy minimization in PHENIX.REFINE[55] using individual isotropic ADP, TLS refinement and imposing the Ca–O or K–O distances to 2.43 Å ± 0.05 $\sigma$ or 2.81 Å ± 0.05 $\sigma$, respectively, during the refinement step. All figures were produced with PyMOL v.1.7 (http://www.pymol.org). C5a sequence alignment was performed in MULTIALIN[58] and the sequence conservation analysis was done in ALINE[59].

**Biacore direct binding assay format.** The Biacore 2000 instrument was set to a constant temperature of 37 °C. The system was cleaned using the DESORB method before the start of each experiment/immobilization of a new chip. After docking a maintenance chip, the instrument was consecutively primed with desorb solution 1 (0.5% SDS), desorb solution 2 (50 mM glycine, pH 9.5) and HBS-EP pH 7.4 buffer. Finally, the system was primed with HBS-EP pH 7.4 buffer (GE Healthcare).

Human and murine C5a and their mutants were immobilized by amine coupling procedure on the CM5 sensor chips (flow cells 2–4, whereas flow cell 1 served as dextran surface control). A volume of 100 μl of a 1:1 mixture of 0.4 M EDC (1-ethyl-3-(3-dimethylaminopropyl) carbodiimide in $H_2O$; GE, BR-1000-50) and 0.1 M NHS (N-hydroxysuccinimide in $H_2O$; GE, BR-1000-50) were injected using the QUICKINJECT command at a flow of 10 μl min$^{-1}$. The C5a proteins were diluted to a concentration of 1 μg ml$^{-1}$ in 10 mM sodium acetate, pH 5.5, with 1 μM NOX-D20 and subsequently C5a was immobilized covalently up to approximately 500 response units (RUs) at a flow of 10 μl min$^{-1}$. The flow cells were blocked by an injection 70 μl of 1 M ethanolamine hydrochloride (GE, BR-1000-50) at a flow of 10 μl min$^{-1}$.

On the day of the sample measurement, the sensor chip was primed twice with degassed physiological running buffer (20 mM Tris pH 7.4; 150 mM NaCl; 5 mM KCl, 1 mM $MgCl_2$ and 1 mM $CaCl_2$) and equilibrated at a flow of 50 μl min$^{-1}$ until the baseline appeared stable. Before sample measurement, the chip underwent at least three injection and regeneration cycles whereby regeneration was performed by injecting 30 μl of 5 M NaCl at a flow of 30 μl min$^{-1}$. Stabilization time of baseline after each regeneration cycle was set to 1 min at 30 μl min$^{-1}$. A concentration series (500; 250; 125; 62.5; 31.3; 15.6; 7.8; 3.9; 1.95; 0.98; 0.49; 0.24; 0.12; 0.06; 0 nM) of NOX-D19 or NOX-D20 was prepared in physiological running buffer and injected starting with the lowest concentration. Regeneration was performed after each measurement. In all experiments, the analysis was performed at 37 °C using the KINJECT command defining an association time of 240 s and a dissociation time of 240 s at a flow of 30 μl min$^{-1}$. The assay was double referenced, whereas flow cell 1 served as (blocked) surface control (bulk contribution) and a series of buffer injections without analyte determined the bulk contribution of the buffer itself on the loaded flow cells. At least one Spiegelmer concentration was injected a second time at the end of the experiment to monitor the regeneration efficiency and chip integrity during the experiments. Data analysis and calculation of dissociation constants ($K_d$) by fitting the data to a 1:1 Langmuir algorithm were done with the BIAevaluation 3.1.1 software (BIACORE AB) with a constant refractive index and an initial mass transport coefficient $k_t$ of $1 \times 10^7$ (RU M$^{-1}$ s$^{-1}$).

**Biacore competitive binding assay format.** For competitive analysis, mC5a or hC5a(Cys704Arg), respectively, was immobilized on a CM5 sensor chip as described above. Co-injection of NOX-D20 with single C5a proteins will result in a competition between the chemokine in solution and the immobilized C5a on the sensor chip for NOX-D20 binding, thus leading to a reduction in signal. The reduction of the signal can be evaluated at a predefined report point, that is, at the end of the dissociation time (see Supplementary Fig. 5a). For detailed analysis, a concentration series of the respective C5a, C5 or C3a protein was injected. Potential unspecific binding of the individual proteins to the dextran matrix or the immobilized C5a was monitored by a control flow cell without the C5a, as well as by an injection of the respective proteins without NOX-D20.

**Influence of cations on NOX-D20 binding.** Murine C5a was immobilized as described above. First 100 nM of NOX-D20 was co-injected with a concentration series (16; 8; 4; 2; 1; 0.5; 0.25; 0.125; 0.0625; 0.0313; 0 (2 ×) mM) of EDTA in physiological running buffer (20 mM Tris, pH 7.4; 150 mM NaCl; 5 mM KCl, 1 mM $MgCl_2$ and 1 mM $CaCl_2$). The $Mg^{2+}$- and $Ca^{2+}$-dependent binding of NOX-D20 to immobilized mC5a was addressed by co-injection of 100 nM NOX-D20 with a concentration series (16; 8; 4; 2; 1; 0.5; 0.25; 0.125; 0.0625; 0.0313; 0 (2 ×) mM) of $MgCl_2$ or $CaCl_2$, respectively, in physiological running buffer without the corresponding divalent ion. To confirm the influence of $Ca^{2+}$ ions also on the stability of the complex, $CaCl_2$ was only titrated in the running/dissociation buffer, while kept at a physiological concentration (1 mM) in the injection/association buffer. To determine the presence of potassium ions, KCl was also titrated down in affinity measurements starting at the physiological concentration of 5 mM (5; 2.5; 1.25; 0.625; 0.313; 0 mM). This was followed by the titration of LiCl (40; 20; 10; 5; 2.5; 1.25; 0.625; 0.313; 0.156; 0 mM) in the absence of KCl to replace potentially present trace amounts of $K^+$ and break the G-quartet if potassium was required for proper folding. Data analysis and calculation of association rate constant ($k_a$) was performed as described above. All final data were analysed using Prism 5 software (GraphPad Software).

**Chemotaxis assay.** Chemotaxis assays were performed in mouse BA/F3 cells stably transfected using a plasmid coding for the human C5a receptor, hC5aR1 (ref. 22). The hC5aR1-expressing BA/F3 cells were stimulated with recombinant mutated mC5a and hC5a proteins at varying concentrations as indicated. Recombinant WT mC5a or recombinant hC5a(Cys704Arg) were used as controls, respectively. Commercially available recombinant WT mC5a and hC5a (R&D Systems) were also tested for comparison. Chemotaxis was followed in Transwell plates with 5 μm pores at 37 °C for 3 h in HBH buffer. For quantification of migrated cells, 50 μM resazurin in PBS was added and incubated at 37 °C for 2.5 h. Fluorescence was measured at 590 nm (excitation wavelength 544 nm). Data were analysed using Prism 5 software (GraphPad Software).

## References

1. Walport, M. J. Advances in immunology: Complement (First of two parts). *N. Engl. J. Med* **344**, 1058–1066 (2001).
2. Janeway, Jr C. A., Travers, P., Walport, M. & Shlomchik, M. J. in *Immunobiology: the Immune System in Health and Disease* 5th edn (Garland Science, 2001).
3. Ricklin, D., Hajishengallis, G., Yang, K. & Lambris, J. D. Complement: a key system for immune surveillance and homeostasis. *Nat. Immunol.* **11**, 785–797 (2010).
4. Carroll, M. C. The complement system in regulation of adaptive immunity. *Nat. Immunol.* **5**, 981–986 (2004).
5. Dunkelberger, J. R. & Song, W. C. Complement and its role in innate and adaptive immune responses. *Cell Res.* **20**, 34–50 (2010).
6. Ember, J. A. & Hugli, T. E. Complement factors and their receptors. *Immunopharmacology* **38**, 3–15 (1997).
7. Klos, A., Wende, E., Wareham, K. J. & Monk, P. N. International Union of Pharmacology. LXXXVII. Complement peptide C5a, C4a, and C3a receptors. *Pharmacol. Rev.* **65**, 500–543 (2013).
8. Zhou, W. The new face of anaphylatoxins in immune regulation. *Immunobiology* **217**, 225–234 (2012).
9. Haas, P. J. & van Strijp, J. Anaphylatoxins: their role in bacterial infection and inflammation. *Immunol. Res.* **37**, 161–175 (2007).
10. Laudes, I. J. *et al.* Expression and function of C5a receptor in mouse microvascular endothelial cells. *J. Immunol.* **169**, 5962–5970 (2002).
11. Klos, A. *et al.* The role of the anaphylatoxins in health and disease. *Mol. Immunol.* **46**, 2753–2766 (2009).
12. Guo, R. F. & Ward, P. A. Role of C5a in inflammatory responses. *Annu. Rev. Immunol.* **23**, 821–852 (2005).
13. Bosmann, M. & Ward, P. A. Role of C3, C5 and anaphylatoxin receptors in acute lung injury and in sepsis. *Adv. Exp. Med. Biol.* **946**, 147–159 (2012).
14. Rittirsch, D., Redl, H. & Huber-Lang, M. Role of complement in multiorgan failure. *Clin. Dev. Immunol.* **2012**, 962927 (2012).
15. Sjoberg, A. P., Trouw, L. A. & Blom, A. M. Complement activation and inhibition: a delicate balance. *Trends Immunol.* **30**, 83–90 (2009).
16. Emlen, W., Li, W. & Kirschfink, M. Therapeutic complement inhibition: new developments. *Semin. Thromb. Hemost.* **36**, 660–668 (2010).
17. Ricklin, D. & Lambris, J. D. Progress and trends in complement therapeutics. *Adv. Exp. Med. Biol.* **735**, 1–22 (2013).
18. Ward, P. A., Guo, R. F. & Riedemann, N. C. Manipulation of the complement system for benefit in sepsis. *Crit. Care Res. Pract.* **2012**, 427607 (2012).
19. Woodruff, T. M., Nandakumar, K. S. & Tedesco, F. Inhibiting the C5-C5a receptor axis. *Mol. Immunol.* **48**, 1631–1642 (2011).
20. Vater, A. & Klussmann, S. Turning mirror-image oligonucleotides into drugs: the evolution of Spiegelmer therapeutics. *Drug Discov. Today* **20**, 147–155 (2015).
21. Ellington, A. D. & Szostak, J. W. *In vitro* selection of rna molecules that bind specific ligands. *Nature* **346**, 818–822 (1990).
22. Hoehlig, K. *et al.* A novel C5a-neutralizing mirror-image (l-)aptamer prevents organ failure and improves survival in experimental sepsis. *Mol. Ther.* **21**, 2236–2246 (2013).
23. Schatz-Jakobsen, J. A. *et al.* Structural and functional characterization of human and murine C5a anaphylatoxins. *Acta Cryst. D* **70**, 1704–1717 (2014).
24. Winn, M. D. *et al.* Overview of the CCP4 suite and current developments. *Acta Cryst. D* **67**, 235–242 (2011).
25. Lech, C. J., Heddi, B. & Phan, A. T. Guanine base stacking in G-quadruplex nucleic acids. *Nucleic Acids Res.* **41**, 2034–2046 (2013).
26. Pyle, A. M. Metal ions in the structure and function of RNA. *J. Biol. Inorg. Chem.* **7**, 679–690 (2002).
27. Campbell, N. H. & Neidle, S. in *Interplay between Metal Ions and Nucleic Acids. Metal Ions in Life Sciences, 10* 119–134 (Springer, 2012).
28. Sen, D. & Gilbert, W. Guanine quartet structures. *Methods Enzymol.* **211**, 191–199 (1992).
29. Kankia, B. I. & Marky, L. A. Folding of the thrombin aptamer into a G-quadruplex with Sr(2 + ): stability, heat, and hydration. *J. Am. Chem. Soc.* **123**, 10799–10804 (2001).
30. Zheng, H., Chruszcz, M., Lasota, P., Lebioda, L. & Minor, W. Data mining of metal ion environments present in protein structures. *J. Inorg. Biochem.* **102**, 1765–1776 (2008).
31. Cook, W. J., Galakatos, N., Boyar, W. C., Walter, R. L. & Ealick, S. E. Structure of human desArg-C5a. *Acta Cryst. D* **66**, 190–197 (2010).
32. Siciliano, S. J. *et al.* Two-site binding of C5a by its receptor: an alternative binding paradigm for G protein-coupled receptors. *Proc. Natl Acad. Sci. USA* **91**, 1214–1218 (1994).
33. Mollison, K. W. *et al.* Identification of receptor-binding residues in the inflammatory complement protein C5a by site-directed mutagenesis. *Proc. Natl Acad. Sci. USA* **86**, 292–296 (1989).
34. Bubeck, P. *et al.* Site-specific mutagenesis of residues in the human C5a anaphylatoxin which are involved in possible interaction with the C5a receptor. *Eur. J. Biochem.* **219**, 897–904 (1994).
35. Hagemann, I. S., Narzinski, K. D., Floyd, D. H. & Baranski, T. J. Random mutagenesis of the complement factor 5a (C5a) receptor N terminus provides a structural constraint for C5a docking. *J. Biol. Chem.* **281**, 36783–36792 (2006).
36. Hagemann, I. S., Miller, D. L., Klco, J. M., Nikiforovich, G. V. & Baranski, T. J. Structure of the complement factor 5a receptor-ligand complex studied by disulfide trapping and molecular modeling. *J. Biol. Chem.* **283**, 7763–7775 (2008).
37. Toth, M. J. *et al.* The pharmacophore of the human C5a anaphylatoxin. *Protein Sci.* **3**, 1159–1168 (1994).
38. Vallazza, M. *et al.* First look at RNA in L-configuration. *Acta Cryst. D* **60**, 1–7 (2004).
39. Rypniewski, W. *et al.* The first crystal structure of an RNA racemate. *Acta Cryst. D* **62**, 659–664 (2006).
40. Tucker, W. O., Shum, K. T. & Tanner, J. A. G-quadruplex DNA Aptamers and their Ligands: Structure, Function and Application. *Curr. Pharm. Des* **18**, 2014–2026 (2012).
41. Padmanabhan, K., Padmanabhan, K. P., Ferrara, J. D., Sadler, J. E. & Tulinsky, A. The structure of alpha-thrombin inhibited by a 15-Mer single-stranded-DNA aptamer. *J. Biol. Chem.* **268**, 17651–17654 (1993).
42. Miyoshi, D., Nakao, A., Toda, T. & Sugimoto, N. Effect of divalent cations on antiparallel G-quartet structure of d(G(4)T(4)C(4)). *FEBS Lett.* **496**, 128–133 (2001).
43. Lee, M. P., Parkinson, G. N., Hazel, P. & Neidle, S. Observation of the coexistence of sodium and calcium ions in a DNA G-quadruplex ion channel. *J. Am. Chem. Soc.* **129**, 10106–10107 (2007).
44. Krauss, I. R., Pica, A., Merlino, A., Mazzarella, L. & Sica, F. Duplex-quadruplex motifs in a peculiar structural organization cooperatively contribute to thrombin binding of a DNA aptamer. *Acta Cryst. D* **69**, 2403–2411 (2013).
45. Gelinas, A. D. *et al.* Crystal structure of interleukin-6 in complex with a modified nucleic acid ligand. *J. Biol. Chem.* **289**, 8720–8734 (2014).
46. Phan, A. T. *et al.* Structure-function studies of FMRP RGG peptide recognition of an RNA duplex-quadruplex junction. *Nat. Struct. Mol. Biol.* **18**, 796–804 (2011).
47. Huang, H. *et al.* A G-quadruplex-containing RNA activates fluorescence in a GFP-like fluorophore. *Nat. Chem. Biol.* **10**, 686–691 (2014).
48. Warner, K. D. *et al.* Structural basis for activity of highly efficient RNA mimics of green fluorescent protein. *Nat. Struct. Mol. Biol.* **21**, 658–663 (2014).
49. Khan, M. A. *et al.* Targeting complement component 5a promotes vascular integrity and limits airway remodeling. *Proc. Natl Acad. Sci. USA* **110**, 6061–6066 (2013).
50. Bajic, G., Yatime, L., Klos, A. & Andersen, G. R. Human C3a and C3a desArg anaphylatoxins have conserved structures, in contrast to C5a and C5a desArg. *Protein Sci.* **22**, 204–212 (2013).
51. Kabsch, W. Automatic processing of rotation diffraction data from crystals of initially unknown symmetry and cell constants. *J. Appl. Crystallogr* **26**, 795–800 (1993).
52. McCoy, A. J., Grosse-Kunstleve, R. W., Storoni, L. C. & Read, R. J. Likelihood-enhanced fast translation functions. *Acta Cryst. D* **61**, 458–464 (2005).
53. Popenda, M. *et al.* Automated 3D structure composition for large RNAs. *Nucleic Acids Res.* **40**, e112 (2012).
54. Doudna, J. A. Chemical biology at the crossroads of molecular structure and mechanism. *Nat. Chem. Biol.* **1**, 300–303 (2005).
55. Adams, P. D. *et al.* PHENIX: building new software for automated crystallographic structure determination. *Acta Cryst. D* **58**, 1948–1954 (2002).
56. Emsley, P. & Cowtan, K. Coot: model-building tools for molecular graphics. *Acta Cryst. D* **60**, 2126–2132 (2004).
57. Davis, I. W. *et al.* MolProbity: all-atom contacts and structure validation for proteins and nucleic acids. *Nucleic Acids Res.* **35**, W375–W383 (2007).
58. Corpet, F. Multiple sequence alignment with hierarchical clustering. *Nucleic Acids Res.* **16**, 10881–10890 (1988).
59. Bond, C. S. & Schuttelkopf, A. W. ALINE: a WYSIWYG protein-sequence alignment editor for publication-quality alignments. *Acta Cryst. D* **65**, 510–512 (2009).
60. Fredslund, F. *et al.* Structure of and influence of a tick complement inhibitor on human complement component 5. *Nat. Immunol.* **9**, 753–760 (2008).

## Acknowledgements

We thank the beamline staffs at MAX-lab and PETRA III for support during data collection and Lotte Vogensen for technical assistance. We also thank Lucas Bethge, Tino Struck and Gabi Anlauf from NOXXON's chemistry group for the synthesis of the highly purified Spiegelmers, Dirk Zboralski, Lisa Bauer and Katrin Schindele from the *in vitro pharmacology* group for providing the cell-based assays with C5a mutants and John Achenbach for helpful discussions. We thank the Lundbeck Foundation for supporting this work through the grant: Lundbeck Foundation Nanomedicine Center for Individualized Management of Tissue Damage and Regeneration. This project was also supported by DANSCATT and by the Novo-Nordisk Foundation through a Hallas-Møller Fellowship to G.R.A.

## Authors contributions

L.Y. prepared all recombinant proteins, performed crystallization, data collection and structure determination/refinement. C.M. performed the SPR measurements. K.H. provided potential two-dimensional structures as starting points for the structural analysis. L.Y., C.M. and A.V. designed the experiments and analysed the data. L.Y., A.V., K.H., C.M., S.K. and G.R.A. wrote the paper.

## Additional information

**Accession codes:** Coordinates and structure factors for the NOX-D20:mC5a and NOX-D20:mC5a-desArg complexes have been deposited in the Protein Data Bank with accession codes 4WB2 and 4WB3, respectively.

**Competing financial interests:** C.M., K.H., S.K. and A.V. are employees of NOXXON Pharma AG, which has filed a patent application on the composition of matter and the use of NOX-D20. The remaining authors declare no competing financial interests.

# 20

# A nanobuffer reporter library for fine-scale imaging and perturbation of endocytic organelles

Chensu Wang[1,2], Yiguang Wang[1], Yang Li[1], Brian Bodemann[2], Tian Zhao[1], Xinpeng Ma[1], Gang Huang[1], Zeping Hu[3], Ralph J. DeBerardinis[3], Michael A. White[2] & Jinming Gao[1]

Endosomes, lysosomes and related catabolic organelles are a dynamic continuum of vacuolar structures that impact a number of cell physiological processes such as protein/lipid metabolism, nutrient sensing and cell survival. Here we develop a library of ultra-pH-sensitive fluorescent nanoparticles with chemical properties that allow fine-scale, multiplexed, spatio-temporal perturbation and quantification of catabolic organelle maturation at single organelle resolution to support quantitative investigation of these processes in living cells. Deployment in cells allows quantification of the proton accumulation rate in endosomes; illumination of previously unrecognized regulatory mechanisms coupling pH transitions to endosomal coat protein exchange; discovery of distinct pH thresholds required for mTORC1 activation by free amino acids versus proteins; broad-scale characterization of the consequence of endosomal pH transitions on cellular metabolomic profiles; and functionalization of a context-specific metabolic vulnerability in lung cancer cells. Together, these biological applications indicate the robustness and adaptability of this nanotechnology-enabled 'detection and perturbation' strategy.

[1] Department of Pharmacology, Simmons Comprehensive Cancer Center, University of Texas Southwestern Medical Center, 5323 Harry Hines Boulevard, Dallas, Texas 75390, USA. [2] Department of Cell Biology, University of Texas Southwestern Medical Center, 5323 Harry Hines Boulevard, Dallas, Texas 75390, USA. [3] Children's Medical Center Research Institute, University of Texas Southwestern Medical Center, 5323 Harry Hines Boulevard, Dallas, Texas 75390, USA. Correspondence and requests for materials should be addressed to M.A.W. (email: michael.white@utsouthwestern.edu) or to J.G. (email: jinming.gao@utsouthwestern.edu).

Endocytic organelles play an essential role in many cell physiological processes and are a primary site of cell–nanoparticle interactions. In cell biology, endosomes/lysosomes act as a nidus for signal transduction events that coordinate cell and tissue responses to nutrient availability and protein/lipid metabolism[1–3]. In drug and gene delivery, endosomes are the first intracellular organelles encountered after nanoparticle uptake by endocytosis[4–6]. Numerous nanocarriers are under development to achieve early endosomal release of therapeutic payloads and avoid lysosomal degradation[7,8]. A ubiquitous biological hallmark that affects all the above processes is the luminal pH of endocytic organelles[9]. For example, along the endocytic pathway, progressive acidification compartmentalizes ligand–receptor uncoupling (for example, low-density lipoprotein receptor) and activation of proteases for protein/lipid degradations into endosomes and lysosomes, respectively[1,2]. Most gene/siRNA delivery systems (for example, polyethyleneimines[10]) behave as a 'proton sponge' to increase osmotic pressure of endosomes for enhanced cytosolic delivery of encapsulated cargo. Although there have been remarkable advances in the effectiveness of these delivery systems, little is known about how perturbations of endosomal/lysosomal pH by these nanoparticles may affect cell homeostasis. Reagents currently used to manipulate and study the acidification of endocytic organelles include lysosomotropic agents (for example, chloroquine (CQ) and NH$_4$Cl), v-ATPase inhibitors (for example, bafilomycin A1) and ionophores (for example, nigericin and monensin)[11]. However, these reagents are broadly membrane permeable and likely simultaneously target multiple acidic organelles (for example, Golgi apparatus with a pH of $\sim 6.5$)[1], presenting significant challenges for discrete analysis of endosome and lysosome/autophagolysosome biogenesis.

In this study, we report a nanotechnology-enabled strategy for operator-controlled real-time imaging and perturbation of the maturation process of endocytic organelles; and application to investigation of the integration of endosomal maturation with cell signalling and metabolism. Previously, we developed a series of ultra-pH-sensitive (UPS) nanoparticles that fluoresce upon contact with a very narrow pH range ($<0.25$ pH units)[12,13]. These nanoparticles are 30–60 nm in diameter and enter cells exclusively through endocytosis. In this study, we report for the first time that these UPS nanoparticles can clamp the luminal pH at any operator-determined pH (4.0–7.4) based on potent buffering characteristics. We demonstrate application of a finely tunable series of these UPS nanoparticles to quantitative analysis of the contribution of endosomal pH transitions to endosome maturation, nutrient adaptation and growth homeostasis.

## Results

**A nanoparticle library with sharp buffer capacity.** We synthesized a series of amphiphilic block copolymers PEO-$b$-P(R$_1$-$r$-R$_2$), where PEO is poly(ethylene oxide) and P(R$_1$-$r$-R$_2$) is an ionizable random copolymer block (Fig. 1a and Supplementary Fig. 1). The molecular composition of each copolymer is shown in Supplementary Table 1. At high pH (for example, 7.4 in phosphate-buffered saline (PBS)), these copolymers self-assemble into core-shell micelle structures (diameter 30–60 nm, surface electrostatic potential $-2$ to $0$ mV, Supplementary Table 1 and Supplementary Fig. 2). At pH below the apparent pK$_a$ of each copolymer, micelles dissociate into unimers because of the protonation of tertiary amines. Our previous studies exploited the sharp pH-dependent micelle transitions for the development of a series of tunable, UPS fluorescence sensors[14].

Here we report the UPS nanoparticles have potent pH-tunable buffer capacity at a narrow pH interval across a broad range of pH (4.0–7.4). Figure 1b shows the pH titration curves of three exemplary UPS$_{4.4}$, UPS$_{5.3}$ and UPS$_{6.2}$ nanoparticles (each subscript indicates the pK$_a$ of the corresponding copolymer, Supplementary Table 1) in the presence of 150 mM NaCl. UPS$_{4.4}$, UPS$_{5.3}$ and UPS$_{6.2}$ (2 mg ml$^{-1}$) buffered the pH at their apparent pK$_a$ at 4.4, 5.3 and 6.2, respectively, when HCl (0.4 M) was added into the polymer solution. In contrast, CQ, a widely used small molecular base in biological studies, showed a broad pH response in the range of pH 6.0–9.0 (pK$_a = 8.3$). Moreover, polyethylenimine (PEI), widely used for nucleic acid delivery, also behaved as a broad pH buffer[15]. Determination of buffer capacity ($\beta = -\text{dn}_{\text{H}}^+/\text{dpH}$, where dn$_{\text{H}}^+$ is the quantity of added H$^+$ and dpH is the associated pH change) from the pH titration curves (Fig. 1c and Supplementary Fig. 3) showed exceptionally strong and selective buffering at specific pHs in the range of 4.0–7.4. In particular, the maximal $\beta$ values for UPS$_{4.4}$, UPS$_{5.6}$ and UPS$_{7.1}$ nanoparticles were 1.4, 1.5 and 1.6 mmol HCl per 40 mg of nanoparticle, which are 339-, 75- and 30-fold higher than CQ at pH 4.4, 5.6 and 7.1, respectively (Fig. 1c). To examine the consequences of the UPS nanoparticles on endo/lysosomal membrane and plasma membrane integrity, we employed recombinant cytochrome $C$ release studies[16] and haemolysis assays[17]. No detectable perturbation of endosomal or plasma membrane lysis, at 200 or 400 µg ml$^{-1}$ of UPS nanoparticles, was detected as compared with positive or negative controls (Supplementary Fig. 4, see Supplementary Methods). This collection of UPS nanoparticles thus provides a unique set of pH-specific 'proton sponges' for the functional range of organelle pH from early endosomes (E.E., 6.0–6.5)[18] to late endosomes (L.E., 5.0–5.5)[18] to lysosomes (4.0–4.5)[9].

**pH buffering of endocytic organelles.** For simultaneous imaging and buffering studies, we established a new nanoparticle design with a dual fluorescence reporter: an 'always-ON' reporter to track intracellular nanoparticle distribution regardless of the pH environment and a pH-activatable reporter (OFF at extracellular medium pH 7.4 and ON at specific organelle pH post endocytosis, see Supplementary Methods). Our initial attempts at conjugating a dye (for example, Cy3.5) on the terminal end of PEO produced an 'always-ON' signal, however, the resulting nanoparticles were unstable because of dye interactions with serum proteins (data not shown). To overcome this limitation, we employed a heteroFRET design using a pair of fluorophores that were introduced in the core of micelles. As an example, we separately conjugated a FRET pair (for example, BODIPY and Cy3.5 as donor and acceptor, respectively) to the P(R$_1$-$r$-R$_2$) segment of the UPS$_{6.2}$ copolymer. Mixing of the two dye-conjugated copolymers (optimal molar ratio of donor/acceptor = 2:1) within the same micelle core allowed the heteroFRET-induced fluorescence quenching of donor dye (that is, BODIPY) in the micelle state (pH > pK$_a$), but fluorescence recovery in the unimer state after micelle disassembly at lower pH (Supplementary Fig. 5a upper panel). To generate the 'always-ON' signal, a low weight fraction of Cy3.5-conjugated copolymer in the micelles was used (for example, 40%) to avoid homoFRET-induced fluorescence quenching for the acceptor dye in the micelle state[13] (Supplementary Fig. 5b). The resulting UPS nanoparticle showed constant fluorescence intensity in the Cy3.5 channel across a broad pH range, whereas achieving UPS activation at specific pH for the BODIPY signal (Supplementary Fig. 5c).

UPS$_{6.2}$, UPS$_{5.3}$ and UPS$_{4.4}$ were chosen for cellular imaging and buffering studies as their apparent pK$_a$'s correspond to early endosomes, late endosomes and lysosomes, respectively[18]. All cell-based experiments were performed in the presence of both

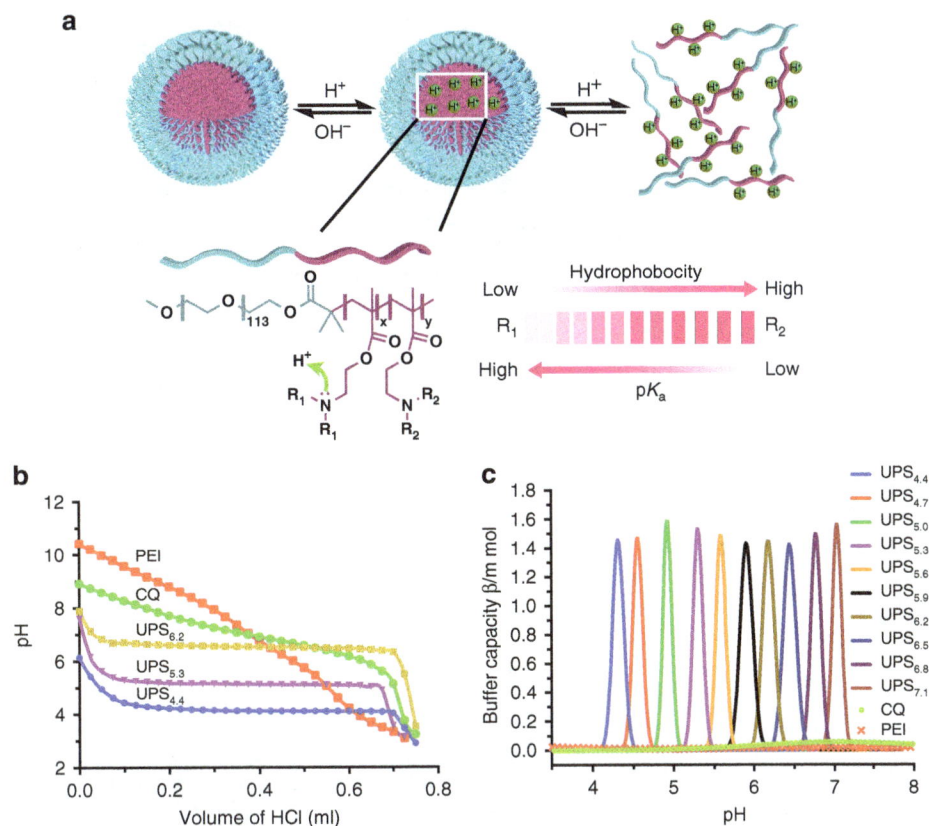

**Figure 1 | A UPS nanoparticle library with sharply defined buffer capacity across a broad physiological pH range.** (a) Schematic illustration of the buffer effect of UPS nanoparticles and the chemical structures of PEO-$b$-P($R_1$-$r$-$R_2$) copolymers with finely tunable hydrophobicity and pK$_a$. The composition for each copolymer is shown in Supplementary Table 1. (b) pH titration of solutions containing UPS$_{6.2}$, UPS$_{5.3}$ and UPS$_{4.4}$ nanoparticles using 0.4 M HCl. The maximum buffer pH corresponds to the apparent pK$_a$ of each copolymer. Chloroquine (CQ, pK$_a$ = 8.3 and 10.4), a small molecular base, and polyethyleneimines (PEI) were included for comparison. (c) Buffer capacity ($\beta$) for each component of the UPS library was plotted as a function of pH in the pH range of 4.0–7.4. At different pH values, UPS nanoparticles were 30- to 300-fold higher in buffer strength over CQ. L.E. and E.E. are abbreviations for late endosomes and early endosomes, respectively.

HEPES (25 mM) and sodium bicarbonate buffers in a 5% CO$_2$-controlled environment. HeLa cells were incubated with an increasing dose (100, 400 and 1,000 µg ml$^{-1}$) of UPS$_{6.2}$, UPS$_{5.3}$ or UPS$_{4.4}$ for 5 min at 37 °C to allow particle uptake via endocytosis[19], then washed with fresh medium (10% FBS in DMEM). At 100 µg ml$^{-1}$, we observed half maximal UPS$_{6.2}$ activation (BODIPY channel) by 30 min, half maximal UPS$_{5.3}$ activation by 60 min and half maximal UPS$_{4.4}$ activation by 90 min (Fig. 2d–f and Supplementary Fig. 6). In contrast, at 1,000 µg ml$^{-1}$, activation of BODIPY signal was delayed by at least 60 min despite clear indication of particle uptake in the HeLa cells by the Cy3.5 signal (Supplementary Fig. 6). *In situ* quantification of the endosomal pH with Lysosensor showed dose-dependent sustained pH plateaus at pH 6.2, 5.3 and 4.4 upon exposure of cells to 400 and 1,000 µg ml$^{-1}$ of UPS$_{6.2}$, UPS$_{5.3}$ and UPS$_{4.4}$ (Fig. 2a–c, Supplementary Fig. 7 and Supplementary Table 2), respectively. For either nanoparticle, 100 µg ml$^{-1}$ was insufficient to delay organelle acidification.

To further quantify the acidification rates, we measured the number of micelle nanoparticles per HeLa cell based on the fluorescence intensity of internalized UPS divided by the cell number (see Supplementary Methods for detials). Particle accumulation appropriately corresponded to the incubation dose (Supplementary Table 2). Based on the number of amino groups per micelle (64,000)[20] and an average of 200 endosomes/lysosomes per cell[21], we measured the acidification rate as ~140–190 protons per second for each organelle. To our best

knowledge, this is the first example of quantitative measurement of proton accumulation rates in endocytic organelles based on the endosome specificity and unique 'buffer and report' design of the UPS nanoparticles. This result is consistent with extrapolations (280–300 protons per second) based on 2 protons per ATP hydrolysed per v-ATPase[22], 3 ATP molecules consumed per rotation[23], 2.4 revolutions per second[24] and an average of 20 v-ATPases per organelle[25].

**pH thresholds exist in nutrient-induced mTORC1 activation.** We examined the consequences of UPS buffering of luminal pH on endosome protein coat maturation and endo/lysosome-dependent signal transduction. For this purpose, we selected UPS nanoparticles that discretely report and buffer at pH 6.2, 5.3, 5.0, 4.7 and 4.4. This range covers established luminal pH values in early endosomes, late endosomes and lysosomes. A discriminating feature of early endosome biogenesis is recruitment of the Rab5 GTPase[26], which corresponds to a luminal pH range of 6.0–6.5 (ref. 18). Fully mature lysosomes are LAMP2 positive with a luminal pH range of 4.0–4.5 (ref. 9). To enable quantification of co-localization of UPS-positive endosomes with endosomal maturation markers, UPS$_{6.2}$–Cy5 and UPS$_{4.4}$–Cy5 were developed with a low dye/polymer ratio that allowed for detectable fluorescence in the micelle state[20] (Fig. 3a–c). We used fluorescent dextran as a temporally synchronized comparator that does not perturb luminal pH. Within 15 min at a concentration of

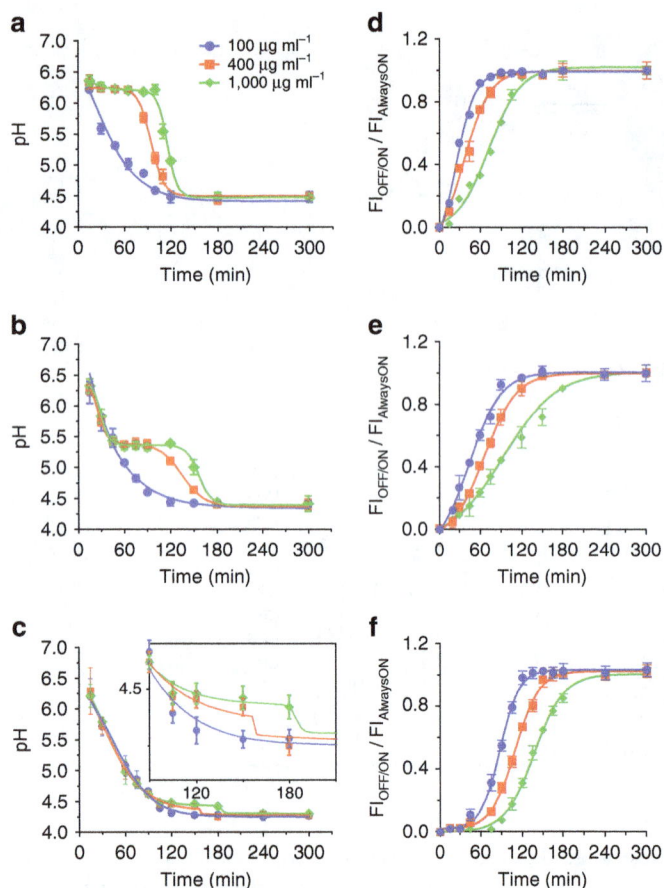

**Figure 2 | pH-sensitive buffering of endocytic organelles in HeLa cells.** Real-time measurement of endo/lysosomal pH in HeLa cells treated with the indicated doses of $UPS_{6.2}$ (**a**), $UPS_{5.3}$ (**b**) and $UPS_{4.4}$ (**c**). The inset is a zoomed-in view of the curve from pH 4.1 to 4.7 and 90 to 210 min in **c**. Lysosensor ratiometric imaging probe was used for *in situ* pH measurement. Quantitative analyses of the activation kinetics of always-ON/OFF-ON $UPS_{6.2}$ (**d**), $UPS_{5.3}$ (**e**) and $UPS_{4.4}$ (**f**). The fluorescent intensity of punctae in BODIPY channel (OFF-ON) was normalized to that of Cy3.5 (always-ON). The error bars represent s.d. from 50 organelles at each time point. In **a** and **b**, the 100 $\mu g\ ml^{-1}$ curves are significantly different from the 400 and 1,000 $\mu g\ ml^{-1}$ ones, with a *P*-value <0.0001, whereas in **c**, the 100 $\mu g\ ml^{-1}$ curve is significantly different from the 400 and 1,000 $\mu g\ ml^{-1}$ ones at 120 and 180 min time points, with a *P*-value <0.05. Two-way analysis of variance and Dunnett's multiple comparison tests were performed to assess the statistical significance. Blue, red and green plots indicate 100, 400 and 1,000 $\mu g\ ml^{-1}$ of UPS nanoparticles in all panels.

1,000 $\mu g\ ml^{-1}$, over 60% of $UPS_{6.2}$-, $UPS_{4.4}$- and dextran-positive endosomes were also Rab5 positive (Fig. 3a,d). $UPS_{4.4}$- and dextran-positive endosomes further transitioned to a Rab5-negative/LAMP2-positive maturation state within 60 min (Fig. 3b,e–f). Notably, $UPS_{6.2}$-positive endosomes also became LAMP2 positive in a similar timeframe despite inhibition of the luminal acidification that normally accompanies this transition (Fig. 3b,d). However, $UPS_{6.2}$ delayed release of Rab5, resulting in transient accumulation of anomalous Rab5/LAMP2-positive endosomes at 60 min (Fig. 3b,f and Supplementary Fig. 8). These observations indicate the presence of a regulatory mechanism that recruits LAMP2 to nascent endolysosomes independent of the luminal pH and the presence of a luminal pH-sensitive Rab5 release mechanism.

To evaluate integration of endosome maturation with cell regulatory systems, we examined signal transduction events coupled to endosomal compartments. The epidermal growth factor receptor, EGFR, has been shown to activate mitogenic signalling cascades following its ligand-dependent internalization in early endosomes[27]. As might be expected from a pH-regulated system, the delayed endosomal maturation induced by $UPS_{6.2}$ also resulted in delayed EGFR degradation and prolonged activation of ERK1/2 and AKT in response to EGF (Supplementary Fig. 9b).

To further examine the consequence of luminal pH clamping on endo/lysosome biology, we investigated a key regulatory system recently reported to be linked to lysosome biogenesis—namely, nutrient-dependent activation of cell growth via mammalian target of rapamycin complex 1 (mTORC1). In mammalian cells, mTORC1 localizes to endo/lysosomal membranes in response to internalized free amino acids[28]. Furthermore, the physical interactions between the v-ATPase and Rag GTPases on endo/lysosomal membranes are essential for mTORC1 activation in response to nutrient availability[29]. To evaluate amino-acid-induced mTORC1 activation, we employed two quantitative reporters of mTORC1 pathway activation: phosphorylation/activation of the mTORC1 substrate p70S6 kinase (p70S6K) and nuclear/cytoplasmic distribution of the mTORC1 substrate transcriptional factor EB (TFEB).

Incubation of HeLa cells for 2 h in a nutrient-free balanced salt solution (Earle's balanced salt solution (EBSS)) was sufficient to inhibit mTORC1 activity as indicated by reduced accumulation of activation site phosphorylation on both p70S6K and its substrate S6. Addition of essential amino acids was sufficient to induce pathway activation within 5 min (Fig. 4a, Supplementary Figs 10a,b and 12a–c). Pretreatment with 1,000 $\mu g\ ml^{-1}$ of $UPS_{4.7}$ or $UPS_{4.4}$ had little to no effect on the mTORC1 response to free amino acids. In contrast, pretreatment with 1,000 $\mu g\ ml^{-1}$ $UPS_{6.2}$, $UPS_{5.3}$ and $UPS_{5.0}$ both delayed and significantly suppressed the mTORC1 pathway response to free amino acids (Fig. 4a,b, Supplementary Figs 10a,b and 12). This correlated with inhibition of amino-acid-induced recruitment of mTOR to endo/lysosomes (Supplementary Fig. 10d). The selective UPS inhibition of the mTORC1 pathway response was mirrored by TFEB nuclear/cytoplasm distribution. Phosphorylation of this transcription factor by mTORC1 results in nuclear exclusion, thereby inhibiting the TFEB transcriptional programme in nutrient replete conditions[30–32]. In Hela cells, with stable expression of GFP-tagged TFEB, pretreatment with $UPS_{6.2}$, $UPS_{5.3}$ and $UPS_{5.0}$ inhibited redistribution of TFEB to the cytoplasm upon addition of free amino acids. In contrast, in cells pretreated with $UPS_{4.7}$ and $UPS_{4.4}$, TFEB redistribution proceeded normally (Fig. 4c,d).

The above data suggest acidification of endosomes below a threshold of pH 5 is necessary for free amino-acid-induced activation of mTORC1. We performed similar experiments employing bovine serum albumin (BSA) as a macromolecular nutrient source rather than free amino acids. Similar to free amino acids, BSA exposure was sufficient to reactivate mTORC1 following nutrient starvation (Supplementary Fig. 11). However, in contrast to free amino acids, $UPS_{4.4}$ delayed mTORC1 activation in response to BSA (Supplementary Fig. 11a,d). Given that cells treated with $UPS_{4.4}$ responded normally to free amino acids, we surmised the delayed response to BSA is the consequence of inhibition of the proteolysis of BSA by acid hydrolases in the lysosome. Consistent with this, we found significant inhibition of cathepsin B activity in the presence of $UPS_{4.4}$ (Supplementary Fig. 10c) as well as inhibition of autophagic degradation of p62/SQSTM1 otherwise induced by serum-deprivation (Supplementary Fig. 9a). Together, these

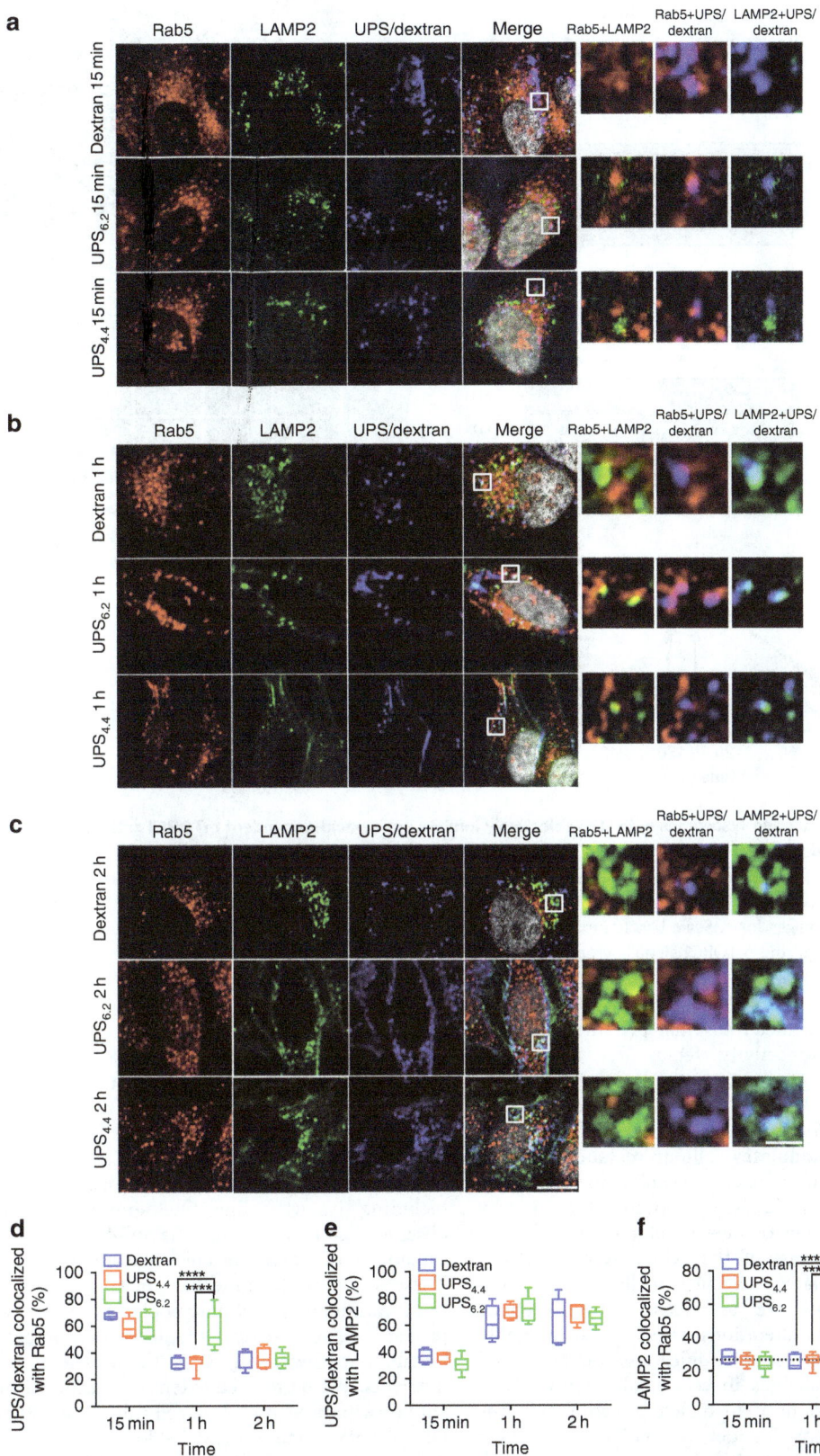

**Figure 3 | Buffering the pH of endocytic organelles affects their membrane protein dynamics.** HeLa cells were treated with $500\,\mu g\,ml^{-1}$ dextran-TMR or $1,000\,\mu g\,ml^{-1}$ $UPS_{6.2}$-Cy5 or $UPS_{4.4}$-Cy5 for 5 min for cell uptake. Then they were fixed after 15 min (**a**), 1 h (**b**) and 2 h (**c**). Immunofluorescence (IF) images show the localization of UPS nanoparticles in early endosomes (Rab5) or lysosomes (LAMP2). Scale bar, 10 and 5 μm (inset). Imaris software was used to analyse co-localization of z-stacked confocal images. The fraction of UPS/dextran co-localized with Rab5 (**d**) and LAMP2 (**e**) and the fraction of Rab5 co-localized with LAMP2 (**f**) were calculated from thresholded Mander's coefficient (see Supplementary Methods), $n = 10$, $\alpha = 0.05$, ****$P < 0.0001$. Two-way analysis of variance and Sidak's multiple comparison tests were performed to assess the statistical significance. The dashed line in (**f**) represents the basal level of Rab 5 and LAMP2 co-localization in HeLa cells without any treatment.

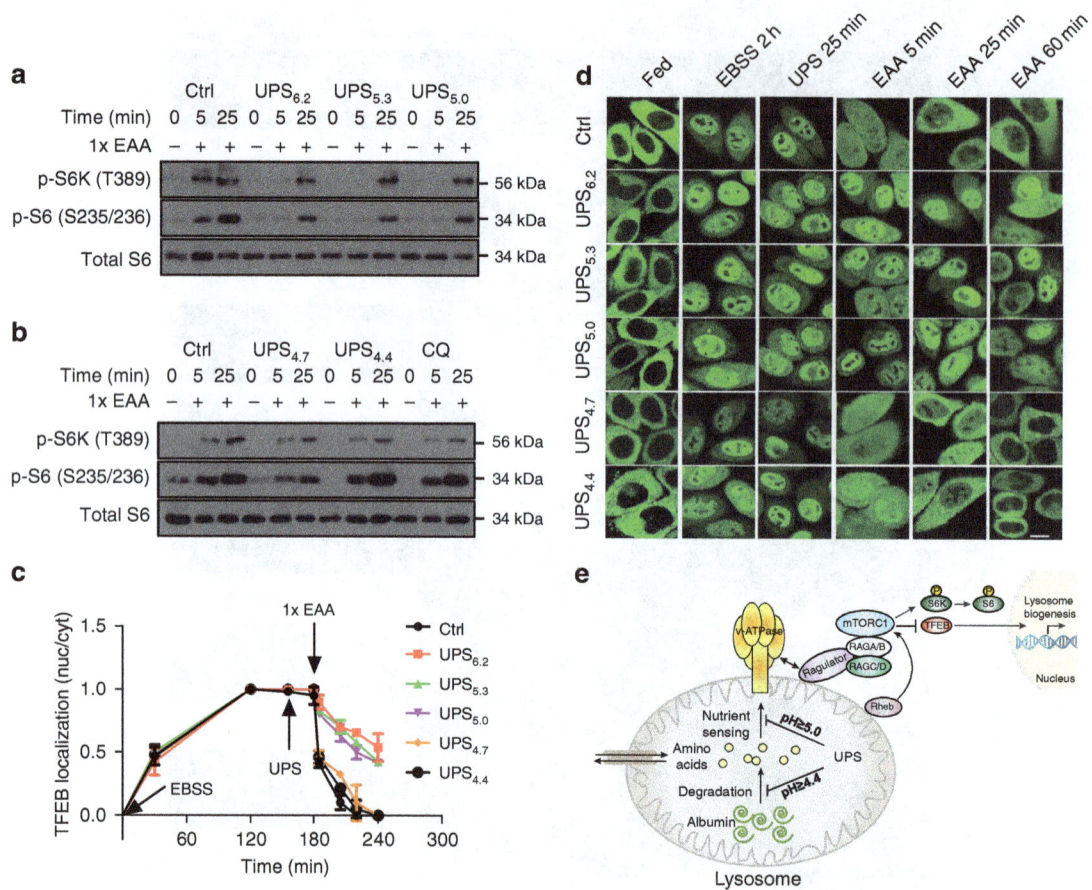

**Figure 4 | Clamping luminal pH of endo-lysosomes with UPS selectively inhibits amino-acid-dependent mTORC1 activation.** HeLa cells were starved in EBSS for 2 h and then stimulated with essential amino acids (EAAs) for indicated time intervals in the presence of (**a**) $UPS_{6.2}/UPS_{5.3}/UPS_{5.0}$ and (**b**) $UPS_{4.7}/UPS_{4.4}$. Water and 50 μM chloroquine (CQ) were used as control. Accumulation of the indicated phosphoproteins was assessed by immunoblot of whole-cell lysates. (**c**) Quantitative analysis of the nuclear/cytosolic distribution of GFP-TFEB following the indicated treatments. Error bars represent s.d., $n = 10$. (**d**) Representative images for **c**. Scale bar, 10 μm. (**e**) Working model of pH transitions required for free amino acid versus albumin-derived amino-acid-dependent activation of the mTORC1 signalling pathway.

observations indicate that distinct lysosomal pH thresholds are required for acid hydrolase activity versus free amino-acid sensing (Fig. 4e).

**Clamping lysosomal pH modulates cellular metabolite pools.** Lysosomes recycle intracellular macromolecules and debris to produce metabolic intermediates deployed for energy production or for construction of new cellular components in response to the nutrient status of the cellular environment[3]. Abnormal accumulation of large molecules, including lipids and glycoproteins, in lysosomes are associated with metabolic disorders. To broadly assess alterations associated with highly selective perturbation of lysosomal acidification, we quantified accumulation of small metabolites in cells loaded with $UPS_{4.4}$ under nutrient-starved versus nutrient-replete growth conditions (see Supplementary Methods for details). Following a 12-h exposure to 0, 200 and 400 μg ml$^{-1}$ of $UPS_{4.4}$, HeLa cells were lysed and intracellular metabolites were quantified using liquid chromatography-triple quadrupole mass spectrometry (LC/MS/MS). Sixty-eight metabolites were quantifiable from $3 \times 10^6$ HeLa cells, revealing a number of dose-dependent and nutrient-dependent consequences of pH arrest at 4.4 in lysosomes (Fig. 5a). Under nutrient replete conditions, as the dose of $UPS_{4.4}$ increased, the relative abundance of most metabolites also increased when normalized to cellular protein content. This

included most amino acids (Fig. 5b upper panel), consistent with an inhibition of the anabolic signals required to use them for protein synthesis and/or defects in lysosomal export of amino acids. In nutrient-deprived conditions, $UPS_{4.4}$ enhanced the relative abundance of nucleotides and their precursors (for example, bottom cluster in Fig. 5a) and massively suppressed the second messenger cAMP. The loss of many essential amino acids including lysine, valine, methionine and arginine was also observed, consistent with the inhibition of starvation-induced catabolism of macromolecules like albumin (Fig. 5b lower panel). These results confirm mechanistic connections between organelle acidification and metabolite pools, and fortify the hypothesis that proper lysosomal acidity is required for homeostasis of numerous metabolic pathways, either in the presence or in the absence of nutrients. We performed extensive additional metabolic profiling with or without exposure of cell cultures to 1,000 μg ml$^{-1}$ $UPS_{6.2}$, $UPS_{5.3}$ $UPS_{4.4}$ and 100 nM bafilomycin A1 (baf A1) under both nutrient-deprived and nutrient-replete conditions. Using the quantitative metabolite profiles as response vectors, we found that the UPS probes clustered nicely according to expectations associated with their pH clamping activity (Fig. 5c). $UPS_{6.2}$, $UPS_{5.3}$ and $UPS_{4.4}$ probes induced concordant drops in key amino acids. Baf A1 clustered with $UPS_{6.2}$, consistent with these agents maximally perturbing endosomal acidification (Supplementary Fig. 13). Notably, baf A1 also induced distinct metabolic changes as compared with all other probes and carrier

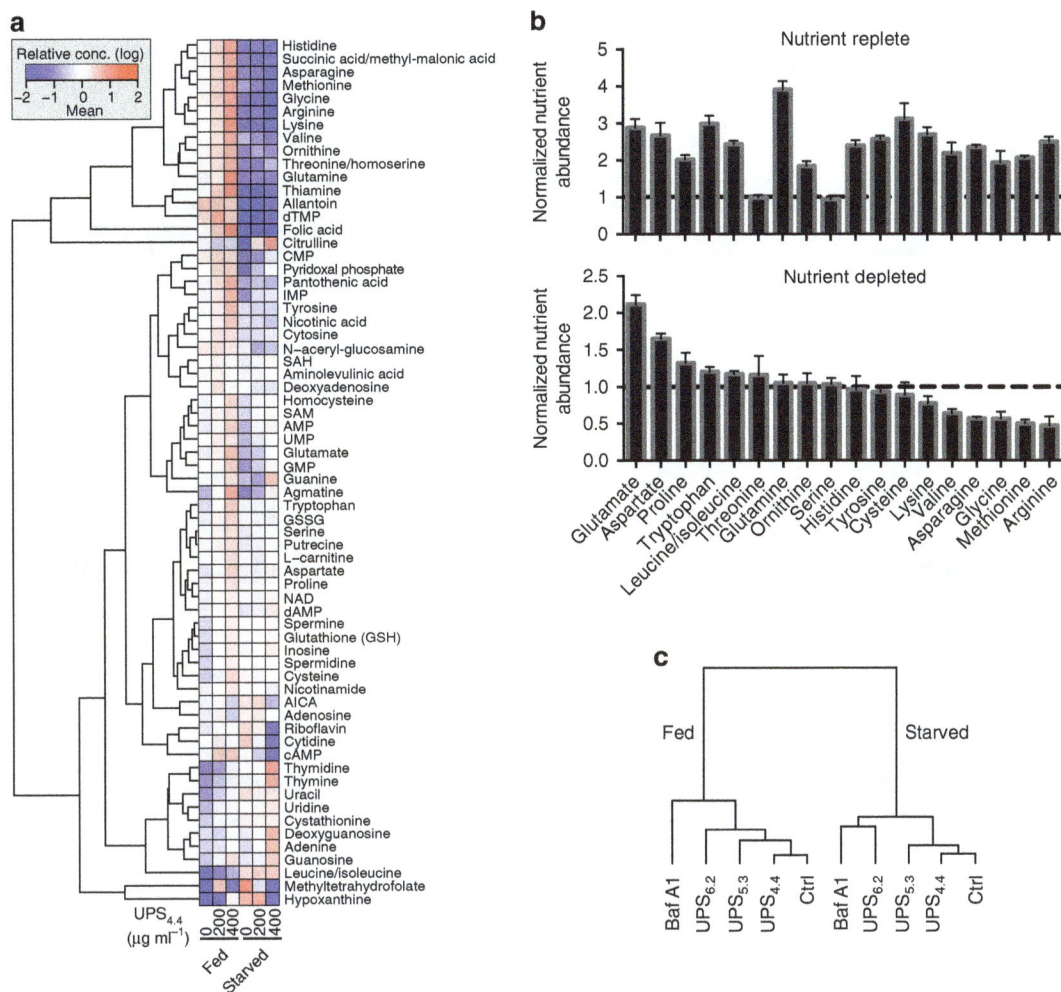

**Figure 5 | Selective buffering of lysosomal pH modulates the cellular metabolite pool. (a)** Dendrogram indicates relative abundance of the indicated metabolites in nutrient replete (fed) or deprived (starved) medium as normalized to the total protein content. Cells were treated with UPS$_{4.4}$ at the indicated doses. **(b)** Normalized abundance of the selected amino acids under nutrient replete and nutrient-deprived conditions. Error bars represent s.d., $n = 6$. **(c)** An unsupervised hierarchical clustering of different treatments including 1,000 µg ml$^{-1}$ UPS$_{6.2}$, UPS$_{5.3}$ and UPS$_{4.4}$ or 100 nM baf A1 under fed and starved conditions. Dentrogram was generated based the abundance of 108 metabolites.

controls (Fig. 5c and Supplementary Fig. 13), which may reflect its broader 'target space' in cells.

**Vulnerability of KRAS$^{mut}$/LKB1$^{mut}$ NSCLC cells to pH arrest.** We recently described a selective metabolic vulnerability in non-small-cell lung cancer (NSCLC) cells, whereby co-occurring mutations in the KRAS oncogene and LKB1 tumour-suppressor result in cellular addiction to lysosomal catabolism for maintenance of mitochondrial health[33]. Genetic or chemical inhibition of v-ATPase activity was sufficient to selectively induce programmed cell death in this oncogenic background. This was proposed to be a direct consequence of inhibition of a lysosome-dependent supply of trichloroacetic acid cycle substrates for ATP production. The UPS library afforded an opportunity to directly test this hypothesis in the absence of confounders associated with the pleiotropic contributions of v-ATPases to cytosolic pH and mTORC1/AMPK protein complexes in cancer cells[29,34]. As a model system, we employed normal (HBEC30KT) and tumour-derived (HCC4017) cell lines from the same patient together with an isogenic progression series in which the KRAS and LKB1 lesions were artificially introduced into the normal cell background (Fig. 6a)[35]. A comparison of cell number and

morphology between HCC4017 and HBEC30KT treated with UPS$_{6.2}$, UPS$_{5.3}$ and UPS$_{4.4}$ at high dose revealed highly selective toxicity of these UPS nanoparticles to HCC4017 (Fig. 6b). The expression of oncogenic KRAS together with inhibition of LKB1 was sufficient to induce sensitivity of bronchial epithelial cells to UPS-induced programmed cell death (Fig. 6c–e). Importantly, this phenotype was rescued in both the tumour-derived cells (Fig. 6f) and the genetically engineered cells (Fig. 6g) upon addition of cell permeable analogues of trichloroacetic acid cycle substrates (methyl pyruvate and α-ketoglutarate). Although UPS nanoparticles are able to clamp endo/lysosomal pH, they did not have detectable effects on cytosolic pH (Supplementary Fig. 14). Thus, selective vulnerability of KRAS/LKB1 co-mutant NSCLC cells to lysosomal function likely arises from addiction to catabolism of extracellular macromolecules.

**Discussion**

Luminal acidification is a hallmark of maturation of endocytic organelles in mammalian cells with pH-selective mechanistic consequences on receptor recycling, organelle trafficking and protein/lipid catabolism[1,2]. Existing tools and reagents employed to manipulate luminal acidification (for example, CQ, NH$_4$Cl, baf

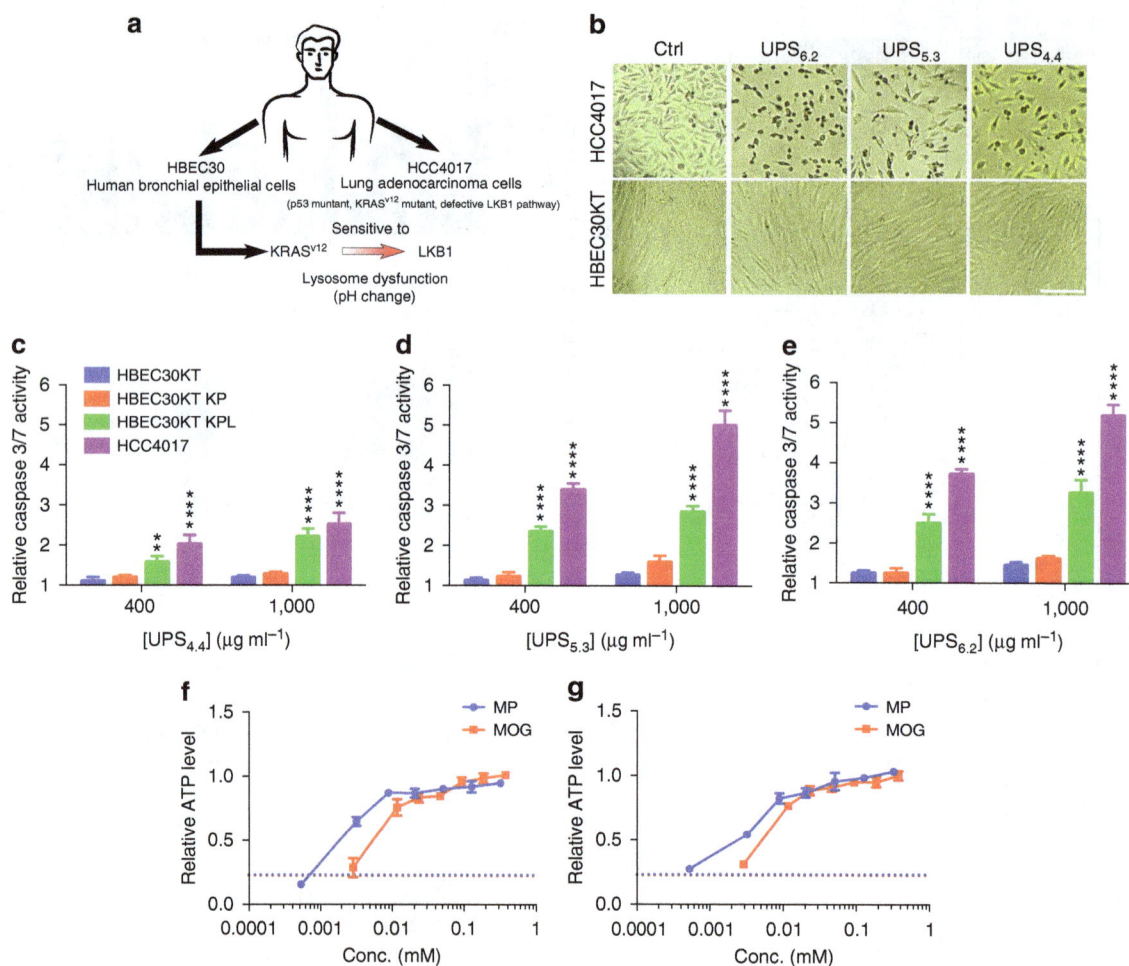

**Figure 6 | UPS nanoparticles selectively kill NSCLC cells that are sensitive to lysosomal stress. (a)** Schematic of the cell models employed and their corresponding vulnerabilities to lysosomal maturation. **(b)** Bright field images indicating the relative viability of HBEC30 KT and HCC4017 cells with and without exposure to UPS at effective doses (UPS$_{6.2}$ and UPS$_{5.3}$ = 400 μg ml$^{-1}$, UPS$_{4.4}$ = 1,000 μg ml$^{-1}$). Scale bar, 100 μm. **(c–e)** Caspase3/7 activity in HBEC30KT, HBEC30KT KP, HBEC30KT KPL and HCC4017 cells was measured 72 h after exposure to the indicated doses of UPS. Two-way analysis of variance and Sidak's multiple comparison tests were performed to assess statistical significance of observed differences between HBEC30KT and HCC4017, and HBEC30KT KP and HBEC30KT KPL, $\alpha$ = 0.05, **$P$ < 0.01, ****$P$ < 0.0001. **(f,g)** Cellular ATP levels were measured after exposure of HCC4017 **(f)** and HBEC30 KT KPL **(g)** to 1,000 μg ml$^{-1}$ UPS$_{6.2}$ for 72 h together with the indicated concentrations of methyl pyruvate (MP), dimethyl-2-oxoglutarate (MOG) or water (dash line). Values were normalized to no treatment (that is, without UPS) controls. Error bars indicate s.d., $n$ = 4.

A1) are membrane permeable and perturb a broad range of pH-dependent cellular activities. Consequently, investigation of endosome/lysosome biology using these agents can suffer from compounded, non-specific effects on multiple acidic organelles (for example, Golgi). In contrast, UPS nanoparticles enter cells exclusively through endocytosis and allow robust and fine-scale buffering of luminal pH at operator-determined thresholds along the endocytic pathway without disrupting cell or organelle membranes. The exceptional potency and specificity of the UPS nanoparticle buffering characteristics, together with previously reported UPS fluorescence response[12,13], holds considerable advantage over reported pH-sensitive probes (for example, small molecular dyes[36], peptides[37,38] or photoelectron transfer nanoprobes[39,40] with ten-fold signal change over 2 pH unit). The ultra-pH responsive property of the UPS system is a unique nanoscale phenomenon for self-assembled systems. The hydrophobic micellization (phase transition) dramatically sharpens the pH transition leading to cooperative protonation of tertiary amines. As a result, the UPS nanoparticles yielded a high-resolution buffer effect within 0.3 pH unit. In contrast to

small molecular pH buffers/sensors that are mostly controlled by electron withdrawing/donating substituents[36], the buffered pH range (centred around apparent pK$_a$) of the UPS platform can be fine-tuned by the hydrophobicity of the PR segment. This chemical simplicity and versatility to achieve fine-tunability by the UPS design is advantageous over reported pH-sensitive non-covalent polymer systems (for example, polyaminoacids[41] such as poly(L-histidines)[42,43]) or covalent polymer systems (for example, pH-labile bonds such as trimethoxy benzylidenes)[44,45]. The unique pH-specific, tunable 'proton sponge' effect is also distinct from other low-resolution polybase buffers (for example, PEI, Fig. 1c). To enhance simultaneous imaging and buffering capability, we constructed an always-ON/OFF-ON dual reporter design employing a heteroFRET strategy.

Detailed evaluation of the UPS library in cells illuminated mechanistic integration of dynamic luminal pH transitions in endosomes with multiple cell physiological processes. For example, the 'perturb and report' characteristics of the library allowed for time-resolved quantitation of endosome maturation, and uncovered previously unappreciated consequences of luminal pH on endosomal coat protein exchange. Notably, we found that recruitment of the 'mature' lysosome marker, LAMP2, occurs independently of luminal acidification. On the other hand, release of the early endosome marker Rab5 is delayed by luminal alkalization, resulting in the *de novo* accumulation of Rab5/LAMP2-positive endosomes. This indicates the presence of currently undescribed, but explorable, pH-sensitive and pH-insensitive mechanisms governing endosome/lysosome biogenesis. The ability to fine-tune UPS buffering capacity also allowed discrimination of distinct pH thresholds required for free amino acid versus albumin-dependent activation of mTORC1 pathway. We speculate that acidification to pH 5.0 or below is required to release free amino acids for 'inside-out' communication with v-ATPase protein complexes, or for induction of conformational changes in v-ATPase during amino-acid sensing[29]. Acidification to pH 4.4 or below is necessary for albumin-dependent activation of mTORC1, most likely due to the need for hydrolase activation and subsequent protein catabolism. Some hydrolases are reported to have pH-sensitive specific activity changes within the ranges spanned by the UPS probes employed here. For example, the *in vitro* activity of cathepsin B shows 10–20% fluctuation from pH 4.0 to 6.0 (ref. 46). We did not observe this range of activity in cells treated with UPS$_{6.2}$, UPS$_{5.3}$ and UPS$_{4.4}$, which may result from the detection limit of the assays we used. The scalability of UPS synthesis enabled broad-spectrum quantification of the cellular metabolite milieu upon inhibition of lysosomal consumption of extracellular macromolecules. The exclusive uptake of UPS within endocytic organelles afforded the opportunity to specifically evaluate the participation of endosomal/lysosomal pH in growth regulatory signalling pathways and cell metabolism.

In summary, we report a new class of biologically compartmentalized, high performance, imageable nanobuffers with high pH precision and resolution at operator-predetermined pH transitions. The combined and controlled perturb and report strategy delivers powerful biophysical tools for quantification of acidification kinetics of endocytic organelles in diverse biological contexts, and for time-resolved perturbation of this kinetics for evaluation of biological relevance. We anticipate these tools will also help generate new insights for biocompatibility and safety assessment of responsive nanomaterials in the rapidly growing fields of nanobiotechnology and drug/gene delivery[47,48].

## Methods

**Chemicals.** The Cy5-NHS, BODIPY-NHS and Cy3.5-NHS esters were purchased from Lumiprobe Corp. Monomers 2-(diethylamino) ethyl methacrylate and 2-aminoethyl methacrylate were purchased from Polyscience Company. Monomers 2-(dibutylamino) ethyl methacrylate (DBA-MA)[13], 2-(dipropylamino) ethyl methacrylate (DPA-MA) and 2-(dipentylamino) ethyl methacrylate[49] were prepared according to the method described in our previous work, as well as the PEO macroinitiator (MeO-PEO$_{114}$-Br)[13]. *N,N,N',N',N''*-Pentamethyldiethylenetriamine (PMDETA) was purchased from Sigma-Aldrich. Amicon ultra-15 centrifugal filter tubes (MWCO = 100 K) were obtained from Millipore. Other reagents and organic solvents were analytical grade from Sigma-Aldrich or Fisher Scientific Inc.

**Cells, culture medium and biological reagents.** The NSCLC cell line HCC4017 and its matched normal bronchial epithelial cell line HBEC30KT were developed from the same patient. The generation of these cell lines and the corresponding HBEC30KT oncogenic progression series was as previously reported[35]. HCC4017 and all HBEC30-derived cell lines were cultured in ACL4 medium (RPMI 1640 with 25 mM HEPES and 2.0 g l$^{-1}$ NaHCO$_3$ supplemented with 0.02 mg ml$^{-1}$ insulin, 0.01 mg ml$^{-1}$ transferrin, 25 nM sodium selenite, 50 nM hydrocortisone, 10 mM HEPES, 1 ng ml$^{-1}$ EGF, 0.01 mM ethanolamine, 0.01 mM O-phosphorylethanolamine, 0.1 nM triiodothyronine, 2 mg ml$^{-1}$ BSA, 0.5 mM sodium pyruvate) with 2% fetal bovine serum (FBS, Atlanta Biologicals) and 1% antibiotics (Gibco). HeLa and GFP-TFEB HeLa cells were cultured in DMEM (Invitrogen, containing 25 mM HEPES and 3.7 g l$^{-1}$ NaHCO$_3$) with 10% FBS and 1% antibiotics (Invitrogen). EBSS (10 ×, Sigma) was diluted to 1 × with Milli-Q water supplemented with 2.2 g l$^{-1}$ sodium bicarbonate (Sigma). All cell-based studies were performed with 25 mM HEPES buffer in a humidified chamber with

5% CO$_2$. Antibodies were from Cell Signaling (p70 S6K-pT389 (catalogue# 9205), total p70 S6K (catalogue# 2708), S6-Ribosomal-Protein-pS235/236 (catalogue# 4858), total S6 Ribsomal Protein (catalogue# 2217), Rab5 (catalogue# 3547), mTOR (catalogue# 2983), p62 (catalogue# 8025), EGFR (catalogue# 4267), EGFR-pY1068 (catalogue# 3777), MEK1/2-pS217/221(catalogue# 9154), total MEK1/2 (catalogue# 9126), Akt-p-S473 (catalogue# 4060), pan Akt(catalogue# 4691), GAPDH (catalogue# 5174)) and Abcam (LAMP2 (catalogue# ab13524)). All primary antibodies used for immunoblot are all diluted in 1:1,000. Secondary antibodies were from Jackson ImmunoResearch Laboratories (Peroxidase-conjugated AffiniPure Goat Anti-Rabbit IgG (catalogue# 111-035-144)) and Invitrogen (Alexa Fluor 488 goat anti-rat IgG (catalogue# A11006), Alexa Fluor 594 goat anti-rabbit IgG (catalogue# A11037), Alexa Fluor 635 goat anti-rabbit IgG (catalogue# A31577)). Other biological agents include Hoechst 33342 (Invitrogen), 70 kDa Dextran-TMR (Invitrogen), LysoSensor Yellow/Blue DND 160 (Invitrogen), Magic Red Cathepsin B Assay Kit (Immunochemistry Technology), Bafilomycin A1 (Sigma), Chloroquine (Sigma) Cytochrome C (Sigma) and BCA Protein Assay Kit (Thermo).

**Syntheses of PEO-*b*-(P(R$_1$-*r*-R$_2$)) block copolymers.** The copolymer PEO-*b*-(P(R$_1$-*r*-R$_2$)) was synthesized using atom transfer radical polymerization method as reported[14] (Supplementary Fig. 1). The molar fractions of R$_1$ and R$_2$ were varied to control the hydrophobicity of the PR segment. In a typical procedure using PEO-*b*-P(DPA$_{60}$-*r*-DBA$_{20}$) (UPS$_{5.9}$) as an example, DPA-MA (1.3 g, 6 mmol), DBA-MA (0.48 g, 2 mmol), PMDETA (21 µl, 0.1 mmol) and MeO-PEO$_{114}$-Br (0.5 g, 0.1 mmol) were charged into a polymerization tube. The monomer and initiator were dissolved in a mixture of 2-propanol (2 ml) and dimethylformamide (DMF) (2 ml). Three cycles of freeze–pump–thaw were performed to remove the oxygen, then CuBr (14 mg, 0.1 mmol) was added into the tube protected by nitrogen, and the tube was sealed *in vacuo*. After 8 h polymerization at 40 °C, the reaction mixture was diluted in 10 ml tetrahydrofuran (THF), and the mixture was passed through a neutral Al$_2$O$_3$ column to remove the catalyst. The organic solvent was removed by rotovap. The residue was dialysed in distilled water and lyophilized to obtain a white powder. The composition and physical properties of UPS library are listed in Supplementary Table 1.

**Titration of UPS nanoparticles.** In a typical procedure, 20 ml micelle solution (2 mg ml$^{-1}$) was first prepared at 150 mM concentration of NaCl. Chloroquine (2 mg ml$^{-1}$, 20 ml) or branched PEI (6.2 mg, 20 ml) solutions in 150 mM NaCl were used for comparison. pH titration was carried out by adding small volumes (25 µl increment) of 0.4 M HCl solution under stirring. The pH decrease in the range of 9 to 3 was monitored as a function of added volume of HCl. The pH values were measured using a Mettler Toledo pH metre with a microelectrode. Figure 1b shows the representative titration curves for the UPS nanoparticles. For each sample, the pK$_a$ value was calculated as the pH in the middle of the two equivalence points in the titration curve. The pK$_a$ values for all the UPS nanoparticles are listed in the Supplementary Table 1. No HCO$_3^-$/H$_2$CO$_3$ or any other buffers were included in the system to avoid possible interference from external systems to nanoparticle alone. However, all cell-based studies were performed with 25 mM HEPES buffer in a humidified chamber with 5% CO$_2$.

**Measurement of endo/lysosomal pH.** HeLa cells were plated in 4- or 8-well Nunc Lab-Tek II Chambered Coverglass (Thermo Scientific) and allowed to grow for 48 h. The cells were then loaded with 25 µM LysoSensor Yellow/Blue DND-160 and 100, 400 or 1,000 µg ml$^{-1}$ UPS nanoparticles in serum-free medium at 37 °C for 5 min. The cells were washed twice and immediately imaged. Imaging was performed using an epifluorescent microscope (Deltavision, Applied Precision) equipped with a digital monochrome Coolsnap HQ2 camera (Roper Scientific). Fluorescence images were collected using SoftWoRx v3.4.5 (Universal Imaging). Data were recorded at excitation/emission wavelengths of 360/460 and 360/520 nm. The single band-pass excitation filter for 4,6-diamidino-2-phenylindole (DAPI; 360 nm) is 40 nm, and the band pass of emission filters for DAPI (460 nm) and fluorescein isothiocyanate (520 nm) is 50 and 38 nm, respectively. Cell fluorescence ratios were determined by image analysis using ImageJ software. For each cell, a region of interest was defined as the punctae in cytosol that emitted fluorescent signals from both UPS nanoparticles and LysoSensor. Fluorescent intensity ratio was calculated for each intracellular punctate as $R = (F_1\text{-}B_1)/(F_2\text{-}B_2)$, where $F_1$ and $F_2$ are the fluorescence intensities at 360/520 and 360/460, respectively, and $B_1$ and $B_2$ are the corresponding background values determined from a region on the same images that was near the punctae in the cytosol. To calibrate the relationship between $R$ and pH, we used a modified protocol established by Diwu et al[50]. Cells were loaded with LysoSensor and then permeabilized with 10 µM monensin and 10 µM nigericin. These cells were treated for 30 min with the equilibration buffers consisting of 5 mM NaCl, 115 mM KCl, 1.2 mM MgSO$_4$ and 25 mM MES (MES buffer) varied between pH 4.0 and 7.4. The cells were kept in the buffer until imaging. The curves for 400 and 1,000 µg ml$^{-1}$ UPS nanoparticles were fit with the bi-dose–response fitting function in OriginLab (v8.0), whereas the curves of 100 µg ml$^{-1}$ were fit with the dose–response function. Two-way analysis of variance and Dunnett's multiple comparison tests were performed to assess the statistical significance using Graphpad Prism (v6.0) software.

**Measurement of intracellular pH.** HBEC30 KT and HCC4017 cells were plated in 4- or 8-well Nunc Lab-Tek II Chambered Coverglass (Thermo Scientific) and allowed to grow for 48 h. The cells were then loaded with 1,000 µg ml$^{-1}$ UPS nanoparticles in ACL4 medium at 37 °C with 5% $CO_2$ for 24 h. 2′,7′-bis-(2-carboxyethyl)-5-(and-6)-carboxyfluorescein, acetoxymethyl ester (BCECF, AM) stock solution was diluted in EBSS (with 2.2 g l$^{-1}$ NaHCO$_3$, without amino acids or buffers containing primary or secondary amines that may cleave the AM esters and prevent loading) to 5 µM. Cells were then loaded with BCECF at 37 °C with 5% $CO_2$ for 20 min. Then cells were washed twice and immediately imaged. A spinning-disk confocal microscope was used to acquire images with excitation at 445 and 488 nm, and a 525/40-nm EMCCD emission wheel. Cell fluorescence ratios were determined by using ImageJ software. Fluorescent intensity ratio in the cytosolic region was calculated for each cell as $R = (F_1 \text{-} B_1)/(F_2 \text{-} B_2)$, where $F_1$ and $F_2$ are the fluorescence intensities at 488/525 and 445/525, respectively, and $B_1$ and $B_2$ are the corresponding background values determined from a region on the same images that was near the cell. To calibrate the relationship between $R$ and pH, we used a modified protocol established by Thomas et al.[51] and others[52,53]. Different pH buffers (135 mM KCl, 5 mM K$_2$HPO$_4$, 20 mM HEPES, 1.2 mM CaCl$_2$, 0.8 mM MgSO$_4$, pH was adjust to 6.0, 6.5, 7.0, 7.5 and 8.0 by adding HCl or KOH) with 10 µM nigericin was used to equilibrate internal and external pH. After 5 min equilibrium, images were taken and analysed as described above to obtain the 488/445 ratio for each calibration pH value. Graphpad Prism (v6.0) software was used to obtain an equation (a sigmoidal plot) that best describes the data.

**Immunofluorescence assays.** HeLa cells were plated on glass coverslips in a 12-well tissue culture dish at 500,000 cells per well. After 24 h, the cells were treated with 500 µg ml$^{-1}$ 70 kDa dextran-TMR, 1,000 µg ml$^{-1}$ UPS$_{6.2}$–Cy5 or UPS$_{4.4}$–Cy5 for 5 min in serum-free DMEM, followed by three washes in PBS and subsequent incubation in DMEM + 10% FBS for the indicated time. Coverslips were then rinsed with PBS and fixed for 10 min with 4% paraformaldehyde in PBS at room temperature. Following fixation, coverslips were rinsed twice with PBS and cells were permeablized with 0.1% Triton X-100 in PBS for 10 min at 4 °C. After rinsing twice with PBS, the coverslips were incubated with 10% normal goat serum block solution for 45 min at room temperature. The primary antibodies (from different species) were diluted 1:100 in the block solution, and were co-incubated with cells overnight in the dark at 4 °C. Secondary antibodies were diluted 1:200 in the block solution. The cells were then washed with PBS and incubated with secondary antibodies at room temperature for 1 h. The coverslips were washed three times before being mounted on glass slides using Vectashield (Vector Laboratories) and imaged using confocal microscopy.

**Image acquisition and analysis.** Confocal laser scanning microscopy was used to investigate the intracellular activation and distribution of UPS nanoparticles. HeLa cells were plated in 4- or 8-well Nunc Lab-Tek II Chambered Coverglass (Thermo Scientific) and allowed to grow for 48 h. Cells were incubated with always-ON/OFF-ON UPS nanoparticles for 5 min in serum-free medium, and washed three times with PBS before imaging. Confocal images in Supplementary Fig. 6 were acquired with a Nikon ECLIPSE TE2000-E confocal microscope with identical settings for each experiment. Data were recorded at excitation wavelengths of 488 nm (BODIPY) and 560 nm (Cy3.5). EZ-C1-free viewer v3.90 (Nikon) and ImageJ software (NIH) were used to convert and analyse the images. The FI$_{\text{OFF-ON}}$ $_{\text{(BODIPY)}}$/FI$_{\text{Always-ON (Cy3.5)}}$ ratio was determined by image analysis using ImageJ (NIH) software. For each cell, a region of interest was defined as the punctae in cytosol that emitted fluorescent signals from both BODIPY and Cy3.5 channels. Fluorescent intensity ratio was calculated for each intracellular punctate as $R = (F_1 \text{-} B_1)/(F_2 \text{-} B_2)$, where $F_1$ and $F_2$ are the fluorescence intensities from BODIPY and Cy3.5 channels, respectively, and $B_1$ and $B_2$ are the corresponding background values determined from a region on the same images that was near the punctae in the cytosol. All the ratios of each nanoparticle were normalized to their end-time-point ratio, and the curves were fit with the dose–response function with Graphpad Prism (v6.0) software. Images from the immunofluorescence assay (Fig. 3a–c) were taken by using a spinning disk confocal microscope (Andor). Z-stack images were used after deconvolution in the co-localization analysis. The data were analysed using the Coloc module of Imaris 7.7 (Bitplane). The thresholded Mander's coefficient was used as an indicator of the proportion of the co-localized signal over the total signal[54,55]. The distribution and volume analysis was also done in Imaris 7.7. Z-stack images were used after deconvolution, the background was subtracted and a 'surface' was built for the DAPI, Rab5 and LAMP2 channels, respectively. The volume of each subject in the Rab5 and LAMP2 channels was calculated based on their voxel numbers, and the shortest distance of each to the nucleus surface were calculated using the MATLAB plugin 'Distance transformation'. The medium distance and volume of Rab5- or LAMP2-positive vesicles in each cell was calculated for each cell. Two-way analysis of variance and Sidak's multiple comparison tests were performed to assess the statistical significance using Graphpad Prism (v6.0) software. A spinning disk confocal microscope (Andor) was used to obtain images in Supplementary Fig. 10d. An average intensity projection was used on the Z-stack images (42 slices) in ImageJ.

**Amino acid or BSA starvation and stimulation of the cells.** The method was adapted from Sancak et al.[28]. Cells were rinsed and incubated with 1 × EBSS for 2 h. For amino-acid starvation, cells were first pretreated with 10 × glutamine (final concentration 1 ×) for 1 h to facilitate the transport of other amino acids into the cells. UPS nanoparticles were added in the last 25 min of glutamine treatment if needed. Tenfold essential amino-acid solution was added to stimulate cells (final concentration 1 ×). After stimulation, the level of essential amino acids and glutamine in EBSS was the same as in DMEM. The mTOR and LAMP2 assay followed the same protocol here and also the immunofluorescence protocol mentioned above. For BSA starvation, cells were pretreated with UPS nanoparticles for 25 min before adding BSA solution with a final concentration of 2 mg ml$^{-1}$ (ref. 56).

**Apoptosis assay.** A Caspase Glo 3/7 assay was used to measure caspase 3/7 activities. A normal bronchiole epithelia-derived (HBEC30KT) cell line, tumour-derived HCC4017, KRAS$^{mut}$ HBEC30KT and KRAS$^{mut}$/LKB1$^{mut}$ HBEC30KT were seeded in 96-well plates (Corning). UPS nanoparticles were added 24 h later. Caspase Glo reagent (Promega) was added after 72 h according to the manufacturer's instructions. Plates were read with PHERAstar FS microplate reader (BMG LABTECH).

## References

1. Maxfield, F. R. & McGraw, T. E. Endocytic recycling. *Nat. Rev. Mol. Cell Biol.* **5**, 121–132 (2004).
2. Yeung, T., Ozdamar, B., Paroutis, P. & Grinstein, S. Lipid metabolism and dynamics during phagocytosis. *Curr. Opin. Cell Biol.* **18**, 429–437 (2006).
3. Settembre, C., Fraldi, A., Medina, D. L. & Ballabio, A. Signals from the lysosome: a control centre for cellular clearance and energy metabolism. *Nat. Rev. Mol. Cell Biol.* **14**, 283–296 (2013).
4. Rajendran, L., Knölker, H.-J. & Simons, K. Subcellular targeting strategies for drug design and delivery. *Nat. Rev. Drug Discov.* **9**, 29–42 (2010).
5. Pack, D. W., Hoffman, A. S., Pun, S. & Stayton, P. S. Design and development of polymers for gene delivery. *Nat. Rev. Drug Discov.* **4**, 581–593 (2005).
6. Tasciotti, E. et al. Mesoporous silicon particles as a multistage delivery system for imaging and therapeutic applications. *Nat. Nanotechnol* **3**, 151–157 (2008).
7. Nel, A. E. et al. Understanding biophysicochemical interactions at the nano-bio interface. *Nat. Mater.* **8**, 543–557 (2009).
8. Dahlman, J. E. et al. In vivo endothelial siRNA delivery using polymeric nanoparticles with low molecular weight. *Nat. Nanotechnol* **9**, 648–655 (2014).
9. Casey, J. R., Grinstein, S. & Orlowski, J. Sensors and regulators of intracellular pH. *Nat. Rev. Mol. Cell Biol.* **11**, 50–61 (2010).
10. Boussif, O. et al. A versatile vector for gene and oligonucleotide transfer into cells in culture and in vivo: polyethylenimine. *Proc. Natl Acad. Sci. USA* **92**, 7297–7301 (1995).
11. Tartakoff, A. M. Perturbation of vesicular traffic with the carboxylic ionophore monensin. *Cell* **32**, 1026–1028 (1983).
12. Zhou, K. et al. Tunable, ultrasensitive pH-responsive nanoparticles targeting specific endocytic organelles in living cells. *Angew. Chem. Int. Ed.* **50**, 6109–6114 (2011).
13. Zhou, K. et al. Multicolored pH-tunable and activatable fluorescence nanoplatform responsive to physiologic pH stimuli. *J. Am. Chem. Soc.* **134**, 7803–7811 (2012).
14. Ma, X. et al. Ultra-pH sensitive nanoprobe library with broad pH tunability and fluorescence emissions. *J. Am. Chem. Soc.* **136**, 11085–11092 (2014).
15. Suh, J., Paik, H.-J. & Hwang, B. K. Ionization of poly (ethylenimine) and poly (allylamine) at various pH's. *Bioorg. Chem.* **22**, 318–327 (1994).
16. Lin, M. L. et al. Selective suicide of cross-presenting CD8 + dendritic cells by cytochrome c injection shows functional heterogeneity within this subset. *Proc. Natl Acad. Sci. USA* **105**, 3029–3034 (2008).
17. Bignami, G. S. A rapid and sensitive hemolysis neutralization assay for palytoxin. *Toxicon* **31**, 817–820 (1993).
18. Weisz, O. A. Acidification and Protein Traffic. *Int. Rev. Cytol.* **226**, 259–319 (2003).
19. Conner, S. D. & Schmid, S. L. Regulated portals of entry into the cell. *Nature* **422**, 37–44 (2003).
20. Wang, Y. et al. A nanoparticle-based strategy for the imaging of a broad range of tumours by nonlinear amplification of microenvironment signals. *Nat. Mater.* **13**, 204–212 (2014).
21. Holtzman, E. *Lysosomes* (Springer, 1989).
22. Deamer, D. W., Kleinzeller, A. & Fambrough, D. M. *Membrane Permeability: 100 Years Since Ernest Overton* (Academic, 1999).
23. Cross, R. L. & Muller, V. The evolution of A-, F-, and V-type ATP synthases and ATPases: reversals in function and changes in the H + /ATP coupling ratio. *FEBS Lett.* **576**, 1–4 (2004).
24. Imamura, H. et al. Evidence for rotation of V1-ATPase. *Proc. Natl Acad. Sci. USA* **100**, 2312–2315 (2003).

25. Rodman, J. S., Stahl, P. D. & Gluck, S. Distribution and structure of the vacuolar H + ATPase in endosomes and lysosomes from LLC-PK1 cells. *Exp. Cell Res.* **192**, 445–452 (1991).

26. Huotari, J. & Helenius, A. Endosome maturation. *EMBO J.* **30**, 3481–3500 (2011).

27. Wells, A. EGF receptor. *Int. J. Biochem. Cell. B* **31**, 637–643 (1999).

28. Sancak, Y. *et al.* Ragulator-Rag complex targets mTORC1 to the lysosomal surface and is necessary for its activation by amino acids. *Cell* **141**, 290–303 (2010).

29. Zoncu, R. *et al.* mTORC1 senses lysosomal amino acids through an inside-out mechanism that requires the vacuolar H(+)-ATPase. *Science* **334**, 678–683 (2011).

30. Pena-Llopis, S. *et al.* Regulation of TFEB and V-ATPases by mTORC1. *EMBO J.* **30**, 3242–3258 (2011).

31. Settembre, C. *et al.* A lysosome-to-nucleus signalling mechanism senses and regulates the lysosome via mTOR and TFEB. *EMBO J.* **31**, 1095–1108 (2012).

32. Roczniak-Ferguson, A. *et al.* The transcription factor TFEB links mTORC1 signaling to transcriptional control of lysosome homeostasis. *Sci. Signal.* **5**, ra42 (2012).

33. Kim, H. S. *et al.* Systematic identification of molecular subtype-selective vulnerabilities in non-small-cell lung cancer. *Cell* **155**, 552–566 (2013).

34. Zhang, C.-S. *et al.* The lysosomal V-ATPase-Ragulator complex is a common activator for AMPK and mTORC1, acting as a switch between catabolism and anabolism. *Cell Metab.* **20**, 526–540 (2014).

35. Ramirez, R. D. *et al.* Immortalization of human bronchial epithelial cells in the absence of viral oncoproteins. *Cancer Res.* **64**, 9027–9034 (2004).

36. Urano, Y. *et al.* Selective molecular imaging of viable cancer cells with pH-activatable fluorescence probes. *Nat. Med* **15**, 104–109 (2009).

37. Viola-Villegas, N. T. *et al.* Understanding the pharmacological properties of a metabolic PET tracer in prostate cancer. *Proc. Natl Acad. Sci. USA* **111**, 7254–7259 (2014).

38. Weerakkody, D. *et al.* Family of pH (low) insertion peptides for tumor targeting. *Proc. Natl Acad. Sci. USA* **110**, 5834–5839 (2013).

39. Diaz-Fernandez, Y. *et al.* Micelles for the self-assembly of 'Off-On-Off' fluorescent sensors for pH windows. *Chemistry* **12**, 921–930 (2006).

40. Uchiyama, S., Iwai, K. & de Silva, A. P. Multiplexing sensory molecules map protons near micellar membranes. *Angew. Chem. Int. Ed.* **47**, 4667–4669 (2008).

41. Bellomo, E. G., Wyrsta, M. D., Pakstis, L., Pochan, D. J. & Deming, T. J. Stimuli-responsive polypeptide vesicles by conformation-specific assembly. *Nat. Mater.* **3**, 244–248 (2004).

42. Lee, E. S. *et al.* Super pH-sensitive multifunctional polymeric micelle for tumor pH(e) specific TAT exposure and multidrug resistance. *J. Control. Release* **129**, 228–236 (2008).

43. Lee, E. S., Na, K. & Bae, Y. H. Super pH-sensitive multifunctional polymeric micelle. *Nano Lett.* **5**, 325–329 (2005).

44. Gillies, E. R. & Frechet, J. M. J. pH-responsive copolymer assemblies for controlled release of doxorubicin. *Bioconjug. Chem.* **16**, 361–368 (2005).

45. Gillies, E. R., Jonsson, T. B. & Frechet, J. M. Stimuli-responsive supramolecular assemblies of linear-dendritic copolymers. *J. Am. Chem. Soc.* **126**, 11936–11943 (2004).

46. Almeida, P. C. *et al.* Cathepsin B activity regulation heparin-like glycosaminoglycans protect human cathepsin B from alkaline ph-induced inactivation. *J. Biol. Chem.* **276**, 944–951 (2001).

47. Dobrovolskaia, M. A. & McNeil, S. E. Immunological properties of engineered nanomaterials. *Nat. Nanotechnol* **2**, 469–478 (2007).

48. Peer, D. *et al.* Nanocarriers as an emerging platform for cancer therapy. *Nat. Nanotechnol* **2**, 751–760 (2007).

49. Li, Y. *et al.* Chaotropic-anion-induced supramolecular self-assembly of ionic polymeric micelles. *Angew. Chem.* **53**, 8074–8078 (2014).

50. Diwu, Z., Chen, C.-S., Zhang, C., Klaubert, D. H. & Haugland, R. P. A novel acidotropic pH indicator and its potential application in labeling acidic organelles of live cells. *Chem. Biol.* **6**, 411–418 (1999).

51. Thomas, J. A., Buchsbaum, R. N., Zimniak, A. & Racker, E. Intracellular pH measurements in Ehrlich ascites tumor cells utilizing spectroscopic probes generated in situ. *Biochemistry* **18**, 2210–2218 (1979).

52. James-Kracke, M. R. Quick and accurate method to convert BCECF fluorescence to pHi: calibration in three different types of cell preparations. *J. Cell. Physiol.* **151**, 596–603 (1992).

53. Kapus, A., Grinstein, S., Wasan, S., Kandasamy, R. & Orlowski, J. Functional characterization of three isoforms of the Na + /H + exchanger stably expressed in Chinese hamster ovary cells. ATP dependence, osmotic sensitivity, and role in cell proliferation. *J. Biol. Chem.* **269**, 23544–23552 (1994).

54. Manders, E., Verbeek, F. & Aten, J. Measurement of co-localization of objects in dual-colour confocal images. *J. Microsc.* **169**, 375–382 (1993).

55. Bolte, S. & Cordelières, F. P. A guided tour into subcellular colocalization analysis in light microscopy. *J. Microsc.* **224**, 213–232 (2006).

56. Commisso, C. *et al.* Macropinocytosis of protein is an amino acid supply route in Ras-transformed cells. *Nature* **497**, 633–637 (2013).

## Acknowledgements

We thank S.M. Ferguson, Yale University, for generously sharing GFP-TFEB-transfected HeLa cells, A. Bugde and K. Phelps from the Live Cell Imaging Facility at UT Southwestern for helping with confocal imaging and analysis, and J. Cooper, S. Mendiratta, M. Potts and B. Eskiocak and all the other members of the Gao and the White laboratories for thoughtful comments and discussions, X. Xie for helping with statistics and E. Macmillan for the support on biostatistics and bioinformatics. This work is supported by the National Institutes of Health (NIH) grants R01EB013149 and R01CA129011 (J.G.) and R01CA71443 and R01CA176284 (M.A.W.), Cancer Prevention and Research Institute of Texas (CPRIT) grants RP120094 (J.G.) and RP121067 and RP110710 (M.A.W.), and the Welch Foundation grant I-1414 (M.A.W.). C.W. is a Howard Hughes Medical Institute (HHMI) International Student Research Fellow.

## Author contributions

C.W., M.A.W. and J.G. are responsible for all phases of the research. Y.W. and G.H. helped with the experimental design of UPS buffering study in HeLa cells, and B.B. contributed to the design of the nutrient-sensing and selective vulnerability experiments. Y.W. and X.M. synthesized the UPS polymers; T.Z. designed and synthesized the always-ON/OFF-ON UPS nanoparticles. Y.L. performed the titration experiments and TEM characterization of UPS nanoparticles. C.W., Z.H. and R.J.D. designed the metabolomics experiments, and C.W. and Z.H. performed the studies and analysed the data. C.W. wrote the initial draft. R.J.D., M.A.W. and J.G. revised the final draft.

## Additional information

**Competing financial interests**: The authors declare no competing financial interests.

# 21

# Microbial metabolomics in open microscale platforms

Layla J. Barkal[1,2,*], Ashleigh B. Theberge[1,2,3,*], Chun-Jun Guo[4,*], Joe Spraker[5], Lucas Rappert[6], Jean Berthier[7], Kenneth A. Brakke[8], Clay C.C. Wang[4,9], David J. Beebe[1,2], Nancy P. Keller[6,10] & Erwin Berthier[1,2]

The microbial secondary metabolome encompasses great synthetic diversity, empowering microbes to tune their chemical responses to changing microenvironments. Traditional metabolomics methods are ill-equipped to probe a wide variety of environments or environmental dynamics. Here we introduce a class of microscale culture platforms to analyse chemical diversity of fungal and bacterial secondary metabolomes. By leveraging stable biphasic interfaces to integrate microculture with small molecule isolation via liquid–liquid extraction, we enable metabolomics-scale analysis using mass spectrometry. This platform facilitates exploration of culture microenvironments (including rare media typically inaccessible using established methods), unusual organic solvents for metabolite isolation and microbial mutants. Utilizing Aspergillus, a fungal genus known for its rich secondary metabolism, we characterize the effects of culture geometry and growth matrix on secondary metabolism, highlighting the potential use of microscale systems to unlock unknown or cryptic secondary metabolites for natural products discovery. Finally, we demonstrate the potential for this class of microfluidic systems to study interkingdom communication between fungi and bacteria.

[1] Department of Biomedical Engineering, University of Wisconsin-Madison, Madison, Wisconsin 53705, USA. [2] Carbone Cancer Center, University of Wisconsin-Madison, Madison, Wisconsin 53705, USA. [3] Department of Urology, University of Wisconsin-Madison, Madison, Wisconsin 53705, USA. [4] Department of Pharmacology and Pharmaceutical Sciences, University of Southern California, Los Angeles California 90089, USA. [5] Department of Plant Pathology, University of Wisconsin-Madison, Madison, Wisconsin 53705, USA. [6] Department of Medical Microbiology and Immunology, University of Wisconsin-Madison, Madison, Wisconsin 53705, USA. [7] Department of Biotechnology, CEA-University Grenoble-Alpes, 17 Avenue des Martyrs, 38054 Grenoble, France. [8] Department of Mathematics, Susquehanna University, Selinsgrove, Pennsylvania 17870, USA. [9] Department of Chemistry, University of Southern California, Los Angeles California 90089, USA. [10] Department of Bacteriology, University of Wisconsin-Madison, Madison, Wisconsin 53705, USA. * These authors contributed equally to this work. Correspondence and requests for materials should be addressed to E.B. (email: erwin.berthier@gmail.com).

Microbial secondary metabolism is an incredibly complex source of bioactive compounds that have important implications for human, animal and plant health. Filamentous fungi, in particular, produce secondary metabolites that are key virulence determinants of human and plant disease[1,2], a prominent threat to food and feed supplies[3,4], and a rich source of therapeutic compounds[5]. As only a small fraction of the potential fungal metabolite pool has been discovered, the identification of novel fungal compounds is the focus of much current interest[6,7]. However, studying fungal secondary metabolites is particularly challenging because they are often produced in response to very specific environmental cues (temperature, available nutrients, signals from nearby organisms) to provide a competitive advantage to the fungus[8-10]. Interactions with adjacent organisms, including bacteria and insects (so-called multikingdom interactions), can also have a significant impact on fungal secondary metabolite production[9,10]. Indeed, genome sequencing of hundreds of fungi has found innumerable cryptic secondary metabolite clusters not expressed under traditional laboratory growth conditions[11-13]. Identification of the environmental inducers of these clusters is, however, limited by current fungal culture and metabolite isolation tools that do not allow a simple and time-efficient exploration of the wide range of culture conditions representative of those found in nature.

Traditional methods of fungal culture use flasks or petri dishes, which require large volumes of reagents, incubator or shaker space and significant processing time. An important and time-consuming step in metabolomic analysis is sample preparation to remove matrix effects from culture media (such as salts, proteins and cell debris) that could impact downstream liquid chromatography-mass spectrometry (LC-MS) results. This is most commonly achieved by either solid-phase or liquid–liquid extraction, but these processes are time consuming and manually intensive as solid cultures must first be homogenized while liquid cultures are typically centrifuged or filtered. Liquid–liquid extraction is also imprecise as the immiscible phase extraction step is a highly serial process, with each pipetting step requiring precise selection of the location of the fluid interface. As such, there is a need for a culture method that has a small physical footprint, is efficient to use and is compatible with a passive, reproducible small molecule extraction process.

Here we address these obstacles by presenting a microscale platform that simplifies and accelerates the workflow of secondary metabolism studies, allows the exploration of a larger spectrum of microenvironmental cues, and brings salient features of microscale platforms (surface/volume ratios, segregated culture chambers, matrix design) to bear on microbial metabolomics research. We leverage microfluidic interfaces[14] to create an open biphasic system in which organic solvent is guided over a microbial culture environment allowing for the integrated and passive extraction of metabolites. These methods build on advances made in open and suspended microfluidics (microfluidics in channels that have any number of open interfaces) that have demonstrated unique advantages for mammalian cell culture and metabolomics[15]. The open microscale culture and extraction technologies presented here are intended to demonstrate simple microbial culture and extraction concepts that can integrate with many other microfluidic methods that provide enhancements in the concentration of secreted factors, the use of rare samples, and the creation of physiologically relevant in vitro models[16-18]. We used these devices to perform multidimensional arrayed experiments, varying culture conditions and metabolite isolation conditions, and demonstrate that efficiency of extraction from microscale is sufficient for LC-MS analysis. Using fungi well known for their rich secondary metabolism, Aspergillus nidulans (A. nidulans) and Aspergillus fumigatus (A. fumigatus), as model systems, we demonstrate the potential of a combinatorial approach to metabolite extraction and microbial culture for microbial metabolomics. Finally, we designed an integrated coculture and extraction platform that is capable of performing cultures of fungi and bacteria and further enables studies of chemical interactions between kingdoms.

## Results and discussion

**Micrometabolomics platform.** We engineered a microscale metabolomics platform that satisfies requirements of microbial culture and solvent flow (known as spontaneous capillary flow[15,19]) in open microfluidic channels. Using this platform, we demonstrate the concept of microscale microbial metabolomics (Fig. 1). In the micrometabolomics device, a micro-agar pad or liquid well is used to culture the microorganisms within an open microfluidic channel (for example, using a teardrop-shaped channel as exemplified in Fig. 2a). Metabolite extraction is simple to perform with a pipette; the open microfluidic channel is designed to direct the flow of solvent over the aqueous culture areas and form stable biphasic interfaces (Fig. 2b). This surface tension-based stability of the aqueous component is confirmed by numerical modelling using the Evolver software[20]

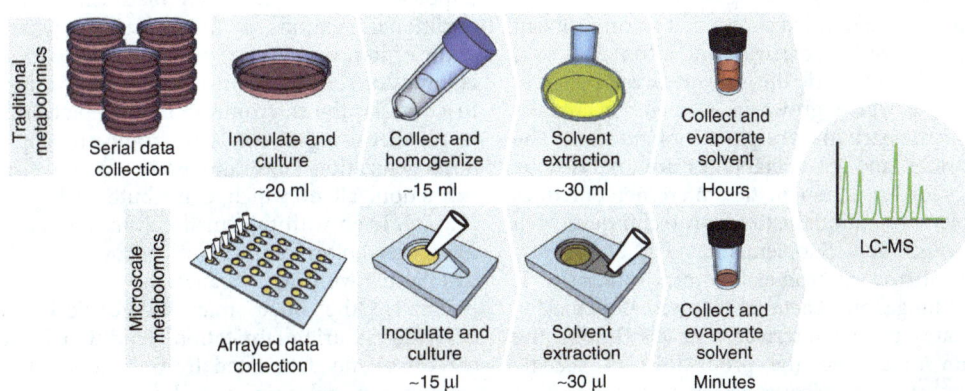

**Figure 1 | The micrometabolomics platform workflow is simpler, faster, and takes up less space than traditional metabolite extraction.** Traditional fungal metabolomics workflows (top) require serial inoculation of cultures, collection and homogenization of the sample, extraction of metabolites with solvent, and finally evaporation of that solvent prior to analysis. The microscale workflow (bottom) allows for arrayed inoculation and on-chip metabolite extraction without the need for culture collection and homogenization. Besides the streamlined process, the micrometabolomics platform also uses ~1000 × less solvent, which cuts down on evaporation time and makes the workflow faster.

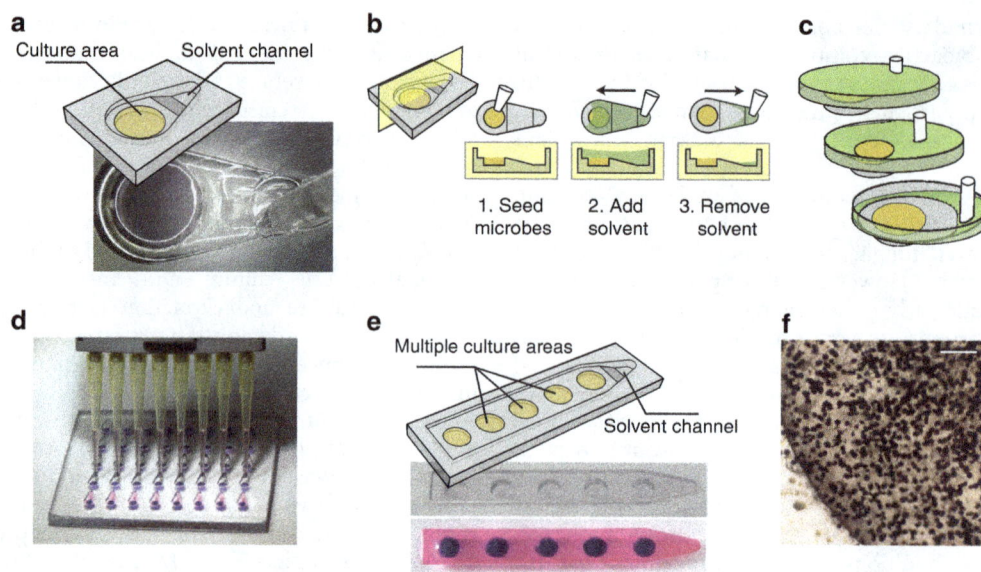

**Figure 2 | Extraction at microscale. (a)** The micrometabolomics device is comprised of a central culture well with an overlying pipet-accessible solvent channel. **(b)** The device is operated in three simple steps. **(c)** Simulations of fluid flow in the platform demonstrate that solvent removal does not disturb the liquid culture underneath. **(d)** The devices are arrayable and compatible with a multichannel pipette. **(e)** The extraction module can also be integrated with a platform for pooled extraction. **(f)** Sporulating culture of *A. fumigatus* overlayed on solid GMM agar grown in the micrometabolomics platform for 2 days. Scale bar, 250 μm.

(Fig. 2c; Supplementary Fig. 1), as well as experimentally (Supplementary Movie), provided the solvent is less dense than the aqueous media. An important factor contributing to the stability is the radially symmetrical covering of the aqueous interface. To facilitate symmetrical covering, we designed a device that drives filaments of solvent around the aqueous well before the solvent–aqueous interface is initiated (Supplementary Fig. 1A); this improves the evenness of solvent covering and prevents destabilizing perturbations. Once formed, the biphasic interface enables the extraction of metabolites based on preferential partitioning into the organic phase (Fig. 2b). At the end of the extraction, the solvent is collected in a second simple pipetting step; because the location of the liquid–solvent interface is precisely controlled, this step is repeatable, arrayable, automatable and does not require an additional processing step (for example, centrifugation). Importantly, the open microfluidic design allows the retrieval of solvent without carrying any aqueous media with it—an essential condition for sample preparation. The stability and integrity of the aqueous fluid was also validated by numerical simulation using the Evolver software (Fig. 2c; Supplementary Fig. 1B), which shows that during solvent retrieval, the solvent level decreases and the aqueous compartment becomes domed until the solvent breaks around an eye of aqueous media which grows in diameter eventually allowing the media to settle back in its original configuration. The solvent is then evaporated and the metabolites are analysed by LC-MS. The efficiency of extraction in the micrometabolomics platform is comparable to traditional extraction techniques such as vortexing or homogenization (Supplementary Fig. 2).

The open nature of the micrometabolomics platform is especially enabling for fungal and bacterial cultures. Biologically, the open surface creates an air interface that conditions the sporulation of certain fungi and the production of specific secondary metabolites[21,22]. *A. fumigatus* grown in the device at the air interface retained its expected morphology (Fig. 2f). In addition, the open design avoids the external pumping methods common among most other microfluidic devices that would make screening experiments nearly impossible and fabrication challenging, effectively preventing widespread integration into biology labs. The concepts of open micrometabolomic methods developed in this work represent the simplest form of open-culture devices (Fig. 2a) and have been used to demonstrate the potential of microscale devices for combined culture and metabolomic analysis. These techniques can also be extended to more complex segregated cocultures (see below) or devices for pooled extraction (Fig. 2e) that would allow for an efficient screening strategy or analyses that require larger amounts of metabolites extracted while maintaining the microscale culture geometries.

The design of the micrometabolomics platform is particularly useful for experimental spaces involving many micro-environmental conditions, time points and strains. The micrometabolomics platform benefits from the small size of each device (Supplementary Fig. 3A; Supplementary Data 1), which makes it possible to array devices for a simpler, more systematic workflow. A set of 30 devices fits easily into the footprint of a double-width microscope slide (Fig. 2d), whereas a stack of 30 petri dishes occupies a shelf of an incubator. Perhaps more important is that the open platform is fast and easy to use; traditional methods of fungal secondary metabolite extraction require samples to be processed individually and use large volumes of solvent, both of which lead to long processing times. In contrast, the micrometabolomics platform processes samples in parallel using a simple micropipette (Fig. 2d), eliminates the homogenization step, and uses ∼1,000× less solvent for the extraction, all of which contribute to a much faster workflow (Fig. 1). Even with the small culture and extraction volumes, the micrometabolomics platform recovers enough material to be compatible with LC-MS analysis.

Finally, the open microfluidic design confers significant advantages during fabrication. Traditional microfluidic platforms are commonly fabricated from materials that are incompatible with many solvents and known to sequester hydrophobic small molecules[23,24]. Open designs allow the use of a wider range of fabrication techniques (for example, injection molding) and of materials as bonding is not required[25]. Furthermore, by remaining completely open on top, the micrometabolomics device can be treated using deposition techniques, such as coating with Parylene C, to render the device material solvent resistant[26]. When similar deposition coating is performed on closed microfluidics, the treated surface is uneven throughout the device[27].

**Metabolite profile depends on extraction solvent.** The ability to use a large range of solvents for the extraction process allows the study of diverse chemical structures[28-30]. To demonstrate the potential of our open microscale extraction system, we tested how varying the extraction solvent affected metabolite profiling from fungal cultures. We cultured *A. nidulans*—a fungal species with ~40% of its secondary metabolome characterized[31]—on glucose minimal media (GMM)[32] both in the micrometabolomics platform and at macroscale (conventional petri dish culture with flask-based extraction). We then extracted metabolites using three different solvents, chloroform, 1-pentanol, and γ-caprolactone, which were chosen to cover a range of polarities (Fig. 3a; Supplementary Table 1). As solvent volatility is less important in the microscale platform, we were able to use high boiling point solvents, pentanol (bp = 138 °C) and γ-caprolactone (bp = 219 °C), not typically employed in fungal secondary metabolite extraction. The chromatograms of microscale cultures extracted with the different solvents had visible differences (Supplementary Fig. 4), and to get at these changes, features extracted from the chromatograms were compared using principal component analysis (PCA). As untargeted metabolomics typically yields a large number of unidentifiable features, we first narrowed the comparison: only

features that could be putatively annotated as secondary metabolites based on exact mass (error <10 p.p.m.) when compared with databases of known *A. nidulans* compounds were used in the PCA (Table 1; Fig. 3b). Of the ~1,000 features observed by LC-MS, 33 features could be putatively annotated based on databases of known *A. nidulans* compounds. These 33 features correspond to 19 putative metabolites; as is commonly observed in metabolomics data, a single metabolite may form multiple adducts (for example, features 14–16 were annotated as three different diorcinol adducts). When considering the reduced data set of 33 features, samples extracted with each of the different solvents are well separated in the PCA indicating the efficiency of secondary metabolite extraction differs by solvent (Fig. 3a). The same separation of solvents is observed at macroscale (Supplementary Fig. 5; Supplementary Table 2), though the time required to run macroscale experiments with low-volatility solvents was markedly longer (requiring 20-fold longer solvent evaporation times) than in microscale. We also performed PCA on the global metabolite profiles (which contain ~1,000 features) and found that the separation based on solvent persists at both micro and macroscale (Supplementary Fig. 6).

The impact of solvent choice is even more apparent at the level of single features (Fig. 3c). The loadings plot shows the weights of each normalized feature in calculating principal component (PC) 1 and PC 2; in general, if a feature is in the same quadrant on the loadings plot as the sample is in the PCA plot, it is enriched in that sample. Feature 4, for example, was putatively identified as asperfuranone, a polyketide[33], and was extracted almost exclusively in γ-caprolactone. This is demonstrated not only by

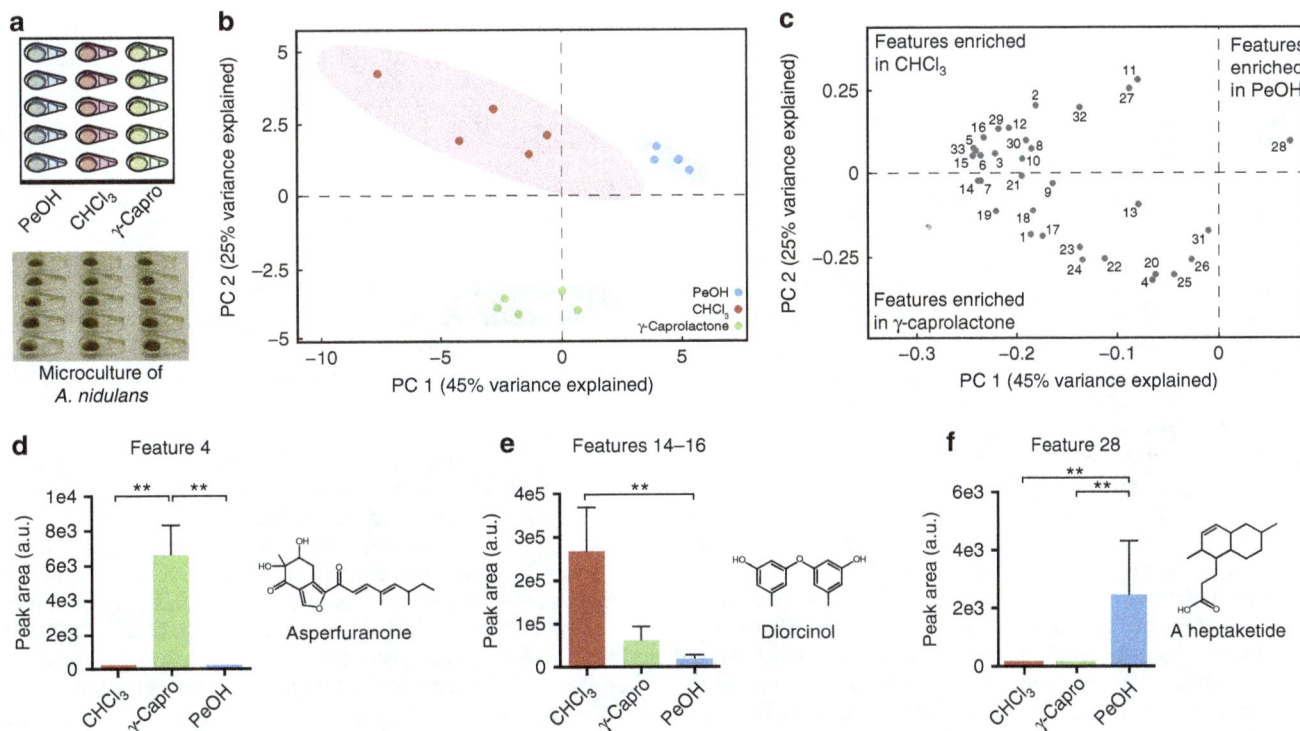

**Figure 3 | Solvent selection impacts extraction of secondary metabolites.** (**a**) Experimental design and culture photo. PeOH is 1-pentanol, CHCl₃ is chloroform and γ-Capro is γ-caprolactone. (**b**) Principal component analysis of *A. nidulans* cultured in the micrometabolomics platform. Only features that could be annotated as known secondary metabolites were used for clustering (Table 1). Each dot represents one of five independent cultures per condition from one experiment and the shaded ellipses represent 95% confidence intervals. Variance explained refers to the amount of total variation observed between samples that can be attributed to segregation along that principal component. (**c**) Loadings plot of individual features for the PCA in **b**. (**d–f**) Peak areas (integrated peak intensities, arbitrary units (a.u.)) of three of the features numbered in **c**. Error bars represent s.d. of the five replicates and statistics were performed using the Kruskal–Wallis test as described in the methods; **P value <0.01. Peak areas for features 14–16 were summed as they are adducts of the same compound. Structures are of the putative annotation for each peak.

**Table 1 | Putative annotations of peaks isolated from *A. nidulans* culture on GMM agar subsequently extracted with chloroform, γ-caprolactone or pentanol.**

| ID | Annotation | m/z | Adduct | Error |
|---|---|---|---|---|
| 1 | Cordycepin | 501.194 | [2M − H] − | 3.88 |
| 2 | Cordycepin | 286.070 | [M + Cl] − | 5.88 |
| 3 | Arugosin G | 537.255 | [M + FA − H] − | 9.73 |
| 4 | Asperfuranone | 377.157 | [M + FA − H] − | 8.43 |
| 5 | Aspoquinolone A/B | 502.165 | [M + K − 2H] − | 2.27 |
| 6 | Austinol intermediate (C25H30O7)* | 441.192 | [M − H] − | 0.49 |
| 7 | Dehydroaustinol | 493.131 | [M + K − 2H] − | 7.81 |
| 8 | Dehydroaustinol | 501.177 | [M + FA − H] − | 1.67 |
| 9 | Dehydroaustinol | 491.149 | [M + Cl] − | 1.76 |
| 10 | Dehydroaustinol | 455.171 | [M − H] − | 0.63 |
| 11 | Dehydrocitreoisocoumarin or 2-acetoacetyl T4HN | 137.024 | [M − 2H] − | 1.60 |
| 12 | Desacetylaustin or austinol | 457.187 | [M − H] − | 1.22 |
| 13 | Desacetylaustin or austinol | 493.164 | [M + Cl] − | 0.65 |
| 14 | Diorcinol | 229.085 | [M − H] − | 6.70 |
| 15 | Diorcinol | 459.178 | [2M − H] − | 6.26 |
| 16 | Diorcinol | 505.182 | [2M + FA − H] − | 9.59 |
| 17 | Emericellamide A | 646.354 | [M + K − 2H] − | 6.66 |
| 18 | Emericellamide A | 644.380 | [M + Cl] − | 0.05 |
| 19 | Emericellamide C/D | 640.393 | [M + FA − H] − | 0.66 |
| 20 | Emericellamide C/D | 630.365 | [M + Cl] − | 0.96 |
| 21 | Emericellamide C/D | 632.339 | [M + K − 2H] − | 6.08 |
| 22 | Emericellamide E/F | 658.395 | [M + Cl] − | 0.17 |
| 23 | Emericellamide E/F | 668.425 | [M + FA − H] − | 1.34 |
| 24 | Emericellamide E/F | 660.375 | [M + K − 2H] − | 0.60 |
| 25 | Emericellin | 393.172 | [M − H] − | 2.16 |
| 26 | Emericellin | 815.386 | [2M − H] − | 7.26 |
| 27 | Emodic acid | 336.977 | [M + K − 2H] − | 2.86 |
| 28 | A heptaketide (C15H24O2)† | 257.154 | [M + Na − 2H] − | 7.21 |
| 29 | Isoaustinone | 471.202 | [M + FA − H] − | 1.88 |
| 30 | Isoaustinone | 463.153 | [M + K − 2H] − | 0.39 |
| 31 | Nidulalin A or B | 301.072 | [M − H] − | 0.73 |
| 32 | Nidulol | 239.057 | [M + FA − H] − | 2.41 |
| 33 | Variecoxanthone A | 679.254 | [2M − H] − | 1.74 |

Compound ID numbers match Fig. 3c. *m/z* is mass to charge ratio detected by the mass spectrometer. Annotations were made by comparing observed masses with databases of known *A. nidulans* secondary metabolites. Annotations were only made if the error (difference between predicted and measured mass) was <10 p.p.m. Predicted adducts are compatible with the observed spectra and adducts annotated as the same compound eluted within 45 s of each other.
*Based on exact mass, the austinol intermediate could be neoaustinone, austinolide or 11β-hydroxyisoaustinone.
†Based on exact mass, the heptaketide formula could take multiple structures.

the areas of the extracted peaks (Fig. 3d), but also the fact that the feature falls clearly within the lower left-hand quadrant of the loadings plot, the same quadrant containing all the γ-caprolactone samples in the PCA (Fig. 3b,c). In contrast, features 14–16 were annotated as various adducts of diorcinol, an antibiotic secondary metabolite involved in fungal development[34], and were collectively extracted best in chloroform (Fig. 3c,e). Feature 28, annotated as a heptaketide[35], was the only known peak best extracted in pentanol which is why it segregates so strongly on the PCA loadings plot towards the quadrant with the pentanol samples (Fig. 3c,f).

These results demonstrate the power of alternative solvents to more fully extract the fungal secondary metabolome. To the best of our knowledge, γ-caprolactone has not been used as an extraction solvent in previous metabolomics studies, likely because its low volatility makes it a challenging solvent to use with traditional extraction methods. The micrometabolomics platform makes it feasible to use γ-caprolactone and other low-volatility solvents to explore segments of the metabolome that simply are not extracted when using more volatile solvents such as chloroform. The geometry of the platform can be simply modified to work with a range of solvents. For example, for solvents of higher volatility, a deeper solvent channel and the addition of a lid make solvents such as ethyl acetate feasible for extractions up to an hour long (Supplementary Fig. 3B;

Supplementary Data 2 and 3). The range of possible volatilities and the open nature of the platform, which allows for the deposition of solvent-protective coatings, render the micro-metabolomics device compatible with a myriad of solvents.

**Culture size impacts the metabolites produced.** In mammalian and bacterial cell culture, confinement in micro and nano-litre volumes of fluid leads to profound changes in soluble factor signalling and corresponding functional changes in cell behaviour[18]. These behavioural effects are typically induced by changes in the concentrations of autocrine and paracrine factors when the cell to culture volume ratio is reduced[16,36]. We thus aimed to identify the potential effect of culture geometry (well diameter and depth) on the fungal secondary metabolome. It is known that fungal secondary metabolism is affected by spatial factors, such as proximity to other fungi, which differ depending on culture geometry[3]. However, this effect is not often investigated due to the predominance of standard-sized cultureware, such as petri dishes.

We designed a panel of devices with different depths and diameters of the central culture well (Supplementary Fig. 3C; Supplementary Data 4) in which we analysed the landscape of metabolites produced by *A. nidulans* (Fig. 4a). The cultures were inoculated and extracted in proportion to the surface area of the

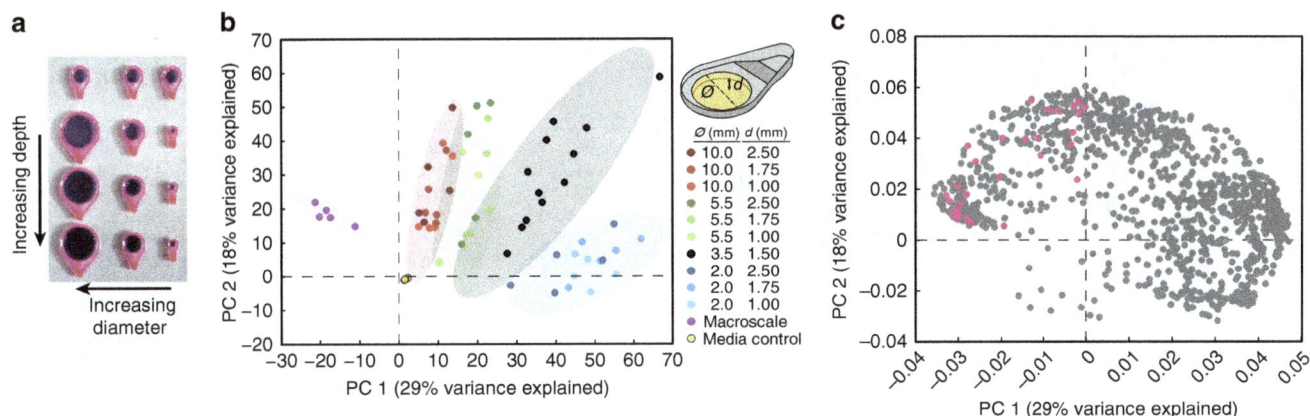

**Figure 4 | Global metabolite profiles segregate by well diameter but not well depth.** (**a**) Panel of devices with varying diameters and depths. (**b**) Principal component analysis of *A. nidulans* grown on GMM agar in wells of varying diameter and depth and extracted with PeOH. Legend values of diameter and depth are given in microns. 'Macroscale' refers to core samples of fungal growth on a 10-cm petri dish while 'Media control' refers to core samples of agar alone. Each dot represents one of five independent cultures per condition from one experiment and the shaded ellipses represent 95% confidence intervals for each of the different diameters or control conditions. (**c**) Loadings plot of individual features for the PCA in **b**. Each dot represents one feature. Features in pink were putatively annotated as *A. nidulans* secondary metabolites based on exact mass <10 p.p.m. error (Supplementary Table 3).

well, which was also used to scale feature intensity after LC-MS analysis. The PCA demonstrates a clear relationship between culture well diameter and global metabolite profile (Fig. 4b). As expected, this trend clusters the largest diameter microwells nearest to the macroscale culture. Interestingly, culture well depth had little impact on the global metabolite profiles. Features coloured pink in the loadings plot (Fig. 4c) correspond to features putatively annotated as *A. nidulans* secondary metabolites by exact mass (Supplementary Table 3). Interestingly, all of these compounds fall in the upper left hand quadrant of the loadings plot. This implies that despite scaling by the surface area of the well, these metabolites are produced to a greater extent in macroscale and large diameter microscale cultures; it is not simply that larger wells have proportionally more compound produced.

The spatial distribution of secondary metabolism throughout a fungal culture may explain the changes observed in different diameter microwells. Some secondary metabolites are made primarily in the spores of the fungus while others are made primarily in the hyphae[22,37,38]. In a point-inoculated culture, spores are produced in the centre of the colony with hyphae radiating towards the edges[39]. The change in the ratio of culture surface area to culture perimeter between the small and large wells could explain the differences in metabolite composition based on developmental stage of the fungus. There are a number of features primarily found in wells of small diameter that are not annotated as secondary metabolites. Some of these could be the uncharacterized *A. nidulans* secondary metabolites (which are estimated to comprise ∼60% of the *A. nidulans* metabolome) not produced in traditional macroscale culture[31]. The micrometabolomics platform moves beyond the limited geometries offered by standard petri dishes and can be adapted to include more complex shapes that could have a significant impact on secondary metabolite production.

**Culture of *A. fumigatus* on blood**. While fungi are typically considered opportunistic pathogens in humans and have not likely experienced evolutionary pressure towards producing metabolites that specifically target the human host, it has been shown that secondary metabolites play an important role during infection[40,41]. There are many characterized *A. fumigatus* secondary metabolites implicated in virulence including verruculogen (interacts with the epithelial lining of the respiratory tract[42]), endocrocin (inhibits neutrophil recruitment[40]), fumagillin (causes neutrophil toxicity[43]), hexahydroastechrome (unknown mechanism[44]) and gliotoxin (induces apoptosis in macrophages and inhibits phagocytosis and ROS production in neutrophils[41]). With the exception of gliotoxin and hexahydroastechrome, macroscale studies have not been published documenting whether these metabolites can be detected during pathogenesis; there is a gap between compound discovery and compound relevance to disease.

Using the micrometabolomics platform, a large panel of microenvironments can be tested, including rare and expensive matrices that better mimic conditions during disease. We performed an experiment to screen secondary metabolite production across three types of fungal inoculum, five types of fungal media, and two modes of culture all in triplicate (Fig. 5a). *A. fumigatus* was selected as it is the best characterized opportunistic fungal pathogen in the context of secondary metabolism. We chose different culture environments that represent chemical landscapes that can be found in the human lung, including the presence of blood cells and eicosanoids—fatty acid derivatives that are key in immune regulation. Synthetic eicosanoids are expensive, typically costing hundreds of dollars per 100 μg or requiring custom synthesis. The *ΔppoA* mutant has one of the three fungal cyclooxygenase-like enzymes deleted and the mutant with all three cyclooxygenase-like enzymes deleted is hypervirulent in a mouse model[45]. The media conditions supplemented with eicosanoids were selected with the thought that they might differentially impact the *ΔppoA* mutant.

When wild type (WT) *A. fumigatus* grown in solid culture on blood is compared with growth on solid GMM, there are evident changes in its global metabolite profile (Fig. 5b). In particular, zone 1 contains 56 features that are produced primarily when *A. fumigatus* is grown on blood. The 66 features in zone 2 are produced by *A. fumigatus* on both blood and GMM, and the 183 features in zone 3 are produced primarily when *A. fumigatus* is grown on GMM. Features in zones 1–3 were compared with databases of known *A. fumigatus* secondary metabolites and annotated by exact mass (error <10 p.p.m.) (Fig. 5c). Of these putative annotations, one metabolite from each zone was selected for further analysis. The peak areas for each feature support its zone classification, with triacetylfusarinine C produced primarily when the fungus is grown on blood, endocrocin (a spore

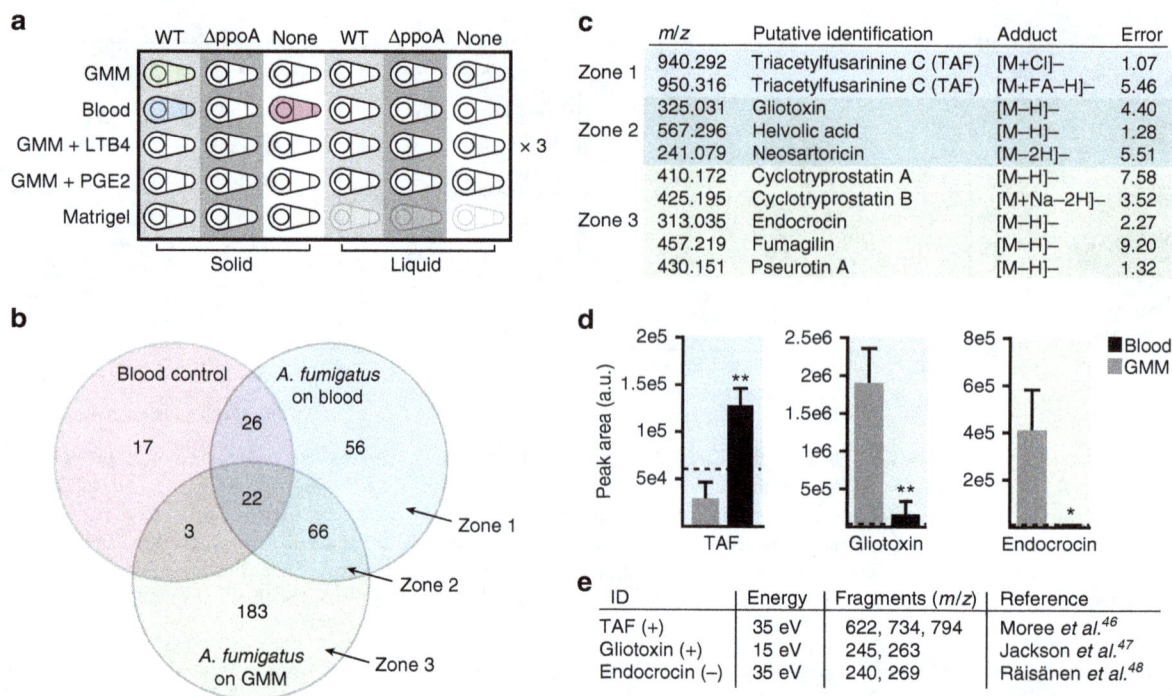

**Figure 5 | A. fumigatus metabolite production varies when grown on blood as compared to GMM. (a)** Experimental conditions tested in the micrometabolomics platform in triplicate. Data for all three independent cultures from one experiment are shown in **b–e**. **(b)** Overlap of global metabolite profiles extracted with PeOH from blood, A. fumigatus grown on blood, and A. fumigatus grown on GMM. Peaks were extracted and aligned using XCMS online. Numbers represent peaks unique in m/z value and retention time. **(c)** Putative annotation of peaks based on exact mass of known A. fumigatus secondary metabolites. Error given in p.p.m. Adducts are compatible with the observed spectrum and colour corresponds to peak location in **b**. **(d)** Peak areas for three putative secondary metabolites produced by A. fumigatus grown on either blood or GMM. The peak areas for both TAF adducts were summed. The dotted line is the peak threshold of 60,000 below which is considered noise. Error bars represent s.d. of three microchannels. P values were calculated using unpaired Student's t-tests: *P value < 0.05, ** P value < 0.005. **(e)** LC-MS/MS of the three peaks with putative IDs in **d**. Replicates were pooled prior to LC-MS/MS. TAF and gliotoxin were analysed in positive mode, endocrocin in negative mode.

metabolite) produced primarily when the fungus is on GMM, and gliotoxin produced on either medium (Fig. 5d). The putative IDs of these three features were confirmed by tandem mass spectrometry (LC-MS/MS) (Supplementary Table 4); retention times and fragments of gliotoxin and endocrocin matched purchased standards (retention time difference < 30 s; fragment m/z difference < 0.05) and fragments of all three compounds matched the published literature (Fig. 5e)[46–48].

Features in zone 1 are produced only when A. fumigatus is grown on blood. These features are of significant interest as they could be more relevant to fungal virulence than features present solely on GMM agar. The only feature that could be annotated in zone 1 is triacetylfusarinine C (Fig. 5d). The microenvironment of blood induced production of this siderophore (iron binding molecule)[49], which was otherwise produced only at low levels. The other treatments in this screen yielded few significant differences: the ΔppoA mutant was similar to WT and the supplemented GMM conditions were similar to GMM alone. The micrometabolomics platform allowed us to efficiently test culture conditions and going forward, the ease of using this platform will enable experiments with more biologically accurate microenvironments to help uncover secondary metabolites with relevance to disease progression, biomarker diagnosis and therapeutic discovery that would otherwise be hidden.

**Coculture of fungi and bacteria.** Interkingdom communication (for example, fungal-bacterial) is prevalent and causes changes in secondary metabolite production as a defence mechanism or in reaction to stress[13]. For example, culture with B. gladioli causes R. microsporus to produce enacyloxin antibiotics[50] and the

coculture of Sphingomonas and A. fumigatus isolates leads to the production of glionitrin A[51]. Traditional cocultures are performed either by completely mixing the culture or by inoculating a plate with the two organisms side by side. In a mixed culture, both organisms must be able to grow in the same media and it can be difficult to separate physical effects from soluble factor effects. In a side by side culture, there is significant spatial heterogeneity; microbes in each culture are of differing distance from the other culture. We extended the micrometabolomics platform to allow culture and extraction of two organisms in soluble communication (Supplementary Fig. 3D; Supplementary Data 5) to address these challenges.

The coculture device is made up of two micrometabolomics platforms placed on each opposing face of a thermoplastic layer and connected via pores in the bottom of each culture well (Fig. 6a). This allows for diffusion of secreted chemical factors between the two cultures while keeping the bulk of each culture separate. The cultures are both exposed to air, have a nearly uniform distance from each other, and can be grown on different media in the same device. To characterize the contact between the two culture wells, diffusion of a fluorescent dye, Alexa 488, across the two agar pads was measured over time. Over 3 h, there is clear movement of the fluorophore from the bottom compartment where it was applied, through the agar, to the top compartment (Fig. 6b). This indicates that the coculture device allows compounds to diffuse between the cultures on a biologically relevant timescale and can be extracted on each side of the platform.

To test the feasibility of using the coculture device to study interkingdom interactions, we cultured fungi together with

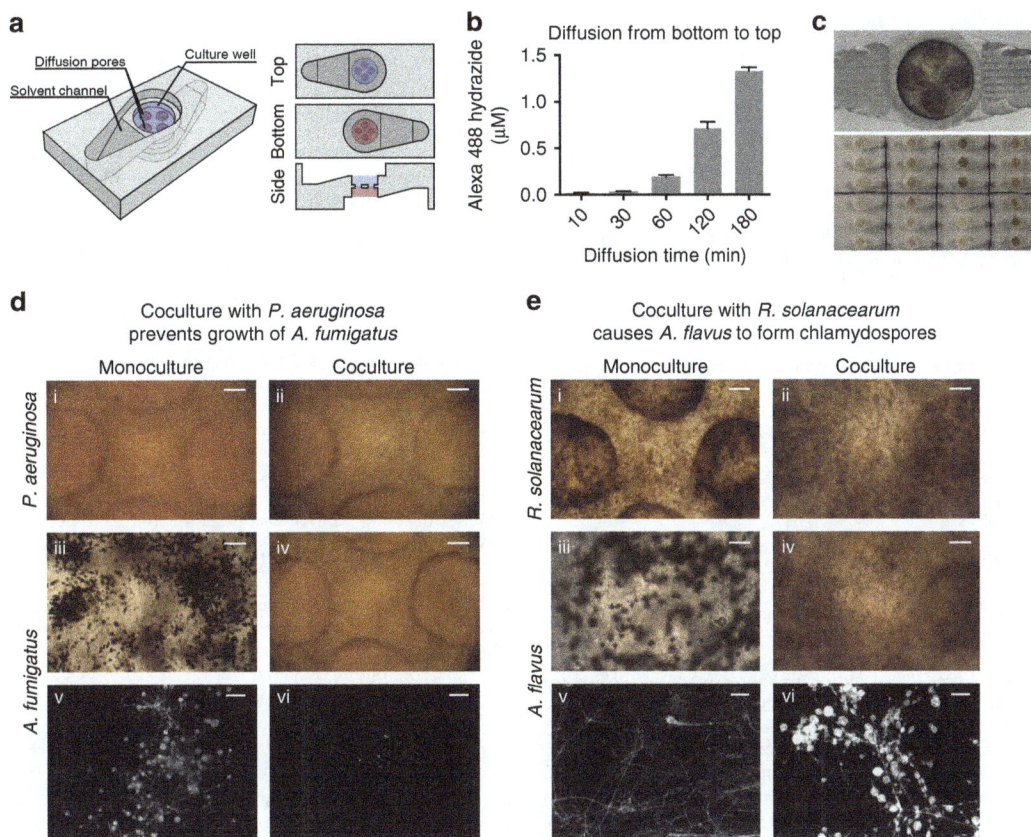

**Figure 6 | A microscale coculture platform that enables segregated analysis of interkingdom communication. (a)** The coculture platform is comprised of two micrometabolomics devices in diffusion contact via four pores in the floor of each culture well. **(b)** Diffusion of a 25-µM solution of fluorescent dye, Alexa 488 hydrazide, through the agar culture pads into the upper solvent well is time dependent. Error bars represent s.d. of three devices. **(c)** Photos of bacterial and fungal growth within the coculture devices. **(d)** Monoculture and coculture of P. aeruginosa and A. fumigatus after 3 days at 30 °C shows that coculture prevents A. fumigatus growth. **(e)** Monoculture and coculture of R. solanacearum and A. flavus for 3 days at 30 °C shows that coculture causes A. flavus to generate chlamydospores. Images were taken at × 4 ((**d,e**) i–iv), scale bar, 250 µm, and a subset of wells were stained with calcafluor white and imaged at × 10 ((**d,e**) v–vi), scale bar, 25 µm. Images are representative of three culture wells.

bacteria, all of which were able to grow successfully in the devices (Fig. 6c). Besides monoculture of each organism, the human pathogen *Pseudomonas aeruginosa* was cocultured with *Aspergillus fumigatus*, and plant pathogen *Ralstonia solanacearum* (*R. solanacearum*) was cocultured with *Aspergillus flavus* (*A. flavus*) and *Fusarium sporotrichioides*, two common plant pathogenic fungi. Despite growing well in monoculture, *A. fumigatus* is unable to grow when in coculture with *P. aeruginosa* (Fig. 6d). In addition, *A. flavus* has a dramatic induction of chlamydospore formation when in coculture with *R. solanacearum* (Fig. 6e) as does *F. sporotrichioides* to a lesser extent (Supplementary Fig. 7). Chlamydospores are large, thick-walled cells that are induced by environmental stressors including bacterial metabolites[52]. The coculture platform enables simple phenotypic screening of multikingdom cocultures while the integrated solvent extraction channels make it possible to do chemical analysis of secreted metabolites in the two cultures. Taken together, the coculture extension of the micrometabolomics device opens up another important microenvironmental factor, signals from surrounding organisms, to ready manipulation.

## Conclusions

We introduce a microscale culture platform to study microbial secondary metabolism in response to engineered chemical, mechanical and geometrical microenvironmental cues as well as coculture with other organisms. Our open microculture platform

is compatible with liquid media or solid agar fungal culture methods, and provides the opportunity to incorporate mammalian or microbial cell types in coculture. Furthermore, the device is easily arrayed and can be used to screen microenvironmental factors that may have an effect on secondary metabolite production, including incorporation of biological samples such as blood, sputum, mucus, extracellular matrix components, etc. The microscale culture well makes it feasible to do large studies with rare samples, such as biological fluids or cells from asthma patients who experience exacerbations from inhaling fungal spores.

In addition, we developed an integrated, open, passive, biphasic metabolite extraction system that leverages surface tension forces to operate at microscale. The device is compatible with a wide range of solvents, giving users access to previously uninterrogated segments of the metabolome. The extraction process can be multiplexed and automated, providing increased efficiency for large parameter studies, while still providing sufficient recovery of metabolites to perform LC-MS. The platform has multiple strengths, including the identification of candidate metabolites and their required culture conditions prior to larger scale structural elucidation studies, the appearance or disappearance of peaks of interest in rare or biologically relevant matrices, or the identification of metabolite differences related to observable physiological changes of the microorganism. Our results demonstrate that the micrometabolomics platform can be used to investigate how the fungal microenvironment influences metabolite production and set the stage to use the platform as a fast and simple way to probe secondary metabolites hidden within cryptic gene clusters.

## Methods

**Device fabrication.** Devices were micromilled as previously described[53]. Briefly, polystyrene sheets of 3 mm thickness were purchased from Goodfellow (#ST313300). Features were cut out of the polystyrene using a CNC micromilling machine (Tormach pCNC 700). Devices used for testing different solvents were coated with Parylene C (CAS #28804-46-8) using a Specialty Coating Systems PDS 2010 vacuum deposition system; 25 g Parylene C dimer was used to get a 15-μm thick coating.

**Coculture device characterization.** Diffusion in the coculture device (Fig. 6b) was assessed by filling the culture compartments with a 7% low melt agarose gel. Alexafluor 488 hydrazide (Invitrogen) was diluted to 25 μM in PBS and applied to the solvent well of the bottom compartment. The device was turned over and PBS was added to the top solvent well. Periodically, the PBS in the top well was removed for fluorescence analysis using a BMG Pherastar Multimode plate reader with excitation at 485 nm and emission at 520 nm.

**Fungal culture.** Strains of *A. nidulans* (WT, RDIT 9.32)[54] and *A. fumigatus* (WT, Af293; *ΔppoA*, TDWC 1.13)[55] were maintained as glycerol stocks. They were activated by culture on solid GMM[32] made with low-gelling temperature agar (CAS #39346-81-1) for 4 days at 37 °C at which point conidia were collected in 0.01% Tween-80.

For control macroscale cultures, 200 μl spores at a concentration of 10e5 spores per ml was spread over the surface of GMM in a 10-cm petri dish. For microscale cultures, the culture well was filled with 20 μl media and 4 μl spores were spread at a concentration of 10e4 spores per ml over the top. Blood agar was made using 96% human blood (Bioreclamation LLC) with 4% low-gelling temperature agar or PBS for solid and liquid cultures, respectively. Matrigel media was made with 50% Basement Membrane Matrix (BD biosciences) and 50% GMM without agar. Where indicated, conditions supplemented with eicosanoids were made by adding 1 μM LTB$_4$ or PGE$_2$ (Cayman Chemical) to the spores prior to inoculation. *A. nidulans* cultures were grown at 30 °C and *A. fumigatus* cultures were grown at 37 °C for 3–4 days in a humidified chamber to prevent evaporation.

**Fungal and bacterial coculture.** *P. aeruginosa* (PA 14, from Dr Yun Wang, Northwestern University) and *A. fumigatus* (Af293, from CBS, Centraalbureau voor Schimmelcultures Fungal Biodiversity Centre of the Royal Netherlands Academy of Arts and Sciences) in both mono- and coculture were grown on potato dextrose agar (PDB from BD, Franklin Lakes, NJ, mixed with low melt agar). *R. solanacearum* (GMI1000, from ATCC, Manassas, VA), *A. flavus* (NRRL 3357, from CBS) and *F. sporotrichioides* (from University of Wisconsin-Madison, Department of Plant Pathology Teaching Lab) were all grown in mono- or coculture on ISP2 media[56]. *P. aeruginosa* and *R. solanacearum* cultures were inoculated with 2 μl overlay of 10e8 bacteria per ml. *A. fumigatus* and *A. flavus* cultures were inoculated with 2 μl overlay of 10e6 spores per ml. *F. sporotrichioides* was inoculated with a punch of macroscale culture; a wide orifice pipet tip was used to transfer consistent amounts of fungal culture from macroscale to the microscale devices. Monoculture studies were done by inoculating the same organism in both the top and bottom culture wells. Cultures were grown at 30 °C for 3 days in a humidified chamber to prevent evaporation.

In coculture experiments, fungal cell walls were stained (Fig. 6d,e v, vi) by adding 10 μl calcofluor white (1 mg ml$^{-1}$) to the wells 1 min prior to microscopy. Cultures were collected from coculture devices by cutting the edges of the agar pads using a sterile 18 g needle and aspirating into 250 μl large orifice pipet tips (USA Scientific). Cultures were wet mounted using another 10 μl ddH$_2$O and imaging was done using a Zeiss Axio Imager A10 microscope (Carl Zeiss, Oberkochen, Germany) equipped with a Zeiss A Plan- × 10 lens, a series 120 X-Cite light source (EXFO), and a DAPI excitation/emission filter set. Images were collected with AxioVision Release 4.7 software (Carl Zeiss).

**Metabolite extraction.** Metabolites in the macroscale samples were extracted by removing 10 cores of 10 mm diameter from the culture and homogenizing them in 0.01% Tween-20. The homogenized suspension was removed to a glass vial and a volume of solvent, 2.449 ml, proportional to the culture surface area of the cores was added. Samples were agitated every 5 min for 30 min after which they were centrifuged at 2,500 r.p.m. (608g) for 10 min. The solvent layer was removed to a fresh vial for evaporation as below. Metabolites were extracted from the microscale cultures by pipetting solvent into the tapered end of the teardrop channel. Solvent volumes used were proportional to the surface area of the culture; for standard 3.5 mm diameter microscale cultures, 30 μl solvent was used. The metabolites were allowed to passively diffuse into the solvent layer for 30 min after which the solvent was collected into glass vials. Samples were evaporated to dryness (1–4 h) in a vacuum concentrator (Thermo Express SC250EXP) without heat, except samples

extracted with γ-caprolactone that were heated to 37 °C for the first hour of evaporation.

**Metabolite analysis.** Extracts were re-dissolved in 50 μl of 20% DMSO/MeOH and centrifuged at 10,000g for 10 min. Sample order was randomized and a 10 μl portion was examined by high performance liquid chromatography with diode-array detection (HPLC-DAD) and MS or MS/MS analysis. HPLC-DAD-MS was done using an Agilent 6210 TOF LC-MS. The same reverse-phase C18 column (Alltech Prevail C18 2.1 mm by 100 mm with a 3-μm particle size) was used for all samples at a flow rate of 125 μl per min. The solvent gradient was 95% acetonitrile (MeCN)/H$_2$O (solvent B) in 5% MeCN/H$_2$O (solvent A), both containing 0.05% formic acid, as follows: 0% solvent B from 0 to 5 min, 0 to 100% solvent B from 5 to 35 min, 100% solvent B from 35 to 40 min, 100 to 0% solvent B from 40 to 45 min and reequilibration with 0% solvent B from 45 to 50 min. Nitrogen was used as auxiliary and sheath gas. Source voltage was set at 4,000 V, nebulizer pressure at 20 psig, drying gas flow rate at 10 l per min and drying gas temperature at 350 °C.

HPLC-MS/MS was done using a ThermoFinnigan LCQ Advantage ion trap mass spectrometer. The solvent gradient was as described above for HPLC-MS. Source heater temperature was set to 0 °C, sheath flow rate was set to 49.22, current was set to 5.52 μA, voltage was set to 5.03 kV, capillary temperature was set to 275 °C and capillary voltage was set to − 15.74 V. The MS/MS fragmentation scheme, including collision energies, for confirmation of three putatively identified compounds (Fig. 5) is described in Supplementary Table 4.

**XCMS analysis.** Agilent.d files were converted to.mzXML files using ProteoWizard[57]. They were then grouped by condition and uploaded to XCMS online[58–60]. Multigroup analyses were performed using default HPLC/ UHD Q-TOF settings that have been optimized for an HPLC run with ∼ 60 min gradient and subsequent analysis using a high-resolution ESI-QTOF-MS. Polarity was specified to be negative. The XCMS multigroup analysis includes retention time correction, peak picking using the centWave algorithm, and peak grouping algorithms.

**Peak identification and PCA.** Data on each of the peaks was downloaded and imported into MATLAB. Peaks were removed from analysis if they had a q value, as calculated by XCMS, of <0.05 or were present in fewer than 66% of replicates of at least one condition in the experiment. Peaks were also removed if present in the solvent control or if their maximum intensity was <5,000 (60,000 for all peaks in the matrix experiment because of higher background in blood). Putative identification of peaks was made if the observed m/z matched the predicted m/z within 10 p.p.m. error when checking against the Reaxys database (version 2.20770.1, Elsevier Information Systems GmbH, Frankfurt, Germany) limited to secondary metabolites of *A. nidulans* or *A. fumigatus* depending on the experiment and an in-house database of secondary metabolites for these two fungal species. Putative identifications were discarded if proposed adducts were incompatible with those present in the spectrum. For adducts annotated as the same compound, retention times agreed within 45 s. Each peak was scaled to have unit variance and additional scaling by surface area was performed for experiments using devices with varying diameters. PCA was done using the PCA function in MATLAB with the singular value decomposition algorithm.

**Statistical analysis of metabolite levels.** Statistical analysis was performed using GraphPad Prism 6 software. To compare metabolite extraction efficiency with different solvents (Fig. 3d–f), we used the non-parametric Kruskal–Wallis test with Dunn's multiple comparison correction. To compare metabolite levels between two different growth conditions, Student's *t*-test was used (Fig. 5d).

## References

1. Woloshuk, C. P. & Shim, W. B. Aflatoxins, fumonisins, and trichothecenes: a convergence of knowledge. *FEMS Microbiol. Rev.* **37**, 94–109 (2012).
2. Friesen, T. L., Faris, J. D., Solomon, P. S. & Oliver, R. P. Host-specific toxins: effectors of necrotrophic pathogenicity. *Cell Microbiol.* **10**, 1421–1428 (2008).
3. Brown, S. H. *et al.* Oxygenase coordination is required for morphological transition and the host–fungus interaction of Aspergillus flavus. *Mol. Plant Microbe Interact.* **22**, 882–894 (2009).
4. Wu, F., Groopman, J. D. & Pestka, J. J. Public Health Impacts of Foodborne Mycotoxins. *Annu. Rev. Food Sci. Technol.* **5**, 351–372 (2014).
5. Evidente, A. *et al.* Fungal metabolites with anticancer activity. *Nat. Prod. Rep.* **31**, 617 (2014).
6. Kusari, S., Hertweck, C. & Spiteller, M. Chemical ecology of endophytic fungi: origins of secondary metabolites. *Chem. Biol.* **19**, 792–798 (2012).
7. Wiemann, P. & Keller, N. P. Strategies for mining fungal natural products. *J. Ind. Microbiol. Biotechnol.* **41**, 301–313 (2014).
8. Yin, W. & Keller, N. P. Transcriptional regulatory elements in fungal secondary metabolism. *J. Microbiol.* **49**, 329–339 (2011).
9. Scherlach, K., Graupner, K. & Hertweck, C. Molecular bacteria-fungi interactions: effects on environment, food, and medicine. *Annu. Rev. Microbiol.* **67**, 375–397 (2013).

10. Rohlfs, M. & Churchill, A. C. Fungal secondary metabolites as modulators of interactions with insects and other arthropods. *Fungal Genet. Biol.* **48**, 23–34 (2011).

11. Grigoriev, I. V. *et al.* Fueling the future with fungal genomics. *Mycology* **2**, 192–209 (2011).

12. Brakhage, A. A. Regulation of fungal secondary metabolism. *Nat. Rev. Microbiol.* **11**, 21–32 (2013).

13. Bertrand, S. *et al.* Metabolite induction via microorganism co-culture: a potential way to enhance chemical diversity for drug discovery. *Biotechnol Adv.* **32**, 1180–1204 (2014).

14. Atencia, J. & Beebe, D. J. Controlled microfluidic interfaces. *Nature* **437**, 648–655 (2005).

15. Casavant, B. P. *et al.* Suspended microfluidics. *Proc. Natl Acad. Sci. USA* **110**, 10111–10116 (2013).

16. Domenech, M. *et al.* Cellular observations enabled by microculture: paracrine signaling and population demographics. *Integr. Biol. (Camb)* **1**, 267–274 (2009).

17. Young, E. W. & Beebe, D. J. Fundamentals of microfluidic cell culture in controlled microenvironments. *Chem. Soc. Rev.* **39**, 1036–1048 (2010).

18. Boedicker, J. Q., Vincent, M. E. & Ismagilov, R. F. Microfluidic confinement of single cells of bacteria in small volumes initiates high-density behavior of quorum sensing and growth and reveals its variability. *Angew. Chem. Int. Ed. Engl.* **48**, 5908–5911 (2009).

19. Berthier, J. *et al.* The dynamics of spontaneous capillary flow in confined and open microchannels. *Sensors Transducers* **183**, 1726–5479 (2014).

20. Brakke, K. The Surface Evolver Version 2.70. Available at http://www.susqu.edu/brakke/evolver/evolver.html(accessed on 25 August 2013).

21. Etxebeste, O., Garzia, A., Espeso, E. A. & Ugalde, U. Aspergillus nidulans asexual development: making the most of cellular modules. *Trends Microbiol.* **18**, 569–576 (2010).

22. Lim, F. Y., Ames, B., Walsh, C. T. & Keller, N. P. Co-ordination between BrlA regulation and secretion of the oxidoreductase FmqD directs selective accumulation of fumiquinazoline C to conidial tissues in Aspergillus fumigatus. *Cell Microbiol.* **16**, 1267–1283 (2014).

23. Lee, J. N., Park, C. & Whitesides, G. M. Solvent compatibility of poly(dimethylsiloxane)-based microfluidic devices. *Anal. Chem.* **75**, 6544–6554 (2003).

24. Toepke, M. W. & Beebe, D. J. PDMS absorption of small molecules and consequences in microfluidic applications. *Lab Chip* **6**, 1484–1486 (2006).

25. Berry, S. M., Alarid, E. T. & Beebe, D. J. One-step purification of nucleic acid for gene expression analysis via immiscible filtration assisted by surface tension (IFAST). *Lab Chip* **11**, 1747–1753 (2011).

26. Fortin, J. & Lu, T. M. *Chemical Vapor Deposition Polymerization* (Springer, 2004).

27. Ramachandran, A., Junk, M., Koch, K. P. & Hoffmann, K. P. A study of parylene C polymer deposition inside microscale gaps. *IEEE T. Adv. Packaging* **30**, 712–724 (2007).

28. Bi, H. *et al.* Optimization of harvesting, extraction, and analytical protocols for UPLC-ESI-MS-based metabolomic analysis of adherent mammalian cancer cells. *Anal. Bioanal. Chem.* **405**, 5279–5289 (2013).

29. Yang, Y. *et al.* New sample preparation approach for mass spectrometry-based profiling of plasma results in improved coverage of metabolome. *J. Chromatogr. A* **1300**, 217–226 (2013).

30. Chen, S. *et al.* Simultaneous extraction of metabolome and lipidome with methyl tert-butyl ether from a single small tissue sample for ultra-high performance liquid chromatography/mass spectrometry. *J. Chromatogr. A* **1298**, 9–16 (2013).

31. Andersen, M. R. *et al.* Accurate prediction of secondary metabolite gene clusters in filamentous fungi. *Proc. Natl Acad. Sci. USA* **110**, E99–E107 (2013).

32. Shimizu, K. & Keller, N. P. Genetic involvement of a cAMP-dependent protein kinase in a G protein signaling pathway regulating morphological and chemical transitions in Aspergillus nidulans. *Genetics* **157**, 591–600 (2001).

33. Chiang, Y. M. *et al.* A gene cluster containing two fungal polyketide synthases encodes the biosynthetic pathway for a polyketide, asperfuranone, in Aspergillus nidulans. *J. Am. Chem. Soc.* **131**, 2965–2970 (2009).

34. Butnick, N. Z., Yager, L. N., Hermann, T. E., Kurtz, M. B. & Champe, S. P. Mutants of Aspergillus nidulans blocked at an early stage of sporulation secrete an unusual metabolite. *J. Bacteriol.* **160**, 533–540 (1984).

35. Fujii, I., Watanabe, A., Sankawa, U. & Ebizuka, Y. Identification of Claisen cyclase domain in fungal polyketide synthase WA, a naphthopyrone synthase of Aspergillus nidulans. *Chem. Biol.* **8**, 189–197 (2001).

36. Domenech, M., Bjerregaard, R., Bushman, W. & Beebe, D. J. Hedgehog signaling in myofibroblasts directly promotes prostate tumor cell growth. *Integr. Biol. (Camb)* **4**, 142–152 (2012).

37. Gauthier, T. *et al.* Trypacidin, a spore-borne toxin from Aspergillus fumigatus, is cytotoxic to lung cells. *PLoS ONE* **7**, e29906 (2012).

38. Fetzner, R., Seither, K., Wenderoth, M., Herr, A. & Fischer, R. Alternaria alternata transcription factor CmrA controls melanization and spore development. *Microbiology* **160**, 1845–1854 (2014).

39. Gifford, D. R. & Schoustra, S. E. Modelling colony population growth in the filamentous fungus Aspergillus nidulans. *J. Theor. Biol.* **320**, 124–130 (2013).

40. Berthier, E. *et al.* Low-volume toolbox for the discovery of immunosuppressive fungal secondary metabolites. *PLoS Pathog.* **9**, e1003289 (2013).

41. Dagenais, T. R. & Keller, N. P. Pathogenesis of Aspergillus fumigatus in invasive Aspergillosis. *Clin. Microbiol. Rev.* **22**, 447–465 (2009).

42. Khoufache, K. *et al.* Verruculogen associated with Aspergillus fumigatus hyphae and conidia modifies the electrophysiological properties of human nasal epithelial cells. *BMC Microbiol.* **7**, 5 (2007).

43. Fallon, J. P., Reeves, E. P. & Kavanagh, K. Inhibition of neutrophil function following exposure to the Aspergillus fumigatus toxin fumagillin. *J. Med. Microbiol.* **59**, 625–633 (2010).

44. Yin, W. B. *et al.* A Nonribosomal Peptide Synthetase-Derived Iron(III) Complex from the Pathogenic Fungus Aspergillus fumigatus. *J. Am. Chem. Soc.* **135**, 2064–2067 (2013).

45. Tsitsigiannis, D. I. *et al.* Aspergillus cyclooxygenase-like enzymes are associated with prostaglandin production and virulence. *Infect. Immun.* **73**, 4548–4559 (2005).

46. Moree, W. J. *et al.* Interkingdom metabolic transformations captured by microbial imaging mass spectrometry. *Proc. Natl Acad. Sci. USA* **109**, 13811–13816 (2012).

47. Jackson, L. C., Kudupoje, M. B. & Yiannikouris, A. Simultaneous multiple mycotoxin quantification in feed samples using three isotopically labeled internal standards applied for isotopic dilution and data normalization through ultra-performance liquid chromatography/electrospray ionization tandem mass spectrometry. *Rapid Commun. Mass Spectrom.* **26**, 2697–2713 (2012).

48. Räisänen, R., Björk, H. & Hynninen, P. H. Two-dimensional TLC separation and mass spectrometric identification of anthraquinones isolated from the fungus Dermocybe sanguinea. *Z. Naturforsch C.* **55**, 195–202 (2000).

49. Nilius, A. M. & Farmer, S. G. Identification of extracellular siderophores of pathogenic strains of Aspergillus fumigatus. *J. Med. Vet. Mycol.* **28**, 395–403 (1990).

50. Ross, C., Opel, V., Scherlach, K. & Herweck, C. Biosynthesis of antifungal and antibacterial polyketides by Burkholderia gladioli in coculture with Rhizopus microsporus. *Mycoses* **57**, 48–55 (2014).

51. Park, H. B., Kwon, H. C., Lee, C. H. & Yang, H. O. Glionitrin A, an antibiotic-antitumor metabolite derived from competitive interaction between abandoned mine microbes. *J. Nat. Prod.* **72**, 248–252 (2009).

52. Li, L. *et al.* Induction of chlamydospore formation in Fusarium by cyclic lipopeptide antibiotics from Bacillus subtilis C2. *J. Chem. Ecol.* **38**, 966–974 (2012).

53. Guckenberger, D. J., de Groot, T. E., Wan, A. M., Beebe, D. J. & Young, E. W. Micromilling: a method for ultra-rapid prototyping of plastic microfluidic devices. *Lab Chip* **15**, 2364–2378 (2015).

54. Tsitsigiannis, D. I., Zarnowski, R. & Keller, N. P. The lipid body protein, PpoA, coordinates sexual and asexual sporulation in Aspergillus nidulans. *J. Biol. Chem.* **279**, 11344–11353 (2004).

55. Dagenais, T. R. *et al.* Defects in conidiophore development and conidium-macrophage interactions in a dioxygenase mutant of Aspergillus fumigatus. *Infect. Immun.* **76**, 3214–3220 (2008).

56. Atlas, R. M. *Handbook of Microbiological Media* 4th edn (CRC Press, 2010).

57. Chambers, M. C. *et al.* A cross-platform toolkit for mass spectrometry and proteomics. *Nat. Biotechnol.* **30**, 918–920 (2012).

58. Smith, C. A., Want, E. J., O'Maille, G., Abagyan, R. & Siuzdak, G. XCMS: processing mass spectrometry data for metabolite profiling using nonlinear peak alignment, matching, and identification. *Anal. Chem.* **78**, 779–787 (2006).

59. Tautenhahn, R., Bottcher, C. & Neumann, S. Highly sensitive feature detection for high resolution LC/MS. *BMC Bioinformatics* **9**, 504 (2008).

60. Tautenhahn, R., Patti, G. J., Rinehart, D. & Siuzdak, G. XCMS Online: a web-based platform to process untargeted metabolomic data. *Anal. Chem.* **84**, 5035–5039 (2012).

## Acknowledgements

We thank Neda Jasemi, Andrew Siedschlag, Benjamin Horman, Kelsie Harris and Aniket Biswas for assistance with device fabrication. We also thank Guillaume Delapierre for his support, Erin Gemperline for the helpful discussions and Sumit Kar for help with photography. This work was supported in part by the National Science Foundation-Emerging Frontiers in Research and Innovation-MIKS: Grant 1136903 (C.C.C.W., D.J.B., N.P.K. and E.B.); in part by the National Institute of Health: R01 AI065728 (N.P.K.), P30 CA014520 (D.J.B.), K12 DK100022 (A.B.T.), T32 ES007015 (A.B.T.); and in part by the National Library of Medicine: 5T15LM007359 (L.B.). L.B. is a student in the UW-MSTP (T32 GM008692). We thank the NIH West Coast Metabolomics Center at the University of California, Davis for helpful discussions and training in metabolomics data collection and analysis (U24 DK097154).

## Author contributions

L.B., A.B.T., C.-J.G., J.S., J.B., C.C.C.W, D.J.B., N.K. and E.B. designed the study. L.B., A.B.T., C.-J.G., J.S., L.R., J.B., K.A.B. and E.B. performed experiments. L.B., A.B.T.,

C.-J.G., J.S., L.R., J.B., K.A.B, N.K. and E.B. analysed data. L.B., A.B.T., N.K. and E.B. wrote the manuscript.

## Additional information

**Accession codes:** Processed mass spectra were deposited in the XCMS repository (https://xcmsonline.scripps.edu/) with the following accession codes: Data set ID 1023001 (data associated with Fig. 3), Data set ID 1026465 (data associated with Fig. 4) and Data set ID 1036638 (data associated with Fig. 5). The mass spectra have undergone retention time alignment and peak picking algorithms as well as basic statistical comparisons, but no further post processing has been done to the stored data.

**Competing financial interests:** D.J.B. has ownership in BellBrook Labs, LLC; Salus Discovery, LLC; Tasso, Inc.; and Stacks to the Future, LLC. E.B. has ownership in Salus Discovery, LLC; Tasso, Inc.; and Stacks to the Future, LLC. A.B.T. has ownership in Stacks to the Future, LLC. The remaining authors declare no competing financial interests.

# 22

# Structural identification of electron transfer dissociation products in mass spectrometry using infrared ion spectroscopy

Jonathan Martens[1], Josipa Grzetic[1], Giel Berden[1] & Jos Oomens[1,2]

Tandem mass spectrometry occupies a principle place among modern analytical methods and drives many developments in the 'omics' sciences. Electron attachment induced dissociation methods, as alternatives for collision-induced dissociation have profoundly influenced the field of proteomics, enabling among others the top-down sequencing of entire proteins and the analysis of post-translational modifications. The technique, however, produces more complex mass spectra and its radical-driven reaction mechanisms remain incompletely understood. Here we demonstrate the facile structural characterization of electron transfer dissociation generated peptide fragments by infrared ion spectroscopy using the tunable free-electron laser FELIX, aiding the elucidation of the underlying dissociation mechanisms. We apply this method to verify and revise previously proposed product ion structures for an often studied model tryptic peptide, $[AlaAlaHisAlaArg + 2H]^{2+}$. Comparing experiment with theory reveals that structures that would be assigned using only theoretical thermodynamic considerations often do not correspond to the experimentally sampled species.

[1] Radboud University, Institute for Molecules and Materials, FELIX Laboratory, Toernooiveld 7c, 6525ED Nijmegen, The Netherlands. [2] Van 't Hoff Institute for Molecular Sciences, University of Amsterdam, Science Park 908, 1098XH Amsterdam, The Netherlands. Correspondence and requests for materials should be addressed to J.O. (email: j.oomens@science.ru.nl).

Mass spectrometry-based analysis of peptides and proteins in the bioanalytical and clinical sciences relies on the gas-phase dissociation of their molecular ions to give sequence fragments from which the original primary structure can be inferred. Sequencing by the increasingly popular electron-induced dissociation methods (ExD, such as electron capture and transfer dissociation, ECD and ETD) has recently seen rapid development and widespread application. ECD in Fourier transform ion cyclotron resonance (FT-ICR) mass spectrometry (MS)[1–3] and more recently ETD in a much broader range of mass spectrometers[4–7] constitute the two primary variations of this method. These methods have shown impressive improvements over collision-induced dissociation tandem MS, primarily in the sense that labile post-translational modifications are not detached during activation revealing their position along the backbone, and that sequence coverage is increased, making the sequencing of intact proteins in top-down strategies possible[8–12].

Electron attachment to multiply protonated peptides or proteins leads to extensive backbone fragmentation, which most often takes the form of backbone N–C$_\alpha$ bond cleavages to provide the c- and z-type ion series[1,2,13,14]. The process of electron-induced N–C$_\alpha$ bond cleavage in peptide cation radicals has been the subject of many recent experimental and theoretical studies[13,15–20]. However, the mechanisms involved in electron attachment to peptide ions, the possible role of excited electronic states, and the structural rearrangements and fragmentation reactions that can follow, remain only partially understood. Furthermore, answers to questions regarding the ratios of fragment ions produced and their structures, especially the odd electron fragments, remain, at least partially, elusive. This is undoubtedly related to the uncertainty in the detailed nature of the (open-shell) dissociation products. For example, ExD product ions are known to undergo hydrogen atom migration reactions that increase/decrease the expected masses of the fragment ions, having direct practical implications for ExD-based sequencing applications[21,22].

Several reaction mechanisms have been proposed for ExD of which the Cornell mechanism from the group of McLafferty[2] and the more recent Utah-Washington (UW) mechanism from the groups of Tureček and Simons[19,23–25] are best known. Frison et al.[26] have discussed the different structures of c-type fragments that would result from different ECD mechanisms and were able to spectroscopically identify an amide c-type ion. While not being able to exclude the possibility that this structure results from a rearrangement after the ECD process, this assignment appears to support the UW mechanism, rather than mechanisms that would directly produce enol-imine c-type ions.

In ExD dissociation, the radical is typically on the z-type fragment, although H-atom migration reactions in the dissociating molecule may transfer the radical to the c-type fragment, resulting in changes in the $m/z$ values of the fragments and complicating the interpretation of ExD MS/MS data. As a result, apparent unit mass shifts of backbone sequence ions in ExD spectra are thus not rare and result in an increased occurrence of ions with overlapping masses. As an illustration, for a peptide with the sequence AlaAlaHisAlaArg, hydrogen atom migration to the open-shell $z_3^{\bullet+}$ fragment would give a closed-shell $z_3^+$ ion having the same chemical formula as the $c_4^+$ fragment and makes this mass peak unassignable for sequencing purposes.

Unfortunately, the detailed information about an ion's molecular structure is scarce and often undecipherable from MS/MS data alone. Infrared spectroscopy allows the structural characterization of both the trapped precursor and fragment ions. Infrared ion spectroscopy (IRIS) has been used extensively to determine the gas-phase structures of molecular ions in MS, and specifically in the effort to elucidate peptide fragmentation mechanisms involved in collisional dissociation[27–34]. Tunable infrared free-electron lasers (FELs), such as FELIX at our institute[35], have played an important role in the recent development of IRIS[36], and are especially useful to obtain fingerprint infrared spectra of ionic species on commercial MS platforms[37]. However, to date, only a single example of infrared spectroscopy on ECD-generated fragments has been reported, in which a small closed-shell c-type ion was examined[26].

Structural characterization of peptidic ExD product ions is something that has been highly sought after for a number of years using a variety of methods, including collisional activation[38,39], ion mobility methods[40] and more recently ultraviolet photo-dissociation studies[41–43]. However, in comparison with infrared spectroscopy these methods provide limited information, regarding the structure and conformation of gas-phase ions. Structural characterization of the dissociation products by infrared spectroscopy and quantum chemistry provides a stringent test to ascertain and confirm uncertain aspects of the fragmentation mechanisms and allows for additional information to be extracted from MS/MS data. Here we present the first direct structural characterization of ETD-generated fragments using IRIS and demonstrate the strengths of this technique for addressing the questions surrounding gas-phase peptide radicals.

## Results

**ETD MS/MS of [AAHAR + 2H$^+$]$^{2+}$.** The ETD MS/MS spectrum of the [AAHAR + 2H]$^{2+}$ (263 $m/z$) ion is presented in red in Fig. 1. We have characterized each of the z$^\bullet$-type fragments (depicted in the peptide sequence shown in Fig. 2) from the ETD MS/MS spectrum using infrared spectroscopy, providing a comprehensive identification of their structures and conformations. This detailed characterization forms a basis for analysing the reactions, leading to their formation.

Tryptic peptides of the AAXAR type and their ETD MS/MS behaviour have been extensively studied previously[39,42,44–47]. Here, consistent with previously reported results, we observe that fragment ion intensity is approximately split over the charge reduced ion (ETnoD) and z$^\bullet$-type sequence ions (Supplementary Fig. 1). For AAHAR, having arginine in the C-terminal position, it is not surprising that C-terminal z$^\bullet$-type fragments primarily retain the proton after dissociation and that they are the principle

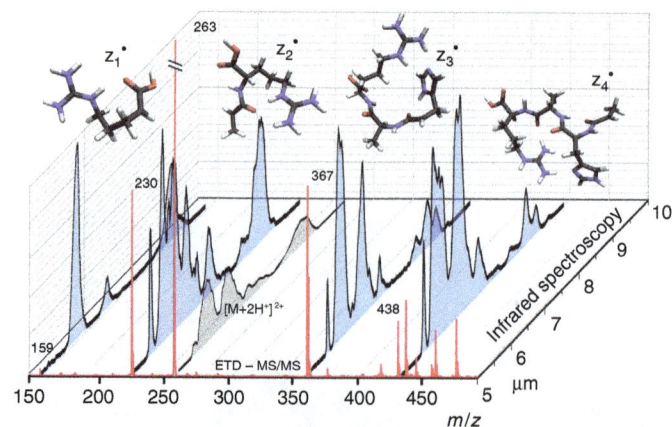

**Figure 1 | The ETD MS/MS spectrum of [AAHAR + 2H]$^{2+}$ with the corresponding infrared spectra.** The infrared spectra of the ETD-generated fragments are shown in black/blue and that of the precursor peptide in black/grey. Supplementary Fig. 1 contains the comprehensive ETD MS/MS results.

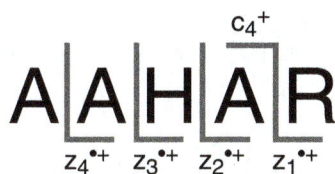

**Figure 2 | Dissociation scheme and notation used for product ions.** c- and z-type peptide fragments typically result from ETD MS/MS. Here we label only the discussed sequence ions from ETD of $[AAHAR + 2H]^{2+}$. Note that fragments carrying a '•' symbol are open-shell radicals and those without are closed shell.

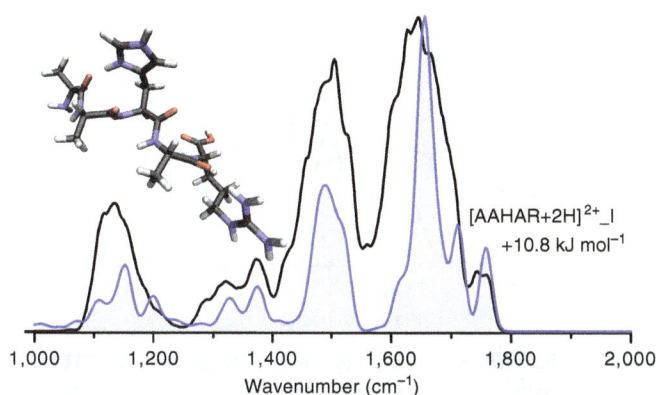

**Figure 3 | The infrared spectrum of $[AAHAR + 2H]^{2+}$ 263 $m/z$.** The experimental spectrum is presented in black and the spectrum of the assigned calculated structure from this study is shown in blue along with the structure and relative free energy at 298 K.

**Table 1 | Summary of structural properties and relative free energies for selected calculated structures.**

|  | $m/z$ | His tautomer | Radical | Rel. $\Delta G$ (kJ mol$^{-1}$) |
|---|---|---|---|---|
| $[AAHAR + 2H^+]^{2+}$ | 263 | | | |
| I | | — | — | +10.8* |
| II | | — | — | 0.0 |
| $z_1^{\bullet+}$ | 159 | | | |
| I | | — | $\alpha$ | +5.4* |
| II | | — | $\delta$ (Arg) | 0.0 |
| III | | — | $\beta$ (Arg) | +10.1 |
| IV | | — | $\gamma$ (Arg) | +14.7 |
| $z_2^{\bullet+}$ | 230 | | | |
| I | | — | $\alpha$ | 0.0* |
| II | | — | $\alpha$ | +10.2 |
| III | | — | $\delta$ (Arg) | +26.5 |
| $z_3^{\bullet+}$ | 367 | | | |
| I | | N3 | $\alpha$ | +22.4* |
| II | | N1 | $\alpha$ | 0.0 |
| III | | N3 | $\beta$ (His) | 0.0 |
| $z_4^{\bullet+}$ | 438 | | | |
| I | | N3 | $\alpha$ | 0.0* |
| II | | N3 | $\alpha$ | +62.5 |
| III | | N3 | — | +54.4 |
| IV | | N3 | $\beta$ (His) | +5.7 |
| $z_3^+$ | 368 | | | |
| I | | N1 | — | +28.5* |
| $c_4$_I | | — | — | — |
| II | | N3 | — | 0.0 |

Rel., relative.
An asterisk (*) indicates the structure assigned spectroscopically.

fragments in the ETD MS/MS spectrum. The electron attachment process is no doubt affected by the protonation sites and conformation of the parent peptide. In its doubly protonated state, this peptide has been shown to protonate on the imidazole group of the histidine side chain and the guanidine group of the arginine side chain[45], and our spectroscopic data and calculations confirm this.

**Structure of $[AAHAR + 2H^+]^{2+}$ precursor peptide at $m/z$ 263.** Figure 3 presents the infrared spectrum obtained for the doubly protonated peptide $[AAHAR + 2H]^{2+}$ at $m/z$ 263. Comparison with the calculated spectrum (blue) allows assignment of the protonation sites (His and Arg sidechains) and the conformation to be made. The band just <1,800 cm$^{-1}$ is consistent with the free (or weakly H-bonded) carboxyl group. Both charged sites are hydrogen bound to backbone carbonyl groups. In terms of hydrogen bonding, involving the charged sites and the C and N termini, this assignment confirms a previously proposed structure[44], although it has an overall more extended conformation, easily distinguished by comparison with the infrared spectra (Supplementary Fig. 2).

For singly charged fragments retaining the His residue, the imidazole ring can tautomerize (having the H on either the N1 or N3 position) and we address this issue for the $z_3^{\bullet+}$ and $z_4^{\bullet+}$ fragments. Note that all fragments carrying a '•' symbol are open-shell radicals and those without are closed shell, where H's and electrons are implicit. Backbone sequence fragments studied here with the structural identification based on our spectroscopic results, as detailed in the following are summarized in Table 1.

**$z_1^{\bullet+}$ fragment structure.** The $z_1^{\bullet+}$ fragment is the smallest C-terminal fragment obtained upon ETD of the precursor peptide. Calculated structure $z_1^{\bullet}$_I was identified to match most closely with experiment, as demonstrated in the top panel of Fig. 4. This structure is the lowest-energy calculated structure from our selection of ~30 structures from the molecular dynamics (MD) procedure. In structure $z_1^{\bullet}$_I, the radical is located at the $\alpha$-carbon adjacent to the C terminus and the carboxyl C=O stretch is found at 1,635 cm$^{-1}$. On the basis of this structure, radicals at other positions along the carbon chain of this fragment were defined and optimized, most of which were higher in energy. H-atom migration to the $\delta$-carbon of the Arg side chain gave a structure 5.4 kJ mol$^{-1}$ more stable; however, a calculated barrier of 115.2 kJ mol$^{-1}$ (Supplementary Fig. 3) must be overcome to reach it and it is not consistent with the experimental infrared spectrum. Barriers of >100 kJ mol$^{-1}$ for H-atom migrations have been reported for different ETD fragments[39,48,49]. With the radical in the $\alpha$-position, conjugation with the carbonyl occurs giving partial double-bond character to the CC-bond, while lowering the bond order of the carbonyl, causing a significant red shift of its stretching mode relative to a carbonyl stretch >1,700 cm$^{-1}$ in the $\beta$-, $\gamma$- and $\delta$-radicals (Supplementary Fig. 4). This is a clear demonstration that the position of the radical can strongly influence the vibrational spectrum, in this case the C=O stretch, highlighting the value of infrared spectroscopy for characterizing open-shell peptide fragments.

**$z_2^{\bullet+}$ fragment structure.** Figure 5 and Supplementary Fig. 5 present infrared spectra for the $z_2^{\bullet+}$ fragment from ETD of

**a**

$z_1^{\bullet +}\_I$
+5.4 kJ mol$^{-1}$

**b**

$z_4^{\bullet +}\_I$
0.0 kJ mol$^{-1}$

1,000   1,200   1,400   1,600   1,800   2,000
Wavenumber (cm$^{-1}$)

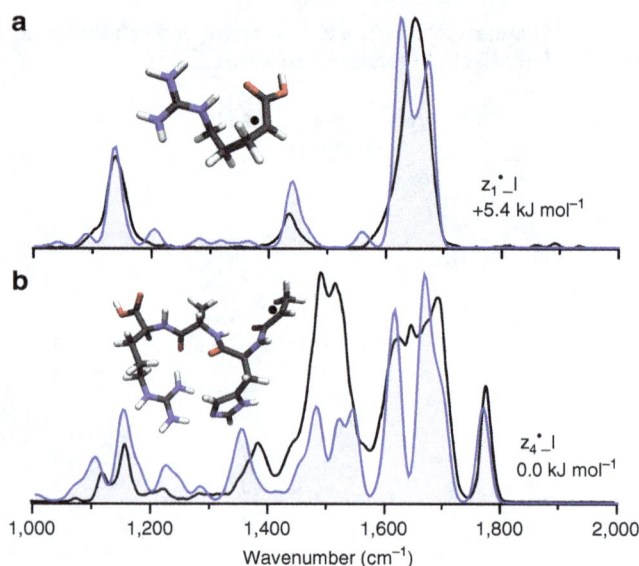

**Figure 4 | The infrared spectra of the $z_1^{\bullet +}$ and $z_4^{\bullet +}$ fragments from ETD of [AAHAR + 2H]$^{2+}$.** The experimental spectra are presented in black in both cases and assigned calculated structures are shown in blue for (**a**) the $z_1^{\bullet +}$ fragment and (**b**) the $z_4^{\bullet +}$ fragment. The associated relative free energies (298 K) and structures are inlayed with the radical sites labelled by '$\bullet$'.

[AAHAR + 2H$^+$]$^{2+}$. The calculated spectrum of $z_2^{\bullet}\_I$, presented in the top panel, matches the experimental spectrum well. This structure is the lowest-energy calculation obtained from our computational procedure. A previously proposed structure[39], labelled here as $z_2^{\bullet}\_II$ and presented in the centre panel, closely resembles $z_2^{\bullet}\_I$. However, the alternate hydrogen bonding orientation of the C terminus is distinguished by comparison with experimental and calculated infrared spectra in the 1,300–1,400 cm$^{-1}$ region and the carbonyl stretching region just < 1,800 cm$^{-1}$. This refinement of the conformation reduces the relative energy by 10.2 kJ mol$^{-1}$. The bottom panel in the figure shows a comparison with the infrared spectrum calculated for the product of hydrogen migration from the α-carbon of the Ala residue to the δ-carbon of the Arg residue, a species 26.5 kJ mol$^{-1}$ higher in energy and readily distinguishable spectroscopically.

$z_3^{\bullet +}$ **fragment structure.** Figure 6 presents the infrared spectrum for the $z_3^{\bullet +}$ fragment from ETD of [AAHAR + 2H$^+$]$^{2+}$ with the spectrum of the assigned calculated structure $z_3^{\bullet}\_I$ in blue and spectra of unassigned alternative structures in red. This assignment was made after considering different conformers, imidazole tautomers and products, resulting from hydrogen atom migration interconverting between the α-radical and the β-radical on the His side chain. A low-energy His N1 tautomer ($z_3^{\bullet}\_II$) was identified to be 22.4 kJ mol$^{-1}$ lower in energy than $z_3^{\bullet}\_I$, a His N3 tautomer; however, the calculated spectrum of this species does not match as well to the experiment, especially in the 1,200–1,400 cm$^{-1}$ region (and the 3400–3600 cm$^{-1}$ region displayed in Supplementary Fig. 6). Furthermore, a β-radical structure ($z_3^{\bullet}\_III$) was found to be 22.4 kJ mol$^{-1}$ more stable, but we do not assign this species on the basis of its spectral mismatch (Fig. 6 and Supplementary Fig. 6). The majority of stable structures we identified for each structure/tautomer of the $z_3^{\bullet +}$ ion features a stabilizing hydrogen bonding interaction between the imidazole side chain of His and the guanidinium side chain of Arg, leaving little flexibility for the orientation of the carboxyl group and giving a free C–OH group and a

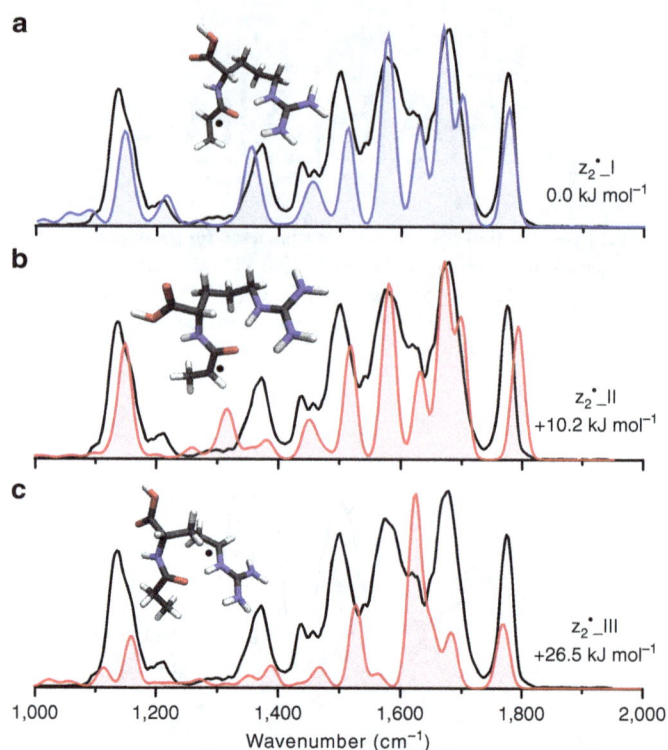

**a**

$z_2^{\bullet}\_I$
0.0 kJ mol$^{-1}$

**b**

$z_2^{\bullet}\_II$
+10.2 kJ mol$^{-1}$

**c**

$z_2^{\bullet}\_III$
+26.5 kJ mol$^{-1}$

1,000   1,200   1,400   1,600   1,800   2,000
Wavenumber (cm$^{-1}$)

**Figure 5 | The infrared spectrum of the $z_2^{\bullet +}$ fragment from ETD of [AAHAR + 2H]$^{2+}$.** The experimental spectrum is presented in black and is compared with computed spectra for different low-energy structures. (**a**) The calculated spectrum for the assigned structure is shown in blue. (**b,c**) The calculated spectra for structures disregarded on the basis of spectral mismatch are shown in red. Calculated structures and relative free energies (298 K) are inlayed for each plot.

C = O weakly hydrogen bonded to the adjacent amide N–H. Being very sensitive to local environment, the position of the carboxyl C = O stretch just < 1,800 cm$^{-1}$ can be used as a diagnostic signature. Structures $z_3^{\bullet}\_II$ and $z_3^{\bullet}\_III$ feature hydrogen bonds between the imidazole nitrogen and hydrogens of the two primary nitrogens of the guanidinium group, while in $z_3^{\bullet}\_I$ the hydrogen bond of the imidazole nitrogen is shared between the hydrogens of the secondary nitrogen and one primary nitrogen.

$z_4^{\bullet +}$ **fragment structure.** The bottom panel of Fig. 4 presents the infrared spectrum for the $z_4^{\bullet +}$ fragment from ETD of [AAHAR + 2H$^+$]$^{2+}$ and the assigned calculated structure (blue). This is the overall lowest-energy structure obtained after an extensive MD-based search over different conformers for various tautomers and structures obtained by hydrogen migration from the α-radical to the His β-radical (see comparison of $z_4^{\bullet}\_IV$ in Supplementary Fig. 7). Similar to $z_3^{\bullet}\_I$, the assigned structure, $z_4^{\bullet}\_I$, has a hydrogen bond between the imidazole side chain and the guanidinium group of the Arg residue. Structure $z_4^{\bullet}\_I$ offers a refinement over a previously proposed structure[39] (here, re-optimized at the currently applied level of theory), giving both a better spectral match (see $z_4^{\bullet}\_II$ in Supplementary Fig. 7) and being ~ 60 kJ mol$^{-1}$ lower in energy.

**Fragment ion at $m/z$ 368 can be $z_3^+$ or $c_4^+$.** While inter- and intramolecular (between c and z fragment pairs) hydrogen migration reactions are commonly observed in ETD MS/MS, their behaviour is still relatively weakly understood. For [AAHAR + 2H$^+$]$^{2+}$, only for the $z_3^{\bullet +}$ cation do we observe an

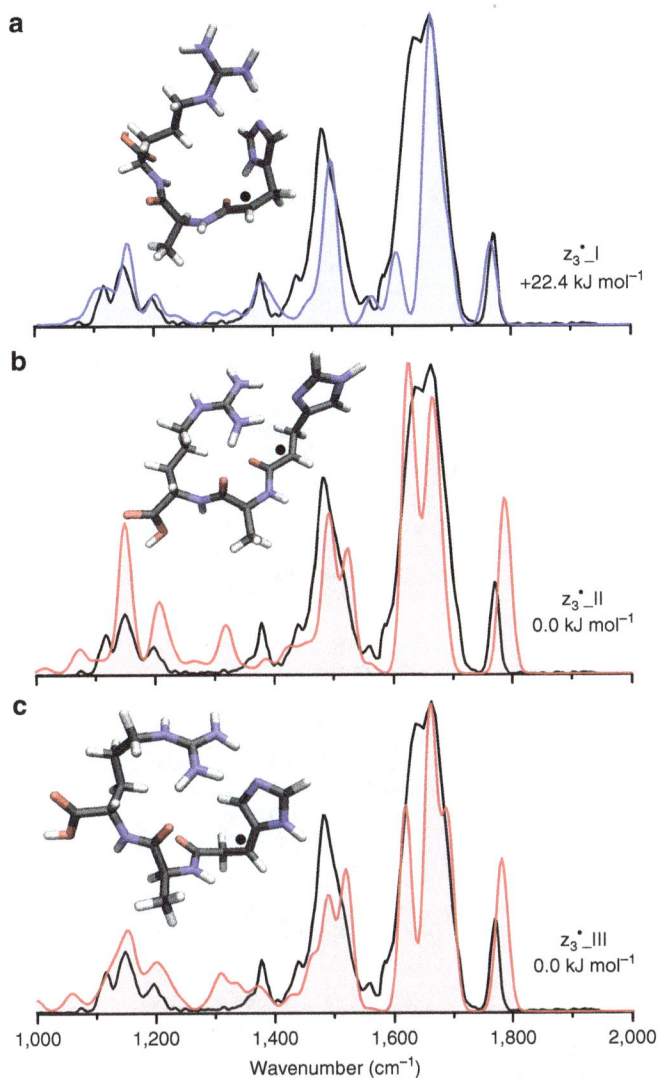

**Figure 6 | The infrared spectrum of the $z_3^{\bullet+}$ fragment from ETD of [AAHAR + 2H]$^{2+}$.** The experimental spectrum is presented in black and is compared with computed spectra for different low-energy structures. (**a**) The calculated spectrum for the assigned structure is shown in blue. (**b,c**) The calculated spectra for structures disregarded on the basis of spectral mismatch are shown in red. Calculated structures and relative free energies (298 K) are inlayed for each plot.

**Figure 7 | The infrared spectrum of the $m/z$ 368 fragment from ETD of [AAHAR + 2H]$^{2+}$.** The experimental spectrum is presented in black and is compared with computed spectra for different low-energy structures. (**a**) The spectrum of the assigned calculated closed-shell $z_3^+$ structure is shown in blue with the structure and relative free energy (298 K) inlayed. (**b**) $c_4$_I is a low-energy $c_4^+$ conformation and its calculated infrared spectrum is presented in red. (**c**) $z_3$_II is the lowest energy calculated $z_3^+$ structure identified in this work.

appreciable extent of such a reaction, where we see both the open-shell $z_3^{\bullet+}$ fragment ($m/z$ 367) and a closed-shell $z_3^+$ fragment ($m/z$ 368). Highlighting the complications that can arise from hydrogen atom migrations in ETD, the $c_4^+$ fragment ($C_{15}N_7H_{25}O_4$) has the same chemical formula and overlaps the closed-shell $z_3^+$ fragment also at $m/z$ 368. Identification and consideration of the hydrogen migration products (loss/gain) are important for assigning fragment ions and correct sequencing[38].

In Fig. 7, we identify the $m/z$ 368 fragment ion as the $z_3^+$ species ($z_3$_I) based on infrared spectral matching. This structure features an alternative hydrogen bonding arrangement in comparison with the open-shell $z_3^{\bullet+}$ and $z_4^{\bullet+}$ fragments described above, most significantly affecting the C terminus (trans configuration) and red shifting the carboxyl $C=O$ stretch for $z_3$_I away from the position just $<1{,}800\,cm^{-1}$ for the open-shell $z_3^{\bullet+}$ and $z_4^{\bullet+}$ fragments. A closed-shell equivalent of the geometry of the open-shell $z_3^{\bullet}$_I structure is defined as $z_3$_II and is 28.5 kJ mol$^{-1}$ lower in energy than $z_3$_I. Calculated spectra

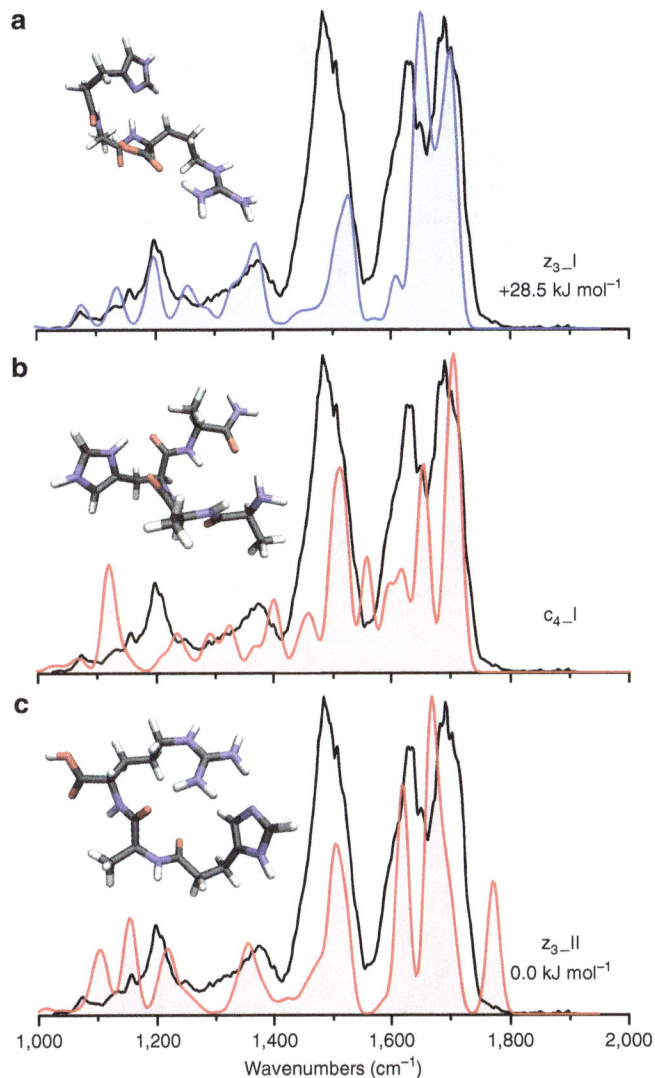

for structure $z_3$_II and $c_4$_I, a low-energy $c_4^+$ conformation, are presented in the bottom two panels of Fig. 7 and support the assignment of $z_3$_I.

## Discussion

These results demonstrate the first use of IRIS to characterize the structures of ETD-generated peptide fragments. Using a model tryptic pentapeptide, precursor ion conformation has been related to the observed ETD fragmentation pattern, and the structures and conformations of the various fragment ions. We show that it is possible to distinguish both conformational details and different radical species, when this approach is combined with routine computational modelling.

We conclude that if structural assignments were made only on the basis of theoretical (thermodynamic) considerations, these assignments would in many cases not match the species observed in experiment—highlighting the potential for IRIS to diagnostically identify gas-phase organic radicals and, more specifically, the mechanisms associated with peptide fragmentation in electron attachment methods. Our results suggest that hydrogen atom transfer necessary for radical migration often does not occur after ETD (without additional activation), leading to the frequent observation of non-equilibrium product ions. Understanding intermolecular hydrogen atom migration is also of practical importance, as it causes shifts in ETD fragment masses and makes sequence ion assignments more complicated.

## Methods

**Ion spectroscopy in a modified ion trap mass spectrometer.** The experiment is based on a commercial quadrupole ion trap mass spectrometer (Bruker, AmaZon Speed ETD) coupled to the infrared beam line of the FELIX FEL. $[M + 2H]^{2+}$ peptide ions are generated by electrospray ionization. AAHAR (GeneCust (Luxemburg), 95% purity) solutions of $10^{-5}-10^{-6} \, mol \, l^{-1}$ (in 50:50 acetonitrile:water, $\sim 1\%$ formic acid) are introduced at $120 \, \mu l \, h^{-1}$ flow rates and desolvated by a pressurized nebulizing gas ($N_2$). The key hardware modifications to the instrument providing optical access to the ion population in the trap were the introduction of a new ring electrode having 3 mm holes at its top and bottom, the installation of mirrors below the trap to direct the beam back out of the instrument and optical windows in the vacuum housing. In ETD experiments, ions were accumulated for 0.1–15 ms in the trap, mass isolated and then reacted with fluoranthene radical anions for $\sim 250$ ms. A fragment ion of interest was mass isolated in a subsequent MS/MS stage and irradiated by the tunable infrared beam from the FEL. In the experiments reported here, the FELIX FEL was set to produce infrared radiation in the form of 5–10 μs macropulses at 5 or 10 Hz and of 30–60 mJ (bandwidth $\sim 0.4\%$ of the centre frequency). Resonant absorption of infrared radiation leads to an increase in the internal energy of the molecule aided by intramolecular vibrational redistribution of the absorbed energy. When a sufficient number of photons is absorbed (here, typically in a single macropulse), unimolecular dissociation occurs and produces frequency-dependent fragment ion intensities in the mass spectrometer (Supplementary Note 1). Relating the parent and fragment ion intensities in the observed mass spectral data (yield = $\Sigma I$(fragment ions)$/\Sigma I$(parent + fragment ions)) generates an infrared vibrational spectrum. The yield at each infrared point is obtained from averaged mass spectra and is linearly corrected for laser power; the frequency is calibrated using a grating spectrometer.

**Computational chemistry.** We have employed a molecular mechanics (MM)/MD approach using AMBER 12 (refs 50,51). Molecular structures manually defined based on chemical intuition where first optimized for each ion at the B3LYP/6-31 + + G(d,p) level in Gaussian09 (ref. 52). Restrained electrostatic potential (RESP) charges from these initial results were used for parameterization of the nonstandard peptide ions in the antechamber program. After minimization within AMBER, a simulated annealing procedure up to 1,000 K was used with a 1 fs step size. Five hundred structures were obtained as snapshots throughout the procedure and after MM minimization were grouped based on structural similarity using *prtraj* in AMBER. Of these, 30–50 unique structures were then each optimized at the B3LYP/6-31 + + G(d,p) level[51,53,54] and vibrational spectra were calculated within the harmonic oscillator model (vibrational frequencies were scaled by 0.975). This computational approach was applied to all structural isomers considered for each ion, except for the small $z_1^{\bullet+}$ ion, where the MM/MD conformational search was only applied once using the alpha-radical species. Calculated line spectra were broadened using a Gaussian function with a full-width at half-maximum of 25 cm$^{-1}$ to facilitate comparison with experiment. Additional calculations using the LC-BLYP and M06 functionals for a selection of $z_1^{\bullet+}$ and $z_2^{\bullet+}$ structures, and ab initio MP2 calculations for the $z_1^{\bullet+}$ structures, were performed to verify the validity of the choice of functional[49,55] and these results are summarized in Supplementary Table 1 and Supplementary Fig. 8. In general, vibrational frequencies were found to be best modelled at the B3LYP/6-31 + + G(d,p) level and calculated free energies are mostly consistent between these levels of theory. Supplementary Data 1–4 contain optimized geometries of the assigned z-type fragments.

## References

1. Kruger, N. A., Zubarev, R. A., Horn, D. M. & McLafferty, F. W. Electron capture dissociation of multiply charged peptide cations. *Int. J. Mass. Spectrom.* **185–187,** 787–793 (1999).

2. Zubarev, R. A., Kelleher, N. L. & McLafferty, F. W. Electron capture dissociation of multiply charged protein cations. a nonergodic process. *J. Am. Chem. Soc.* **120,** 3265–3266 (1998).

3. Zubarev, R. A. *et al.* Electron capture dissociation for structural characterization of multiply charged protein cations. *Anal. Chem.* **72,** 563–573 (2000).

4. Syka, J. E. P., Coon, J. J., Schroeder, M. J., Shabanowitz, J. & Hunt, D. F. Peptide and protein sequence analysis by electron transfer dissociation mass spectrometry. *Proc. Natl Acad. Sci. USA* **101,** 9528–9533 (2004).

5. Pitteri, S. J., Chrisman, P. A., Hogan, J. M. & McLuckey, S. A. Electron transfer ion/ion reactions in a three-dimensional quadrupole ion trap: reactions of doubly and triply protonated peptides with SO2•. *Anal. Chem.* **77,** 1831–1839 (2005).

6. Xia, Y. *et al.* Implementation of ion/ion reactions in a quadrupole/time-of-flight tandem mass spectrometer. *Anal. Chem.* **78,** 4146–4154 (2006).

7. Coon, J. J. *et al.* Protein identification using sequential ion/ion reactions and tandem mass spectrometry. *Proc. Natl Acad. Sci. USA* **102,** 9463–9468 (2005).

8. Swaney, D. L., McAlister, G. C. & Coon, J. J. Decision tree-driven tandem mass spectrometry for shotgun proteomics. *Nat. Meth.* **5,** 959–964 (2008).

9. Breuker, K. & McLafferty, F. W. Native electron capture dissociation for the structural characterization of noncovalent interactions in native cytochrome c. *Angew. Chem. Int. Ed.* **42,** 4900–4904 (2003).

10. Breuker, K., Oh, H., Lin, C., Carpenter, B. K. & McLafferty, F. W. Nonergodic and conformational control of the electron capture dissociation of protein cations. *Proc. Natl Acad. Sci. USA* **101,** 14011–14016 (2004).

11. Oh, H. *et al.* Secondary and tertiary structures of gaseous protein ions characterized by electron capture dissociation mass spectrometry and photofragment spectroscopy. *Proc. Natl Acad. Sci. USA* **99,** 15863–15868 (2002).

12. Breuker, K. & McLafferty, F. W. Stepwise evolution of protein native structure with electrospray into the gas phase, 10 − 12 to 102s. *Proc. Natl Acad. Sci. USA* **105,** 18145–18152 (2008).

13. Zubarev, R. A. Reactions of polypeptide ions with electrons in the gas phase. *Mass. Spectrom. Rev.* **22,** 57–77 (2003).

14. Horn, D. M., Zubarev, R. A. & McLafferty, F. W. Automated de novo sequencing of proteins by tandem high-resolution mass spectrometry. *Proc. Natl Acad. Sci. USA* **97,** 10313–10317 (2000).

15. Tureček, F. & Julian, R. R. Peptide radicals and cation radicals in the gas phase. *Chem. Rev.* **113,** 6691–6733 (2013).

16. Simons, J. Mechanisms for S–S and N–Cα bond cleavage in peptide ECD and ETD mass spectrometry. *Chem. Phys. Lett.* **484,** 81–95 (2010).

17. Anusiewicz, I., Skurski, P. & Simons, J. Refinements to the Utah–Washington mechanism of electron capture dissociation. *J. Phys. Chem. B* **118,** 7892–7901 (2014).

18. Li, X., Lin, C., Han, L., Costello, C. E. & O'Connor, P. B. Charge remote fragmentation in electron capture and electron transfer dissociation. *J. Am. Soc. Mass. Spectrom.* **21,** 646–656 (2010).

19. Syrstad, E. A. & Tureček, F. Toward a general mechanism of electron capture dissociation. *J. Am. Soc. Mass. Spectrom.* **16,** 208–224 (2005).

20. Zhurov, K. O., Fornelli, L., Wodrich, M. D., Laskay, U. A. & Tsybin, Y. O. Principles of electron capture and transfer dissociation mass spectrometry applied to peptide and protein structure analysis. *Chem. Soc. Rev.* **42,** 5014–5030 (2013).

21. Savitski, M. M., Kjeldsen, F., Nielsen, M. L. & Zubarev, R. A. Hydrogen rearrangement to and from radical z fragments in electron capture dissociation of peptides. *J. Am. Soc. Mass. Spectrom.* **18,** 113–120 (2007).

22. Liu, J., Liang, X. & McLuckey, S. A. On the value of knowing a z• ion for what it is. *J. Proteome. Res.* **7,** 130–137 (2008).

23. Sobczyk, M. *et al.* Coulomb-assisted dissociative electron attachment: application to a model peptide. *J. Phys. Chem. A* **109,** 250–258 (2004).

24. Anusiewicz, I., Berdys-Kochanska, J. & Simons, J. Electron attachment step in electron capture dissociation (ECD) and electron transfer dissociation (ETD). *J. Phys. Chem. A* **109,** 5801–5813 (2005).

25. Chen, X. & Tureček, F. The arginine anomaly: arginine radicals are poor hydrogen atom donors in electron transfer induced dissociations. *J. Am. Chem. Soc.* **128,** 12520–12530 (2006).

26. Frison, G. *et al.* Structure of electron-capture dissociation fragments from charge-tagged peptides probed by tunable infrared multiple photon dissociation. *J. Am. Chem. Soc.* **130,** 14916–14917 (2008).

27. Lucas, B. *et al.* Investigation of the protonation site in the dialanine peptide by infrared multiphoton dissociation spectroscopy. *Phys. Chem. Chem. Phys.* **6,** 2659–2663 (2004).

28. Erlekam, U. *et al.* Infrared spectroscopy of fragments of protonated peptides: direct evidence for macrocyclic structures of b5 ions. *J. Am. Chem. Soc.* **131,** 11503–11508 (2009).

29. Bythell, B. J., Erlekam, U., Paizs, B. & Maître, P. Infrared spectroscopy of fragments from doubly protonated tryptic peptides. *Chemphyschem* **10,** 883–885 (2009).

30. Polfer, N. C., Oomens, J., Suhai, S. & Paizs, B. Spectroscopic and theoretical evidence for oxazolone ring formation in collision-induced dissociation of peptides. *J. Am. Chem. Soc.* **127,** 17154–17155 (2005).

31. Polfer, N. C., Oomens, J., Suhai, S. & Paizs, B. Infrared spectroscopy and theoretical studies on gas-phase protonated leu-enkephalin and its fragments: direct experimental evidence for the mobile proton. *J. Am. Chem. Soc.* **129**, 5887–5897 (2007).

32. Polfer, N. C. & Oomens, J. Vibrational spectroscopy of bare and solvated ionic complexes of biological relevance. *Mass. Spectrom. Rev.* **28**, 468–494 (2009).

33. Yoon, S. H. *et al.* IRMPD spectroscopy shows that agg forms an oxazolone b2 + ion. *J. Am. Chem. Soc.* **130**, 17644–17645 (2008).

34. Perkins, B. R. *et al.* Evidence of diketopiperazine and oxazolone structures for ha b2 + ion. *J. Am. Chem. Soc.* **131**, 17528–17529 (2009).

35. Oepts, D., van der Meer, A. F. G. & van Amersfoort, P. W. The free-electron-laser user facility FELIX. *Infrared Phys. Technol.* **36**, 297–308 (1995).

36. Oomens, J., Sartakov, B. G., Meijer, G. & Von Helden, G. Gas-phase infrared multiple photon dissociation spectroscopy of mass-selected molecular ions. *Int. J. Mass. Spectrom.* **254**, 1–19 (2006).

37. Bakker, J. M., Besson, T., Lemaire, J., Scuderi, D. & Maître, P. Gas-phase structure of a π-allyl – palladium complex: efficient infrared spectroscopy in a 7T Fourier transform mass spectrometer. *J. Phys. Chem. A* **111**, 13415–13424 (2007).

38. Hamidane, H. B. *et al.* Electron capture and transfer dissociation: peptide structure analysis at different ion internal energy levels. *J. Am. Soc. Mass. Spectrom.* **20**, 567–575 (2009).

39. Ledvina, A., Chung, T., Hui, R., Coon, J. & Tureček, F. Cascade dissociations of peptide cation-radicals. part 2. Infrared multiphoton dissociation and mechanistic studies of z-ions from pentapeptides. *J. Am. Soc. Mass. Spectrom.* **23**, 1351–1363 (2012).

40. Moss, C. L. *et al.* Assigning structures to gas-phase peptide cations and cation-radicals. an infrared multiphoton dissociation, ion mobility, electron transfer, and computational study of a histidine peptide ion. *J. Phys. Chem. B* **116**, 3445–3456 (2012).

41. Shaffer, C. J., Marek, A., Pepin, R., Slovakova, K. & Turecek, F. Combining UV photodissociation with electron transfer for peptide structure analysis. *J. Mass. Spectrom.* **50**, 470–475 (2015).

42. Nguyen, H. T. H., Shaffer, C. J. & Tureček, F. Probing peptide cation–radicals by near-uv photodissociation in the gas phase. structure elucidation of histidine radical chromophores formed by electron transfer reduction. *J. Phys. Chem. B* **119**, 3948–3961 (2015).

43. Nguyen, H. T. H., Shaffer, C. J., Pepin, R. & Tureček, F. U. V. Action spectroscopy of gas-phase peptide radicals. *J. Phys. Chem. Lett.* **6**, 4722–4727 (2015).

44. Tureček, F., Moss, C. L. & Chung, T. W. Correlating ETD fragment ion intensities with peptide ion conformational and electronic structure. *Int. J. Mass. Spectrom.* **330–332**, 207–219 (2012).

45. Tureček, F. *et al.* The histidine effect. electron transfer and capture cause different dissociations and rearrangements of histidine peptide cation-radicals. *J. Am. Chem. Soc.* **132**, 10728–10740 (2010).

46. Zimnicka, M., Moss, C., Chung, T., Hui, R. & Tureček, F. Tunable charge tags for electron-based methods of peptide sequencing: design and applications. *J. Am. Soc. Mass. Spectrom.* **23**, 608–620 (2012).

47. Chung, T., Hui, R., Ledvina, A., Coon, J. & Tureček, F. Cascade dissociations of peptide cation-radicals. part 1. scope and effects of amino acid residues in penta-, nona-, and decapeptides. *J. Am. Soc. Mass. Spectrom.* **23**, 1336–1350 (2012).

48. Chung, T. W. & Tureček, F. Backbone and side-chain specific dissociations of z ions from non-tryptic peptides. *J. Am. Soc. Mass. Spectrom.* **21**, 1279–1295 (2010).

49. Riffet, V., Jacquemin, D. & Frison, G. H-atom loss and migration in hydrogen-rich peptide cation radicals: The role of chemical environment. *Int. J. Mass. Spectrom.* **390**, 28–38 (2015).

50. Case, D.A. *et al.* AMBER 12 (University of California, San Francisco, CA, 2012).

51. Martens, J. K., Grzetic, J., Berden, G. & Oomens, J. Gas-phase conformations of small polyprolines and their fragment ions by IRMPD spectroscopy. *Int. J. Mass. Spectrom.* **377**, 179–187 (2015).

52. Frisch, M. J. *et al.* Gaussian 09 (Gaussian, Inc., Wallingford, CT, 2009).

53. Halls, M. D. & Schlegel, H. B. Comparison of the performance of local, gradient-corrected, and hybrid density functional models in predicting infrared intensities. *J. Chem. Phys.* **109**, 10587–10593 (1998).

54. Kapota, C., Lemaire, J., Maître, P. & Ohanessian, G. Vibrational signature of charge solvation vs salt bridge isomers of sodiated amino acids in the gas phase. *J. Am. Chem. Soc.* **126**, 1836–1842 (2004).

55. Riffet, V., Jacquemin, D., Cauët, E. & Frison, G. Benchmarking DFT and TD-DFT functionals for the ground and excited states of hydrogen-rich peptide radicals. *J. Chem. Theory. Comput.* **10**, 3308–3318 (2014).

## Acknowledgements

We gratefully acknowledge the FELIX staff, particularly Dr B. Redlich and Dr A.F.G. van der Meer. In addition, the authors thank Professor F. Tureček and Dr C. Schaffer for discussion. Finally, the authors express their gratitude for the technical assistance from staff at Bruker Daltonics in Bremen, Germany, while implementing this experiment, in particular Dr C. Gebhardt. Financial support for this project was provided by the Chemical Sciences division of the 'Nederlandse organisatie voor Wetenschappelijk Onderzoek' (NWO) under VICI project no. 724.011.002. We also thank NWO Physical Sciences (EW) and the SurfSARA Supercomputer Center for providing the computational resources. This work is part of the research program of FOM, which is financially supported by NWO.

## Author contributions

J.M., G.B. and J.O. conceived and designed the experiments. J.M. and J.G. performed the experiments and computations. The manuscript was co-written by all authors.

## Additional information

# Permissions

# List of Contributors

**Raphael Böhm, Andrea Maggioni, Mark von Itzstein and Thomas Haselhorst**
Institute for Glycomics, Griffith University, Gold Coast Campus, Southport, Queensland 4222, Australia

**Fiona E. Fleming, Vi T. Dang, Gavan Holloway and Barbara S. Coulson**
Department of Microbiology and Immunology, The University of Melbourne at the Peter Doherty Institute for Infection and Immunity, Melbourne, Victoria 3000, Australia

**Igor Pavlovic, Divyeshsinh T. Thakor, Laurent Bigler, Gilles Gasser and Philipp Anstaett**
Department of Chemistry, University of Zurich, Winterthurerstrasse 190, Zurich 8057, Switzerland

**Jessica R. Vargas, Colin J. McKinlay and Paul A. Wender**
Departments of Chemistry and Chemical and Systems Biology, Stanford University, Stanford, California 94305, USA

**Sebastian Hauke and Carsten Schultz**
European Molecular Biology Laboratory (EMBL), Cell Biology & Biophysics Unit, Meyerhofstrasse 1, 69117 Heidelberg, Germany

**Rafael C. Camuña**
Departamento de Química Orgánica, Facultad de Ciencias, Universidad de Málaga, Malaga 29071, Spain

**Henning J. Jessen**
Department of Chemistry and Pharmacy, Albert-Ludwigs University Freiburg, Albertstrasse 21, 79104 Freiburg, Germany

**Joan J. Soldevila-Barreda, Isolda Romero-Canelón, Abraha Habtemariam and Peter J. Sadler**
Department of Chemistry, University of Warwick, Gibbet Hill Road, Coventry CV4 7AL, UK

**Jos J.A.G. Kamps, Bas J.G.E. Pieters and Jasmin Mecinović**
Institute for Molecules and Materials, Radboud University, Heyendaalseweg 135 , 6525 AJNijmegen, The Netherlands

**Jiaxin Huang and Haitao Li**
Department of Basic Medical Sciences, Center for Structural Biology, School of Medicine, Tsinghua University, Beijing 100084, China

**Jordi Poater**
Department of Theoretical Chemistry and Amsterdam Center for Multiscale Modeling, VU University, De Boelelaan 1083, 1081 HV Amsterdam, The Netherlands

**F. Matthias Bickelhaupt**
Institute for Molecules and Materials, Radboud University, Heyendaalseweg 135 , 6525 AJNijmegen, The Netherlands
Department of Theoretical Chemistry and Amsterdam Center for Multiscale Modeling, VU University, De Boelelaan 1083, 1081 HV Amsterdam, The Netherlands

**Aiping Dong, Jinrong Min and Chao Xu**
Structural Genomics Consortium, University of Toronto, 101 College Street, Toronto, Ontario, Canada M5G 1L7

**Woody Sherman and Thijs Beuming**
Schrödinger, Inc., 120 West 45th Street, New York, New York 10036 USA

**Ruili Huang, Menghang Xia, Srilatha Sakamuru, Jinghua Zhao, Sampada A. Shahane, Matias Attene-Ramos, Tongan Zhao, Christopher P. Austin and Anton Simeonov**
Division of Pre-clinical Innovation, National Center for Advancing Translational Sciences, National Institutes of Health, 9800 Medical Center Drive, Rockville, Maryland 20850, USA

**Hartmut Jahns, Martina Roos, Jochen Imig, Fabienne Baumann, Yuluan Wang and Jonathan Hall**
Department of Chemistry and Applied Biosciences, ETH Zürich, Vladimir-Prelog-Weg-4, CH-8093 Zürich, Switzerland

**Ryan Gilmour**
Institute for Organic Chemistry, Westfälische Wilhelms-Universität Münster, D-48149 Münster, Germany

**Chul-Jin Lee, Qinglin Wu, Javaria Najeeb and Jinshi Zhao**
Department of Biochemistry, Duke University Medical Center, Durham, North Carolina 27710, USA

**Xiaofei Liang and Ramesh Gopalaswamy**
Department of Chemistry, Duke University, Durham, North Carolina 27708, USA

**Eric J. Toone and Pei Zhou**
Department of Biochemistry, Duke University Medical Center, Durham, North Carolina 27710, USA

Department of Chemistry, Duke University, Durham, North Carolina 27708, USA

**Marie Titecat, Florent Sebbane and Nadine Lemaitre**
Inserm, Univ. Lille, CHU Lille, Institut Pasteur de Lille, CNRS, U1019-UMR 8204-CIIL-Center for Infection and Immunity of Lille, F-59000 Lille, France

**Rainer Müller, Frank Stein, Suihan Feng and Carsten Schultz**
European Molecular Biology Laboratory, Cell Biology and Biophysics Unit, Meyerhofstraße 1, 69117 Heidelberg, Germany

**André Nadler**
European Molecular Biology Laboratory, Cell Biology and Biophysics Unit, Meyerhofstraße 1, 69117 Heidelberg, Germany
Max Planck Institute of Molecular Cell Biology and Genetics, Pfotenhauerstraße 108, 01307 Dresden, Germany

**Dmytro A. Yushchenko**
European Molecular Biology Laboratory, Cell Biology and Biophysics Unit, Meyerhofstraße 1, 69117 Heidelberg, Germany
Institute of Organic Chemistry and Biochemistry, Academy of Sciences of the Czech Republic, Flemingovo náměstí 2, 16610 Prague 6, Czech Republic

**Christophe Mulle and Mario Carta**
Institut Interdisciplinaire de Neurosciences, CNRS UMR 5297 Université Bordeaux 2, 146, rue Léo-Saignat, 33077 Bordeaux, France

**Marco Grossi and Marina Morgunova**
Department of Pharmaceutical and Medicinal Chemistry, Royal College of Surgeons in Ireland, 123 St Stephen's Green, Dublin 2, Ireland

**Shane Cheung and Donal F. O'Shea**
Department of Pharmaceutical and Medicinal Chemistry, Royal College of Surgeons in Ireland, 123 St Stephen's Green, Dublin 2, Ireland
School of Chemistry and Chemical Biology, Conway Institute, University College Dublin, Belfield, Dublin 4, Ireland

**Dimitri Scholz, Emer Conroy, Marta Terrile and William M. Gallagher**
School of Biomolecular and Biomedical Science, Conway Institute of Biomolecular and Biomedical Research, University College Dublin, Belfield, Dublin 4, Ireland

**Angela Panarella and Jeremy C. Simpson**
School of Biology and Environmental Science, Conway Institute of Biomolecular and Biomedical Research, University College Dublin, Belfield, Dublin 4, Ireland

**Tobias Klein**
Discovery Sciences, AstraZeneca R&D, Alderley Park, Macclesfield SK10 4TG, UK
Bayer Healthcare, GP Grenzach Produktions GmbH, Postfach 1146, D-79629 Grenzach-Wyhlen, Germany (T.K.)

**Navratna Vajpai**
Discovery Sciences, AstraZeneca R&D, Alderley Park, Macclesfield SK10 4TG, UK
Biological E. Ltd, ICICI Knowledge Park, Shameerpet, Ranga Reddy District, Hyderabad, Telangana 500078, India (A.D.S.)

**Gareth Davies, Geoffrey A. Holdgate, Chris Phillips, Julie A. Tucker, Richard A. Norman and Andrew D. Scott**
Discovery Sciences, AstraZeneca R&D, Alderley Park, Macclesfield SK10 4TG, UK

**Jonathan J. Phillips**
MedImmune, Granta Park, Cambridge CB21 6GH, UK
Astbury Centre for Structural Molecular Biology, Faculty of Biological Sciences, University of Leeds, Leeds LS2 9JT, UK (A.L.B.)

**Daniel R. Higazi and David Lowe**
MedImmune, Granta Park, Cambridge CB21 6GH, UK

**Gary S. Thompson**
Astbury Centre for Structural Molecular Biology, Faculty of Biological Sciences, University of Leeds, Leeds LS2 9JT, UK

**Alexander L. Breeze**
Discovery Sciences, AstraZeneca R&D, Alderley Park, Macclesfield SK10 4TG, UK
Astbury Centre for Structural Molecular Biology, Faculty of Biological Sciences, University of Leeds, Leeds LS2 9JT, UK

**Fange Liu**
Department of Chemistry, Georgia State University, Atlanta, Georgia 30303, USA
Department of Chemistry, University of Chicago (F.L.)

**Lu Huo**
Department of Chemistry, Georgia State University, Atlanta, Georgia 30303, USA
Molecular Basis of Disease Area of Focus Program, Georgia State University, Atlanta, Georgia 30303, USA
Department of Pharmaceutical Sciences, University of Connecticut (L.H.)

**Ian Davis, Shingo Esaki and Aimin Liu**
Department of Chemistry, Georgia State University, Atlanta, Georgia 30303, USA
Molecular Basis of Disease Area of Focus Program, Georgia State University, Atlanta, Georgia 30303, USA

**Babak Andi**
Photon Sciences Directorate, Brookhaven National Laboratory, Upton, New York 11973, USA

**Hiroaki Iwaki and Yoshie Hasegawa**
Department of Life Science and Biotechnology and ORDIST, Kansai University, Suita, Osaka 564-8680, Japan

**Allen M. Orville**
Photon Sciences Directorate, Brookhaven National Laboratory, Upton, New York 11973, USA
Biosciences Department, Brookhaven National Laboratory, Upton, New York 11973, USA

**Erwan Poivet, Narmin Tahirova, Lu Xu, Clara Altomare, Anne Paria, Dong-Jing Zou and Stuart Firestein**
Department of Biological Sciences, Columbia University, New York, New York 10027, USA

**Zita Peterlin**
Corporate Research and Development, Firmenich Incorporated, Plainsboro, New Jersey 08536, USA

**Vanessa Delfosse, Tiphaine Huet and William Bourguet**
Inserm U1054, Montpellier 34090, France
CNRS UMR5048, Centre de Biochimie Structurale, Montpellier 34090, France
Université de Montpellier, Montpellier 34090, France

**Béatrice Dendele, Marina Grimaldi, Abdelhay Boulahtouf, Vincent Cavaillès and Patrick Balaguer**
Université de Montpellier, Montpellier 34090, France
IRCM, Institut de Recherche en Cancérologie de Montpellier, Montpellier 34298, France
Inserm, U1194, Montpellier 34298, France
ICM, Institut régional du Cancer de Montpellier, Montpellier 34298, France

**Bertrand Beucher and Jean-Marc Pascussi**
Université de Montpellier, Montpellier 34090, France
Inserm U661, Montpellier 34094, France
CNRS UMR5203, Institut de Génomique Fonctionnelle, Montpellier 34094, France

**Sabine Gerbal-Chaloin**
Université de Montpellier, Montpellier 34090, France
Inserm U1040, Montpellier 34295, France

**Martine Daujat-Chavanieu**
Université de Montpellier, Montpellier 34090, France
Inserm U1040, Montpellier 34295, France
CHU de Montpellier, Institut de Recherche en Biothérapie, Montpellier 34295, France

**Dominique Roecklin, Christina Muller and Valérie Vivat**
NovAliX, Illkirch 67400, France

**Roger Rahmani**
INRA UMR 1331, TOXALIM, Sophia-Antipolis 06903, France

**Michael K. Fenwick, Yang Zhang and Steven E. Ealick**
Department of Chemistry and Chemical Biology, Cornell University, 120 Baker Lab, Ithaca, New York 14853, USA

**Angad P. Mehta, Sameh H. Abdelwahed and Tadhg P. Begley**
Department of Chemistry, Texas A&M University, College Station, Texas 77843, USA

**Xiangzhao Ai, Junxin Aw, Jing Mu and Edwin K.L. Yeow**
Division of Chemistry and Biological Chemistry, School of Physical and Mathematical Sciences, Nanyang Technological University, 637371 Singapore, Singapore

**Chris Jun Hui Ho, Amalina Binte Ebrahim Attia and Malini Olivo**
Singapore Bioimaging Consortium, Agency for Science Technology and Research (A*STAR), 138667 Singapore, Singapore

**Yu Wang**
Department of Chemistry, National University of Singapore, 117543 Singapore, Singapore

**Xiaogang Liu**
Department of Chemistry, National University of Singapore, 117543 Singapore, Singapore
Institute of Materials Research and Engineering, A*STAR (Agency for Science, Technology and Research), 117602 Singapore, Singapore

**Xiaoyong Wang, Xiaoyuan Chen and Gang Liu**
State Key Laboratory of Molecular Vaccinology and Molecular Diagnostics, Center for Molecular Imaging and Translational Medicine, School of Public Health, Xiamen University, 361102 Xiamen, China

**Yong Wang, Huabing Chen and Mingyuan Gao**
School of Radiation Medicine and Protection, Soochow University, 215123 Suzhou, China

**Bengang Xing**
Division of Chemistry and Biological Chemistry, School of Physical and Mathematical Sciences, Nanyang Technological University, 637371 Singapore, Singapore
Institute of Materials Research and Engineering, A*STAR (Agency for Science, Technology and Research), 117602 Singapore, Singapore

**Anupam Bandyopadhyay, Kelly A. McCarthy, Michael A. Kelly and Jianmin Gao**
Department of Chemistry, Merkert Chemistry Center, Boston College, 2609 Beacon Street, Chestnut Hill, Massachuetts 02467, USA

**Ricardo M.P. da Silva**
Simpson Querrey Institute for BioNanotechnology (SQI), Northwestern University, Chicago, Illinois 60611, USA
Laboratory of Macromolecular and Organic Chemistry and Institute for Complex Molecular Systems, Eindhoven University of Technology, Eindhoven MB 5600, The Netherlands
Craniofacial Development & Stem Cell Biology, King's College London, London, SE1 9RT, UK

**Daan van der Zwaag and E.W. Meijer**
Laboratory of Macromolecular and Organic Chemistry and Institute for Complex Molecular Systems, Eindhoven University of Technology, Eindhoven MB 5600, The Netherlands

**Lorenzo Albertazzi**
Laboratory of Macromolecular and Organic Chemistry and Institute for Complex Molecular Systems, Eindhoven University of Technology, Eindhoven MB 5600, The Netherlands
Nanoscopy for Nanomedicine Group, Institute for Bioengineering of Catalonia (IBEC), Barcelona 08028, Spain

**Sungsoo S. Lee**
Department of Materials Science and Engineering, Northwestern University, Evanston, Illinois 60208, USA

**Samuel I. Stupp**
Simpson Querrey Institute for BioNanotechnology (SQI), Northwestern University, Chicago, Illinois 60611, USA
Department of Materials Science and Engineering, Northwestern University, Evanston, Illinois 60208, USA
Department of Chemistry, Northwestern University, Evanston, Illinois 60208, USA
Department of Medicine, Northwestern University, Chicago, Illinois 60611, USA
Department of Biomedical Engineering, Northwestern University, Evanston, Illinois 60208, USA

**Andrew C. Larsen and Sujay P. Sau**
The Biodesign Institute, Arizona State University, Tempe, Arizona 85287-5301, USA

**Matthew R. Dunn**
The Biodesign Institute, Arizona State University, Tempe, Arizona 85287-5301, USA
School of Life Sciences, Arizona State University, Tempe, Arizona 85287-5301, USA

**Andrew Hatch and Cody Youngbull**
School of Earth and Space Exploration, Arizona State University, Tempe, Arizona 85287-5301, USA

**John C. Chaput**
The Biodesign Institute, Arizona State University, Tempe, Arizona 85287-5301, USA
Department of Chemistry and Biochemistry, Arizona State University, Tempe, Arizona 85287-5301, USA
Department of Pharmaceutical Sciences, University of California, 147 Bison Modular, Building 515, Irvine, California 92697

**Laure Yatime and Gregers R. Andersen**
Department of Molecular Biology and Genetics, Aarhus University, Gustav Wieds Vej 10C, DK-8000 Aarhus, Denmark

**Christian Maasch, Kai Hoehlig, Sven Klussmann and Axel Vater**
NOXXON Pharma AG, Max-Dohrn-Strasse 8-10, 10589 Berlin, Germany

**Yiguang Wang, Yang Li, Tian Zhao, Xinpeng Ma, Gang Huang and Jinming Gao**
Department of Pharmacology, Simmons Comprehensive Cancer Center, University of Texas Southwestern Medical Center, 5323 Harry Hines Boulevard, Dallas, Texas 75390, USA

**Chensu Wang**
Department of Pharmacology, Simmons Comprehensive Cancer Center, University of Texas Southwestern Medical Center, 5323 Harry Hines Boulevard, Dallas, Texas 75390, USA
Department of Cell Biology, University of Texas Southwestern Medical Center, 5323 Harry Hines Boulevard, Dallas, Texas 75390, USA

**Brian Bodemann and Michael A. White**
Department of Cell Biology, University of Texas Southwestern Medical Center, 5323 Harry Hines Boulevard, Dallas, Texas 75390, USA

**Zeping Hu and Ralph J. DeBerardinis**
Children's Medical Center Research Institute, University of Texas Southwestern Medical Center, 5323 Harry Hines Boulevard, Dallas, Texas 75390, USA

**Layla J. Barkal, David J. Beebe and Erwin Berthier**
Department of Biomedical Engineering, University of Wisconsin-Madison, Madison, Wisconsin 53705, USA
Carbone Cancer Center, University of Wisconsin-Madison, Madison, Wisconsin 53705, USA

**Ashleigh B. Theberge**
Department of Biomedical Engineering, University of Wisconsin-Madison, Madison, Wisconsin 53705, USA
Carbone Cancer Center, University of Wisconsin-Madison, Madison, Wisconsin 53705, USA
Department of Urology, University of Wisconsin-Madison, Madison, Wisconsin 53705, USA

**Chun-Jun Guo**
Department of Pharmacology and Pharmaceutical Sciences, University of Southern California, Los Angeles California 90089, USA

**Joe Spraker**
Department of Plant Pathology, University of Wisconsin-Madison, Madison, Wisconsin 53705, USA

**Lucas Rappert**
Department of Medical Microbiology and Immunology, University of Wisconsin-Madison, Madison, Wisconsin 53705, USA

**Jean Berthier**
Department of Biotechnology, CEA-University Grenoble-Alpes, 17 Avenue des Martyrs, 38054 Grenoble, France

**Kenneth A. Brakke**
Department of Mathematics, Susquehanna University, Selinsgrove, Pennsylvania 17870, USA

**Clay C.C. Wang**
Department of Pharmacology and Pharmaceutical Sciences, University of Southern California, Los Angeles California 90089, USA
Department of Chemistry, University of Southern California, Los Angeles California 90089, USA

**Nancy P. Keller**
Department of Medical Microbiology and Immunology, University of Wisconsin-Madison, Madison, Wisconsin 53705, USA
Department of Bacteriology, University of Wisconsin-Madison, Madison, Wisconsin 53705, USA

**Jonathan Martens, Josipa Grzetic and Giel Berden**
Radboud University, Institute for Molecules and Materials, FELIX Laboratory, Toernooiveld 7c, 6525ED Nijmegen, The Netherlands

**Jos Oomens**
Radboud University, Institute for Molecules and Materials, FELIX Laboratory, Toernooiveld 7c, 6525ED Nijmegen, The Netherlands
Van 't Hoff Institute for Molecular Sciences, University of Amsterdam, Science Park 908, 1098XH Amsterdam, The Netherlands

# Index

www.ingramcontent.com/pod-product-compliance
Lightning Source LLC
Chambersburg PA
CBHW080530200326
41458CB00012B/4387